Civil Engineer's Handbook of Professional Practice

Civil Engineer's Handbook of Professional Practice

SECOND EDITION

Karen Lee Hansen, PhD

Kent E. Zenobia, PE, BCEE

ASCE PRESS

WILEY

Library of Congress Cataloging-in-Publication Data applied for:

Hardback ISBN: 9781119739791

Cover Design: Wiley
Cover Images: © Maarten Zeehandelaar/Shutterstock, Gorodenkoff/Shutterstock, GreenOak/Shutterstock, Sirikunkrittaphuk/Shutterstock

Set in 10.5/13pt STIXTwoText by Straive, Pondicherry, India

SKY10090213_110724

Contents

Contributing Authors

Keith A. Bisharat, BS, MS is an emeritus professor in the Department of Construction Management at California State University, Sacramento. Keith has a Bachelor of Science in Architecture and a Master of Science in Engineering and Project Management from the University of California, Berkeley He is also a licensed general contractor with more than 35 years of experience in construction as a sole proprietor, partner, forensic construction consultant, developer, building designer, project manager, superintendent, project engineer, carpenter, and laborer. He is author of Construction Graphics: A Practical Guide to Interpreting Working Drawings, a book that shows how construction graphics "translate" into construction methods and practices.

Mary Balogh, MS obtained her BS degrees in Natural Resource Management & Water Quality Management and a MS degree in Remote Sensing from the University of Wisconsin. She worked for the US Fish & Wildlife Service (USFWS) where she provided GIS, Global Positioning System (GPS), and Remote Sensing support to the regional office and field offices. She also mapped vegetation at National Wildlife Refuges in support of Comprehensive Conservation Plans. While at the USFWS Mary has written training documentation for GIS, GPS, and Remote Sensing and has published and edited articles for Photogrammetric Engineering & Remote Sensing, a peer reviewed scientific journal.

Dr. Tim Brady has been researching innovation and innovation management since 1980. He is a Principal Research Fellow at the Center for Research and Innovation Management (CENTRIM), at the University of Brighton, United Kingdom. He joined CENTRIM in 1994 to work on a study of the management of innovation within complex product systems (CoPS) and later became Deputy Director of the Economic and Social Research Council (ESRC)-funded CoPS Innovation Centre. His current research interests include learning and capability development in project- based business, and the emergence of integrated solutions. He was a member of the Engineering and Physical Sciences Research Council (EPSRC) network: Rethinking Project Management and organized the eighth International Network on Organizing by Projects (IRNOP) research conference, which took place in Brighton in September 2007. He previously worked at the Science Policy Research Unit (SPRU), University of Sussex, and at the University of Bath. Dr. Brady's Ph.D. dissertation examined business software 'make-or-buy' decisions.

Jody Bussey, DBIA, PMP has worked for architects, general contractors, and construction management firms since 2000. She graduated magna cum laude from California State

University, Sacramento with a BS in Construction Management with a minor in Business Administration. Her involvement on a LEED Gold high rise construction project introduced her to sustainable design and construction. Jody worked at PMA Consultants, acting as a senior engineer assisting with construction management services on the San Francisco Water System Improvement Program, involving multiple pipelines, a water treatment facility, and crossover valve facility projects totaling $300M. The projects include the $85M Tesla UV Water Treatment Plant, a LEED-certified facility that at the time was the third largest in the country and the largest in California. These projects are part of a $4B overall program utilizing state of the art construction management software and award-winning best practices procedures. Jody currently works at McCarthy Building Companies, Inc. as a project executive.

E.J. Koford is a biologist and project manager with 20 years of experience preparing environmental permitting documents, wildlife, and fisheries investigations, threatened and endangered species surveys, EIS/EIRs, water quality evaluations, and environmental regulatory compliance with requirements of CEC, FERC, SMARA, CERCLA, RCRA, NEPA, and CEQA. He has performed field surveys in 18 states and countries. Mr. Koford has an M.S. in Ecology from the University of California at Davis, an A.B. in Zoology from the University of California at Berkeley and is a Certified Wildlife Biologist of the Wildlife Society.

Dr. Iain A. MacLeod, a Chartered Member and Fellow of both the Institute of Civil Engineers (ICE) and Institution of Structural Engineers (IStructE), was a Professor Emeritus in the Department of Civil Engineering, Strathclyde University. He worked as a design engineer and consultant in the United States and Canada and in design research with the Portland Cement Association in the United States. He was Professor of Structural Engineering at the University of Strathclyde in Glasgow for 23 years and Professor and Head of Department at Paisley University. He was a former Lecturer at the University of Glasgow. His research work has spanned a range of topics in the design of buildings, including the analysis of tall buildings, the use of information technology (IT) in design and studies in design process. He is author of Modern Structural Analysis: Modelling Process and Guidance, published by Thomas Telford Ltd., a book that redresses the imbalance in risk between computer models based around generally determinate calculation outputs and possibly non-de- terminate understandings of the actual modeling process.

Dr. Jane E. Millar, principal of Jane Millar & Associates in Brighton, United Kingdom, consults in Policy Research. She has been a Senior Research Fellow at the Migration Research Unit (MRU), University College London; at the Institute for Public Policy Research in London; and at the Policy Research Unit (SPRU), University of Sussex. She holds a Ph.D. in Cognitive and Computing Sciences from the University of Sussex and managed a wide range of projects in both industry and academia.

Brian Neale, a Chartered Engineer, a Fellow of both the Institution of Civil Engineers (ICE) and the Institution of Structural Engineers (IStructE), and an Honorary Fellow of the Institute of Demolition Engineers (IDE) in the United Kingdom, where he is a former independent consultant and former Executive Secretary of the UK based Hazards Forum.

Previously he worked for various Government departments and other professional Civil Engineering organizations. He has been active at various stages with each of those three professional bodies as well as in national and international standards-making. An example includes chairing the drafting committee of the BS6187:2011 Code of Practice for Full and Partial Demolition standard and also its 2010 predecessor. A further example is as a European Committee for Standardization (CEN) convenor, where Mr. Neale oversaw the drafting of one of the Structural Eurocodes as well as contributing to others. He initiated and chaired the Organizing Committees of the four International Conferences on Forensic Engineering from 2001 organized by the ICE and supported by the American Society of Civil Engineers (ASCE) and was editor of the associated published proceedings. His published papers include an international dimension, and his consultancy includes a training element.

Danny Nguyen, E.I.T. has a Bachelor of Science degree from California State University, Sacramento in Civil Engineering and is currently a CSUS graduate student in Transportation/ Civil Engineering. With a leadership background, experience in field work, and high activity in both student and professional organizations, Danny gives frequent presentations as part of the civil engineering educational curriculum. He is bilingual and biliterate in Vietnamese and is bilingual in Spanish and Japanese. He is a current active member of the Institute of Transportation Engineers (ITE) and was the past President and is a member of the University Chapter ASCE.

Rob Nixon, PE, GE, is a licensed civil and geotechnical engineer, and has been working in the engineering consulting industry since 1994. He has served in roles including project manager, group manager, contract manager, client account manager, capture manager, department manager, regional market sector leader, growth leader, and division leader. His experience includes design, construction, and supervision of staff and resources, to support development of large-scale public works civil infrastructure. Mr. Nixon received his Bachelor of Science degree in Civil Engineering from California Polytechnic State University, San Luis Obispo, and most recently serves as Vice President at AECOM.

Greg Oslund, P.E. has more than 22 years of experience in the planning, approval, design, management and oversight of transportation projects. He has spent his entire career developing a comprehensive understanding of the project development phases required for these projects including project initiation, planning, programming, project approval and environmental (PA&ED), design (PS&E), utility coordination, permit- ting, R/W acquisition and engineering support during construction. He has served as project engineer, project manager and/or principal in charge for more than 25 large transportation projects. In addition, Mr. Oslund has more than 15 years business development experience involving major transportation project pursuits as the prime consul- tant. He has served as client service manager, pursuit manager and regional business development manager responsible for setting and implementing the business develop and marketing strategy for a large engineering and construction firm.

George T. Qualley, P.E. is a licensed professional engineer with 40 years of civil engineering design, construction, operation, and maintenance experience for the State of California. He served for 13 years as Flood Management Division Chief for the California Department of Water Resources, responsible for a staff of over 300 carrying out an integrated statewide program with a balance of structural ("controlling" floodwaters) and non-structural (reducing exposure to flood damage through improved floodplain management) approaches to flood risk reduction. He grew up on a grain and a cattle farm near Fargo, North Dakota. George holds a Bachelor of Science Degree from North Dakota State University.

Tony Quintrall, P.E. is a geotechnical project engineer with HDR Engineering, Inc. in Folsom, CA. At HDR he has been involved in numerous geotechnical investigations and design and construction activities for levees and small dams throughout Northern California. He has been involved with all aspects of the design process, from preliminary investigations and analysis to construction management, functioning as a technical specialist performing analysis as well as providing oversight and quality control.

Dr. Tarek Salama, P.E., PMP is an Assistant professor at California State University, Sacramento. He earn ed a BSc. (Construction Management) and a MSc. (Structural Engineering) at Alexandria University, Alexandria, Egypt. He also obtained a MEng. (Civil Engineering) and PhD. (Building Engineering) from Concordia University, Montreal, Canada. From 2003 to 2012, Dr. Salama worked in several multinational organizations related to offshore and onshore oil and gas industries. His research interests include structural analysis, modular construction, project planning, and scheduling. Dr. Salama is an associate member at the American Society of Civil Engineers (ASCE) and a member of the Egyptian Engineers Syndicate and holds a project management professional certificate (PMP) from the Project Management Institute (PMI). He received Hydro Quebec financial award from the faculty of engineering and computer science in Concordia University from 2014 to 2016, and the Stephen G. Revay Award from the Canadian Society for Civil Engineering (CSCE) in 2018 as well as the Research and Creative Activity (RCA) faculty award from California State University, Sacramento in 2021. Also, he is a reviewer of international top-tier scholarly journals.

Dr. Matthew Salveson, P.E. is a licensed civil engineer and has been working in the transportation engineering field since 1991. His project experience includes the planning and design of various transportation facilities in California, including bridges, freeways, local roads, and interchanges. He has also managed the construction, retrofit and repair of numerous bridges. Dr. Salveson received his Bachelor of Science, Master of Science, and Doctor of Philosophy in Civil Engineering from the University of California, Davis. He is currently an Adjunct Professor of Civil Engineering at California State University, Sacramento.

Michael A. Sanchez, P.E. is a licensed civil engineer with 27 years of experience focused on improving transportation for communities and local agencies throughout the State of California. He has focused much of his career leading engineering teams and coordinating roadway improvement projects including corridors, highways, interchanges, intersections, roundabouts, along with bridge replacement and rehabilitation projects. Michael received his Bachelor of Science from the University of California, Berkeley as a Civil Engineer with a focus in transportation design. Michael leads a team of specialized transportation design engineers in his current role with Consor (formerly Quincy Engineering). Beyond project delivery, Michael enjoys leadership roles working directly with civil engineering students, mentoring younger staff, and delivering projects on expedited schedules. Michael contributed by sharing his professional leadership acumen to the Leadership portions of this book.

Michael A. Turco, P.E., BCEE is a licensed professional engineer and certified project manager, with 40 years of engineering, design, and management experience in and for the oil, chemical, hazardous waste management and environmental consulting industries. He is board certified by the American Academy of Environmental Engineers in hazardous waste management and holds a BS in Chemical Engineering, an MS in Environmental Engineering and an MBA, all from Drexel University.

Dr. Jorge A. Vanegas, has a B.S. in Architecture from the Universidad de Los Andes, Bogatá, Colombia, and is a registered Architect in Colombia. He also holds a M.S. in Construction Engineering & Management and Ph.D. in Civil Engineering from Stanford University. He is a tenured Professor in the Department of Architecture (ARCH) in the School of Architecture (COA) at Texas A&M University, where he served as Dean for over twelve years. Jorge's primary areas of scholarly and professional activities in research, education, and service have been, among others: (1) creativity, innovation, design, and entrepreneurship; (2) sustainability of the built environment – urban, infrastructure, facilities, and housing; and (3) advanced strategies, technologies, methods, and tools for integrated delivery and management of capital projects in the Architecture, Engineering, and Construction (AEC) industry.

Scott D. Woodland, P.E., M. ASCE is a licensed professional engineer in the State of California. With experience in design and construction, operations and maintenance and planning for the California Department of Water Resources he is an 18-year veteran of California's on-going struggles to deliver water and protect the State's citizens from floods. He currently is helping with the implementation of the California FloodSAFE and Integrated Regional Water Management Programs. Scott has a BS in Civil Engineering from the University of California, Davis. Scott contributed to portions of this book related to executing a professional commission, engineer's role in project development, and professional engagement.

Phil Welder, P.E., PMP is a chemical/environmental engineer with nearly 20 years of experience managing complex large-scale toxic and hazardous waste remediation projects for both the private and public sector, particularly the federal government. He is a certified project management professional (PMP), and is an Associate at GeoEngineers, Inc., where he monitors and assists project managers with their daily project oversight activities. Phil has a BS in Chemical Engineering from Trinity University, Texas. Phil contributed to portions of this book related to executing a professional commission, products that engineers deliver, and professional engagement.

Contributing Editors

Dr. Sandra M. Benedet holds a Ph.D. in Spanish from Stanford University, a BA from San Francisco State University, and has taught at Stanford University, Roosevelt University, Northwestern University, and the University of Iowa. She has taught a wide range of courses, including language, composition, and literature, including a course on urban literature that examines the way in which the Latin American city has been imagined in the 20th century. She has worked extensively on questions of modernity as they relate to the avant-garde. Her work has appeared in "La palabra y el hombre: Revista de la Universidad Veracruzana," and "Contratiempo," a Chicago-based publication.

Phil Brozek, P. E. is a Professional Engineer in the State of California and has more than 30 years of professional experience in contract management, construction management, and project management on large US Army Corps of Engineers projects. Phil is currently a partner in Brozek & Associates providing project leadership for natural resource conservation projects.

Gareth Figgess, BS, MS is the Chair of the Construction Management Department at California State University, Sacramento (CSUS). Gareth has a Bachelor of Science in Construction Management and a Master of Science in Business Administration. He has received an impressive grant from the US Environmental Protection Agency (EPA) for "Net-Zero Residential Construction," where he brought students together from several disciplines across campus to build a family home that produces more energy than it consumes. His work has won awards from the US EPA and the Sacramento Municipal Utility District (SMUD) and will advance the current methods of residential construction to a more-efficient standard. Gareth holds a General Contractor's license, has been a project engineer with a large general engineering contractor, and is considered a University expert in construction management.

Marina Gamez is an International Business Bachelors graduate from Zuyd University in Maastricht, Netherlands. She is currently working in the parking industry as an International Account Manager in Maastricht, Netherlands. In her spare time, Marina enjoys playing the guitar, drawing in her sketchbook, and jogging through the "hills" of Limburg when it is not raining. Marina is honing her artistic talents and integrating them into her new career. She agreed to apply these artistic skills and developed several original illustrative figures in this book to help clarify and enhance the points of discussion.

Becky Flegel Hansen is a retired English teacher. She holds a BA in English (summa cum laude) from California State University, Hayward/East Bay. She worked nearly 20 years in the communication industry and then began her teaching career in the Fremont Unified School District. While there, she was instrumental in establishing a local program for easing student transition from elementary to junior high school. After twenty-one years teaching seventh and eighth grades, she retired to Rocklin, CA. where she has volunteered at several elementary school libraries in support of her grandchildren. Becky has assisted local authors in editing and refining their works. She recently participated with Pearson Education in evaluating video submissions as part of the application for secondary English credentials.

Dr. Janis E. Hulla, D.A.B.T., has worked with the U.S. Army Corps of Engineers (USACE) since 2002. She provides environmental health and toxicological expertise to the USACE the Department of Defense. She identifies and frames national issues at the intersection of policy, science, and field practice to resolve both longstanding and emerging issues. She serves as an advisor to, and project manager for, the Physical Sciences and Life Sciences Divisions of the Army Research Office located in Research Triangle Park, NC. Prior to moving to Sacramento, Dr. Hulla was a senior fellow at the National Institute of Environmental Health Sciences, RPT, NC. A former faculty member of the University of North Dakota and North Dakota State Toxicologist, Dr. Hulla earned her B.S. in Microbiology and M.S. in Biochemistry from Mon- tana State University. Her Ph.D. was earned in Pharmacology from the University of Washington School of Medicine. Dr. Hulla is certified as a Diplomate of the Ameri- can Board of Toxicology (ABT) and currently serves on its Board of Directors.

Dr. John Johnston, P.E. is professor of environmental engineering in the Department of Civil Engineering at California State, Sacramento (CSUS) and Technical Advisor in the CSUS Office of Water Programs where he has guided stormwater research for all Caltrans projects. He served as Senior Environmental Engineer, Camp Dresser and McKee, Inc., in Boston, MA, where he managed EPA-sponsored technology evaluation of in-vessel composting sys- tems for municipal sludge, and a study of sludge dewatering system options for the City of Fall River, MA. Dr. Johnston also was a Civil Engineer with U.S. Army Corps of Engineers, Sacramento District, where he designed water and wastewater systems, roads, and facilities at Corps reservoirs in California.

Thomas J. Kelleher, Jr. is an attorney and Senior Partner with Smith, Currie, & Hancock LLP, a nationally recognized firm that practices in the areas of construction law, government contracts, and environmental law. He graduated cum laude from Harvard University and graduated from the University of Virginia School of Law. He served in the U.S. Army from 1968 through 1973 including positions as the Assistant Chief and Instructor in the Procurement Law Division at the U.S. Army Judge Advocate General's School, Charlottes- ville, Virginia. Mr. Kelleher has extensive government and construction contract experience on the spectrum of issues in- volving bidding, changes, differing site conditions, delays, and terminations. He has represented clients on hospital projects, airport facilities, research laboratories, convention facilities, prisons, federal and state courthouse and office complexes, and resort hotels and has practiced before the various federal government boards of contract appeals, as well as federal and state courts. In addition, he has represented clients in mediations, as well as arbitration proceedings. Mr. Kelleher is co-editor of Common Sense Construction Law: A Practical Guide for the Construction Professional.

Peter Ouchida, BS, MS, P.E., PMP is an adjunct professor in Civil Engineering at California State University, Sacramento. Peter holds his Project Management Professional (PMP) certification from the Project Management Institute. He specializes in engineering economics and the professional practice of Civil Engineering. Peter previously enjoyed a successful career at the California Air Resources Board.

Dr. Debra Larson, P.E. is Associate Dean of the College of Engineering, Forestry and Natural Sciences at Northern Arizona University (NAU). She joined in 1994 as an Associate Professor after completing a Ph.D. in Civil Engineering from Arizona State University and working in industry as a civil and structural engineer for ten years. Her research interests have included alternative building materials and techniques, value-added wood products, low-rise structures, and engineering pedagogy. Dr. Larson has designed and managed numerous American Society of Civil Engineers (ASCE)-sponsored Excellence in Civil Engineering Education (ExCEEd) Teaching Workshops for civil engineering educators and participated actively as a member of the ASCE's Body of Knowledge (BOK) Educational Fulfillment Committee. She also has lead ABET, Inc.—formerly Accreditation Board for Engineering and Technology—specialized evaluation teams in reviewing academic institutions and programs to ensure that they are meeting established standards of educational quality.

Todd Kamisky, PE, GE, is a licensed civil and geotechnical engineer, and has been working in the geotechnical engineering field since 1994. His project experience includes all geotechnical aspects of residential subdivisions, detention basins, bridges, communication towers, schools, and commercial/industrial developments. Mr. Kamisky received his Bachelor of Science degree in Civil Engineering from California State University, Chico and a Master of Science degree in Civil Engineering with emphasis in Geotechnical Engineering, from University of California, Davis.

Bridget Crenshaw Mabunga is an Assistant Adjunct Professor of English in the Los Rios Community College District and a Writer/Editor. She also volunteers as an Assistant Editor for Narrative Magazine. She holds a BA in English (cum laude) from CSU, Chico and an MA in English (emphasis Creative Writing) from CSU, Sacramento.

Janet Riser, MBA, CFM, CRPC obtained her undergraduate degree from the University of Pittsburg, and an MBA from Drexel University before entering the financial investment community as a financial advisor for over 25 years with Merrill Lynch and now with Janney,

Montgomery, Scott LLC as a First Vice-President. Janet earned her Chartered Retirement Planning Counselor designation from the College of Financial Planning in 2007 and in 2009 received Five Star Wealth Manager Award in the Delaware Valley. Janet specializes in the financial planning process, helping her clients deal with life cycle and market transitions. One of Janet's greatest pleasures in her work is the long-term relationships working with and growing extended families through multiple generations. Janet contributed to portions of this book related to the client relationship, communication, and professional engagement.

Preface

The American Society of Civil Engineers (ASCE) has made a concerted effort to work with ABET (formerly named the Accreditation Board for Engineering and Technology) in order to assure that civil engineering education anticipates and responds to the profession's evolving needs. The ASCE has formed several task forces over the last two decades not only to address these needs in the present but also to foresee significant trends.

The ASCE has incorporated these findings in multiple reports and policy statements, including: Policy 465—Academic Prerequisites for Licensure and Professional Practice; the vision articulated by the Summit on the Future of Civil Engineering— 2025; and the *Civil Engineering Body of Knowledge for the 21st Century* (BOK1-2004, BOK2-2008, and BOK3/CEBOK-2019). Policy 465 supports the concept of the master's degree or equivalent as a prerequisite for licensure and the practice of civil engineering at the professional level. The attendees of the Summit on the Future of Civil Engineering—2025 articulated a vision that sees civil engineers as being entrusted by society to be leaders in creating a sustainable world and enhancing the global quality of life. (More information is available at: www.asce.org/raisethebar).

Each of the current *CEBOK's 21* outcomes could command its own textbook. The goal of this book is to provide an easily understood and readily usable resource for civil engineering educators, students, and professional practitioners that develops overall understanding and points readers to additional resources for further study.

The *Civil Engineer's Handbook of Professional Practice* targets both academia and industry. The book can be used as a textbook for Professional Practice, Senior Project, Infrastructure Engineering, and Engineering Project Management courses. It is appropriate for upper division and graduate level students in the major. Additionally, the book is a helpful reference for practicing civil engineers.

The information contained in the ASCE's CEBOK provides a vision for a civil engineering body of knowledge. The *Civil Engineer's Handbook of Professional Practice* builds on that vision by providing illuminating techniques, quotes, case examples, problems and information to assist the reader in addressing the many challenges facing civil engineers in the real world. This book:

- Focuses on the business and management aspects of a civil engineer's job, providing students and practitioners with sound business management principles
- Addresses contemporary issues, such as permitting, globalization, sustainability, and emerging technologies

- Offers proven methods for balancing speed-quality-price with contracting and legal issues in a client-oriented profession
- Includes guidance on juggling career goals, life outside work, compensation, and growth

Additionally, the authors and publisher have established a website: www.wiley.com/go/cehandbook. Wiley and the Authors wish to support this book and to enable communication between the readers and authors and offer this website address as a convenient mechanism to do so.

While every effort has been made to ensure the accuracy and completeness of the information provided in this handbook, readers are advised to consult with a qualified professional before applying any of the practices or techniques discussed to specific projects or situations.

Acknowledgments

This book was born through our involvement with the students of the Departments of Civil Engineering and Construction Management at California State University, Sacramento (CSUS) and a desire to help them become highly functioning, competent, ethical, and successful Civil Engineering and Construction professionals. We have been guided by the vision established by the American Society of Civil Engineers (ASCE) in the *Bodies of Knowledge 1 (2004) and 2 (2008), and 3 (2019) Civil Engineering Body of Knowledge for the 21st Century: Preparing the Civil Engineer for the Future*, and other ASCE policy statements. We would like to acknowledge both our students and the many professional Civil Engineers and Constructors, both past and present, who have inspired us.

We have relied heavily on the insights and professional experience of our many expert contributing authors and technical reviewers and are most grateful for their participation. To engage with these professionals, who are part of an engineering community that is dedicated to continuous improvement, mentoring, public health and safety, was a pleasure. The contents of this book truly reflect a national and international flavor and represent the diversity of our fellow engineers in academia, public service, and the private sector. These dedicated professionals are acknowledged and listed with their credentials in the following pages.

The authors also thank our colleagues in the CSUS Department of Civil Engineering and the Construction Management Department for their assistance with this project and for helping to provide an environment that is both stimulating and nurturing. Specifically, we thank Dr. Ramzi Mahmood, Director of the Office of Water Programs, for his support. Keith Bisharat is thanked for great leadership and insight into the initial mystery of book publishing. Keith was able to show us the true end product, his book titled *Construction Graphics*, and often made himself available for consulting and coaching. Civil Engineering (CE) Senior Project class faculty are acknowledged for their leadership and contributions in helping students solve actual engineering problems prepared by graduating CE students under the tutelage of volunteer professional Civil Engineers. We also are grateful for additional guidance and encouragement provided by Dr. John Johnston, Mikael Anderson, P.E., and current College of Engineering and Computer Science Dean and Civil Engineer, Dr. Kevan Shafizadeh.

On a personal level, Karen Hansen wishes to thank all of those who have assisted in this book-writing-publishing odyssey. Several good friends and relatives have provided warmth as well as homes away from home. I am forever indebted to Susan Padilla, her sister Maxine Padilla-Selby, her brother Mark Padilla, and to my aunts and uncles, Gordon and Peggy Winlow and Blanche and Herbert Jensen, for their support. These friends and family used all of their considerable collective creative powers to help me keep on track. My parents, Barbara Lee Winlow and Robert W. Hansen, have given me the curiosity and drive required to see this project through to completion. How fortunate I have been to have these people in my life!

There are many others, who have offered intellectual counterpoints, good humor, and strong shoulders. Among these are Sandra Benedet, my cousin Kristie Denzer, Carole Hyde, Ben Langhout, Marion Lee, Iain MacLeod, Irene McNay, Noel (Bill) Stewart, Dr. Jorge Vanegas, and my cousins Dave, Dale, and Doug Winlow. Also of tremendous influence have been my professors at Stanford University – Prof. Donald S. Barrie, Prof. John W. Fondahl, Dr. James G. March, Dr. Raymond E. Levitt, Prof. Clarkson H. Oglesby, Prof. Henry W. Parker, Dr. Boyd C. Paulson, and Dr. C.B. (Bob) Tatum). Thank you all!

Kent Zenobia wishes to thank several people who helped immensely with the production of this book. I would like to thank my life partner and wife, Sylvia, for her love, support, and patience during the past three or so years it has taken me to research and produce this work. She demonstrated great patience and understanding throughout the process. She helped with subject–matter presentation and editing. I am so fortunate to have her as a partner in life and love. I thank my two children, Taylor and Jack, and stepdaughter, Marina, for their love, support, and patience.

I am treated to an additional dimension of engineering by my fellow colleagues at CSUS. Working as an adjunct professor at California State University, Sacramento, has provided me with another family of associates for which I am truly grateful.

Creating this handbook has been stimulating, numbing, satisfying, frustrating, and always challenging. Each author wishes to thank the other for their patience, grace under pressure, and insights we anticipate our readers will find constructive. Together, we hope our multi-dimensional views from academic, public service, industry, and national/international perspectives enhance readers' professional practice of Civil Engineering and Construction.

Finally, we thank John Wiley and Sons, Inc. for their efforts producing this handbook. We whole-heartedly thank Kalli Schultea, Senior Editor, who helped initiate this project; Isabella Proietti, Editorial Assistant; Jayashree Saishankar, Managing Editor; and Kavin Shanmughasundaram, Content Refinement Specialist, for their patience, craftsmanship, and experience in the actual publication of this work. We would also want to thank Donna Dickert, ASCE Acquisitions Editor, and Tara Hoke, Aff. M. ASCE as the General Counsel at ASCE and a member of the Virginal bar for review of this Second Edition manuscript.

Karen Lee Hansen and Kent E. Zenobia
October 2024

List of Abbreviations

A

AA	Affirmative Action
AAA	American Arbitration Association
AACE	American Association of Cost Engineers
AAEO	Affirmative Action and Equal Employment Opportunity
AAP	Affirmative Action Program
AASHTO	American Association of State Highway and Transportation Officials
ABET, Inc.	Accreditation Board for Science and Technology (formerly)
ACEC	American Council of Engineering Companies
ACEEE	American Council for an Energy-Efficient Economy
ACI	American Concrete Institute
ACLC	Administrative Civil Liability Complaint
ACM	Asbestos Containing Material
A.D.	Anno Domini, also Common Era (CE)
ADA	Americans with Disabilities Act
ADR	Alternative Dispute Resolution
A / E	Architect / Engineer
AEA	Atomic Energy Act
AEC	Architectural / engineering / construction
AEI	Architectural Engineering
AGC	Associated General Contractors
AHA	Activity Hazard Analysis
AI	Artificial Intelligence
AIA	American Institute of Architects
AICCP	Spanish Asociación de Ingenieros de Caminos, Canales, y Puertos
AISI	American Iron and Steel Institute
ANSI	American Nation Standards Institute
AP	Accredited Professional
APN	Assessor's parcel number
APP	Accident Prevention Plan
ARMA	Asphalt Roofing Manufactures Association
ASCE	American Society of Civil Engineers

ASDS	Advanced Systems Development Services
ASHRAE	American Society of Heating, Refrigerating, and Air-Conditioning Engineers
ASPE	American Society of Professional Estimators
ASTM	American Society for Testing and Materials (formerly)
AutoCAD	Automated Computer-aided Design
AWI	Architectural Woodwork Institute
AWS	American Welding Society

B

BAA	British Airports Authority
BAC	Budget at Completion
B.C.	Before Christ, also Before the Common Era (BCE)
BCEE	Board Certified Environmental Engineer
BD	Business Development
BD&A	Big Data and Analytics
BEES	Building for Environmental and Economic Sustainability
BHMA	Builders Hardware Manufacturers Association
BIM	Building Information Modeling
BLISS	Building Life-cycle Information System
BOK1	*Civil Engineering Body of Knowledge for the 21st Century* (ASCE, 2004)
BOK2	*Civil Engineering Body of Knowledge for the 21st Century* (ASCE, 2008)
BOK3	*Civil Engineering Body of Knowledge for the 21st Century* (ASCE, 2019)
BREEAM	Building Research Establishment Environment Assessment Method
B.S.	Bachelor of Science degree
BS EN ISO	British Standard Euro Norm International Standard

C

CAA	Clean Air Act
CAD	Computer-Aided Design
CAM	Computer-Aided Manufacturing
CATIA	Computer Aided Three-Dimensional Interactive Application
CBD	*Commerce Business Daily*
CBS	Cost Breakdown Structure
CCPM	Critical Chain Project Management
CCR	California Code of Regulations
CD	Construction Documents

CDFG	California Department of Fish and Game
CFR	Code of Federal Regulation
CEBOK	*Civil Engineering Body of Knowledge* (ASCE, 2019)
CEEQUAL	Civil Engineering Environmental Quality Assessment and Award Scheme
CEN	European Committee for Standardization
CEng	Chartered Engineer (United Kingdom)
CEO	Chief Executive Officer
CEQA	California Environmental Quality Act
CERCLA	Comprehensive Environmental Response, Compensation and Liability Act
CI	Collective Intelligence
CICE	Chinese Institute of Civil Engineering
CIFE	Center for Integrated Facility Engineering
CIH	Certified Industrial Hygienist
CIPP	Continuous Improvement of the Project Process
CII	Construction Industry Institute
CM	Construction Manager or Management
CNRA	California Natural Resources Agency
CO_2	Carbon Dioxide
COPR	Coasts, Oceans, Ports, and Rivers
COR	Contracting Officer Representative
CPD	Continuing Professional Development
CPI	Cost performance index
CPM	Critical path method
CSA	County Service Area
CSI	Construction Specifications Institute
CSP	Certified Safety Professional
CSWP	Code of Safe Work Practice
CURT	Construction Users Roundtable
CV	Cost Variance
CVRWQCB	Central Valley Regional Water Quality Control Board
CWA	Clean Water Act

D

D	Duration
DA	Design Assist

DASHO	Designated Agency Safety and Health Official
DB	Design Build
DBA	Doing Business As
DBB	Design-Bid-Build
DBIA	Design Build Institute of America
DD	Design Development
DER	Distributed Energy Resources
DFOW	Definable Features of Work
DL	Design (Team) Leader
DODI	Department of Defense Instruction
DPM	Design Performance Measure
DRB	Dispute Review Board
DUNS	Dun & Bradstreet Number
DWR	Department of Water Resources (California)

E

EAC	Estimate at Completion
EAP	Emergency Action Plan
EC2000	Engineering Criteria 2000
EEOC	Equal Employment Opportunity Commission
EF	Early Finish
EIR	Environmental Impact Report
EIS	Environmental Impact Statement
EIT	Engineer in Training
EJCDC®	Engineers Joint Contract Documents Committee®
EM	Engineering Manual
EM	Engineering Mechanics
EO	Equal Opportunity
EO	Executive Order (Presidential)
EPA	Environmental Protection Agency (United States)
EPCRA	Emergency Planning and Community Right-to-Know Act
ES	Early Start
ESA	Endangered Species Act
ESG	Environmental, Social, and Governance
ETC	Effort (or Estimate) to Complete

F

FAR	Federal acquisition regulation
FARS	Federal Acquisition Regulation System

FF	Free Float
FFDCA	Federal Food, Drug, and Cosmetic Act
FIATECH	Fully Integrated and Automated Technology (formerly)
FIDIC	Fédération Internationale Des Ingénieurs-Conseils (International Federation of Consulting Engineers)
FHWA	Federal Highway Administration
FIFRA	Federal Insecticide, Fungicide, and Rodenticide Act
FS	Feasibility Study
FWPCA	Federal Water Pollution Control Amendments

G

GC	General Contractor
GDA	Government Designated Authority
GINA	Genetic Information Nondiscrimination Act
GMP	Guaranteed Maximum Price
GPS	Global Positioning Systems
GreenLITES	Green Leadership in Transportation and Environmental Sustainability
GIS	Geographic Information Systems

H

H&S	Health and Safety
HAZWOPER	Hazardous Waste Operations and Emergency Response
HCP	Hazard Communication Plan
HIPP	Heat Illness Prevention Plan
HMO	Health Maintenance Organization
HR	Human Relations
HSP	Health and Safety Plan
HSP/APP	Health and Safety Plan and Accident Prevention Plan
HUBZONE	Historically Underutilized Business Zone
HVAC	Heating, Ventilating, and Air Conditioning

I

IAI	International Alliance for Interoperability
ICC	International Code Council
ICE	Institution of Civil Engineers (United Kingdom)
IDW	Integrated Design Workflow
IFCE	International Federation of Consulting Engineers
IIPP	Injury and Illness Prevention Plan

I-LAST	Illinois-Livable and Sustainable Transportation Rating System and Guide
IPCC	Intergovernmental Panel on Climate Change
IPD	Integrated Project Delivery
ISE	Institute of Structural Engineers
ISO	International Organization for Standardization
ISS	International Space Station
IT	Information Technology

J

| JHA | Job Hazard Analyses |
| JSCE | Japan Society of Civil Engineers |

K

| KO | Contracting Officer |

L

LBMS	Location Based Management System (Scheduling)
LCC	Lifecycle Cost
LCCA	Lifecycle Cost Analysis
LCM	Lead Containing Material
LEED	Leadership in Energy and Environmental Design
LF	Late Finish
LLC	Limited Liability Company
LOB	Line of Balance (Scheduling)
LOE	Level of Effort
LS	Late Start
LSM	Linear Scheduling Method

M

MCM	Mercury Containing Material
ME	Mentored Experience
MEP	Mechanical, Electrical, Plumbing
MILCON	Military Construction
MP	Multiple Prime
MSA	Master Services Agreement

N

| NAICS | North American Industry Classification System |
| NBIC | Nano-Bio-Info-Cognizance |

NBIMS	National BIM (Building Information Modeling) Standards
NCEES	National Council of Examiners for Engineering and Surveying
NCS	National Computer Aided Design (CAD) Standard
NECA	National Electrical Contractors Association
NEMA	National Electrical Manufacturers Association
NEPA	National Environmental Policy Act
NIBS	National Building Information Modeling (BIM) Standard
NIST	National Institute of Standards and Technology
NOA	Naturally Occurring Asbestos
NOAA	National Oceanic and Atmospheric Administration
NMFS	National Marine Fisheries Service
NPDES	National Pollutant Discharge Elimination System
NRCA	National Roofing Contractors Association
NSPE	National Society of Professional Engineers

O

O&M	Operation and Maintenance
OBS	Organizational Breakdown Structure
OFCCP	Office of Federal Contract Compliance Programs
OPA	Oil Pollution Act
OSHA	Occupational Safety and Health

P

P&L	Profit and Loss
PCB	Polychlorinated Bisphenols
PDRI	Project Definition Rating Index
PDT	Project Delivery Team
PERT	Performance Evaluation Review Technique
PG	Post Graduate (Education)
PgMP	Program Management Plan
PHA	Position Hazard Analyses (PHA)
PHCC	Plumbing, Heating, Cooling Contractors Association
PM	Project Manager
PMBOK	Project Management Body of Knowledge
PMI	Project Management Institute
PMP	Project management plan
PPA	Pollution Prevention Act
PPE	Personal Protective Equipment
PPO	preferred provider organization
PS&E	Plans, Specifications, And (Cost) Estimates

Q

Q	Quantitative Scheduling
QA	Quality Assurance
QAP	Quality Assurance Plan
QBS	Qualifications Based Selection
QC	Quality Control
QCP	Quality Control Plan

R

RAC	Risk Assessment Code
R&D	Research and Development
RCRA	Resource Conservation and Recovery Act
RF	Radio Frequency
RFI	Request for Information
RFP	Request for Proposal
RFQ	Request for Qualifications
RP	Responsible Party
RTC	Response to Comment
R / W	Right of Way

S

SAE	Society of Automotive Engineers
SAICE	South African Institution of Civil Engineers
SAM	System for Award Management
SARA	Superfund Amendments and Reauthorization Act
SBA	Small Business Administration
SB	Small Business
SD	Self-Developed
SDI	Steel Door Institute
SDG	Sustainable Development Goal
SDVOSB	Service-Disabled Veteran-Owned Small Business
SDWA	Safe Drinking Water Act
SF330	Standard Form 330
SMACNA	Sheet Metal and Air Conditioning Contractors National Association
SMD	Sewer Maintenance District
SOH	Safety and Occupational Health
SOQ	Statement of Qualifications
SOP	Standard Operating Procedure
SOW	Statement, or Scope, Of Work
SPCC	Spill prevention, containment, and contingency
SPI	Schedule Performance Index

SSHP	Site Safety and Health Plan
SSPC	Systems and Specifications for Painting and Coatings
STARS	Sustainable Transportation Analysis Rating System
SUNRA	Sustainability National Road Administrations
SV	Schedule Variance
SWP	Safe Work Practices

T

T&DI	Transportation and Development Institute
TBD	To Be Determined
TF	Total Float
TOC	Theory of Constraints
TQM	Total Quality Management
TSCA	Toxic Substances Control Act

U

UESI	Utility Engineering and Surveying Institute
UK	United Kingdom
UN	United Nations
USACE	United States Army Corps of Engineers
USC	United States Code
USFWS	US Fish and Wildlife Service

V

VAC	Variance at Completion
VOSB	Veteran-Owned Small Business
VR	Virtual Reality

W

WBS	Work Breakdown Structure
WOSB	Woman-Owned Small Business
WWT	Wastewater Treatment
WWTP	Wastewater Treatment Plant

Numerical

4D	Four Dimensions (CAD plus time/schedule)
5D	Five Dimensions (CAD plus money)

About the Companion Website

This book is accompanied by a companion website:

www.wiley.com/go/hansen/CivilEngineersHandbook

This website includes

- Instructor's Manual
  - Lecture notes
  - Topics for discussion
  - Sample assignments with rubrics
  - Quizzes and examinations
- Architecture, Engineering, and Construction (AEC) Industry contracts
- Sample Power Point presentations
- Other relevant information

CHAPTER 1

Introduction

Big Idea

"Entrusted by society to create a sustainable world and enhance the global quality of life, Civil Engineers serve competently, collaboratively, and ethically as: master planners, designers, constructors; stewards of the natural environment and its resources; innovators and integrators; managers of risk and uncertainty; and leaders in discussions and decisions shaping public environmental and infrastructure policy."

—ASCE Body of Knowledge 2

Key Topics Covered

- The Need for Accreditation
- American Society of Civil Engineers (ASCE)
- 21st Century Engineer
- Goal of This Book
- Reader's Guide

Civil Engineer's Handbook of Professional Practice, Second Edition. Karen Lee Hansen and Kent E. Zenobia.
© 2025 John Wiley & Sons, Inc. Published 2025 by John Wiley & Sons, Inc.
Companion website: www.wiley.com/go/hansen/CivilEngineersHandbook

---

Related Chapters in This Book

- Chapters 2 through 21 and Appendices A, B, C, D, E, F, G

1.1   BACKGROUND

The *Civil Engineer's Handbook of Professional Practice* is a professional practice guide for civil engineers. The first two decades of the 21st century have afforded many opportunities to reflect on the role civil engineers will play in coming years. The global economy and world banking system, national security, climate change, dwindling natural resources, technological advances, and societal changes have provided sufficient food for thought. In retrospect, the 2001 American Society of Civil Engineers (ASCE) report, titled *Engineering the Future of Civil Engineering*, which acknowledged that civil engineering must respond proactively to increasingly complex challenges related to public health, safety, and welfare, appears prophetic.

As a university program, civil engineering has been growing in the 21st century. Enrollment in most universities across the nation continues to increase, partially due to shrinking opportunities in other technical fields as a result of outsourcing. Civil engineers work very closely with government agencies and on projects requiring significant local knowledge, making outsourcing of their work difficult.

According to the US Bureau of Labor Statistics, civil engineers are expected to experience an 8 percent employment growth during the projection's decade [2020–2030] (www.bls.gov/oco/ocos027.htm#outlook, 2009). Related to the need to improve the Nation's infrastructure, more civil engineers will be required to design and construct or expand transportation, water supply, flood control, pollution control systems, buildings, building complexes, and other major civil engineering projects. They also will be needed to repair or replace existing roads, bridges, public structures and to respond to the increasing pressures of climate change and the demand for resilient infrastructure.

The ASCE persists in sending danger signals regarding the condition of the nation's infrastructure through report cards, the first of which was issued in 1998 (ASCE, 2021). For many years the country's infrastructure has been given a grade of "D" on the ASCE's infrastructure report card (ASCE, 2019, 2020); in 2021 the ASCE Report Card displayed some infrastructure categories moved out of the "D," and into the C-range. According to the ASCE, "The most recent analysis reveals that while we've made incremental immediate gains in some of the infrastructure categories, our long-term investment gap continues to grow. We're still just paying about half of our infrastructure bill—and the total investment gap has gone from $2.1 trillion over 10 years to $2.59 trillion over 10 years."

"The Global Infrastructure Construction Market was valued at USD 2,242.3 billion in 2021 and is expected to reach USD 3,267.3 by 2027, registering a growth rate of 6.48% during the forecast period."

Source: Modor Intelligence, Global Infrastructure Construction Market – Growth and Trends (2022–2027)

1.2   THE NEED FOR ACCREDITATION

The ASCE has made a concerted effort to work with ABET, Inc., formerly named the Accreditation Board for Engineering and Technology, to assure that civil engineering education anticipates and responds to the profession's evolving needs. ASCE has formed several task forces not only to address these needs in the present but also to foresee significant trends.

ABET, Inc. accredits civil engineering programs within universities and plays a significant role in determining the development of the profession. University Departments of Civil Engineering undergo extensive, periodic reviews by ABET to maintain their accreditation (ASCE.Org. Policy 465, 2006, 2019), (ABET.Org, 2022).

ABET, Inc. was established more than 90 years ago as the Engineers' Council for Professional Development (ECPD). A survey of multiple engineering societies revealed the need for quality control, and in 1932, seven societies founded ECPD. These societies included: the American Society of Civil Engineers (ASCE); the American Society of Mining and Metallurgical Engineers (now the American Institute of Mining, Metallurgical, and Petroleum Engineers); the American Society of Mechanical Engineers (ASME); the American Institute of Electrical Engineers (now IEEE); the Society for Promotion of Engineering Education (now the American Society for Engineering Education–ASEE); the American Institute of Chemical Engineers (AIChE); and the National Council of State Boards of Engineering Examiners (now NCEES). Today 35 member societies help set professional standards for accreditation. By 2022, ABET had accredited approximately 4,400 programs at more than 850 universities and colleges in 41 countries (Galloway, P. 2008), (ABET.Org, 2022).

1.2.1   ABET Outcomes

Following a long period of development, in 1997, ABET adopted engineering criteria 2000 (ec2000), which took a completely new approach to engineering education. By defining outcomes of engineering education, ec2000 focused on what is learned rather than what is taught (ABET.Org, 2022).

ABET'S 2022–2023 *Criteria for Accrediting Engineering Programs* identifies seven student outcomes of civil engineering education. To achieve ABET accreditation, each civil engineering program must document student outcomes that support the program's educational objectives. Attainment of these outcomes is meant to prepare baccalaureate graduates to enter the professional practice of engineering. These outcomes are:

1. The ability to identify, formulate, and solve complex engineering problems by applying principles of engineering, science, and mathematics.
2. an ability to apply engineering design to produce solutions that meet specified needs with consideration of public health, safety, and welfare, as well as global, cultural, social, environmental, and economic factors.
3. an ability to communicate effectively with a range of audiences.
4. an ability to recognize ethical and professional responsibilities in engineering situations and make informed judgments, which must consider the impact of engineering solutions in global, economic, environmental, and societal contexts.
5. an ability to function effectively on a team whose members together provide leadership, create a collaborative and inclusive environment, establish goals, plan tasks, and meet objectives.
6. an ability to develop and conduct appropriate experimentation, analyze and interpret data, and use engineering judgment to draw conclusions.
7. an ability to acquire and apply new knowledge as needed, using appropriate learning strategies. (ABET.Org. 2022)

1.3   AMERICAN SOCIETY OF CIVIL ENGINEERS

Meanwhile, the American Society of Civil Engineers has made a concerted effort to work with ABET to assure that civil engineering education anticipates and responds to the profession's evolving needs.

The ASCE has formed several task forces not only to address these needs in the present but also to foresee significant trends. ASCE *Policy Statement 465 – Civil Engineering Body of Knowledge and the Practice of Civil Engineering* states:

The American Society of Civil Engineers (ASCE) supports the attainment of the Civil Engineering Body of Knowledge (CEBOK) as a requirement for exercising responsible charge in the practice of civil engineering. The CEBOK is defined as the knowledge, skills, and attitudes necessary to exercise responsible charge in the practice of civil engineering and is attained through undergraduate and post-graduate engineering education, mentored experience, and self-development. Licensure constitutes a legal authority to practice engineering, however, the requirements for licensure do not ensure attainment of the CEBOK.

ASCE, Adopted by the Board of Direction on October 14, 2019

As depicted in Figure 1.1, Policy 465 supports the concept of the master's degree or equivalent as a prerequisite for licensure and the practice of civil engineering at the professional level.

For civil engineers to maintain leadership in the infrastructure and environmental arena, an implementation master plan was needed; and the basis of this master plan is a document called the Body of Knowledge. The Body of Knowledge 1 (BOK1), published in 2004, defines categories of knowledge and recommends 15 outcomes that collectively prescribe a "substantially greater depth and breadth of knowledge, skills, and attitudes required of an individual aspiring to the practice of civil engineering at the professional level (licensure) in the 21st Century." The BOK1 also emphasized the importance of attitude: "knowledge and skill, while necessary, are not sufficient to be a fully functioning civil engineer."

The ASCE published the second edition of BOK1, the Body of Knowledge 2 (BOK2), in 2008. The BOK2 also used the "outcomes" approach developed by ABET to define the knowledge, skills, and attitudes necessary to enter civil engineering practice at the professional level in the 21st century. The BOK2 further adopted Bloom's Taxonomy to indicate the desired level of achievement for each outcome. Bloom's Taxonomy of Educational Objectives, which is a collaborative effort to establish an outcome-based way of measuring achievement of

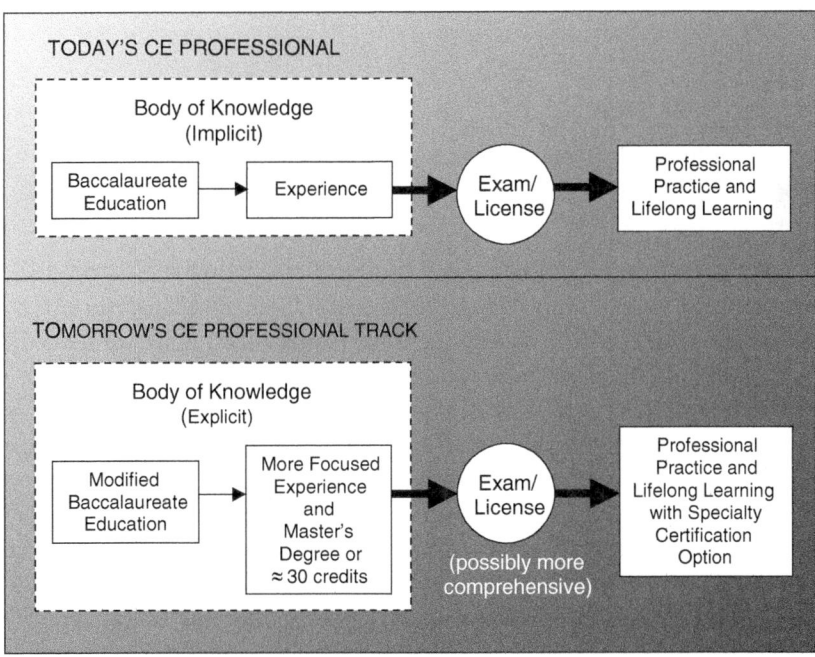

FIGURE 1.1   ASCE's vision of preparation for a career in civil engineering. *Source*: Adapted from ASCE Policy Statement 465.

TABLE 1.1   Civil engineering body of knowledge 3 (BOK3) outcomes.

Foundational	Engineering Fundamentals	Technical	Professional
1) Mathematics (c)	5) Materials Science (c)	9) Project Management (c)	16) Communication (c, a)
2) Natural Sciences (c)	6) Engineering Mechanics (c)	10) Engineering Economics (c)	17) Teamwork and Leadership (c, a)
3) Social Sciences (c)	7) Experimental Methods & Data Analysis (c)	11) Risk and Uncertainty (c)	18) Lifelong Learning (c, a)
4) Humanities (c)	8) Critical Thinking and Problem Solving (c)	12) Breadth in Civil Engineering Areas (c)	19) Professional Attitudes (c, a)
		13) Design (c)	20) Professional Responsibilities (c, a)
		14) Depth in Civil Engineering Areas (c)	21) Ethical Responsibilities (c, a)
		15) Sustainability (c, a)	

Note: (c) indicates "Cognitive" and (a) indicates "Affective" Outcomes
Source: ASCE BOK3 (2019)/American Society of Civil Engineers.

educational goals, includes three domains. Handbook 1: Cognitive, first published in 1956, focuses on knowledge-based domains. Handbook 2: Affective, published in 1964, concentrates on awareness and growth in attitudes and feelings. Handbook 3: Psychomotor, illuminating change and/or development of behavior/skills, was never fully published.

ASCE published the third edition of the BOK in 2019, BOK3. As depicted in Table 1.1, the BOK3 includes 21 outcomes organized into four categories: foundational outcomes, engineering fundamentals, technical, and professional. According to the BOK3:

> *The foundational outcomes provide the knowledge on which all other outcomes are built, both for civil engineers and those in most other learned professions. The engineering fundamentals outcomes form a bridge between the foundational and technical outcomes for all civil engineers, and notably for many other disciplines of engineering as well. Both the foundational and engineering fundamentals typically would be fulfilled as part of the undergraduate education. The technical outcomes specify knowledge more specific to civil engineering, and the professional outcomes focus on interpersonal and professional skills needed to be successful in the practice of civil engineering at the professional level*
>
> (ASCE, 2019).

In creating the BOK3, the ASCE also conducted an extensive literature review to capture leading educational theories. Additionally, the task force responsible for finalizing the BOK3 engaged constituents through a series of structured surveys. The BOK3 expands on the use of Bloom's Taxonomy by indicating that some outcomes are both "Cognitive" and "Affective," meaning that these outcomes require thinking and feeling. The resultant outcomes and levels of achievement are shown in Table 1.2, depicting outcomes in the cognitive domain levels of achievement, and Table 1.3, showing outcomes in the affective domain levels of achievement.

TABLE 1.2   Entry into the practice of civil engineering at the professional level requires fulfilling 21 outcomes to the appropriate cognitive domain level of achievement.

	Cognitive Domain Levels of Achievement					
Outcomes	1 Remember	2 Comprehend	3 Apply	4 Analyze	5 Synthesize	6 Evaluate
Foundational						
1. Mathematics	UG	UG	UG			
2. Natural Sciences	UG	UG	UG			
3. Social Sciences	UG	UG	UG			
4. Humanities	UG	UG	UG			
Engineering Fundamentals						
5. Materials Science	UG	UG	UG			
6. Engineering Mechanics	UG	UG	UG			
7. Experimental Methods and Data Analysis	UG	UG	UG	PG		
8. Critical Thinking and Problem Solving	UG	UG	UG	ME	ME	
Technical						
9. Project Management	UG	UG				
10. Engineering Economics	UG	UG				
11. Risk and Uncertainty	UG	UG	ME			
12. Breadth in Civil Engineering Areas	UG	UG	ME			
13. Design	UG	UG	ME	ME		
14. Depth in Civil Engineering Areas	UG	UG	PG	ME		
15. Sustainability	UG	UG	ME			

(Continued)

TABLE 1.2   *(Continued)*

		Cognitive Domain Levels of Achievement					
	Outcomes	1 Remember	2 Comprehend	3 Apply	4 Analyze	5 Synthesize	6 Evaluate
	Professional						
16.	Communication	UG	UG	UG	ME	ME	
17.	Teamwork and Leadership	UG	UG	UG	ME	ME	
18.	Lifelong Learning	UG	UG	UG	ME	ME	
19.	Professional Attitudes	UG	UG	ME	ME		
20.	Professional Responsibilities	UG	UG	ME	ME	ME	
21.	Ethical Responsibilities	UG	UG	UG	ME	ME	

Source: Adapted from Table F-5 of the ASCE Body of Knowledge 3 (2019).

TABLE 1.3   Entry into the practice of civil engineering at the professional level requires affective domain level of achievement.

		Affective Domain Levels of Achievement				
	Outcomes	1 Receive	2 Respond	3 Value	4 Organize	5 Characterize
	Technical					
15.	Sustainability	UG	UG	ME	SD	
	Professional					
16.	Communication	UG	UG	ME	SD	
17.	Teamwork and Leadership	UG	UG	ME	SD	
18.	Lifelong Learning	UG	UG	ME	SD	
19.	Professional Attitudes	UG	UG	ME	SD	
20.	Professional Responsibilities	UG	UG	ME	SD	
21.	Ethical Responsibilities	UG	UG	ME	ME	SD

TABLE 1.3 *(Continued)*

Outcomes	Affective Domain Levels of Achievement				
	1 Receive	2 Respond	3 Value	4 Organize	5 Characterize
Key:	**UG**	Undergraduate Education leading to a bachelor's degree in civil engineering, or a closely related engineering discipline, accredited by the Engineering Accreditation Commission (EAC) of ABET			
	PG	Postgraduate education equivalent to or leading to a master's degree in civil engineering, or a closely related engineering discipline, generally equivalent to one year of full-time study			
	ME	Mentored Experience, early career experience under the mentorship of a civil engineer practicing at the professional level, progressing both in complexity and level of responsibility			
	SD	Self-Developed through formal or informal activities and personal observation and reflection			

Source: Adapted from Table F-6 of the ASCE Body of Knowledge 3 (2019).

ASCE Has Developed a Global Vision of the Profession:

Entrusted by society to create a sustainable world and enhance the global quality of life, Civil Engineers serve, competently, collaboratively, and ethically as master:

- Planners, designers, constructors, and operators of society's economic and social engine, the built environment
- Stewards of the natural environment and its resources
- Innovators and integrators of ideas and technology across the public, private, and academic sectors
- Managers of risk and uncertainty caused by natural events, accidents, and other threats
- Leaders in discussions and decisions shaping public environmental and infrastructure policy

Source: Civil Engineering Body of Knowledge for the 21st Century (BOK2).

All three editions of the Civil Engineering Body of Knowledge stress the need for change in the way civil engineers practice their profession and in the way they are educated. Though not strictly prescriptive, BOK3 offers guidance to academia in helping to educate future engineers. Summary findings are highlighted below.

Key issues facing engineering education

BOK1 identifies the key issues facing civil engineering as:
- Escalated complex risks and challenges to public safety, health, and welfare
- Vulnerability to human-made hazards and disasters (such as terrorism, and climate change)
- Globalization
- Four-year bachelor's degree inadequacy in providing formal academic preparation for the practice of civil engineering at the professional level

BOK2 adds further concerns:
- Sustainability
- Emerging technology

BOK3 introduces:
- Fourth new category, engineering fundamentals, to the list of outcomes

Teaching/learning modes

BOK1 identifies four teaching/learning modes:
- Undergraduate study typically leading to a BSCE
- Graduate study or equivalent
- Cocurricular and extracurricular activities
- Post-B.S. engineering experience prior to licensure

BOK1 also concludes that distance learning will increasingly improve accessibility to high-quality formal education.

BOK3 adds self-development—refining/extending student character or abilities through self-study and personal observation and reflection—as a pathway to outcome fulfillment (ASCE, 2004, 2008, 2019).

Faculty member characteristics

BOK1 identifies characteristics of the model full- or part-time civil engineering faculty member:
- Scholars having and maintaining expertise in the subjects they teach
- Teachers who effectively engage students in the learning process
- Professionals with practical experience, preferably with professional engineering licenses
- Positive role models for the profession

Measurement of success in achieving learning outcomes

BOK3 presents extensive rubrics for measuring student success in achieving the Body of Knowledge outcomes in the cognitive domain and, where applicable, the affective domain.

ASCE Education Summit: Mapping the Future of Civil Engineering Education, 2020

OBJECTIVE 1: Reexamine, and potentially redefine, the domain of Civil Engineering.

- A clear consensus among Summit participants is that the world is becoming increasingly complex—thus, the challenges faced by engineers are becoming increasingly complex. One aspect of this complexity relates to the interconnected nature of infrastructure, environmental, political, and social systems. Such interconnectedness is a major driver of the dissolution of traditional "boundaries" that define a particular engineering discipline.

OBJECTIVE 2: Elevate professional skills to a truly equal footing with technical skills.

- Certainly, the need for strong professional skills has long been recognized by both civil engineering educators and practitioners. ... Summit participants placed significant emphasis on this topic; of the 20 prioritized Opportunity Statements, seven (7) address professional skills and abilities. Moving forward, topics related to professional skills should be elevated in importance within curricula—to be thought of not as "desirable," but "required," on an equal basis with the various technical/design skills currently emphasized in undergraduate programs.

OBJECTIVE 3: Develop a diverse, inclusive, equitable, and engaging culture within the civil engineering profession.

- Although it may be tempting to place these topics and discussions within the realm of professional skills and attitudes, this subject rose to represent a major theme of the event. At least four (4) of the "top 20" prioritized Opportunity Statements address the concept of civil engineering culture. Participants explored the distinct yet interconnected nature of diversity, inclusion, and equity; the need to engage students at all levels; and the concept of permeating the student educational experience with these concepts.

OBJECTIVE 4: Implement a regular schedule of national/ international civil engineering education events and dedicate resources to address findings.

- Planners of the 2019 Civil Engineering Education Summit consulted a significant body of literature to explore topics and themes arising from previous assessments of engineering education. ... Summit participants agreed that the 2019 Civil Engineering Education Summit was highly worthwhile, and expressed both hope and confidence that real and significant change could result from the work accomplished during the Summit (ASCE.Org, 2019, 2020).

1.4   21ST CENTURY ENGINEER

Aspiring civil engineers face challenges posed by the unique attributes and characteristics of facilities and civil infrastructure systems, as well as the complexities of the current processes and the diverse set of resources required for both their delivery and their use (futureworldvision.org, 2022), (National Academy of Engineering, 2004). Please refer to Figure 1.2 – The Civil Engineering body of knowledge study guide.

THE *CIVIL ENGINEERING* BODY OF KNOWLEDGE

STUDY GUIDE FOR FACULTY

PREPARE THE CIVIL ENGINEER OF TOMORROW

ASCE
AMERICAN SOCIETY OF CIVIL ENGINEERS

asce.org/engineer-tomorrow | engineertomorrow@asce.org

FIGURE 1.2   The Civil Engineering body of knowledge study guide.

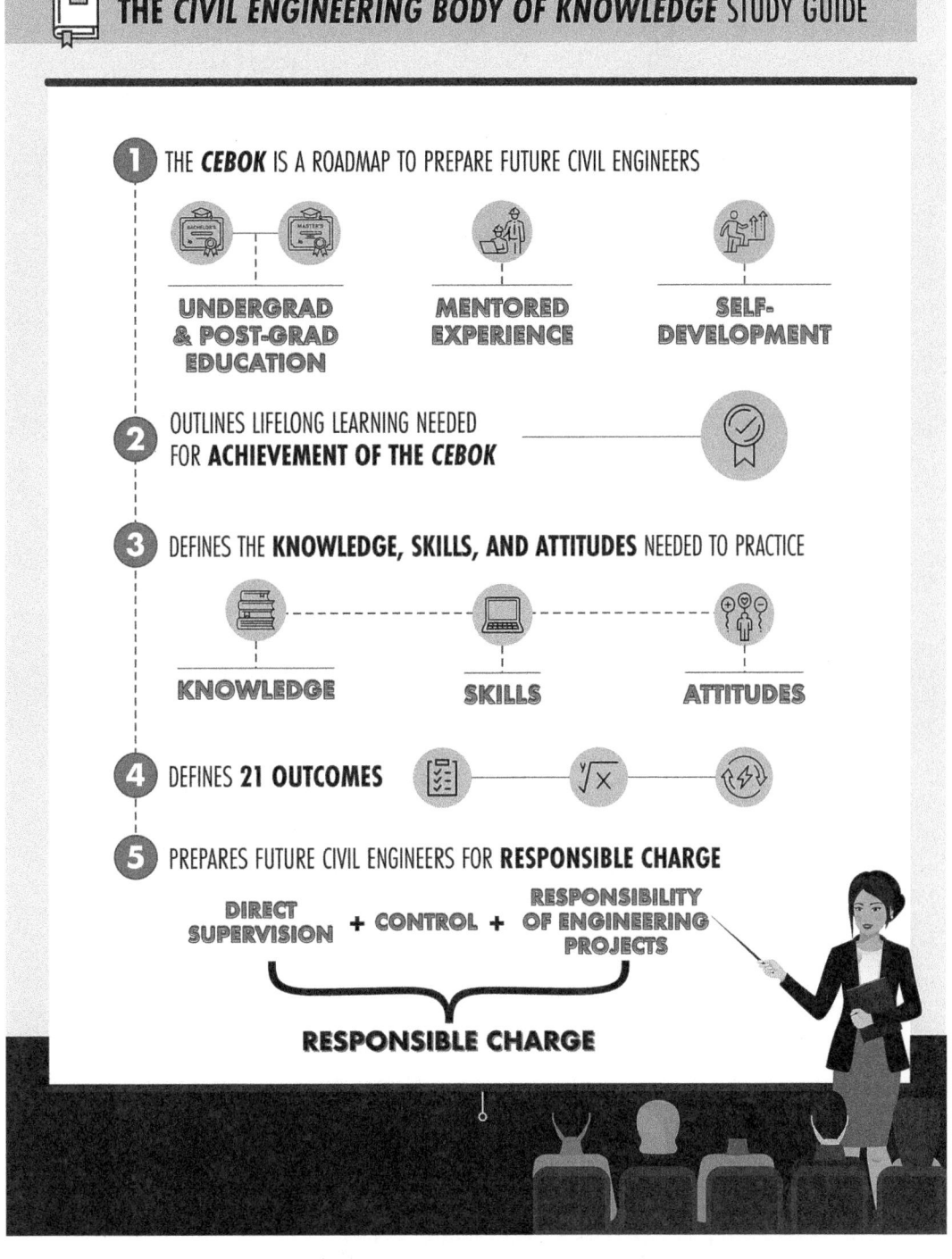

FIGURE 1.2 *(Continued)*

HELP **STUDENTS** CHART THEIR COURSE > HERE'S ONE TYPICAL PATH...

Make sure to introduce all 21 *CEBOK* outcomes

EVERYTHING STARTS WITH
UNDERGRADUATE EDUCATION

Take the **FUNDAMENTALS OF ENGINEERING** (F.E.) exam

Earn a degree from an
ABET ACCREDITED program

Take the **PRINCIPLES AND PRACTICE IN ENGINEERING** (P.E.) exam

POST-GRAD EDUCATION
AND **MENTORED EXPERIENCE**

ATTAINING THE *CEBOK* IS A JOURNEY,
AND YOU ARE THEIR FIRST GUIDE

LIFELONG LEARNING AHEAD

 GET **STUDENTS** STARTED ON THEIR PATH

FIGURE 1.2 (*Continued*)

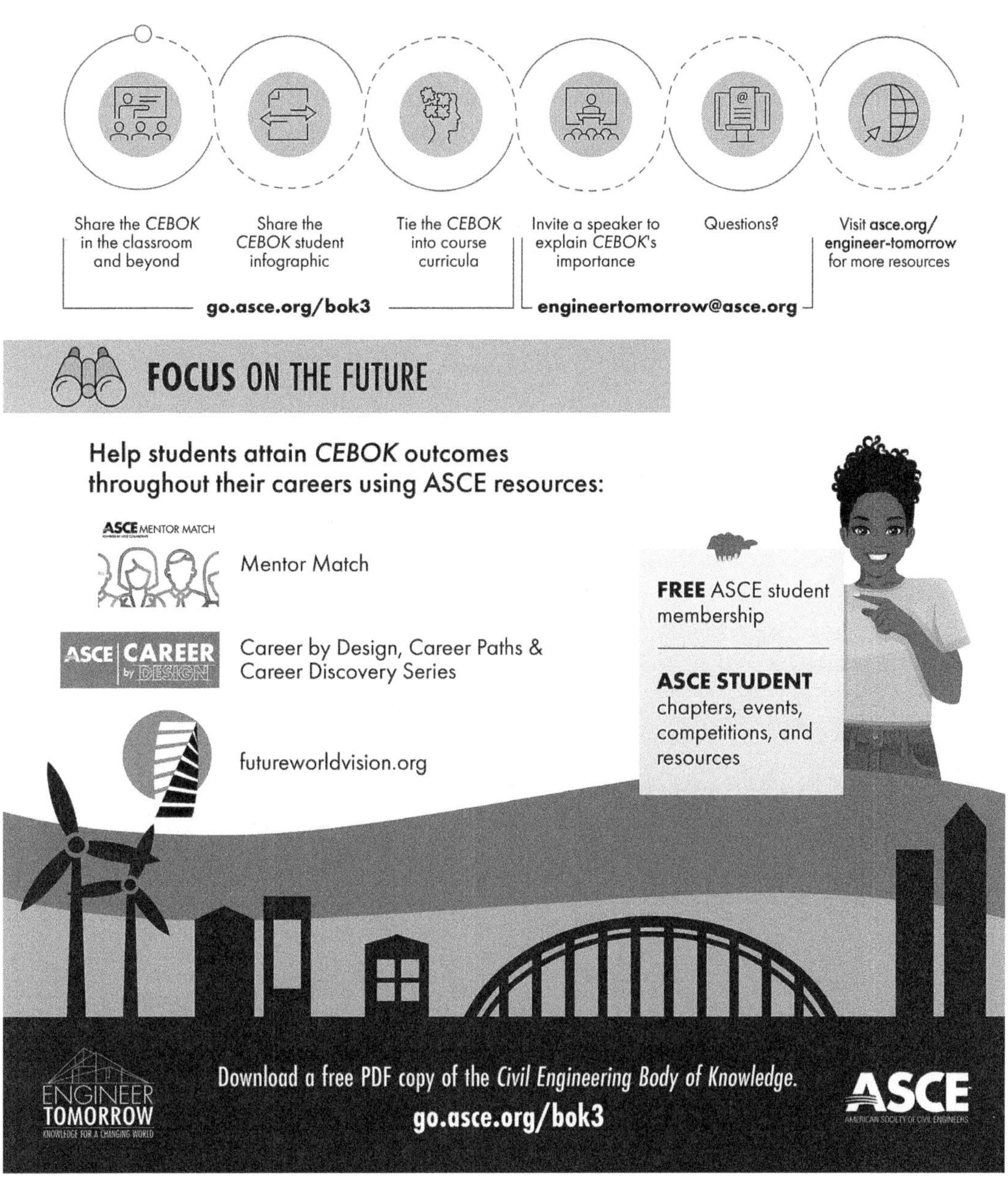

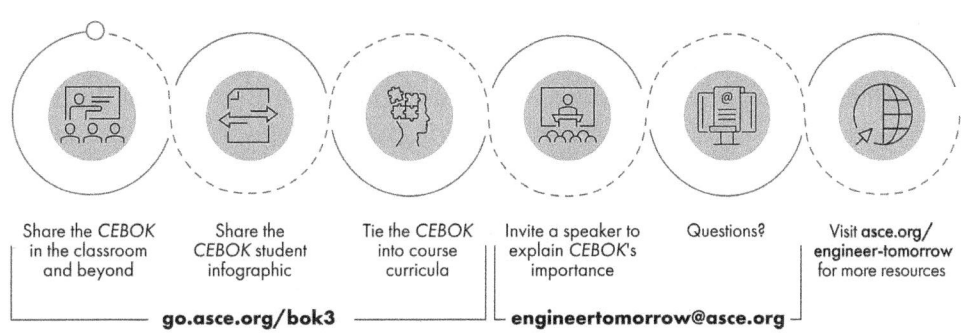

Share the *CEBOK* in the classroom and beyond

Share the *CEBOK* student infographic

Tie the *CEBOK* into course curricula

Invite a speaker to explain *CEBOK*'s importance

Questions?

Visit asce.org/ engineer-tomorrow for more resources

go.asce.org/bok3

engineertomorrow@asce.org

FOCUS ON THE FUTURE

Help students attain *CEBOK* outcomes throughout their careers using ASCE resources:

ASCE MENTOR MATCH

Mentor Match

ASCE | CAREER by DESIGN

Career by Design, Career Paths & Career Discovery Series

futureworldvision.org

FREE ASCE student membership

ASCE STUDENT chapters, events, competitions, and resources

Download a free PDF copy of the *Civil Engineering Body of Knowledge*.
go.asce.org/bok3

ENGINEER TOMORROW
KNOWLEDGE FOR A CHANGING WORLD

ASCE
AMERICAN SOCIETY OF CIVIL ENGINEERS

FIGURE 1.2 (*Continued*)

BOK3 gives some structure to what is a large educational challenge. These new outcomes and approaches have raised the bar substantially for civil engineering educators. Twentieth-century civil engineering education focused on learning about engineering mechanics; doing calculations; writing essays and lab reports; acquiring knowledge; and working with determinant processes. Twenty-first-century civil engineering practice requires innovative thinking and relies heavily on tacit knowledge—understanding, judgment, associativity, and intuition (MacLeod, 2009), (Rockefeller Foundation's 2050 Forum, 2008). (See Figure 1.3.)

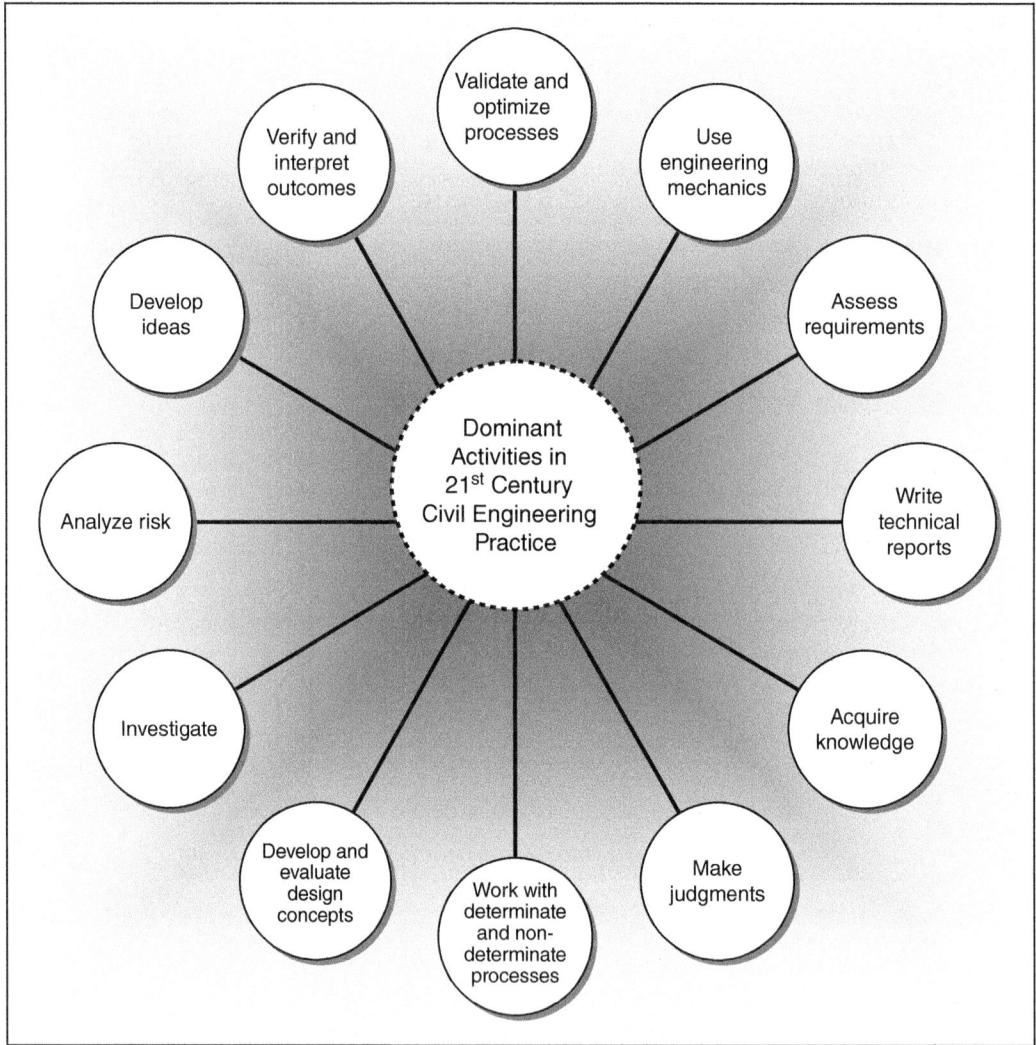

FIGURE 1.3   Dominant activities in 21st century practice.
Source: Dr. Iain A. MacLeod, Professor Emeritus, Department of Civil Engineering, Strathclyde University, Glasgow, Scotland.

1.5 GOAL OF THIS BOOK

Given these complexities, the question is: How can the new BOK outcomes be achieved? Clearly, each of the BOK3's 21 outcomes could command its own textbook. The goal of this book is to provide an easily understood and readily usable resource for civil engineering educators, students, and professional practitioners that develops overall understanding and points readers to additional resources for further study. The book distills 14 of the BOK3's outcomes (one engineering fundamentals outcome, seven technical outcomes, and all six professional outcomes) as well as other relevant issues.

The *Civil Engineer's Handbook of Professional Practice* targets both academia and industry. The book can be used as a textbook for Professional Practice, Senior Project, Infrastructure Engineering, and Engineering Project Management courses. It is intended for junior, senior, and graduate-level students. As the issues addressed in the 2019 BOK3 are disseminated and better understood by educators, all Civil Engineering Departments will need to offer a course on Practice Management, if they do not do so already.

Additionally, the book is a helpful reference for practicing civil engineers. The information imbedded in the 156-page BOK3 provides a vision for a civil engineering body of knowledge. The Civil Engineer's Handbook of Professional Practice builds and supplements that vision and is essential for the rapidly evolving practice of civil engineering.

1.6 READERS' GUIDE

Of the 21 outcomes discussed in BOK3, this book addresses the following:

8. Critical Thinking and Problem Solving	14. Depth in Civil Engineering Areas
9. Project Management	15. Sustainability
10. Engineering Economics	16. Communication
11. Risk and Uncertainty	17. Teamwork and Leadership
12. Breadth in Civil Engineering Areas	18. Lifelong Learning
13. Design	19. Professional Attitudes
	20. Professional Responsibilities
	21. Ethical Responsibilities

The Civil Engineer's Handbook of Professional Practice offers additional relevant information such as: the design professional's role in the project development process; the legal infrastructure in the United States; the fundamental contents of contracts; the origin of conflicts; the various roles that the civil engineer plays in construction projects; how the legal world views construction disputes; the basic economics of civil engineering practice; and emerging technologies relevant to civil engineering. Each chapter concludes with references for further reading and or study.

The book presents information in three levels of increasing detail through the use of graphics (photographs, illustrations, line drawings, graphs, text boxes, and cartoons) and text. These illustrations form one level of information, the commentary that accompanies the illustrations forms another, and the third is the actual text. The first page of each chapter outlines the key concepts presented and contains a unique graphic that helps to orient the reader.

The chapters of the Civil Engineer's Handbook of Professional Practice can be read in the order that best suits the reader. Following is a brief summary of the chapters and appendices:

- Chapter 1—Introduction
  This chapter addresses the overall issues outlined in the ASCE's Body of Knowledge, first, second, and third editions (BOK1, BOK2, and BOK3) and the need for a new approach to civil engineering.
- Chapter 2—Background and History of the Profession
  This chapter addresses BOK3 Outcome 12–Breadth in Civil Engineering Areas and gives an overview of the Architectural/Engineering/Construction (AEC) industry.
- Chapter 3—Ethics
  This chapter covers BOK3 Outcomes 19–Professional Attitudes and 21–Ethical Responsibilities.
- Chapter 4—Professional Engagement
  This chapter covers BOK3 Outcomes 8–Critical Thinking and Problem Solving and 20–Professional Responsibilities.
- Chapter 5—The Engineer's Role in Project Development
  This chapter covers BOK3 Outcome–8 Critical Thinking and Problem Solving 13–Design.
- Chapter 6—What Engineers Deliver
  This chapter covers BOK3 Outcomes 8–Critical Thinking and Problem Solving and 13–Design.
- Chapter 7—Executing a Professional Commission
  This chapter covers BOK3 Outcomes 7–Project Management and 20–Professional Responsibilities.
- Chapter 8—Permitting
  This chapter covers BOK3 Outcome 19–Professional Attitudes and 14–Depth in a Civil Engineering Area.
- Chapter 9—The Client Relationship
  This chapter covers BOK3 Outcomes 16–Communication and 19–Professional Attitudes.
- Chapter 10—Leadership
  This chapter covers BOK3 Outcomes 17–Teamwork and Leadership and 19–Professional Attitudes.

- Chapter 11—Legal Aspects of Professional Practice
  This chapter covers BOK3 Outcomes 11–Risk and Uncertainties and 20–Professional Responsibilities, as well as the additional legal aspects.
- Chapter 12—Managing the Civil Engineering Enterprise
  This chapter covers BOK3 Outcomes 10–Engineering Economics and 20–Professional Responsibilities.
- Chapter 13—Communicating as a Professional
  This chapter covers BOK3 Outcome 16–Communication.
- Chapter 14—Balancing Life, Family, and Career
  This chapter covers BOK3 Outcomes 18– Lifelong Learning and Outcome 19–Professional Attitudes.
- Chapter 15—Globalization
  This chapter addresses BOK3 Outcome 12–Breadth in Civil Engineering Areas.
- Chapter 16—Sustainability
  This chapter covers BOK3 Outcome 15–Sustainability.
- Chapter 17—Emerging Technologies
  This chapter addresses BOK3 Outcome 14–Depth in a Civil Engineering Area.
- Chapter 18—Human Relations Policies and Employment Laws
  This chapter covers BOK3 Outcomes 19–Professional Attitudes and 20–Professional Responsibilities.
- Chapter 19—Construction Management for Civil Engineers
  This chapter covers BOK3 Outcomes 9–Project Management and 11–Risk and Uncertainty.
- Chapter 20—Health and Safety Knowledge for Civil Engineers
  This chapter covers BOK3 Outcome 20–Professional Responsibilities.
- Chapter 21—What Civil Engineers Need to Know
  This chapter covers BOK3 Outcomes 8–Critical Thinking and Problem Solving, 12–Breadth in Civil Engineering Areas, 14–Depth in Civil Engineering Areas, 19–Professional Attitudes, 20–Professional Responsibilities.

1.7 SUMMARY

This *Civil Engineering Handbook of Professional Practice*, 2nd edition can provide guidance and resources for a fruitful and successful career in civil engineering.

The demands of society and the related high standards required by both ABET and ASCE present civil engineers and civil engineering educators with numerous challenges. Practicing civil engineers need to know about design; sustainability and other contemporary issues; risks and uncertainties; project management; communication; public policy; business;

leadership; teamwork; human resources laws/requirements; Occupational Safety and Health Administration (OSHA) laws/requirements; contract law; effective communications; and professional and ethical responsibility.

The Bachelor's degree in civil engineering represents the beginning of the journey in the civil engineering profession. The graduate civil engineer likely will encounter many high-risk, legal, ethical, and/or potentially dangerous situations requiring more knowledge, information, mentoring, and experience. The authors hope that the *Civil Engineer's Handbook of Professional Practice*, 2nd edition will provide both aspiring and practicing civil engineers, as well as civil engineering educators, with useful information that assists them in meeting the needs of society and achieving their own personal goals.

While every effort has been made to ensure the accuracy and completeness of the information provided in this handbook, readers are advised to consult with a qualified professional before applying any of the practices or techniques discussed to specific projects or situations.

BIBLIOGRAPHY

American Society of Civil Engineers. (2019). *Civil Engineering Body of Knowledge for the 21st Century*, 3rd edition. ASCE report, ASCE, Reston, VA.

American Society of Civil Engineers. (2008). *Civil Engineering Body of Knowledge for the 21st Century*, 2nd edition. ASCE report, ASCE, Reston, VA.

American Society of Civil Engineers. (2004). *Civil Engineering Body of Knowledge for the 21st Century*, 1st edition. ASCE report, ASCE, Reston, VA.

American Society of Civil Engineers. (2020). *Civil Engineering Education Summit: Mapping the Future of Civil Engineering Education*. ASCE report, ASCE, Reston, VA.

American Society of Civil Engineers. (2021). *Infrastructure Report Card*. ASCE report, ASCE, Reston, VA.

American Society of Civil Engineers. (2019). *Policy 465*. ASCE report, ASCE, Reston, VA.

American Society of Civil Engineers. (2006). *Policy 465*. ASCE report, ASCE, Reston, VA.

Galloway, Patricia D. (2008). *21st Century Engineer: A Proposal for Education Reform*. American Society of Civil Engineers, Reston, VA.

www.abet.org/about-abet/history.html (accessed July 20, 2022).

www.bls.gov/oco/ocos027.htm#outlook (accessed November 7, 2009).

https://www.futureworldvision.org/why-future-world-vision (accessed July 17, 2022).

MacLeod, Iain A. (2009). The Education of Innovative Engineers. Presentation made.

National Academy of Engineering. (2004). *The Engineer of 2020: Visions of Engineering in the New Century*. National Academies Press, Washington, D.C. ISBN-10: 0–309-09162–4.

Rockefeller Foundation's 2050 Forum. (2008). *Rebuilding and Renewing: 21st Century Infrastructure Agenda*, May 9, 2008, Lincoln Institute of Land Policy, Washington, D.C., May 2008, https://www.lincolninst.edu/es/news/press-releases/forum-national-plan-infrastructure-may-9-washington-dc.

CHAPTER 2

Background and History of the Profession

Big Idea

". . . lessons learned from the behavior and especially the failure of even ancient designs are no less relevant today . . . good design practice of engineers in centuries past can serve as models for the most sophisticated designs of the modern age."

—Henry Petroski

Key Topics Covered

- Civil Engineering's Historical Inheritance
- The Ancient Engineers
- Engineering in Medieval Times
- Engineering in the Renaissance and the Age of Enlightenment
- The Industrial Revolution
- Modern Civil Engineering

Civil Engineer's Handbook of Professional Practice, Second Edition. Karen Lee Hansen and Kent E. Zenobia.
© 2025 John Wiley & Sons, Inc. Published 2025 by John Wiley & Sons, Inc.
Companion website: www.wiley.com/go/hansen/CivilEngineersHandbook

2.1   BACKGROUND

Chapter 2 examines civil engineering as a profession and the significant contributions civil engineers have made to civilization. The chapter explores civil engineering's historical inheritance, provides examples of outstanding projects—from ancient to modern times—and profiles several legendary civil engineers. Knowledge of civil engineering history and culture helps civil engineers communicate the importance of their profession to the world.

Noted engineering historian, Henry Petroski, posits that engineering history is both history and engineering. Additionally, familiarity with civil engineering history can assist with the practice of the profession.

"The lessons of the past are not only brimming with caveats about what mistakes should not be repeated but also are full of models of good engineering judgment." [Henry Petroski, 1994]

2.2   CIVIL ENGINEERING'S HISTORICAL INHERITANCE

Much of the material in this section is derived from L. Sprague de Camp's seminal work, *The Ancient Engineers*. To begin learning about civil engineers' rich historical inheritance, we have to turn the clock back 6,000 years to the dawn of civilization. Mr. de Camp observes:

The first engineers were irrigators, architects, and military engineers. The same man was usually expected to be an expert at all three kinds of work. This was still the case thousands of years later, in the Renaissance, when Leonardo, Michelangelo, and Dürer were not only all-around engineers but outstanding artists as well. Specialization within the engineering profession has developed only in the last two or three centuries. [p. 9]

After 4000 B.C., when humans began to abandon the nomadic way of life, the need for water, permanent shelter, religious monuments and burial sites, and fortification emerged. Early river valley civilizations, such as those around the Tigris and Euphrates (Mesopotamia), Nile (Egypt), Indus (India), and Hwang-ho (China), required canal systems to irrigate surrounding land so that farmers could raise sufficient food to support the population

(Scarborough, Vernon L., 2003). Kings or rulers desired houses larger than huts of stone, clay, or reed; and priests wanted homes for the gods at least as grand (Morgan, William N., 2008). To protect the growing wealth of these early settlements, walls and moats needed to be constructed. These were the challenges that occupied the first engineers (Encyclopedia Britannica, Civil Engineering, 2009), (James, William, 1998).

Studying the Past Yields Valuable Lessons

". . . any lessons learned from the behavior and especially the failure of even ancient designs are no less relevant today, and the good design practice of engineers in centuries past can serve as models for the most sophisticated designs of the modern age. Indeed, ignoring wholesale the lessons and practices of the past threatens the continuity of engineering and design judgment that appears to be among the surest safeguards against recurrent failures."

—(Henry Petroski, Design Paradigms: Caase Histories of Error and Judgement in Engineering, p. 143)

2.3   THE ANCIENT ENGINEERS

Some early writing on stone and brick in Mesopotamia and Egypt has survived, but other written accounts of ancient engineering in those areas have perished. The same can be said about the documentation of the ancient engineering feats of the Persians, Indians, and Chinese. Because of the limited number of written accounts, relatively more is known about ancient Greek and Roman engineering. Around 100 B.C., several Greek writers created lists of the seven most magnificent engineering feats of which they were aware. Shown in Figure 2.1, the typical list included:

1. Great Pyramid at Giza, Egypt
2. Hanging Gardens of Babylon, Mesopotamia
3. Statue of Zeus at Olympia, Greece
4. Temple of Artemis at Ephesus, modern Turkey
5. Tomb of King Mausolos of Karia at Halikarnassos, Greece
6. Colossus of Rhodes, Mediterranean
7. Pharos Lighthouse of Alexandria, Egypt

Of the ancient wonders included on these lists of, the Pyramids of Egypt (circa 2700−1600 B.C.) alone survive in a recognizable form today. (See Figure 2.2 for more information on the Egyptian Pyramids.) The Greek writers could list only the wonders they had heard of, so the

FIGURE 2.1   Seven wonders of the ancient world.
Source: en:User:Slof - en:Image/Wikipedia Commons/CC BY-SA 3.0.

Great Wall of China, the dam at Ma'rib, which furnished water to a valley in southwest Arabia for about 1,000 years, the Buddhist stûpas of Sri Lanka, enormous domed structures over religious relics, and other feats of civil engineering are missing from the Greeks' lists (de Camp, L. Sprague, 1993).

Around 5,000 years ago, the typical Egyptian king or noble was buried in a rectangular, mud brick structure made of inward sloping walls and set over an underground chamber. Although unfired mud brick is a poor building material, these very early Egyptians learned that if they tapered the walls of their tombs inward from bottom to top, the walls did not crumble as quickly as if constructed vertically. Eventually these mud brick tombs gave way to stone structures; and although the reason for the sloping walls no longer existed, the stone tombs maintained the same form.

Around 2,700 B.C., the first engineer/architect known to us by name – Imhotep – emerged. Imhotep was a genius who was known as a builder, physician, statesman, writer, and overall sage. Together with his ruler, he embarked on improving the traditional tomb so that raiders would be less successful at entering. He and the king started with a stone tomb that was square rather than rectangular, 200' on a side and 26' high. Changes were made several times and the tomb was enlarged by adding stone to the sides. Before the second enlargement was finished, the king decided to build another level on top of the first. He changed his mind several times more, and the tomb that Imhotep finally built was comprised of six stages of decreasing size, or levels, over a burial chamber – the first step pyramid.

Succeeding generations of Egyptian kings also built step pyramids and eventually filled in the steps, creating true straight-sided pyramids. The largest of these was built around 2,500 B.C. near Giza, a town on the west bank of the Nile River, just upstream from Cairo. King Khufu, or Cheops as the Greeks called him, built his pyramid 756 feet square and about 480 high. The Great Pyramid is made of approximately 2,300,000 blocks of limestone, each weighing an average of two and one half tons, and is faced with a higher quality limestone.

Most of the stone for the pyramids was cut from local stone outcrops. Early scholars theorized that the stone was dragged on sleds to the building sites. Based on existing evidence – tomb paintings, ancient tools found in modern times, tool marks on stones, and quarries with blocks partially detached – rollers under sleds were not used. Rather, workers may have poured liquid, possibly milk, on the soil in front of the sled to improve slipping. More recently researchers have found through experimentation that large stones fixed inside ingenious wood crates can be rolled by several workers, not the many required to pull sleds. Finer limestone used for exterior facing and granite used to line chambers had to be moved down the Nile on barges and then lugged to sites. To position these stones, agricultural workers conscripted during off-seasons, not slaves, used elaborate levers and ramps.

As a pyramid rose, workers built a large earthen mound surrounding the structure. After one course was laid, the mound and accompanying ramp were raised to a new level. When the pyramid was complete, workers had to haul away this vast amount of soil; and masons standing on the ramp removed any irregularities in the facing stone.

The last Egyptian pyramids were built approximately 1,600 B.C., possibly because of the prohibitive cost of construction or liberalization of religious doctrines. But Egyptian engineers learned much about quarrying, shaping, and moving heavy stones, and this knowledge became part of the world's collective technological wisdom.

—Adapted from L. Sprague de Camp, *The Ancient Engineers.* pp. 20 – 36.

FIGURE 2.2   Origins of the Egyptian Pyramids.
Source: (a) Hajor/Wikimedia Commons. (b) Wikimedia Commons.

Though civilization in Mesopotamia, "the land between the rivers" in Greek, may have begun several hundred years before Egypt's, little remains of its monumental architecture. Mesopotamia comprised most of the area that is modern-day Iraq. In ancient Babylon, this land was predominately desolate and barren except where water from the Tigris and Euphrates rivers provided irrigation. According to de Camp:

In southern Mesopotamia, at the beginning of recorded history [5,000 to 6,000 years ago], the Sumerians—a people of unknown origins—built the city walls and temples and dug the canals that comprised the world's first engineering works. Here, for over two thousand years, little city-states bickered and fought over water rights.

[de Camp, L. Sprague. (1993)]

Unfortunately for those interested in history, the Mesopotamian plain contained no stone suitable for building; and the only timber available had to be brought down the Tigris River from the Assyrian hills. Kiln-dried or burnt brick was expensive because of the lack of fuel (wood) for kilns. Consequently, the predominate building material was sun-dried mud brick, which was strong when dry but crumbled when wet. Mesopotamian temples and palaces were faced with kiln-dried brick but interiors were sun-dried. Consequently, when cracks developed and were left untended, sharp winter rains penetrated the mud brick within and the buildings eventually disintegrated. The upper part of Mesopotamian public structures has disappeared but their rubble has protected the foundations beneath.

One marvelous Mesopotamian invention was paving. On processional ways that were regular features of cities, Mesopotamian engineers lay flat bricks in a bed of mortar made from lime, sand, and asphalt. Sandstone flagstones were placed on top of the bricks. Special rules governed these sacred streets; parking the odd chariot or other vehicle along such a road could result in impalement. Eventually pavement was applied to major thoroughfares and then to heavily traveled roads outside main cities.

Mesopotamia is also home of the earliest recoded stone bridge over a river. Previous bridges had been made of tree trunks, reeds, or inflated goatskins. This first stone bridge over the Euphrates River was 380 feet long and rested on seven piers constructed of fired brick, stone, and timber. Due to shifts in the river channel over centuries, the sizes of the bridge and its large piers, 28 feet by 65 feet in plan, are known because the piers have been excavated in modern times. Most large, ancient bridges were constructed in a similar way—the piers took up half the width of the river (Landel, John Gray, 2000).

Contemporary with the construction of the Egyptian Pyramids at Giza and the urbanization of Mesopotamia was the construction of Stonehenge. Stonehenge, located in what is now modern Britain, is a magnificent feat of ancient engineering and organization of human labor. Its real function and meaning are not yet clear, but the scale of effort and command of physical principles necessary to build it can be admired today. (See Figure 2.3 for more information on Stonehenge.)

Based on radio carbon dating, Stonehenge was constructed more than 5,000 years ago. Its ancient British builders were working on it at the same time the Egyptians were constructing the Great Pyramids at Giza. But what was it and what purpose did this collection of giant stones fulfill?

Part of a collection of remarkable stone circles in northwestern Europe, Stonehenge attracts a wide array of the curious interested in diverse topics like archeology, astronomy, meteorology, sacred geography, geomancy, and shamanism. In truth, knowledge from all of these fields is necessary to begin to explain something created by people who lived in the Neolithic (Stone) Age.

Stonehenge was the centerpiece of a culture that flourished on the Salisbury plain in what is now western England. The builders moved Sarsen Stones, the tallest uprights and lintels weighing 50 tons each, from Fyfield Down over 20 miles away. The smaller blue stones came from the Preseli Mountains, Wales, over 150 miles. Evidentially, this was a very special site!

The Sarsen Stones are a type of sandstone harder than granite. These stones were dressed with mauls weighing up to 65 pounds. Five trilithons [two uprights spanned by one lintel] originally stood in the center of the circle at a height of 17 to 25 feet. An approximately 100 foot diameter Sarsen Circle surrounded the trilithons. The bluestones formed an inner circle approximately 75 feet in diameter, as well as a horseshoe around the trilithons.

Early researchers strove to understand the geometry of the stones, and the position of the stones is linked to the rising and setting positions of the Sun, Moon, and stars. One stone has been called the *Heel Stone* for centuries; some thought this because there is an indentation on the stone resembling a heel print. However, in the Old Welsh language *ffriw yr haul* is phonetically very similar and means, "appearance of the sun." Students of Stonehenge do not always agree on what the *ghost in the machine* is. What is known is that people very much like ourselves prepared, surveyed, and marked-out the site; transported the megaliths [very large stones]; and erected them.

FIGURE 2.3 Stonehenge: An ancient mystery.
Source: Wikipedia Commons, Stonehenge: Ancient Temple of Britain.

2.3.1 Persian Engineers

Around 550 B.C., Cyrus the Great founded the Persian Empire, modern Iran, which ruled the Near and Middle East for more than 200 years. Persian kings did not rule from a single capital but maintained four capitals among which they moved. Darius, his son Xerxes, and his grandson Artaxerxes labored for decades in the 5th century B.C. to create a magnificent royal center at what the Greeks named Persepolis, "Persian City," one of the four capitals. In 331 B.C. Alexander the Great burned Persepolis and it was abandoned. Alexander's action strangely conserved more to be appreciated in our times because the other three capitals continued as great cities, and what existed has been demolished or built upon.

The Persians spread ideas about building, such as a system of irrigation, far and wide. One of their innovations, a ghanat, was adopted widely. As shown in Figure 2.4, a ghanat

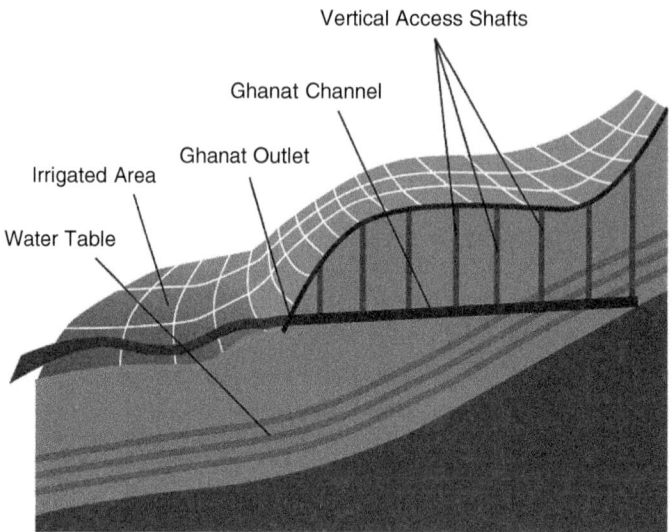

FIGURE 2.4   The Persian ghanat: Groundwater distribution.
Source: Wikimedia commons, Persian qanat.

is a sloping tunnel that conveys water from an underground source in a range of hills to a dry plain below. Less water is lost to evaporation than in an open-air aqueduct. To construct a ghanat, a line of vertical shafts is dug. The bottoms of these shafts are connected by the continuous tunnel (Vali-Khodjeini, Ali., 1995). To allow workers to maintain the tunnel or to draw water, other shafts are dug at an angle from the surface to the tunnel. The water is distributed into irrigation channels when the tunnel reaches its destination. Ghanats were a family effort that could take several generations to complete. The ghanat system is still in use in the Middle East and North Africa today (James, William, 1998)

The Persians also accomplished major feats in military bridge-building. In 480 B.C. in a campaign involving the famous battle at Thermopylae that pitted the Persian King Xerxes and his forces against Leonidas and his 300 Spartans, Persian engineers constructed a pontoon bridge over the Hellespont, a narrow straight dividing Europe and Asia Minor. The price of failure for many early engineers was quite high—the first bridge was torn apart in a storm and the engineers were beheaded. The new engineers built a bigger bridge with larger factors of safety. The new bridge was constructed of 674 galleys anchored in a double row connected by two enormous flax cables and four cables of papyrus. Planks were laid at right angles over the cables, brush was piled on the planks, and finally soil was piled on the brush. Xerxes's army of perhaps 150,000 soldiers and an equal number of noncombatant camp followers marched over the bridge. (Early historians may have exaggerated this figure; nonetheless the number of troops was enormous.)

Other ancient civilizations around the Mediterranean—the Phoenician, Carthaginian, Lydian, Hebrew, Assyrian, Minoan—accomplished major feats of engineering. Over 3,000 years

FIGURE 2.5   Sewers: Knossos, Crete, 3000–100 B.C.
Source: Unknown author/Wikipedia Commons/CC BY-SA 3.0.

ago on the island of Crete, seafaring kings built exquisite unfortified palaces with stone walls and post-and-lintel colonnades. Ceramic drain pipes carried water away from baths and toilets, creating some of the first sanitary sewer systems (see Figure 2.5). The palaces were destroyed by earthquakes and were rebuilt more splendidly. The Minoan civilization collapsed around 400 B.C. and these magnificent buildings with their advanced sewer systems fell into disrepair.

2.3.2   Greek Engineers

Around 1000 B.C., when Kings David and Solomon ruled in Israel, Aryan invaders began to attack the Aegean coastline and Mediterranean islands. Three to four centuries later, the aggressors and locals had mingled to form a new people, the Greeks. The Greeks were influenced by other people—the Egyptians, Babylonians, and Phoenicians (seafaring people whose home was the coast of current-day Lebanon). But the Greeks began doing something different—they were starting to connect engineering and pure science, freed for the first time from the supervision of priests, who until then had controlled intellectual life. One of the earliest examples of a scientific approach to physical and mathematical problems applicable to civil engineering is the work of Archimedes in the 3d century B.C. To this day Archimedes' Principle continues to inform our understanding of buoyancy and offers practical solutions, such as Archimedes' screw, which can be used in irrigation to transfer water from a lower source (lake, creek, river) to a higher level such as a ditch or canal (Hart, Ivor B., 1928).

In the 5th century B.C. during Greece's Golden Age, the leader of Athens—Pericles—commissioned leading artists, architects, and engineers to cover the Acropolis with temples, shrines, and statues. The Acropolis is a huge rock outcropping whose top was reached through a winding, processional path. At the top of this path, worshippers entered the site through a monumental gateway, the Propylaia. The Propylaia is notable for the wrought iron bars used to reinforce marble ceiling beams, among the first known use of metal structural members in a building. One of the temples constructed on the Acropolis was the Parthenon, a temple to the goddess Athena. The Greek temple style spread all over the Mediterranean and lasted for centuries. It was revived in Renaissance Europe and again in the 19th and early 20th centuries. The style often has been used in the design of art museums, banks, churches, and memorials.

In the 4th century B.C., Alexander of Macedonia—Alexander the Great—subdued all of Greece and later conquered much of the Middle East, including Egypt, Mesopotamia, and Persia. This was a period of intellectual fervor, travel and tourism, scholarship and research, invention, and intermarriage of people and cultures, not unlike our own time. Aristotle and Plato lived then. In Egypt, Alexander ordered the construction of the city of Alexandria. Alexander's successor created a splendid harbor and erected a magnificent lighthouse on a rocky islet called Pharos. The Pharos Lighthouse of Alexandria is one of the seven ancient Wonders of the World. A later ruler created an enormous library; and while it endured, the Library of Alexandria was the intellectual capital of the Mediterranean world.

Ancient Structural Systems and Success Factors

In the ancient world, building styles depended on locally available materials: clay, stone, and wood. Buildings of antiquity utilized one or a combination of four devices to support roofs or upper stories:

1. Corbel—an "arch" that requires no falsework or shoring. Stones are layered in courses from two sides, overhanging each previous course until the two sides meet in the middle.
2. Post and lintel—a system of vertical columns crossed by horizontal beams.
3. Arch and vault
4. Truss

Mesopotamia had lots of clay but no stone or wood and, thus, preferred the corbel or arch and vault construction. Egypt had stone and clay, while Greece and China had stone, clay, and wood; these civilizations favored post-and-lintel construction. Europe had abundant sources of wood and consequently developed the truss.

Underpinning the success of ancient engineers were three factors:

1. Intensive and careful use of existing principles and tools, such as the water level and astronomical observation
2. Unlimited labor and the power to organize and command it
3. A different perspective of time

Perhaps the most important is the last factor—the Ancients seemed to have infinite patience.

—Adapted from L. Sprague de Camp, The Ancient Engineers. pp. 26–31.

2.3.3   Roman Engineers

The founding of ancient Rome is traced to the 8th century B.C., and the fall of the Roman Empire dates to the 5th century. Roman engineering, like that of other defining empires, relied on intensive application of existing principles and tools, cheap labor, and time. Rome possessed plenty of raw materials in the form of clay for brick, stone, and timber; and because of its rapid expansion, it also had an abundance of slave labor. Romans devoted more resources to constructing useful public works than their predecessors and developed civil structures throughout their empire, including aqueducts, harbors, bathhouses, markets, bridges, dams, and roads. Some scholars argue that Rome contributed little to pure science; but Roman genius had more to do with pragmatism. Remarkable Roman statesmen, soldiers, administrators, and jurists built on others' scientific findings and artistic creations.

Principally, Roman engineering is civil engineering. Romans themselves developed new building methods, which continued from the early years of the Empire's expansion in the 4th century B.C. for nearly 800 years. By the 1st century B.C., Vitruvius—author of the only surviving book on engineering and architecture from classical antiquity—wanted his ideal architect (engineer) to be a scholar, a skillful draftsperson, a mathematician, a student of philosophy, familiar with historical studies, acquainted with music, not ignorant of medicine, knowledgeable of jurisconsults' responses, and familiar with astronomy—a point of view that resonates with the American Society of Civil Engineer's (ASCE) 2008 Body of Knowledge 2 (ASCE, 2008).

Although Vitruvius gives sound reasons for having all these skills—for instance, a knowledge of music is useful in tuning catapults by striking the tension skeins—the difficulty is the same as in all professions in all ages. Vitruvius' requirements are a counsel of perfection, because nobody lives long enough to learn everything that might be useful to him.

[de Camp, L. Sprague. (1993)]

Perhaps this is not dissimilar to the situation in which the 21st century civil engineer finds himself or herself.

Vitruvius's vision was grand; however, the scale of invention and innovation was small in ancient times. When a city was sacked, some creations could be destroyed and then they vanished for centuries. Innovations were created, lost, and recreated. For example, a palace in Beycesultan, southwestern Anatolia (Turkey), built in 1200 B.C. and excavated in 1954, had indirect central heating. Ducts beneath the floor suggested a central heating plant. Then no evidence of this innovation appeared for a thousand years. By the 1st century B.C., Roman engineers had harnessed the power of geothermal springs for use in public baths. The concept also was applied to baths in houses, which were equipped with under-floor ceramic ducts through which air heated by fire passed. Eventually, Romans applied this system to whole buildings.

About 300 B.C., Romans discovered concrete. They found that when sandy volcanic ash was mixed with lime mortar, a cement formed and dried rock-hard, even under water. Concrete was created by mixing this cement with sand and gravel. After centuries of exposure, some examples of Roman concrete are harder than many natural rocks. At first, Romans only used concrete in limited applications—like a superior mortar. Then it began to replace the brick and stonework it was helping to bond in walls and fences.

The Romans not only invented new construction materials and ways of combining new and old materials, they also created new architectural forms. They excelled in building secular rather than religious edifices. Romans improved the arch and vault, making public buildings adaptable due to large, clear spans and giving these buildings a feeling of spaciousness. They developed methods of erecting huge, well-constructed buildings in a fraction of the time and for far less expense than other ancient engineers. They also were able to adapt circular dome ceilings to square or rectangular buildings. This was accomplished through the use of *pendentives*, triangular sections of masonry leaning in from the rectilinear base to connect with the circular base of the dome. Later in Byzantium, pendentives were used extensively. Their temples, however, followed the Greek style of post-and-lintel structures. Yet a Roman innovation was to make columns solid rather than a series of "drums" erected one on top of the other to form pieced columns, as the Greeks had done.

Romans also were master road builders. Only since the advent of the car have road-building standards returned to anything close to Roman engineering criteria. Today the Appian Way, constructed in the 4th century B.C., still exists southeast of Rome. Roman engineers designed their roads to require little maintenance and to last a minimum of 80 to 100 years; obviously, some lasted longer. Primarily, Roman roads were intended to enable an army to move swiftly. During most of the Roman Empire, the army was composed of heavy infantry. So Roman engineers were more concerned with providing a firm footing for marching soldiers than hoofed animals. Important roads were paved, and most secondary and provincial roads were graveled. An unpaved strip might be included on either side of a paved road. Romans preferred to build roads as straight as land contours allowed and to go over hills rather than around, even if this meant grades of 20 percent.

Roman surveyors laid out routes using simple instruments, such as water levels and plumb-bobs, to establish horizontal lines. To determine right angles, they used a pair of boards nailed to make a right-angled cross, which they mounted on a post. Then they leveled and sighted along the cross pieces. A Roman paved road has been likened to a massive wall lying on its side. They started with a trench several feet deep; and if the soil was not firm at that depth, they drove piles. The rest of the road construction depended on the importance of the route and the availability of local materials. A major, fully paved road in Italy might be made of five layers totaling 4 feet thick and 6 to 20 feet wide. First was a layer of sand, mortar, or both. Then was a layer of small squared stones set in cement or mortar. Next was a layer of gravel set in clay or concrete, followed by a layer of rolled concrete made with sand aggregate. On top of everything were large blocks of hard rock, dressed on their upper surfaces, set in concrete.

The Romans also excelled in bringing water to their cities (see Figure 2.6). They were not the first to build aqueducts (Scarborough, Vernon L., 2003). The Mesopotamians, Greeks, and Phoenicians all had constructed them, but the Romans built more and bigger aqueducts. Rows of arches remaining from Roman aqueducts can be seen today in Italy, France, Spain, North Africa, Greece, and other locations in Asia Minor. Parts of several, those around Rome, Segovia (Spain), and Athens still function. Why did the Romans build so many? Most large cities are built on rivers; but even without the knowledge of bacteria, the ancients knew that

FIGURE 2.6 Roman aqueduct: Pont du Gard, Nîmes, France.
Source: Unknown author/Wikipedia Commons/CC BY-SA 3.0.

spring water was better than river water. Additionally, Romans had inherited Greek ideals of civilized living that called for public fountains, baths, and gardens, all of which required water.

Roman engineers built aqueducts on a simple pattern consisting of a series of small round arches bearing on tall piers of stone or brick. Above was a water channel of concrete covered with an arched or gabled roof. When the aqueduct had to cross an exceptionally deep gorge, two or three rows of arches were erected on top of each other. Sometimes two or three channels shared the same arcade. Because the water flowing in open channels was moved by gravity all the way from the source to the point of distribution, the channels needed a fairly consistent downgrade of two to three feet per mile. The famous Pont du Gard in Nîmes, France, has three superimposed arcades. While arcades were the most conspicuous part, the vast majority of the Roman aqueduct system was in conduits and tunnels.

A Roman Civil Engineer's Field Report

Nonius Datus on the Difficulties of Building a Tunnel for an Aqueduct Saldae, Algeria, 152 A.D.

I found everybody sad and despondent. They had given up all hopes that the opposite sections of the tunnel would meet, because each section had already been excavated beyond the middle of the mountain. As always happens in these cases, the fault was attributed to me, the engineer, as though I had not taken all precautions to ensure the success of the work. What could I have done better? For I began by surveying and taking the levels of the mountain, I drew plans and sections of the whole work, which plans I handed over to Petronius Celer, the Governor of Mauretania; and to take extra precaution, I summoned the contractor and his workmen and began the excavation in their presence with the help of two gangs of experienced veterans, namely, a detachment of marine infantry and a detachment of alpine troops. What more could I have done? After four years' absence, expecting every day to hear good tidings of water at Saldae, I arrive; the contractor and his assistants had made blunder upon blunder. In each section of tunnel they had diverged from the straight line, each towards right, and had I waited a little longer before coming, Saldae would have possessed two tunnels instead of one.

—Ivor B. Hart, The Great Engineers. p. 24.

2.3.4   Indian Engineers

Further to the east of the Mediterranean, kings kept up roads and irrigation systems, especially the great canal network in Babylonia. Iranians built many bridges, including one constructed in the 8th century near Susa that had abutments made of iron slag and lead, using these materials as concrete. East of Iran lies India, but much less is known about the ancient history of India. When the Persian King Darius conquered the Punjâb in approximately 515 B.C., Persian construction techniques were introduced into India; but the Indians continued to prefer to build in wood than stone. Darius's arrival corresponded with the life of Buddha, and Buddhism eventually brought changes to Indian building. Previously, Indian religious structures had been unpretentious, wooden religious shines. With the advent of Buddhism came monasteries, stupâs, and rock temples. These new structures required different building materials, and wood gave way to brick and later stone.

Indian monasteries either consisted of numerous cells built around a compound or were stepped in tiers up mountainsides. Stupâs housed relics of Buddhist saints. As can be seen in Figure 2.7, a stupâ's main feature was a domed structure over an actual relic. Also, each of the four sides of the base had a symbolic gateway. Some stupâs, such as those in Sri Lanka, were very grand, their domes having 300-foot diameters and the entire monuments' height reaching 250 feet. In the 2d century B.C., the foundations of one such stupâ were constructed of layers of stone, clay, and iron, all compacted by elephants wearing leather boots. Rock temples were built into the sides of rocky hills, similar to the one built by Rameses II at Abu Simbel, Egypt. Over many centuries, Hinduism and Buddhism vied for dominance and Hindus also constructed many magnificent temples carved from rock hillsides. Other Hindu temples were large compounds accessed through monumental gateways.

FIGURE 2.7   Buddhist Stupâ, Polonnaruwa, Sri Lanka.
Source: Bernard Gagnon/Wikimedia Commons/CC BY-SA 3.0.

Indians used post-and-lintel construction with domes and arches. Rather than mortar, Indian builders preferred to use iron dowels to join large stones. In fact, ancient Indians knew the secret of good steel. In Roman times, Indian steel was exported widely. Then and later, Indian steel found its way to Damascus, where it was made into sword blades that became famous for their strength and durability. The Iron Pillar of Dehli, dating approximately 415 A.D., was a 24-foot shaft that bore a manbird statue, the steed of the Hindu god Vishnu. The figure has been lost, but the column remains. Around the same time, Indians were making suspension bridges supported with iron chains.

2.3.5   Chinese Engineers

Due to great barriers such as jungles, mountains, deserts, forests, and seas, China remained largely cut off from the activities of the ancient Mesopotamians, Egyptians, Greeks, and Romans to the west. Based on archeological remains, a civilization similar to that of ancient Sumer existed in northwest China around 2000 B.C., located near a narrow band of passable territory that became a trade route. Another civilization arose approximately 500 years later in Hunan, also connected to a trade route. Once established, these trade routes had more or less continuous traffic. However, traders usually carried goods from point to point, perhaps over a few hundred miles. Traders traveled several days from home and then returned. The route acted as a kind of filter, through which goods passed more easily than ideas. Therefore, Chinese engineers largely were cut off from western influences.

China had many building materials at its disposal and was not limited to clay bricks as were the ancient inhabitants of Mesopotamia. Stone foundations with wood and occasional brick superstructures topped by clay tile roofs were usual. The Chinese also knew about the barrel vault and used wood post-and-lintel construction; but they did not use the truss. By the 4th century B.C., the Chinese had discovered how to make cast iron. By the 10th century A.D. continuing to the 15th century A.D., the Chinese constructed pagodas, memorial towers adjacent to temples that were derived from the tupa form, entirely of cast iron. They also supported suspension bridges with cables made of bamboo fiber. Through communication with India via Buddhist monks, bridges suspended from iron chains began to appear in China in the 8th century A.D.

Ancient China was not as large as it is today; most development took place in the north-central part of modern China. There were periods of relative unity and also of great division, with lesser rulers opposing the dynasty in power and imposing a sort of local feudal power. For centuries, nomadic peoples—Huns, Avars, Turks, Uighurs, Tartars, Mongols, and Uzbeks—invaded and conquered parts of China. Finally, around 220 B.C. the king of the Tsin people conquered all other contending states and founded the first centralized, autocratic rule. This emperor, Ch'in Shih Huang Ti, undertook the largest single engineering work of the ancient period—the Great Wall of China—in order to hold back the invaders. If the ancient Greeks, who compiled lists of the Seven Wonders of the World, had known of this magnificent creation, they certainly would have included the Great Wall.

As the crow flies, the Great Wall is 1,400 miles long; taking into consideration curves, branches, and loops, it stretches over 2,000 miles (3,200 kilometers). Under the direction of General Meng T'ien, construction of the wall began by first laying out farms along the route to supply workers with food. Ancient engineers had to think a lot like military generals planning for food and supplies to care for the large labor force needed to accomplish such enormous tasks. Construction varied in different sections depending on the existence of previously constructed wall sections and availability of local building materials. In most places the wall was 30 feet high, 25 feet wide at the base, and 15 feet wide at the crest. The paved road that capped the structure had a 6-foot crenellated parapet (wall with a zigzag top) on the invader side and 3-foot crenellated parapet on the homeland side. There were watchtowers at regular intervals, averaging 35 feet square and 45 feet high. In most sections, the core of the wall was rammed earth or rubble faced with cut stone, fieldstone, or brick and mortar (see Figure 2.8). The masonry was excellent—the emperor ruled that any worker who left a crack into which a nail could fit should be beheaded instantly. Quality problems are generally handled with a little more relaxed attitude today. An inexperienced autocratic manager might go as far as firing (beheading) a noncompliant engineer, but a more experienced manager would likely counsel the engineer about the need for improvement and quality control.

FIGURE 2.8   Great Wall of China at Jinshanling.
Source: Georgio/Wikimedia Commons/CC BY-SA 3.0.

2.3.6   African Engineers

Nearly every civil engineer is aware of the Egyptian pyramids. The inspiration for pyramids may have come from the practice of burying dead under a mound of earth and stones (tumulus, burial mound) in parts of Africa and Asia. The Great Pyramid of Giza, created almost 5,000 years ago, is the oldest and only enduring edifice listed as one of the original Seven Wonders of the World. It remained the tallest human-made structure on earth for thousands of years. Other impressive pyramids, such as the Pyramids of Meroë built at the time of the Kushite Kingdom, developed to the south of Giza (see Figure 2.9).

Bordering the Mediterranean, the later Roman Empire extended across the top of the African continent. Roman-engineered cities existed in Tunisia, Algeria, and Morocco. Though smaller than the Colosseum in Rome, the Amphitheater of El Jem, Tunisia (Roman Thysdrus), could seat 35,000 people. However, the sociopolitical arrangement of mainland Africa contrasted sharply with that of the Romans. Hundreds of self-reliant groups lived in the arid sub-Saharan zone, tropical savannahs, coastal forests, and in quiet river basins (Gates, Henry Louis, et. al, 2017).

Because of the wide variation in climate and landscape, there was no pan-African style of building. Construction methods included *banco*, a wet-clay process similar to coil pottery, and other readily available building materials—stones, wood, grass, animal skins. Groupings of homesteads and villages reflected social structures and functions. Each family unit had a

FIGURE 2.9   Pyramids of Meroë, Sudan; Great Mosque of Djenné, Mali; Sankore Mosque, Timbuktu, Mali; Kasbah Gate, Marrakesh, Morocco; Bete Giyorgis, Lalibela, Ethiopia (Clockwise from upper left). *Source*: (a) Georgio/Wikipedia Commons/CC BY-SA 4.0. (b) Andy Gilham/Wikipedia Commons/CC BY-SA 3.0. (c) Anne and David/Wikipedia Commons/CC BY-SA 4.0. (d) Bernard Gagnon/ Wikipedia Commons/ CC BY-SA 3.0. (e) calflier001/Wikipedia Commons/CC BY-SA 2.0.

grinding house and granary, stable, and beer store; these buildings were grouped and linked by straight or enclosing walls.

Eventually, many cities and hubs developed around the Sahara. The use of domesticated camels enabled expansion of extensive trade routes, linking these trade centers. According to Harvard professor, Dr. Henry Louis Gates, Jr.:

From the 10th to 15th centuries, north and west Africa underwent a dramatic transformation. Celebrated cities arose, including Timbuktu [Mali], Marrakech [Morocco], and Ile-Ife [Nigeria]. Great scholars wrote books that filled libraries. Wealth from trans-Saharan trade grew at a spectacular speed, drawing attention from afar to the wonders of the African continent . . . Farmers, traders, and nomads from the barren fringe of the Sahara Desert would build some of the most advanced civilizations in history (see Figure 2.9).

Other major sites developed in East Africa. To the north of the River Jordan, in a mountainous region at the center of current day Ethiopia, lies a 12th century complex of eleven monolithic churches carved out of rock. According to the United Nations Economic, Scientific and Cultural Organization (UNESCO):

The churches were not constructed in a traditional way but rather were hewn from the living rock of monolithic blocks. These blocks were further chiselled out, forming doors, windows, columns, various floors, roofs etc. This gigantic work was further completed with an extensive system of drainage ditches, trenches, and ceremonial passages, some with openings to hermit caves and catacombs.

Biete Medhani Alem, with its five aisles, is believed to be the largest monolithic church in the world, while Biete Ghiorgis has a remarkable cruciform plan. Most were probably used as churches from the outset, but Biete Mercoreos and Biete Gabriel Rafael may formerly have been royal residences. Several of the interiors are decorated with mural paintings (see Figure 2.9).

Further to the south in modern-day Zimbabwe on a high plateau between the Zambezi and Limpopo rivers, lie the Great Zimbabwe ruins. The Great Zimbabwe ruins are what is left of a once thriving religious and civic center. Starting in the 11th century for approximately 300 years, the ancestors of today's Shona people, who currently inhabit that area, enjoyed a thriving cattle-based economy and traded luxury goods with outsiders—beads and pottery fragments have been found from China, Persia, and other Middle Eastern countries. The Great Zimbabwe site offered abundant resources: reliable rainfall, freedom from tsetse flies and malaria, ample woodlands for timber and fuel, arable soil for crops, open grasslands for grazing cattle, minerals, and gold.

From north to south, the Great Zimbabwe site is approximately one half mile (800 meters). It can be divided into three distinct zones: structures atop a kopje (ridge of granite); a Great

FIGURE 2.10   Great Zimbabwe, Zimbabwe.
Source: (a) Jan Derk/Wikimedia Commons/Public Domain. (b) Macvivo/Wikipedia Commons/CC BY-SA 3.0.

Enclosure on a flat granite shelf across a valley from the kopje; and a number of smaller structures on the shallow slopes of the valley. The Great Enclosure, or Elliptical Enclosure, is the largest ancient structure in southern Africa. The builders used stone blocks occurring naturally and made others by heating and then quenching granite. They also made use of puddled clayey soil mixed with daga (a fine aggregate), which set hard in the sun. This material functioned as wall plaster as well as flooring covering (see Figure 2.10).

2.3.7   American Engineers

Although war and trade connected Europe, Asia, and Africa, the American continents stood alone from the rest of the ancient world. Of course, the earliest Americans came from Asia across what is now the Bering Strait approximately 30,000 years ago. Once these inhabitants had settled in North, Central, and South America, they adopted a wide variety of building materials and systems. The earliest Americans built stone domes braced with whale bone, igloos made from snow and ice, gabled cedar houses, and partially sunken pit houses, for protection against the cold and wind. In the Southwest of the United States there is evidence of flood irrigation systems as early as several centuries B.C.

By the 5th century B.C., in central Mexico and the Gulf Coast the social order involved a ruling class with priests who were experts on the calendar and weather. The priests interceded with the divine on behalf of the agriculturally based population. Large monumental structures began to be constructed that magnified the importance of religion and the state. On the island of La Venta, located in the mangrove wetlands in the Mexican state of Tabasco, a rounded pyramid was constructed, perhaps, earlier than the 5th century B.C. Apparently, La Venta was a civic and religious center where a small number of priests and possibly a related labor force were housed.

La Venta foreshadowed the developments at Teotihuacán, a magnificent religious center and premier market town located 25 miles (40 kilometers) northeast of modern-day Mexico City. One can visit Teotihuacán today and experience its principal characteristics, which were developed between 100 B.C. and 200 A.D. Teotihuacán's axis, now called the "Avenue of the Dead," is approximately 3 miles (5 kilometers) long. At the north end of this axis is the Pyramid of the Moon and along the east side are the Temple of Quetzalcóatl, with its famed feathered serpent heads, and the towering Pyramid of the Sun. Originally, there were hundreds of smaller platforms along this axis. As with many ancient sites, the building at Teotihuacán seeks to capture the nature of the cosmic order.

As shown in Figure 2.11, both the Pyramid of the Sun and the Pyramid of the Moon are terraced. The Pyramid of the Sun was made of horizontal layers of clay faced with unshaped stones. The newer Pyramid of the Moon was built with a core of tufa (volcanic stone) piers and rubble, which filled the shafts between the piers. Angled walls buttressed the core and determined the slope of the main terraces.

To the south, in the Yucatán Peninsula of Mexico, Guatemala, San Salvador, and Honduras, the Maya culture reached its zenith between 600 A.D. and 900 A.D. Maya land is exemplified by temples—stepped stone pyramids with temples set atop, ball courts, and clusters of one-story buildings sometimes referred to as "palaces," but whose actual function is not

FIGURE 2.11   Teotihuacán, Mexico D.F.
Source: Selefant/Wikipedia Commons/CC BY-SA 3.0.

clear. Occasional burials have been found inside these temple pyramids, but their primary purpose was different from Old Kingdom Egyptian pyramids. The temple on top of Mayan pyramids contained a two-room sanctuary where human sacrifices were made. The temple used post-and-lintel construction topped with corbel vaults. Steel hard sapodilla wood formed the lintels and the walls were made of rubble, lime mortar, and a casing of cut stone. Several million Mayans built ritual centers at Palenque, Chichén Itzá, Uxmal, Tikal, and Copan, and many lesser sites.

Further to the south and many centuries later, the Incas flourished in an immense territory including all of modern Peru, Ecuador, and northern Chile. With their capital in Cuzco, Peru, the Incas had been active in the Andean highlands since 1200 A.D. When their toughest adversaries—the Chimu—submitted, the Incas found themselves the uncontested masters of an enormous domain. The Incas and the Romans had much in common.

At the apex of their imperial power, the Incas ruled a superbly organized and well-administered domain, which was connected by a vast network of roads. All roads started in Cuzco and spread out to the four quarters into which the empire was divided, not on the basis of compass points but on the land's topography. The 3,100-mile (5,000-kilometer) royal road cut through the Andes. Draft animals and the wheel were not known, so the roads were not paved. Incan roads accommodated people and llamas by tunneling through spurs, becoming stairs at sharp ridges, and providing stone-lined causeways in swamps. Suspension bridges across valleys were made of enormous plant-fiber cables anchored by stone towers. At regular intervals, there were posts for runners carrying official messages and resthouses for bureaucrats and merchants. Masonry walls of perfectly matched polygonal blocks characterized Inca building, and examples can be seen in Cuzco today. See Figure 2.12 for more information about Incan civil engineering.

Back in the Valley of Mexico, the Aztec federation was the strong power. Contemporaries of the Inca, the Aztecs had a dazzling metropolis the size of London. Tenochtitlán is the site of modern-day Mexico City. The Aztecs were latecomers from the north of Mexico who emerged as a cohesive group in the area around 1200 A.D. They established Tenochtitlán as their capital in about 1325 A.D. on an island in Lake Texcoco, a salt lake. By 1450 A.D. they occupied a position of primacy in central Mexico. By filling large containers of wickerwork (chinampas) with mud, they reclaimed wetlands and turned them into arable land. Three causeways that doubled as levees lead from terra firma to the central plaza of Tenochtitlán, where twin temple pyramids replaced earlier shrines in the 1480s A.D. (Wright, Kenneth R., 1999). A fourth causeway stopped at the eastern bank of the island. Fresh water was brought to the city from Chapultepec and Coyoacán via aqueducts. Of course, most of us know the story of Cortés and the Spanish invasion that occurred in 1519:

Bernal Díaz del Castillo, who was there, left a vivid description of what they saw. He speaks of the communities along the shoreline, of boats on the lake bringing food-stuffs and carrying out merchandise, of terrace houses, of the aqueduct coming in from Chapultepec . . . .Of the market of Tlatelolco, he writes: "There were among us soldiers

Machu Picchu, the most well-known Inca archeological site, was home to a permanent population of 300 residents; but the inhabitants grew to 1,000 when the Inca emperor was in residence. Situated near the headwaters of the Amazon River in the Peruvian Andes Mountains, Machu Picchu was inhabited primarily between 1450 – 1540 AD, prior to the arrival of the Spanish Conquistadores.

Due to its effectively engineered foundation and drainage systems, the royal retreat survived – abandoned in a South American rainforest – for over 400 years. Hiram Bingham, a professor from Yale University, 'discovered' Machu Picchu in 1911. The city that he and other 20th century scientists investigated was nearly in the same condition as it had been four centuries earlier.

Macho Picchu's engineered drainage system and foundations are the secret of its longevity. Without good foundations and drainage, many of its buildings would have crumbled and its agricultural terraces would have been unrecognizable due to high levels of rainfall, sheer slopes, slide-prone soils, and subsidence.

The engineers and builders of Machu Picchu gave serious consideration to both surface and subsurface water drainage. Extensive excavations have shown a deep subdrainage system under the agricultural terraces, as well as urban and agricultural drainage channels integrated with stairways, walkways, and temple interiors. Additionally, there are over 100 strategically placed drain outlets in numerous stone building and retaining walls. Like many other early engineers, the Inca builders constructed Machu Picchu to last an eternity, perhaps the original version of sustainable engineering.

FIGURE 2.12   Machu Picchu–An Inca Engineering Marvel.
Source: Wright et al. (1999)/from American Society of Civil Engineers.

who had been to many parts of the world, to Constantinople, to the whole of Italy and to Rome, and they said they had never seen a market so well organized and orderly, so large, so full of people."

[Kostof, Spiro. (1995).]

No wonder the Spaniards found Tenochtitlán so impressive. At the time of the Spanish invasion of the Americas, Europe was just awakening from the medieval era and the Renaissance was preparing to make its debut.

The Power of Ancient Building

To chart a place on earth—that is the supreme effort of the built environment in antiquity . . . . To mediate between cosmos and polity, to give shape to fear and exorcise it, to affect a reconciliation of knowledge and the unknowable—that was the charge of ancient architecture.

—Spiro Kostof, *A History of Architecture: Settings and Rituals*, 2d edition. pp. 240–241.

2.4   ENGINEERING IN MEDIEVAL TIMES

The term "medieval" literally means "between ages" and is used to describe the time in Western Europe between the end of the Roman era and the beginning of the Renaissance in the 15th century. Of course, the people living then had no concept that they were between anything—except perhaps a rock and a hard spot.

Much has been said of the fall of the Roman Empire, usually dated 476. While civilization continued in the eastern Mediterranean, Iran, Iraq, India, and China as before, the fall of the Western Roman Empire was no small event. Due to the lack of a strong central government, Roman roads, aqueducts, and harbors fell into ruin over a vast area. In the West, communities demolished Roman buildings to make fortifications and dismantled roads and bridges to slow down marauding Goths, Germans, and Vikings. Literacy almost vanished, science became superstition, and engineering deteriorated to rule-of-thumb craftsmanship.

In the Byzantine Empire, which was an extension of the Roman Empire in Asia Minor, certain trends continued that had started when the Roman Empire was united; but these had to do more with governance and religion than Roman engineering. In the late 11th century when the Crusaders arrived in Byzantium, the capital of which was Constantinople or modern-day Istanbul, they had little in common with their allies. The Byzantines had also spent centuries battling invaders. Intellectual activity was the domain of churchmen, some of whom did scientific work; the general attitude toward science, however, was one of indifference or hostility.

Meanwhile, in the 7th century a religious revolution led by Muhammad ibn-Abdallah took place in Arabia. Within one century Islam had spread from Spain to Turkestan. These invaders were quicker to master the arts of those whom they conquered than the Germanic people who overran the Western Roman Empire. Starting in approximately 750 for a century and a half, the caliphs (rulers) in Baghdad employed scholars to translate Western wisdom into Arabic. For the two previous centuries the Persians had done the same at Jundishapur, translating Greek and Sanskrit into their language. Thus, the Middle East became the intellectual center of the Mediterranean-facing world.

In terms of building, the Arabs continued using the system of fortifications, walls with battlements and towers, developed by the Romans and Byzantines. The mosque was a distinctly Muslim style of building that used domes and arches. Another uniquely Muslim development was the minaret—a tall, slender tower from which the public are called to prayer. The need for and interest in irrigation and canal building continued.

In Europe during what was once called the Dark Ages, between the 6th and 10th centuries, engineering and architecture stopped being recognized as professions. Design and construction were carried out by artisans, such as stone masons and carpenters, rising to the role of master builder. Knowledge was retained in guilds and advances in technology came slowly. For many centuries in Western Europe, construction in stone became rare while wood and plaster were common, resulting in the half-timbered medieval building style. Churches were constructed in the Romanesque style. These were rather plain, massive stone buildings with small windows and many round arches.

The 12th and 13th centuries were a period when conflicts between the major monotheistic religions, Christianity and Islam, and schisms within them were an everyday reality. There were frenzied outbreaks of religious hysteria and fanaticism, including massacres of "heretics." European feudal lords fought incessantly. However, engineering began to regain some of the ground lost after the fall of Rome. Scholars pondered the nature of motion, force, and gravity; and Medieval builders made advances in structural forms. In addition to the semicircular arch of the Romans, the Islamic pointed arch was introduced. Another advance was the use of the truss to support roofs. Unfortunately, no one could analyze these structures so Medieval roof trusses had unnecessary members that contributed to visual clutter but nothing to the trusses' load-carrying capacity.

The most significant engineering achievement of the time, however, was the development of the Gothic cathedral. The word "Gothic" meant *barbarous* to the Italians (due to the name of one of the early invading ethnic groups, the Goths), but the style spread over most of Europe. As shown in Figure 2.13, Gothic cathedrals were characterized by soaring vaulted interiors and large stained-glass windows. In anticipation of modern skyscrapers, the structure of the Gothic cathedral was a skeleton, represented by piers and flying buttresses. The walls were used to keep out the weather, not as structural support. Vaults were developed that enabled clear spaces of over 100 feet high. Lacking scientific principles, Medieval builders relied on trial and error. The roof of Beauvais Cathedral with a ceiling of 154 feet, the tallest of all Gothic cathedrals, collapsed twice. These massive undertakings could take several generations to complete.

The other noteworthy building type of this period was the fortified castle. Feudal warfare encouraged castle building. Until the advent of gunpowder, these edifices were so successfully engineered that they could withstand sieges for months and often were captured only through treachery. One of the best preserved European-style castles, Kerak des Chevaliers, was built in modern-day Syria for the Knights Hospitallers of St. John in the 12th century A.D. Ironically, the finest Medieval Muslim palace remaining today is the Alhambra, in Granada, Spain.

Medieval times also saw advances in the use of water wheels. The ancients had used water wheels for raising water and for milling grains. The notebook of a 13th-century craftsman shows a water-powered sawmill. In the later Middle Ages, water power also was applied to the bellows of smelting furnaces, to trip hammers for crushing ore or bark in tanneries, and to grinding and polishing armor and other metal wares.

Improvements also were made in canal building. Canals enabled people and goods easier movement than did the existing rutted, unpaved roads; and the development of the lock

FIGURE 2.13   Chartres Cathedral, France: Gothic masterpiece.
Source: Benutzer:Honge/Wikimedia Commons/CC BY-SA 3.0.

changed everything. The origins of the canal lock are uncertain, but this innovation dates to the late 14th century in The Netherlands or Italy. In the 1450s the engineer Bertola da Novate put forward-looking ideas about locks into practice:

> *The dukes and republics of North Italy kept Bertola, the ablest canal builder of his time, busy all his life digging canals for them. Sometimes they quarreled over who should have priority on his services. His only trouble was that his workmen sometimes could not understand his advanced concepts.*

[de Sprague, p. 381]

2.5   ENGINEERING IN THE RENAISSANCE AND THE AGE OF ENLIGHTENMENT

The term "Renaissance," which means *rebirth*, applies to Western Europe in the 15th through 16th centuries. In a narrow sense, the name refers to the revival of learning that took place in that period. Fashionable people had at least a veneer of scholarship. Study of classical antiquity, the writing and architecture of Greece and Rome, became vogue. However, many other sweeping changes also were taking place: the Reformation, world exploration, the downfall of the old astronomy that put Earth at the center of the universe, and the creation of the first patent systems for encouraging innovation.

Engineering again grew to be respected, and engineers became famous and, sometimes, well paid. They were no longer anonymous craftsmen; they promoted themselves and were not shy about arguing with employers or rivals. One of the earliest engineers of the Renaissance was the Florentine Filippo Brunelleschi. He mastered perspective drawing and competed for and won the commission to build the famous dome on Florence's cathedral, Santa María del Fiore (shown in Figure 2.14), among other accomplishments. Brunelleschi first competed for the award in 1407, received the order to build in 1419, and finished the task in 1436. The entire cathedral is 351 feet high, and the dome is 105 feet high (approximately ten stories) and 143 feet in diameter. The City of Florence also gave Brunelleschi the first known patent, for a canal boat fitted with cranes capable of moving heavy cargo. Like others of the same period— Leonardo da Vinci, Michelangelo Buonarroti, Andrea Palladio—Brunelleschi had to serve as both a civil and military engineer.

Most early Renaissance engineers achieved fame through word of mouth. Later in the 15th century, the printing press helped to disseminate engineering knowledge. An Italian engineer/architect/painter/philosopher/musician/poet, Leon Battista Alberti, wrote a book in Latin on rules of thumb for the proportions of structures, such as bridges. This work

FIGURE 2.14   Florence Cathedral: Brunelleschi's Renaissance dome.
Source: trialsanderrors/Wikimedia Commons/CC BY 2.0.

originally was published in 1452 and circulated in manuscript form among Alberti's friends. Later, however, it was translated into Italian, French, Spanish, and English. In 1472, Roberto Valturio published a book that surveyed the state of military engineering. In the 1580s, Palladio, who had perfected the bridge truss, wrote about that subject and others in *I quattro libri dell' architectura* (*The Four Books of Architecture*).

Through the Spanish Inquisition (starting in the 15th century and lasting several hundred years) and Counter-Reformation (16th to mid-17th centuries), a dim light shone on science, as demonstrated by the threats of torture to which Galileo Galilei was subjected for proposing that the Earth did indeed rotate around the sun, rather than the other way around. Italy did not fare well during this period because in addition to changing religious views, armies from France, Spain, and the Holy Roman Empire (centered in Vienna, Austria), kept waging war there. But technical progress did continue elsewhere. About the 16th century people began to write about "modern discoveries"; and by the 18th century, engineering schools appeared in France.

This was the Age of Enlightenment (18th century) and many unforeseen changes were taking place. In an Enlightened Europe there was a strong appetite for attack on the church. The church began to lose power to nations; the Jesuits were expelled from Portugal, France, Spain, and Naples. As in the early Renaissance when Henry VIII of England seized monastic property, many of the Jesuits' holdings became "available." Access to this property, social unrest, capitalism, and the notion that existing structures could be replaced with more up-to-date ones helped to establish land as a liquid, negotiable commodity.

2.6   THE INDUSTRIAL REVOLUTION

At the close of the 18th century, the first stirrings of the Industrial Revolution were beginning to be felt. In England, earlier than in the rest of Western Europe, the transition from an agrarian, handcraft-based economy to a machine-dominated economy was underway. The trend had earlier roots, but mechanized labor, inanimate power—particularly steam—and inexpensive raw materials accelerated dramatic changes. Workers were moving away from home-based (cottage) industry and shops to mills and factories. In England the countryside was under assault as scores of towns emerged around country plants making anything from cast iron to cotton cloth. In the country, industry could flourish away from the influence of guilds and government regulations.

Up until the late 18th century, military engineers had undertaken the construction of public infrastructure in support of expanding industry. However, in 1768 an Englishman named John Smeaton is credited with being the first person to call himself a civil engineer. By describing himself as a "civil engineer," Smeaton identified a new and distinct profession that encompassed all nonmilitary engineering. Smeaton's work was backed by thorough research, and he became a member of the prestigious Royal Academy of Engineering.

In 1771, he founded the Society of Civil Engineers (now known as the Smeatonian Society). His objective was to bring together engineers, entrepreneurs, and lawyers to promote the building of large public works, such as canals (and later railways). These new professionals also recognized that they needed to obtain parliamentary approval necessary to execute their schemes.

The Industrial Revolution brought with it new materials and methods for producing and using them. Cast and wrought iron are good examples. As early as 1780, cast iron columns began to be substituted for wood posts supporting the roofs of cotton mills in England. Bricks and timber (lumber) were produced using industrial methods and glass began to replace oiled paper as window coverings. Structural innovations accompanied these developments enabling spectacular early applications in bridges and railroad tracks.

Iron Bridge, designed by Thomas Farnolls Pritchard, is an outstanding monument to both civil engineering and the Industrial Revolution. In 1779, Iron Bridge, the world's first cast iron bridge, opened for traffic over the River Severn in Coalbrookdale, Shropshire, England. The bridge was cast in the local foundries by a man named Abraham Darby III. His grandfather, Abraham Darby, was the first to use less-expensive iron, rather than brass, to cast strong thin pots for the poor. Under his son and grandson, the Coalbrookdale foundry flourished. In 1777, Abraham Darby III began erecting 378 tons of cast iron to build the bridge, which spans 100 feet (30 meters) (see Figure 2.15).

The development of mills and factories in the countryside attracted workers by tens of thousands. Because good roads and rail systems did not yet exist, canals connecting

FIGURE 2.15   Iron bridge over the River Severn, Coalbrookdale, Shropshire, England: First use of cast iron in a bridge.
Source: Archivist/Adobe Stock.

locks, wharves, boatyards, limekilns, and warehouses were constructed at a frantic pace. The first public railroad opened in 1825. The race was on to shrink distance and speed up time.

The use of iron and glass continued to shake up traditional construction methods. According to Kostof, "Not since the Roman invention of concrete had a building technology so radicalized architecture" [Kostof, Spiro, 1995]. Actually, Kostof continues to say that architects were not so thrilled about the appearance of cast iron and tended to conceal or decorate it in the sort of public buildings they specialized in designing. Its characteristics, however, were impossible not to appreciate. Iron was less expensive than stone and possessed exciting mechanical properties: It withstood fire better than wood and it could be prefabricated, shipped to the site, and assembled with relative ease.

As early as 1813 an iron and glass dome was built over a granary in Paris. Fifteen years later iron and glass roofs were used to span commercial arcades and shopping streets for Parisian pedestrians. England's most innovative uses of iron were railroad stations and bridges. Civil engineers embraced these new materials and created magnificent, awe-inspiring new structural forms.

For a time, the Scot Thomas Telford, first president of the Institution of Civil Engineers (ICE) in the United Kingdom, lived near Iron Bridge; he must have been fascinated by what he saw. He later used cast iron in many innovative bridge designs, including a chain suspension bridge over the Menai Straight in Wales (see Figure 2.16). French immigrant to the United Kingdom, Marc Brunnel, and his son, Isambard Kingdom Brunnel, also pushed the limits of civil engineering design and construction with projects such as the first tunnel under the River Thames for the new underground rail system in London. Isambard Kingdom Brunnel went on to design railroads, bridges (see Figure 2.17), train stations, and a ship—he also owned the Great Western Railroad. Brunnel's design for Paddington Station in London (1849–1854) resulted in a flexible covered space without columns. New railways were regarded as sources of future prosperity for provincial cities and towns, and the public took intense interest in Brunnel's daring schemes.

As the Industrial Revolution rolled along, many social changes were taking place. One significant development was the rise of the professions. New societal needs, commerce, educational opportunities, and exciting developments in technology converged. Institutions and societies were created to lend credibility, codify conduct, and provide a place where meetings of minds could occur. The following years are important in the development of civil engineering and architecture as professions:

- Institution of Civil Engineers (ICE)—launched 1818
- Royal Institute of British Architects (RIBA)—launched 1834
- American Society of Civil Engineers (ASCE)—launched 1852
- American Institute of Architects (AIA)—launched 1857

FIGURE 2.16 Thomas Telford: Menai chain suspension bridge, Wales; Craigellachie cast iron bridge, Scotland; elevation and view of construction.
Source: (a) Dumelow/Wikimedia Commons/Public Domain. (b) Mick Knapton/Wikimedia Commons/CC BY-SA 3.0. (c) Leithp/Wikimedia Commons/Public Domain. (d) High school engo teacher/Wikimedia Commons/Public Domain.

FIGURE 2.17 Isambard Kingdom Brunnel; Clifton suspension bridge, Bristol, England.
Source: (a) Robert Howlett/Wikimedia Commons/Public Domain. (b) Arpingstone/Wikimedia Commons/Public Domain.

ASCE's Profile

The American Society of Civil Engineers, a professional organization representing more than 146,000 [150,00 in 2022] civil engineers, celebrated its 150th anniversary in 2002 [170th in 2022]. When the 12 founders gathered at the Croton Aqueduct on November 5, 1852, and agreed to incorporate the American Society of Civil Engineers and Architects, one can only wonder if they dreamed the profound significance and long-lasting impact ASCE would have on the overall development of society. They laid a foundation for what proves to be one of the most prominent engineering societies in the world (American Society of Civil Engineers, 2019).

Today ASCE is a worldwide leader for excellence in civil engineering. With a mission to advance professional knowledge and improve the practice of civil engineering, ASCE is a focal point for the development and transfer of research results, and technical policy and managerial information. Through strategic emphasis in key areas, including infrastructure renewal and development, policy leadership and professional development, ASCE delivers the highest quality publications, programs, and services to its worldwide membership, demonstrating a daily commitment to sustaining the profession.

As civil engineering enters a new millennium, the American Society of Civil Engineers not only reflects on the profession's rich heritage, but equipped with this knowledge, ASCE continues to develop flexible, forward-thinking plans for the future of the society and the civil engineering profession.

—ASCE website, www.asce.org

In the United States, other civil engineers were designing and building canals, railroads, municipal water systems, and bridges. The Croton Aqueduct (Figure 2.18) was a 41-mile (66-kilometer) water distribution system constructed for New York City between 1837 and 1842. It brought water from the Croton River into reservoirs in Manhattan. During the 1830s, New York City desperately needed a fresh water supply to combat both disease and fire. After numerous proposals and a plan abandoned after two years, construction began in 1837 under the expertise of John Bloomfield Jervis.

The field of civil engineering grew with the times. A German immigrant to the United States, John Roebling, designed the first suspension bridge using steel cables—the Brooklyn Bridge. Planning for the bridge began in 1867 and construction was completed in 1883. The Brooklyn Bridge stretches 5,989 feet (1,825 meters) over the East River and connects the New York City boroughs of Manhattan and Brooklyn (see Figure 2.19). At the time of its completion, it was the longest suspension bridge in the world.

FIGURE 2.18 Croton Aqueduct: Clean water for New York City.
Source: Kristen Herde/Wikimedia Commons/CC BY-SA 2.5.

2.7 MODERN CIVIL ENGINEERING

Civil engineering has continued to evolve. The 20th century saw increasing specialization and advancements in theoretical understanding, materials and methods, and technologies (see Figure 2.20).

Just as the Greeks compiled a list of The Wonders of the Ancient World, the American Society of Civil Engineers has compiled a list of wonders of the modern world, which are summarized in Table 2.1, Seven Wonders of the Modern World and are shown in Figure 2.21.

Other innovative projects continue to excite the imagination. The Millau Viaduct (shown in Figure 2.22), a large cable-stayed road-bridge spanning the valley of the River Tarn in southern France, was completed in 2004. Designed by structural engineer Michel Virlogeux and British architect Norman Foster, it is the tallest vehicular bridge in the world. One mast's summit is 1,125 feet (343 meters), only 125 feet (38 meters) shorter than the Empire State Building. The bridge won the 2006 International Association for Bridge and Structural Engineering (IABSE) Outstanding Structure Award.

Taipei 101, completed in 2005 in Taipei, Taiwan, was the world's tallest building until being surpassed by Burj Dubai. Designed by C.Y. Lee & Partners and constructed by Samsung Engineering & Construction, Taipei 101 incorporates many innovations necessary to build skyscrapers in earthquake and high wind zones (see Figure 2.23). The building is 101 stories

FIGURE 2.19   John August Roebling; Niagara suspension bridge; Cincinnati suspension bridge; Brooklyn Bridge.
Source: (a, c) Wikimedia Commons/Public Domain. (b) William Notman/Wikimedia Commons/Public Domain. (d) Postdlf/Wikimedia Commons/CC BY-SA 3.0.

above ground (1,670 feet, 509 meters) and five stories underground. A steel-tuned mass damper (TMD) weighing 662 metric tons and consisting of 41 layered steel plates welded together to form a 5.5-meter diameter sphere is suspended from the 92d and 88th floors. The TMD acts like a giant pendulum to counteract the building's movement, reducing sway by 30 to 40 percent.

Burj Khalifa, formerly called Burj Dubai, has held the record for the world's tallest building at 2,717 feet (828 meters) since 2010. A collection of the world's tallest man-made structures is depicted in Figure 2.24.

From Left to Right:

Karl Terzaghi: Published his theory of consolidation in 1923, which taken together with his earlier theories on earth pressures and piping, established modern soil mechanics.

Arthur Casagrande: As a professor at Harvard University, continued and expanded upon Terzaghi's work through experimentation and development of soil testing techniques.

Ralph Peck: Taught at the University of Illinois, conducted research and an active international consulting practice, co-authored *Soil Mechanics in Engineering Practice* with Terzaghi in 1948; emphasized judgment in engineering practice.

FIGURE 2.20 Founders of modern geotechnical engineering.
Source: (a) foto fra/Wikimedia Commons/Public Domain. (b) Norwegian Geotechnical Institute, NGI. (c) Wikipedia Commons.

TABLE 2.1 Seven wonders of the modern world.

	Modern Wonder	Started	Finished	Location
1	Channel Tunnel	1987	1994	Strait of Dover, between the United Kingdom and France
2	CN ower Tallest freestanding structure in the world 1976–2007	1973	1976	Toronto, Ontario, Canada
3	Empire State Building Tallest structure in the world 1931–1967 First building with 100+ stories	1930	1931	New York, NY, US
4	Golden Gate Bridge	1933	1937	Golden Gate Strait, north of San Francisco, California, US
5	Itaipu Dam	1970	1984	Paraná River, between Brazil and Paraguay
6	Delta Works/Zuiderzee Works	1950	1997	The Netherlands
7	Panama Canal	1880	1914	Isthmus of Panama, Panama

Source: American Society of Civil Engineers website: www.asce.org.

FIGURE 2.21   ASCE's seven wonders of the modern world.
Source: (a) Herbert Ortner/Wikimedia Commons/CC BY-SA 3.0. (b) Unknown author/Wikimedia Commons/ CC BY-SA 3.0. (c) Daniel Schwen/Wikimedia Commons/CC BY-SA 4.0. (d) Christian Mehlführer/Wikimedia Commons/CC BY-SA 3.0. (e) Stan Shebs/Wikimedia Commons / CC BY-SA 3.0. (f) Donar Reiskoffer/Wikimedia Commons/CC BY 3.0. (g) Martin St-Amant/Wikimedia Commons/CC BY-SA 3.0.

FIGURE 2.22 Millau Bridge.
Source: Mike Switzerland/Wikimedia Commons/CC BY-SA 2.5.

FIGURE 2.23 Taipei 101, Taiwan.
Source: Wikimedia Commons.

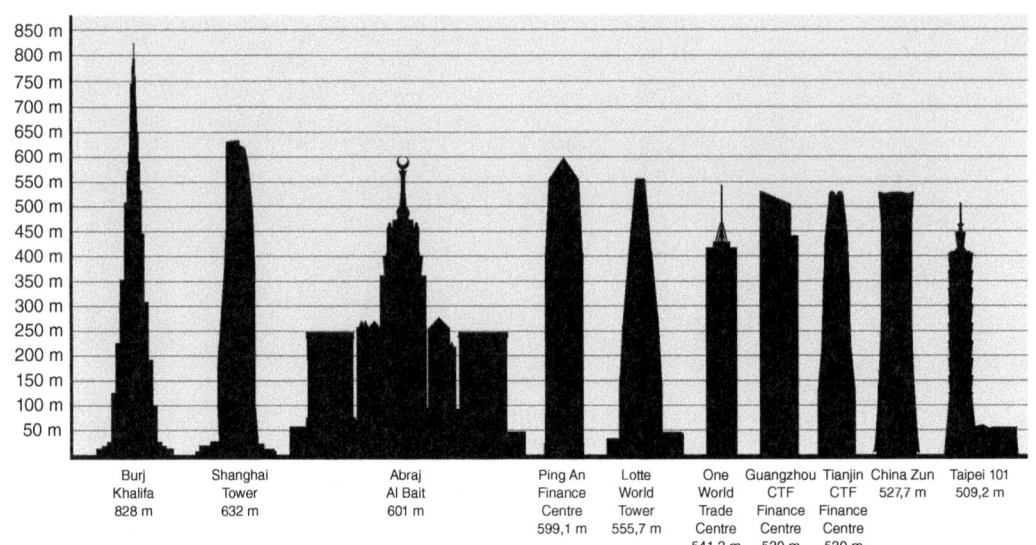

FIGURE 2.24   The world's tallest buildings.
Source: Phoenix CZE/Wikimedia Commons/CC BY-SA 4.0.

2.8   SUMMARY

Between 4000 and 2000 B.C. in Mesopotamia and ancient Egypt, humans started to abandon a nomadic existence, requiring increasingly complex structures such as fortifications, temples, and residences. During the medieval period, craftsmen such as masons and carpenters, carried out most building. Knowledge was retained in guilds that were not compelled to make changes to their well-guarded knowledge.

The history of civil engineering is linked to developments in mathematics and science and other fields. Civil engineers now have specialized educations involving diverse topics that enable them to recognize and solve problems. In addition to mathematics and science, practicing civil engineers need to know about design; sustainability and other contemporary issues; risks and uncertainties; project management; communication; public policy; business; leadership; human relations, mandatory health and safety requirements, contract law and liability, teamwork; and professional and ethical responsibility. Although the average citizen may not recognize the role of civil engineers in society, civil engineers continue to shape the quality of our lives. As the ASCE puts it:

> . . . *historic civil engineers used their creativity and ingenuity to lead the way to innovative civil engineering design, serving as an inspiration for today's practitioners. The legacy of their accomplishments remains a lasting tribute to their passion for the civil engineering profession.*

> **—ASCE History and Heritage website,**
> *http://content.asce.org/history/index.html*

"We shape our buildings, and afterwards our buildings shape us."
    —*Winston Churchill, address in the House of Parliament, London, October 28, 1943*

BIBLIOGRAPHY

American Society of Civil Engineers. (2008). *Civil Engineering Body of Knowledge for the 21st Century*, 2d edition. Prepared by the Body of Knowledge Committee on the Academic Prerequisite for Professional Practice. ASCE, Reston, VA. ISBN-13: 978-0-7844-0965-7.

American Society of Civil Engineers. (2019). *Civil Engineering Body of Knowledge for the 21st Century*, 3rd edition. ASCE report, ASCE, Reston, VA. ISBN (print): 9780784415221ISBN (PDF): 9780784481974.

Bachner, John Philip. (1991). *Practice Management for Design Professionals: A Practical Guide to Avoiding Liability and Enhancing Profitability*. John Wiley & Sons, New York. ISBN 0-471-52205-8.

"civil engineering." Encyclopædia Britannica. (2009). Encyclopædia britannica online. Nov. 5, 2009 www.britannica.com/EBchecked/topic/119227/civil-engineering.

de Camp, L. Sprague. (1993). *The Ancient Engineers*. Ballantine Books, New York. (First published 1960.) ISBN 0-345-48287-5.

Gates, Henry Louis, Jr., McGee, Dyllan, and Kunhardt, Peter. (2017). *PBS Africa's Great Civilizations*. McGee Media LLC and Inkwell Films, Inc., Produced by the Corporation for Public Broadcasting and PBS.

Hart, Ivor B. (1928). *The Great Engineers*. Methuen & Co., London.

James, William. (1998). *Professor of Water Resources Engineering, "A Historical Perspective on the Development of Urban Water Systems,"* University of Guelph, Guelph, Ontario, Canada.

Kostof, Spiro. (1995). *A History of Architecture: Settings and Rituals*, 2nd edition. Oxford University Press, New York. ISBN-13 978-0-19-508378.

Landel, John Gray (2000). *Engineering in the Ancient World*, rev edition. University of California Press, 1978, Berkeley and Los Angeles.

Morgan, William N. (2008). Earth architecture: from ancient to modern. ISBN-13 978-0-8130-3207-8. ISBN-10: 0-813032075.

Petrowski, Henry. (1994). *Design Paradigms: Case Histories of Error and Judgment in Engineering*. Cambridge University Press, Cambridge, ISBN 0-521-46108-1 (hardcover), ISBN 0-521-46649-0 (paperback).

Scarborough, Vernon L. (2003). *The Flow of Power—Ancient Water Systems and Landscapes. A School of American Research Resident Scholar Book*. SAR Press, Santa Fe, New Mexico. ISBN 1-930618-32-8.

Vali-Khodjeini, Ali. (1995). Human impacts on groundwater resources in Iran. Man's Influence on Freshwater Ecosystems and Water Use Proceedings of a Boulder Symposium, July 1995. IAHS Publication No. 230.

Wright, Kenneth R., Zegarra, Alfredo Valencia, and Lorah, William L. (1999). "Ancient Machu Picchu drainage engineering." *Journal of Irrigation and Drainage Engineering*. ASCE Press, November/December, Vol 125, No. 6.

https://www.bls.gov/ooh/architecture-and-engineering/civil-engineers.htm#tab-5 (accessed July 20, 2022).

https://whc.unesco.org/en/list/Africa (accessed July 20, 2022).

https://whc.unesco.org/en/list/18 (accessed June 01, 2024).

CHAPTER 3

Ethics

Big Idea

"The adoption of a code is significant for the professionalization of an occupational group, because it is one of the external hallmarks testifying to the claim that the group recognizes an obligation to society that transcends mere economic self-interest."
— Illinois Institute of Technology—Center for the Study of Ethics, Jan 2016

Key Topics Covered

- Defining the Engineer's Ethical Code
- ASCE's New Code of Ethics—Summary (Part 1 or 2)
- ASCE's New Code of Ethics—Summary (Part 2 or 2)
- The American Society of Civil Engineer's Code of Ethics
- The American Council of Engineering Companies (ACEC) Ethical Conduct Guidelines
- The International Federation of Consulting Engineers (IFCE) Code of Ethics
- Importance and Relevant Policy Statements by ASCE and National Society of Professional Engineers (NSPE)

Civil Engineer's Handbook of Professional Practice, Second Edition. Karen Lee Hansen and Kent E. Zenobia.
© 2025 John Wiley & Sons, Inc. Published 2025 by John Wiley & Sons, Inc.
Companion website: www.wiley.com/go/hansen/CivilEngineersHandbook

Related Chapters in This Book

- Chapters 2 through 21 and Appendices A, B, C, D, E, F

3.1   INTRODUCTION

Ethical conduct is essential to a civil engineer's successful career. The information presented here provides civil engineers with the knowledge and tools to follow the civil engineering practice boundaries and guidelines. These ethics apply to the conduct and practice of CE design, client/public and agency interactions, contracting, communications, construction, business, and all aspects of civil engineering practice. The practice of proper ethical conduct is essential for a successful CE career.

One of the basic components of high ethical standards is trust, the ability to do "the right thing" independent of profit, pressure, or personal feelings. It seems ethics and laws may have some gray areas, which could be subject to interpretation or presentation. However, "laws" pretty clearly define acceptable actions. However, ethics can be thought of as a "code," crudely defined as what a person would do if nobody were looking and nobody would ever know. So, where do engineers fall on the public's view of ethical professions?

For the 18th year in a row, Americans rate the honesty and ethics of nurses highest among a list of professions that Gallup asks US adults to assess annually. Currently, 85 percent of Americans say nurses' honesty and ethical standards are "very high" or "high," essentially unchanged from the 84 percent who said the same in 2018. Engineers are next with a 66 percent honesty and ethical standards are "very high" or "high," essentially unchanged from the 65 percent who said the same in 2016 (engineers were not rated in 2017 and 2018). Medical doctors and pharmacists follow engineers and seem to be consistently within one or two percentage points (Gallup, Politics, 2020).

In the UK, the engineering profession makes up about 19 percent of their workforce. This article stated engineers were deemed trustworthy by 89 percent of the population. Engineering closely follows nurses (93 percent) and doctors (91 percent). This marks engineering's highest placing to date since its inclusion in their MORI Veracity Index in 2018 (The Engineer, 2020).

Like several other chapters in this book, the challenge with writing a chapter on ethics is that there are many volumes of references for this subject. The challenge then becomes how to sort through thousands of pages of text to produce reference material for practical use by the engineer. It is logical then, to begin with the definition of ethics.

Ethics are referred to as moral philosophy and recommend concepts of right and wrong behavior for professionals practicing within a profession (Davis, M., 1991). Professional ethics are considered as moral philosophy as applied to what's right and wrong for professionals.

Contemporary ethical theories can be divided into three general subject areas (Copp, 2006):

- Metaethics
- normative ethics
- applied ethics

Metaethics investigates the origin of ethical principles, their meaning, and source within society. There is some question about whether metaethics involve more than expressions of our individual emotions (Kant, 2002). Metaethical subjects include issues focusing on universal truths such as the reality and the will of God and logical reasoning in ethical judgments.

Normative ethics include practical issues like moral standards society sets to regulate right and wrong conduct. Sometimes this causes global or international conflict when one nation's standards appear to be violated by another nation's practice (Copp, 1995). A contemporary example of this could be human rights, an appropriate age to marry, or how males treat females in a particular society. It could also involve good habits, ethical or moral duties like caring for our family elders, or the consequences our behavior has on others like smoking in public places.

Finally, applied ethics involve examining how moral principles inform our judgments on debatable, controversial issues, such as the death penalty, environmental concerns, same sex marriage, and animal rights, among others (Harris, 1995).

For the purpose of ethics related to professional engineering, the focus is on *normative ethics* where the practicing professionals set standards to regulate right and wrong conduct (Copp, 2006). The civil engineer can imagine ethical considerations might be like the inputs when negotiating any ethical issue as depicted in Figure 3.1. The following references, codes, and discussions are presented to assist with these considerations.

FIGURE 3.1 Engineering ethics inputs. *Source:* Anaya Learning PTE LTD/https://www.ozassignments.com/solution/engineering-ethics-case-studies-oz-assignments/(accessed June 01, 2024).

3.2   DEFINING THE ENGINEER'S ETHICAL CODE

Before defining the code it's important to discuss why it even exists. An interesting study from the Illinois Institute of Technology—Center for the Study of Ethics in the Professions notes (Illinois Institute of Technology—Center for the Study of Ethics, Jan 2016):

> *The adoption of a code is significant for the professionalization of an occupational group, because it is one of the external hallmarks testifying to the claim that the group recognizes an obligation to society that transcends mere economic self-interest (p. 138).*

Michael Davis makes a strong positive case for professional codes of ethics. Davis argues that codes of ethics should be understood as conventions between professionals. Davis writes,

> *The code is to protect each professional from certain pressures (for example, the pressure to cut corners to save money) by making it reasonably likely . . . that most other members of the profession will not take advantage of her good conduct. "A code protects members of a profession from certain consequences of competition. A code is a solution to a coordination problem." (ibid, p154)*

Davis goes on to suggest that having a code of ethics allows an engineer to object to pressure to produce substandard work not merely as an ordinary moral agent, but as a professional. Engineers (or doctors, clergy, and other professionals) can say "As a professional, I cannot ethically put business concerns ahead of professional ethics."

Davis gives four reasons why professionals should support their professions code:

> *First . . . supporting it will help protect them and those they care about from being injured by what other engineers do. Second, supporting the code will also help assure each engineer a working environment in which it will be easier than it would otherwise be to resist pressure to do much that the engineers would rather not do. Third, engineers should support their profession's code because supporting it helps make their profession a practice of which they need not feel . . . embarrassment, shame, or guilt. And fourth, one has an obligation of fairness to do his part . . . in generating these benefits for all engineers. (p. 166)*

Harris and colleagues summarize Stephen Unger's analysis of the possible functions of a code of ethics:

> *First, it can serve as a collective recognition by members of a profession of its responsibilities. Second, it can help create an environment in which ethical behavior is the*

norm. Third, it can serve as a guide or reminder in specific situations . . . Fourth, the process of developing and modifying a code of ethics can be valuable for a profession. Fifth, a code can serve as an educational tool, providing a focal point for discussion in classes and professional meetings. Finally, a code can indicate to others that the profession is seriously concerned with responsible, professional conduct. (p. 35).

To better understand the actual description of the language, reference, and detail in an engineer's code of ethics we can visit the code as accepted by several engineering organizations including:

- The American Society of Civil Engineers (ASCE)
- The American Council of Engineering Companies (ACEC)
- The International Federation of Consulting Engineers (FIDIC)
- The National Society of Professional Engineers (NSPE)

Remember, **engineers** regulate the ethical standards for the profession. If these standards appear to need revising, there is an amendment process in place. The codes can be revised within the organization and agreed to by the members of the profession. Ethical standards should be *usable and live* codes accepted and followed by all engineers. Adherence to ethical standards makes choosing the right road for civil engineers obvious.

Terminology Used in Codes of Ethics

- *Principles* refer to a fundamental and comprehensive doctrine (morality) regarding behavior and conduct.
- *Canons* are broad principles of conduct.
- *Standards* are more specific goals toward which individuals should aspire in professional performance and behavior.
- *Rules of Conduct* are mandatory; violation of a Rule usually is grounds for disciplinary action. Rules can implement more than one Canon or Standard.

As depicted in Figure 3.2, engineering ethics is a combination of many important concepts some of which have other deeper meanings, such as "integrity." This concept includes honesty, truth, honor, and possessing strong moral principles.

For a clear and succinct definition, we can begin with the ASCE. (The ACEC, NSPE, and FIDIC codes are presented later on.)

- Negotiating Ethical Situations and Decisions in Your Career is Like Driving a High-Performance Sports Car on a Mountain Road. Be Careful!

- Ethical decisions in your Civil Engineering Career involve legal implications, risk, intelligent thinking, and reward.

- Use caution when considering ethical situations and decisions. There are a lot of resources and research available through organizations like ASCE. Know your limits of knowledge and experience and know when "legal counsel" is required.

FIGURE 3.2 Choosing the right road. Artist: Marina Gamez (2024).

3.3   ASCE'S NEW CODE OF ETHICS—SUMMARY PART 1 OF 2

QUESTION: How does the new Code of Ethics compare with its predecessor?

DISCUSSION: Looking first to the overall structure of the new Code of Ethics, one of the most significant changes in the new code is its abandonment of the "canon model" in favor of what has been termed a "stakeholder model." Whereas the prior code grouped its ethical guidelines into a set of eight canons—each outlining a broad ethical principle such as competence, fidelity, or integrity—the new code organizes ethical duties under the specific stakeholder to which the duty is owed. The goal of this change is to make the code's structure more intuitive to engineers seeking to resolve ethical concerns. For example, an engineer with qualms about critiquing another professional's work might not immediately look for guidance from a canon on "unfair competition" but would quickly recognize the applicability of a section outlining the engineer's ethical duty to peers.

1. LOOKING FIRST TO THE OVERALL STRUCTURE OF THE NEW CODE OF ETHICS, ONE OF THE MOST SIGNIFICANT CHANGES IN THE NEW CODE IS ITS ABANDONMENT OF THE "CANON MODEL" IN FAVOR OF WHAT HAS BEEN TERMED A "STAKEHOLDER MODEL."

A second important change is the ordering of the five stakeholders: Society; Natural and Built Environment; Profession; Clients and Employers; and Peers. One common criticism of the previous Code of Ethics was that it lacked a clear hierarchy among ethical principles, except for the express obligation to "hold paramount" the public health, safety, and welfare. As such, the previous code offered little assistance to engineers in resolving dilemmas raised by two conflicting ethical principles, such as a conflict between truthfulness (the central precept of the old code's Canon 3) and protection of a client's confidences (an element of faithful service under Canon 4). Conversely, the new code expressly states that in cases of conflict, the stakeholders are listed in order of priority, meaning that an ethical obligation to the profession takes precedence over a conflicting obligation to clients and employers, which in turn takes precedence over duties to peers. While not a complete solution to the problem of competing ethical obligations—conflicts may still arise within a category of stakeholders, such as conflicting duties to an employer and client—the new direction is yet another illustration of the new code's aim of easing the application of its ethical guidance to the professional practice.

Perhaps the most visually evident change in the new code is its significant reduction in size. Including its Fundamental Principles, Canons, and Guidelines to Practice, the previous ASCE Code of Ethics totaled some 2,200 words, while the new code is less than a third of that length. While this reduction means some loss of specificity in the ethical guidance provided, this change is not necessarily a detriment to users. For example, an engineer grappling with a potential conflict of interest can still be guided by the general language on conflicts regardless of the specific facts of the case; plus, there may be less temptation for the engineer to assume a questionable action is permitted if it is not one of the specific scenarios directly addressed in the code.

2. PERHAPS THE MOST VISUALLY EVIDENT CHANGE IN THE NEW CODE IS ITS SIGNIFICANT REDUCTION IN SIZE.

Moving to the text itself, the new Code of Ethics begins with a brief Preamble. Like the Fundamental Principles that served as a preface to the previous Code of Ethics, the new

Preamble is not intended to be an enforceable statement of ethical obligations itself; rather, it is an aspirational statement of the moral aims that underpin the code's provisions. While both the new code and its predecessor identify four underlying principles, the specific principles themselves have substantial differences.

The new Preamble states that engineers:

- Create safe, resilient, and sustainable infrastructure.
- Treat all persons with respect, dignity, and fairness in a manner that fosters equitable participation without regard to personal identity.
- Consider the current and anticipated needs of society.
- Utilize their knowledge and skills to enhance the quality of life for humanity.

Though the fourth bullet roughly mirrors one of the prior code's Fundamental Principles, the remaining principles of that code had a distinctly business-centric focus: the commitment to serve clients and employers faithfully, increase the stature of the profession, and support professional societies. This difference suggests a substantial shift in emphasis between the old and new codes. While sustainability and inclusivity were important parts of the prior code and fidelity and reputation are certainly not lost in the new code, the new Preamble elevates the engineer's role in shaping the lives of others over more pragmatic and internal goals of business needs and professional advancement.

3. YET WHILE PROTECTING THE PUBLIC REMAINS "FIRST AND FOREMOST" OVER ALL OTHER ETHICAL DUTIES, THE NEW CODE OF ETHICS REFLECTS AN UNDERSTANDING THAT THIS IS BY NO MEANS THE ENGINEER'S ONLY ETHICAL DUTY TO SOCIETY.

The first stakeholder in the new Code of Ethics is Society, and the nine provisions in this section describe the engineer's ethical obligations to the public at large. As the foremost ethical duty defined in the new code, this section roughly takes the place of the prior code's Canon 1, and indeed, two provisions in particular are direct corollaries to this canon. Section 1(a) of the new code states that engineers must "first and foremost, protect the health, safety, and welfare of the public," while Section 1(i) directs engineers to "report misconduct to the appropriate authorities where necessary to protect the health, safety, and welfare of the public."

Yet while protecting the public remains "first and foremost" over all other ethical duties, the new Code of Ethics reflects an understanding that this is by no means the engineer's only ethical duty to society. Truthfulness and objectivity play vital roles in preserving the public's trust in engineers; thus, Section 1(c) of the new code directs engineers to "express professional opinions truthfully and only when founded on adequate knowledge and honest conviction."

The concept of equity and inclusion supports communities and professionals with diverse needs and objectives, so Section 1(f) instructs members to "treat all persons with respect, dignity, and fairness, and reject all forms of discrimination and harassment."

Much of this section of the new code closely approximates language in the prior Code of Ethics, but there are a few notable additions. Section 1(d) incorporates the old Canon 6 mandate of "zero tolerance for bribery, fraud, and corruption," but it also directs engineers to report violations to the proper authorities—a significant change from the previous code, which imposed a reporting obligation on engineers only in cases of threats to the public health, safety, and welfare.

In addition, section 1(h) directs engineers to "consider the capabilities, limitations, and implications of current and emerging technologies when part of their work." Projecting that the coming years will continue to see rapid advancements in artificial intelligence and other technological innovations, this new provision incorporates both an encouragement for engineers to embrace new tools and methods for delivering services and a caution to use good engineering judgment and relentless attention to safety when using such innovations.

Though the structure and style of the new code are dramatically different from the canon-based model of ASCE's past Code of Ethics, the text of the first stakeholder category—Society—suggests that many of the prior code's core ethical principles remain fundamentally unchanged. The January column will offer a similar overview of the remaining stakeholder categories: Natural and Built Environment; Profession; Clients and Employers; and Peers.

—Tara Hoke, General Counsel for the American Society of Civil Engineers
"ASCE Adopts New Code of Ethics." ASCE Civil Engineering Magazine (December 2020)

3.4   ASCE'S NEW CODE OF ETHICS—SUMMARY PART 2 OF 2

QUESTION: How does the new Code of Ethics compare with its predecessor?

DISCUSSION: The second of five stakeholders in the new Code of Ethics is the Natural and Built Environment, a category that corresponds to the sustainability provisions under the prior code's Canon 1. But whereas the previous canon asked only that engineers "strive" to incorporate sustainability into their engineering practices, the new Code of Ethics is much more prescriptive. Section 2a requires that engineers "adhere to the principles of sustainable development," while the remaining provisions expand on the nature of this adherence; engineers must "consider and balance societal, environmental, and economic impacts" of their work, "mitigate adverse societal, environmental, and economic effects," and "use resources wisely while minimizing resource depletion."

Also notable is section 2b's directive for engineers to seek "opportunities for improvement" when considering the impacts of their work. While sustainability is often expressed in terms of the engineer's role in reducing negative impacts, this language is a crucial reminder that engineers should use their knowledge and judgment to create positive impacts as well, either by remediating harms caused by past development or by introducing technological solutions to problems arising from the natural environment.

Third in the new code's hierarchy of stakeholders is Profession, and the provisions in this section focus on the role of individual engineers in maintaining the reputation of the engineering profession as a whole. Much like the first stakeholder category, Society, this section represents an amalgam of ethical concepts from throughout the previous Code of Ethics. Included in this section is language on unfair competition (much like the prior code's Canon 5), continuing education (mirroring the old Canon 7), and the duty to "uphold the honor, integrity, and dignity of the profession" (a near-exact replication of the old Canon 6).

A new addition to this code is section 3b's requirement that engineers "practice engineering in compliance with all legal requirements in the jurisdiction of practice." While financial crimes such as bribery and fraud were highlighted in the prior code, and certainly most unlawful acts would run afoul of at least one of the past code's canons, it is interesting that none of the past iterations of ASCE's Code of Ethics contained an overall requirement for engineers to comply with the law. This new language could be broadly interpreted to make ethical imperatives of everything from licensure requirements to tax reporting, and its inclusion in the code's third section reflects a belief that legal compliance is essential to preserve the public trust in engineering professionals.

One noteworthy absence in the new Code of Ethics is the previous code's extensive treatment of advertising conduct. In what some might consider the stuffiest section of the previous code, guideline f under the old Canon 5 devoted some 200 words to what it deemed as "permissible" ways for engineers to advertise their professional services; engineers could place listings "in rosters or directories published by responsible organizations" or display ads in "recognized dignified business and professional publications" but could not use their names in "public endorsement(s) of proprietary products." While the new code's language on truthfulness, proper credit, and fair competition would apply to advertising conduct, the new code makes no express reference to advertising. This is perhaps the clearest example of the new code reflecting a change in ethical philosophy, with today's engineers placing greater value on commercial speech and autonomy than on advertising decorum.

Fourth among the stakeholders is Clients and Employers. Though the placement of clients and employers near the bottom of the ethical hierarchy may seem startling, this should not be read to suggest that an engineer's duty to clients and employers is unimportant.

Rather, this structure is an illustration of how strong the engineer's commitment is to professional integrity and the public good; in cases of ethical conflict, those values take precedence over even the engineer's basic duty of service to clients and employers.

Unlike other sections of the new code, this stakeholder category derives most of its content from a single canon of the prior code. Section 4a instructs engineers to "act as faithful agents of their clients and employers," in a close corollary to the previous code's Canon 4, while other provisions mirror the old Canon 4's guidelines on conflicts of interest, confidentiality, and communication with clients and employers.

The fourth stakeholder section also includes two provisions originating from the prior code's Canon 2, namely, the requirements for engineers to "perform services only in areas of their competence" and to "approve, sign, or seal only work products that have been prepared or reviewed by them or under their responsible charge." Given how closely these two requirements are linked to public safety, one might question whether these two provisions should have been elevated to the top stakeholder category. While the best placement for this language was debated extensively by the drafters of the new code, ultimately it was felt that safety considerations were covered in the Society section's language on protecting the public health, safety, and welfare, and that it was important also to recognize competence as an ethical obligation owed to clients and employers.

The final stakeholder in the new Code of Ethics is designated as Peers, although the provisions that follow make it clear that this section defines an engineer's ethical obligations not only to engineering peers but also to subordinates, colleagues from other technical or nontechnical disciplines, and others in the engineer's professional sphere. As might be expected in a section on engineering peers, several provisions echo guidance from the old Canon 5 on unfair competition; sections 5a and 5b note that engineers must "only take credit for professional work they have personally completed" and "provide attribution for the work of others," while 5h warns engineers to "comment only in a professional manner on the work, professional reputation, and personal character of other engineers."

While elements of Canon 8's stricture to "treat all persons fairly" may be found in nearly all sections of the new code, inclusivity is a resounding theme in this category. Section 5d directs engineers to "promote and exhibit inclusive, equitable, and ethical behavior in all engagements with colleagues," and 5g instructs engineers to "supervise equitably and respectfully." This section also includes a unique focus on the engineer's role in building a safe and constructive work environment; engineers are expected, for example, to "foster health and safety in the workplace," "act with honesty and fairness on collaborative work efforts," and "encourage and enable the education and development of other engineers."

Last in order among the provisions of the new code is another new mandate; section 5i requires engineers to "report violations of the Code of Ethics to the American Society of Civil Engineers." While ASCE's bylaws have long included a requirement for members to report observed unethical conduct to the Committee on Professional Conduct, the inclusion of this language in the code puts a new spotlight on the obligation to share knowledge of members whose conduct does not embody the ethical values of the Society and the civil engineering community.

While the new Code of Ethics features dramatic structural changes and a new hierarchy of ethical obligations, a review of the text of this new code reveals more similarities than differences with its predecessor. While the history of ASCE's codes is one of slow but regular generational shifts in ethical values, it is clear that moral precepts such as competence, diligence, and integrity remain at the heart of today's engineering practice.

—Tara Hoke, General Counsel for the American Society of Civil Engineers
"ASCE Adopts New Code of Ethics." ASCE Civil Engineering Magazine
(January/February 2021)

The articles written by Tara Hoke, General Counsel for ASCE, summarize the recent changes made to the ASCE Code of Ethics and the rationale behind those changes. The new code underscores many aspects of the civil engineering profession, its history, and applied business practices. ASCE Code of Ethics connection to other chapters of the *Civil Engineer's Handbook of Professional Practice* is briefly outlined in Table 3.1.

TABLE 3.1   Connection of ASCE's ethical standards to the chapters in the civil engineer's handbook of professional practice.

ASCE's Ethical Standards	Chapter Relationship and the Direct Connection
Create safe, resilient, and sustainable infrastructure	Related Chapter: "Sustainability"—Civil Engineers and ASCE recognizes the critical importance to design, construct, operate, and function in a sustainable fashion to safeguard our resources and environment for our future.
Address five stakeholders and their Hierarchy: Society; Natural and Built Environment; Profession; Clients and Employers; and Peers	Related Chapters: Background and History; Engineer's Role in Project Development; What Engineers Deliver; The Client Relationship; Leadership; Managing the CE Enterprise; Communicating as a Professional Engineer; Globalization; Sustainability; and Emerging Technologies. These chapters are related to our interactions with the clients; the public health, safety, and welfare; the design, construction and operation of infrastructure projects; leading project teams and clients effectively; communicating alternatives and emerging technologies and more.
Treat all persons with respect, dignity, and fairness	Related Chapters: The Client Relationship and Business Development; Leadership; Legal Aspects of Professional Practice; Managing the Civil Engineering Enterprise; Communicating as a Professional Engineer. These chapters all discuss interactions and relationships with Clients, Stakeholders, Permitting Agencies and Staff, and specifically include legal requirements for employment and professional requirements under law.
Consider the current and anticipated needs of society	Related Chapters: Background and History; Engineer's Role in Project Development; What Engineers Deliver; The Client Relationship; Leadership; Managing the CE Enterprise; Communicating as a Professional Engineer; Globalization; Sustainability; and Emerging Technologies: These chapters discuss CE interactions with the Clients; Project Stakeholders and Partners; the design, construction and operation of infrastructure projects; communicating alternatives and emerging technologies and more.
Utilize their knowledge and skills to enhance the quality of life for humanity	Related Chapters: This Handbook's entire "Table of Contents" virtually discusses how civil engineer's apply their skills to communicate and crystallize the owner's and project's needs and alternatives to enhance the quality of life for humanity.

3.5   THE AMERICAN SOCIETY OF CIVIL ENGINEERS CODE OF ETHICS

The American Society of Civil Engineers Code of Ethics was approved by the ASCE Board of Direction on October 26, 2020. More details on the ASCE Code can be found in the "Final Notes[1-5] on the ASCE Code of Ethics" at the end of this chapter, and at: https://www.asce.org/publications-and-news/civil-engineering-source/civil-engineering-magazine/issues/magazine-issue/article/2020/12/part-1-of-2-asces-new-code-of-ethics-explained

3.5.1   Preamble

Members of The American Society of Civil Engineers conduct themselves with integrity and professionalism, and above all else protect and advance the health, safety, and welfare of the public through the practice of Civil Engineering.
Engineers govern their professional careers on the following fundamental principles:

- create safe, resilient, and sustainable infrastructure;
- treat all persons with respect, dignity, and fairness in a manner that fosters equitable participation without regard to personal identity;
- consider the current and anticipated needs of society; and
- utilize their knowledge and skills to enhance the quality of life for humanity.

All members of The American Society of Civil Engineers, regardless of their membership grade or job description, commit to all of the following ethical responsibilities. In the case of a conflict between ethical responsibilities, the five stakeholders are listed in the order of priority. There is no priority of responsibilities within a given stakeholder group with the exception that 1a. takes precedence over all other responsibilities.[1]

3.5.2   Code of Ethics

1. **SOCIETY**
   Engineers:
   a. first and foremost, protect the health, safety, and welfare of the public;
   b. enhance the quality of life for humanity;
   c. express professional opinions truthfully and only when founded on adequate knowledge and honest conviction;
   d. have zero tolerance for bribery, fraud, and corruption in all forms, and report violations to the proper authorities;
   e. endeavor to be of service in civic affairs;

    f. treat all persons with respect, dignity, and fairness, and reject all forms of discrimination and harassment;

    g. acknowledge the diverse historical, social, and cultural needs of the community, and incorporate these considerations in their work;

    h. consider the capabilities, limitations, and implications of current and emerging technologies when part of their work; and

    i. report misconduct to the appropriate authorities where necessary to protect the health, safety, and welfare of the public.

2. NATURAL AND BUILT ENVIRONMENT

Engineers:

    a. adhere to the principles of sustainable development;

    b. consider and balance societal, environmental, and economic impacts, along with opportunities for improvement, in their work;

    c. mitigate adverse societal, environmental, and economic effects; and

    d. use resources wisely while minimizing resource depletion

3. PROFESSION

Engineers:

    a. uphold the honor, integrity, and dignity of the profession;

    b. practice engineering in compliance with all legal requirements in the jurisdiction of practice;

    c. represent their professional qualifications and experience truthfully;

    d. reject practices of unfair competition;

    e. promote mentorship and knowledge-sharing equitably with current and future engineers;

    f. educate the public on the role of civil engineering in society; and

    g. continue professional development to enhance their technical and non-technical competencies.

4. CLIENTS AND EMPLOYERS

Engineers:

    a. act as faithful agents of their clients and employers with integrity and professionalism;

    b. make clear to clients and employers any real, potential, or perceived conflicts of interest;

    c. communicate in a timely manner to clients and employers any risks and limitations related to their work;

    d. present clearly and promptly the consequences to clients and employers if their engineering judgment is overruled where health, safety, and welfare of the public may be endangered;

    e. keep clients' and employers' identified proprietary information confidential;

     f. perform services only in areas of their competence; and

     g. approve, sign, or seal only work products that have been prepared or reviewed by them or under their responsible charge.

5. PEERS

Engineers:

     a. only take credit for professional work they have personally completed;

     b. provide attribution for the work of others;

     c. foster health and safety in the workplace;

     d. promote and exhibit inclusive, equitable, and ethical behavior in all engagements with colleagues;

     e. act with honesty and fairness on collaborative work efforts;

     f. encourage and enable the education and development of other engineers and prospective members of the profession;

     g. supervise equitably and respectfully;

     h. comment only in a professional manner on the work, professional reputation, and personal character of other engineers; and

     i. report violations of the Code of Ethics to the American Society of Civil Engineers.

Following are the American Council of Engineering Companies' (ACEC) published "Ethical Conduct Guidelines."

3.6 THE AMERICAN COUNCIL OF ENGINEERING COMPANIES ETHICAL CONDUCT GUIDELINES

3.6.1 The ACEC Guideline Preamble

Consulting engineering is an important and learned profession. The members of the profession recognize that their work has a direct and vital impact on the quality of life for all people. Accordingly, the services provided by consulting engineers require honesty, impartiality, fairness and equity and must be dedicated to the protection of public health, safety and welfare. In the practice of their profession, consulting engineers must perform under a standard of professional behavior which requires adherence to the highest principles of ethical conduct on behalf of the public, clients, employees and the profession.

I. Fundamental Canons

Consulting engineers, in the fulfillment of their professional duties, shall:

    1. Hold paramount the safety, health and welfare of the public in the performance of their professional duties.

    2. Perform services only in areas of their competence.

3. Issue public statements only in an objective and truthful manner.
4. Act in professional matters for each client as faithful agents or trustees.
5. Avoid improper solicitation of professional assignments.

II. Rules of Practice

1. Consulting engineers shall hold paramount the safety, health and welfare of the public in the performance of their professional duties.

   a. Consulting engineers shall at all times recognize that their primary obligation is to protect the safety, health, property and welfare of the public. If their professional judgment is overruled under circumstances where the safety, health, property or welfare of the public are endangered, they shall notify their client and such other authority as may be appropriate.

   b. Consulting engineers shall approve only engineering work which, to the best of their knowledge and belief, is safe for public health, property and welfare and in conformity with accepted standards.

   c. Consulting engineers shall not reveal facts, data or information obtained in a professional capacity without the prior consent of the client except as authorized or required by law or these Guidelines.

   d. Consulting engineers shall not permit the use of their name or firm nor associate in business ventures with any person or firm which they have reason to believe is engaging in fraudulent or dishonest business or professional practices.

   e. Consulting engineers having knowledge of any alleged violation of these Guidelines shall cooperate with the proper authorities in furnishing such information or assistance as may be required.

2. Consulting engineers shall perform services only in the areas of their competence

   a. Consulting engineers shall undertake assignments only when qualified by education or experience in the specific technical fields involved.

   b. Consulting engineers shall not affix their signatures to any plans or documents dealing with subject matter in which they lack competence nor to any plan or document not prepared under their direction and control.

   c. Consulting engineers may accept an assignment outside of their fields of competence to the extent that their services are restricted to those phases of the project in which they are qualified and to the extent that they are satisfied that all other phases of such project will be performed by qualified registered or otherwise associates, consultants or employees, in which case they may then sign the documents for the total project.

3. Consulting engineers shall issue public statements only in an objective and truthful manner.

   a. Consulting engineers shall be objective and truthful in professional reports, statements or testimony. They shall include all relevant and pertinent information in such reports, statements or testimony.

    b. Consulting engineers may express publicly a professional opinion on technical subjects only when that opinion is founded upon adequate knowledge of the facts and competence in the subject matter.

    c. Consulting engineers shall issue no statements, criticisms, or arguments on technical matters which are inspired or paid for by interested parties, unless they have prefaced their comments by explicitly identifying the interested parties on whose behalf they are speaking and by revealing the existence of any interest they may have in the matters.

4. Consulting engineers shall act in professional matters for each client as faithful agents or trustees.

    a. Consulting engineers shall disclose all known or potential conflicts of interest to their clients by promptly informing them of any business association, interest or other circumstances which could influence or appear to influence their judgment of the quality of their services.

    b. Consulting engineers shall not accept compensation, financial or otherwise, from more than one party for services on the same project, or for services pertaining to the same project, unless the circumstances are fully disclosed to, and agreed to, by all interested parties.

    c. Consulting engineers in public service as members of a governmental body or department shall not participate in decisions with respect to professional services solicited or provided by them or their organizations in private engineering practices.

    d. Consulting engineers shall not solicit or accept a professional contract from a governmental body on which a principal or officer of their organization serves as a member.

5. Consulting engineers shall avoid improper solicitation of professional assignments.

    a. Consulting engineers shall not falsify or permit misrepresentation of their, or their associates', academic or professional qualifications. They shall not misrepresent or exaggerate their degree of responsibility in or for the subject matter of prior assignments. Brochures or other presentations incident to the solicitation of assignments shall not misrepresent pertinent facts concerning employees, associates, joint ventures or past accomplishments with the intent and purpose of enhancing their qualifications and their work.

    b. Consulting engineers shall not offer, give, solicit or receive, either directly or indirectly, any political contribution in an amount intended to influence the award of a contract by public authority, or which may be reasonably construed by the public of having the effect or intent to influence the award of the contract. They shall not offer any gift or other valuable consideration to secure work. They shall not pay a commission, percentage or brokerage fee to secure work except to a bona fide employee or bona fide established commercial or marketing agencies retained by them.

Following is the International Federation of Consulting Engineers Code of Ethics.

3.7   THE INTERNATIONAL FEDERATION OF CONSULTING ENGINEERS (FIDIC) CODE OF ETHICS

According to the International Federation of Consulting Engineers' website:

> *The International Federation of Consulting Engineers (Fédération Internationale Des Ingénieurs-Conseils) recognizes that the work of the consulting engineering industry is critical to the achievement of sustainable development of society and the environment.*

To be fully effective not only must engineers constantly improve their knowledge and skills, but also society must respect the integrity and trust the judgment of members of the profession and remunerate them fairly.

All member associations of FIDIC subscribe to and believe that the following principles are fundamental to the behavior of their members if society is to have that necessary confidence in its advisors.

The FIDIC Code of Ethics follows:

3.8   FIDIC CODE OF ETHICS

3.8.1   Responsibility to Society and the Consulting Industry

The consulting engineer shall:

- Accept the responsibility of the consulting industry to society.
- Seek solutions that are compatible with the principles of sustainable development.
- At all times uphold the dignity, standing and reputation of the consulting industry.

3.8.2   Competence

The consulting engineer shall:

- Maintain knowledge and skills at levels consistent with development in technology, legislation and management, and apply due skill, care and diligence in the services rendered to the client.
- Perform services only when competent to perform them.

3.8.3   Integrity

The consulting engineer shall:

- Act at all times in the legitimate interest of the client and provide all services with integrity and faithfulness.

3.8.4 Impartiality

The consulting engineer shall:

- Be impartial in the provision of professional advice, judgement or decision.
- Inform the client of any potential conflict of interest that might arise in the performance of services to the client.
- Not accept remuneration which prejudices independent judgement.

3.8.5 Fairness to Others

The consulting engineer shall:

- Promote the concept of "Quality-Based Selection" (QBS).
- Neither carelessly nor intentionally do anything to injure the reputation or business of others.
- Neither directly nor indirectly attempt to take the place of another consulting engineer, already appointed for a specific work.
- Not take over the work of another consulting engineer before notifying the consulting engineer in question, and without being advised in writing by the client of the termination of the prior appointment for that work.
- In the event of being asked to review the work of another, behave in accordance with appropriate conduct and courtesy.

3.8.6 Corruption

The consulting engineer shall:

- Neither offer nor accept remuneration of any kind which in perception or in effect either a) seeks to influence the process of selection or compensation of consulting engineers and/or their clients or b) seeks to affect the consulting engineer's impartial judgement.
- Co-operate fully with any legitimately constituted investigative body which makes inquiry into the administration of any contract for services or construction.

—https://fidic.org/sites/default/files/fidic_codeofethics_201510.pdf, accessed March 5, 2023.

Reference: International Federation of Consulting Engineers
FIDIC, Box 311—CH–1215, Geneva 15, Switzerland
SKYPE fidic.secretariat, Tl +41–22-799 49 00, Fx +41–22-799 49 01
www.fidic.org

3.9   IMPORTANT AND RELEVANT POLICY STATEMENTS BY ASCE AND NSPE

Referenced below are several of ASCE's very interesting and relevant "policy statements" of which engineers should be aware. These policies are shown with the policy statement title, adoption date, the actual written adopted policy and the issue. The policies regard:

- Continued education requirements for annual "ethics training," as stated in Policy Statement 376
- Engineer's judgment and adherence to the ASCE Code of Ethics, as stated in Resolution 502
- Use of the term "civil engineering professional" as stated in Policy 535.

3.9.1   ASCE Policy Statement 376—Continuing Education in Ethics Training

Approved by the National Engineering Practice Policy Committee on March 8, 2007
Approved by the Policy Review Committee on March 9, 2007
Adopted by the Board of Direction on April 24, 2007

3.9.1.1   Policy

The American Society of Civil Engineers (ASCE) encourages all state boards of engineering licensure to institute a minimum professional development requirement consisting of at least one (1) hour per year on professional ethics for professional licensure which would be reciprocal with other states. The one hour per year should be based upon the fundamental canons of professional conduct and other appropriate administrative rules or regulations, and designed to demonstrate a working knowledge of professional ethics.

3.9.1.2   Issue

Professional ethics is the cornerstone of engineering practice. Adherence to a Code of Ethics encourages engineers to practice in areas in which they are competent and that 'they will hold the safety, health and welfare of the public as their highest duty. The majority of complaints referred to state boards of licensure for investigation and possible penalty action involve ethics and, often, a lack of understanding of the Fundamental Canons of Professional Conduct.

3.9.1.3   Using a Code of Ethics

Codes of ethics are created in response to actual or anticipated ethical conflicts. Considered in a vacuum, many codes of ethics would be difficult to comprehend or interpret. It is only in the context of real life and real ethical ambiguity that the codes take on any meaning.

Codes of ethics and case studies need each other. Without guiding principles, case studies are difficult to evaluate and analyze; without context, codes of ethics are incomprehensible. The best way to use these codes is to apply them to a variety of situations and see what results. It is from the back and forth evaluation of the codes and the cases that thoughtful moral judgements can best arise.

3.9.2 ASCE Resolution 502—Professional Ethics and Conflict of Interest

Approved by the Engineering Practice Policy Committee on March 26, 2009
Approved by the Policy Review Committee on March 27, 2009
Adopted by the Board of Direction on July 25, 2009
First Approved in 2003

3.9.2.1 Policy

The American Society of Civil Engineers (ASCE) believes that:

- The engineer's judgment and adherence to the ASCE Code of Ethics must be above reproach and beyond the influence of competing interests. Even the appearance of a conflict of interest is to be avoided.
- The ability to exercise the independent judgment required of engineers to protect the public health, safety, welfare and environment should not be compromised in any way by the rules of any organization to which the engineer belongs.
- Laws, regulations, conditions of employment and collective bargaining agreements must permit engineers to maintain their independence and avoid potential conflicts of interest to protect the public health, safety, welfare, and environment.
- Engineers should not be subject to disciplinary or demeaning actions for holding the public interest above all others.

3.9.2.2 Issue

Engineering is a learned profession that has a direct impact on the environment and the safety, health and welfare of the public. Accordingly, the services provided require high standards of honesty, integrity and fairness.

ASCE's Code of Ethics recognizes the unique employment aspects of the engineer, regardless of the employer, public or private. Employment conditions for engineers must support their duty to hold paramount the health, safety, welfare and environment of the public in their engagements. To fulfill their duty, engineers must apply responsibly their independent judgment in design and construction matters. This duty to the public supercedes any actual or perceived obligations engineers have to the owners of their projects, their employers, or any organizations to which they belong.

3.9.2.3   Rationale

Engineers must adhere to ASCE's Code of Ethics and operate under the jurisdiction of state licensure laws and are subject to discipline for violation of these laws. Engineers are also subject to discipline from the professional societies of the engineering profession for violation of the public trust. These laws and standards include the responsibility for properly preparing design documents or performing field observation and testing to document construction.

An engineer relies on a variety of resources, including non-professional personnel, in rendering professional engineering services. An engineer must oversee the performance of those resources for public health, safety, welfare and the environment.

Since ASCE is composed of individual members, the Society is concerned about matters that affect its members and will voice its concerns relative to the employment conditions of its professional members while simultaneously striving to protect the health, safety, welfare and the environment of the public it serves.

3.9.3   ASCE Policy Statement 535—Use of the Term "Civil Engineering Professional"

First Approved 2011; this version:
Approved by the Engineering Practice Policy Committee on March 20, 2014
Approved by Public Policy Committee on May 9, 2014
Adopted by the Board of Direction July 13, 2014

3.9.3.1   Policy

The American Society of Civil Engineers (ASCE) recognizes the Civil Engineering Professional, the Civil Engineering Technologist, and the Civil Engineering Technician as important members of the civil engineering project team. ASCE defines each as follows:

- Civil Engineering Professional (CE Professional)—A person who holds a professional engineering license. A person initially obtains status as a CE Professional by professional engineering (PE) licensure obtained through the completion of requisite formal education, experience, examination, and other requirements as specified by an appropriate Board of Licensure. A person working as a CE Professional is qualified to be professionally responsible for engineering work through the exercise of direct control and personal supervision of engineering activities and can comprehend and apply an advanced knowledge of widely applied engineering principles in the solution of complex problems.
- Civil Engineering Technologist (CE Technologist)—A person who exerts a high level of judgment in the performance of engineering work, while working under the direct control and personal supervision of a CE Professional. A person initially obtains status as a

CE Technologist through the completion of requisite formal education and experience and may include examination and other requirements as specified by a credentialing body. A person working as a CE Technologist can comprehend and apply knowledge of engineering principles in the solution of broadly defined problems.

- Civil Engineering Technician (CE Technician)—A person typically performing task-oriented scientific or engineering related activities and exercising technical judgments commensurate with those specific tasks. A person working as a CE Technician works under the direct control and personal supervision of a CE Professional or direction of a CE Technologist. A person initially obtains status as a CE Technician through the completion of requisite formal education, experience, examination(s), and/or other requirements as specified by an appropriate credentialing body. A person working as a CE Technician is expected to comprehend and apply knowledge of engineering principles toward the solution of well-defined problems.

3.9.3.2   Issue

Civil engineering, like other learned professions, consists of a work continuum withvarying complexities that is most effectively accomplished by individuals with differentranges of responsibilities, qualifications, and work experience. The civil engineeringcontinuum of work can be segmented into three broad categories: engineering work;technology work; and technician work. However, the roles and titles for CE Professional, CE Technologist, and CE Technician are not well defined in the civil engineering community, making proper assignment of work difficult. Currently CE firms use individuals performing work in these three categories, but assign them many differenttitles and roles. A lack of definition for roles and titles of the team members also makes their support and recognition more difficult.

Additionally it is recognized that not all members of the civil engineering workforce can be characterized as civil engineering professionals, civil engineering technologists, or civil engineer technicians. These include engineering interns (EIs), engineering-in-training (EITs), individuals in the developmental stages of becoming technologists or technicians, and specialists who are critical to the success of the civil engineering enterprise, but who do not necessarily have an engineering education. This latter group may include individuals such as contract specialists, resource managers, marketing, and other business-related specialties.

3.9.3.3   Rationale

To effectively provide civil engineering services, the proper use of the entire civilengineering project team will become increasingly important in the future. ASCE believes it is essential to improve the utilization, recognition and support of technologistsand technicians within the civil engineering project team. Advances in software continue to change how engineering work is accomplished, but more importantly, who within the team is doing the work. Therefore, it is critical that the roles of each team member be properly defined and that appropriate requirements for entry into the CE workforce, continuing education, career advancement, licensure and/or certification are defined and implemented. The health, safety, and welfare of the public is best assured by assigning the segments of civil engineering work to the members of the CE workforce most qualified to complete them.

3.9.4   NSPE Position on Potential Incidents of the Unlicensed Practice

NSPE has issued guidance to NSPE State Societies ("State Societies") and to NSPE members on reporting potential incidents of the unlicensed practice, or offer to practice, of engineering.

The practice of engineering by unlicensed practitioners potentially places the public health, safety and welfare at risk. For this reason, it is of interest to State Societies to encourage members to report potential unlicensed practice, or offers to practice, to State Licensing Boards. Note that efforts to prevent unlicensed practice are intended solely to protect the public health, safety, and welfare, and are not intended to improperly restrict lawful activities or practices.

In many jurisdictions, a Professional Engineer has an ethical and legal obligation to report unlicensed practice to the State Licensing Board.

A recent National Council of Examiners for Engineering and Surveying (NCEES) survey of State Professional Engineering Licensing Boards ("State Licensing Boards") requested information from each Board on the categories of violations indicated for cases opened during a two-year period. Responses from 43 boards were received, with information on the reported violations for 3,369 disciplinary cases. The frequency of categories of violations reported in that survey is as follows, beginning with the most frequent:

- Incompetence/negligence
- Unlicensed practice/offer
- Ethics/professional conduct/misconduct
- Fraud, deceit, misrepresentation
- Sealing of work not prepared under the direct supervision and control of the licensee

Some case studies of actual violations posted on the State of California Board for Professional Engineers and Land Surveyors and NSPE websites follow:

Case Studies

Case studies provide valuable insight and enlightenment to civil engineers regarding licensing issues and ethics. The California Business and Professions Code is presented below in a text box for reference. Each state has a similar code by which the Licensing Board may receive and investigate complaints against registered professional engineers. Ethical violations of the code may also be investigated and acted upon by these state boards.

California Business and Professions Code 6775*

The board may, upon its own initiative or upon the receipt of a complaint, investigate the actions of any professional engineer licensed under this chapter and make findings thereon.

By a majority vote, the board may publicly reprove, suspend for a period not to exceed two years, or revoke the certificate of any professional engineer licensed under this chapter on any of the following grounds:

(a) Any conviction of a crime substantially related to the qualifications, functions, and duties of a licensed professional engineer, in which case the certified record of conviction shall be conclusive evidence thereof.

(b) Any deceit, misrepresentation, or fraud in his or her practice.

(c) Any negligence or incompetence in his or her practice.

(d) A breach or violation of a contract to provide professional engineering services.

(e) Any fraud, deceit, or misrepresentation in obtaining his or her certificate as a professional engineer.

(f) Aiding or abetting any person in the violation of any provision of this chapter or any regulation adopted by the board pursuant to this chapter.

(g) A violation in the course of the practice of professional engineering of a rule or regulation of unprofessional conduct adopted by the board.

(h) A violation of any provision of this chapter or any other law relating to or involving the practice of professional engineering.

(Amended by Stats. 2013, Ch. 178, Sec. 2. (SB 152) Effective January 1, 2014.)

*Note: 6775.1 states that the board may, upon its own initiative or upon the receipt of a complaint, investigate the actions of any engineer-in-training and make findings thereon.

Several case studies are presented for reference below and show how many common issues are interpreted and acted upon with details on some fines to the licensee.

3.9.4.1   Citations Issued to Board Licensees

Citations are issued to licensed engineers and land surveyors when the severity of a violation may not warrant suspension or revocation of the licensee's right to practice. When a fine is levied with a citation, payment of the fine represents satisfactory resolution of the matter. Summaries of the citations, including each licensee's name and license number, remain on the website for five years after the citation is final, unless further action is taken against the licensee. All citations issued by the Board are matters of public record.

Case 1   Expired License, Mr. W.

Civil Engineer C 3xxxx
Citation 5L
Final: October 27, 2002
Action: Order of Abatement; $7,500 fine

   An investigation revealed that Mr. W., whose Civil Engineer License, C 3xxxx, expired on September 30, 1998, violated Business and Professions Code Sections 6733 and 6737(a) and (e) by performing civil/geotechnical engineering on several projects in California during the period his license was expired. The citation ordered Mr. W. to cease and desist providing civil engineering services in California until such time as his delinquent license is renewed and reinstated and to pay administrative fines to the Board in an amount totaling $7,500.00. The administrative fines have been paid. In accordance with Section 125.9(d) of the Business and Professions Code, payment of an administrative fine does not constitute admission of any violation(s) charged but represents a satisfactory resolution of the matter.

Case 2   False, Misleading, and Deceitful Information, Mr. D.

Civil Engineer C 1xxxx
Citation 52-L
Final: March 26, 2005
Action: Order of Abatement; $250 fine

   A citation was issued to Mr. D. on May 15, 2000, alleging that Mr. D. had provided false, misleading, and deceitful information on reference forms for an applicant for licensing as a civil engineer. The references were dated July 20, 1999, and October 12, 1999. An investigation, including a review by at least one licensee of the Board competent in civil engineering, determined Mr. D. had signed and sealed reference forms which contained incorrect information. At an informal conference following service of the citation, which was reissued February 25, 2005, Mr. D. admitted that he failed to adequately check the dates of employment on the reference forms he signed and that he had taken the applicant's word that the information on the forms was correct. The Board ordered Mr. D. to cease and desist from violating Business and Professions Code Section 6775(f) and to pay an administrative fine of $250. The administrative fine has been paid in full. In accordance with Section 125.9(d) of the Business and Professions Code, payment of an administrative fine does not constitute admission of any violations charged but represents a satisfactory resolution of the matter.

Case 3   Failing to Sign and Stamp a Feasibility Study Report, Ms. R.

Civil Engineer C 2xxxx
Citation 56-L
Final: July 22, 2005
Action: Order of Abatement; $250 fine

Investigation determined that Ms. R. violated Business and Professions Code Section 6735 by failing to sign and stamp a feasibility study report that was released to her client, who then released it to the public during a public meeting. When questioned about who was responsible, Ms. R. and her colleagues signed and stamped the report but failed to include the date of signing and stamping. Ms. R. was ordered to obey all laws by properly signing and stamping all final engineering reports and include both the date of expiration of her license and the date reports are signed and stamped. Additionally, she was ordered to pay an administrative fine of $250. The fine has been paid. In accordance with Section 125.9(d) of the Business and Professions Code, payment of an administrative fine does not constitute admission of any violations charged but represents a satisfactory resolution of the matter.

Case 4   Providing Structural Engineering Services Without a Contract, Mr. M.

Civil Engineer C 4xxxx
Citation 59-L
Final: October 7, 2003
Action: Order of Abatement; $500 fine

The Board found that Mr. M. violated Business and Professions Code Section 6749 by providing structural engineering services for a room addition to a residence without entering into a written contract. Mr. M. stated he was hired by an unlicensed designer to provide services on the project but was paid directly by the homeowner. The unlicensed designer is not legally authorized to provide civil engineering services to his clients unless he has a partner who is a licensed civil engineer or is part of a business that is owned or co-owned by a licensed civil engineer. The homeowner stated the project was never completed; however, Mr. M. provided the client with a refund of all of the fees paid to him concerning the project. The Board ordered Mr. M. to enter into written contracts as required by Section 6749 when providing civil engineering services and to pay an administrative fine of $500. The administrative fine has been paid. In accordance with Section 125.9 (d) of the Business and Professions Code, payment of an administrative fine does not constitute admission of any violation(s) charged but represents a satisfactory resolution of the matter.

3.9.5   NSPE Ethics Case Study

Case 5   Sustainable Development, NSPE Board of Ethical Review,
Case No. 0X-X, 4/8/08—FINAL

Sustainable Development—Threatened Species Facts: Engineer A is a principal in an environmental engineering firm and is requested by a developer client to prepare an analysis of a piece of property adjacent to a wetlands area for potential development as a residential condominium. During the firm's analysis, one of the engineering firm's biologist reports to Engineer A that in his opinion, the condominium project could threaten a bird species that inhabits the adjacent protected wetlands area. The bird species in not an "endangered species," but it is considered a "threatened species" by federal and state environmental regulators. In subsequent discussions with the developer client, Engineer A verbally mentions the concern, but Engineer A does not include the information in a written report that will be submitted to a public authority that is considering the developer's proposal.

Question

Was it ethical for Engineer A not to include the information about the threat to the bird species in a written report that will be submitted to a public authority that is considering the developer's proposal?

References

Section I.3.—NSPE Code of Ethics: Engineers, in the fulfillment of their professional duties, shall issue public statements only in an objective and truthful manner.

Section I.5.—NSPE Code of Ethics: Engineers, in the fulfillment of their professional duties, shall avoid deceptive acts.

Section II.3.a.—NSPE Code of Ethics: Engineers shall be objective and truthful in professional reports, statements, or testimony. They shall include all relevant and pertinent information in such reports, statements, or testimony, which should bear the date indicating when it was current.

Section III.2.d.—NSPE Code of Ethics: Engineers are encouraged to adhere to the principles of sustainable development in order to protect the environment for future generations.

Section III.4.—NSPE Code of Ethics: Engineers shall not disclose, without consent, confidential information concerning the business affairs or technical processes of any present or former client or employer, or public body on which they serve.

Discussion

In January 2006, the NSPE Board of Directors approved a change to the NSPE Code of Ethics to add Section III.2.d. to the NSPE Code. The new section stated that "engineers shall strive to adhere to the principles of sustainable development in order to protect the environment for future generations." A footnote (Footnote 1) was also included at the end of the NSPE Code of Ethics. The footnote further clarified and defined the term "sustainable development." It stated that "sustainable development" is "the challenge of meeting human needs for natural resources, industrial products, energy, food, transportation, shelter, and effective waste management while conserving and protecting environmental quality and the natural resources base essential for future development." Thereafter, in July 2007, the NSPE House of Delegates voted to modify the language in NSPE Code Section III.2.d. to state that "engineers are encouraged to adhere to the principles of sustainable development in order to protect the environment for future generations. With this added language and further clarification, the NSPE Board of Ethical Review will review this language as a matter of first impression and in the context of other language in the NSPE Code and earlier NSPE Board of Ethical Review Opinions.

Unlike earlier NSPE Board of Ethical Review cases of this type, the facts in this case present a situation that often raises very difficult issues for engineers in dealing with clients. Engineering practice sometimes places the engineer in the position where the interests of a client and the interests of the public are in open and serious conflict.

As this Board has noted on several occasions, engineers play an essential role in society by taking steps and actions to see that products, systems, facilities, structures, and the land surrounding them are reasonably safe. Sometimes engineers are placed in situations where they must balance the extent of their obligations to their employer or client with their obligations to protect the public health and safety. NSPE Code Section III.2.d. places some additional responsibilities on engineers for the protection of environment.

At the same time, as noted in ABC Case No. ZZ-Y, there are various rationales for the nondisclosure language contained in NSPE Code Section III.4. Engineers, in the performance of their professional services, act as "agents" or "trustees" to their clients. They and the members of their firms are privy to a great deal of information and background concerning the business affairs of their client. The disclosure of confidential information could be quite detrimental to the interests of their client and, therefore, engineers as agents or trustees are expected to maintain the confidential nature of the information revealed to them in the course of rendering their professional services.

3.10 SUMMARY

Ethics are referred to as moral philosophy, which recommends concepts of right and wrong behavior for professionals practicing within a profession. Professional ethics are considered as moral philosophy as applied to what is right and wrong for professionals. It is important for engineers to be aware of potential ethical issues and acceptable responses under the code of ethics. Your license to practice depends upon it.

Organizations with Useful Information about the Professional Practice of Engineering

- ACEC—http://www.acec.org
- American Institute of Architects (AIA)—http://www.aia.org/index.htm
- ASCE—http://www.asce.org
- Association of General Contractors (AGC)—http://www.agc.org
- Design Build Institute of America (DBIA)—http://www.dbia.org
- NSPE—http://www.nspe.org/index.html
- NCEES—http://www.ncees.org
- International Federation of Consulting Engineers -fidic@fidic.org—FIDIC.org

FINAL NOTES ON THE ASCE CODE OF ETHICS

1. This Code does not establish a standard of care, nor should it be interpreted as such.
2. The ASCE's Code of Ethics was adopted on September 2, 1914 and was most recently amended on July 23, 2006, and 2020. Pursuant to the Society's Bylaws, it is the duty of every Society member to report promptly to the Committee on Professional Conduct any observed violation of the Code of Ethics.
3. In April 1975, the ASCE Board of Direction adopted the fundamental principles of the Code of Ethics of Engineers as accepted by the Accreditation Board for Engineering and Technology, Inc. (ABET).
4. In November 1996, the ASCE Board of Direction adopted the following definition of Sustainable Development: "Sustainable Development is the challenge of meeting human needs for natural resources, industrial products, energy, food, transportation, shelter, and effective waste management while conserving and protecting environmental quality and the natural resource base essential for future development."
5. "Sustainable development" is the challenge of meeting human needs for natural resources, industrial products, energy, food, transportation, shelter, and effective waste management while conserving and protecting environmental quality and the natural resource base essential for future development.

BIBLIOGRAPHY

Reinhart, R.J. (January 6, 2020). Continue to rate highest in honesty, ethics. GALLUP, Politics.

Copp, David. (1995). *Morality, Normativity, and Society*. Oxford University Press, New York.

Copp, David. (June, 2006). Metaethics and normative ethics. The Oxford Handbook of Ethical Theory.

Davis, Michael. (1991). "Thinking like an engineer: the place of a code of ethics in the practice of a profession." *Philosophy and Public Affairs*. Vol. 20, No. 2. 150–167.

The Engineer. (26th November, 2020). "Engineers rated one of UK's most trusted professions".

Harris, Charles E., Jr., Pritchard, Michael S., and Rabins, Michael J. (1995). *Engineering Ethics: Concepts and Cases*. Wadsworth Publishing, Belmont, CA.

Hoke, Tara. (December 2020). " ASCE adopts new code of ethics." ASCE Civil Engineering Magazine.

Hoke, Tara. (January/February, 2021). "ASCE adopts new code of ethics". ASCE Civil Engineering. Magazine.

Illinois Institute of Technology—Center for the Study of Ethics (IIT CSEP). (January, 2016). http://ethics.iit.edu.

Kant, Immanuel. ([1785] 2002). *Grounding for the Metaphysics of Morals*. Trans. and ed., T.E. Hill Jr. and A. Zweig. Oxford University Press, Oxford.

CHAPTER 4

Professional Engagement

Big Idea

Understanding how Civil Engineers obtain work is essential for those in private practice as well as their clients. Mastering the process of Professional Engagement enhances the civil engineer's probability of success.

"Chance favors only the prepared mind."

—Louis Pasteur

Key Topics Covered

- Contracting with the Government
- Fee-Based Selection
- Writing Engineering Proposals
- The Contract
- Budgeting
- Enhancing the Engineering Firm's Probability for a Successful Professional Engagement

Civil Engineer's Handbook of Professional Practice, Second Edition. Karen Lee Hansen and Kent E. Zenobia.
© 2025 John Wiley & Sons, Inc. Published 2025 by John Wiley & Sons, Inc.
Companion website: www.wiley.com/go/hansen/CivilEngineersHandbook

- Working Example of a "Request for Proposals" (RFPs)
- Typical Civil Engineering Proposal
- Typical Engineering Feasibility Study Report

Related Chapters in This Book

- Chapter 3: Ethics
- Chapter 5: The Engineer's Role in Project Development
- Chapter 6: What Engineers Deliver
- Chapter 7: Executing a Professional Commission, Project Management
- Chapter 8: Permitting
- Chapter 9: The Client Relationship and Business Development
- Chapter 10: Leadership
- Chapter 11: Legal Aspects of Professional Practice
- Chapter 12: Managing the Civil Engineering Enterprise
- Chapter 13: Communicating as a Professional
- Chapter 14: Balancing Life, Family, and Career
- Chapter 15: Globalization
- Chapter 16: Sustainability
- Chapter 17: Emerging Technologies
- Chapter 18: Human Relations Policies and Employment Practices
- Chapter 19: Construction Management
- Chapter 20: Critical Health and Safety Knowledge for Civil Engineers
- Chapter 21: What Engineers Need to Know

4.1   INTRODUCTION

In this reference book "professional engagement" is defined as:

To secure professional services; to hire.

One of the immediate issues recognized by someone requiring engineering services is the difficulty in defining and communicating the specific needs and/or tasks required to arrive at a solution. While potential clients may realize they have a problem to be solved, most do not understand engineering or what it takes to define and communicate their problem to the engineer.

Engineers recognize this situation and refer to it as the need to create a scope, or statement, of work (SOW). One of the challenges for engineers is that sometimes learning and understanding the client's operation and/or objectives takes quite a bit of time, and there may be several ways to address the problem with a variety of capital and/or expense scenarios.

> *The situation may be analogous to a patient telling a medical doctor that they have a headache and then asking, "How will you fix it and (by the way) how much will it cost?" The doctor usually has no idea without at least examining the patient—it might be as simple as prescribing a pain reliever or it might require surgery.*

In the course of assessing the client's needs, the engineer may spend quite a bit of their own (or their company's) time and expenses to provide a detailed SOW. The SOW needs to include a cost estimate, project schedule, and tabulated labor categories, which may include subcontractors and more, so that the client may choose an appropriate path forward. Occasionally the engineer may find that the client is surprised at the depth of their own needs, or the costs associated with resolving them. The client may then provide the engineer's SOW and proposal package to a competing engineer for an alternate approach or lower cost. The original engineer may then be left "holding the bag" with respect to sunk costs for the initial services and scoping effort.

To increase the chances of preparing a winning proposal and securing the professional engagement, engineering firms often create and follow a business development process as illustrated in Figure 4.1.

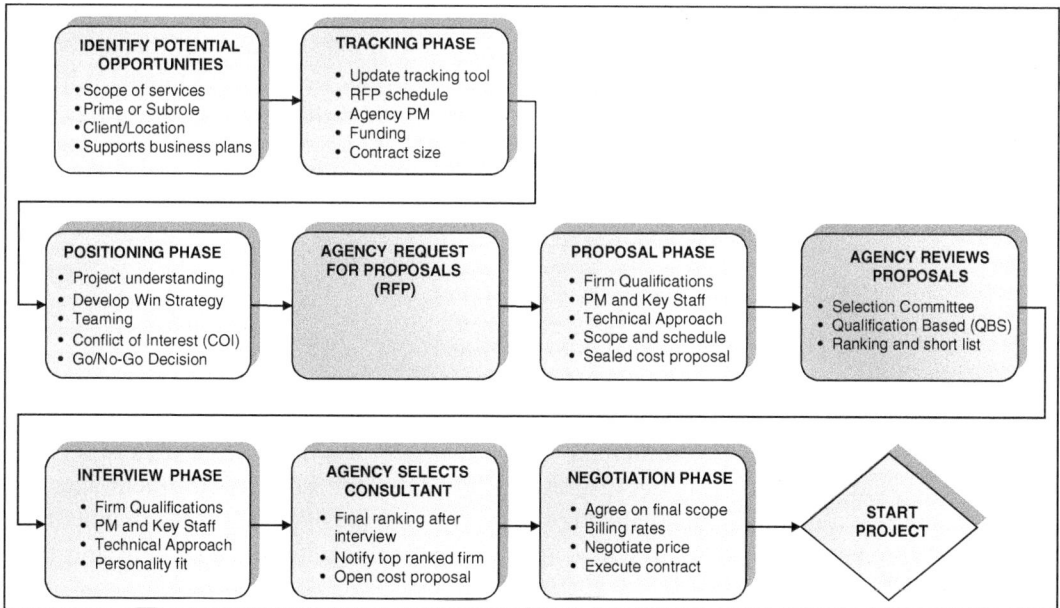

FIGURE 4.1 Business development process.

This process involves the civil engineering firm identifying potential leads early in the project initiation phases and following these projects through to the RFP and proposal phases. Savvy firms work to position themselves strategically with exceptional project experiences, talented staff, and other differentiating factors that can give them distinct advantages to win these projects.

Professional engineering services are best provided when the client knows and trusts the engineer. We always (or should always) act in the best interest of the client. This is why large clients (companies, facilities, or organizations) may have their own engineering staff. These staff engineers know the mission and requirements of their own employer and can act in the latter's best interest. Sometimes, however, the engineering staff may become overwhelmed or may require specialized help; in these cases, the staff engineers may require assistance just as companies that doesn't have engineering staffs might. The engineering staff may be in a good position to prepare a SOW for outside engineering support. Alternately, a firm could establish a support contract with an engineering firm to provide these services as the need arises.

We've now envisioned a scenario where a need for engineering services arises, and the client is in a position to request these services. But, the key question is "how" does the client request these services and specifically "what" services do they request? This situation is not unique to private industry; federal, state, local governments, as well as nonprofit organizations and members of the general public, can experience the same situation. One way to arrange for engineering services is for the client to prepare a request for qualifications (RFQ) that may or may not include a specific SOW. The client then solicits responses and selects an engineer based upon their general qualifications. This arrangement is referred to as qualification-based selection, or QBS. The benefit of QBS is that the client selects an engineer because of specific experience and capabilities relevant to the client's need.

The only part of the equation that may be missing is the working relationship and the "trust factor" between the client and the engineer. However, professional engineers are bound by ethics, business law, and contracts so the trust factor is not usually a problem, although it's human nature to hire or work with someone you know rather than a perfect stranger. (More on this subject is discussed in Chapter 3, Ethics; Chapter 5, The Engineer's Role in Project Development; Chapter 9, The Client Relationship and Business Development; and Chapter 13, Communicating as a Professional Engineer.)

4.2   CONTRACTING WITH THE GOVERNMENT

There are many reasons to want to obtain a contract with the government to provide engineering service. The federal government is itself the largest purchaser of goods and services in the world. And government is not just one entity. State, and Local government agencies each have their own unique policies and buying practices. It is critical to success in obtaining work for government to know as much as you can about this potential customer and their professional engagement practices.

Although each government agency is unique in many ways, there are important similarities. Generally, a company must become registered with that agency which includes meeting the requirements for registration. Registration facilitates tracking what opportunities are available. A firm should decide to compete only for contracts for which your company is highly qualified and can provide a compelling proposal in the format used by that agency. Becoming selected or recognized by an agency as highly qualified for the services they need is the first major step. The next steps are to continue with that agency in the process they use to further evaluate those companies they deem to be highly qualified. If selected for award of work, the contracting process may continue on and involve scoping, cost estimating, fees or cost negotiation, and contract terms finalization.

4.2.1 Registering with the Federal Government

The federal government uses on-line databases to manage procurement data. Engineers wishing to bid on federal government proposals must become familiar with the use of the System for Award Management (SAM) database. This database is under a program of modernization to become a unified one stop for all of the older databases previously used for procurement data management. This modernization is currently ongoing. The name has changed to "SAM. gov" and this terminology is used in the remainder of this chapter. It's up to the CE engineering firms to stay up to date with these ever-changing requirements and database details.

4.2.2 DUNS Number

To start the registration process, you or your company will need a Dun & Bradstreet (DUNS) number. A DUNS number is a unique nine-digit identification number for each physical location of your business.

When registering for your DUNS number, you'll need to have the following on hand:

- Legal name
- Headquarters name and address for your business
- Doing Business As (DBA) or other name by which your business is commonly recognized
- Physical address, city, state, and ZIP Code
- Mailing address (if different from headquarters and/or physical address)
- Telephone number
- Contact name and title
- Number of employees at your physical location
- Whether you're a home-based business

To apply for a DUNS number, visit DUNS Request Service.

4.2.3   NAICS Code

It is important to become familiar with the North American Industry Classification System (NAICS) code. NAICS codes classify businesses based on the particular product or service they supply. A business will generally have a primary NAICS code, but it can also have multiple NAICS codes if it sells multiple products and services. The NAICS code for Engineering Services is 541330.

To find your NAICS code, view the NAICS code list at the US Census Bureau.

4.2.4   Register

With the above information completed a firm will then go ahead and register with SAM. Remember that SAM is the database that government agencies search to find contractors they want to work with. It's important the registration is accurate and complete. To participate in government contracting, a firm must comply with all laws and regulations. The federal government's contracting process is governed by the Federal Acquisition Regulations (FARs). SAM also tracks firms that may perform poorly on federal contracts or violate the FARs. Poor ratings can make winning new work difficult or even lead to disbarment from future work for the government.

4.2.5   Small Business Program

A civil engineering firm seeking work with the federal government will want to understand and (if applicable) qualify for the small business program. The Federal Government has substantial goals for award of contract dollars to Small Businesses. Many contracts are "set-aside" and generally will only be awarded to small businesses. If the engineering firm is seeking work with the federal government, the firm will be greatly assisted by taking advantage of the opportunity to be registered in the SAM database as a small business. Small business may also qualify for additional preferences. The Federal Small Businesses are:

4.2.5.1   Federal Small Businesses

SMALL BUSINESS (SB)—Located in US, organized for profit, independently owned and operated, not dominant in field of operations in which it is bidding on Government contracts, AND meets Small Business Administration (SBA) size standards. Size standard is based upon the North American Industrial Classification Standard (NAICS) assigned to the specific procurement dependent upon product/service purchased. More information can be found at www.sba.gov.

WOMAN-OWNED SMALL BUSINESS (WOSB)—Small Business, at least 51% owned by a woman or women. Management and daily business operations must also be controlled by the women owners.

8(a) SMALL DISADVANTAGED BUSINESS—The 8(a) Business Development Program is available to socially and economically disadvantaged individuals. If accepted into the program, 8(a) certification can only last for a period of 9 years but allows agencies to make sole-source awards.

HISTORICALLY UNDERUTILIZED BUSINESS ZONE (HUBZONE)—Small Business, owned and controlled at least 51% by US citizens, SBA-certified as a HUBZone concern (principal office located in a designated HUBZone and at least 35% of employees live in a HUBZone).

SERVICE-DISABLED VETERAN-OWNED SMALL BUSINESS (SDVOSB) and VETERAN-OWNED SMALL BUSINESS (VOSB)—Small Business, veteran-owned as defined in 38 USC 101(2), at least 51% owned by 1 or more veterans, with management/daily operations controlled by 1 or more veteran owners. For SDVOSB, veteran owner(s) must have a service-connected disability as defined in 38 USC 101(16) and documented on DD 214 or equivalent. In the case of veteran with permanent and severe disability, the spouse or permanent caregiver of such veteran may also qualify.

4.2.5.2   Certifying as a Small Business

To be eligible for government contracts reserved for small businesses, the business must meet size requirements set by the Small Business Administration (SBA). These size standards define the maximum size that a business can be to qualify as a small business for a particular contract. The size standards are based on your NAICS Code. For example, for engineering services, the firm would be a small business with annual revenues over a three-year average of less than. $16.5 million.

Using SAM, a firm must be able to certify that the business is eligible for contracts that are reserved for small businesses. The firm should also be able to represent if the business is eligible for contracts under an SBA contracting program because it is disadvantaged, women owned, veteran owned, or located in an underutilized area.

The small business' profile in SAM should read like a résumé. Creating a profile that's accurate and appealing is important to winning government contracts. It's important to use accurate, descriptive terms about the business so the contracting officials will be able to find the firm in search results.

4.2.6   Searching for Business Opportunities

Federal contract opportunities over $25,000 in general must be listed in the SAM database. The total number of opportunities can be huge. When looking for a prime contracting opportunity, the database of opportunities must be aggressively filtered to only those opportunities for which the firm has a reasonable chance of being selected. As an example of the filtering process, the NAICS code for the primary services needed must match the NAICS code of the services the firm can provide. Location is another key filter. Where will the services be performed and does the firm have or need experience in that location to be selected for contracting.

The opportunity for a prime contract should match the business size. If the opportunity is open to any registered bidder, it is unlikely a small business will be successful. Similarly, if the firm is a disadvantaged small business, the firm's marketing time may be best spent on opportunities set aside for Disadvantaged Small Businesses.

Many other filters are typically applied by successful federal providers to find key opportunities. Successful filters may be in the form of key words to find agencies that need skills for which the firm is especially proficient and experienced. Some engineers seek contracts only with particular government agencies for which they have past experience.

4.2.7   Teaming and Subcontracting

Government agencies may seek to procure contractors to provide a wide range of disciplines and services and over a wide geography. The government's needs must be clearly stated in the contract opportunity synopsis published in SAM by the government's contracting officer. The possibility of being evaluated as the most qualified company for the contract award is often enhanced by adding teaming partners or subcontractors to the firm's proposal. Team members may provide added geography, disciplines, or depth of available labor or specialized experience to meet contract capacity.

Large businesses using subcontractors must find a means to award certain percentages of the subcontracted work to small businesses, often in multiple categories.

4.2.8   Federal Proposal Preparation

The Qualifications-Based Selection process for the federal government starts with the solicitation of proposals from firms interested in the contract. These firms are asked to submit a completed Standard Form 330 (SF330). The SF330 replaced the Standard Form 255 and Standard Form 254. All Federal agencies are supposed to use the SF330 when soliciting for architecture and engineering services. Many state agencies have also adopted this form.

It is important as an engineer interested in public sector work to understand how to complete this form. Even if one is working for a large company with other people to complete this form the engineer will need to maintain their own resume in SF330 format and adapt that resume for a particular solicitation. The key is showing one has relevant experience on projects similar to the new work needed by the client.

Sections A, B, and C of the SF330 are easy and routine. Section D is the organizational chart. The org chart should convey a single point of contact for the team, enough people for the job, staff for all necessary disciplines and skills, a clear management structure including quality assurance personnel, the role of subcontractors, and any specialists and subject matter experts on the team. The org chart should evolve as the rest of the proposal is prepared.

Section E is composed of resumes of the proposed key staff including the key staff from subconsultants. Resumes should be kept to one page. Some clients will allow certain resumes to go over a page. Each resume should include five projects. There is a requirement to check whether or not the project was completed while with the person's current firm. In general, the SF330 requires that experience listed in sections E and F to have occurred within the last 10 years. There are two fields after the project name. The first is "services" and the next is "construction." The date services were completed is the year this person stopped providing services for this project. The construction year is when the person stopped working on the construction. The person's title for his/her projects should match up with the role identified in the proposal.

Section F includes ten relevant projects selected to show past experience in the services the client is planning to hire. The projects should be within the last 10 years. For this and any section the client may require something other than the norm, such as requiring projects that are no more than five years old. The most effective proposals have ten projects that are very highly relevant to the client's needs. The relevance might be enhanced with photos and call outs of particularly relevant elements. One or two of the ten projects can be from a subcontractor.

Effective preparation of an SF330 can often start with completion of Section G. Section G will list the key staff to be used on the contract and the ten relevant projects from Section F. Then an "X" is placed on each project the staff member worked on. This shows visually the connection between the key staff proposed and the experience provided. Logic would dictate that the client wants to see as many "x marks" as possible on this form so everybody needs an x. This includes showing that the team has worked with the subcontractor personnel in the past.

Section H is space for additional relevant information the firm may wish to highlight about the team.

SF 330 Part II is the replacement for the Standard Form 254 and provides general information about the firm. SF330 Part II should be included in the submission unless directed by the client not to. In some cases, Part II should be for each one (location) that will be working on the project. Forms are constantly being updated and it is critical to check that the proposal to the government uses the latest version of the forms in addition to meeting the specific instructions for that specific contract.

4.2.9   Mentor-Protégé Programs and Joint Ventures

The federal government has extensive programs known as mentor/protégé which engineers should consider if seeking contracted work. The programs allow large businesses to identify as mentors to assist small business with business development in several areas such as navigating the contract bidding, acquisition, and performance processes. Small businesses can identify as protégés and select a mentor firm to receive this assistance. The largest program is administered by the federal Small Business Administration (SBA). Many agencies also administer their own programs, notably the Department of Defense. Many states and some larger local governments have similar programs.

The mentor and the protégé will negotiate a mentor-Protégé agreement which must be approved by the SBA. For contracting purposes this team can go further and form a joint venture. An approved joint venture is allowed to propose on contracts set aside for small businesses. The government is able to contract to a small business to help with meeting their goals, the protégé has the support of the experienced large business, and the mentor firm obtains some work with the government agency that would otherwise be unavailable to them.

A joint venture is a separate company and must be properly listed in SAM. The proposal is allowed to use experience and personnel from either of the companies enhancing the evaluation and chances for selection. Performance of the work must adhere to the mentor-protégé agreement and SBA guidelines.

4.2.10   Government Evaluation of Proposals and Contractor Selection

Following the evaluation of the statements of qualifications, the board prepares a report that recommends the firms to be included on the short list. Short-listed firms are those that the evaluation board has chosen to interview. The evaluation board's report generally evaluates and ranks at least three of the firms for the purpose of discussing the project with them in another meeting referred to as a "short-list interview." In the event that only two firms submitted qualifications, the client may elect to only evaluate these two firms but may retain the option of re-advertising with a re-written Scope of work (SOW). Evaluation boards are not limited in the number of firms that they can select for these "interviews"; it is left to the discretion of each board.

4.2.11   Interviews/Discussions with Firms

The interviews usually involve discussions on project concepts and the relative utility of alternative methods of furnishing the required services. Before the interview, some agencies send detailed selection criteria and other information about the project to the firms recommended for further consideration. Although conceptual alternatives may be presented, under the system established by QBS, the architect-engineer designer does not produce any design product in competing for the project.

Usually, these interviews are held at the agency's office. Occasionally, and in special circumstances, phone interviews are conducted. The interviews are brief, usually lasting only 30 to 60 minutes. The interview is the best chance for the engineer to display the firm's knowledge, unique strengths, project experience, and to present the project team to the decision-makers. For more information on this topic, the reader should also review Chapter 6, What Engineers Deliver; Chapter 7, Executing a Professional Commission; Chapter 8, Permitting; Chapter 9, The Client Relationship and Business Development; and Chapter 13, Communicating as a Professional Engineer.

4.2.12   Ranking of the Top Three Firms

Following the interviews, the board's report is presented to the agency head or a person who is designated to act on behalf of the agency head. The report lists, in order of preference, at least three firms that are considered to be the most highly qualified to perform the services. This is considered to be the final selection of the competing firms. If the firm listed as the most preferred is not the firm that was recommended as the most highly qualified by the evaluation board, the head of the agency will provide a written explanation for the reason for the preference. The head of the agency, or that person's designate, may not add names of other firms to the final report. The report reviews the recommendations of the evaluation board, and from that, the agency head makes the final selection. Samples of federal government architect/engineer evaluation forms employed by the Veteran's Administration are illustrated in Figure 4.2—Example Short List Form; Figure 4.3—Example A/E Interview Scoresheet; and Figure 4.4—Example A/E Performance Form. A detailed evaluation of these example forms will enlighten the competing A/E firms on the criteria for independent judgments and allow potential time for the firm to prepare their presentation materials in concert with the reviewing materials.

SHORT-LIST CRITERIA UTILIZING THE SF330 FORM

Department of Veterans Affairs – Architect/Engineer Evaluation Board

1. Specialized experience and technical competence of the firm (including a joint venture or association) with the type of services required
   Assignable point rang ....(0 to 40)

2. Specific experience and qualifications of personnel proposed for assignment to the project and record of working together as a team
   Assignable point range ....(0 to 40)

3. Professional capacity of the firm in the designated geographic area of the project to perform work (including any specialized services) within the time limitations. Unusually large existing workload that may limit A/E's capacity to perform project work expeditiously
   Assignable point range ....(0 to 20)

4. Past record of performance on contracts with the Department of Veterans Affairs. This factor may be used to adjust scoring for any unusual circumstances that may be considered to deter adequate performance by an A/E. (Firms with no previous VA experience receive a +5 rating)
   Assignable point range ....(−20 to 20)

5. Geographic location and facilities of the working office(s) which would provide the professional services and familiarity with the area in which the project is located
   Assignable point range ....(0 to 20)

6. Demonstrate success in prescribing the use of recovered materials and achieving waste reduction and energy efficiency in facility design
   Assignable point range ....(0 to 20)

7. Inclusion of small business consultant(s) (1 point), and/or minority-owned consultant(s) (1 point), and/or women-owned consultant(s) (1 points), and/or veteran owned consultant(s) (1 point), and/or disadvantage veteran owned consultant(s) (1 point), and/or HUBZone consultant(s) (1 point)
   Assignable point range ....(0 to 6)

SCORING KEY						
SCORING FACTORS	RANGE	POOR	MARGINAL	ACCEPTABLE	VERY GOOD	OUTSTANDING
1 and 2	0–40	0	5–10	15–25	30–35	40
3,5,6	0-20	0	5	10	15	20
4	(−20)–(+20)	(−20)	(−10)	0	10	20
7	0–6	0	1–2	3–4	5	6

Revised: July 13, 2004

FIGURE 4.2   Example short list form.

A/E INTERVIEW SCORESHEET

Project Title:		A/E APPLICANTS								

Project Location:	Raw Score Key										
	0.9 to 1.0	Excellent									
Project #:	0.7 to 0.8	Very Good									
	0.4 to 0.6	Acceptable									
Date:	0.2 to 0.3	Marginal									
	0.0 to 0.1	Poor									

FACTORS	WEIGHT	RAW SCORE	WEIGHTED SCORE	RAW SCORE	WEIGHTED SCORE	RAW SCORE	WEIGHTED SCORE	RAW SCORE	WEIGHTED SCORE
I - TEAM PROPOSED FOR THIS PROJECT									
Background of the personnel									
1. Project Manager									
2. Other key personnel									
3. Consultants									
II - PROPOSED MANAGEMENT PLAN									
Team organization									
1. Design Phase									
2. Construction Phase									
III- PREVIOUS EXPERIENCE OF PROPOSED TEAM									
Project Experience									
IV - LOCATION AND FACILITIES OF WORKING OFFICES									
A. Prime firm									
B. Consultants									
V - PROPOSED DESIGN APPROACH FOR THIS PROJECT									
A. Proposed design philosophy									
B. Anticipated problems and potential solutions									
VI - PROJECT CONTROL									
A. Techniques planned to control the schedule and costs									
B. Personnel responsible for schedule and cost control									
VII - ESTIMATING EFFECTIVENESS									
Ten most recently bid projects									
VIII - SUSTAINABLE DESIGN									
Team design philosophy and method of implementing									
IX- MISCELLANEOUS EXPERIENCE & CAPABILITIES									
A. Interior Design									
B. CADD & Other Computer Applications									
C. Value Engineering & Life Cycle Cost Analyses									
D. Environmental & Historic Preservation Considerations									
E. Energy Conservation & New Energy Resources									
F. CPM & Fast Track Construction									
X - AWARDS									
A. Awards received for design excellence									
XI - INSURANCE AND LITIGATION									
A. Type and amount of liability insurance carried									
B. Litigation involvement over the last 5 years & its outcome									
TOTALS									

Remarks: ....
....
....

Signature of Chairman	Signature of Member
Signature of Member	Signature of Member
Signature of Member	Signature of Member

08-INT 2/12/98   Medical - General

FIGURE 4.3 Example A/E interview scoresheet.

A/E PERFORMANCE

Project #:
Project Title:
Location:

A/E Name:
Architect:
Interior Designer:
Structural Engineer:
HVAC Engineer:
Plumbing Engineer:
Civil Engineer:
Fire Protection Engr:
Electrical Engineer:
Landscape Architect:
Estimator:

Stage of Service: ☐ Schematic Design ☐ Design Development ☐ Contract Documents

Performance

Rating Factors

Legend:	Rating Symbol	Score	Discipline	Not Applicable	Accuracy	Completeness	Cooperation	Coordination	Management	Meeting Schedule	Personnel Ability	Work Quality	SCORE [(-10) to +10]	Reviewer Initials	Date
Excellent	E	10	Architectural												
Very Good	VG	5	Interiors												
Acceptable	A	0	Structural												
Marginal	M	–5	HVAC												
Poor	P	–10	Plumbing												
			Civil												
			Fire Protection												
			Electrical												
			Landscape Arch.												
			Estimating												
			OVERALL												

Remarks:

1421a-12/94

FIGURE 4.4 Example A/E performance form.

4.2.13   Negotiation with the Top-Ranked Firm

The final process of being selected for a government contract varies considerably between agencies. The process described below is provided as an example. The engineer may be required to complete or at least participate with the proposal team for preparation. The final documents may become attachments to a contract and are prepared by the top ranked firm from the proposal process described above. If the government is not satisfied with the documents the contracting officer can decide to move to the second ranked firm, etc.

4.2.13.1   Quality Assurance Plan

The firm will be required to submit and quality assurance plan (QAP). The plan will describe overall how each scope of work to be performed under the contract is to be maintained on schedule and on an agreed budget for that scope (task order, etc.). The QAP will describe the system that will be maintained in place to assure the government that all work will be performed with a high standard of professionalism and that any mistakes are discovered and corrected before the work is deemed complete. The proposed QAP must state that the quality assurance program will be applicable to subcontractors.

4.2.13.2   Health and Safety Plan

The firm will prepare a plan that describes how an acceptable company-wide health and safety program will be staffed and activated for the work performed for the government. The plan must demonstrate the implemented health and safety plan will conform to the safety standards published by the client and applicable other government agencies. The health and safety plan must apply to subcontractor personnel as well as the prime contractor.

4.2.13.3   Cost Volume

The firm will prepare a cost volume which is based on audited financial data. The labor costs may be the actual salaries for the personnel proposed for use during the life of the contract and also calculated for the company's standard job titles. Job titles are based on education and years of experience. In addition to labor costs the cost volume will list the rates that will be charged for equipment items maintained in stores for use on projects. Rates may be provided with justification for combined equipment/labor such as the performance of laboratory tests.

The highest cost proposed to the government and part of the cost volume is the overhead rate for the company. The overhead are costs such as office space, daily supplies, communications, computing to name a few. The government will take great care to ensure that the overhead does not include some costs that the government considers "unallowable." Most marketing costs for example, are "unallowable."

The cost volume must include costs the government will incur if subcontractors are used in portions of the work. It is the prime contractor's duty to review the costs submitted by subcontractors for conformance to the Federal Acquisition Regulation System FARs. The cost volume will propose and provide justification for the profit markup to be applied to labor, subcontractors, overhead, and purchases. Some contracts have a fixed percentage markup while other types of contracts may pay a fixed fee.

4.2.13.4 Initial Scope of Work

Some government agencies will request preparation of the first technical proposal based on a scope of work. The technical proposal will include the list of tasks to be performed to complete the scope of work in sufficient detail that each task can later be broken into detailed activities with time and budget components. The government is looking for appropriate solution to their need as described in the scope of work statement. They will judge if the selected firm is working with expertise, innovation, and efficiency.

4.2.14 SUMMARY—The Final Selection

When the final selection is made by the agency head, the contracting officer is authorized to begin negotiations with the top-ranked firm. The negotiations are conducted pursuant to the procedures set forth in the Federal Acquisition Regulation (FAR). Usually, the firm is requested to submit a fee proposal listing direct and indirect costs as the basis for contract negotiations. Contract negotiations are conducted following an evaluation of the fee proposal and an audit when the proposed design fee is more than $100,000.

If a fee is not agreed upon within a reasonable time, the contracting officer will conclude negotiations with the top-ranked firm and initiate negotiations with the second-ranked firm. If a satisfactory contract cannot be accomplished with this firm, then this procedure will be continued until a mutually satisfactory contract is negotiated. If negotiations fail with all selected firms, the contracting officer may return to the original contractors' submittal or re-advertise the project. The negotiation process will then continue until an agreement is reached and a contract awarded. As a practical note, it is rare that a contract is not successfully negotiated with the top-ranked firm. For more information, the reader should also review Chapter 12, Managing the Civil Engineering Enterprise.

4.3 FEE-BASED SELECTION

An alternate method of selection of an engineer of professional services is referred to as "fee-based selection." This selection method is based upon the client (alone) or the client and an engineer creating a unilateral scope of work and negotiating a fee for these services. This fee may be a flat rate or a percentage of the overall cost of the project.

This method, which can be used by public agencies and private clients to select an A/E firm, requires that the client prepare either an RFQ or a request for proposal (RFP). While an RFQ describes the project in general terms, the RFP includes a detailed SOW. These RFQs or RFPs are then either advertised or sent to potentially qualified A/E firms and selection is made from responding firms.

A detailed description of the 6 Percent Fee Limitation has been presented by American Council of Engineering Companies (ACEC) and appears as follows:

Fee Limitations on Design Contracts

Since 1939, federal construction agencies have been required by law to limit the fee payable to an architect or engineer to 6 percent of the estimated construction cost. Presently, there are at least four statutes that prescribe limitations on architect-engineer fees and apply to all civilian and military construction agencies with the exception of the US Department of State.

Federal agencies have interpreted the statutory fee limitations as applying only to the part of the fee that covers the production and delivery of "designs, plans, drawings, and specifications." The agencies, therefore, consider that the 6 percent fee limitation does not apply to the cost of field investigations, surveys, topographical work, soil borings, inspection of construction, master planning, and similar services not involving the production and delivery of designs, plans, drawings, and specifications. Most direct federal awarding agencies have, as a part of their supplement to the FAR, a list of those items exempt from the 6 percent fee limitation. American Council of Engineering Companies (ACEC), 2023.

4.4   WRITING ENGINEERING PROPOSALS

A critical component of the civil engineer's tool box is problem solving. Clients sometimes request engineering and technical support without a clear understanding of their needs. Conversely, some clients have a very clear understanding of their needs but are unsure how to address them. Regardless of the specific client situation, it is imperative that the engineer have a clear understanding and demonstrated skill to communicate problem solving. These key components to problem solving include:

- Identifying the client's particular problem/s
- Possessing background knowledge, the ability to work as a team member, and preparing a clear and comprehensive SOW
- Understanding the client's requirements and constraints
- Having the ability to communicate clearly
- Formulating technical alternatives
- Providing the client with alternative evaluation and/or selection
- Performing engineering design including engineering plans, specifications, and cost estimates
- Offering construction assistance, construction monitoring, or construction management,

- Providing start-up assistance and/or operations and maintenance assistance
- Creating a realistic project schedule

"Nothing is so inspiring as seeing big works well laid out and planned and a real engineering organization."

—Sir Frederick Handley Page, English Industrialist and Pioneer in the Aircraft Industry

More details on these components appear below.

4.4.1 Problem Identification

An engineering project is typically born as a problem or challenge to the client or operating group. Often, the client does not clearly understand the problem and it is rarely articulated well in the form of clearly explained subtasks, tasks elements, schedule, or budget. For the purposes of these discussions, we will use the terms "client" and "operating group" synonymously, since the operating group within an industry or agency will actually be the client anyway.

The engineer generally has some idea of the problem from initial communication with the client, or possibly an RFP, and then gains more understanding after a site tour. A client site tour is often referred to as a "site walk." If the client intends to solicit outside commercial engineering support, a site walk may be accomplished with representatives from competing engineering firms, potentially in groups. The client can then address the questions from all the competitors, so the competition is fair for all parties. If, by chance, the site walk is performed with the representative(s) from only one firm this could be good or bad news. A long-term client may perform singular-firm site walks with an engineering firm that is trusted. Otherwise, singular site walks can be an indication that the client is simply seeking another proposal or an alternate idea from a competing engineering firm. Experience shows that the success rate, sometimes referred to as the "hit rate" for this scenario, is lower than the "hit rate" for a site walk performed with groups.

Once the engineer has a basic understanding of the problem, they can begin to articulate the "problem identification." The problem identification statement should generally appear early in the introductory section of the engineering proposal. This statement should be constructed in the engineer's own language and could be enhanced with relevant practical knowledge from other similar projects in the engineer's portfolio. The problem statement will form the foundation of the proposal and should demonstrate to the client that the engineer has a clear understanding of the client's situation and impact on the business activity and operations.

Our experience has shown that a problem identification statement can be greatly enhanced by including a short description of the client's primary objective and any secondary/tertiary objectives if applicable. This "objective" statement should be clear and concise, approximating a paragraph or less. It is also desirable to include an "approach" statement immediately following the objective. This statement will very briefly announce the engineer's overall approach to solving the problem. Again, it is recommended that the statement be clear and concise, approximating a paragraph or less. These objective and approach statements serve as a brief announcement of the overall direction of the proposal.

So far, the engineer has launched the proposal effort and described their understanding of the problem, the objective, and approach. Now is an excellent opportunity to integrate background knowledge into the direction of the scope of work task elements. It may also be an opportune time to suggest optional tasks, beyond the SOW, if they would be applicable and valuable to the client's overall objectives. Background knowledge can consist of other similar projects and/or the application of engineering principles the client is unaware of or did not consider. This is where the engineer demonstrates that they have truly comprehended the client's problem/ objective and has analyzed the condition and synthesized potential solutions for the client.

Once the problem analyses and synthesis are complete the engineer should incorporate the client's requirements and other related constraints. The constraints may be as simple as a client choice of color, security requirements, operating hours, or as complex as the presence of a radiological species on the site. However, these requirements and constraints should be regarded as hard boundaries for the client's project.

A complete well-written proposal should include proposal assumptions. Proposal assumptions are a very important tool for the engineer. In the course of calculating the potential solutions and alternatives to the client's problem the engineer makes numerous assumptions. These assumptions are likely related to the site conditions, complexity of the client's original problem, weather, client contract review periods, client report review periods, resource availability and costs, site accessibility, number of meetings, meeting times, permitting requirements/costs/conditions, and a myriad of other project-related conditions. It is critical to capture these assumptions in the event it becomes necessary to show how task/subtask costs were developed if the project SOW changes or the project schedule is revised.

4.4.2   Background Knowledge, Teamwork, and Scope of Work

The application of the engineer's background knowledge can be weaved throughout three sections of a proposal for work which include the scope of work, the "qualifications" section, and the "project team and personnel resumes" section.

The scope of work is a logical set of tasks and subtasks that will accomplish the client's objective and solve the problem described in the problem identification statement. The work breakdown structure (WBS) is a fundamental tool used in project management and systems engineering. It is likened to a structure resembling a tree that describes the summation of subordinate costs for tasks, materials, and so forth, into their successively higher-level "parent" tasks, materials, and so on. Each element of the WBS includes a description of the parent task to be performed. This technique is used to describe, define, and organize the total scope of work for a project (Norman et al. 2008). More details on task and subtask identification and the work breakdown structure appear in Chapter 7, Executing a Professional Commission.

A typical SOW often includes a task for synthesizing technical alternatives for the problem. If it's not a task that the client specifically included then the engineer may have an opportunity to add a task that considers alternate, cost-effective solutions.

If technical alternatives are included in the SOW, another typical task element is the "alternative evaluation" of these alternatives. These analyses will most likely include a detailed evaluation of the advantages/disadvantages of each alternative with a corresponding summary on the capital outlay, operation/maintenance costs, permitting requirements, sustainability evaluation, implementability, constructability, and any other criteria important to the client or stakeholders.

Other follow-on tasks can include construction monitoring (or construction management) of the actual construction and implementation of the project. The objective of this task is to have the original designer (the A/E firm) review and verify that the construction conforms to the intended design concepts. Another benefit of this service is to review the contractor's work and be present to comment on any potential change orders the contractor may request. In addition, the A/E firm can also be available to comment on the resource requirements and commitments the contractor provides to the job. The benefit is related to the overall impact to the project schedule and the project delivery date. The engineer generally has a great deal of design and construction experience. If the engineer is on-site for this task, it would be possible to provide a credible opinion on the resource commitments to the project before incurring a potential delay in the final delivery of the project.

Another critical element construction monitoring accomplishes is verification of material specifications. The contractor/builder may use the specifications of the construction materials to their advantage for price reduction to maximize their profits. Substitution of lesser quality materials is usually a disadvantage for the owner with regard to life cycle and/or performance. In addition, if a contractor uses materials of lesser quality this practice may cause a potential liability for the engineer in the overall performance (or failure) of the finished product. Construction monitoring is usually a win/win scenario for the owner and engineer. There will be more on this subject later in this chapter.

4.4.3   Client Requirements and Constraints

A clear understanding of the client's requirements and constraints is an essential building block in the client relationship. The requirements will have a major impact on the scope of services. Comprehending the constraints and explaining how the engineer's proposal will address these constraints will demonstrate critical thinking to the client.

4.4.4   Clear Communication

Clear communications provide the conduit for transmitting the engineer's knowledge and experience of practical application to the client's project. Clear communications are a key component in the client relationship and include verbal, nonverbal, and written skills. More information on this subject may be found in Chapter 13, Communicating as a Professional Engineer.

4.4.5   Formulating Technical Alternatives

Clients employ engineers to apply technical knowledge and critical thinking to complex problems and projects. Most projects have many different alternative solutions and these solutions come with a myriad of advantages and disadvantages from project initiation through construction. Civil engineering firms with technical expertise matching the client's needs are best positioned to develop viable technical alternatives to address their client's requirements.

4.4.6   Alternative Evaluation

Many technical alternatives have different cost elements in a specific practical application. For example, one alternative may be significantly more expensive, more reliable, and require much less maintenance versus a simpler alternative that is less expensive. An engineer can present these alternatives in a concise fashion to help the client choose between the alternative that best fits their needs. This is where an understanding of the client's requirements and comprehension of their constraints will enable the engineer to provide superior client service.

4.4.7   Design, Plans, Specifications, and Cost Estimates

Engineering design usually culminates with a set of engineering plans, specifications, and cost estimates sometimes referred to as P, S & E. The level of detail in the engineering plans should be described in detail in the scope of services, assumptions, limitations, and corresponding contract. Depending upon the engineering plans, the set may include civil drawings accompanied with structural, mechanical, electrical, process, and architectural drawings, among others. The specifications may provide details on the materials of construction, preferred vendors or suppliers, interfaces, quality, quantity, type, and compliance with other specifications and specific codes for the general contractor (GC) to procure and install in accordance with the intent of the design. The engineering plans and specifications often include an engineer's estimate for the entire project. The cost estimate may include a large variable: labor costs. In fact, the project should clearly state whether the project requires union labor. The engineer usually has a good idea about material, labor, and installation costs. However, a GC's labor rate can vary according to the local demand for labor or specific classifications of labor. Therefore, the engineer's estimate can only provide general guidance on the total cost of a project. After the project is sent out for bidding, the corresponding construction bids show the real costs.

4.4.8   Construction Assistance, Monitoring, and Management

The engineer may also be invited to provide construction assistance or construction monitoring during the GC's installation. The objective of this task is to provide the owner with the materials of construction installed in accordance with the intent of the design.

One critical note for the engineer performing construction monitoring is that the GC (usually) works for the owner. If the engineer observes an inconsistency in materials compared to the specification or a potential installation flaw, the engineer should promptly notify the owner. Frankly, GCs often don't like to have engineers monitor their construction. Extreme caution should be employed so the engineer does not give instructions to the GC. There are many instances where an engineer innocently provided guidance or instruction to the GC in a true effort to provide client service. The GC may have intended (and bid) the project in a different way or provided alternative materials and, therefore, feels a construction claim or change order is warranted since the engineer directed the GC to perform a task differently. These construction claims will come as a surprise to the owner. Remember, owners don't like surprises and claims associated with the engineer's presence on the job can tarnish the engineer's reputation with the client.

4.4.9 Start-Up and/or Operations and Maintenance Assistance

Once the construction of a facility is complete, an owner may have a need for start-up assistance or operations and maintenance assistance. These services may be optional services in the original contract or may be recognized later by the client as necessary. The engineer as the designer is recognized and accepted as the expert and may have an opportunity to provide these services to the client. This is another example where clear communications and client service will "pay off" for the engineer as a reward in the form of additional work and as a win/win scenario for the engineer and the owner.

4.4.10 Scheduling

Work Breakdown Structure—The WBS is a comprehensive classification of the project scope of work that concentrates on the planned project outcomes in a hierarchical fashion. In summary, the WBS is a purposeful family tree that captures all the work of a project into smaller increments in an organized way. Preparing the project WBS is an important step toward managing the project's inherent complexity and should be developed before the project schedule is prepared (Norman et al. 2008).

Internal Reviews, Client Reviews—It is important to include internal reviews and client review periods in the overall project schedule. These review cycles are an important component of the project quality system and generally improve the integrity and accuracy of the delivered project. Review periods vary with the complexity of the engineer's product and vary from several days to weeks depending upon the project elements and complexity. In addition, incorporating key client reviews of the deliverable product provides the engineer with an opportunity to achieve a high degree of client satisfaction and allows the owner to have "ownership" in the process.

Response to Comment (RTC) Tables and Resolutions—An RTC table is a tool commonly used by the engineer to communicate a complete understanding of the client's review comments and review comments from others in a comprehensive and organized fashion. The RTC table summarizes all these comments and states "how" they were handled in a column labeled "accepted," "rejected," or "revised," and usually include the final disposition of the comment. This table is an important tool because it provides a summary of every reviewer's comments, some of which may contradict one another and becomes part of the project file documents. It also simplifies the final checking process since the reviewers do not have to go into the document and track the final disposition of their comments. Additional details on the RTC table and a sample format appear in Chapter 13, Communicating as a Professional Engineer.

Final Production Time—The engineer should consider the final project production time and include this time period in the project schedule. The final product may be a report or data, but it does take time to assemble, and it should be checked for accuracy and completeness before delivery.

Delivery Time—Upon completion of the final product the engineer can arrange for delivery. Depending upon the ultimate location of the client(s), at least a one-to-two-day period should be reserved for final delivery. Electronic delivery of engineering reports or data can be accomplished much more quickly but time should still be reserved for this activity. It is also recommended to follow-through that delivery was achieved, and the client can open or read the final product (in the event of electronic delivery).

Reserve Time—An experienced engineer will usually place a one-to-two-day reserve period in the project schedule. Experience shows that despite superior planning, events beyond our control can disrupt the project schedule and impact the delivery of the finished product.

Software Programs—Software products should be verified for accuracy before final delivery to the client. It is also recommended to have a brief introductory session where the engineer can present the product and perform an initial demonstration of its utility and capabilities.

4.5 THE CONTRACT

The contract is the glue that binds the client and the engineer to the SOW. It is a formal, legal document that usually cites the tasks from the original RFP, the engineer's tasks, schedule, assumptions, and budget. It includes numerous clauses and terms that usually trump any discussions and verbal agreements made during the entire process. Sometimes clients expect the Engineer to begin the project and tasks for the project before the contract is executed. If a client insists that work begins immediately, it is highly recommended that the Engineer be certain that the contract is in full force and effect and that the owner's representative has the authority to approve the contract. Otherwise, the Engineer may not be compensated for any work performed without a contract. Most contracts include a clause that states "the terms of the contract supersede any other agreements including oral agreements."

Contract review and approval is not simply a formality. Typical contracts are incredibly detailed and complicated legal instruments. For example, if disputes were to occur between the client and the engineer, the case would be tried in the city where the last party signed the contract. Clients usually ask the engineer to review and approve a contract before sending it back to the home office or facility location. Therefore, the last party to sign and date the contract did so at their home office where they likely have legal representation and recognition in their own city. If the engineer were to be challenged and had to appear in court, the travel, meals, lodging, any attorney's fees and preparation would be borne by the engineer. This fact alone might enter into the final decision whether to attempt to challenge a dispute because of the great expense incurred.

Key elements to the contract are the clauses that typically appear in small print on the back side of the contract. These clauses describe the details and conditions for the contract. A standard contract will generally begin with a statement and reason for the contract and list the names of the parties involved in the agreement. There will likely be some additional information such as:

1. The original "request for services" or definition of the scope of services from the originator (client)
2. The original proposal from the consulting engineer and any addenda, meeting minutes, or telephone records relevant to the proposal
3. Any other relevant agreements such as a "master services agreement" (referred to as an MSA)

The contract terms are generally complex, detailed, and are prepared to serve as a legal document governing the arrangement for the parties involved in the contract. The engineer may be tempted to skip over these terms because they are anxious to perform the services, but this could be a costly mistake. These terms are generally listed in numerical order and may range from a few clauses to 50 or more. They outline the business elements of the legal arrangement ranging from the terms for payment to insurance requirements required to receive payment for the professional services rendered. Each client and respective contract is unique so the contract terms should be read carefully by the engineer. The engineer should be confident that they will be able to comply with the contract terms while performing the services, after completion of the work, and when receiving payment for the services.

Some of the typical terms and their meanings are listed for reference as follows:

1. Scope of Services: This clause is one of the most critical ones in the contract. It defines the actual tasks to be performed, the deliverable product(s) resulting from the effort, the budget, the schedule, the project location, the fees, and any other relevant statements describing the services the consulting engineer will provide. The engineer should take great care describing the scope of work and reaching an agreement with the client on an SOW that meets their needs.
2. Standard of Care: The standard of care refers to "how" the consulting engineer will perform their services. This clause describes that standard and generally limits this standard to the level of care and skill normally exercised by similar engineering professionals in the same locale, under similar conditions (such as time requirements, budgets, or task

elements in the SOW), at the date the services were performed. This clause is important because it's directly related to the overall quality of the deliverable product that can be produced by the engineering professionals in this area at this time for a given price.

3. Engineer's Responsibility: This clause describes "how and what" the engineer is responsible for and includes a statement that the engineer is an independent contractor which is directly related to providing an independent opinion and professional ethics. The clause may also state that the consulting engineer will provide qualified staff to perform the work; will employ safe work practices for these staff; will not be responsible for the safety or liability of other staff outside the organization that may be working on the same project; will work cooperatively with the client's employees, consultants, subcontractors, or other staff; and will retain a record copy of the project file and deliverable product for a given period of time.

4. Client's Responsibilities: This clause generally states the client will provide all the material, information, reports, and data pertaining to these services without limitation. It may also state that the client will disclose information regarding potentially hazardous situations, chemical compounds, and underground conditions including utility information; past or present information on the general and environmental compliance with local/state and federal regulations; any potential or pending court actions; or any other relevant information on the project or site location. Another client responsibility includes informing the engineer whether any other regulations or requirements should be included in the engineer's scope of services and whether other labor-related conditions may be applicable such as trade union representation or prevailing wage regulations. The clause may also state that the client's employees, consultants, subcontractors, or other staff will cooperate with the engineer. Many companies often require that this new work described in the contract be completed while the company remains in operation. So, the consulting engineer must evaluate if this requirement may add unanticipated costs to the contract to comply with this requirement.

5. Insurance: There are different types and amounts of insurance including automobile, general liability, and more. The insurance clause is important because these requirements may not be available or may impose greater cost that impacts the engineer's fee and profit.

6. Revisions or Contract Changes: This clause states that contract changes or revisions may be recommended or required by either party after the project begins by altering, deleting, or adding tasks to the scope of services. If the scope of services is altered, both parties are responsible to renegotiate in good faith to assess an equitable adjustment in the project budget, schedule, and deliverable product and to prepare and approve a project change order or work order reflecting this new adjustment. The clause will also state that if both parties cannot agree and approve a change order that the work will be suspended without penalties to either party and that the engineer is entitled for just compensation for the services performed to date on the project.

7. Contract Term and Termination: This clause generally states that the contract will begin at the approval (signature) of the agreement and that the agreement will be in force for a period of time or until the project is complete. It will also likely state that either party may cancel the contract at any given time without cause by providing an advance written notice with a time period varying from two to ten days. The clause also states that the client will compensate the engineer for reasonable expenses and labor charges up to the cancellation of the contract including any relevant demobilization fees and that the engineer will provide any related file information and partial deliverable products to the client.

8. Force Majeure: If, in the course of performing services, the engineer encounters conditions or causes beyond their control then "force majeure" is declared. Force majeure includes acts of God; acts of a legislative or judicial office; acts by the client's subcontractors; labor disturbance or strikes; floods; hurricanes; fires; war; or severe weather.

9. Site Access: This clause states that the client will provide unimpeded site access to the engineer and subcontractors; space for equipment, materials, or vehicles; utility access including utility services and relevant utility clearances; and any relevant permits.

10. Warranty and Ownership of Waste Products: This clause generally has two subject components. Warranty refers to an overall warranty for the deliverable product produced by the engineer. The engineer is cautioned to limit the deliverable product(s) and the subject clause to the accuracy and timeliness of the verbal and written information provided by the client (or client's contractors, agents) and to also limit the warranty to the general limitations clause of the contract.

    The other major subject component regarding this clause is related to any waste products and/or testing materials provided by the client to the engineer. The engineer should be certain to limit the risk of loss of sample materials and the responsibility for disposal of residual sample materials to the owner. The owner is considered the original generator of said materials and is ultimately responsible for proper manifesting and disposal in accordance with federal, state, and local regulations. The engineer should be clear that final return of waste materials and ultimate disposal is not included in the original scope of services unless, of course, this task is included in the scope of work. A typical construction project can generate a great deal of waste materials requiring proper transport and disposal.

11. Subcontracting: The subcontracting clause notifies the client that subcontractors may be employed to perform specific tasks for the project and that these staff may require site access or information access as part of the project team. The clause may also mention a mark-up for these services to the client if the engineer contracts with this subcontracting firm. Finally, the clause may also state that the contract agreement cannot be assigned to a third party without the written consent of both originating parties. This clause protects both parties from having to execute a contract with parties other than the original ones in the agreement.

12. Dispute Resolution: This clause outlines the procedure for resolving disputes that may arise out of the interpretation, enforcement, implementation, or performance of the services in the scope of work and contract. Generally, both parties agree that they will attempt to resolve the dispute in a meeting within the existing upper management structure of both parties within a relatively short period of time. If this is unsuccessful, the next level of resolution is generally by third party or mediation. One mediation procedure often cited is the architect's Construction Industry Mediation Rules or an alternate association. Finally, if this level of resolution is unsuccessful the clause may mention legal claims by filing suit in court in the jurisdiction of one of the corporate headquarters. The caution here is that if a lawsuit is filed both parties may want the suit to be tried in their hometown because the courts may favor the local entity.

    The losing party in the lawsuit may have to pay travel expenses for the plaintiff and/or attorneys.

13. The clause on dispute resolution is helpful to provide a roadmap for resolving disputes. This is particularly important because a dispute can often stop or slow the progress on the project which can lead to an array of other problems. It is also helpful to resolve project-related problems at the lowest level of authority possible to keep the client-CE relationships amicable and positive.

14. Governing Laws: This clause generally states that the contract is subject to the laws of the United States and specifically to the laws where either the engineer's or client's home office is located.

15. Severability: This clause usually states that each contract term is separate from one another. For example, it's acceptable for both parties to choose to ignore a specific clause. However, the other clauses of the contract are still in effect.

16. Entire Agreement: This clause generally states that the agreement applies to the immediate task order and any future task orders. It also usually states that any amendments to the agreement shall be in writing. Verbal agreements outside of the written agreement will not be recognized. This clause is important to the consulting engineer because it clearly states that the written agreement is the legal controlling document and that any verbal agreements between the engineer and project-related clients are not recognized. If a client insists on a verbal agreement the engineer is cautioned to prepare a revision clause to the written agreement and have it approved by the client before proceeding with the verbal direction.

17. Risk Allocation: This clause attempts to discuss risks and rewards and tries to balance claims against the engineer to a reasonable proportion for responsibility for the actual cause of any losses. The clause often has multiple parts including:

    a. Limitation of Liability: This subpart clause generally states that the engineer will only be responsible for potential damage or loss payments up to a preset dollar limit (in fees ranging from $10,000 to $100,000) or the limit of the actual contract. The limit usually applies to any potential losses or claims in connection to the contract, and it usually has some time period specified when it's possible to seek these claims generally varying from three months to a few years. The clause may also state that claims may only be filed against the organization and not filed against any specific individuals providing some security to the engineering employees for the company. In addition, more recent clauses sometimes limit the percentage of fees that may be based on an estimated percentage of the engineer's contribution toward the resulting problem or loss. This means that if an engineer had a very small role on a project and a corresponding small, estimated contribution toward a failure of a large construction project, then the engineer would be responsible for a similar small percentage of the fees to correct the problem or loss. In the recent past, some consulting engineers have performed very limited technical services (with correspondingly small fees, such as a few thousand dollars) and have been sued for large sums, thinking the engineering firm had large assets or large insurance coverage. This clause attempts to level the playing field and potentially expose the engineer to a fair and reasonable amount of exposure when working on large, intricate, or expensive projects for small portions of the actual project. The final contract may have a similar clause protecting the client in the event the engineer suffers losses from a client or subcontractor-related cause.

A Note on Liability: The most important concepts of this contract clause are for the engineering firm to limit the monetary extent of liability in the project and to attempt to limit the liability to potential causes of damage to the scope of services performed by the engineering firm or some fee like $10,000, if possible. Typically contracts present reasonable liability coverage to the monetary limits of the engineering services or $50,000, whichever is less. Limiting liability exposure to an originating cause within the firm allows the firm to be protected by their own insurance.

b. However, some clients will only accept contracting terms using their own preapproved contracts. Under this condition it is especially important for the engineer to carefully review the liability clauses. We have seen liability clauses by large public and private clients where this clause asks the engineering firm to accept liability for their own subcontractors, and other contractors and subcontractors working on the site. These clauses are particularly onerous because the engineer has no oversight for these other firms and may risk the viability of the entire engineering firm for mistakes made by others.

c. Indemnification for engineer clause: This subpart clause generally states that the client indemnifies the engineer and their employees, officers, shareholders, or agents from any lawsuits, damages claims including attorney's fees potentially caused by the client's negligent performance of services under the contract.

d. Indemnification for client clause: This subpart clause generally states that the consulting engineer indemnifies the client, and their employees, officers, shareholders, or agents from any lawsuits, damages claims including attorney's fees potentially caused by the consulting engineer's negligent performance of services under the contract.

18. Ownership of Instruments of Service: This clause generally states that the client owns the documents, but the consulting engineer retains the right to retain a copy for their files. This clause is sometimes challenged when there may be an occasion when the client may not pay for the engineering consulting services in a timely manner and the engineer feels that it is necessary to hold back the finished product until payment is received. In addition, it is advised that both the A/E firm and the client work closely on this clause to avoid misunderstandings in data packages, usability, readability, availability, and presentation.

19. Late Payment/Assessments: A late payment and assessment clause in the contract is important because of the time value of money and the potential capitalization of (potentially) many simultaneous projects within the engineering firm at any given time. The typical engineering firm prepares and sends invoices monthly. This process captures all the project-related costs for the previous month including labor, project expenses, and overhead. Once these costs are computed a draft invoice is typically prepared for the engineering project manager's review and approval which in total may take an additional two to three weeks. Often the total time elapsed by the time the client receives the invoice from the previous month is six to eight weeks. Now the invoice has to proceed through the client's review and approval process which in turn can take an additional three to six weeks.

Often the total time elapsing from a professional at the engineering firm working on the project on Day 1 until the firm receives payment from the client is 8 to 12 *weeks under optimum circumstances. Under these conditions, the typical engineering* firm is paying at least 8 to 12 weeks' interest for the firm's entire portfolio of projects which can lead to increased overhead expenses thereby increasing the firm's cost and making the firm less competitive. A late payment clause can be a good tool to remind the client about the importance of paying promptly and hopefully the assessment component of this clause will not need to be accessed.

See *Chapter 11—Legal Aspects of Professional Practice* for additional information on contracts.

4.6   BUDGETING

The project budget is like the fuel for the jet aircraft with the client and the engineer onboard working together. Proper budgeting and budget tracking is critical for both parties—the project and the careers depend upon it. Some key points to remember:

- Budget by tasks, subtasks, and staff codes
- Include subcontractors, oversight and management costs, and any mark-ups

- Include project-related expenses like mileage, vehicles, equipment, per diem, or other related items
- Review assumptions for hidden costs like permitting fees that might be missed
- Include other related items like delivery expenses or report binders, color copies, and the like

4.7   ENHANCING THE ENGINEERING FIRM'S PROBABILITY FOR A SUCCESSFUL PROFESSIONAL ENGAGEMENT

"ASCE's Body of Knowledge Technical Outcome 8—Problem Recognition and Solving" is one of the most appropriate examples of engineering and business applications for the engineering firm. As part of the client proposal process the engineer and the firm clearly need to apply the six levels of cognitive achievement shown in the following list (American Society of Civil Engineers, www.ASCE.org).

ASCE's Six Levels of Cognitive Achievement Presented in the *Civil Engineering Body of Knowledge for the 21st Century (ASCE.Org, 2023)*:

- Knowledge
- Comprehension
- Application
- Analysis
- Synthesis
- Evaluation

These are the key variables in crafting an effective proposal for the client except for two additional variables that the civil engineer may find him- or herself working on for his or her entire career:

- Client relationship and
- Effective project budgeting and project management

This concept can be further explained by the graphic depicted in Figure 4.5.

The civil engineering firm has an opportunity to enhance their chance for winning the client's project and securing a professional engagement by completing the proposal in an effort that matches ASCE's six levels of cognitive achievement discussed above. Reaching this high level of cognitive achievement, building positive client relationships, and including effective project budgeting will usually "seal the deal" for the engineer and the client.

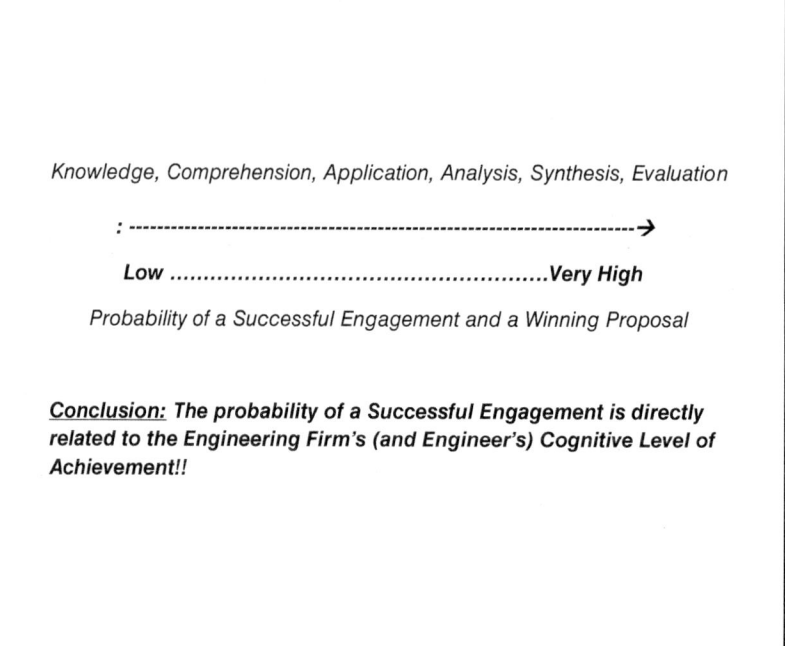

Knowledge, Comprehension, Application, Analysis, Synthesis, Evaluation

Low ....**Very High**

Probability of a Successful Engagement and a Winning Proposal

<u>**Conclusion:**</u> **The probability of a Successful Engagement is directly related to the Engineering Firm's (and Engineer's) Cognitive Level of Achievement!!**

FIGURE 4.5   Levels of cognitive achievement for the CE professional and CE firm.

4.8   WORKING EXAMPLE OF A REQUEST FOR PROPOSALS (RFPs)

Request for Proposals (RFPs) are often published by public agencies, private companies, or other owners asking for engineering assistance to solve a specific problem or to design and build a capital improvement. The RFP is a vehicle that owner's use to publicly request civil engineering firms to view and consider their request for engineering services. The RFPs can be published as a public document or simply provided to select engineering firms the owner is acquainted with. **Appendix A** presents a Draft RFP for a real problem that a local Wastewater Treatment (WWT) Authority published requesting engineering services for the decommissioning of an inadequate WWT plant and the design of a new conveyance line to an existing downstream regional facility.

Typically, an engineering firm would evaluate this RFP for an assessment on how the scope of work fits with the firm's capability, schedule, and business objectives as discussed in Chapter 12, Managing the CE Enterprise, Chapter 9, The Client Relationship and Business Development, other related chapters herein, and other related considerations like staffing and risk.

If the engineering firm decides to respond to the RFP, the firm will prepare a proposal for this project.

4.9   TYPICAL CIVIL ENGINEERING PROPOSAL

Assuming the engineering firm decides to proceed with the time, effort, and expense to prepare a proposal for the RFP discussed above and depicted in Appendix A, the firm would assemble a qualified team that could best complete the project cost-effectively for the client. The firm will then likely perform the following activities:

- Review the Draft RFP, select a project team led by a project manager,
- Meet with the client to discuss the project and gain valuable information that may not appear on the RFP,
- Conduct a site visit,
- Conduct an internal proposal kick-off meeting,
- Prepare a proposal outline including objective and approach statements consistent with the proposal,
- Prepare a draft proposal including task descriptions, schedules, level of effort (LOE work hours) budgets, assumptions, and qualifications. Integrate any outside sub-contractors that may be required,
- Depending upon the size and complexity of the client's project it might be necessary to upgrade the draft proposal to a 90 percent draft final proposal,
- Review, comment on, and revise the 90 percent proposal,
- Prepare a 100 percent proposal,
- Submit the final proposal.
- Meet with the client on the final proposal and respond to client comments, questions, or comments,
- The client may request a draft 30-minute presentation on the final proposal,
- Internally, Review, comment on, and revise the draft presentation,
- Present the final presentation to the client,
- Follow-up with the client on their thoughts about the presentation,
- Hopefully win the project and move into the contract and then the work phases.

An example "engineering firm's proposal" to the RFP for the typical CE problem discussed above appears in **Appendix B**. This problem was an actual one experienced by a small county Wastewater Collection and Treatment Authority in northern California. The actual engineering proposals were similar to this mock proposal.

4.10   TYPICAL ENGINEERING FEASIBILITY STUDY

The RFP depicted in Appendix A and the engineering proposal presented in Appendix B are both for the same project with a client who is a small northern California Wastewater Collection and Treatment Authority. The engineering Feasibility Study performed for this client's needs includes:

- Cover Letter
- Introduction
  - Project Overview
  - Purpose and Limitations of the Study
  - Summary of Recommended Alternative
- Project Constraints
  - Design Standards and Permitting Requirements
  - Physical Constraints
  - Environmental Constraints
- Development of Alternatives
  - Strategy for Identifying Alternatives
  - Reasons for Disqualification
- Primary Alternatives
  - Common Elements and Design Considerations
  - Three Alternatives Presented
- Evaluation of Alternatives
  - Ranking Criteria
  - Alternative Rankings
- Recommendation
  - Reason for Final Recommendation
  - Benefits
  - Cost Ranges
- List of Tables
- List of Figures
- References
- Appendices

The Feasibility Study for this project appears in **Appendix C.**

4.11   SUMMARY

This chapter describes the request for proposal process, contracting with the government, engineering proposal content, selection processes, followed by contracting details for performing the project, and enhancing the engineering firm's probability for a successful professional engagement. In addition, a sample RFP is included in Appendix A for reference. This RFP is then followed by a sample proposal displayed in Appendix B for the project stated in the RFP presented in Appendix A. Finally, a sample Feasibility Study Report in presented in Appendix C for the project included in Appendices A and B for an comprehensive, complete picture of this process.

Overall, this chapter emphasizes that the engineer can increase the probability of submitting winning proposals and winning more jobs by thoroughly applying the six levels of cognitive achievement: knowledge, comprehension, application, analysis, synthesis, and evaluation to the client's problem. The consulting engineer may also find that they will continually improve two other variables for their entire career: building client relationships, and effective project budgeting and project management.

BIBLIOGRAPHY

American Council of Engineering Companies, (ACEC). (2023). www.acec.org.

American Society of Civil Engineers (ASCE). (2023). www.asce.org.

Norman, Eric S. et al. (2008). *Work Breakdown Structures: The Foundation for Project Management Excellence*. John Wiley and Sons, Hoboken, NJ. ISBN: 978-0-470-17712-9.

CHAPTER 5

The Engineer's Role in Project Development

Big Idea

Civil engineers play many different roles in the project delivery process.

> Engineering problems are under-defined, there are many solutions, good, bad and indifferent. The art is to arrive at a good solution. This is a creative activity, involving imagination, intuition and deliberate choice.
>
> —Ove Arup, engineer and entrepreneur

Key Topics Covered

- Participants in the Process—The Players
- The Flow of Work
- Predesign
- Design
- Design During Bid and Construction
- Post-construction Activity

Civil Engineer's Handbook of Professional Practice, Second Edition. Karen Lee Hansen and Kent E. Zenobia.
© 2025 John Wiley & Sons, Inc. Published 2025 by John Wiley & Sons, Inc.
Companion website: www.wiley.com/go/hansen/CivilEngineersHandbook

Related Chapters in This Book

- Chapter 2: Background and History of the Profession
- Chapter 3: Ethics
- Chapter 4: Professional Engagement
- Chapter 6: What Engineers Deliver
- Chapter 7: Executing a Professional Commission
- Chapter 8: Permitting
- Chapter 11: Legal Aspects of Professional Practice
- Chapter 15: Globalization
- Chapter 16: Sustainability
- Chapter 17: Emerging Technologies

5.1   BACKGROUND

The architectural, engineering, and construction (AEC) industry always has operated on the "virtual" organization principle and is infamous for its fragmentation. Constructed products involve a staggering number of players. These include private owners, developers, government agencies, engineers and architects, other designers, builders, product and material suppliers, real estate agents, lending institutions, the general public, and inspectors among others. In the United States, where the AEC industry historically has made up 9 to 10 percent of the gross national product (GNP), the majority of firms employ fewer than 20 people.

The industry is design-intensive because most projects are one-of-a-kind. Often with limited local knowledge, AEC professionals must produce unique products with stringent cost, schedule, and quality standards. Much of the design for any particular constructed product is performed by separate individuals and firms. Furthermore, project team members may not be focused on a shared goal. Clients and owners come in a variety of types and sizes, some with considerable design and construction experience and, more frequently, others with none. The owner typically thinks in terms of quality, as well as short- and long-term costs. Architects and engineers have a different perspective; often they are motivated by the desire to avoid mishaps and to minimize their costs relative to billable hours.

Civil engineers must work with a staggering number of determinant and nondeterminate processes, a vast array of participants, and the need to evaluate outcomes and manage risk to develop and evaluate prospective designs. This chapter examines the civil engineer's role in project development. The chapter discusses the people involved in moving projects from ideas and needs, through design and permitting, and through to completion. It explores the many roles civil engineers play in the design and the project delivery process, as well as the deliverables connected to that process.

5.2   PARTICIPANTS IN THE PROCESS—THE PLAYERS

As discussed in Chapter 2, Background and History of the Profession, the first civil engineers had to develop knowledge and skill in a wide range of fields: irrigation, palace and tomb building, weapons, and fortification. Today's civil engineer also must develop considerable expertise, but this knowledge tends to be more technically specialized. What has remained constant for millennia is a need for civil engineers to be problem solvers, innovators, analysts, critical thinkers, and communicators.

Another uninterrupted theme throughout the history of civil engineering is the involvement of three key players in the project delivery process:

- Owner or client
- Designer (engineer or architect)
- Builder or contractor

For a description of the roles that these participants play, see the accompanying textbox—-*Participants in the Design Process.*

These three main groups—owner or client, designer (engineer or architect), and builder or contractor—coexist and interact with a fourth group composed of legislative and statutory bodies, regulatory and permitting agencies, licensing organizations and professional societies, and the public and special interest groups. (See Figure 5.1.) Chapter 8, Permitting, describes the important role played by these participants in the design process.

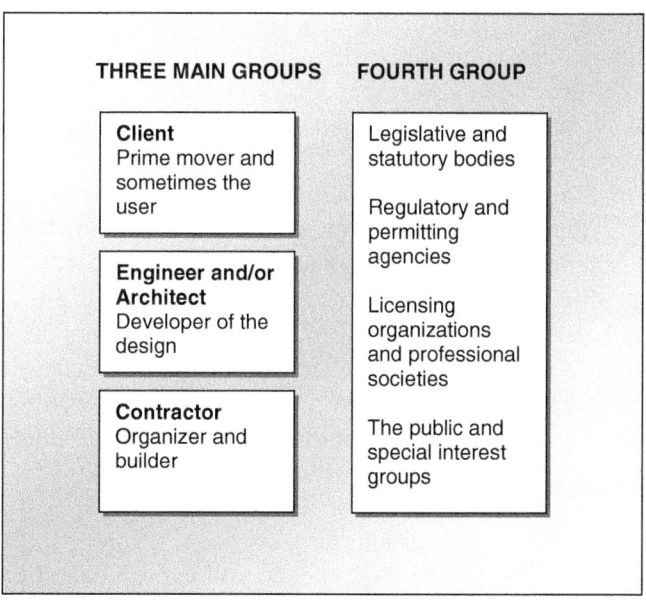

FIGURE 5.1   Players involved in project delivery.

Participants in the Design Process: Owners, Design Professionals, and Contractors

Designing and constructing projects frequently involves hundreds of participants; however, there are three principal players in every project: the owner, the design professional, and the contactor.

Owners

Although the list of potential construction project owners is nearly infinite, the short list includes governments (federal, state, and local), districts (school, irrigation, and reclamation), for-profit and nonprofit corporations, partnerships, and individuals. Construction projects occur when a representative of one of these groups seeks to mitigate a need or realize an idea. The owner's role generally is to provide the site; finance both the design and construction of the improvements; give timely, accurate feedback to the design professional; and operate the facility. Owners commonly engage the services of a design professional to conceive the design and produce the construction documents. Owners place contractors under contract to execute the work described in the construction documents.

The Design Professional and Design Consultants

Design professionals—primarily architects and engineers—offer a wide variety of services to project owners. Traditionally, they have created projects and produced construction documents and contract administration on behalf of owners, under service agreements called design contracts. The tumult in the design profession in the last couple of decades has prompted architects and engineers to diversify the services they offer for a fee. Many now include facility lifecycle analysis, recycling and management, [sustainable design and energy efficiency], construction management, as well as practice management in the range of services they provide.

The number of different design professionals involved in producing construction drawings varies according to the type and complexity of the project. Individuals or very small organizations generate most drawings for homes. In some states, laypersons may design homes and duplexes without a design professional's license. Developing the design for a hospital, performance center, [high rise building], or manufacturing facility, in contrast, may require many highly specialized design professionals who, after rigorous examination, have been licensed by the states in which they do business.

The core participants commonly responsible for the design of building construction projects include architects and landscape architects, and geotechnical, civil, structural, mechanical, and electrical engineers.

Architects, whose authority to design projects derives from state licensing boards, have the daunting task of identifying their clients' problems during the predesign phase and describing their solutions to them, using pictures and words, during the design phase. Architects conceive the physical attributes of a project and incorporate local land-use ordinances and applicable building code requirements into their designs. Their interests and professional responsibilities are focused primarily on how a project looks (aesthetics) and how it works as a product (that is, will it protect its users from the elements and from injury during catastrophic events such as earthquake and fire? Does it fit effectively into its environment? Does it fulfill the owner's needs?).

The number of specialists and the variety of services that design consultants offer is substantial; however, architects commonly hire structural, mechanical, and electrical engineers for significant portions of building work—areas of specialty for which they frequently do not have the training, license, or personnel. Large design firms, however, frequently have in-house engineering capability, which gives them more market share, greater efficiency, and more control over the design process. Such organizations are commonly referred to as architect/ engineer (AE) firms. Regardless of the size and organizational structure of the office, the overall responsibility and liability for the design of a project reside with the architect, who becomes known as the prime design professional (the "prime" designer or contractor is the term given to the entity that signs a contract with the project owner).

Although many civil engineers are qualified to prescribe the treatment required to prepare soil for a project, geotechnical engineers are registered professional engineers who are required to devote several more years to practice and/or additional education after becoming licensed civil engineers before they can legally call themselves geotechnical engineers. They are hired by owners to investigate a project site and produce a comprehensive evaluation of its soil conditions, which are recorded in a geotechnical report. Geotechnical engineers [and in some cases environmental engineers] commonly investigate the past uses of the site and its hydrology, identify its soil types, determine whether and to what extent a site is contaminated, and delineate any procedures that the contractor must follow to prepare the soil for its intended role. For example, soils must be made stable and competent to bear the weight of structures and vehicular traffic for years, and soils may be used to encapsulate solid waste and to line excavations and earthen structures that will contain water.

A host of participants in the design and building process use the geotechnical report. The structural engineer uses the report to design the foundation of a structure; the landscape architect uses the report to develop the specifications for the planting and irrigation of landscaped areas; and the contractor and subcontractors use the report to determine the costs of earthwork (such as site preparation, excavation, soil preparation, pile-driving, and foundation work) and evaluate the risk associated with it.

The principal concern of geotechnical engineers is how the soil will perform over time with the planned activities imposed on it. Their contracts with the owner normally

(Continued)

require them to prepare the geotechnical report, and monitor, inspect, and approve earthwork while it is being performed. Additionally, the geotechnical engineer resolves issues that arise in the course of construction, such as the mitigation of contaminated soil that might not have been apparent during the site investigation. Beyond these functions, they do not typically get involved in design.

Civil engineers typically produce most of the construction documents related to engineering construction (streets and highways, sewer and water treatment plants, harbors, dams, levees, bridges, and utilities). They must be licensed by the state in which the work is performed. On commercial building projects, the civil engineer plays a relatively limited design role, normally taking responsibility for on-site grading, drainage, and paving plans and specifications; for off-site improvements (driveways, gutters, curbs, and sidewalks along a public thoroughfare); and for the design of certain on-site underground utilities (sewer lines, fire system supply, storm drainage, domestic water supply). Civil engineers often cite the standard specifications of the city, county, or state in which the project is located, particularly in the design of off-site improvements. These specifications are frequently tried-and-true specifications that are developed by the state departments of transportation (which invest considerable funding in research) and are often wholly adopted by public works departments at the local level.

Structural engineers specialize in the design of foundations (piles, caissons), substructures (habitable portions of a structure that are below ground, such as basements), and superstructures (the portion of the project above grade, or above the water in the cases of bridges built across bays, lakes, and rivers). Like civil engineers, structural engineers are licensed by the state in which they do business, but they are frequently required to have specialized education and training beyond that of a civil engineer. Structural engineers—frequently hired by architectural firms for their expertise—are focused on the performance of the structural system under various loading conditions that fall into two classifications: static and dynamic loading. Static loading comprises dead loads (gravitationally imposed loads resulting from the weight of the structure and its permanent equipment) and live loads (mobile loads that are not necessarily present at all times). Furniture, snow, hydrostatic pressure (the pressure at any point exerted on a surface by a liquid at rest), and a building's occupants are examples of live loads. Dynamic loads, such as seismic activity and wind can occur suddenly, and vary in intensity, duration, and location.

Structural engineers are responsible for protecting the lives and property of project users in a cost-effective way. Although their focus is on the performance of a structure under the loading conditions just mentioned, they should also be aware of the aesthetics of the project.

Mechanical engineers involved in building project design are responsible for plumbing, sewage and piping systems, and for heating, ventilating, and air conditioning systems (HVAC). Mechanical engineers commonly form consultant agreements with the A/E to develop and describe the plumbing and HVAC systems in buildings, which are designed to ensure the comfort and health of building occupants. Plumbing and sewage systems provide an adequate source of water for human consumption and sanitation, and effectively dispose of wastes generated in the building. The heating, ventilating, and air conditioning equipment is used to control environmental comfort factors such as the temperature of the ambient air in a building, the mean radiant temperature of the surrounding surfaces, the relative humidity of the air, the pureness of the air, and air motion. HVAC and plumbing systems in building projects present a significant design challenge, particularly in the distribution of conditioned air and piping through the structure. The involvement of mechanical engineers in the design process increases dramatically when they are involved in industrial construction projects, such as refineries, manufacturing facilities, chemical plants, and waste and water treatment plants. Indeed, they may hold the prime design professional role on these projects. Mechanical engineers concentrate on the performance of the systems they design.

Electrical engineers are involved in the design of a variety of construction projects, including massive power generation and distribution systems for state and federal governments, cogeneration power plants, and building construction projects, to name a few. As with the other engineers, electrical engineers must be licensed by the state in which they conduct business. In building construction projects, these engineers design the electrical service and communications systems on the site, as well as the site lighting, usually at the request of the A/E. They also design the service and distribution systems inside the structures. In addition, electrical engineers must design and clearly spell out the type and location of the electrical equipment and cabinetry and the means of distribution and controlling the power. Those engineers who work for the local utility company frequently control the design of the off-site system (the portion found in public utility easements). Electrical engineers focus on the proper sizing of the system, the location of the equipment, the distribution of the power, and the safety of the end user.

Landscape architects, also licensed by the state, specialize in developing ornamental landscaping plans, which includes selecting trees, shrubs, ground cover, and grasses, and designing the irrigation system required to support them. The landscape architect's work may also include some site improvements (such as walkways, garden structures, screens, fencing, and water features, all of which are referred to generally as *hardscape*.) Landscaping plays an important role—not only for the visual beauty it brings to a project, but for the beneficial effects that well-conceived and -executed design can have on the energy consumption of a building, as well as on air and water pollution.

Contractors

Although the term "contractor" is loosely applied to anyone who earns income from constructing things, sole proprietorships, partnerships, corporations, and joint ventures are the common legal entities that assume responsibility and liability for constructing projects under contract with the owner. Many states regulate contractors through licensing

(Continued)

boards, which assure the health, welfare, and safety of the public through education, testing, and, where applicable, the enforcement of state license laws and building cor.
There are distinct categories of contractor:

- *Engineering contractors* construct engineering projects such as highways, bridges, and industrial construction projects.
- *General building contractors* produce residences, multiple-family projects, commercial and civic buildings, and/or retail spaces.
- *Specialty contractors* focus on one portion of a project, such as plumbing, sheet metal and air conditioning, roofing, insulation, tile, floor coverings, and elevators.

The contractor who signs a construction contract with an owner is called the *prime contractor*. The prime contractor, for a variety of reasons, frequently hires specialty contactors for portions of the work, who become subcontractors under the construction contract. Plumbing, mechanical, and electrical specialty contractors are commonly hired in this fashion.
—Keith Bisharat (2008). *Construction Graphics*, John Wiley & Sons, pp. 5–7.

5.3   THE FLOW OF WORK

The typical project moves through several phases: predesign, design, bid, construction, occupancy, and eventually adaptive-reuse, and decommissioning and/or demolition. Civil engineers can be involved in any of these phases. (See Figure 5.2.)

5.4   PREDESIGN

In predesign, or planning, clients enter a "discovery" phase where needs and wishes are explored. If a client is large, client staff may be responsible for preparing a general plan of action and outlining requirements. Often large clients have constructed previous projects and have a wealth of experience on which to draw. A client without in-house capacity may hire a consultant, usually an architect or civil engineer, to help evaluate the need to build. These consultants may hire additional consultants to support their efforts. Table 5.1 gives examples of professional services available in predesign.

At this initial stage, the entity responsible for the evaluation forms a working organization and identifies the information needed. Data should include:

- History of the events leading up to the decision to build
- Purpose and function of the project

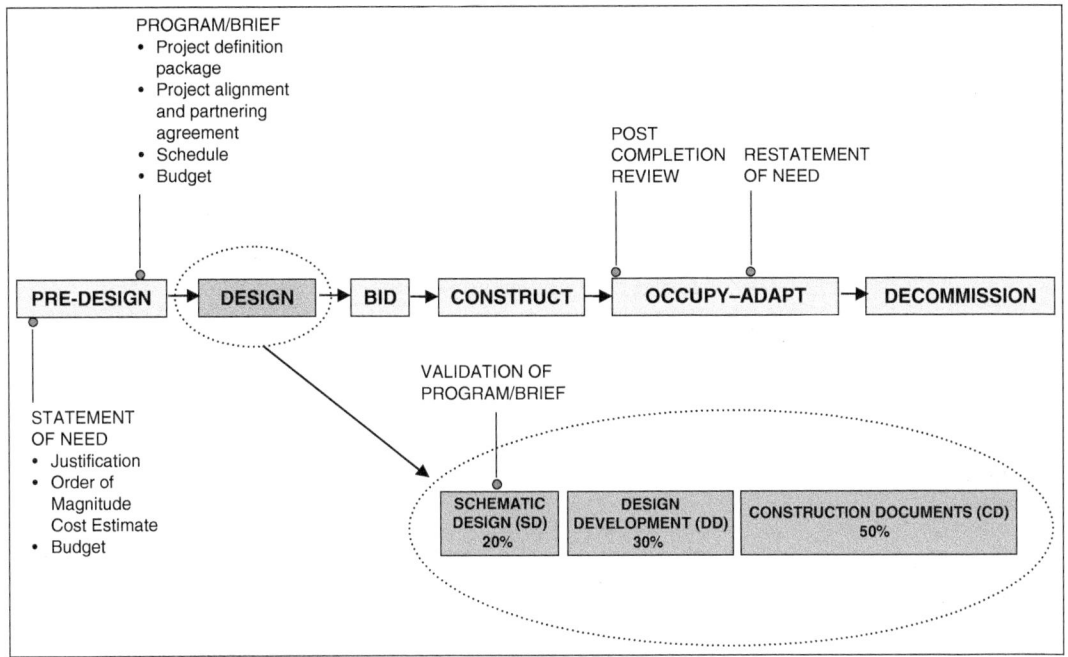

FIGURE 5.2   The flow of work.
Source: Adapted from Blyth and Worthington (2001).

- Policy decisions
- Timescale for the project
- Cost limit, or budget, of the project
- Details of the site and services
- Basic details of building requirements
- Comparable best practice

The result of this effort is a *Statement of Need*, which identifies the need for a new or remodeled facility based on business objectives or public policy and enables the client to gain internal approval for the project. In private corporations, upper management grants this approval. In public organizations, governing bodies, such as boards or councils, give the go-ahead.

The Statement of Need concisely states the problem, not the solution. Even after organizational approval has been gained, the client continues to assess needs. Either through the efforts of in-house personnel, with the assistance of consultants, or by a combination of the two, the client evaluates the needs and resources of the organization and generates options to meet those needs. All options should be considered, including the "do nothing" option. Options that do not involve new construction, such as leasing additional capacity or

TABLE 5.1   Professional services available in predesign.

Profession	Special Skills
General Civil Engineer	• Conducts site assessment • Highlights opportunities, potential problems, and building permit requirements • Coordinates other professionals and prepares design criteria, program, or a "brief"
Environmental Engineer	• Assesses need for environmental impact reports and other permitting requirements • Researches project site's history in relationship to previous uses and hazardous materials (Phase I Environmental Assessment)
Geotechnical Engineer	• Investigates and reports on soils conditions • Proposes initial approach for design of foundations
Structural Engineer	• Advises other professionals regarding load-carrying capacity of existing structures • Evaluates feasibility of conceptual designs
Transportation Engineer	• Performs transportation planning • Investigates rights of way, current/proposed/and use • Observes existing traffic conditions • Conducts traffic impact analysis, travel demand forecasts
Water Resources Engineer	• Evaluates flood plain analysis, channel design, stormwater conveyance and treatment, water supply
Land Surveyor	• Provides legal description and topography of site
Architect	• Conducts site and/or building assessment • Highlights opportunities and potential problems • Coordinates other professionals and prepares design criteria, program, or brief • Assesses Americans with Disabilities Act (ADA) requirements of existing building and/or new improvements
Landscape Architect	• Surveys condition of existing trees and planting • Performs Americans with Disabilities Act (ADA) assessment of existing site improvements
Urban Planner	• Interprets planning regulations • Advises on probable outcome of development proposal, changes of use, or new development
Contractor	• Reviews conceptual design for constructability (buildability) • Evaluates availability of materials and labor • Establishes initial construction schedule
Cost Estimator	• Forecasts project costs including design, construction, and other fees and expenses

remodeling existing facilities, also need to be contemplated. Relative benefits, drawbacks, and risks need to be analyzed. Feasibility studies may be conducted on particular aspects of the proposed options. Alternatives may be tested to determine their financial, economic, technical, or other advisability.

Workshops with users may be conducted to establish whether the project represents value for money and meets both organizational and functional needs. Key activities include:

- Confirming that options have been identified
- Agreeing upon which option(s) to pursue
- Identifying potential problems with items in the budget and agreeing on a total budget
- Establishing a timeline
- Carrying out a risk assessment
- Defining clear objectives
- Preparing a *Program* or *Brief*

The *Program, Statement of Need, Basis of Design (BOD)* or *Brief* (as it is called in the United Kingdom) captures the essence of the project. It converts organizational and business language into building terms and fixes functional relationships and major elements of the design. During the proposal phase discussed in Chapter 4, the program or brief forms a key component of the Request for Proposal (RFP). It sets out project parameters used by the client to instruct and select the design or design-build team. It also forms the basis on which to judge the relative merits of proposals submitted by these Teams (Hansen et al. 1996). The program or brief frequently is written with the help of a designer. Table 5.2 depicts a process for developing this document.

Each requirement included in the completed document should be described as being:

- Complete: fully describes the functionality desired
- Correct: is compatible with larger project (system) objectives and accurately reflects users' needs
- Feasible: possible to implement given the overall project and its environment
- Necessary: is really needed for conformance to overall projects, requirements, or a standard
- Prioritized: assigned a priority that indicates its relative importance
- Unambiguous: is written in simple, straightforward language so that all readers arrive at a single interpretation
- Verifiable: can be shown to be accomplished

TABLE 5.2   Steps used in creating a program or brief.

Elicit	Analyze	Document	Verify
• Write vision and scope	• Consider need to build	• Write purpose and functions of project	• Inspect requirements document
• Identify project objectives	• Evaluate physical context of project	• Record business case	• Cross-check functional performance requirements
• Define procedure for developing requirements	• Evaluate nonphysical context of project	• Establish cost and schedule limits	• Define user acceptance criteria
• Identify key users	• Prioritize requirements	• Adopt Statement of Need template	• Define project team
• Select champions	• Model requirements in terms of cost and schedule	• Identify sources of requirements	• Confirm requirements are being met
• Establish focus groups	• Apply Quality Function Deployment	• Create requirements traceability matrix	• Confirm Owner's approval hierarchy and communnications
• Set up support organization	• Establish roles for project team members	• Prepare job duty statements	

Source: Adapted from Blythe and Worthington.

Getting requirements right early results in significant payoffs: improved product quality, savings of time and budget, and better client relations. The concept depicted in Figure 5.3 is known to most firms involved in delivering complex products and is sadly familiar to clients who have experienced cost overruns and schedule delays. Project team members can exercise maximum control over the project's final outcome during the earliest phases. With the passage of time, the ability to exert a positive influence over the end product diminishes. On the other hand, mistakes and omissions in design briefs can lead to higher costs, increased litigation, schedule delays, and lower quality of the final constructed product (Salisbury 1990).

The importance of early project definition has been recognized for many years. As early as the 1970s, Preiser and Peña in the United States made important contributions to the field of *facility programming*. Even earlier in the United Kingdom, a 1960s investigation at the Tavistock Institute addressed briefing from the perspective of communication within the construction industry (Higgin and Jessop 1963). Others in well-established professional institutions, academia, and client organizations have continued to investigate the subject:

- The Royal Institute of British Architects (RIBA) divides the briefing process into four phases: Stage A, Inception; Stage B, Feasibility; Stage C, Outline Proposals; and Stage D, Scheme Design.

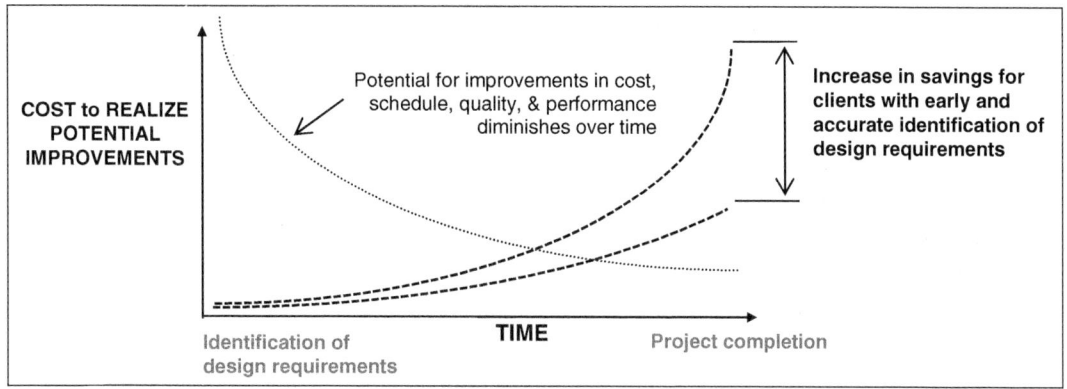

FIGURE 5.3 Importance of getting requirements right.
Source: Adapted from Boyd C. Paulson, Stanford University.

TABLE 5.3 Key programming (briefing) issues.

Common Client Concerns	Conditions Necessary for Success	Attributes of an Effective Brief	
• Early indication of cost • Early indication of schedule • Functional constraints – spaces and services • Environmental issues – planning and site	• Client and design team clear points of contact • Decision-making authority of client rep. • Effective client/ design team organization • Communication media clearly understood by all	• Limited number of key objectives expressed • Layered and iterative, reflecting design stages • Fluid until the last possible moment • Fixed and frozen once complete	• Requirements backed-up with hard data • Performance measured • Innovation balanced with established use • Users and building demands balanced • Related to chosen procurement process • Concise and clear

Compiled from Murray et al.   Compiled from Blythe and Worthington

- A monograph produced by the Building Research Establishment (BRE), *Better Briefing Means Better Buildings*, presents an outline that serves as a checklist of important considerations.
- John Worthington and others at the Institute of Advanced Architectural Studies (IoAAS) at the University of York have conducted professional level courses on design briefing and have gathered a large amount of empirical data drawn from its close association with DEGW, an international firm that consults on the planning, design, and management of workspace. See Table 5.3.

- An investigation conducted by Professor James Murray at the University of Reading viewed the brief as a communication tool. Case studies indicated that clients shared four project-related concerns and additional areas crucial to the success of the briefing process (Murray, J. 1993). See Table 5.3.
- The University of Salford's Construct I.T. Centre of Excellence have benchmarked the effectiveness of information technology (IT) and identified best practice in briefing and design. This study found that most briefing activities use terminology very specific to the AEC industry, thereby making communication with clients difficult. Additionally, the study noted that a significant gap exists between the leading-edge IT and best practice.
- Dr. Iain MacLeod of the Department of Civil Engineering at the University of Strathclyde has indicated that establishing a *Requirements File* assists in managing the briefing process. At the outset of the design work a requirements file is opened which contains: (1) a statement of what the client requires the design team/consultant to achieve; (2) a statement by the client delineating performance criteria and constraints; (3) a *design requirements statement*, a comprehensive list of the project requirements drawn up by the design team; and (4) a requirements checklist, based on the design requirements statement, which then is used in the design review process.
- Edward T. White, while on sabbatical from Florida A&M University, conducted a rich study called *Design Briefing in England*. His study showed that a good brief does not ensure good design, but a good design is very difficult to produce with a poor brief.

Clearly, the subject of design definition continues to gain the attention of academics and practitioners alike. The emergence of international standards such as ISO 9000 and ISO 14000 and increasingly complex design requirements assure sustained interest. As depicted in Figure 5.2, the final program or brief should be validated by the design team selected in the request for proposals/qualifications process (RFP/RFQ) to assure that client requirements have been understood. This graphic illustration shows the criticality of establishing the "basis of design" in the early phases of the project when it's cost-effective to make revisions conceptually, on paper (O'Reilly, J.J.N. 1987). Conversely, as the project timeline progresses toward completion it becomes significantly more expensive to integrate design revisions once construction is well underway. For more information, see the textbox *Design Programming Primer*.

Design Programming Primer

At the beginning of any project, design requirements are identified. The program or basis of design (in the United States) or design brief (in the United Kingdom) is a statement of requirements that ideally should contain everything a designer needs to know about a client's proposed project. It anticipates functionality, aesthetics, project costs, schedule, quality, safety, and so forth. It also sets the tone for communication among project participants.

One of the early practitioner/authors to address programming in the United States was William Peña of Caudill, Rowlett and Scott (CRS) Architects. In 1969, he directed the first publication of his *Problem Seeking* to clients and planning officials within institutions, corporations, and various public bodies. Soon practicing architects and architectural students discovered the booklet, and in the late 1970s the second edition of *Problem Seeking* joined a multitude of other new publications on programming methods. In 1994, Hellmuth, Obata + Kassabaum, Inc. (HOK) acquired CRS (then CRSS) and eventually undertook publication of the fourth edition of Peña's work. The fourth edition, published in 2001, represents a range of principles developed by Peña as well as other practitioners at CRS and HOK.

Peña purports that "Programming IS analysis. Design IS synthesis." The program, or problem statement, is the last step in problem seeking (programming) and the first step in design. The problem seeking method clearly was a breakthrough more than 30 years ago, and Peña's book continues to be read by most practitioners. However, as discussed in a later section, the possibilities introduced by digital design and a faster-paced world may necessitate the expansion of these original concepts.

Another well-respected design methodology innovator is Wolfgang Preiser, whose initial work, *Facility Programming*, appeared in 1978. Preiser contributed to the development of the emerging field of facility programming by introducing a generic programming process. This facility programming process included the major actors in the process, the principal beneficiaries of the process, and a discussion of who would pay for the process. The book's last chapter anticipates the potential for a computer-driven database that provides decision support and links facility programming to design and post-occupancy evaluation. Through the use of this database, this 1978 edition also identifies the possibility for designers to apply lessons learned from successes and failures in building performance to future buildings.

Preiser's next book *Programming the Built Environment* appeared in 1985. A third book, *Professional Practice in Facility Programming*, published in 1993, expands on the possibility of using databases as part of knowledge-based, expert systems shells interfacing with computer-aided design (CAD). Preiser speculates that facilities management may dominate the creation of appropriate software systems and modules and that these systems will add modules for facility programming and post-occupancy evaluation. Though more than 20 years old, this vision has yet to be realized.

There are many recent US books addressing briefing (Cherry 1998; Duerk 1997; Hershberger 1999; Kumlin 1995). These books have been written by experienced practitioners for practitioners. Both Cherry and Hershberger present a general approach to programming that adopts the best of current methods and also offers a text to be used in an educational context. The books by Duerk and Kumlin similarly focus on methods for eliciting information from clients and creating programs that lead to satisfactory design solutions.

Numerous research projects sponsored by the Construction Industry Institute (CII, 2000a) at various universities across the United States also have demonstrated early

(Continued)

and accurate project definition to be crucial for successful project outcome. The findings from these research efforts can be found in a variety of publications (CII, 2000b) that address topics such as:

- Scope Definition and Control (Pub. #RS6-2);
- Project Objective Setting (Pub. #RS12-1);
- Input Variables Impacting Design Effectiveness (Pub. #SD-26);
- Work Packaging for Project Control (Pub. #SD-28);
- Adaptation of Quality Function Deployment to Engineering and Construction Project Development (Pub. #SD-97);
- Pre-Project Planning (Pub. #SP39-2);
- Project Definition Rating Index (PDRI) for Industrial and Building Construction (Pub. #IR155-2 and IR113-2);
- Alignment During Pre-Project Planning—Key to Project Success (Pub. #IR113-3); and
- Framework and Practices for Cost-Effective Engineering in Capital Projects in the A/E/C Industry (Pub. #IR113-3).

An analysis of these CII publications, and of other sources found in the literature, reveals the importance of project definition.

The programming methodology that Peña and Preiser pioneered has now become "mature," but a strong need for innovation in the identification and management of design requirements remains. Clients exert an ever-increasing influence over the way design professionals perform their tasks (Hansen and Tatum 1996). A new importance is being placed on collaboration with the client. The American Institute of Architects' (AIA's) document, *The Client Experience*, urges architects to reach beyond traditional roles both as a profession and for their client base. The document identifies the genesis or predesign phase of a new built environment as one of the areas of greatest growth potential for designers. The AIA's current *Architect's Handbook of Professional Practice* devotes the more than 40 pages to project definition.

5.5 DESIGN

After the client has developed a program or brief and has selected a designer, the client and designer enter into a contract for professional services. (See Chapter 4, Professional Engagement and Chapter 11, Legal Aspects of Professional Practice.) In addition to being a legal document, the contract is a communication tool. It spells out the:

- Design tasks to be performed
- Parties' (client's and designer's) specific responsibilities during design

- Client approvals required
- Schedule, including start date, end dates, and major milestones
- Budget, including any contingencies

In some public projects, such as water and sewage treatment plants, private civil engineering firms may work collaboratively with public utility departments to produce a design. In other public infrastructure projects, such as highways and bridges, departments of transportation may develop all of the plans and specifications internally. Many contracts divide the design effort into several discrete phases: schematic design, design development, construction documents, bidding, and construction as depicted in Figure 5.2. Frequently, the client's payment of the designer's invoices is linked to successful completion of these design phases. The percentages of the design effort that these various stages (schematic design, design development, construction documents) represent can vary slightly. Depending on the type of project, the three phases may be referred to as:

1. Schematic Design, expressed as "per cent complete": 20% + 30% + 50%,
2. Design Development, expressed as "per cent complete": 30% – 60% – 90%, or
3. Construction Documents, expressed as "per cent complete": 35% – 65% – 95%.

Schematic design involves establishing the general project scope, relationships among project components, basic geometry, and client understanding and acceptance. As part of schematic design, the designer also validates the program or brief that the client has provided.

During *design development*, the design concept is elaborated. In other words, major systems are defined, important decisions are documented, and a clear, coordinated description of the project is developed. Gaining the client's understanding and acceptance is extremely important so that preparation of construction documents can proceed smoothly (CIB World Building Congress 2001 2001).

Construction documents provide the contractor with sufficient information to build the project and delineate the responsibilities of the two parties who sign the construction contract—the client (owner) and the contractor. They also provide information about the role of the designer, who is not a party to this contract but who has responsibilities during the bidding and construction phases. The construction documents are comprised of drawings and a project manual, made up of bidding requirements and technical specifications.

See Table 5.4 for a summary of the purpose, activities, and deliverables associated with the various project phases.

5.5.1 Design Process

Though the three phases of design often are depicted in an organized, linear manner for the sake of clarity, in reality the actual design process is far more iterative. There are many books written on the creative design process. Sometimes describing the actual act of design seems like trying to capture lightning in a bottle. Christopher Alexander, a British architect who taught for years at the University of California, Berkeley, wrote two classics that address the involvement

TABLE 5.4 Design in project phases.

	PREDESIGN	DESIGN			BID	CONSTRUCTION	POST-CONSTRUCTION
		Schematic Design (20%)	Design Development (30%)	Construction Documents (50%)			
Purpose	Establish project scope, budget, and schedule Obtain necessary permits Select prime designer	Establish basic geometry and relationships among project components	Develop clear, coordinated description of project, including major systems	Provide contractor with sufficient information to build the project	Assist client in selecting contractor	Assist client in assuring that contractor is building per Contract Document	Work to improve design/construct process and position firm to acquire new work
Activities	Assist client in: analyzing project requirements, gaining funding approval, performing environmental impact report/s obtaining geotechnical reports and preparing request for proposal (RFP)	Verifying information contained in RFP Organizing info and synthesizing possibilities Involving other designers and subconsultants Securing client understanding and acceptance	Defining major systems Conducting workshops: Value engineering Life-cycle analysis Sustainability review Documenting important decisions Freezing client's design changes	Delineating responsibilities of the client (owner) and the contractor Supplying an appropriate level of detail in plans and specifications Conducting Constructability review	Conducting prebid conference Answering potential bidders' questions Assisting client in evaluating bids	Making field observations that fulfill requirements for level of attention and testing cited in contract and specifications Resolving problems and discrepancies as they arise	Conducting Post-Occupancy Review
Deliverables	Design criteria, program, or brief Order of magnitude cost estimate Schedule RFP and scope of work (SOW)	Schematic plans Outline specifications Schematic cost estimate	Design development plans Design development specifications Design development cost estimate	Complete plans Complete Project Manual including Procurement Requirements, Conditions of the Contract, and Technical Specifications Opinion of probable construction cost	Addendum/a	Responses to Requests for Information (RFI's) Reviews of submittals and shop drawings	Post-Occupancy Review Report

of users (end users and clients/owners) in the design process: *A Pattern Language* (Oxford University Press, 1977) and *The Timeless Way of Building* (Oxford University Press, 1979).

Involving users can reap significant rewards. Most civil engineering projects involve considerable stakeholder participation, especially during the early phases of design. Frequently owners hire facilitators to help manage large public meetings where stakeholders express their concerns. Involving stakeholders early is advantageous because their apprehensions are known and mitigations can be developed in a timely manner. Discovering late in the design process that stakeholders may mount a campaign against a project can become extremely expensive, both to the owner and their engineers. (See textbox, *Collaborative Design*.)

Collaborative Design

In order to become more responsive to clients, design professionals have recognized the need for better communication among the disciplines. In the late 20th century, "constructability" or "buildability" became a way of bringing useful insights from builders into the design process. As ideas about collaboration evolve, the locus and extent of collaboration is expanding to include greater participation not only of professionals but also of stakeholders (Murray, James, 1993).

Arriving at an approved design that actively involves stakeholders requires several key components, including:

- A high-level person in the client organization who champions "buy-in"
- Clear guidelines regarding scope, budget, and schedule
- A project organization with explicit roles and responsibilities
- Stakeholder-focused building site committee(s) committed to bringing closure to sometimes difficult issues
- Management and project teams willing to embrace the collaborative process
- Recognition up front by all parties of the impact in time and cost for rework and changes.

As a prelude to beginning actual design, a design workshop should be held to introduce the building site committee(s) to the design team. A diagram that explains the process that will be used should be presented. The diagram should convey the idea that the building site committee(s) will be making most of the big decisions during the schematic design and design development phases and that committee involvement will taper off during the construction documents phase (Preiser, Wolfgang F.E. 1993). Additionally, an easily read Gantt (bar) chart should be presented that depicts critical review points and presentations to be made to the client and regulatory organizations.

Finally, open communication is key to collaborative design. "I never saw that before!" is a stakeholder comment that strikes fear in the hearts of design project managers and signals potentially serious, negative impacts to the project schedule.

"While the word *design* is used to signify the individual act of conceptualization that puts an idea on the back of an envelope, the same word is used to signify the often long and unusually collaborative process of carrying out the detailed calculations that flesh out the first sketch and thus make it possible to put specific dimensions and manufacturing instructions on formal drawings."

> —Henry Petroski, (1997). Remaking the World: adventures in engineering, Vintage Books

The design process works with information as well as "flashes of insight" on many levels. In pursuit of appropriate and acceptable solutions, designers must process:

- Client requirements
- Technical variables
- Physical, budgetary, and schedule constraints
- Permitting and code issues
- Political realities

Design Analysis

Throughout an iterative process of examination and criticism, the design emerges. An effort to understand thoroughly the problems to be solved speeds this process. A careful analysis should include:

- Program analysis: convert the information the client has provided into understandable and usable information
- Site analysis: visit the site and organize information into a common scale and format
- Zoning and code analysis: concurrently with site analysis, translate zoning and code issues into building form
- Documentation of existing conditions: establish clear and accurate documentation of existing conditions
- Scheduling: examine the need for project phasing, fast-track sequencing, and time required for permits
- Cost: analyze the project budget and allocate funds to aspects of the project necessary for overall success
- Construction industry practice: evaluate local building practices to know availability of materials and labor and to understand standard processes
- Design precedents: assess previous projects for relevant precedents and similar program, site, context, size, cost, and other design issues

> —AIA. (2007). Architect's Handbook of Professional Practice, John Wiley & Sons, pp. 522–523.

If design begins with analysis, it proceeds with synthesis. Whatever the method, designers must move past the base data they have amassed. Through a combination of sketching, talking, calculating, and thinking, designers must reach sufficient understanding to form a concept. Tim Brown, CEO of IDEO, the firm known for such innovative designs as the iMAC offers a methodology for making this transition that he calls "design thinking." See Figure 5.4, which outlines the three phases of design thinking. The three phases of design thinking are *inspiration*, *ideation*, and *implementation*. Designers are encouraged to:

- Explore the circumstances
- Generate, develop, and test ideas
- Chart the path to completion

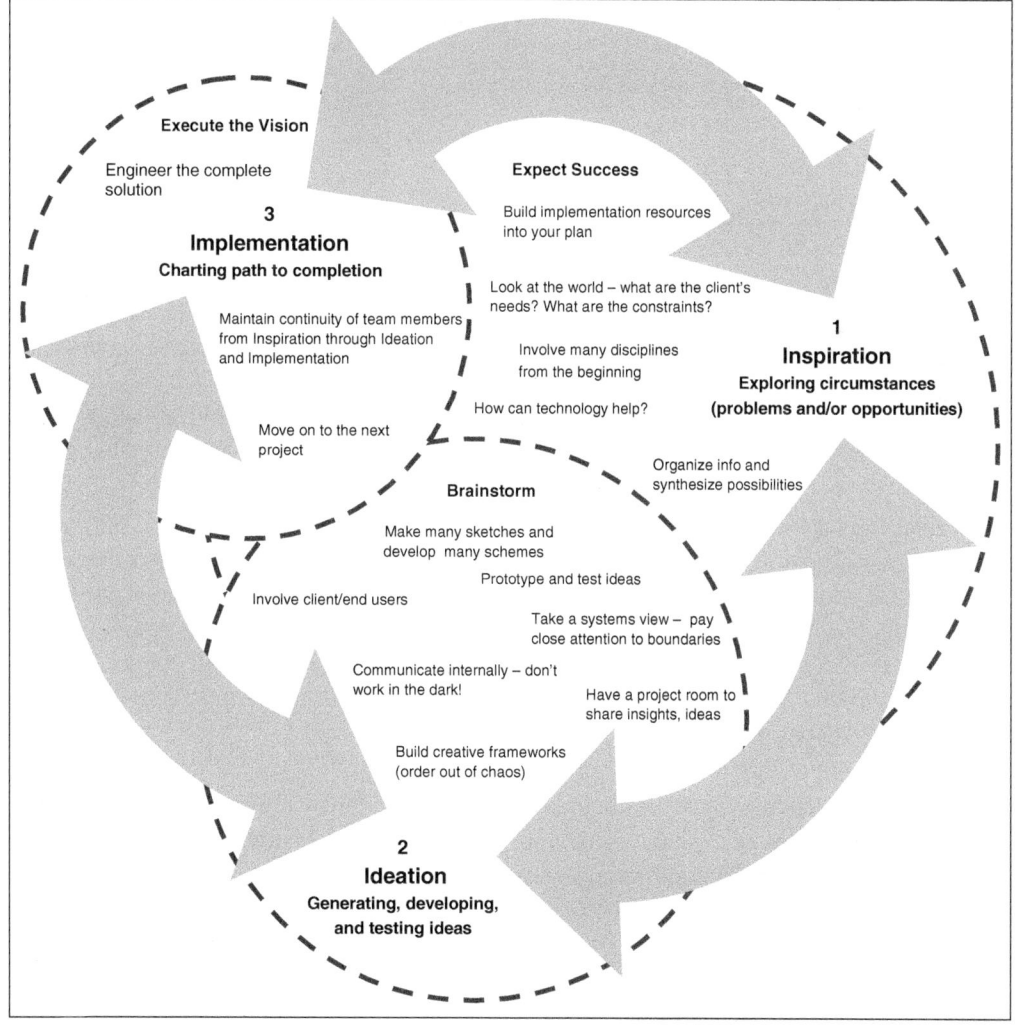

FIGURE 5.4 Design thinking.
Source: Adapted from Tim Brown, "Design thinking," Harvard business review, June 2008.

For information on design thinking and project-based learning, see the following section on *Design Thinking*.

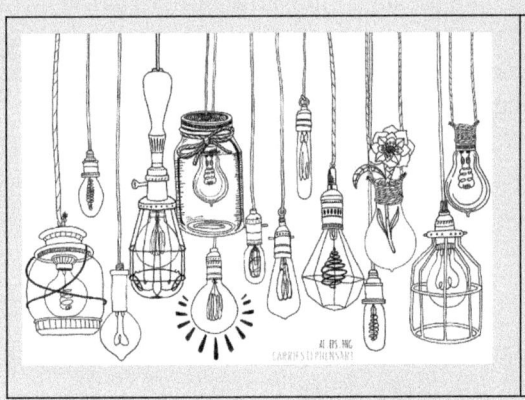

Design thinking utilizes elements from the designer's toolkit like empathy and experimentation to arrive at innovative solutions. By using design thinking, you make decisions based on what future customers really want instead of relying only on historical data or making risky bets based on instinct instead of evidence.

–David Kelly, Founder of IDEO and Stanford University's Hasso Plattner Institute of Design, known as the d.school

(Illustration: https://carriestephensart.com)

5.5.2   Design Thinking

"**Design thinking** is a process to stimulate innovation, drawing on methods from engineering and design and combining them with ideas from the arts, social sciences, and the business world." (http://scpd.stanford.edu/ppc/design-thinking-courses-workshops-and-programs)

Colleges and Schools of Architecture have been teaching design thinking before it was a "thing." As stated on the Texas A&M (TAMU) Dept. of Architecture's website:

The faculty is committed to the studio [laboratory] education method that employs project-based learning to model professional behavior and motivate students through meaningful application of the knowledge that they acquire . . . In the studio projects, a shared emphasis is placed on the technical and expressive content of the design work, the process by which students research, synthesize, and document their design ideas, and the creation of tangible products that achieve high quality of graphic and physical craft. (https://dept.arch.tamu.edu)

This **studio culture** creates ample opportunity for student–faculty and student–student interaction. It also spawns a progressive approach to solving increasingly complex design challenges. In the College of Architecture, studio space abounds. As part of the studio operational procedure:

Students and faculty in every design studio will embody the fundamental values of optimism, respect, sharing, engagement, and innovation. Every design studio will therefore encourage

the rigorous exploration of ideas, diverse viewpoints, and the integration of all aspects of architecture (practical, theoretical, scientific, spiritual, and artistic), by providing a safe and supportive environment for thoughtful innovation.

Every design studio will increase skills in professional communication, through drawing, modeling, writing, and speaking . . .

Every design studio will establish opportunities for timely and effective review of both process and products. Studio reviews will include student and faculty peer review. Where external reviewers are introduced, the design studio instructor will ensure that the visitors are aware of the Studio Culture Statement and recognize that the design critique is an integral part of the learning experience. (https://dept.arch.tamu.edu/about/studio-culture/index.html)

The majority of studio assignments are project-based. **Project-based learning** (PBL) is an instructional methodology that encourages students to learn and apply knowledge and skills through engaging in "real life" experiences. PBL presents opportunities for deeper, in-context learning and for the development of important skills tied to university and professional success.

PBL activities involve:

a. **An extended time frame**—rather than many individual assignments, projects have multiple phases/increments
b. **Collaboration**—students often work in teams and bounce ideas off of each other, faculty, and industry professionals
c. **Inquiry, investigation, and research**—in addition to a textbook, university and community resources are used for exploration
d. **Construction of an artifact or performance of consequential tasks**—students' final assignments typically include proposals, physical and/or digital models, drawings/renderings, posters, photographs, cost estimates/schedules, and/or presentations

In addition to project-based learning, the College of Architecture has created two busy maker spaces to encourage innovation. Tom and David Kelley, authors of the book *Creative Confidence* state that creative confidence is a way of "experiencing the world that generates new approaches and solutions." (Tom Kelley and David Kelley, *Creative Confidence*.)

In a makerspace, the learning happens through making based on an individual's interest. Students engage in playful ways to approach and solve problems through experimentation and discovery. Ideally, this tinkering builds a bridge between intuition and mechanical and/or scientific principles. It also develops practical skills that can build students' creative confidence.

As noted in the Summer 2018 issue of the MIT Sloan Management Review:

The maker movement is a cultural phenomenon that celebrates shared experimentation, iterative learning, and discovery through connected communities that build together, while always emphasizing creativity over criticism. (https://sloanreview.mit.edu/article/lessons-from-the-maker-movement)

Both the Stanford University Graduate School of Business (GSB) and the Massachusetts Institute of Technology (MIT) Sloan School of Management offer short, executive programs in design thinking. IDEO, the company that initiated the design thinking movement, awards a Certificate of Design when participants successfully complete two online classes. The goal of these courses is to teach creativity, brainstorming, and a concept generation process in designing solutions that meet client needs.

Throughout the design process, and particularly in the construction documents phase where most work is in "production" mode, a *quality control plan* should be implemented. This could include internal quality control review, independent technical review, and in important cases, peer review. For more information about managing work quality, see Chapter 7, Executing a Professional Commission and the textbox, *Importance and Value of a Comprehensive Quality Control Plan.*

Importance and Value of a Comprehensive Quality Control Plan

Tony Quintrall, HDR

Quality Control Plans (QCP) are an important aspect of any successful project. The first step in creating a comprehensive QCP is reviewing previous project experience and coordinating with the project manager, principal in charge, and other appropriate staff members. Important elements of a QCP include: a knowledgeable project manager adept at implementing a QCP, an experienced QC Team capable of reviewing contract documents, and an extensive and successful QCP outline.

Once the QCP is drafted, the next step is implementation. Implementation of a comprehensive QCP includes the following: a clear and concise organizational chart outlining roles and responsibilities of the QC Team, proper scheduling of review, including ample time for each review, and good project management practices to ensure the reviews are completed.

A comprehensive QCP can reduce the risks associated with incomplete or poorly completed work products. It can increase the overall quality of the work performed by providing an additional step between the completion of the work and the delivery to the client, which enables a final review for any missing items or incorrect standards. Both of these attributes will lead to a better end product, which will increase client satisfaction and maximize the client's desire to solicit the designer for future work. Repeat business is paramount for sustaining a successful and profitable organization, and any steps that can be taken to ensure this should be implemented. Here's another example of the importance and relevance of Figure 5.3 – Importance of Getting Requirements Right. An effective QCP and the process of quality control needs to be applied throughout the project.

5.6   DESIGN DURING BID AND CONSTRUCTION

The work of civil engineers typically does not end with the completion of the construction documents. Most clients rely on their prime designers to help them through the bid phase. As part of the bid process, civil engineers may be responsible for including Division 00—Procurement and Contracting Requirements in the project manual. (See Chapter 6, What Engineers Deliver.) Division 00 includes:

- Advertisement for bids
- Invitation to bid
- Instructions to bidders (contractors)
- Prebid meetings
- Land survey information
- Geotechnical information
- Bid forms
- Owner-contractor agreement forms
- Bond forms
- Certificate of substantial completion form
- Certificate of completion form
- Conditions of the contract
- Procedure for answering bidders' questions

Whether acting in the capacity of prime designer or subconsultant, civil engineers usually attend prebid meetings to acquaint prospective bidders with the project. They also answer bidders' questions during the time allotted for the bid and group those responses and clarifications in an *addendum* or *addenda*, if more than one installment is required.

Following contract award (the owner and contractor enter into a contract), civil engineers may be responsible for:

- Attending a preconstruction conference
- Responding to field questions, called requests for information (RFIs)
- Making field observations
- Reviewing submittals, including shop drawings

Making field observations fulfills the requirements for level of attention and testing cited in contract and specifications and encourages quality. The civil engineer's jobsite presence also can head-off problems, because contract documents are never perfect. Additionally, owners may request civil engineers to validate contractors' payment requests (invoices) for accuracy by comparing work or materials in place with percent complete invoiced.

Submittal and shop drawing review is the final element of design review. Reviewing submittals, such as concrete mix design, and shop drawings, such as steel fabrication drawings, assures that the design detailing by the contractor conforms with the intent of the design. However, due to time and budgetary constraints, time to review is limited. The process used by civil engineers for reviewing submittals and shop drawings should be referenced in the general conditions of the construction contract and discussed at the prebid and preconstruction conferences.

Shop Drawing Review Process

1. A/E identifies shop drawings required and establishes schedule for review and resubmission
2. Contractor reviews each shop drawing and establishes acceptability as to:
   - Means
   - Methods
   - Techniques
   - Operations and sequences of construction
   - Safety precautions
3. A/E reviews each shop drawing and determines conformity to:
   - Design intent
   - Compliance with contract documents (plans and specifications); may ask contractor to "revise and resubmit"
4. Contractor appraises A/E of any changes from what was specified in contract documents—A/E may or may not accept
5. Contractor should pay A/E if shop drawings vary considerably from original design (difficult to accomplish because contractor does not have contract with A/E)
6. A/E returns any shop drawings that they have not required

—John Bachner (1991). Practice Management for Design Professionals: A Practical Guide to Avoiding Liability and Enhancing Profitability

In order to reduce liability, architects and engineers use special language when reviewing submittals and shop drawings. Typical language might include:

Review is limited solely to the purpose of checking for conformance with Civil Engineer's design intent and conformance with information contained in the Contact Documents. Review is not conducted to determine the accuracy and completeness of other information such as dimensions, completeness, installation instructions, or performance of equipment or systems supplied by the contractor, all of which remain the contractor's responsibility. Review neither extends nor alters any contractual obligations and shall not relieve the contractor of responsibility for deviation from the requirements of the Contract Documents.

Additionally, when returning the submittals and shop drawings to the contractor, civil engineers can select among several options, which conclude:

- Accepted as Noted
- Revise and Resubmit
- Rejected
- Not Reviewed, Submittal not Required by Contract Documents
- Reviewed for Project Closeout Requirements Only

5.7 POST-CONSTRUCTION ACTIVITY

Some sophisticated client organizations conduct their own design reviews throughout the design process. They also document how closely architectural and engineering firms design to budget by comparing the opinion of probable construction cost provided at the end of design with the bid submitted by the successful contractor. They also track the number of RFIs issued by the contractor as an indication of the quality and completeness of the plans and specifications. As Figures 4.2, 4.3, and 4.4 in the previous chapter indicate, the Veterans Administration (VA) scores the prime designer and subconsultants at the end of schematic design, design development, and construction document design phases; the VA then uses this information when selecting architects and engineers to perform new work.

Though not always done, most design organizations could benefit from a Postoccupancy Review with the client and end users.

Finally, though difficult to achieve, many organizations—both academic and commercial—are exploring the use of *design performance measures* (DPMs). The belief is that measurement of design performance will lead to improved designs. See the textbox, *Design Performance Measures* for additional information.

Design Performance Measures

Numerous research projects at various US universities over the last two decades have demonstrated that early and accurate project definition is crucial for successful project outcome. The findings from these research efforts can be found in a variety

(Continued)

of Construction Industry Institute publications (CII 2000a) (CII 2000b) that address topics such as: Scope Definition and Control; Project Objective Setting; Input Variables Impacting Design Effectiveness; Work Packaging for Project Control; Adaptation of Quality Function Deployment to Engineering and Construction Project Development; Pre-Project Planning; Project Definition Rating Index (PDRI) for Industrial and Building Construction; Alignment During Pre-Project Planning—Key to Project Success; and Framework and Practices for Cost-effective Engineering in Capital Projects in the AEC Industry.

An analysis of these CII publications and of other sources found in the literature reveals the importance of project definition. However, most publications stop short of actual linkage of design quality to measurement of performance. When addressed, design performance measures (DPMs) are usually cost-based. Such measures can be product-oriented, such as lifecycle cost analysis, or process-oriented, such as construction change orders as a percentage of total project cost. Few DPMs address qualitative aspects of design, such as client and user satisfaction, innovation, or aesthetic appeal. This may be the case because practitioners do not perceive the need, or they find the problem too difficult to solve, or they do not wish to enable discussions regarding subjective design factors with clients. Additionally, there is limited funding available for academics to pursue this line of investigation.

Within the US architectural, engineering, and construction industry, development of DPMs is nascent. However, researchers have identified performance measures that explicitly represent project objectives, such as those that might appear in a brief. A guiding principle in defining DPMs is the identification of a critical variable that measures, reflects, or significantly influences a particular performance objective. In most instances, a high-level performance objective will need to be delineated by multiple metrics that influence its overall satisfaction. The following discussion of current research on developing DPMs is divided into two sections: cost-based methods and other approaches.

Cost-Based DPMs

Life-cycle cost is a relatively straightforward performance objective to delineate. However, others, such as energy efficiency, may be more difficult. A group at the Lawrence Berkeley National Laboratory recognizes that metrics cannot stand on their own as they are linked to design assumptions and/or operating conditions. In order to evaluate performance against a benchmark, a Building Life-cycle Information System (BLISS) has been developed. During design, data from this model can be used to simulate performance. Later, simulated performance can be compared to actual building performance. An aggregate Life-Cycle Cost performance objective is shown below (Hitchcock et al. 1998):

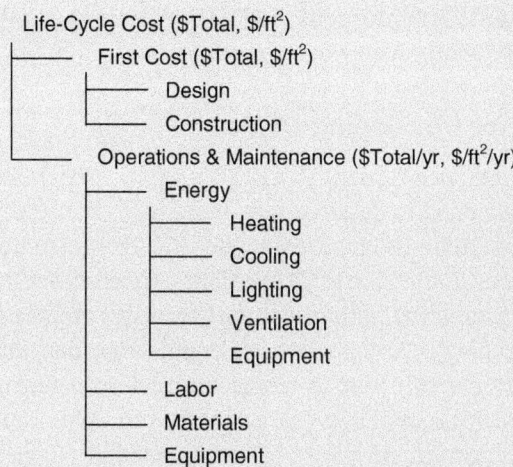

Life-Cycle Cost ($Total, $/ft^2)
— First Cost ($Total, $/ft^2)
  — Design
  — Construction
— Operations & Maintenance ($Total/yr, $/ft^2/yr)
  — Energy
    — Heating
    — Cooling
    — Lighting
    — Ventilation
    — Equipment
  — Labor
  — Materials
  — Equipment

At least two organizations at the National Institute of Standards and Technology (NIST) concern themselves with DPMs. The Building and Fire Research Laboratory (BFRL) focuses on developing measurement methods, fundamental data, simulation models, and life-cycle environmental and economic analysis tools to support sustainability in design. The BFRL has created software called BEES (Building for Environmental and Economic Stability) for designers to select among alternative building materials and products based on environmental and economic performance (NIST 2024). Following the attacks of September 11, 2001, NIST's Office of Applied Economics has developed financial models to be applied in design to optimize investments in "protective," that is, anti-terrorist, strategies (Marshall 2002).

At Stanford University's Center for Integrated Facility Engineering (CIFE), investigators acknowledge that the AEC industry is experiencing a profound change that brings with it the need for productivity improvements, leaner organizations, and more consistent and rigorous performance metrics. Several near-term metrics have been proposed, some cost-based and others having different quantitative measures. Schwegler et al. (2001) describe these as: quality of design documentation (ratio of drawings or 3D objects to dollar value) of work; individual team task performance (transactional data provided by project extranets); assembly complexity (number of simultaneous activities occurring during construction); design iterations (experience in manufacturing has shown the benefits of increasing design iterations but AEC industry practitioners frequently "freeze" design at a low number of design iterations in order to control soft costs); and

(Continued)

response time for requests for information (RFIs) and shop drawing review (extremely long response times frequently have been shown to result in major unplanned changes).

Other Approaches for Developing DPMs

In 1999 the High Performance Structures (1999) group at the Massachusetts Institute of Technology was commissioned to design a new building to house Civil and Environmental Engineering (CEE). Both students and faculty rated the design solutions on a set of predetermined design measures that included: (1) flexibility (upgradeability, adaptability, expandability); (2) aesthetics (character, human comfort/security, proportion and scale, material palette, lighting, landscaping); (3) high level of engineering performance (showcase, HPS components); (4) environmentally friendly (impact to environment, energy efficient, material flow, choice of materials); (5) accessibility (ease of entry, multiple circulation paths, controlled access, location of space, communication); (6) constructability (construction efficiency, low impact); and (7) maintainability (cleanability, reparability, and site maintenance). These elements are weighted and then graded on a scale of 1, 2, or 3. Scores on individual elements were aggregated to create an overall score for the design.

Within the Construction Engineering and Management Group in the College of Civil and Environmental Engineering at the Georgia Institute of Technology, a Project Definition Matrix has been developed over time and has been used successfully in workshop settings with industrial clients. This three-dimensional model is based on: six layers of stakeholder perspectives (owner, vendor/suppliers, construction, design, user/operator, external parties); twelve performance parameters (contextual compatibility and response, functional performance, physical performance, cost, time, quality/reliability, safety/security, risk, constructability, maintainability, health, sustainability); and six types of internal and external influences (project characteristics, project objectives, project scope, physical context of the project, non-physical context of the project, project risks). These largely qualitative considerations assist in aligning client and designer expectations (Vanegas 2001).

A US Defense Department goal is for all military construction to use principles of sustainability, addressing issues such as siting, water and energy efficiency, minimization of pollution, indoor environmental quality, etc. Consequently, the US Army Civil Engineering Research Laboratory (CERL) has developed the Sustainable Project Rating Tool (SPRT). SPRT will be used as a standard measure to rate the sustainability of Army building and infrastructure designs (Flanders et al. 2000).

Researchers at Stanford's Civil and Environmental Engineering Project Based Learning Lab (PBL2) have used metrics to measure cross-disciplinary learning in distributed AEC teams that can relate to improved design quality. The methodology involves a four-tiered classification based on cognitive and situative learning theories. The four tiers are: island of knowledge, awareness, appreciation, and understanding. The approach used has relevance for planning deployment of automated briefing systems (Fruchter and Emery 2000).

Given the fragmented nature of the US construction industry, the lack of DPMs is not surprising. In a single project, design usually is performed by a panoply of consulting firms. However, the subject of design definition and quality will continue to gain the attention. The emergence of international standards such as ISO 9000 and ISO 14000 and increasingly complex design requirements assure sustained interest of academics and practitioners alike in improving the briefing process. At the 2001 CIB Congress in New Zealand, 33 papers focused on the "demand side," or stakeholders' perspective, of the construction delivery process—the highest number ever.

Need for a Different Approach

The publications and studies emphasize how crucial good design definition is and clearly point to the necessity for better tools and strategies for handling design requirements. Classifications and checklists are helpful in conceptualizing what a brief is and what it should include. However, by definition, they are generic and cannot make allowances for each project's unique nature. These approaches make a very iterative process look predictable and linear.

5.8 SUMMARY

This chapter details the many roles civil engineers play in pre-design, design, and construction. Civil engineers can be involved from the very initial stages of the project inception through project closeout, owner occupancy, and later adaptive reuse or decommissioning. Co-players in the project development process—clients, civil engineers and other design professionals, contractors, and regulatory agencies—possess diverse perspectives and agendas. Much of design, especially the early phases, requires civil engineers to take abstract ideas and convert them into tangible deliverables. Because of the complexity of today's projects, encouraging involvement of professionals and stakeholders through a collaborative design process can reap significant rewards. Developing and implementing a quality control plan (QCP) also yields tangible benefits.

The next chapter—Chapter 6, What Engineers Deliver—discusses the deliverables connected to—the project delivery process.

BIBLIOGRAPHY

Alexander, Christopher. et al. (1977). *A Pattern Language*. Oxford University Press, New York.

Alexander, Christopher. et al. (1979). *The Timeless Way of Building*. Oxford University Press, New York.

American Institute of Architects. (2007). *Architect's Handbook of Professional Practice*, J.A. Demkin, ed. John Wiley & Sons, New York.

Bachner, John. (1991). *Practice Management for Design Professionals: A Practical Guide to Avoiding Liability and Enhancing Profitability*. John Wiley & Sons, New York.

Bisharat, Keith. (2008). *Construction Graphics*. John Wiley & Sons.

Blyth, Alastair and Worthington, John. (2001). *Managing the Brief for Better Design*. SPON Press, London.

Brown, Tim. (2008). "Design Thinking," Harvard Business Review. June 1, 2008.

Building and Fire Research Laboratory. (February, 2000). National Institute of Standards and Technology, Technology Administration, U.S. Department of Commerce. http://www.bfrl.nist.gov/goals_programs/EBP_goal.htm.

Cherry, Edith. (1998). *Programming for Design: From Theory to Practice*. John Wiley & Sons, New York.

CIB World Building Congress 2001. (2001). "Construction Industry Board". Conference Proceedings, 2–6 April, Wellington, New Zealand.

Construction Industry Institute (CII). (2000a). The University of Texas at Austin, Austin Texas. http://construction-institute.org.

Construction Industry Institute (CII). (2000b). The University of Texas at Austin, Austin Texas. http://construction-institute.org/services/catalogue/catframe.htm.

Duerk, Donna P. (1997). *Architectural Programming: Information Management for Design*. John Wiley & Sons, New York.

Flanders, Stephen N., Schneider, Richard L., Fournier, Donald, and Stumpf, Annette. (2000). http://www.cecer.army.mil/earupdate/nlfiles/2000/sustainable2.cfm.

Fruchter, Renate and Emery, Katherine. (2000). "CDL: Cross-Disciplinary Learning Metrics and Assessment Method." *Proceedings of ASCE – ICCCBE-VIII Conference*, Stanford University, August 2000.

Hansen, Karen Lee, MacLeod, I.A., Tulloch, I.M., and McGregor, D.R. (1996) "*BriefMaker*: A design briefing tool developed on the Internet," International Conference on Trends in Civil and Structural Engineering Design at Strathclyde University, Glasgow. Civil Comp Press, Edinburgh. August 1996.

Hansen, Karen Lee and Tatum, C.B. (1996). "How strategies happen: A decision making framework. *ASCE Journal of Management in Construction* 12 (1): January/February 1996. 40–48.

Hershberger, Robert G. (1999). *Architectural Programming and Predesign Manager*. McGraw-Hill, New York.

Higgin, Gurth and Jessop, Neil (1963). *Communications in the Building Industry: The Report of a Pilot Study*. Tavistock Institute of Human Relations Publishing, Great Britain.

High Performance Structures Group. (1999). http://www.moment.mit.edu/Hps/98-99/designfiles/documents.

Hitchcock, Robert J., Ann Piette, Mary, and Selkowitz, Stephen E. (1998). "Documenting performance metrics in a building life-cycle information system." Proceedings of the ACEEE '98 Summer Study on Energy Efficiency in Buildings, Lawrence Berkeley National Laboratory (LBNL-41940), June 1998.

Kumlin, Robert R. (1995). *Architectural Programming: Creative Techniques for Design Professionals*. McGraw-Hill, New York.

Marshall, Harold E. (2002). "Economic approaches to homeland security for constructed facilities." Keynote address, Tenth Joint W055-W065 International Symposium on Construction Innovation and Global Competitiveness, University of Cincinnati, September 2002.

Murray, James P., Gameson, R.M., and Hudson, J. (1993). *Creating Decision-Support Systems, Professional Practice in Facility Programming*, W.F.E. Preiser, ed. Van Nostrand Reinhold, New York, NY. 427–452.

O'Reilly, J.J.N. (1987). *Better Briefing Means Better Buildings*. Building Research Establishment (BRE), Garston, UK.

Peña, William M. and Parshall, Steven A. (2001). *Problem Seeking: An Architectural Programming Primer*. John Wiley & Sons, New York.

Petroski, H. (1997). Remaking the World: adventures in engineering, Vintage Books, a Division of Random House, NY.

Preiser, Wolfgang F.E. (1993). *Professional Practice in Facility Programming*. Van Nostrand Reinhold, New York.

Rutherford, James H. and Maver, Thomas W. (1994). Knowledge-based design support. In: *Knowledge-Based Computer-Aided Design*. G. Carrara and Y.E. Kalay, ed. Elsevier, New York, NY.

Salisbury, Frank. (1990). *Architect's Handbook for Client Briefing*. Butterworth Architecture, London.

Schwegler, Benedict R., Fischer, Martin. et al. (2001). *Near-, medium-, and long-term benefits of information technology in construction*. Center for Integrated Facilities Engineering (CIFE) Working Paper #65, July 2001.

Vanegas, J. (2001). "The project definition package: A cornerstone for enhanced capital project performance." *Proceedings of the 2001 World Congress of the International Council for Research and Innovation in Building and Construction (CIB)*, Wellington, New Zealand (paper in conference CD ROM).

White, Edward T. (1991). *Design Briefing in England*. Architectural Media Ltd, Tuscon.

https://www.nist.gov/services-resources/software/bees# (accessed 01 June, 2024)

CHAPTER 6

What Engineers Deliver

Big Idea

Civil engineers convert abstract ideas into physical realities through their efforts. The output of these engineering services includes engineering reports, feasibility studies, plans and specifications, and construction administration.

> We know where most of the creativity, the innovation, the stuff that drives productivity lies—in the minds of those closest to the work.
> —Jack Welch, Former Chairman and CEO of General Electric

Key Topics Covered

- Contract Documents
- Drawings
- Specifications
- Drawings and Specifications—Final Thoughts
- Technical Memos and Reports

Civil Engineer's Handbook of Professional Practice, Second Edition. Karen Lee Hansen and Kent E. Zenobia.
© 2025 John Wiley & Sons, Inc. Published 2025 by John Wiley & Sons, Inc.
Companion website: www.wiley.com/go/hansen/CivilEngineersHandbook

- Calculations
- Other Deliverables

Related Chapters in This Book

- Chapter 2: Background and History of the Profession
- Chapter 3: Ethics
- Chapter 4: Professional Engagement
- Chapter 5: The Engineer's Role in Project Development
- Chapter 7: Executing a Professional Commission—Project Management
- Chapter 8: Permitting
- Chapter 11: Legal Aspects of Professional Practice
- Chapter 15: Globalization
- Chapter 16: Sustainability
- Chapter 17: Emerging Technologies

6.1   BACKGROUND

As the design solution evolves through pre-design and the schematic design, design development, and construction documents phases, the final form of the contract documents begins to take shape as discussed in Chapter 4, Professional Engagement; Chapter 5, The Engineer's Role in Project Development; and Chapter 12, Managing the Civil Engineering Enterprise. In the traditional design-bid-build process represented by Figure 5.2—The Flow of Work, the prime designer is responsible for developing the majority of the documents that form the basis of the construction contact between the owner and the contractor. In building projects, the architect is usually the prime designer, and in large civil projects, the civil engineer is usually the prime designer. These prime designers typically hire numerous sub-consultants to assist with the design. Contractors also hire many subcontractors and material suppliers in order to perform the construction. Figure 6.1 depicts the typical contractual arrangements in design-bid-build project delivery. Chapter 11, Legal Aspects of Professional Practice, discusses other forms of project delivery.

Aside from the sheer number of contracts required, the most notable aspect of the contract relationships depicted in Figure 6.1 is that there is no contract between the prime designer and the contractor. Thus, the prime designer and subconsultants must prepare documents,

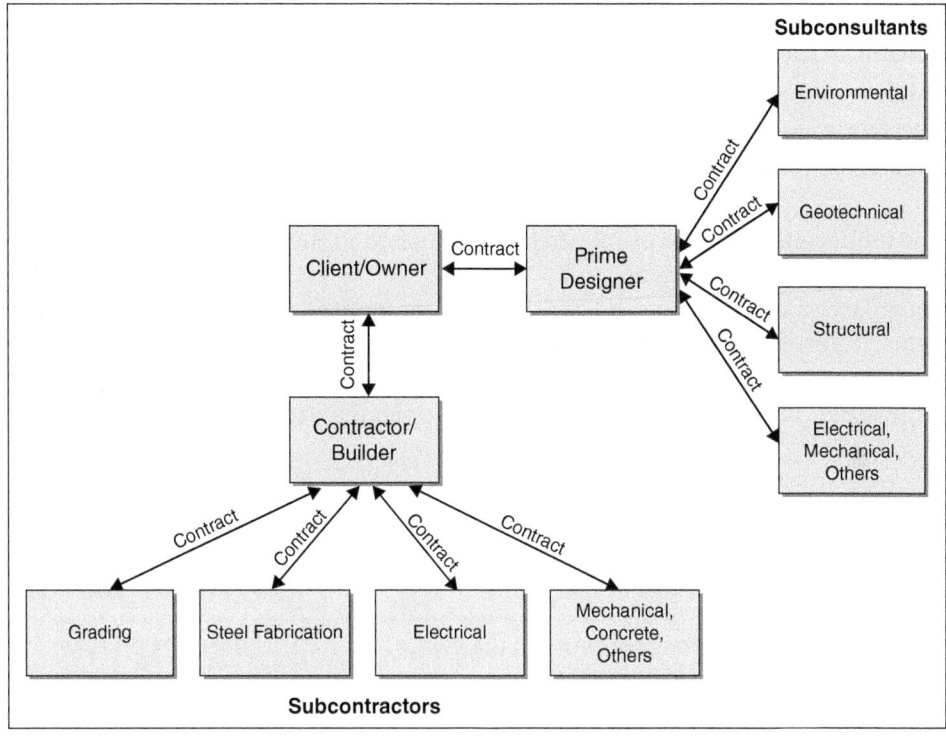

FIGURE 6.1   Contractual relationships in design-bid-build project delivery.

paid for by the client, that the client and stakeholders can understand, and that the contractor can use to build the project. This may seem more than a little challenging, and it is!

This chapter examines the kinds of deliverables for which civil engineers are responsible. These include drawings or plans, specifications, and technical memos and reports, as well as other documents such as calculations, meeting minutes, construction reports, reviews of various submittals, and schedules.

6.2   CONTRACT DOCUMENTS

The term "contract documents" is used widely in the Architecture, Engineering, and Construction (AEC) industry. These are the documents on which the contract for construction is based. Contract documents include more than the drawings and technical specifications. They consist of:

- Agreement/Contract Forms
- Conditions of Contract (General and Supplementary Conditions)

- Drawings
- Technical specifications and any required calculations
- Addendum/a (changes made during the bidding process)
- Modifications to the contract (changes made after the owner and contractor have signed the contract for construction)

Table 6.1 depicts the kinds of information contained in these documents and the parties responsible for their development.

Though technically not considered part of the contract documents, another set of documents, referred to collectively as *procurement and contracting requirements*, are necessary for selecting a contractor using competitive bid methods. The procurement and contracting requirements, general and supplementary conditions, and technical

TABLE 6.1   Contract documents (used by owner and contractor).

Document Name	Contents	Responsible for Development
• Agreement/ Contract Forms	• Contract between the owner and contractor	• Owner's Attorney with input from Prime Designer—Civil Engineer or Architect
• Conditions of Contract	• General and Supplementary Conditions	• Prime Designer—Civil Engineer or Architect
• Drawings	• Information in graphical format depicting location, size, shape, and dimensional relationships of design elements and materials	• Prime Designer—Civil Engineer or Architect
• Technical specifications and any required calculations	• Information in text format describing requirements for materials, equipment, systems, standards and workmanship, and performance of related services	• Prime Designer—Civil Engineer or Architect
• Addendum/a	• Changes made during the bidding process, usually stemming from questions raised by the contractor	• Prime Designer—Civil Engineer or Architect
• Modifications to the contract	• Changes made after the owner and contractor have signed the contract for construction	• Prime Designer—Civil Engineer or Architect and Owner

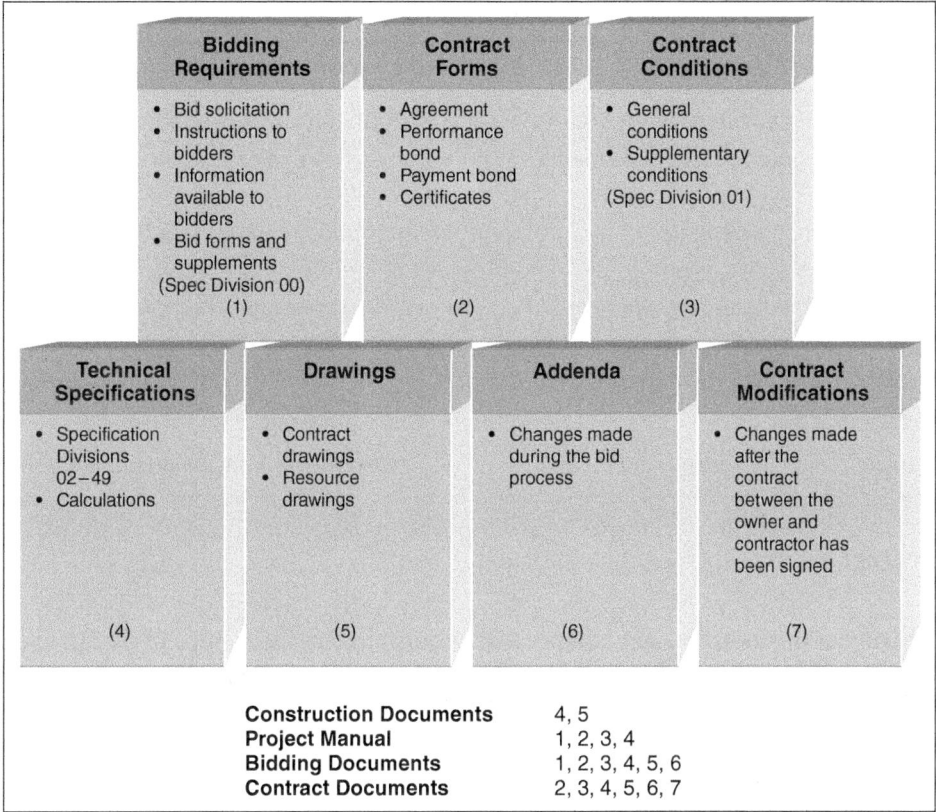

FIGURE 6.2   Documents are the formal building blocks of project delivery.

specifications are collected in a single document assembled by the prime designer. This document commonly is referred to as the *Project Manual*. The agreement/contract forms, project manual, drawings, and addenda comprise the *Bidding Documents*, as shown in Figure 6.2. More information regarding the agreement/contract forms and the different approach used in design-build project delivery is included in Chapter 11, Legal Aspects of Professional Engagement.

There is an important difference in the information that is included in drawings (or plans) and technical specifications—see Table 6.2 for a comparison. The most obvious difference is that drawings largely contain information in graphical/geometric format and specifications contain information in text format. The level of detail included in the drawings and specifications corresponds to the needs of the client, of permitting and regulatory agencies, and of the contractor. The following sections discuss further the content and organization of drawings and specifications.

TABLE 6.2   Information contained in construction drawings vs. technical specifications.

Construction Drawings	Technical Specifications
• Design requirements represented graphically	• Design requirements represented verbally
• Size, shape, and relationship of elements provided	• Properties and characteristics of elements provided
• Location of elements depicted	• Installation requirements for elements established
• Products or materials shown wherever located	• Products or materials described once
• Products or materials shown generically	• Products or materials identified specifically
• Quantity indicated	• Quality indicated
• Few requirements for testing noted	• Requirements for testing clearly spelled out

6.3   DRAWINGS

Drawings depict the location, size, shape, and dimensional relationships of design elements, in addition to materials. A drawing set typically includes site and building plans, elevations, sections/profiles, details, and schedules (matrices or tables, not time-based schedules). Each drawing should include sufficient information to orient the user, including scale, a north arrow on plans, and key plans that locate partial plans within the whole. Each sheet in the drawing set also should contain:

- Designer information
  Names and addresses of consultants
  Seal and signature of engineer or architect, as needed in most states
- Project information
  Title
  Project address
  Frequently, owner's name and address
- Sheet title
  Title of the sheet
  Copyright information, if applicable
- Drawing management information
  Names of those who worked on sheet
  Names of those who checked sheet
- Issue information
  Date(s) of issue, including revisions
  Purpose of issue—bid, permit, construction

TABLE 6.3 Standardized drawing (sheet) sizes.

Designation	Architectural Drawing Sizes		Manufacturing Drawing Sizes*	
	(mm)	(inches)	(mm)	(inches)
A		9 × 12		8.5 × 11
B		12 × 18		11 × 17
C		18 × 24		17 × 22
D		24 × 36		22 × 34
E		36 × 48		34 × 44

*American National Standards Institute (ANSI)

- Drawing identification
  Sheet letter
  Sheet number

Drawings can be produced in a variety of sizes. Standardized sheet sizes are shown in Table 6.3.

A uniform method for formatting sheets has been adopted by various organizations, such as the American Institute of Architects (AIA) (American Institute of Architects 2008). For example, each sheet is divided into three modules: (1) sheet title block, the information listed above, usually on the right side of the sheet; (2) graphic area, based on a hidden modular grid; and (3) perimeter, or border, with alpha numeric grid coordinates.

There also are standards for organizing the types of information contained on individual sheets. The US National Computer Aided Design (CAD) Standard (NCS) recognizes nine sheet types:

0. *General*—symbols, legends, notes
1. *Plans*—site, building, generally horizontal views
2. *Elevations*—vertical views of surfaces
3. *Sections*—vertical cuts across plans
4. *Large-Scale Views*—plans, elevations, stair sections, or sections that are not details
5. *Details*—plans, elevations, sections of smaller-scale components
6. *Schedules and Diagrams*—tables, matrixes
7. *User Defined*—for types that do not fall in other categories, including typical detail sheets
8. *User Defined*—for types that do not fall in other categories, including typical detail sheets
9. *3D Representations*—isometrics, perspectives, photographs

Additionally, drawing sets are organized by disciplines, each with a letter prefix, such as "C" for civil engineering, "S" for structural engineering, "A" for architecture, and so forth.

Information in each discipline is ordered by grouping like information into subdivisions, such as the ten listed above. Thus, drawing organization within each discipline progresses from the general to the specific. The textboxes *Typical Drawing Numbering System* and *Content in Drawing Sets* shown in the following figure contain more information about the organization and the content of drawing sets.

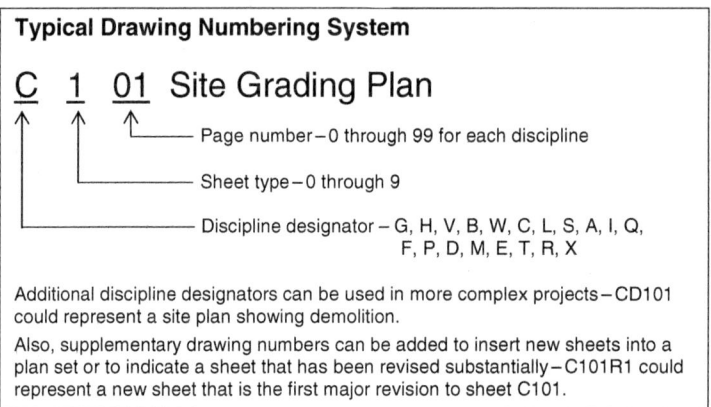

G, H, V, B, W, C, L, S, A, I, Q,
F, P, D, M, E, T, R, X

Additional discipline designators can be used in more complex projects – CD101 could represent a site plan showing demolition.

Also, supplementary drawing numbers can be added to insert new sheets into a plan set or to indicate a sheet that has been revised substantially – C101R1 could represent a new sheet that is the first major revision to sheet C101.

Content in Drawing Sets

While the following order reflects the National CAD Standard, much of what is described pertains to drawings that follow other organizations.

Cover Sheets

Cover sheets are, simply, title pages, with the project title, the owner's name, the names of the design professionals involved, a pictorial of the project (frequently a perspective drawing), and similar content.

G: General Information

Site data, location map, energy compliance calculations, building code summary, project square-foot calculations, key plans, general notes, abbreviations, and an index of sheets all belong in the general information category.

H: Hazardous Materials

Hazardous materials occur in a variety of goods and materials, primarily in older structures. Asbestos, for example, occurs in a variety of materials (predating the 1970s), including floor coverings; plaster; piping insulation; ceiling tiles; plaster, floor, wall, and ceiling insulation; and myriad other materials, thus making sheet organization somewhat problematic in this discipline. The way the NCS is currently established, demolition of an

HVAC or piping system in an older building would be described in mechanical drawings using the Level 2 discipline designator (see sheet identification paragraphs upcoming). The likelihood that asbestos would be encountered in the duct or piping insulation and sealed joints in older buildings is high. Sitework involving hazardous materials—hydrocarbons, for example—are likely to be described in the civil drawings. Just how this discipline will develop remains to be seen; however, the idea is to identify the location of hazardous materials so that their potential for harm is mitigated.

V: Survey/Mapping

These pages contain relevant survey and map information. It is normally the owner's responsibility to provide the design professional with accurate information related to the real estate being developed. Vicinity maps, common on drawings sets, and general layout information are recorded in this sheet set.

B: Geotechnical Information

Providing information in a drawing on site soil conditions makes sense from several viewpoints. A number of people make use of the geotechnical report in their analysis of work requirements, including the owner, the prime contractor, and subcontractors. Though including this information in a drawing set is cumbersome, it is perhaps justified by the ready availability of the information.

W: Civil Works

The civil works category is, for all practical purposes, the same as civil drawings (the next one); however, it was added at the behest of government agencies that are responsible for civil work that encroaches upon the property of multiple landowners. A municipal pipeline project that impinges upon numerous landowners, for example, would be the appropriate project type to describe in this category.

C: Civil Drawings

In a drawing set that describes civil construction projects, such as highway and street improvement projects, it is not necessary to distinguish the discipline from others—the entire project is "civil drawings." When the discipline is distinguished, however, it is generally when civil drawings form a part of a commercial building set. Civil engineers design and describe the off-site improvements (curbs, gutters, and sidewalks along public thoroughfares), on-site grading and paving requirements, and underground utilities for building construction projects. Their work is recorded in the "C" sheets—civil sheets.

(Continued)

L: Landscape Drawings

Landscape drawings generally include planting plans and schedules (lists of plants, shrubs, and grasses), the irrigation system required to support the plants, and hardscape (trellises, fences, site benches, patios, walkways, etc.) described in plan, elevation, section views, and details.

S: Structural Drawings

These drawings describe the elements, components, and assemblies of structural systems, and the manner in which they are connected, for a variety of projects. They are the construction equivalent of the skeleton, tendons, and muscle matter of the human body; in fact, "tendons" is a term used to describe the steel cables that are inserted in components, such as precast, prestressed concrete piles and girders, and in cast-in-place post-tensioned concrete beams, bridge decks, and floor slabs. Architects frequently determine the basic structural system, since the functional arrangement of a project and the required aesthetic treatment may dictate column spacing and therefore spans; however, the calculations and detailed structural design parameters are the bailiwick of the structural engineer.

A: Architectural Drawings

Architectural drawings are the heart and soul of a building project—virtually all other disciplines act in support of the architect's design, which is described in these drawings. To some extent this is due to the architect having overall control as the prime design professional and the uniqueness of building projects; however, in highway projects, for example, where design standards for construction are common and projects are co-developed and managed by district and central DOT offices, the various engineers involved act more as equals. There are as many approaches to design as there are architects—some conceive projects from the exterior and fit the functions within a shell; others determine the appropriate functional relationships of a building and develop the shell from them. No matter the origin of the design, the other disciplines take their cues from the architect's drawings, which are the most wide-ranging drawing set. Depending on the charges to the architect, the drawings can include master project planning and building design (frequently for multiple buildings) from basic systems or shells to complete buildings with interior details, furniture, and even fabric design.

I: Interiors

As just noted, architects might be given the responsibility to design a complete building or building shell for an owner. When the latter occurs, it is often because a developer or owner has anticipated that there will exist a demand for space within the building when it is complete and has undertaken to construct it on a speculative basis. Commercial real estate brokers monitor the construction and lease activity of buildings, and earn fees for facilitating lease agreements between building owners and tenants. Under certain lease agreements, tenants have the responsibility to design and construct their office space, within parameters established by the building owner, and will commission interior or building architects to produce the necessary design documents. It is these drawings, as well as drawings used in subsequent lease activity in the project, that are inserted in a drawing set under the interiors category.

Q: Equipment

Equipment drawings run the gamut of equipment that might be used in a project, from bank vaults, teller equipment, and ATMs to library, theater, videoconferencing, and commercial cooking, bakery, and laundry equipment. Some of the drawings required in this division can be complex, as, for example, in the case of bank vaults, which are subject to compliance with federal legislation governing their construction.

F: Fire Protection

To aid fire departments in their fire-fighting efforts, fire codes require comprehensive fire protection plans from project owners. Access and egress to the site; on-site street widths and radii; the location of hydrants, water mains, trees, overhead power lines, utility service disconnects; and anything else that could affect the success of fire-fighting efforts are subject to review by fire districts. Fire suppression systems, which are systems that actively fight fires (automatic sprinkler systems of a variety of types), as opposed to systems that simply detect or prevent fires, belong in this division as well. In addition to having some of the problems associated with other mechanical systems, fire suppression systems are complicated hydraulic systems whose performance is sensitive to minor changes in design. They are carefully reviewed in the design phase and are actively monitored during and after construction by the fire districts having jurisdiction in the community.

P: Plumbing

Plumbing systems are designed and described by mechanical engineers and recorded in plans, elevations, and sections, as well as in isometric schematics and fixture schedules. The basic parts of the system include drain waste and vent piping, hot and cold-water supply, and fixtures.

D: Process

Process refers to systems that support the conversion of raw materials into a commercial product. Complicated forests of piping, controls, and storage facilities, process facilities are worthy of a distinct division in drawing sets. Refineries, canneries, and wineries are examples of process facilities—the latter being an example of projects for which building construction drawings and process facilities might be combined.

(Continued)

M: Mechanical

Mechanical drawings describe the location, size, and type of equipment for distributing, filtering, humidifying/dehumidifying, cooling, and heating air, as well as the distribution and control systems required in a project.

E: Electrical

Electrical drawings describe the electrical service (utility-provided wiring, metering, main switches, and grounding), distribution (panelboards, switchgear, and wiring emanating from the boards), branchwork (circuitry), and devices used in a project. As with other drawings, the electrical engineer uses plans, sections, details, and schedules to describe the project.

W: Distributed Energy

Distributed energy resources (DERs) are small-scale units of power generation that operate locally and are connected to a larger power grid at the distribution level. Distributed energy drawings describe the location, size, and type of equipment for DERs, such as solar, wind, and batteries.

T: Telecommunications Drawings

Changes resulting primarily from widespread computer use, as well as developments in telecommunications technology, have resulted in a dramatic increase in the attention given to telecommunications systems. The NCS has provided room for additional developments by creating a separate division for these systems.

R: Resource

Resource drawings consist of any drawings that are created prior to and sometimes during construction, as well as "measured" drawings—drawings that describe existing conditions that are used in the development of remodeling plans, among other types. As to subject matter, these drawings contain whatever information might be required for a remodeling or refurbishing project—structural, mechanical, and other plans are among the possibilities.

X: Other Disciplines

This category is a miscellaneous division. Any participant—an acoustical consultant, for example—could produce the necessary drawings for atypical kinds of work.

Z: Contractor Drawings

Shop or fabrication drawings are among the types of drawings that are the responsibility of the contractor, hence, the division "contractor drawings." Subcontractors or manufacturers use shop drawings to demonstrate to their shop personnel, the contractor, and to the design professional how an assembly or component—described in general terms by the architect or engineer—will be produced. Structural steel, trusses, fire suppression systems, and vertical transportation are examples of the kinds of work that are detailed in shop drawings. Shop drawings are the responsibility of the contractor; however, it is generally the architect who provides the list of work items that require shop drawings as a part of the submittal process. The drawings are reviewed initially by the contractor, and then are sent to the design professional, who reviews them for compliance with design intent. Although some controversy has arisen as to the timing (prior to permit approval) of certain submittals, as well as the liability associated with shop drawing approval, the design professional is interested in understanding generally how the contractor plans to execute portions of the work.

O: Operations

This category exists for the benefit of facilities management personnel, who have the responsibility for maintaining the facilities of a company or institution as well as for modifying facilities to suit changing needs. Drawings generated by facilities management employees or by design firms that describe proposed changes to the facility find a home in this category.

—Keith Bisharat (2008), *Construction Graphics*, John Wiley & Sons, pp. 26–28.

6.4 SPECIFICATIONS

The AIA states that specifications are written requirements for materials, equipment, systems, standards and workmanship for the work, and performance of related services. In their book, *Construction Specification Writing: Principles and Procedures*, Rosen and Regener list the following topics best covered in specifications:

- Type and quality of every product, from simple material to system
- Quality of workmanship during manufacturing, fabrication, application, installation, and finishing
- Requirements for fabrication, erection, application, installation, and finishing
- Regulatory requirements, including applicable codes and standards
- Overall and component dimensional requirements for specified materials, manufactured products, and equipment
- Specific descriptions and procedures for product alternates and options
- Specific requirements for administration of the contract for construction

Those who use drawings and specifications—owners/clients, plan checkers, engineer/architect field representatives, estimators, contractors, subcontractors, material suppliers, and inspectors—benefit from standardized formats. Standardization enables AEC professionals

TABLE 6.4   US and Canadian drawing and specification standards.

Drawings	Specifications
• US National Computer Aided Design (CAD) Standard (NCS)	• Construction Specifications Institute (CSI) MasterFormat™
• Construction Specifications Institute (CSI) Uniform Drawing System (UDS)	• Construction Specifications Institute (CSI) SectionFormat™
• American Institute of Architects (AIA) CAD Layer Guidelines	• Construction Specifications Institute (CSI) PageFormat™
• National BIM Standard Project Committee— National Institute of Building Sciences (NIBS)	• American Institute of Architects (AIA) MASTERSPEC®

to communicate more easily with one another, minimizes confusion, and saves time. Some large public owners, such as the US Army Corps of Engineers, and private companies, such as The Boeing Company, develop and adopt their own standards. In such cases, designers and contractors must familiarize themselves with owner-defined standards.

Table 6.4 lists the organizations largely responsible for creating the standardized systems of drawings and specifications used throughout the United States and Canada.

6.4.1   Specification Format

The Construction Specifications Institute (CSI) has developed a specification numbering and formatting system, called MasterFormat™, which is used widely (Construction Specifications Institute 2005). Prior to 2004, CSI-standardized specifications had 16 divisions, 01 through 16. Because of increased project complexity and the desire to address the needs of the market, CSI's 2004 MasterFormat™ was increased to 50 divisions, 00 through 49. It also changed the basic specification numbering system from five to six digits. See the text boxes in the following figure *Typical Specification Numbering System* and *MasterFormat™ Division Numbers and Titles.*

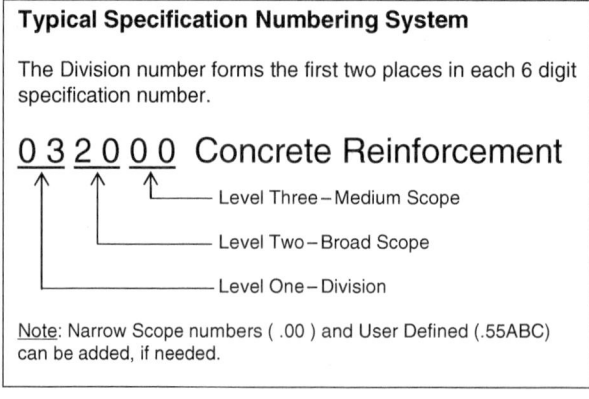

MasterFormat™ Division Numbers and Titles

PROCUREMENT AND CONTRACTING REQUIREMENTS GROUP
Division 00   Procurement and Contracting Requirements

SPECIFICATIONS GROUP
GENERAL REQUIREMENTS SUBGROUP
Division 01   General Requirements

FACILITY CONSTRUCTION SUBGROUP
Division 02   Existing Conditions
Division 03   Concrete
Division 04   Masonry
Division 05   Metals
Division 06   Wood, Plastics, and Composites
Division 07   Thermal and Moisture Protection
Division 08   Openings
Division 09   Finishes
Division 10   Specialties
Division 11   Equipment
Division 12   Furnishings
Division 13   Special Construction
Division 14   Conveying Equipment
Division 15   Reserved
Division 16   Reserved
Division 17   Reserved
Division 18   Reserved
Division 19   Reserved

FACILITY SERVICES SUBGROUP
Division 20   Reserved
Division 21   Fire Suppression
Division 22   Plumbing
Division 23   Heating, Ventilating, and
             Air Conditioning
Division 24   Reserved
Division 25   Integrated Automation
Division 26   Electrical
Division 27   Communications
Division 28   Electronic Safety and Security
Division 29   Reserved

SITE AND INFRASTRUCTURE SUBGROUP
Division 30   Reserved
Division 31   Earthwork
Division 32   Exterior Improvements
Division 33   Utilities
Division 34   Transportation
Division 35   Waterway and Marine Construction
Division 36   Reserved
Division 37   Reserved
Division 38   Reserved
Division 39   Reserved

PROCESS EQUIPMENT SUBGROUP
Division 40   Process Integration
Division 41   Materials Processing and Handling
             Equipment
Division 42   Process Heating, Cooling, and Drying
             Equipment
Division 43   Process Gas and liquid Handling,
             Purification, and Storage Equipment
Division 44   Pollution Control Equipment
Division 45   Industry-Specific Manufacturing
             Equipment
Division 46   Reserved
Division 47   Reserved
Division 48   Electrical Power Generation
Division 49   Reserved

Source: Construction Specifications Institute (csinet.org)

The specifications numbering systems developed by the Construction Specification Institute (CSI) and the American Institute of Architects (AIA) provide for inclusion of *Procurement and Contracting Requirements* and *General Requirements (General Conditions of the Contract)*. Table 6.5 lists typical information included in Divisions 00 and 01.

TABLE 6.5 Information contained in divisions 00 and 01.

Division 00 Procurement and Contracting Requirements	Division 01 General Requirements
• Advertisement for bids	• Summary of work
• Invitation to bid	• Price and payment procedures
• Instructions to bidders	• Product substitution procedures
• Prebid meetings	• Contract modification procedures
• Land survey information	• Project management and coordination
• Geotechnical information	• Construction schedule and documentation
• Bid forms	• Contractor's responsibility
• Owner-contractor agreement forms	• Regulatory requirements (codes, laws, permits, etc.)
• Bond forms	• Temporary facilities
• Certificate of substantial completion form	• Product storage and handling
• Certificate of completion form	• Owner-supplied products
• Conditions of the contract	• Execution and closeout requirements

CSI also has created SectionFormat™, which presents a unified way of depicting information contained in each specification section. Each specification section is divided into three parts:

- Part 1—General

  An extension of Division 01—General Requirements unique to this specification. Describes work covered by the specification, as well as administrative and procedural requirements such as submittals and quality assurance.
- Part 2—Products

  Details regarding materials, products, equipment, systems, and quality control.

  Describes products to be incorporated into project, such as mix design and off-site fabrication.
- Part 3—Execution

  Preparatory and on-site actions to be taken.

  Describes erection/application/installation, as well as field quality control and manufacturer's field services.

Typical information included under these headings is included in Figure 6.3. A complete package of typical construction documents including the project manual, drawings, and specifications is depicted in Figure 6.4.

```
                           SECTION XXXXXX

                            SECTION TITLE

    PART 1– GENERAL

    1.1     SECTION INCLUDES
            A.   Element of Work ⎤
            B.   Element of Work ⎦  Next level of detail
    1.2     RELATED SECTIONS
    1.3     ALLOWANCES
    1.4     UNIT PRICES
    1.5     ALTERNATES
    1.6     REFERENCES
    1.7     DEFINITIONS
    1.8     PERFORMANCE REQUIREMENTS
    1.9     SUBMITTALS
    1.10    QUALITY ASSURANCE
    1.11    DELIVERY, STORAGE, AND HANDLING
    1.12    PROJECT CONDITIONS
    1.13    SEQUENCING AND SCHEDULING
    1.14    WARRANTY
    1.15    MAINTENANCE

    PART 2– PRODUCTS

    2.1     MANUFACTURERS
    2.2     MATERIALS
    2.3     [MANUFACTURED UNITS][EQUIPMENT][COMPONENTS][ELEMENT OF WORK]
    2.4     ACCESSORIES
    2.5     MIXES
    2.6     FABRICATION
    2.7     SOURCE QUALITY CONTROL

    PART 3– EXECUTION

    3.1     EXAMINATION
    3.2     PREPARATION
    3.3     [ERECTION][APPLICATION]INSTALLATION]
    3.4     FIELD QUALITY
    3.5     MANUFACTURER'S FIELD SERVICES
    3.6     ADJUSTMENT AND CLEANING
    3.7     DEMONSTRATION
    3.8     PROTECTION
    3.9     SCHEDULE

                            END OF SECTION

                              XXXXXX - #
```

FIGURE 6.3 Abridged version of CSI's sectionformat™.

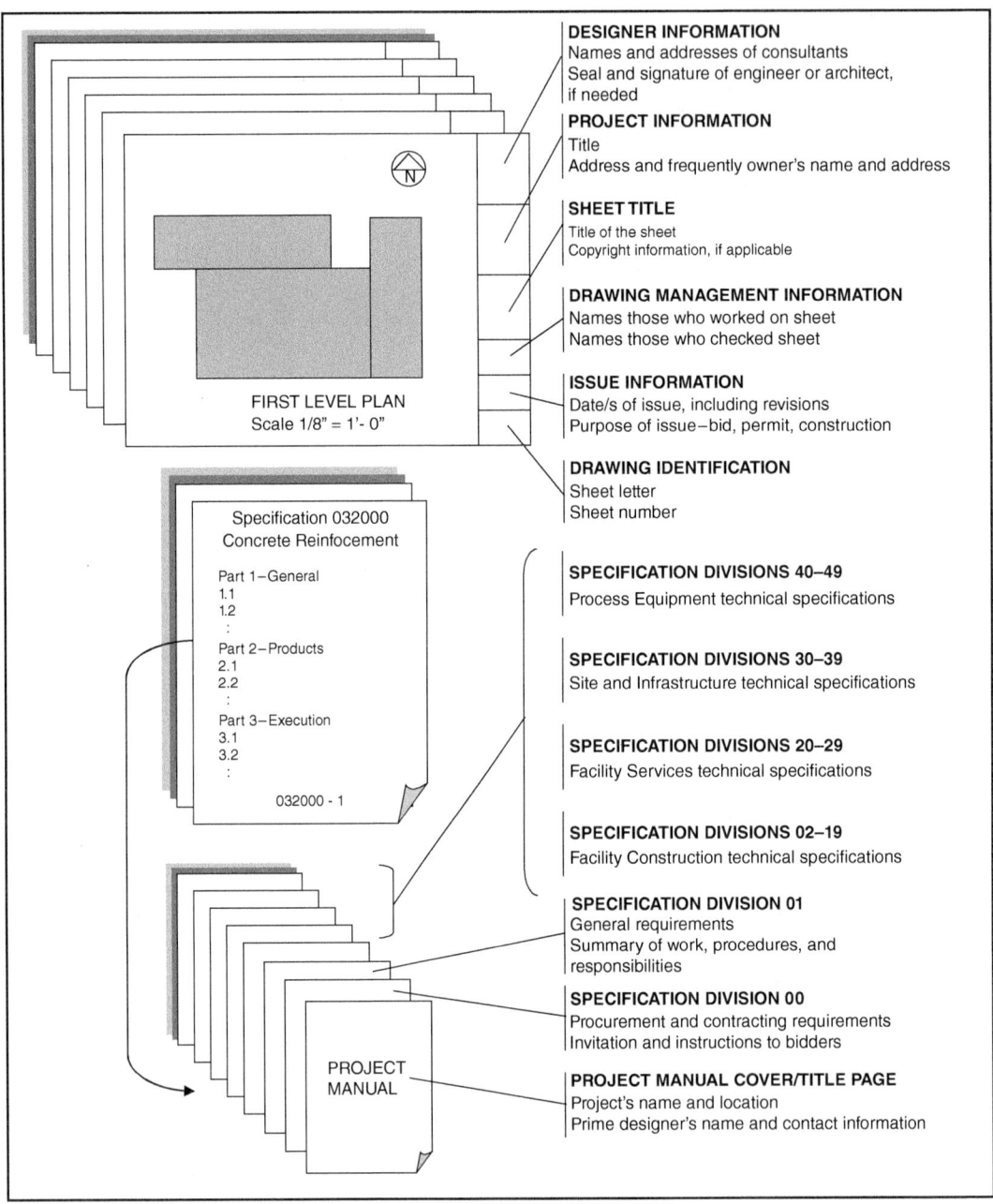

DESIGNER INFORMATION
Names and addresses of consultants
Seal and signature of engineer or architect,
if needed

PROJECT INFORMATION
Title
Address and frequently owner's name and address

SHEET TITLE
Title of the sheet
Copyright information, if applicable

DRAWING MANAGEMENT INFORMATION
Names those who worked on sheet
Names those who checked sheet

ISSUE INFORMATION
Date/s of issue, including revisions
Purpose of issue–bid, permit, construction

DRAWING IDENTIFICATION
Sheet letter
Sheet number

SPECIFICATION DIVISIONS 40–49
Process Equipment technical specifications

SPECIFICATION DIVISIONS 30–39
Site and Infrastructure technical specifications

SPECIFICATION DIVISIONS 20–29
Facility Services technical specifications

SPECIFICATION DIVISIONS 02–19
Facility Construction technical specifications

SPECIFICATION DIVISION 01
General requirements
Summary of work, procedures, and
responsibilities

SPECIFICATION DIVISION 00
Procurement and contracting requirements
Invitation and instructions to bidders

PROJECT MANUAL COVER/TITLE PAGE
Project's name and location
Prime designer's name and contact information

FIRST LEVEL PLAN
Scale 1/8" = 1'- 0"

Specification 032000
Concrete Reinfocement

Part 1–General
1.1
1.2
:

Part 2–Products
2.1
2.2
:

Part 3–Execution
3.1
3.2
:

032000 - 1

PROJECT
MANUAL

FIGURE 6.4   Typical construction documents (drawings and project manual) format.

6.4.2 Methods of Specifying

There are four widely recognized methods of specifying. These are:

- *Descriptive Specifying*: Exact properties of materials and methods are described in detail, without referring to specific manufacturers or suppliers.
- *Reference Standard Specifying*: Standards developed by trade organizations, institutions, and government organizations are cited within the three other specification methods.
- *Proprietary Specifying:* Actual brand names, model numbers, and other unambiguous information define the materials, methods, or systems; may be made less restrictive by naming two or three manufacturers or by providing "or equal" terms.
- *Performance Specifying*: The end result desired is described and the contractor, manufacturer, and/or fabricator supplies the solution that meets the defined criteria; designer must include a provision for appropriate tests to measure performance.

Each type of specifying has advantages and disadvantages. These are summarized in Table 6.6.

Some Authors and Publishers of Reference Standard Specifications

AISI (American Iron and Steel Institute)

ANSI (American National Standards Institute)

ARMA (Asphalt Roofing Manufacturers Institute)

ASHRAE (American Society of Heating, Refrigerating, and Air-Conditioning Engineers)

AASHTO (American Association of State Highway and Transportation Officials)

ASTM International (formerly American Society for Testing and Materials)

AWI (Architectural Woodwork Institute)

AWS (American Welding Society)

BHMA (Builders Hardware Manufacturers Association)

ACI (American Concrete Institute)

ICC (International Code Council)

NECA (National Electrical Contractors Association)

(Continued)

NEMA (National Electrical Manufacturers Association)

NFPA International (formerly National Fire Protection Agency)

NIST (National Institute of Standards and Technology)

NRCA (National Roofing Contractors Association)

OSHA (Occupational Health and Safety Administration)

PHCC (Plumbing, Heating, Cooling Contractors Association)

SAE (Society of Automotive Engineers)

SDI (Steel Door Institute)

SMACNA (Sheet Metal and Air Conditioning Contractors' National Association)

SSPC (Systems and Specifications for Painting and Coatings)

TABLE 6.6   Advantages and disadvantages of specifying methods.

DESCRIPTIVE SPECIFYING		REFERENCE STANDARD SPECIFYING	
Advantages	Disadvantages	Advantages	Disadvantages
• Describes exactly what the designer intends	• Requires designer to describe design intent carefully; "wordsmithing" necessary	• Clearly states which standards of production, workmanship, and quality apply	• Can be cited incorrectly, causing confusion at best or errors at worst
• Is applicable to all conditions and circumstances	• Results in long specification documents	• Is based on well-tested and accepted work developed by experts	• Can be used inappropriately in place of descriptive specifications
• Permits free competition—omits brand names	• Is time consuming to produce and may require more time in evaluating submittals	• Saves designer time by not having to "reinvent the wheel"	• May be difficult to enforce; contractor must interpret the standard
• Provides good basis for bidding; desired work results clear	• May be too elaborate for small projects	• Can be used on most projects, except very small projects	• May be obsolete or based on too low a standard

TABLE 6.6 *(Continued)*

PROPRIETARY SPECIFYING		PERFORMANCE SPECIFYING	
Advantages	Disadvantages	Advantages	Disadvantages
• Controls product selection strictly; contractors know exactly what is expected	• Prefers some manufacturers and suppliers over others	• Spells out design intent only; requires contractor to deliver systems that work	• Requires definition of all attributes, requirements, criteria, and testing
• Bases details closely on data supplied by manufacturers or suppliers	• Reduces or eliminates competition	• Delegates technical responsibilities to the contractor	• Delegates technical responsibilities to the contractor
• Reduces specification production time	• May use products or materials with which the contractor has no or bad experience	• Can result in shorter documents	• Can be time consuming to produce
• Simplifies bidding by narrowing the competition	• May create the possibility of increased cost due to lack of competition	• Encourages the development of new technologies and permits free competition	• May be too elaborate for small projects

Source: Adapted from Harold J. Rosen and John R. Regener (2005), *Construction Specifications Writing: Principles and Procedures,* 5th edition.

6.5 DRAWINGS AND SPECIFICATIONS—FINAL THOUGHTS

Division 01—General Requirements typically spells out whether the drawings or specifications take precedence if the information in one conflicts with that of the other. In a dispute, however, specifications often are given greater significance than drawings for several reasons. First, specifications may show design intent more clearly than drawings. And second, those involved in the legal system—attorneys, judges, and juries—as well as construction managers are more familiar with interpreting text-based rather than graphics-based documents.

The best way to avoid conflicts between drawings and specifications is to limit duplication of information. Drawings and specifications work together to tell the whole story, but their purposes are different. Making global changes to drawings and specifications is difficult because they are created using different software; there is no "search and replace" command to replicate changes through both sets of documents. Many designers request the reader/bidder to ask questions if there's an apparent discrepancy between the drawings and the specifications.

Finally, the information presented in both types of documents should be clear and concise. Many different people use the drawings and specifications. The language selected by civil engineers and other consultants should be comprehensible to owners, technical specialists, construction field personnel, and government agencies alike. Like any professional deliverables, drawings and specifications also should be correct and complete.

6.6   TECHNICAL MEMOS AND REPORTS

Engineers are often given assignments in problem solving that may be part of an overall larger project. However, some problem solving may be on a large scale and may be associated with an operating component of a company or an organization. The size of the problem and its potential impact usually dictate whether the assignment will produce a technical memo or a technical report. A technical memo is usually less formal than a report but still has much of the same content. A technical report will generally include several additional sections beyond a technical memo including an executive summary, references, and/or appendices. In addition a technical report typically has greater detail and depth of the information than a technical memo. However, a technical memo is often requested when there's a time crunch or an immediate project need for this information since it can usually be prepared more quickly. In these cases, a technical report may be requested after the technical memo is quickly prepared.

Probably the most important difference between a technical memo and a technical report is the intended audience. A technical memo can be shorter and include less detail because the audience may be colleagues or peers within the organization, senior management or other related departments within the organization. In such cases, the audience is usually familiar with the problem, technical approaches, and analytical details.

An example of a technical report format is included in Chapter 13, Communicating as a Professional Engineer. In addition, a sample short technical report titled "The Benefits of Green Roofs" may also be found in Appendix E. Additional sample technical reports may be found in the Appendices. For example, Appendix A includes a sample Request for Proposal for a Pipeline Routing Study. Appendix B includes an example engineering proposal to accomplish this requested scope of services in Appendix A. And, Appendix C includes a sample feasibility report to address this RFP found in the Appendix A.

The United States Environmental Protection Agency (US EPA) has several excellent reference documents for engineers including a guidance document for conducting feasibility studies (FSs). This guidance document is titled, "Guidance for Conducting Remedial Investigations and Feasibility Studies under CERCLA, Interim Final" (October 1988) EPA 540/G/89/004, OSWER 9355.3–01. This is an excellent reference for conducting FSs and includes good graphics and tables as examples.

Another useful tool for engineers is a cost estimating guidance document. This US EPA reference can provide useful information but the engineer will need to update specific details relevant to the location, and contemporary material and labor rates. The reference document is titled:

"A Guide to Developing and Documenting Cost Estimates During the Feasibility Study" Provides information about developing and documenting cost estimates during the feasibility study. US Army Corps of Engineers US Environmental Protection Agency EPA 540-R-00–002 OSWER 9355.0–75 www.epa.gov/superfund July 2000

6.7   CALCULATIONS

Engineers are also given assignments in problem solving which may require engineering calculations. The calculations are usually associated with an operating component of a company or an organization or could be part of a repair or replacement for a necessary component. Engineering calculations are often requested by a professional colleague from a technical branch like "plant operations" of a company or agency. The requestor generally has a great deal of knowledge on the overall performance and utility of the system and simply needs some technical assistance from an engineer.

Calculations should be set up much like engineers are taught in their degree programs. The problem statement should be clearly defined and a deliverable product should be agreed upon with the requestor. The original assignment may be given to the engineer in a hallway conversation or in a more formal kick-off meeting. If the engineer is unclear about the assignment or problem, some additional communications will be necessary to ferret out the specific information needed to proceed. A typical format for "engineering calculations" is outlined below:

- Problem Statement
- Objective and Approach
  - The objective statement should be carefully worded to reflect the originator's request in the engineer's own words. The objective statement is usually a sentence, or maybe several, and can include a primary objective, secondary objective, or others as necessary.
  - The approach statement should reflect "how" the engineer plans to approach the problem. It can include details on required tools, any required testing or analyses, required resources, materials, restrictions on ongoing operations, required health and safety details (the H&S plan will likely be included later but mentioned here), and any other pertinent details related to the approach. State whether the operations (if this is part of an operating system) will need to be shut down or restricted in any way during the assessment, maintenance, and repair period.

- Known Information
  - A description and relationship of the specific component within the system should be included, as well as photos or sketches with dimensions and impacts if possible.
  - It is suggested that the engineer include a sketch or diagram if it will help crystallize the problem and related details.
  - For clarity, the "problem statement" should be connected with the objective/approach and known information.
- Technical References
  - Technical references on materials of construction, reference documents, existing engineering drawings, manufacturer's catalogue information, performance objectives, specifications, operating requirements, and other related information also should be included. This information should be made as detailed as necessary to meet the objectives, health and safety requirements for personnel, and operations requirements.
- General Scope of Work Section
  - This section should show the general tasks that are required to accomplish the task.
  - There may be an assessment task to gather information related to the component failure, replacement, or upgrade.
  - If there is an assessment task, an evaluation task, design task, or calculation task will likely follow.
  - Depending upon the complexity of this problem, completing this series of tasks with product evaluation, specification details, ordering, pricing, or custom development and manufacture may be necessary.
  - A clear statement should be made regarding the requestor's desired product, how it should be transmitted, and how it should be delivered.
  - The task effort should include some time for development and planning of a schedule and budget (or time estimate). The engineer is highly cautioned to include assessment time, reference time, and QC time in the overall estimated time to complete the effort. Often the requestors do not understand the interrelationships of components within an overall system and can possibly "oversimplify" the problem statement in an effort to speed up the delivery of the required product. Conversely, the engineer is highly cautioned not to "overcomplicate" the problem when communicating with the requestor. An experienced engineer will find that there is a delicate balance here and may request some mentoring or coaching from more experienced colleagues when addressing this issue.
- Calculations
  - The calculations should be clear and concise showing the reason for the calculation related to the problem statement, objective/approach, known information including drawings or photos, and general scope of work.

- The engineer should write clearly, include references, show detail, include assumptions, verify conditions, initial (or sign) and date each sheet. The engineer should have these calculations peer-reviewed and all calculations should be quality checked (QC checked) by a competent professional in the field of endeavor.
- If the document will be released for public files, the specific state where the work is accomplished may require the engineer to stamp the calculations.
- Transmitting the Final Deliverable
  - The engineer should prepare a letter of transmittal to the requestor that includes the relevant information agreed upon in the kick-off meeting. Sample letters and transmittals are included for reference in Chapter 13, Communicating as a Professional Engineer.
  - The engineer should keep file copies of the information for future reference especially if the laws governing professional engineers in the state where the work is accomplished require the engineer to do so.

6.8   OTHER DELIVERABLES

The civil engineer often has numerous responsibilities during the construction phase of a project and frequently is responsible for: (1) facilitating the permit process; (2) responding to requests for information (RFIs) from the contractor; (3) reviewing fabrication (shop) drawings and other submittals; (4) attending meetings and writing meeting minutes; (5) making recommendations to the owner regarding construction progress payments; and (6) documenting installation instructions. Table 6.7 contains further information related to construction terms.

TABLE 6.7   Useful construction phase terms.

Term	Meaning
Submittal	Shop drawings, material data, and sample required primarily for the engineer and architect to verify that the contractor has purchased the products required in the plans and specifications. Concrete mix design calculations are an example.
Shop drawing	Drawing or set of drawings submitted to the prime designer for review by the general contractor. Produced by the contractor, supplier, manufacturer, subcontractor, or fabricator. Typically required for prefabricated components, such as structural steel, trusses, precast, elevators, etc.
Steel fabrication drawings	Shop drawings showing the fabrication of structural steel components. Depict steel joint connections and detailed dimensional information for all parts and assemblies, including tolerances. Calculations generally included.
Request for information (RFI)	Questions directed to the prime designer by the general contractor to gain clarification or to confirm the interpretation of a detail, specification, or note on the construction drawings. Used to secure a documented directive and often results in a change to the scope of requiring changes in project budget and/or schedule.

6.9 SUMMARY

The civil engineer and other designers are responsible for a dizzying array of deliverables. The challenging and rewarding aspect of the project delivery process is that these designers ultimately create something from nothing. As later chapters show, client relations, project management, and communication are key to this process. Ultimately, the delivery of the civil engineer's stock in trade—studies, reports, calculations, plans, specifications—enables a sort of alchemy to take place. These project deliverables result in build/no-build decisions and have been used to construct almost all the built environment that surrounds us. The need for their accuracy and completeness and clear communication cannot be overstated.

BIBLIOGRAPHY

American Institute of Architects. (2008). *The Architect's Handbook of Professional Practice*, 14th edition. Joseph A. Demkin, AIA, executive editor. John Wiley & Sons, Inc., Hoboken, New Jersey. ISBN 978-0-470-00957-4.

Bisharat, Keith A. (2008). *Construction Graphics: A Practical Guide to Interpreting Working Drawings*, 2nd edition. John Wiley & Sons, Inc., Hoboken, New Jersey. ISBN 978-0-470-13750-5.

Construction Specifications Institute. (2005). *The Project Resource Manual—CSI Manual of Practice*, 5th edition. McGraw-Hill, New York. ISBN 0-07-137004-8.

Madan, Mehta, Scarborough, Walter, and Armpriest, Diane. (2010). *Building Construction: Principles, Materials, and Systems—2009 Update*. Pearson Prentice Hall, Upper Saddle River, New Jersey. ISBN-13: 978-0-13-506476-4.

Rosen, Harold J. and Regner, John R. (2005). *Construction Specifications Writing: Principles and Procedures*, 5th edition. John Wiley & Sons, Inc., Hoboken, New Jersey. ISBN 0-471-43204-0.

CHAPTER 7

Executing a Professional Commission—Project Management

Big Idea

Effective project management knowledge and techniques are essential for conducting project operations. The project manager is at the heart of the project and must be aware of all activities related to the initiation, planning, execution, monitoring and control, and closure of the project.

"Of all the things I've done, the most vital is coordinating the talents of those who work for us and pointing them toward a certain goal."

—Walt Disney

Key Topics Covered

- The Basics of Project Management
- The Major Parties on a Project
- Project Sectors
- Project Teams

Civil Engineer's Handbook of Professional Practice, Second Edition. Karen Lee Hansen and Kent E. Zenobia.
© 2025 John Wiley & Sons, Inc. Published 2025 by John Wiley & Sons, Inc.
Companion website: www.wiley.com/go/hansen/CivilEngineersHandbook

- Project Initiation
- Project Estimates
- Project Management Plan Components
- Staff Selection Guidelines for the PM
- The Project Manager's Responsibilities
- Project Risk Management
- Design Coordination

Related Chapters in This Book

- Chapter 3: Ethics
- Chapter 4: Professional Engagement
- Chapter 5: The Engineer's Role in Project Development
- Chapter 6: What Engineers Deliver
- Chapter 8: Permitting
- Chapter 9: The Client Relationship and Business Development
- Chapter 10: Leadership
- Chapter 11: Legal Aspects of Professional Practice
- Chapter 12: Managing the Civil Engineering Enterprise
- Chapter 13: Communicating as a Professional
- Chapter 14: Balancing Life, Family, and Career
- Chapter 15: Globalization
- Chapter 16: Sustainability
- Chapter 17: Emerging Technologies
- Chapter 18: Human Relations Policies and Employment Practices
- Chapter 19: Construction Management
- Chapter 20: Critical Health and Safety Knowledge for Civil Engineers
- Chapter 21: What Engineers Need to Know.

7.1   INTRODUCTION

7.1.1   Project Management Background

It's appropriate to begin a discussion on project management with a brief description of the historical development of the management of projects. Peter Morris has provided a detailed account of the development of project management as a distinct discipline (Morris 1994). Other authors (Kerzner 1998) and Morris postulate that project management as a discipline was spawned in the mid-20th century.

However, mega-projects date to a distant time; the Roman Coliseum was built by four different contractors (Morris 1994). Early undertakings by the Celts, Egyptians, Greeks, Romans, and Chinese involved the entire community and served to cement secular and religious authority. During the great Gothic period, expression of the devotion to God was more important than timely project delivery but, as economies and technologies became more sophisticated, so did the use of contracts to realize projects. By the 18th century, those who designed projects were separate both contractually and organizationally from those who built them.

The economic development of the Victorian era led to huge infrastructure projects and industrialization. Authors whose early theories on scientific management emerged at the beginning of the 20th century are still quoted. These include: Taylor and Gilbreth (time and motion studies) and Gantt (production scheduling). Weber also established his theories on bureaucracy during this period. But it was not until the 1930s that an academic writer proposed the use of a coordinator who might be used to administer a task involving several functional areas (Morris 1994). Morris views this addition of a separate mechanism to integrate the various entities making up a project as *the inception of modern project management*. He further proposes that the rise of modern project management between the 1930s and 1950s is related to:

- Development of systems engineering in the US defense and aerospace industry
- Engineering management practices in process engineering
- Developments in management theory, particularly in organization design
- Evolution of the computer, enabling many project management tools

Kerzner, on the other hand, holds the more typical view that project management began in the 1950s and 1960s (Kerzner 1998). A major advancement came in 1958 during the development of the POLARIS missile program by the US Navy, helped by the Lockheed Missile Systems division and the consultant firm of Booz-Allen & Hamilton when "program evaluation review technique" (PERT) was developed.

During that period the literature abounds with journal articles that proposed various models of organizational redesign which would lead to better control over resources, and thus, better project control. The result was an organizational structure with multiple layers of management. The 1970s saw a focus on organizational behavior. Firms and researchers pondered how to get desired productivity from these elaborate command and control structures. With advances in information technology, by the 1980s the emphasis shifted to project management quantitative tools, such as "performance evaluation and review technique," or PERT, and "critical path method," or CPM (Render and Stair 1982, Wiest and Levy 1974). As a result of business process re-engineering (BPR) in the 1990s and the changing nature of technology, many firms have fewer layers of management and relatively flat organizational structures. The presence of multidirectional, cooperative work flow necessitates better communications. Additionally, the advent of the "virtual" organization, for instance, a multifirm

TABLE 7.1   Evolution of design toward collaboration and integration.

Modern	Postmodern
Taylorism	Systems Engineering
hierarchical/vertical	flat/layered/horizontal
specialization	interdisciplinary
rational order	messy complexity
centralized control	distributed control
experts	meritocracy
tightly coupled system	networked system
micro-management	blackboxing
hierarchical decision making	consensus-reaching
bureaucratic structure	collegial community
incremental	discontinuous
closed	open

Source: Adapted from Hughes, 1998.

organization (an example would be a prime contractor and key subcontractors) formed around a project that will be disbanded upon project completion, strongly underscores the importance of integration in project management.

Literature on project management reflects the need for flexibility and the changing context in which projects exist. The phrase "management by projects" may best capture the situation today. Thomas Hughes, in *Rescuing Prometheus: The Story of Mammoth Projects* (1998), has captured the essence of these changes by comparing the modern and postmodern approaches to design (Table 7.1).

7.1.2   A Discipline, But Not a Theory

Although books on project management abound, their primary focus is prescriptive. That is, authors outline what to do to optimize cost, schedule, quality, profitability, and so forth. Works on project management *theory* are much less abundant. In fact, project management appears to be a blend of ideas drawn from other disciplines. The Project Management Institute's guide to accepted knowledge and practices of the profession includes nine project management "knowledge areas," each of which has a robust theoretical underpinning (PMBOK Guide, 2021). These areas are:

- Integration management
- Scope management

- Time management
- Cost management
- Quality management
- Human resources management
- Communications management
- Risk management
- Procurement management

To this list can be added:

- Organization structure
- Organizational behavior

Shehnar and Dvir approach project management theory from a different perspective, that of the innovation literature (Shehnar and Dvir 1996). These authors observe that one of the basic deficiencies of project management theory may be that little distinction is made between project type and project managerial problems. They propose a two-dimensional typology of projects that maps system (project) scope against technological uncertainty as a way of distinguishing project types as illustrated in Figure 7.1.

System scope refers to the notion that there are different hierarchies inside a system or product with different levels of activities. Shehnar and Dvir define these categories of project scope as: *assembly* (components and/or modules combined into a single unit); *system* (collection of interactive elements, performing independent functions and fulfilling a specific need); and *array* (widely dispersed collection of different systems functioning together to achieve a common purpose). The technology axis ranges from *low-tech* to *super high-tech*. At one end of the spectrum is technology that exists and is readily available and acceptable (e.g., that used to build roads); at the other end is technology that does not exist at project initiation (e.g., the Apollo moon landing).

Shehnar and Dvir's typology appears to offer a useful way to move toward an integrated theory of project management. The authors suggest other typologies that might be developed. Possible dimensions include: the fit between project type, project management style, and project effectiveness; project environment, such as economic, political, social, geographic, and cultural; and the degree of difficulty in articulating user/customer requirements and the point in the project lifecycle at which these requirements are identified.

Given the factors mentioned above, most projects vary significantly along several critical dimensions rendering a universal typology meaningless. From a broader perspective, acknowledging that all projects have typology with varying phases of complexity can be quite useful. When the Project Manager recognizes this fact, he or she can gain insight into timing and location of potential *hotspots*, such as interface between players, integration, communication, and so forth that may manifest itself later in the project.

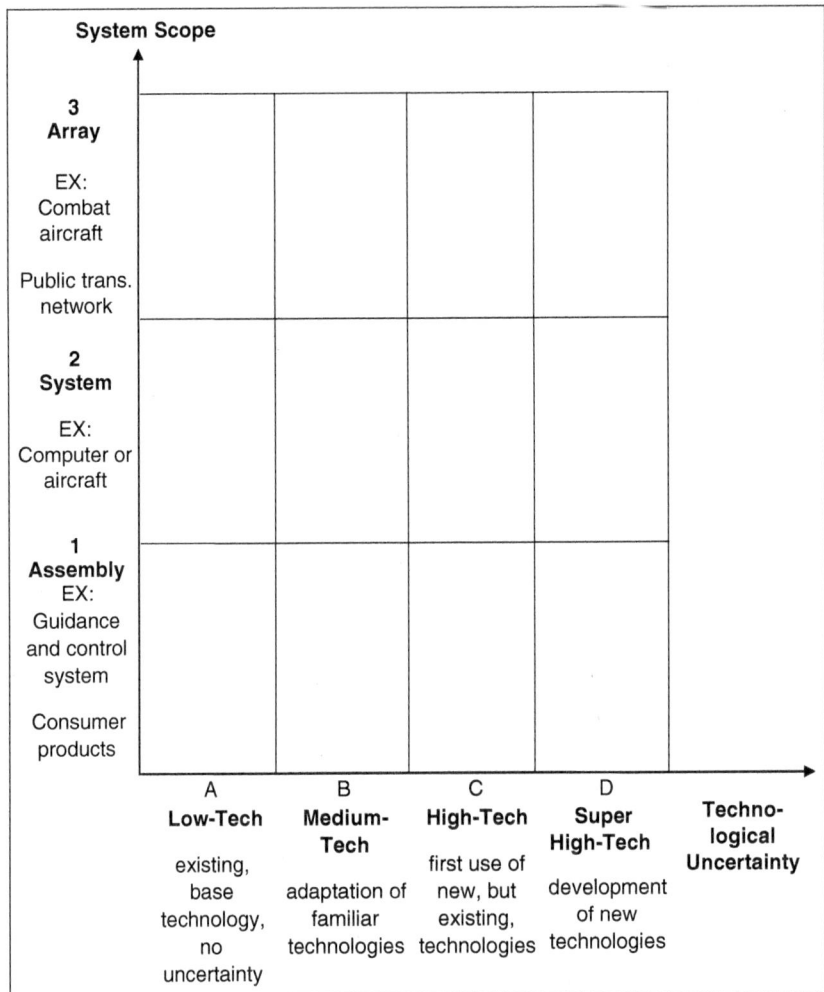

FIGURE 7.1   Two-dimensional typology.

A Glimpse into the Life of a Project Manager

Charlene just had a full morning of budget and projection meetings for her business unit and now she was exhausted as it approached 12:30 p.m. She actually got up almost an hour earlier than usual to get bagels and snacks for the management team since it was her turn. Her Engineering Manager, Steve, needed a complete rundown of the previous quarter's invoicing for her four clients (and all the other client invoices in the unit) plus the anticipated staff level of effort (LOE) resource projections and subcontractor needs for the next quarter. She stayed up late the night before to get the figures into the new spreadsheet requested by her boss through the Denver corporate office. Charlene left the meeting with seven priority tasks on her "to do" list that needed attention by the end of the day. On the way to her office, she ran into Frank who reminded her about the overdue deadline for her comments and recommendations for two new engineer candidate employees she recently interviewed. The interviewees were anxiously awaiting a response from the firm since they both now had competitive offers from other companies and a public agency.

As Charlene returned to her office, she rested her eyes on the pleasant photo on her wall calendar, but she suddenly remembered it's month end and she also needed to enter her time into the time sheet database or her paycheck would be delayed. She hadn't had a free moment all week to enter her time and she was now seriously out of compliance with the policy. She knew she would receive the computer automated reminders and a special "spanking" phone call from the Engineering Manager on this one, since Federal regulations require daily time entries. Her main concern now was the three urgent e-mail alerts on her screen and the blinking light on her voicemail suggesting a missed call from one of her clients. She normally would have seen the e-mails come in on her smart phone, but Steve did not permit cell phones in the quarterly meetings because of the disruptions. Charlene quickly scrolled to the urgent e-mails as she bit her lip unintentionally.

Apparently, Jim, the field crew lead technician, had a minor health and safety incident in the field. Jim hit an obstacle on the construction site with the new truck and had a flat tire and bent wheel rim. In the process of changing the tire the new staff engineer nicked his thumb on the bent rim and required five stitches. Unfortunately, Charlene, as the PM for this project, had to fill in the Accident Form 3592 by the end of the shift and report it to the Engineering Manager and the Director of Corporate Health and Safety. She began to feel her neck ache indicating the first sign of one of her dreaded headaches.

Charlene pulled up Form 3592 on her computer as she called Jim on his cell phone. After four rings and no answer, she left a message on Jim's voicemail to check in and see if he needed any assistance. She then called Sara, his colleague and trainee. Sara answered after one ring anxiously waiting Charlene's call. Sara stated they were driven to the emergency room by the client since their new truck was now disabled. The client's PM, Wendy, was really nice and understanding over the past 18 months on this project—they even exchanged birthday cards. However, Sara said there was one minor problem as she mentioned that Jim's blood stains on the client's car seat. Charlene sounded surprised and slightly irritated about this mishap but quickly recalled they did not have a chance to equip the new truck with a standard company first aid kit. The Engineering Manager would not be happy about this situation and would likely be a lot more irritated than the client with the stained car seat. Charlene hung up the phone and went to find Steve to inform him of the disabled vehicle and stranded technicians at the hospital.

(Continued)

As Charlene rounded the corner, she noticed Steve's lights were out signaling his likely lunch appointment. She phoned him immediately to tell him the news. The Engineering Manager picked up after the second ring and was very quiet on the other end as she recanted the events thus far. Steve apologized to his clients at the lunch table as he called for the bill. He understood that he was likely going to the project site or the hospital to pick up the field crew, Jim and Sara, as he quickly headed back to the office. Jim chose to go to the project site to arrange for a tow truck and thank the client, Wendy, for taking Jim and Sara to the emergency room. He was thinking about what solvent might take Jim's blood stains out of the seat fabric in Wendy's personal car.

Charlene took her personal car to the emergency room to pick up Sara and Jim. They were waiting by the exit door as she pulled up and gave a hug to Jim showing her delight that he was okay after the stitches and unanticipated tetanus shot. Jim apologized for the mistake and took the blame for not having the first aid kit in the new truck. It really wasn't Jim's fault, or Charlene's fault either, since the kit was on backorder and expected in next week. Charlene actually passed within a few blocks of her house as she longed for a long hot bath, a snack, and a glass of wine. When they arrived back at the office, Charlene told them both to go home for the day. In her mind, she thought the minor accident would likely be decided as "avoidable" by the Corporate Health and Safety Manager and that Jim would have to retake the defensive driving course as a reminder and sort of a punishment. She also recalled a recent presentation she heard about "how we pay" for mistakes in this morning's long meeting.

Charlene then returned to her office to tackle the unopened e-mails and voicemails as it approached 4:00 p.m. Her headache was in full force accompanied by a stomach pain now, long after lunch as she searched for the Tylenol and an emergency snack bar in her bottom desk drawer. She scanned here-mail origination time, titles, original sender, and distribution and chose the second unopened e-mail to open immediately as she dialed her voicemail box for that news. The deep voice on the voicemail message confirmed that the message was from her largest telecommunications client Civil Construction Vice President, Jonathan. He said he was surprised that there was a new proposed ASTM testing technique required on the construction materials for all new facilities under construction in the eastern region. He was concerned that it could potentially cost over $25,000 in unanticipated expenses to his firm. He ended the concise message with an irritated tone and requested a callback soon.

Charlene quickly composed a brief e-mail response (knowing Jonathan's personal habit of checking fresh e-mails over unopened ones) and then picked up the phone for Jonathan's cell. They shared the same cell provider, Jonathan's company. Charlene wondered whether this could have had any impact on her firm winning the multi-million-dollar design and construction monitoring contract. Jonathan picked up the call on the first ring knowing it was Charlene while he simultaneously received confirmation of the pending e-mail reply from her. Charlene opened with a warm greeting and then apologized for missing Jonathan's earlier call and previous e-mail. Jonathan returned the greeting warmly and stated he only had a few minutes since he was at the airport waiting for his flight home. Charlene didn't bother providing Jonathan excuses on "why" she could not respond earlier and simply got right to the point and asked whether he had a chance to read yesterday's e-mail notification and pick up her previous phone message. Jonathan stated he had been in strategic planning meetings for the last two days regarding funding for his company's corporate challenge and the current flurry of construction. Charlene mentioned that she had just learned about these new testing requirements, and she had already planned to equip her field staff with the new testing equipment and that, overall, this test will replace a similar but slower technique thereby saving the client money and that construction may be able to proceed a bit faster. Charlene thought to herself how she really appreciated her CE QA team for this information. Jonathan laughed and thanked Charlene for being prepared and promised more good news when he had more time at their regular bimonthly meeting next week. As Jonathan finished his final words Charlene heard a flight announcement in the background. Taking the nonverbal cue Charlene said she looked forward to the upcoming meeting. She quickly said "happy anniversary" and closed with "stay well," a little saying they said to each other often. Jonathan said thanks again and closed with "you stay well, too." Jonathan wondered how he found such an intelligent, responsive PM and one who could keep track of his anniversary date almost better than he could. Charlene thought her new "client key information" database was working well, knowing that two weeks ago Jonathan let it slip out that his ten-year wedding anniversary was at the end of the month.

Charlene took a deep breath, opened the second e-mail as she prepared to call Steve to check on the progress of the disabled new truck. She saw the second e-mail was related to a civil design question for her third largest client. She composed a brief reply on her understanding of the design aspect, included her lead designer on the reply message, and suggested that more recent information could be provided by the lead designer. She knew this client well and suspected her information was sufficient and was going into the client's month-end report. However, she thought it was better to provide more information than less and marked this client as a follow-up phone call in two days. Meanwhile, Steve picked up the call on the second ring, said the new truck went to the dealer's shop and he was on his way home. She communicated her brief accomplishments and that she would be at work a bit longer finishing up. She saved Jonathan's promise of a "surprise" for her next face-to-face meeting with Steve when she needed to explain "how" the truck incident happened.

Charlene addressed the three urgent e-mail alerts and the urgent VMX and now turned to her time sheet, the required Health and Safety Form 3592, Frank's second request for her recommendation on the two interviewees, and the seven things on

(Continued)

her "to do" list. The time sheet was the easy one and only required ten minutes and then she pulled up Form 3592 on her screen. She filled out the urgent part knowing that some information would not be available until the injured party, Jim, returned to work. The accident was fully covered, and Jim had not had an accident in his last five years with the business unit so she wasn't too concerned. Completing these activities lifted a weight off her mind as she thought the Tylenol was kicking in as her headache receded. She remembered the blood stain in her client's (Wendy's) car as she researched a solvent for blood and a car detail company that could accomplish the task. She placed a quick call to Wendy's cell to ask her if she would like a rental car for a day. Wendy turned down the nice request, but Charlene did get a commitment for a date and time to get Wendy's car in for a detail cleaning and a special wash and wax treatment. Wendy was so nice. Charlene thought this fix would require some flowers, too. Charlene thought of the "ways a PM pays when a team member makes a mistake on a project."

Charlene finished a brief message to four of the seven items on her "to do" list as she forwarded them off to her technical team for finishing up the information. She addressed the other three items quickly, and sent off the e-mail and attachments to Steve, as was requested of her. Charlene thought to herself that she actually was quite efficient when nobody was around, the phone didn't ring, and e-mails didn't come in. The quietness of the office was not surprising as it approached 6:30 p.m. She realized she had to leave for her dinner appointment with her best friend. She put on some soft rock music on the quick trip home before changing and heading out for the evening.

Charlene thought to herself that her original plan of becoming invaluable to her employer and clients was a good one. She recognized that nobody could foresee all the potential project impacts and she was proud of her accomplishments for the day. On the way to dinner, she placed one more call to Jim to check on his hand injury and empathize with his pain. Time out: 7:30 p.m., off duty. But Charlene's cell phone is still on.

> **"Mistakes are painful when they happen, but years later a collection of mistakes is what is called experience."**
> —Denis Waitley, motivational consultant

An engineering project manager (PM) may have to pay for a mistake, even if the mistake is not the fault of the PM or the firm:

1. Money: When a mistake or a problem comes between a client and an engineer, the most obvious form of payment is money. The payment can take the form of additional services, materials, or supplies paid by the engineer or shared with the client.

2. Time: The time element is usually experienced with a slip in the schedule or a delay in delivery of the final product or services. The schedule slip will likely be accompanied with additional costs borne by the engineer and/or client

3. Pain: In a professional work atmosphere the third element that may accompany a mistake, depending upon the seriousness and circumstances, is emotional pain. In this instance, the emotional pain may be embarrassment or a sting to one's character or reputation. If the mistake is in the form of a health and safety incident, it may also be accompanied with actual physical pain as well.

7.2   THE BASICS OF PROJECT MANAGEMENT

7.2.1   Definition of a Project

A project is an endeavor that is undertaken to produce results expected from the requesting party. To be a bit more specific, a project is an endeavor that will:

- Accomplish a specific client need or goal
- Include related activities guided by a leader or project manager
- Be composed of cross-departmental personnel or unique experts capable of providing unique services to the project
- Be performed for a fixed duration of time

Generally, the requestor or owner does not have the expertise or the time (or both) to take on the endeavor on their own. The project will likely have a fixed or estimated life of its own which can vary from days to years. For example, a project to repair a software program may take days and a project to build some of the world-renowned civil engineering projects depicted in Chapter 2, may take decades. Once the project is complete, the operating component and maintenance will likely revert to the owner's operations and maintenance personnel as the project team is disbanded. However, some contemporary engineering project owners offer the designers and/or builders a chance to provide services for short-term start-up operations and occasionally long-term facility operations.

7.2.2   Scope/Schedule/Budget Triangular Relationship

A project's scope of work, often referred to as the "scope," is a definition of the tasks required to complete the project to meet the client's needs. The complete scope of work for a project requires a schedule and budget. If the owner does not supply a schedule as part of the scope, an experienced PM will include a schedule for a project and an experienced negotiator will know that the schedule is part of the negotiations. For example, if an owner wants a 40-story hotel and convention center designed and built in 48 weeks, this effort will include a significantly more concentrated effort for managing the design team, materials procurement/delivery, and the construction team than an identical project built over a two-year period.

A project like this was built in Las Vegas—on a fast-track basis with the construction crews framing and placing concrete for each floor over a one-week period and then moving to the next floor. The casino owner had evaluated the increased construction costs for the accelerated construction compared to the standard construction cost. The owner then considered the casino revenue lost on the longer 100-week standard construction period and the costs associated with the shorter 48-week construction period. The casino owner realized that the additional hotel and casino revenue significantly exceeded the costs for the accelerated construction. The owner determined they would attempt building one floor a week for the high-rise hotel and convention center. This example illustrates how the budget is uniquely tied to the scope of work and schedule for each project, as illustrated in Figure 7.2.

This relationship of scope, schedule, and budget can be thought of as a connected triangle where each side represents an essential component of the "project" managed by the project manager (PM). Remembering the connected sides of this triangle enables the PM to see that these three components are interrelated, and that one side cannot move or be adjusted without affecting either of the other two components Kerzner (1997).

Quality during this stage of project development is another important issue that needs to be addressed. The client will likely have an idea about the quality of the project and will include their thoughts on quality in the scope details. The important point is that quality should connect the scope, schedule, and budget. If the final project does not meet or exceed the owner's expectation for quality, the project likely will be rejected by the client. There will be more discussions on quality later in this chapter (PMBOK Guide, 2021).

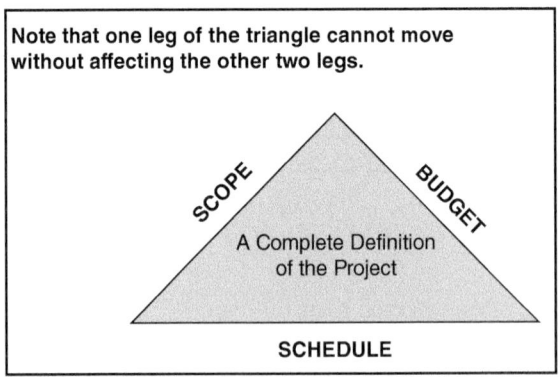

FIGURE 7.2　Scope, schedule, and budget relationship.

7.3   THE MAJOR PARTIES ON A PROJECT

The most important parties in a project are the owner, designer, and builder/contractor. These parties have the most invested in the project and stand to gain or lose the most if the project is performed correctly on time and within budget. A cooperative team approach is best when performing projects although the builder/contractor is often brought into the project at a later stage. It is recommended to have several builder/contractors comment on the early design phases of the project to get valuable feedback on the constructability of the proposed design.

A cooperative team approach with the owner, designer, and builder/contractor will often avert adversarial relationship problems later in the project.

7.3.1   The Owner's Role

The owner's role in the project is critical to setting operational criteria for the details on the scope of work. The owner should also identify:

- Their preferred level of involvement
- Their proposed review process
- Levels of approval based on dollar amounts, scheduled completion, or budget completion
- Any special equipment
- Any important company policies regarding safety, security, labor requirements, and so forth

In addition, the owner should also set parameters on total cost, payment of costs, major milestones, and the required project completion date.

7.3.2   The Designer's Role

The engineering designer should specify the number, type, and details of the:

- Proposed design alternatives
- Number, type, and amount of computations
- Number, type, and amount of drawings
- Level of detail for the specifications

The engineering designer can reduce the level of ambiguity and increase the level of understanding with the owner and builder/contractor by providing a clear understanding in written form with exhibits, tables, and figures on the design detail.

7.3.3   The Contractor's Role

The contractor should clearly state (or be asked to state) that they will:

- Perform all the work in accordance with contract documents prepared by the engineering designer
- Furnish all labor, equipment, material, and know-how necessary to build the project as designed
- Prepare regular budget and status reports for the owner
- Develop project quality reports for the engineer and owner
- Develop accurate budget and schedule controls

In summary, the contractor should act as if they are an owner and report information clearly and effectively to keep the project on schedule and within budget.

7.3.4   A Brief Summary of the Basics

Project management is defined as the art and science of coordinating people, equipment, materials, money, and schedules to complete a specified project on time and within approved cost, and includes the functions of planning, organizing, staffing, directing, and controlling.

7.4   PROJECT SECTORS

Project sectors relate to the market sector the project will service. The building sector generally includes the residential, commercial, or industrial market depending upon the client. The infrastructure sector includes private or, more likely, public owners and involves transportation, water, and other systems needed for a functioning society. The process sector includes the industrial and/or commercial markets. For example:

7.4.1   Building Market

Projects will likely be composed of:

- Commercial, educational, office, hospital, residential buildings
- Prime designer: Architect

7.4.2 Infrastructure Market

Projects will likely be composed of:

- Transportation (roads, bridges, airports, waterways, and water treatment facilities)
- Prime designer: Engineer

7.4.3 Process Facilities Market

Projects will likely be composed of:

- Chemical plants, oil refining, pharmaceutical, pulp and paper, electrical generating
- Prime designer: Engineer

7.4.4 Project Delivery Methods

There is a large array of approaches for delivering projects to the owner in these diverse sectors. Project delivery methods are influenced by whether a project is competitively bid or negotiated. Many engineering firms will not even submit proposals for competitively bid projects, where the focus is mainly on lowest price rather than the engineering firm's qualifications.

The amount of risk a client is willing to accept also greatly influences the choice of a project delivery method. There are risks associated with rapid construction (fast-track). If the client desires an accelerated service, with design solutions in the field, the client must be willing to accept a fairly high level of risk. In fast-track delivery methods, a detailed schedule showing critical dates for design packages and procurement is essential.

Project size is another import factor in selecting the appropriate project delivery method. For example, more elaborate methods, such as integrated project delivery outlined below, may not be appropriate for small projects; but larger, more complex projects may benefit greatly from such approaches.

Chapter 11 discusses project delivery methods in detail; however, a brief list of available methods is included here for reference:

- Design-bid-build (DBB) requires two separate contracts: owner-prime designer and owner-contractor. If the prime designer and contractor enjoy a solid working relationship, the owner's risk exposure is moderate; but if multiple conflicts arise between the prime designer and contractor (who do not have a contract with each other), the owner could be exposed to a high level of risk, including cost over-runs and schedule delays. This situation inevitably leads to a high level of risk for the prime designer also.

- Design-build (DB) requires only one contract from the owner's perspective: owner-design build entity. The design build entity can be a partnership or joint venture between the prime designer and builder. Alternatively, either the designer or builder can take the lead, putting the other in a subordinate position. The DB project delivery approach is meant to reduce the owner's risk and to shift that risk to the design build entity.
- Integrated project delivery (IPD) requires one contract: a multiparty agreement. In IPD, the owner, prime designer, builder, and possibly other significant project participants (a key subcontractor, for example) sign a single contract. In IPD risk is meant to be shared equally by all signatories to the contract.
- Construction management (CM) at risk requires two contracts: owner-prime designer and owner-CM. In this delivery method the construction manager signs contracts on the owner's behalf with the general contractor and key subcontractors. In CM at risk the CM is assuming some of the owner's risk.
- Construction management (CM) as agent typically requires three contracts: owner-CM, owner-prime designer, and owner-contractor. In this project delivery method, the CM acts as the owner's agent and may manage both design and construction; however, the owner assumes the primary risk for design and construction.

7.5 PROJECT TEAMS

Project teams are composed of members vital to the success of the project and must include a leader to guide overall efforts. The team leader is the Project Manager (PM) and the PM must rely on the team for technical expertise. The PM also acts as a coach, answer questions, clear the way for progress, and makes sure desired outcomes are understood by all team members.

The PM influences a diverse set of individuals with sometimes competing goals, needs, and perspectives. The PM needs to motivate these individuals since often the members may be assigned to the project from other departments within the firm. The PM often deals with multiple teams within the organization including administrative, budget office, health and safety, and various engineering disciplines. The design team organizations may include architects, subconsultant engineers, and CAD specialists. Construction teams have a different culture and may have a shorter-term perspective on the project.

The PM must possess team management skills, since these various subteams need to be an integral part of the project organization, referred to as the "project team." The project team must have a well-defined mission with goals and trust instilled by the PM. As part of the team management skill, the PM will likely conduct team building exercises, such as the project kick-off meeting, where they will stress that:

- All participants need to use effective communications and should confirm that information given and received is clearly understood
- All participants have a common customer

- Team success will allow continuity of project team
- Remembering key words "responsibility," "honesty," "kindness," "respect," and "communications" is critical
- The positive aspects of the project lead to success and perhaps future work for the same client

"Teamwork builds trust and trust builds speed."
                                —Lieutenant General Russel Honore, American military leader

With regard to the project team, the PM likely will request or pick members in accordance with the client's requirements for the scope, budget, and schedule; acquire resources for the project; develop processes for decision-making; and develop a leadership style that is respected and accepted by the whole project team.

7.6   PROJECT INITIATION

Once an engineering business unit wins a project after completing the initial proposal effort, interviews, and contracting, then the actual phases of work for CE projects begin, as depicted on Figure 7.3.

The phases of delivering the project include:

- Project definition, referred to as "predesign," which includes an approximate 10 percent design that shows the layout and footprint and possibly an elevation.
- Design phase, which is composed of three phases referred to as schematic design, design development, and construction documents. The schematic design will be about 10 to 20 percent and include input from the owner, the PM, and the design engineer. The design further develops to 60, then 90 percent completion, sometimes referred to as a "check print," for the client's review and comments and possibly the builder/contractor for a constructability review.
- Once the 100 percent design, which becomes part of the contract documents. The Construction Documents typically include the plans, owner's statement, a complete set of reviewed and stamped engineering plans, and a set of specifications including general specs, technical specs, contract requirements, a bid sheet, and a proposed construction schedule. With the exception of the plans, this information frequently is bound into a "book" called the Project Manual.
- Occupancy, which refers to occupancy of the project site by the owner.
- Adaptive re-use and decommissioning, which are activities that can take place years after project completion, but which may be addressed in the current design.

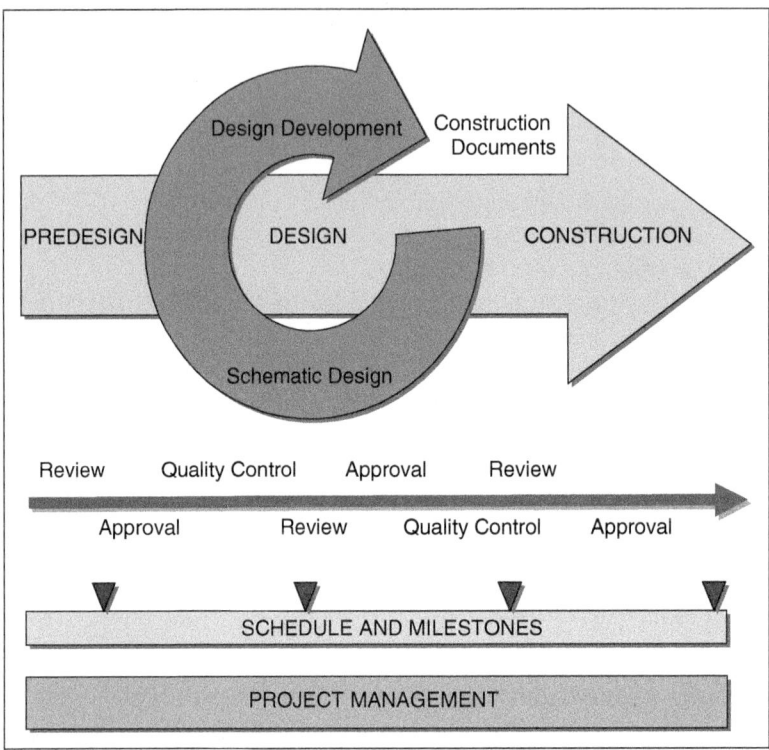

FIGURE 7.3   Delivering the project.

7.7   PROJECT ESTIMATES

Project managers frequently are involved in creating a series of project cost estimates, or "opinions of probable construction cost." The amount and type of information available at the time of the estimate influence the approaches taken and levels of accuracy.

7.7.1   Early Estimates

An early estimate is an estimate prepared before completion of detailed design Early estimates are an important tool for the PM because these very preliminary cost estimates are often the basis for business decisions. These early estimates are also used for:

- Asset development strategies
- Screening of potential projects
- Committing resources for further project development

Inaccurate or careless early cost estimates can lead to:

- Lost opportunities
- Wasted development effort
- Lower-than-expected returns or profit

There are also classifications of estimates/re-estimates including:

- Sponsor's (client's) study (conceptual design)
- Preliminary engineering (preliminary design)
- Schematic engineering (Schematic Design)
- Detailed engineering (design development)
- Final design (construction documents)

There are several primary factors that should be considered when preparing and reviewing estimates, or comparing them to one another, including:

- Standardization of the process to allow for comparison of items
- Alignment of objectives between client and designer so there's an early and common understanding on the cost of a project
- Selection of appropriate methodology for building size, location, material selection, or other methodologies
- Collection of project data and historical costs
- Organization of estimate into desired formats because some funds will likely come from different sources or even organizations
- Standard and/or special material costs
- Determine whether union or non-union labor costs should be used and/or if any specialized trades and associated labor cost are required
- Documentation of basis costs and accuracy of the estimate are important qualifiers
- Review and checking to verify content and accuracy
- Feedback from project implementation for future efforts to perform continuous improvement

There are benefits of alignment with cost estimates throughout the life of a project including:

- Establishing an understanding of the product or service received for the cost paid
- Determining the level of effort associated with the project and can establish a budgetary target for various components of the project
- Establishing work processes and a staffing plan
- It's recommended the cost estimates should be in relation to an approximate project schedule. For example, the cost of excavating in the winter months in the northern latitudes are much different than the associated costs for the same work being performed the summer months
- Highlighting critical issues early in the project so alternatives can be arranged, or high-cost elements can be assessed more closely
- Improving and documenting scope definition for the record as the project progresses
- Assisting the client's understanding of what is included in the estimate and what is not included. Here's another example of where the cost-estimating team should include written assumptions.

- Establishing responsibilities of parties involved and respective budget allocations and responsibilities
- Creating cohesiveness between project team and client in a way that builds trust and commitment

There are some critical questions that should be considered when preparing early estimates, such as:

1. What is the current level of definition for the scope of work? For example, is it enough to know there will be a 10,000-square-foot building or would an estimate be more accurate if the definition included more information such as two floors, one standard elevator and a freight elevator, two pairs of bathrooms, steel roof, masonry construction, two 400-square-foot kitchen units, and two 300-square-foot conference rooms?
2. What level of detail is the customer expecting, a five-line early estimate or a five-page estimate?
3. What resources are required for this deliverable, when is it due, and what is the appropriate format?
4. How will this early estimate be used and what decisions will be made from it?

There's one additional tool that can be very helpful for the engineer regarding the preparation of cost estimates. Cost estimating likely will involve several specialists on the project team and there will likely be a cost estimate kick-off meeting. Here's a checklist of issues for the estimate kick-off meeting:

1. What are the client's main issues, cost, quality, time, or other concerns?
2. What level of accuracy and what format is the client expecting?
3. When is the estimate due, when will the materials be purchased, and when will the work be performed?
4. What is the budget for the preparation of this estimate and how much effort is required?
5. Are there any customer-furnished items such as land, grading, security, utilities, materials, fencing, or other items?
6. Are there any special tax issues or credits, or funding requirements that could have an impact on the project cost?
7. Are there similar projects in the nearby vicinity that this project may be compared to?
8. What is the availability for labor and any specialty trades required for this project?
9. Are there any special permitting issues for this project that could affect the cost or schedule?
10. Does the client have any specific information sources the project team should contact or avoid?

This checklist is project specific, client and time dependent, and includes only the basic checklist items. However, the checklist will provide the engineer with a starting point for cost estimation discussions with the client.

7.7.2   Project Budget Estimates

Project budget estimates levels of accuracy significantly vary with the level of completion of design. As depicted in Figure 7.3, the project moves through the predesign phase, through schematic and design development, on to the construction documents, and to construction. The project becomes increasingly defined and refined. Each level of design generally has a client review component where there is an opportunity to conduct detailed discussions on the project thereby crystallizing the client's image and expectation in the engineers' and project team's minds. Therefore, it follows that if there's little or no design the "level of accuracy" for the budget estimate will be much higher than with more refined engineering design and project detail. The general level of accuracy for project budget estimates appears below:

- No design work: Accuracy $\pm$ 50 percent
- Preliminary design: Accuracy $\pm$ 30 percent
- Design development: Accuracy $\pm$ 15 $-$ 20 percent
- Construction documents: Accuracy $\pm$ 5 $-$ 10 percent

Of course, an owner may have other techniques for estimating their project costs, especially if they perform those projects on a regular basis, such as with small restaurants, commercial buildings, or civil structures. These owner techniques are often evaluated by the following:

- Parametric techniques such as square footage, cubic footage, linear footage, or some other common comparative measure, or
- Historical costs based on past projects, especially if they are in the general vicinity and time frame.

From an engineer's point of view design budgets and the business unit's compensation for these projects can be estimated by several means including:

- Lump sum based upon similar projects or experience
- Salary cost multiplied by a fixed multiplier
- Cost plus a fixed fee (payment) for the service
- Cost plus a variable fee (incentive-based payment) for the service
- A fixed percentage of construction

From a contractor's point of view construction budgets and the contractor's compensation for these projects can be estimated by several means including:

- Fixed price if the contractor has a solid database to work from and a high level of confidence
- Cost reimbursable plus a fixed fee

Work Plans and Engineering Services Proposals

A young engineer may ask a simple question: "So, how does the work plan differ from the proposal we sent the client"?

Pipeline Routing Study—Proposal for Engineering Services, Outline

- Cover Letter Title
- Page
- Table of Contents
- List of Figures
- List of Tables

This is front matter to the business proposal for the Pipeline Routing Study.

1.0 Project Description

    1.1 Background

    1.2 Project Details

    1.3 Objective and Approach

    1.4 Site Description

This outline shows the components of the business proposal for the Pipeline Routing Study from the cover letter to the supportive information, as shown in Sections 1.0 – 7.0.

2.0 Scope of Work
(As shown in the Global Hydraulic Proposal)

    2.1 Task 1—Project Management

    2.2 Task 2—Project Research

    2.3 Task 3—Development of Alternatives

    2.4 Tasks 4—Preparation of Cost Estimates

    2.5 Task 5—Evaluation of Alternatives

    2.6 Task 6—Engineering Feasibility Report

    2.7 Task 7—Oral Presentation to Clients

3.0 Project Schedule

4.0 Project Team

5.0 Project Budget
(Not shown in the Global Hydraulic Proposal)

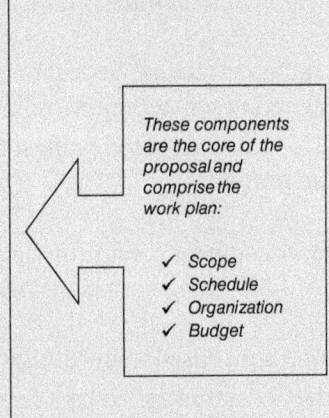

These components are the core of the proposal and comprise the work plan:

✓ Scope
✓ Schedule
✓ Organization
✓ Budget

6.0 Qualifications

7.0 Supportive Information

    Resumes

    Site Photographs

    Color Exhibits

    References

The core components of a proposal that actually discuss the work to be done, the organization and team to do it, the schedule of activities, and the budget to accomplish the effort become part of the work plan. The work plan is contained within the engineering firm's proposal to do the work. In essence, the initial work plan is the proposal minus the introduction and the qualifications and reference materials (marketing materials) to sell the work to the client. The marketing materials included in the proposal usually demonstrate that the proposer has completed similar projects successfully, has the qualified staff to accomplish the work—on time and within budget. So, the marketing material in the proposal has a very important role in helping the client(s) visualize a successfully completed project with a capable team to perform the work, while the proposal's schedule of activities, the team and team organization, schedule, and budget are the starting points for developing the work plan.

The PM begins preparing the work plan first as a "scope of work" within the proposal by analyzing previously gathered material the firm has been collecting since its business development personnel began following the proposed project. This preliminary work plan should reflect the scope of work outlined in the client's proposal. The PM will combine this information with all the background material prepared by the client. The PM will then:

- Become familiar with the owner's objectives and overall project needs
- Formulate an objective and approach
- Identify additional information required
- Organize the review process into three categories:
  - Scope
  - Budget
  - Schedule

The work plan can be strengthened by adding the owner's perspective and specifically adding the authorized representatives as an integral part of the project team. The owner's representatives serve two purposes:

1. To provide information and clarity to the project requirements. Optionally these representatives may actually review and approve all team decisions, and
2. To define quality in addition to scope, cost, and schedule. Optionally, if the owner does not have the expertise in-house to perform this function, they may hire an independent third party to act on their behalf. A construction manager can also accomplish this task for the owner.

If representatives are a part of the project team their detailed role and level of involvement needs to be articulated and well understood to avoid problems later.

7.8   PROJECT MANAGEMENT PLAN COMPONENTS

7.8.1   Plan Purpose

Once the project scope of work, schedule, and budget are defined, it's time to get to work. A very useful document to guide the efforts and provide necessary procedures for the project team is the Project Management Plan (PMP). The PMP is like a roadmap that provides important and general information and project procedures to the project team and stakeholders. The PMP is generally distributed to all team members at the project kick-off meeting, discussed in detail and updated as necessary as the project progresses. The PMP should be a live document, updated on a regular basis and controlled by a revision number and date. PMP revisions should be an agenda item discussed at each project meeting.

7.8.2   PMP Components

- General project information:
  - Purpose
  - Objective
  - Approach
- Major client, stakeholder, and general information:
  - General contract information
  - Contract type (prime contractor, subcontractor, design-build, or multiple prime)
  - General contact information including e-mail and phone
  - Preferred method of contact
- Organizational structure
  - Organization chart
  - Project manager or project engineer
  - Key Disciplines (technical disciplines such as civil engineering, planning, environmental sciences, civil design, and administrative services such as document production, contracting, accounts receivable, and so forth)
  - Key team members and leaders in each discipline area, as well as their roles within the project organization, especially as it relates to client and vendor or subcontractor interaction
- Scope of work summary, or work plan
  - Summarize key tasks and subtasks
  - Show relationships of tasks to key disciplines
  - Show subcontractors, other services, materials, and other resource requirements

- Project schedule
  - Present project schedule
  - Present flow chart of activities that relate disciplines to tasks and schedule, generally shown as an accompanying CPM or PERT chart
- Project budget
  - Present a summary spreadsheet displaying tasks, project team member names, summary of labor hours, subcontractors, materials, or other resource needs. Relate this information to the overall project budget.
- Communication plan
  - Establish project communication protocols for client, project team, stakeholder, and partner contact. Protocols should describe the preferred method of contact and frequency for the client, stakeholders, and key partners (written, teleconference, e-mail contacts, and frequency such as daily, weekly, or monthly).
  - Set up regular schedule and reserve dates and time slots for periodic work sessions and project review meetings. The importance of project progress meetings cannot be overstated. These meetings are vital to ensure that a cross-feed of project information is occurring so that problems can be identified and addressed, or even anticipated and eliminated before they become impediments to the performance of work related to the project.
  - Describe the project filing system, including hard copy and electronic file formats.
- Project health and safety plan
  - Minimum plan content is described below. Plan should be upgraded as required by complexity of the field tasks, capability of the staff, and overall risk assessment.
  - Plan should include information such as:
    - Task activities including soil borings with a drill rig, site recon, enclosed spaces, chemical exposure, and so forth.
    - Potential hazards such as heat stress, poisonous plants, slip, trip and falls, ladder use, and the like.
    - Routes of exposure such as inhalation, skin exposure, puncture wounds, and so on.
    - Suggested safety equipment such as safety shoes, hard hats, safety glasses, sunscreen, or appropriate clothing.
  - Site map and routes to emergency care
    - Name and phone number for police and emergency care
    - Map to first aid facility
  - Suggested field staff partnering system (minimum of two staff members working together as partners if possible). For single staff activities it is recommended to have a phone call in procedure such as one phone call in the morning and one in the afternoon to report progress and safety status.

- Miscellaneous: First aid kit in vehicles, cell phone or radio availability, flashlights, flares, and so forth
- Health and safety plan to be reviewed and approved by senior engineering staff, certified safety professional (CSP), or certified industrial hygienist (CIH) professionals
- Project quality plan
  - Engineering tasks should be checked and verified by other qualified professionals in the discipline area.
  - Quality assurance (QA) personnel should be listed for each major discipline in the organizational chart mentioned above.
  - Quality assurance personnel should be involved from the initial proposal or scoping activities and throughout the project. QA personnel should not be expected to review and verify project information at the conclusion of the project.
  - Engineering documents that include conclusions and recommendations are generally required to be conducted by a registered engineer, stamped and dated by the registered professional. The recommended view should reflect requirements by the Board of Professional Engineers in the state in which the work is contracted.

In summary, the PMP should be a useful document often consulted by the PM, project team members, client, and other project stakeholders for project guidelines and procedures. This document will be a live document that should be discussed in the periodic project meeting and updated as necessary, tracked with a revision number and date so members can keep abreast of developments.

7.9   STAFF SELECTION GUIDELINES FOR THE PM

At some point in the project lifecycle the PM will likely be engaged in the process of staff selection for the project. Staff selection will usually involve numerous individuals, all with a vested interest, including the PM, the individual staff members, each discipline manager, the client or client's representative(s) among others. The experienced PM will often begin soliciting interest for staff before the project contract is signed or the kick-off meeting is conducted. Generally, the goal for the PM will be to select staff for the project based on criteria as illustrated in Figure 7.4. These criteria include:

- Project and client needs
- Staff availability
- Previous experience and qualifications
- Staff development
- Project budgets and/or staff rates

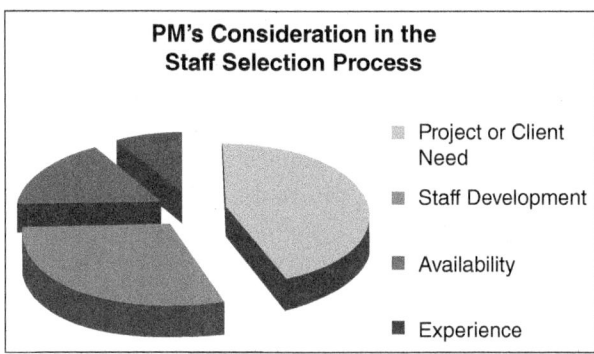

FIGURE 7.4   Staff selection process.

Once the PM has completed the project management plan, the PM can begin the negotiation process for specific staff individuals.

7.9.1   Project and Client Needs

Of course, one of the key elements in this decision will be the client's needs and the best choice for the project staff within each discipline that the project requires. For example, if the project has complex civil design features it would be a good choice to select very experienced civil designers rather than more junior staff. Conversely, if there were common drafting or AutoCAD requirements, then the PM might consider a staff engineer who was eager to learn new AutoCAD skills under the tutelage of another experienced engineer willing to mentor the junior engineer.

7.9.2   Staff Availability

The PM often encounters one important but basic question when negotiating for staff resources on a project: Which staff is "available" to work on a particular project? In the private sector, many companies often desire to operate with the minimum staff for the general market conditions to get the work done. In other words, the private sector generally operates a little light on staff with a higher work load per individual to maintain billability and cost competitiveness. Private sector managers usually like this scenario because they prefer to periodically place overtime requirements on their staff than the opposite condition where they might have to reduce work hours or reduce the number of staff on payroll. This general market condition can fluctuate over time and if long work hours accompany a high work load with corresponding higher profits, then these managers will consider hiring additional people.

In the public sector, the engineering managers generally find the workloads a little more stable unless there are specific conditions like new regulations or new programs that generate a significantly higher work load. So, in summary, the PM should keep abreast of the general market condition and demand for experienced staff and remain flexible in a buyer's or seller's market for the staff to conduct the project.

7.9.3   Previous Experience and Qualifications

The PM's goal is usually to work with the appropriate discipline manager (such as the Planning Group, or Surveying Department) to review the scope of work, task requirements/deliverables, and establish experience criteria for staffing the project. The PM and discipline manager will generally work together to prequalify the staff that have previous experience while also considering individual staff schedules or other commitments, staff career development goals, and/or cross-training between other clients and similar projects. For example, it might make sense to train a junior engineer on a particular project if they were particularly interested in working for a specific client.

7.9.4   Staff Development

Most public agencies and private engineering consulting firms perform annual appraisal and development evaluations on their staff. The first part of these evaluations is generally the appraisal portion that communicates the staff person's performance to the individual. The second portion of these evaluations generally includes the staff's development goals, indicating where the individual would like to direct their career. During these evaluation discussions, the staff individual generally expresses the type of work or experience they would like to obtain over the next year or two to develop their career. This process is a good one because it allows the discipline managers to learn the individual staff's desires while the PM learns where some staff might want specific experience.

7.9.5   Project Budgets and/or Staff Rates

A new PM quickly learns how to accomplish tasks and prepare deliverables for the client or finds alternate work. As part of this education, the PM gets exposed to different ways to accomplish the same goal. For example, the PM may have an option to place a very experienced engineer on a specific task and compare this level of effort and cost versus using a lesser experienced engineer with some senior mentoring and comparing this level of effort and overall cost to the project. There are some more specific examples of this situation under "Tracking Work/Work Breakdown Structure" in this chapter. The point here is that the PM can create alternate scenarios using a mix of staff to accomplish the same goals but come up with different cost impacts to the project. This particular skill can come in very handy for the experienced PM where this applied knowledge may allow the PM to cut costs, create development or coaching scenarios for eager staff, or increase profits for the firm or agency as needed.

In summary, the PM and discipline managers work together to discuss project assignments and corresponding deliverables, confirm staff interest and availability, discuss staff development goals and client requirements, and obtain firm commitments. If the project is delayed for some reason or the staff member is not available, then the PM and discipline managers generally will work together to find alternative solutions. Regular communication between the PMs, discipline managers, department managers, and staff members are required to keep one another updated on developments that may affect project commitments or staff availability.

7.10 THE PROJECT MANAGER'S RESPONSIBILITIES

The most important task of the PM is to maintain the Scope/Schedule/Budget triangular relationship depicted in Figure 7.2. The PM must follow the contract terms with the client (PMBOK Guide, 2021). Of course, the responsibilities do not stop there when there are so many interrelated tasks to manage to accomplish this main goal. These other tasks include:

- Maintaining ethical conduct, as discussed in Chapter 3
- Conducting problem recognition and solving, as discussed in Chapter 4
- Maintaining a high-quality product, as discussed in Chapter 4
- Maintaining the engineer's role in project development, as discussed in Chapter 4 and Chapter 5
- Producing quality engineering deliverables, as discussed in Chapter 6
- Managing staff, effective project management, and business practices to produce profits to maintain a healthy engineering firm, as discussed in Chapter 7.
- Managing the permitting requirements, as discussed in Chapter 8
- Maintaining the client relationship, as discussed in Chapter 9
- Conducting oneself as a leader, as discussed in Chapter 10
- Conducting the project in accordance with the contractual and legal aspects, as discussed in Chapter 11
- Preparing and reviewing the invoices, as described in Chapter 12
- Communicating as a professional, as discussed in Chapter 13
- Balancing life, family, and career, as discussed in Chapter 14
- Possessing and practicing the basic understanding of how globalization affects competition, utility of engineering services, design, cost-effectiveness, construction, decision-making and potential impacts to the civil engineer's own firm and the client, as discussed in Chapter 15.
- Implementing responsible sustainable growth and development practices as recommended and described by the ASCE, in Chapter 16.
- Adopting a personal goal for continuous improvement, becoming aware and presenting potential applicable opportunities for the application of emerging technologies, as described in Chapter 17.

- Following the required human relations policies and employment practices, as discussed in Chapter 18.
- Managing scope, budget, schedule, and quality, and implementing the construction management practices for civil engineers, as discussed in Chapter 19
- Employing the health and safety knowledge and requirements for civil engineers and staff, as discussed in Chapter 20

7.10.1   The PM's Time Commitment

The PM function can require an hour or two per week or can actually be managed by a project management office with numerous staff tracking the key components, depending upon the size and complexity of the project. If the project is relatively simple or has a small budget, the PM may actually perform the PM functions as well as all other task functions. These simple projects may be something as routine as a Preliminary Site Investigative Report or a Spill Prevention, Containment and Contingency (SPCC) Plan, where the engineering budget hours may only require 10 to 100 engineering/technical support staff hours. These relatively small project budgets don't really allow adequate time or budgets for a distinct PM task.

For medium to large projects, the PM is likely just to provide PM functions, depending upon the client and the business unit that employs the engineer. PM services and corresponding budgets typically require from about 2 to 10 percent of the overall project budget, depending upon the client and the business unit. As the projects get more complex and the project budgets increase, the PM budgets will likely become a smaller proportion of the overall project budget.

Complex or large projects typically providing technical services greater than 2,000 engineering/technical support staff hours may be managed by an entire PM office. This office includes staff services such as project management and client services, accounts receivable/payable, quality control, legal support, health and safety, technical writing, project tracking and/or other specific project staff as necessary. Medium-sized engineering projects ranging from 100 to 2,000 engineering/technical support staff hours may require the same services; but these services may be provided by shared matrix-managed staff or staff capable of performing more than one function, such as accounts receivable/payable, technical writing, and project tracking. The *project definition* is then completed by assessing the detailed technical services required to perform the project. For example, a water resources management project may require services like civil engineering, hydraulic modeling, geotechnical engineering, environmental permitting by environmental support staff, site restoration and biological services, landscape design services, technical writing, editing, mapping, geographic information system services, project tracking, accounts payable, contracting and legal services. These technical services comprise the core disciplines that the PM will integrate into the staff matrix.

The PM, usually in combination with the business unit's management, will then assess how the project will be conducted, the staff services that the business unit can provide, and any outside resources that may be required. The key point is that the PM, in conjunction with the business unit management staff, should evaluate the appropriate technical and client services for the project and then assess how these services can be best provided to fit the client's needs and budget. These aims sometimes are addressed in the organization breakdown structure (OBS).

This discussion leads us to the work breakdown structure and the business unit/organizational support.

7.10.2   Work Breakdown Structure

The *work breakdown structure* (WBS) defines the work to be accomplished and divides it into identifiable components that can be managed. The WBS does not yet define responsible parties (RPs) accountable for performing the work. But people are a necessary and important component of this process. After the WBS has been prepared, the next step is to identify the RPs, sometimes referred to as "resources" in the business unit/organization to perform the work. This component of the project is a critical one because the PM and/or the business unit's management team usually assess the overall schedule and availability of the staff and their capabilities to accomplish the tasks identified. For example, an engineering task might be accomplished by 100 staff support hours with senior engineering oversight of 10 to 20 support hours. Alternatively, this same task might be accomplished by about 70 senior engineering staff hours with 8 hours of senior engineering oversight. This evaluation becomes one that considers budget efficiency, staff training, coaching/mentoring, or even staff availability (more on this later).

One of the other input decisions for selecting the business unit's staff is to identify technical disciplines and individual staff capabilities responsible for the WBS functions. Selecting the individual staff members links the WBS to the organization and completes the *project framework*. Costs then can be linked to individual functions within the WBS through a cost breakdown structure (CBS).

Once the project framework is complete, the project schedule can be prepared. A tool useful to the PM who needs to develop a project schedule is the *critical path method* (CPM) (PMBOK Guide, 2021). CPM was originally developed by the DuPont Company in combination with Remington Rand as consultants in the mid-1950s. A similar method, referred to as the *performance evaluation review technique* (PERT), was developed by the US Navy in 1957. Both scheduling techniques are referred to as a network analysis that defines interrelationships of activities with corresponding scheduling of cost elements and resource availabilities. The CPM and PERT techniques require a detailed understanding of the identified tasks and their interrelationships to one another in a logical sequence (Anderson et al. 2009).

After the PM generates the WBS, this network diagram can be linked to a schedule using either the CPM or PERT technique. This handbook will not go into the details of network diagramming since there are many references available on this subject already. However, the PM should understand the basics of these techniques. Their primary purpose is to develop a baseline for the project and measure project progress. Project scheduling techniques can be as simple as bar charts, Gantt charts, Microsoft project scheduling software, or much more involved software programs. In engineering consulting, it is important to pick a scheduling software program that best fits the client's needs, the capability of the PM and the project team, the complexity of the project, and the project budget (Kerzner 2010).

In summary, the phases for the development of project plan are:

- Project Definition
- Project Framework, Organization Breakdown Structure
- Work Breakdown Structure
- Cost Breakdown Structure
- Project Scheduling (CPM and PERT)
- Project Tracking, Evaluation, and Control
- Project closeout and recordkeeping

Some Definitions

- OBS—Organization(al) breakdown structure: identifies organizational, rather than task-based, relationships
- WBS—Work breakdown structure: defines and groups a project's discrete work elements
- CBS—Cost breakdown structure: links to WBS and classifies costs into cost units, elements, and types
- CPM/PERT—Critical path method/performance evaluation review technique: links WBS, and possibly the CBS, to a schedule

Concepts Used in Project Progress Tracking

Project progress can be tracked using various indices. These tracking methods vary but basically compare the work completed in relation to the original budget and original schedule. Some definitions follow:

- Earned Workhours = (budgeted workhours) × (percent complete)
- Percent Complete = $\dfrac{\text{Sum of Earned Workhours for Tasks Included}}{\text{Sum of Budgeted Workhours for Tasks Included}}$

- Cost Performance Index (CPI)
  - $CPI = \dfrac{\text{Sum of Earned Workhours of Tasks Included}}{\text{Sum of Actual Workhours for Tasks Included}}$
  - Only tasks for which budgets have been established are included
  - Change order would need to be prepared for new work
- Schedule Performance Index (SPI)
  - $SPI = \dfrac{\text{Sum of Earned Workhours to Date}}{\text{Sum of Scheduled Workhours to Date}}$
  - Compares amount of work performed to the amount scheduled to a point in time
- If CPI or SPI is greater than 1.0, then:
  - Trend is favorable and can be "cumulative" or for a "defined period"
  - Trend can be plotted on a graph
  - Trend can be applied to design and construction
- If CPI or SPI is less than 1.0, then:
  - Trend is unfavorable, i.e., poor project performance
  - Trend can be plotted on a graph
  - Trend can be applied to design and construction
- Earned Value System

  BCWS = budgeted cost of work scheduled (Original Plan)

  ACWS = actual cost of work scheduled (Actual)

  BCWP = budgeted cost of work performed (Earned)

  ACWP = actual cost of work performed (Actual)

  CV = cost variance

  CV = Earned – Actual

  SV = schedule variance

  SV = Earned – Planned

  BAC = Budget at completion (Total cost for the project)

  EAC = Estimate at completion

  EAC = ACWP + estimated effort still needed to complete the work

  VAC = Variance at completion

  VAC = BAC – EAC
- If a positive number: Indicates an increased potential for profit or accelerated schedule.
- If negative number: Indicates potential cost overrun or schedule slippage.

7.10.3   Tracking Methods

The *percent complete* method is probably the most common method for typical civil engineering project tracking. This method simply compares the percentage of work completed and plots this figure on a graph and compares it to the percentage of time for the task to be completed (Wiest and Levy 1974). The *y*-axis portrays the percentage of work completed and the *x*-axis portrays the percentage of time for the task to be completed. Both axes should be scaled the same so that the point representing 25 percent complete versus 25 percent of time completed forms a 45-degree line bisecting the 90-degree angle of the *x*- and the *y*-axis. A typical graph representing project percentage completion is illustrated in Figure 7.5.

This particular example is very typical and illustrates how project teams often start out rather slowly where the percentage of work completed lags behind the percentage of time on the schedule (PMBOK Guide, 2021). At some point the project team either gets a time extension or significantly ramps up construction to catch up to the percentage of time completion on the *x*-axis. These graphs can be created for each task or collectively by calculating the total project completion percentage data. One caution about this method of project tracking is a tendency for the project team members to overestimate their percentage of work completed. A good cross-check on this estimate is to follow-up this question with:

"What is your estimate to complete the task?"

The answer to this query should provide a better indication of the remaining work to be completed for each task.

An alternate project tracking method is called the *CPI*, which calculates the sum of earned workhours of tasks included divided by the sum of actual workhours of tasks included. This index simply shows that if a project team member completes the assignment for less than,

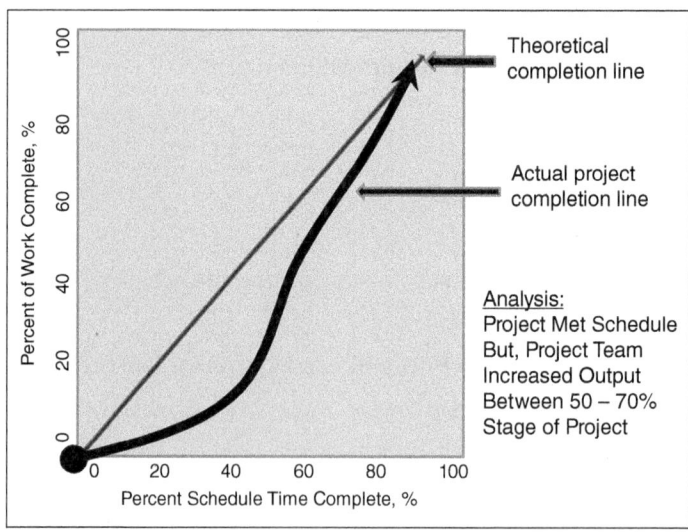

FIGURE 7.5   Typical percent complete graph.

or equal to, the number of hours assigned the ratio will be just greater than or equal to 1.0. A ratio of 1.0 or greater indicates a positive performance and completion within the assigned task budget. If the ratio is less than 1.0, this indicates a negative performance and completion over the assigned task budget with a likely negative impact on the budget.

The *SPI* is a similar tracking method which calculates the sum of earned workhours to date divided by the sum of scheduled workhours to date. This index shows the amount of workhours performed to date compared to the amount of workhours scheduled to be completed to date. If a project team member completes more hours on the project assignment than the amount of scheduled workhours to date, the ratio will be greater than or equal to 1.0. A ratio of 1.0 or greater indicates a positive performance and completion of the amount of hours scheduled or greater than the amount of workhours scheduled. If the ratio is less than 1.0, this could indicate a negative performance and likely incomplete assignment. This answer will require a more definitive answer to how much work had actually been completed. It is possible that efficient team members could have completed the assignment in less than the amount of scheduled hours thereby surprising the PM with a positive budget impact.

Using the earned value system, a PM can calculate a cost variance or a schedule variance for the project. This method compares the budgeted cost of work scheduled (in the original plan) compared to the actual cost of work scheduled (actual cost). A cost variance is calculated by simply subtracting the actual cost from the earned cost. Of course, the PM will have to be sure the task assignment is on schedule and compares these costs for the same stage of task completion for this calculation to be accurate. A schedule variance is calculated in much the same way, by subtracting the actual cost of work scheduled from the budgeted cost of work scheduled.

7.11   PROJECT RISK MANAGEMENT

Another way of looking at the project tracking methods discussed above is to consider them as tools used to analyze, anticipate, and address the risks inherent in any project undertaking. Project Risk Management, however, is a separate process unto itself, and is responsible for responding to both negative (adverse) and positive events in the lifecycle of a project. When defining risk from a health and safety perspective, it could be defined in terms of the possibility of suffering some type of harm or loss. Project risk, however, involves not just negative outcomes, but potentially positive outcomes. Opportunities that may arise from how a project is conducted (Winch 1997).

The first step in managing risk is identifying it, usually by performing a cause and effect analysis of the phases of the project to evaluate the "what if" scenarios—both worst and best case—if project tasks are not performed according to plan. The way to obtain this information is often very prosaic—interview or ask the people performing the work what most worries them about what could go wrong within a given task or portion of the project.

The next step involves quantifying the risk, by using statistical tools or techniques to identify what risks have a fairly high likelihood or occurring, and which ones are not likely to ever happen at all. Once this list has been developed, additional quantification of the potential costs to the project, both to schedule and budget, must also be performed. The outcome from this quantification exercise should result in a short list of manageable opportunities (positive risk) and negative threats.

The approach to responding to these risks is straightforward—either avoid the risk, mitigate for it, or make the management decision to accept the consequences. How to respond to them is a bit more complex. To achieve the desired outcome for a given project risk, the project manager needs to work with his project team to develop contingency plans and alternative strategies or "workarounds" that directly address the risk. Upper management may become involved in the form of deciding to purchase insurance or bonds, if the risk of the project rises to such a level that it warrants it. Ideally, especially for large, complex projects that may have multiple and diverse risks entailed in the performance of the work, a risk management plan should be prepared that identifies, at least conceptually, the corrective actions that will be performed in the project to respond to a risk as it occurs. This may seem a bit akin to "crystal ball gazing," but thinking about the risks a given project may face prior to the actual performance of the work can go a long way toward heading off these threats before they ever can affect the project in the first place—it's referred to as "mitigation of risk."

7.12 DESIGN COORDINATION

For purposes of this discussion, the PM and the design team leader (DL) are two different functions and individuals. The DL will be a responsible member of the project team working for the PM. After the DL develops a work plan that includes the scope, schedule, budget, deliverables (reports, drawings, specs, and so forth) and a work breakdown structure, there are numerous other management functions that should occur including:

- Managing scope growth (creep): This is commonly referred to as "scope creep," which includes incremental changes that alter the original scope. The DL should have a process for controlling these generally small incremental changes. The DL has the control to record these changes and decide when these revisions actually alter the scope with a schedule impact or cost impact. This will be a delicate balance because the PM and the DL are trying to develop a relationship with the client while trying to hold his company's cost to the original effort estimated. The problem is that the project won't function effectively without controls for scope creep. In addition, the contract for the work will have legal implications if the schedule and scope of work aren't controlled.

- Project team meetings: Team meetings should occur at regular intervals throughout the project—a typical frequency is weekly, throughout the project.
- The DL usually sends out "Meeting Agendas" before the meeting and defines the interface of the various disciplines on the project. The DL or an alternate party should record meeting notes. (Sample forms for these meeting notes are included in Chapter 13, Figure 13.3 for reference). It is the DL's responsibility to assure that meetings stay on track, record key decisions, record action items, record responsible individuals to perform those actions with deadlines and finally stay productive.
- Weekly/monthly reports: Regular weekly or monthly reports may actually be part of the contract documents. These reports can act as a record of change as the design progresses through the project lifecycle.
- Distribution of documents: Document distribution must be timely and conducted with a sense of urgency. If document distribution is not conducted effectively, it will increase work load and bog down the project.
- Authority/responsibility checklist: Tracking action items and the responsible parties for accomplishing these actions is also the responsibility of the DL. It is also wise to track the tasks and the responsible individual identified for accomplishing these tasks. The DL may also use a checklist of duties for design (Hansen, K. 1998).

7.12.1 Team Management

The DL is responsible for team management in three distinct areas:

- Within project team
- Between team and client
- Between team and engineering organization's management

Communication between and among the design team, other project team members, and client is critical to achieve a clear understanding of goals and objectives. These communication problems seem to become amplified as clients emphasize the need to stay on schedule. We understand "why" some clients want to keep on track. It's because of the rate of expenditure. Some clients believe in the "open window" concept, which is: The longer the window is open the more dollars fly out. Often there's no real deadline except for the client's desire to cut back on the rate of spending and move closer to construction and overall completion. But, even with effective communication, problems can arise. Some typical problems include:

1. Differing outlooks, perspectives, priorities
2. Fear of completing one job because of the unknown afterward

3. Scope creep by the owner
4. Lack of understanding or coordination among team members
5. Lack of confidence in the PM or DL
6. Unsure of path forward for some team members

To reduce these potential problems the DL or PM should continually emphasize commitment, clarity, and unity. This is where the leader's perception and people skills become very important.

7.12.2   Evaluation of Design Effectiveness

Evaluating the effectiveness of an overall design is difficult to quantify. The opinions of the effectiveness would likely be judgmental and subjective depending on whether the evaluator was an owner, user, regulator, or engineer. One way to quantify design effectiveness is through the number of design revisions or adherence to the original schedule or design budget. There are subcriteria that may also be evaluated such as:

- Constructability: The ease with which the contractor can build the project reflects, at least partially, the quality of the final construction documents. Constructability can be evaluated from the perspective of construction managers (CM) specialists, or contractors. Often, constructability is quantified based on the number of requests for information (RFIs) coming from the field and on the resultant change orders in a given project (Hobday, M. 1998).
- Flexibility: This refers to the capability of using alternate materials that generally meet the same specification and criteria but possibly offer a more attractive price or availability.
- Labor skill/availability: The largest component of most construction jobs is labor. Elaborate or intricate design concepts will likely require a higher-skilled labor force transcending into a much higher cost to the owner. The designers should consider this component in the overall design process to keep the overall construction price within expected ranges.
- Substitution: Experienced designers likely have experience with substituting materials due to temporary shortages or price spikes. Importing steel or decorative tile from a foreign country may have unexpected impacts on the schedule or project budget. Performing a design that can easily handle substitutions can provide the owner with potential negotiation power when the time comes to place the order for these special materials.

In summary, evaluating the effectiveness of a design is judgmental. But creating a design that includes criteria such as constructability, flexibility, skill, and compensation rates for labor and substitution will provide the owner with a degree of comfort like a hidden insurance policy.

From Design Engineer to Project Manager

Bright and motivated engineers ultimately are faced with a dilemma: continue on their path of growing technical expertise or jump to the professional fast track of business development and project management. For many young engineers, the concept of a project manager is quite nebulous. Business and management classes are not typically part of the engineering curriculum (which is unfortunate), and project management involves skills that are not common to the engineering practitioner. The transition from design engineer to project manager can be challenging.

So, what exactly does an engineering project manager do? Simply put, the project manager is responsible for everything. That by no means implies that the managers do everything, or even that they are an expert at everything. It does mean that when anything goes wrong, the responsibility lies with the manager. Responsibility for "everything" includes: business development (one has to get the project in the first place), project delivery (on time, within budget, and with the satisfaction of all project goals), internal staff performance, consultant and subconsultant performance, human resources (staffing, teamwork, dispute resolutions, hiring, and firing), client success (the ultimate measure of project delivery), and time management. Because of the breadth of these responsibilities, project managers are given a tremendous amount of "rope" within most firms. The trick, of course, is to avoid hanging one's self.

Project Management Skills

A manager must have a team, and the project team is often quite diverse, even for relatively small projects. An engineering project will often require specialists in multiple branches of engineering as well as planners and architects. Beyond the immediate engineering challenges that must be resolved, a project manager must deal with business issues, political issues, interpersonal issues, and public relations—all requiring skills that engineers are less than famous for.

Formal engineering training includes a study of mathematics, the physical sciences, computer programming, specific engineering fields, and general education. Often, the only opportunity for early formal training in the soft skills associated with project management comes with "general education." Candidly, engineers tend to avoid coursework that involves the intensive practice of written and oral communication, business, and accounting. Postgraduate training of engineers in these fields is typically not encouraged by employers or embraced by employees. Engineering project management, therefore,

(Continued)

requires that an individual be enthusiastic about deriving personal and professional growth from less formal types of education.

The mentality of an engineer, although useful for the design of a bridge or a machine, is not immediately conducive to project management. Engineers think in absolutes. A system either meets the design criteria or it does not. They take pride in their work and are meticulous about it. Engineers are paid to provide the best possible solution, but often interpret this as a charge to provide the one correct solution. They are not paid to be flexible (allowing a building to collapse "only a little bit" is not an option). Politics and bureaucracy are anathema to an efficient engineering project. Unfortunately, all of these tendencies are counterproductive to effective project management. Flexibility is, of course, critical. There are always better ways of doing things, and the need for a timely project delivery often outstrips the need for a perfect project. Moreover, the engineer is not the only important project stakeholder. Politics and bureaucracy are unavoidable because engineering projects often involve large amounts of public money. Effective communication with non-engineers becomes critical.

Project management requires skills in business development, leadership, and persuasion, among others. Business development means marketing. Successful business development comes from knowledgeable project managers that are familiar with their clients' needs, wants, and constraints. Marketing is actually quite straightforward if one's past project performance is excellent. A successful manager needs to inspire confidence and project competence. All commitments should be met without excuses, and solutions should be presented to a client, never problems. If one can honestly stand behind a track record that demonstrates these qualities, then repeat business will always follow.

The leadership needed to be a project manager does not come from a promotion and it is not granted from above. It comes from leading by example, not by pushing from behind. Leadership is manifested in project management through interactions with the project team, flexibility, discipline, and advocacy for the team. A leader must be both flexible, for nothing ever goes exactly as planned, and, at the same time, disciplined. Certain standards, such as the project schedule, must be set and maintained, or a project will not succeed. Finally, leadership means advocating for the needs of the project team. The client is not always right. Ignoring that fact can mean alienating the people who have the ability to keep the client well satisfied in the long term.

More than anything, project management is about persuasion. A mentor frequently reminds me that his job is to get people to do what they really did not want to do. The need to be persuasive is far reaching: We must persuade our project team to get the job done correctly and on time; we must persuade company principals to provide needed resources; we must persuade our clients that our solutions are effective; and we must persuade agencies and companies to approve our clients' projects. This requires excellent communication skills as well as leadership and mutual trust.

Training

Formal training for engineering project management can be hard to come by. Large firms and agencies sometimes have established internal programs, but they often amount to a stack of manuals and a few seminars. Most successful project managers learn through the school of hard knocks. A few schedules and budgets are blown, lucrative contracts are lost, and fellow employees are alienated. If an individual can avoid making major mistakes more than once, they succeed. If not, they go back to designing, but often at a new firm.

From Design Engineer to Project Manager

Assuming that one did not have the foresight to take extra business and communication courses in college (and most of us did not), other training resources must be utilized. First, one should obtain the appropriate tools. Project management software ranges from terrific to terrible. A project manager needs tools that provide all the relevant project and staffing data, without overwhelming the user with irrelevant information. Proper support from accounting personnel is important. Second, there is no replacement for voracious reading and self-study. Everyone has an opinion about how to manage, and many have written books on the topic (including, or course, the author of this text). At best, one can glean some useful ideas from each of these opinions. Project management style needs to vary with the individual project manager, the particular engineering field, and the particular firm involved, and therefore must be carefully crafted by each individual.

There are no design manuals for how to manufacture a project manager. Finally, a project manager must learn to stop talking and listen. Watch carefully at your next big project meeting—the most impressive person in the room will likely sit quietly through most of the meeting. Take note when they do talk, as it will probably be in regard to something of substance.

Is It Worth It?

Project management can be rewarding, but the profession offers its own set of challenges. For example, from a client's perspective, projects are either great or terrible, and they are not always great. The responsibility for a failed project lies entirely with the project manager. Economies fluctuate, and that means that a manager must occasionally fire and layoff coworkers and sometimes friends. Because of these responsibilities, a good project manager is respected, but a certain emotional distance is necessarily maintained from most coworkers. Project managers (and companies, for that matter) that forget that they are working in a business cease to exist.

(Continued)

For some ambitious engineers, this discussion is discouraging, for project management involves a lot of skills that require additional development and training and are outside the typical engineer's comfort zone. Despite the challenges, a career in management is worth considering because it means career and personal advancement. One becomes responsible for business development, and truly has a major stake in the project ownership. Project managers are typically involved with a project from its inception through its completed construction. This long-term personal stake in a job generates tremendous professional satisfaction. Some engineers even enjoy practicing the interpersonal skills that are required. Finally, great satisfaction is derived from doing a job that most of one's peers cannot do.

What Now?

For those passionate about leading their profession, engineering project management offers a unique opportunity. Ready to make the leap? The first step in preparing yourself is to broaden your skills. In particular:

- Practice writing and public speaking
- Take a business class and read everything you can find
- Take on a leadership role somewhere (it doesn't have to involve engineering)
- Broaden your engineering resume
- Most importantly, spend some time working in construction or manufacturing. Learn to build what you are designing.

> —Matt Salverson, PhD, PE, Wood Rodgers, Inc., lecturer at Sacramento State University

7.13 SUMMARY

Effective project management skills are essential for conducting efficient project operations. The project manager is at the heart of the project much like a captain of a ship is often on the bridge. The PM must maintain close communication with the client and also be aware of all activities related to the initiation, planning, execution, monitoring and control, and closure of the project. Infusing the project team and task efforts with the client's vision of the final deliverable product requires skill and experience. The PM helps to assure a final quality product delivered on time and within budget. A bonus award for the PM, the project team, and the organization would be the client's next job, a letter of appreciation, and/or a good recommendation based on a project well done.

"When a thing is done, it's done. Don't look back. Look forward to your next objective."
—General George Marshall, American Military Leader

BIBLIOGRAPHY

Anderson, David R. et al. (2009). *Quantitative Methods for Business*. South-Western Cengage Learning, Mason, OH. ISBN: 13: 978-0-324-65181-2.

Hansen, Karen Lee. (1998). "Integration of methods for the effective administration of processes in the construction industry." Presented at Expo Construction '98. February 1998, Mexico City. Vanir Construction Management, San Jose, California.

Hobday, Michael. (1998). "*Product complexity, innovation and industrial organisation: cops working paper*." June 1998. CoPS Publication No 52.

Hughes, Thomas P. (1998). "*Creating Open Systems from Edison to the Internet*." Lecture 150, Sloan School of Management, Massachusetts Institute of Technology, Cambridge, Massachusetts, USA.

Kerzner, Harold. (1997). *Project Management: A Systems Approach to Planning, Scheduling, and Controlling*. John Wiley & Sons, New York.

Kerzner, Harold. (1998). *Project Management: A Systems Approach to Planning, Scheduling, and Controlling*. John Wiley & Sons, New York.

Kerzner, Harold. (2010). *Project Management Best Practices: Achieving Global Excellence*. (The IIL/Wiley Series in Project Management). John Wiley and Sons, Inc., Hoboken, NJ. ISBN: 978-0-470-52829-7.

Morris, Peter W. G. (1994). *The Management of Projects*. Thomas Telford, London.

Render, Barry and Stair, Jr., Ralph M. (1982). *Quantitative Analysis for Management*. Allyn & Bacon Inc., Massachusetts.

PMBOK Guide. (2021). Guide to the Project Management Body of Knowledge (PMBOK® Guide) – Seventh Edition and The Standard for Project Management Body of Knowledge. Project Management Institute Publisher, Newtown Square, PA.

Shenhar, Aaron J. and Dvir, Dov. (1996). "Toward a typological theory of project management." *Research Policy*, vol. 25. 4.

Wiest, Jerome D. and Levy, Ferdinand K. (1974). *A Management Guide to PERT/ CPM*. Prentice-Hall of India Private Limited, New Delhi.

Winch, Graham. (1997). "Thirty years of project management: what have we learned?" Presented at British Academy of Management, Aston, 1996 (revised March 1997). Bartlett School of Graduate Studies, University College, London.

CHAPTER 8

Permitting

Big Idea

Permitting is a "key element" in the design process and is part of the "basis of design" for engineering projects. A civil or environmental engineering construction project could have potentially tens or hundreds of applicable federal, state, and local regulations or Presidential Executive Orders. Civil engineers should recognize and understand these regulations and incorporate the essential elements of compliance into the project statement and scope of work.

"What good is a house, if you haven't got a decent planet to put it on?"
— Henry David Thoreau

Key Topics Covered

- Accept Requirements for Permits
- Respect the Staff Implementing the Permits
- Initiate the Permitting Process Early

Civil Engineer's Handbook of Professional Practice, Second Edition. Karen Lee Hansen and Kent E. Zenobia.
© 2025 John Wiley & Sons, Inc. Published 2025 by John Wiley & Sons, Inc.
Companion website: www.wiley.com/go/hansen/CivilEngineersHandbook

- Identifying Permits with the United States (US) Environmental Protection Agency
- Managing Permits
- Streamlining Permits
- Sample Permit Table
- Sample United States Army Corps of Engineers (USACE), Section 408 Permit

Related Chapters in This Book

- Chapter 3: Ethics
- Chapter 5: The Engineer's Role in Project Development
- Chapter 10: Leadership
- Chapter 11: Legal Aspects of Professional Practice
- Chapter 13: Communicating as a Professional
- Chapter 14: Balancing Life, Family, and Career
- Chapter 15: Globalization
- Chapter 16: Sustainability
- Chapter 19: Construction Management
- Chapter 21: What Engineers Need to Know

8.1   INTRODUCTION

Addressing environmental considerations early in the project takes more time up front, but will avoid having costly schedule delays and rework later in the process. To engage the environmental agencies CEs must possess the ability to develop emotional intelligence, employ people skills, clear communication, innovative creativity, and work with and enhance open minds.

> Most people spend more time and energy going around problems than in trying to solve them.
>
> —Henry Ford

Project permitting hardly was addressed in previous engineering handbooks, much less made a chapter title. But ask any engineer today what their major headaches are and environmental permitting will be in the top five. There is no more frustrating problem, and one for which current handbooks offer so little practical advice. Avoiding frustration and delay associated with project permitting requires three important tools:

1. An attitude that accepts the requirement for permits and the expertise and opinions of the agencies and staff that are trained to implement them.
2. An approach that includes project permitting as an important early step in the project schedule.
3. A plan to streamline permits by preparing them in a comprehensive and consolidated manner.

Engineers traditionally were trained to believe that they held all responsibility and authority for a project's implementation and success. In practice, engineers experience confusion, conflict, and frustration when confronted with the need to obtain environmental permits. The conflict arises because: first; a regulatory agency (not the engineer) has authority over the project approval and implementation; second, the regulatory agency (not the engineer) can control (and substantially delay) the project schedule; third, agency staff are often from a land use, cultural resources, or biological background and may seem to have personality and communication methods that contrast with those of engineers. The engineer may perceive a regulator to be less technically competent. (Liu,David, 1974), (Venables, Roger, 2000)

> "We all need people who will give us feedback. That's how we improve."
> —Bill Gates, Businessman and philanthropist

The common reaction is for the engineers to try to deny the permitting authority and proceed forcefully. Some will blame the agency for delaying the schedule and endangering the project. Some will rail against the subjective nature of the permit requirements and complain how difficult the agency staff can be. Some will spend time attempting to rationalize the directions of the regulatory agency. None of these responses will lead to a successful project, a successful relationship, or reduce the stress and anxiety of an engineering career.

A more successful approach to permitting includes the following steps:

- Accept the requirement for permits and actively seek out what may be required
- Respect the agency staff implementing the permits
- Identify the permitting requirements and initiate the permitting processing early in the project—at least 12 months in advance of implementation for a simple project.

8.2   ACCEPT THE REQUIREMENTS FOR PERMITS

Environmental permitting prior to 1965 was practically nonexistent. In the later 1960s the public became increasingly concerned with environmental pollution and enacted a series of laws to reduce further degradation and restore previous conditions. Most of the current environmental permitting requirements stem from the following foundation laws:

- The National Historic Preservation Act 1966
- The National Environmental Policy Act 1969
- The California Environmental Quality Act 1970
- The Clean Air Act 1970
- The Clean Water Act 1972
- Federal Endangered Species Act of 1973
- Archaeological Resources Protection Act of 1979

From these foundations sprung rules and regulations designed to codify and detail necessary compliance at both the state and federal level. Different levels of government may have different permitting requirements, some of which may be similar, but many of which will be stricter than the underlying federal law. The laws describe the public health and trust; the permit process is the mechanism used to implement these definitions of public welfare and environmental protection.

It is important to acknowledge that these laws were created by a process that included input by the people of the United States to protect the public trust. If the project engineer believes the permit requirements are unnecessary, or ignorant, then they should work to revise or change the law. Engineers, clients, and the public should not forget that the laws were enacted to represent the will of the majority in an effort to preserve our heritage, history, environmental quality, endangered species, and quality of life. These concepts are similar to the engineer's ethics as presented in Chapter 3, Ethics, and Chapter 11, Legal Aspects of Professional Practice.

In reality, permitting requirements are a significant project input for the design process. The permitting conditions and requirements are as important as other design requirements because the project may not be built until these requirements are fulfilled. The challenge for the engineer is to identify the permit requirements and schedules, and incorporate this information into the data collection, requirements, and design basis for the project to keep it on schedule and within budget. This information is also presented in more detail in Chapter 5, The Engineer's Role in Project Development.

8.3   RESPECT THE STAFF IMPLEMENTING THE PERMITS

Engineers require years of college education, an advanced degree, and continuing education to gain a professional license. The engineering curriculum includes advanced classes in calculus, physics, statics, chemistry, and more. Engineers deal in facts. Engineers are paid more than planners, cultural resources specialists, and biologists. Therefore, engineers are smarter, right? This attitude becomes a dangerous stumbling block for many projects.

Generally, the permitting staff has as many years of advanced education, continuing education, and professional licenses as engineers. In many cases they were required to take the same advanced classes in calculus, physics, and chemistry as engineers, in addition to training in planning principles, application of the law, economics, and principles of theoretical ecology. In other words, agency staffs have the same level of education, training, and experience in their profession as an engineer does in theirs. Therefore, engineers should show the same respect to non-engineering professionals as they would an engineering colleague.

> "He who speaks without modesty will find it difficult to make his words good."
> —Confucius, *The Sayings of Confucius*

One area to show constructive respect is to listen and read carefully the agency requirements. Understanding the underlying purpose of the permit or agency can be a positive way to show the regulator that you have "done your homework." Be cautious in interpreting or reinterpreting the requirements. Permit requirements may seem subjective, rather than factual, which puts the engineering mindset at an immediate disadvantage. Agency staffs generally have the authority and responsibility to interpret the rules, and it is wise to listen to them carefully.

Interacting with the agency staff is a key area to employ collaboration techniques to establish a win-win scenario. Distrust between project proponents and engineers and the regulatory staff is common for many of the reasons described above. The tendency is for engineers to minimize their interaction with the regulatory staff. On the contrary, using every opportunity to increase time with the agency staff will generally increase trust and improve the permitting relationship. This is one area where social engineering is as important as technical skill in reducing potential schedule and cost impacts. More information on this topic is presented in Chapter 13, Communicating as a Professional Engineer.

8.4   INITIATE THE PERMITTING PROCESSING EARLY

Building permits are generally a "back end" process. Once the project design and schedule are complete, applications are filed for building authorization in a process that extends less than 30 days. Environmental permits, by contrast, may require 12 to 18 months to apply for and obtain approval, *and likely will require changes in the project design or schedule*. Permits may require field investigations or surveys over multiple years before permit applications can be prepared. The project engineer should anticipate that a complex civil or environmental engineering construction project may need to comply with tens or hundreds of applicable federal, state, and local regulations or Presidential Executive Orders. The engineer needs to recognize and understand these regulations, incorporate the essential elements of compliance into the project statement, scope, and schedule of work. Some examples of permits that require extended preapplication times follow:

- **The Endangered Species Act** is intended to prevent the extinction of species by the actions of the federal government. The Act prohibits the "take" of endangered species. "Take" is defined very broadly to include both direct mortality and modification of the habitat supporting the species. All species identified and listed are covered. These can include widely known species like eagles and wolves, and also not so well-known species such as the elderberry beetle, the San Francisco garter snake, and the desert tortoise. The habitat features that support endangered species can be a patch of sand, a rainwater pond that fills once in ten years, or some plant community like coastal scrub. In order to determine that a project will not endanger a species, the applicant needs to determine if the species is present, if the habitat characteristics that support the species are present, and in some cases survey the locations and health of populations that are outside the project area. For species that are only present during a short period of the year or perhaps only every other year, surveys nearly always will require at least two years of field work. An example of this is vernal pool fairy shrimp that occurs in temporary rainwater pools every two to three years. Other fairy shrimp species occur across the United States and more are likely to be listed as endangered. For projects that occur where fairy shrimp may be present, one to two years of survey work and application preparation, followed by at least 135 days of consultation by agency will be necessary. (US EPA, December, 2022)

- "There is no such thing as failure, there's just giving up too soon."
  —Jonas Salk, virologist, medical researcher

- **The Historic Resources Preservation Act** requires that projects not damage or disturb historic resources. Historic resources can include bone, stone, burial grounds, artifacts, or other evidence of previous human inhabitants. Historic resources can now also include historic structures as little as 50 years old. Another more recent extension of the law includes prehistoric resources, such as fossils and fossil evidence. Cultural and prehistoric resources evaluations follow a sequential path of reviewing existing records and literature, performing reconnaissance-level surveys, geomagnetic surveys or potholing, and at the most extreme, excavation, recovery, and curation of resources prior to developing the project. A literature search and reconnaissance survey can take as little as four weeks, but excavations, recovery, and curation can require years to complete. (US EPA, December, 2022)

- *The Clean Water Act* requires that projects not degrade surface waters. Under Section 404 of the CWA, regions identified as "waters" and "wetlands" are also protected. A formal determination of the location and extent of wetlands is required before allowing project impacts. Some wetlands determinations are complex and may require multiple wet or flowering seasons to complete. Section 401 prohibits discharges or changes to water courses that can degrade water. (US EPA, December, 2022)

- *The National Environmental Policy Act (NEPA)* requires that federal agencies evaluate the potential environmental impacts of a project before implementing, approving, or funding them. Any project that uses, for example, **Federal Highway Administration (FHWA)** or **US Army Corps of Engineers** funding will require an Exemption, an Environmental Assessment, or Environmental Impact Statement (EIS) to evaluate the potential impacts of the project and propose methods to avoid, minimize, or compensate for them. In some cases, impacts may be unavoidable, but necessary to achieve a more important objective. NEPA documents generally evaluate 13 different potential impact areas including air quality, cultural resources, biological impacts, water quality, land use, noise, and socioeconomic impacts. An EIS can require one to two years to prepare and publish. An EIS is made available for public comment and if impacts are identified that were not evaluated in the Environmental Impact Report (EIR), the process may need to be repeated one or more times. The final decision based on the NEPA document can be legally challenged on the basis of incomplete information, unsupported conclusions, or a high degree of uncertainty on the analysis, extending the time needed for approval. For further information it is suggested that the reader read "A Citizen's Guide to the NEPA, *Having your Voice Heard*". (Council on Environmental Quality, 2007) It provides an excellent, summary plus appendices and may be found by researching "NEPA" and the Council of Environmental Quality or at the following web address: https://digital.library.unt.edu/ark:/67531/metadc31141/m1/2 (Schmidt and Rosenberg, 2017)

- *California Environmental Quality Act (CEQA)* is California's equivalent of NEPA and requires evaluation for projects authorized, permitted, or funded by state agencies be analyzed for environmental impacts. The CEQA document is an EIR In some cases where both federal and state agencies are expected to authorize the project, a combined EIR/EIS may be prepared. Some agencies such as the California Energy Commission may prepare a "CEQA-equivalent" document instead of a CEQA document. The necessary time and cost of a CEQA document are similar to the NEPA document and can be likewise challenged by the public. (CEQA, 2023)

8.5   IDENTIFYING PERMITS WITH THE US ENVIRONMENTAL PROTECTION AGENCY (USEPA)

Identifying Permits with the USEPA

The EPA can be a valuable resource for the engineer to identify environmental permits required for a proposed engineering project. The EPA's foundation is protecting public health and the environment. Most federal laws proposed and accepted by Congress and signed into law by the President do not have sufficient detail for immediate implementation. Congress authorizes the EPA to write the regulations that explain the details necessary to implement the environmental laws. Also, some laws come into effect as Presidential Executive Orders (EO). These EOs can establish additional conditions or performance standards, or influence the regulatory schedule.

For example, EO 12898 (signed by President William J. Clinton in 1994) is titled "Federal Actions to Address Environmental Justice in Minority Populations and Low-Income Populations." Its goal is to focus federal attention on the environmental and human health effects of federal actions on minority and low-income populations with the goal of achieving environmental protection for all communities. The EO directs each agency to develop a strategy for implementing environmental justice to avoid disproportionately high and adverse human health and/or environmental effects on minority and low-income populations. The Order is also intended to promote nondiscrimination in federal programs that affect human health and the environment, as well as provide access to public information and public participation. The Major Laws and Executive Orders promulgated by the US EPA are presented below:

—US EPA website, 2022: https://www.epa.gov/

Many federal or state agency permits may specify that a CEQA or NEPA document be completed *before* permit authorization is effective. For example, a NEPA document is required before the United States Army Corps of Engineers can consult with the United States Fish and Wildlife Service to extend Section 7 authorization for a state or privately lead project. The California Department of Fish and Game requires that a CEQA document be completed before authorizing work under a Streambed Alteration Agreement.

Some Major Laws and Executive Orders (EOs) that Should Be Considered Are Listed Below:

- Atomic Energy Act (AEA)
- Clean Air Act (CAA)
- Clean Water Act (CWA)—(Originally The Federal Water Pollution Control Amendments of 1972)
- Comprehensive Environmental Response, Compensation and Liability Act (CERCLA, or Superfund)
- Emergency Planning and Community Right-to-Know Act (EPCRA)
- Endangered Species Act (ESA)
- Energy Policy Act
- EO 12898: Federal Actions to Address Environmental Justice in Minority Populations and Low-Income Populations
- EO 13045: Protection of Children from Environmental Health Risks and Safety Risks
- EO 13211: Actions Concerning Regulations That Significantly Affect Energy Supply, Distribution, or Use
- Federal Food, Drug, and Cosmetic Act (FFDCA)
- Federal Insecticide, Fungicide, and Rodenticide Act (FIFRA)
- Federal Water Pollution Control Amendments—see Clean Water Act
- Marine Protection, Research, and Sanctuaries Act (MPRSA, also known as the Ocean Dumping Act)
- National Environmental Policy Act (NEPA)
- National Technology Transfer and Advancement Act (NTTAA)
- Nuclear Waste Policy Act
- Occupational Safety and Health (OSHA)
- Ocean Dumping Act—see Marine Protection, Research, and Sanctuaries Act
- Oil Pollution Act (OPA)
- Pollution Prevention Act (PPA)
- Resource Conservation and Recovery Act (RCRA)
- Safe Drinking Water Act (SDWA)
- Superfund—see Comprehensive Environmental Response, Compensation and Liability Act
- Superfund Amendments and Reauthorization Act (SARA)—see Comprehensive Environmental Response, Compensation and Liability Act
- Toxic Substances Control Act (TSCA)

Laws and EOs that Influence Environmental Protection

Another way for the engineer to cross-check the environmental laws and EOs that might apply to a specific engineering project is to check the EPA list by business sector. In this case the engineer would simply look up the specific industry that is considering engineering services.

Most business sectors are affected by a number of major environmental statutes and regulations. EPA's website offers a regulation's text, history, statutory authority, supporting analyses, compliance information, or related guidance for numerous industry sectors. The nature and scope of activities vary across facilities in any single sector. The information provided on the EPA website will not necessarily apply to all facilities within that specific sector. The key industry sectors are:

REGULATORY INFORMATION BY BUSINESS SECTOR

- Aerospace
- Agriculture
- Automotive
- Chemicals
- Computers/ Electronics
- Construction
- Dry Cleaning
- Electronics/Computers
- Energy
- Federal Facilities
- Fishing
- Food Processing
- Forest Industry
- Furniture
- Garment/Textiles
- Healthcare
- Leather Tanning and Finishing
- Local Governments/Municipalities
- Lumber/Pulp/Paper
- Metals
- Mineral/Mining/Processing
- Paper/Pulp/Lumber
- Pesticides
- Petroleum
- Pharmaceuticals
- Plastics/Rubber
- Power Generation
- Printing
- Pulp/Paper/Lumber
- Retail
- Rubber/Plastics
- Shipping, Shipbuilding, and Repair
- Solid Waste
- Textiles/Garment
- Transportation
- Tribes

Identification of Applicable Permits

The successful engineer will assign experienced project staff or permitting experts when identifying applicable permits for a specific task effort or project. This permitting function has become a specialist's job and the engineer should not be tempted to simply consult the Internet for simple answers where the employer's or client's best interest, legal liability, and funds are at stake. For example, would it be responsible for an inexperienced, energetic individual to design a bridge by using the Internet? United States Environmental Protection Agency (December, 2022).

8.6   MANAGING PERMITS

Recognizing that environmental permits are not a "back-end" process, a successful engineer will list project permitting as a first order task, beginning with proposal or contract review. Following are some suggestions for specific considerations during the proposal, contract review, construction, and postconstruction stages of the project.

During Proposal or Contract Review—Review carefully:

- Any reference to client or project-required environmental policies or regulations
- Any requirement to appoint an individual with specific responsibilities or qualifications (Environmental Manager) who will be responsible to implement and monitor environmental compliance
- Any limitations or approvals needed for specific materials. For example, it is increasingly common in California for agencies to specify that only ultra-low sulfur fuel vehicles be used and that machine idling times be limited. Agencies increasingly specify that no pesticides or herbicides be used without prior agency approval. The potential consequences of these limitations on cost or schedule need to be identified in the proposal and cost estimate.

Upon Project Implementation:

- Consider appointing a dedicated Environmental Permitting Manager from the kick-off of the project.
- Make Environmental Permit Review and Compliance a standing part of project meetings.
- Incorporate your corporate environmental policy into the requirements of the project.
- Monitor and record implementation of environmental policies and procedures that show compliance with permits. For example, important data to show permit compliance include separating waste streams and implementing recycling programs, minimizing waste streams through recycling, documenting regular inspections, worker awareness trainings, ride-sharing programs, and similar initiatives.
- Remain aware of materials, services, and subcontractors that provide environmental benefits for consideration in permits. While not explicitly required in many contracts, such measures are part of many permits, and can be a crucial discriminator or "value added" feature of an engineering or construction agency.
- Remain aware of and train workers so that project participants know they are representatives of the client and the project and that behavior and environmental actions both on and off the job site are important to project success.
- Invite the regulatory agency staff to inspect the site or review project procedures. In fact, this can be the engineer's first step in a collaborative process. It may also seem like inviting enforcement, but in the event of an enforcement action, the judge will generally look with favor on a project proponent that shows a willingness and desire to comply. Any fines or enforcement actions are likely to be waived or reduced as a result. As a practical matter,

agency staff rarely has the time to perform additional reviews or site inspections. This is another opportunity to "reach out" to your agency counterpart to share the information on the current condition of the project. These little actions will build rapport and the reputation for the engineer as a team player in this complicated process.

The Collaborative Process:

Inviting the regulatory agency staff to inspect the site or review the draft work plan or project procedures can be the engineer's first step toward working with the agency in a collaborative process. Engineers should consider every interaction with agency staff an opportunity to establish a closer working relationship with the agency. Building collaborative relationships with regulatory agencies will benefit the client and the engineer!

Communicating with Permitting Agencies:

Nearly all permits require a final report, and some require periodic monitoring and reporting for one, five, or ten years. Communicate with the regulating agencies on the progress of these requirements and submit these reports in a timely manner. If there is an established format, the engineer should consider asking the agency staff person if they would like to view a working draft. If there is no established outline or format, then the engineer should consider submitting a draft outline for agency review and comment.

- Nearly all permits require a letter of contract completion or closure to be filed.
- The project engineer should seek a letter of concurrence or other evidence that all project permit requirements have been completed.

Table 8.1 titled, "How Permitting Is Integrated with Project Phases," describes permitting activities for large civil engineering projects. This summary applies for phases of the project from initiation and proposal efforts through project implementation, construction, completion, and closeout. The general activities are described for each major phase of the project accompanied by the general timeframe and deliverable product from each phase of activity. It's noteworthy that environmental permitting can require from one to five years or more, design from one to three years, and construction can take months to many years to reach completion and closeout. The next section describes how to streamline permits for effective project management. Please recognize that this list of activities and corresponding time frames is a typical example for discussion and learning purposes. The civil engineer/project manager should investigate, list, and verify detailed permitting requirements for their project(s).

TABLE 8.1 How permitting is integrated with project phases (medium to large civil engineering projects).

1	2	3	4	5
3 – 6 months	6 – 12 months	3 to 60 months	1 to 3 years	1 to ?? years
Initiation	Funding	Environmental Review	Design	Construction
• Project Need Identification • Project Scoping Study • Feasibility Study • Preliminary Design • Multiple Alternates • Identify Required Permits, Agencies, Time Frames • Cost Estimate • Schedule	• Confirm Permit Requirements • Cost (Total) Engineering Construction Real Estate Right of Way(s) Utilities Mitigation(s)	• NEPA • CEQA (State EIR) • Refined Preliminary Design • Environmental, Technical, State Reviews Biological Air Noise • Public Meetings • Agency Review/s and Approval/s • Re-Confirm Permit Requirements	• Final Technology Decision, e.g., Selection of Bridge Type • Cost Estimate(s) • Design Submittals for Owner's Approval 30% 60% 90% 100%	• Advertisement • Bid • Award • Construction • Fulfillment of Final Permit Requirements • Close-out
Approval to Proceed	Funding Authorization	Permits	Contract Documents	Completed Project

Source: Adapted from Conrad Bridges, Vice President HDR (retired).

8.7   STREAMLINING PERMITS

There are ways to streamline permits. As noted above, most complex civil and environmental engineering projects will require multiple permits from multiple agencies. Preparing the permit materials together, with close collaboration between design engineers and the permitting staff, can make the process more efficient.

To streamline project environmental permitting:

1. Prepare one combined project description listing the facts and figures of the project. This should include the purpose and objectives of the project, the proposed project footprint (including electrical transmission, gas, water supply, sewer, or other utilities) with clearly established project boundaries, city and county boundaries, and other key landmarks. An access plan will need to be created that shows transportation routes, proposed ingress/egress, and parking locations. The project description should include a proposed schedule, estimates of construction staffing and effort, as well as estimated water use, waste generation, and plans for recycling, disposal, and waste management. The project owner, operator, and construction contractor should be identified. Generally, the construction contractor will not be known and can be entered in permit applications as "TBD" (to be determined). However, the need to identify these parties for the permits forces a timely determination of who will be responsible, and in whose name permits should reside. For example, it may be desirable to have construction waste permits and construction stormwater permits in the name of the contractor *if the regulating agency allows it.*

2. Do not attempt to skip Step 1. Attempting to initiate project permitting without a *written* project description is likely to result in frustration and substantial rework and delay. This is not to ignore that projects will change during design. However, it is generally more efficient to amend or revise the project description subsequently, then to proceed with permit applications that lack a comprehensive project description.

3. Attach the comprehensive project description to each permit application, in lieu of describing details in an application form. In this manner the same information is submitted to all the regulating agencies and can be tracked and amended efficiently. One useful technique is to amass hard facts and figures (area of disturbance, expected megawatts of generation, estimates of waste generation). Using a separate table for factual information allows quick revisions without the need to re-do word processing and document formatting.

8.8   SAMPLE PERMIT TABLE

Table 8.2 lists common regulations and permits required for a hypothetical gas-fired energy project in California as an example of the environmental permits required for a typical project. Natural gas-fired energy projects involve the engineering design, permitting, and installation on industrial or private properties. The projects include construction of access roads to the facilities, engineered foundations for the facilities, installation of the structures as well as

TABLE 8.2 Natural gas-fired energy project—state permit requirements (California).

Permit	Agency	Time Frame	Cost	Requirement
PRECONSTRUCTION				
Application for certification pursuant to Warren Alquist Act	California Energy Commission (CEC)	18–30 months	Cost reimbursement, generally $1–$3M	Required to allow construction of energy generation facilities in California. CEC takes "one stop" authority before all other agencies.
Determination of compliance and authority to construct	Regional Air Pollution Control District	Just after AFC filing	Initial fee + application fee based upon heat input (~$160,000)	Required to allow discharge of air pollutants (oxides of nitrogen, sulfur, carbon particulate, etc.).
401 Water Quality Certification	Regional Water Quality Control Board (RWQCB)	4–6 weeks	$500 filing fee. Based on capital cost of improvements	Required to demonstrate no adverse effect to water quality and beneficial uses.
Section 10 Consultation and Habitat Conservation Plan	US Fish and Wildlife Service (USFWS)	At least 120 days	No filing fee	Required to demonstrate that endangered species will not be adversely affected, and prepare compensation plan for any impacts. Potential cost can be 2% of project cost.
Section 2081 Consultation with Memorandum of Understanding	California Department of Fish and Game (CDFG)	At least 120 days	No filing fee	Required to demonstrate no impacts to species declared endangered by the state. Potential mitigation costs of 1% of project cost.
Section 7 Consultation with Biological Opinion	National Marine Fisheries Services (NMFS)	At least 120 days	No filing fee	Required to demonstrate no adverse impacts to anadromous threatened and endangered species, where a federal agency has permitting authority. Section 7 is required for the USACE to authorize fill of wetlands projected by federal law.

(Continued)

TABLE 8.2 *(Continued)*

Permit	Agency	Time Frame	Cost	Requirement
Clean Water Act Section 404 Permit	US Army Corps of Engineers	45–120 days	No filing fee	Required for activities that would fill waters or wetlands, including vernal pools, creeks, drainages or ditches.
Historic Preservation Act, Section 106 Permit	State Historic Preservation Office	30–120 days	No filing fee	Because issuing a Section 404 authorization is a <u>federal</u> action, the USACE will require compliance with Section 106. The Historic Preservation Act now generally includes both prehistoric (fossils), cultural (native American artifacts and burials) and historic (buildings or structures more than 50 years old) evaluations.
Streambed Alteration Agreement (California Fish and Game Code Section 1600)	California Department of Fish and Game (CDFG)	45–120 days	Generally $500 filing fee	Required for construction or operation activities that would affect the area within the "bed and banks" of any "stream" as defined by CDFG
General Plan Amendment	Local county	45–120 days	$500 filing fee, based on capital cost of project	Required if the proposed project changes the designated use of the property (e.c. from agricultural to industrial or residential)
Conditional Use Permit	Local county	45–120 days	$500 filing fee, based on capital cost of project	In lieu of a General Plan Amendment; required if the proposed project does not conform to the allowed uses of the property.
Encroachment Permit for pipelines	Caltrans	Within 60 Days	$70 per hour for review and inspection	Required for water, gas, or electrical lines to cross or parallel roadways.
Encroachment Permit for pipelines	Local county	2 weeks (work with project engineer)	Varies depending on type of roadway and work	Required for water, gas, or electrical lines to cross or parallel roadways.
Encroachment Permit for pipelines	Railroad	(1)	(1)	Required for water, gas, or electrical lines to cross or parallel railroads.

CONSTRUCTION PERMITS

Permit	Agency	Time	Fee/Cost	Notes
Trenching and Excavation Permit	Cal/OSHA	24 hours	(1)	
Permit for the erection of a fixed tower crane	Cal/OSHA	24 hours		(1)
Single Trip Transportation Permit	Caltrans	2 hours	$16	
Single Trip Transportation Permit	Local county	Same day	$16	
Annual Trip Transportation Permit	Caltrans	2 weeks	$90	
Transportation Permit	Local county	Same day		
Construction Stormwater Permit	State Water Resources Control Board	1 week		Required to demonstrate that measures are implemented to prevent stormwater contaminating surface or groundwater
Construction Permit	City or county	4–6 weeks	Review and inspection costs	

OPERATIONS PERMITS

Permit	Agency	Time	Fee/Cost	Notes
Risk Management Plan (RMP)	Local County Emergency Management			Needed to demonstrate that project materials (compressed gases, lubricants, fuels, explosives) will not create an unacceptable hazard to the local public
Hazardous Material Permit	California Highway Patrol		$100 initial fee/$75 annually	

(Continued)

TABLE 8.2 *(Continued)*

Permit	Agency	Time Frame	Cost	Requirement
Hazardous Materials Business Plan (HMBP)	Local County Emergency Management	2–4 months	$10,000–$20,000 to prepare the HMBP	Obtained prior to commencement of operations
Hazardous Materials Storage Permit (Must be submitted with HMBP)	Local County Emergency Management	2–4 months	~$1,000–$2,000 plus nominal filing fee	Obtained prior to storage of Hazardous Material (HM)
Risk Management Plan (RMP)	CUPA (Local County)	4–6 months	$30,000–$35,000 to prepare document	EPA requires an RMP be prepared prior to project operation.
Industrial Wastewater Permit	Local county	(1)	(1)	(1)
Solid Waste Permit	Local county			
Spill Prevention Control and Countermeasures Plan (SPCC)	CUPA (Local county)	2–4 months	$10,000–$20,000 to prepare the plan	Needed prior to storage of oil at the site
Hazardous Waste Generator Permit	CUPA (Local county)	1 month	Nominal filing fee	Obtained prior to construction
Industrial Stormwater Permit	RWQCB	(1)	(1)	(1)
User Agreement	Local city	(1)	(1)	(1)
Sewer Discharge Permit	To local wastewater jurisdiction	(1)	(1)	(1)
Title 22 Engineering Report	State Department of Health Services	(1)	(1)	(1)

(1) A "Blank Field" on this table indicates that the user must research the specific detailed information for their project and area.

transmission lines to and from the facilities. There are numerous federal, state, and local permits required through all phases of project initiation through construction and operation (as shown in Table 8.1).

Projects involving roads, buildings, bridges, outfalls, or levee improvement will require similar or analogous permits, for which this table may be a useful beginning. Projects involving hazardous materials or remediation of toxic waste sites will have a similar list, predicated on the Comprehensive Environmental Response, Compensation and Liability Act (Superfund) or Resource Conservation and Recovery Act (RCRA). The engineer should prepare an analogous table for each project early in project scoping.

8.9 SAMPLE UNITED STATES ARMY CORPS OF ENGINEERS (USACE), SECTION 408 PERMIT

USACE Section 408 Permit

Following is a brief description of the typical United States Army Corps of Engineers (USACE) permit process required for any type of land disturbance in the vicinity of waters of the United States, including civil projects that build upon, alter, improve, move, occupy, or otherwise affect the usefulness, or the structural or ecological integrity of a USACE project. Typically, Geotechnical and Water Resources Engineers prepare documentation for USACE 408 Permission.

Section 408 Permission from US Army Corps of Engineers

If your project is in the vicinity of the federal flood control project, you probably need a Section 408 Permission from the US Army Corps of Engineers (USACE). It can be a daunting task to secure a Section 408 Permission within your project's construction schedule due to extensive requirements. If your project is in the Central Valley Flood Protection Board's (CVFPB) jurisdiction, then you will also be required to obtain a permit from the CVFPB with the USACE Permission.

Section 408 Program—What is Section 408 Permission?

In order to obtain a Section 408 Permission with the least amount of difficulty, it is very helpful to have a good understanding of Section 14 of the Rivers and Harbors Appropriation Act of 1899, as amended and codified in 33 USC. § 408. A portion of the code is summarized below:

"...*Section 14 of the Rivers and Harbors Appropriation Act of 1899, as amended and codified in 33 USC 408 (Section 408) provides that the Secretary of the Army may,*

(Continued)

upon the recommendation of the Chief of Engineers, grant permission to other entities for the permanent or temporary alteration or use of any US Army Corps of Engineers (USACE) Civil Works project. An alteration refers to any action by any entity other than the Corps that builds upon, alters, improves, moves, occupies, or otherwise affects the usefulness, or the structural or ecological integrity of a USACE project. Section 408 permission requires a determination that the requested alteration is not injurious to the public interest and will not impair the usefulness of the project. This means USACE has the authority to review, evaluate, and approve all alterations to federally authorized civil works projects to make sure they are not harmful to the public and still meet the project's intended purposes mandated by congressional authorization. Routine operations and maintenance do not require 408 permissions . . ."

Applying for Section 408 Permission from USACE and CVFBP Permits

Engineering Circular 1165–2-220 provides the policies and procedural guidance that the USACE follows in processing requests to alter or modify civil works projects that were constructed by the Corps of Engineers. The processing of a Section 408 Permission request begins with a written request from an applicant. Engineering Circular 1165–2-217 lists the USACE requirements for their Safety Assurance Review program. According to USACE, the following are the key steps for each Section 408 permit action:

Section 408 Permit Request Step-by-Step Procedures

Requestors are responsible for preparing and submitting a completed Section 408 Permit request accompanied by the required technical and environmental documents. The request should: (1) confirm the technical soundness of the proposed alteration/modification; (2) provide a basis for environmental acceptability; and (3) formally request that a Section 408 Permit be granted for the proposed alteration or modification.

The following decision process flow chart, depicted in Figure 8.1, by the USACE shows the various steps involved in obtaining the Section 408 permission.

Step 1: Pre-Coordination

Early coordination between USACE, the requester, and/or a non-federal sponsor (if applicable) may aid in identifying potential issues, focusing efforts, minimizing costs, and protecting sensitive information. Requestors should review the information and requirements shown on the USACE website for incorporation into the proposed request. Pre-application conferences are encouraged for projects of a unique nature or those that involve extensive or critical modifications.

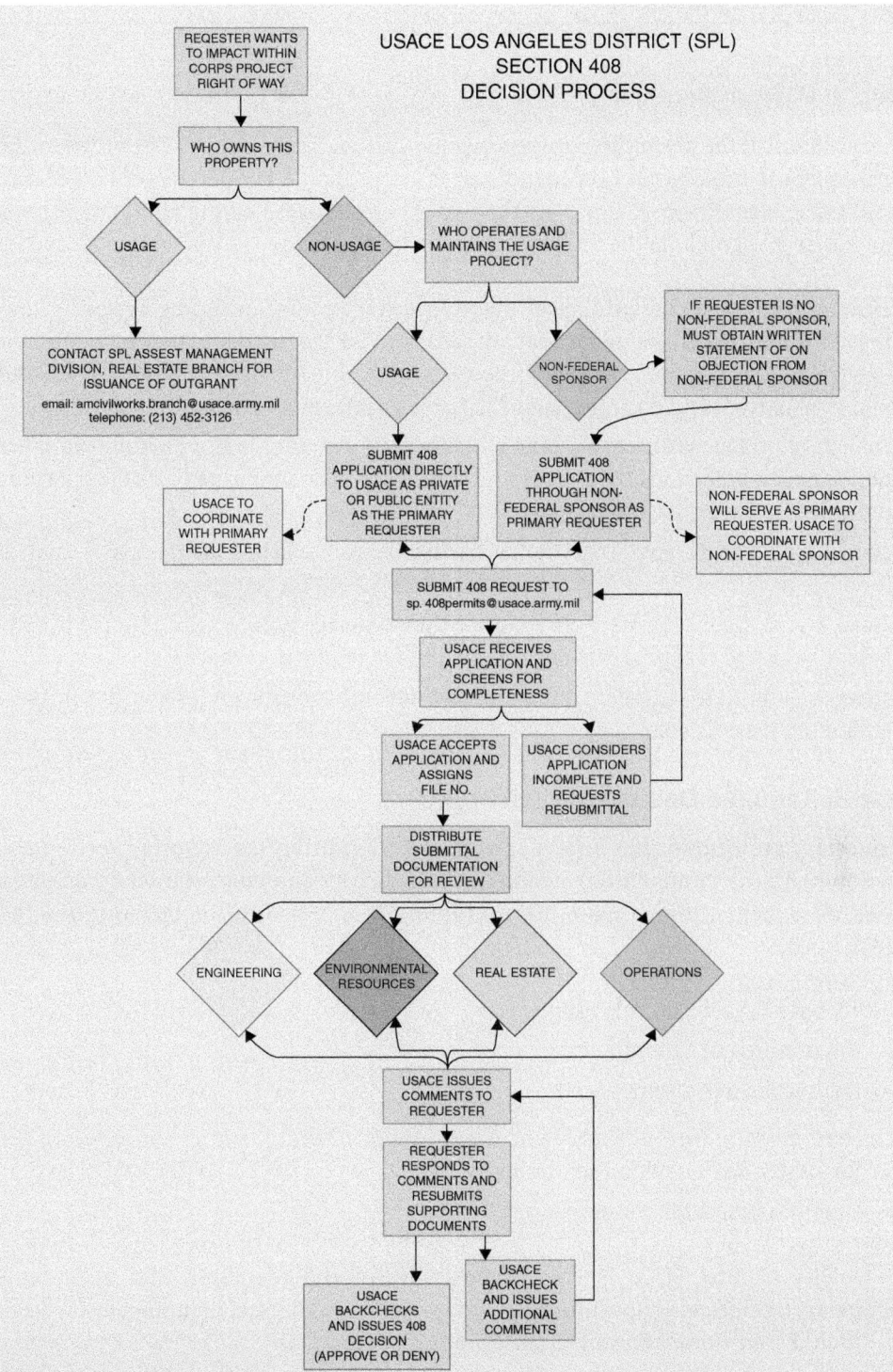

FIGURE 8.1 Section 408 Decision Process Flow Chart.

(*Continued*)

Step 2: Written Request

The initiation of the permitting process begins with a written request. The written request is submitted to the District Commander of the appropriate USACE district office having jurisdiction over the USACE project that would be impacted by the alteration. The written request must include the following five components:

Item 1: A complete description of the proposed alteration including necessary drawings, sketches, maps, and plans that are sufficient for the district to make a preliminary determination as to the location, purpose and need, anticipated construction schedule, and level of technical documentation required to inform its evaluation.

Item 2: A written statement regarding whether the requester is also pursuing authorization pursuant to Sections 10/404/103 and, if so, the date or anticipated date of application/pre-construction notification submittal.

Item 3: Information regarding whether credit is being or will be sought under Section 221 (or other law), or whether approval under Section 204(f) is being or will be sought.

Item 4: A written statement of whether the requester will require the use of either federally owned real property or property owned by the non-federal sponsor; and,

Item 5: A written statement from the non-federal sponsor endorsing the proposed alteration, if applicable.

Step 3: Required Documentation

Requestors are responsible for preparing and submitting the required technical and environmental documentation needed by the district to determine whether the proposed alteration would impair the usefulness of the project or be injurious to the public interest.

- Technical Analysis and Design
- Environmental Compliance
- Real Estate Requirements
- Discussion of Executive Order 11988 considerations
- Requester Review Plan Requirement
- Operations and Maintenance

Quality Control (QC) is the responsibility of the requester, and all submitted documents (including supporting data, analyses, environmental compliance documents, etc.) should have undergone a QC review prior to submittal.

Step 4: Decision-Making Process

The District evaluates each request for an alteration on a case-by-case basis. The technical staff will evaluate proposed alterations against Corps standards (using current USACE guidance, criteria, and staff experience). Additionally, the technical staff will ensure that the proposed alteration will not be injurious to the public, will not impair the usefulness of the project, and is not in conflict with any known laws and/or regulations, as set forth by the policies and procedures in EC 1165-2-220. Upon completion of the review, notification is provided with one of the following: (1) the issuance of the Section 408 Permission for the proposed action; (2) a denial of the request; or (3) an order requesting additional information or revisions to the proposal.

  Final Step: Post-Permission Oversight. Once the District approves the proposed alterations and issues the Section 408 Permission, the permittee is then responsible for the Post-Permission oversight, which includes the following:

(a) *Construction Oversight*: The Section 408 permittee shall oversee the conduct of the work and ensure that construction is in accordance with the issued Section 408 permit and the approved plans and specifications.

(b) *As-Builts*: Drawings showing alterations as finally constructed shall be furnished to the district by the Section 408 permittee after completion of the work. As-builts must be provided within 180 days of construction completion.

(c) *Operations and Maintenance (O&M) Manual Updates*: The Section 408 permittee and/or non-federal sponsor is required to provide an update/supplemental appendix for the O&M Manual. At a minimum, the update should include a description of the new features, a reference to the Section 408 approvals, as-builts, and instructions regarding O&M of any new features not included in the existing manual.

(d) *Post-Construction Closeout*: Upon construction completion, the Requester shall notify the USACE in writing that all construction is complete. The Requester shall also provide electronic copies of the as-built plans to the USACE, as well as any other documents as required by the "Section 408 approval."

—Jay S. Punia, PE, MS, Senior Water Resources Engineer,
jpunia@woodrodgers.com

8.10 SUMMARY

Thirty years ago, the typical civil engineering practice was to design the project, and then initiate environmental permitting just before the planned construction date. This approach is likely to have a high failure rate today. Environmental permitting agencies want to be more involved in the design process so they can include features that will avoid or minimize environmental impacts. Even if none of the environmental features are included in final design, many agencies are obligated to show that design alternatives were considered in developing the project.

Environmental permitting for civil and environmental engineering projects has become a much larger part of project management than was traditionally the case. For this reason, the experience of senior engineers may not be as helpful as an open collaborative attitude to comply with requirements that were nearly absent historically. The primary sources of frustration and project delay associated with environmental permitting are a lack of understanding of the need for permits, a lack of having a project management plan to address them, and not implementing methods to acquire permits in a streamlined and efficient manner. Permitting should be regarded as a "key element" in the design process and be part of the "basis of design" for engineering projects. The permitting task should be delegated to qualified staff and addressed at an early phase of the project to allow time to complete preapplication investigations and modifications to the design, if required.

By using the tools described in this chapter, the successful engineer can distinguish himself/herself by maintaining constructive working relationships with the regulatory agencies, avoiding and minimizing environmental permitting delays and maintaining cost control on projects.

BIBLIOGRAPHY

CEQA: The Californis Environmental Quality Act. (2023). https://www.opr.ca.gov/ceqa.

Council on Environmental Quality (U.S.). (December 2007). *A Citizen's Guide to the NEPA: Having Your Voice Heard.* University of North Texas Libraries, UNT Digital Library, Washington D.C., https://digital.library.unt.edu/ark:/67531/metadc31141. (accessed August 1, 2023), https://digital.library.unt.edu.

Liu, David HF and Liptak, Bela, G., eds. (1974). *Environmental Engineer's Handbook.* Lewis Publishers – A CRC Company, New York.

Schmidt, Eric. and Rosenberg, Jonathan. (March 21, 2017). *How Google Works (Paperback).* Grand Central Publishing, New York, NY.

United States Environmental Protection Agency. (December, 2022). *Regulatory and Guidance Information by Topic.* Laws and Regulations, https://www.epa.gov/regulatory-information-topic.

Venables, Roger. (March 1, 2000). *Environmental Handbook for Building and Civil Engineering Projects.* Volume 528 of CIRIA, Special Publication 58.

The Client Relationship and Business Development

Big Idea

A customer is the most important visitor on our premises, he is not dependent on us. We are dependent on him. He is not an interruption in our work. He is the purpose of it. He is not an outsider in our business. He is part of it. We are not doing him a favor by serving him. He is doing us a favor by giving us an opportunity to do so.

—Mahatma Gandhi

Key Topics Covered

- The Foundation of a Lasting Relationship
- Building upon the Relationship—The Superstructure
- Maintaining the Relationship
- Cultivating Business Opportunities
- Business Development
- Conflict Management

Civil Engineer's Handbook of Professional Practice, Second Edition. Karen Lee Hansen and Kent E. Zenobia.
© 2025 John Wiley & Sons, Inc. Published 2025 by John Wiley & Sons, Inc.
Companion website: www.wiley.com/go/hansen/CivilEngineersHandbook

Related Chapters in This Book

- Chapter 3: Ethics
- Chapter 4: Professional Engagement
- Chapter 5: The Engineer's Role in Project Development
- Chapter 7: Executing a Professional Commission—Project Management
- Chapter 10: Leadership
- Chapter 11: Legal Aspects of Professional Practice
- Chapter 12: Managing the Civil Engineering Enterprise
- Chapter 13: Communicating as a Professional Engineer
- Chapter 14: Balancing Life, Family, and Career
- Chapter 18: Human Relations Policies and Employment Practices

9.1   INTRODUCTION

Almost all engineering projects begin with a client who has a need to be filled with a budget and schedule in mind. Building the relationship with the client is the foundation of business development.

In Chapter 1 the fact that ABET adopted Engineering Criteria 2000 (EC2000) and took a completely new approach to engineering education was discussed. The new focus was to identify *outcomes* of engineering education focusing on what is learned rather than what is taught. ABET's 2022–2023 Criteria for Accrediting Engineering Programs identifies seven student outcomes of civil engineering education.

To achieve ABET accreditation, each civil engineering program must document student outcomes that support the program's educational objectives. Attainment of these outcomes is meant to prepare baccalaureate graduates to enter the professional practice of engineering. These outcomes are:

1. the ability to identify, formulate, and solve complex engineering problems by applying principles of engineering, science, and mathematics.
2. an ability to apply engineering design to produce solutions that meet specified needs with consideration of public health, safety, and welfare, as well as global, cultural, social, environmental, and economic factors.
3. an ability to communicate effectively with a range of audiences.
4. an ability to recognize ethical and professional responsibilities in engineering situations and make informed judgments, which must consider the impact of engineering solutions in global, economic, environmental, and societal contexts.

5. an ability to function effectively on a team whose members together provide leadership, create a collaborative and inclusive environment, establish goals, plan tasks, and meet objectives.

6. an ability to develop and conduct appropriate experimentation, analyze and interpret data, and use engineering judgment to draw conclusions.

7. an ability to acquire and apply new knowledge as needed, using appropriate learning strategies.

A review of these criteria from the professional practice perspective quickly reveals a key missing element. This element is "the client." The client may be a paying customer if the engineer is in private practice, a member of the public if the engineer is in public service, or a commander if the engineer is in the military. Regardless of the engineer's business relationship with their respective "clients," there are key components to successful relationships that can enhance the professional's career, improve the engineer's enjoyment of the practice, and help gain more respect for the profession.

The key components to a successful client relationship are:

- Trust
- Respect
- Managing Expectations—It's all about relationships
- Follow-through
- Effective Communication
- Scope/Schedule/Budget—Maintaining this triangular relationship by applying effective project management skills
- Understanding the client's business model and/or their perspective and his/ her stakeholders
- Anticipating the client's needs before their request. Help crystallize the client's initial thought process in an effort to comprehend their needs.
- Quality
- Going above and beyond the competitors
- Personalize your delivery

Engineers should recognize that clients create the need for engineering services regardless of whether you are in private engineering consulting, in public service, or in the military. "Your" client may be an industry, a paying private customer, or an army General but, it will be someone, numerous people, a community, a state, or the Nation that will depend upon the needed engineering service. Client service will distinguish a good engineer from an excellent engineer. And, the client relationship is at the heart of client service.

9.2   THE FOUNDATION OF A LASTING RELATIONSHIP

The client-engineer relationship is "at will" because both parties may leave the relationship at any time provided there was no express contract for a definite term governing the relationship. In the professional world there is a standard of conduct where patience, courtesy, and professionalism are always expected. An illustration of the relationship appears in Figure 9.1.

Trust, respect, commitment, follow-through, and communication are the foundational elements of the client relationship:

- *Trust*: In the course of the relationship with the Client, the engineer will learn a lot about the client's business. This information about the client's business model may include:
  1. Key partners or subcontractors
  2. Raw materials used in their process
  3. The client's clients and/or competitors
  4. Market segment pricing and/or profit margins
  5. Regulations governing their business activities including commerce and operations
  6. Other business elements for a successful profitable business

One can imagine that any client would be cautious and guarded sharing this valuable information. Often the contractual agreement will have clauses restricting the engineer to confidentiality and limiting the sharing of this information. Therefore, trust is a very important element in the foundation of the client-engineer relationship.

For example, in environmental engineering and particularly the hazardous waste management practice, the engineer learns about the waste produced from an industry's manufacturing operations. Working backward and applying reverse engineering, the engineer could

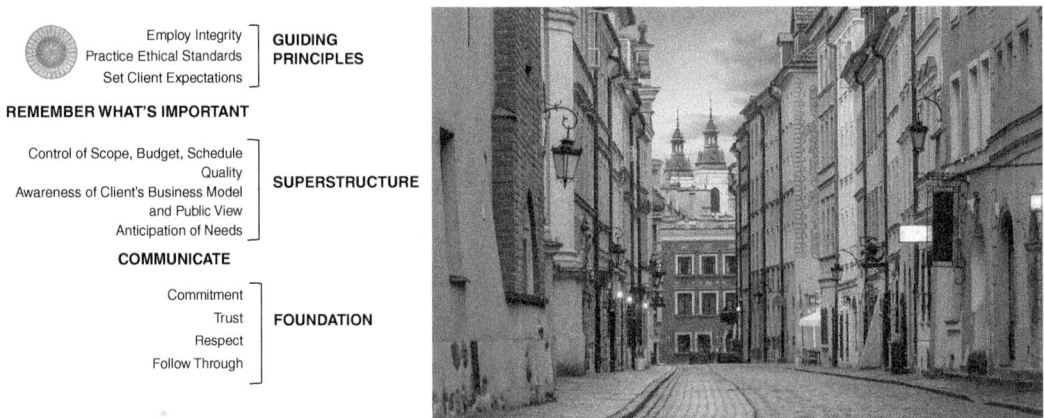

FIGURE 9.1   Building relationships. *Source*: f11photo/Adobe Stock Photos.

learn about the raw materials used in a process, the actual process operations, labor requirements, and potentially marketing information. This information could all be provided to a competitor or potentially used in some way against the client if trust were absent in the relationship. And, in fact, some industrial and chemical manufacturing clients do require confidentiality agreements before engineering tasks proceed. Of course, engineers are also bound by professional ethics regarding sharing private information regarding the Client's business.

- *Respect*: Respect for the individual is a critical component to successful interactions and building relationships. It is the foundation of strong relationships. Respect is demonstrated by recognizing a person's abilities or qualities especially as a component of an organization or company. If an engineer is in public service, respect is essential when interacting with the public and acknowledging their feelings and/or requirements. Respect toward another individual demonstrates your recognition of them as having a sense of worth and a valuable place in their organization and society.
- *Commitment*: Commitment is a promise or a pledge that you provide to someone. People, in general and clients in particular, enjoy hearing commitments but are often cautious about believing or trusting someone who gives a verbal commitment. The caution comes about because almost everyone has learned the adage that "words are cheap and actions are valuable." Commitments are usually verbally stated to clients first and are truly valued after the commitment has been demonstrated. Commitment is an essential component to the successful foundation of a positive relationship. An example of a written commitment is a proposal or contract where promises and expectations are formalized legally, in writing.
- *Follow-through*: In baseball, when hitting the ball, good "follow-through" makes the difference between a pop fly in the outfield and a home run into the bleachers. Follow-through is directly related to commitment because in the engineering profession commitments are usually made in a meeting or a proposal and follow-through is demonstrated by the engineer in the field after the proposal has been accepted and the contract has been signed. Follow-through is the process by which the commitment is driven home to the client (or to the public) and the initial promise is demonstrated in real-time delivery. Follow-through, like commitment, is essential to the successful foundation of a positive relationship.
- *Communication*: There is an entire chapter devoted to communication in this book but, let's just say that to achieve effective communication it is essential to send and receive "the complete message." We believe the strength of the relationship between the "engineer and the client" is directly proportional to the efficiency of their communication.

The Fundamental Rule of Client Relationships: **NO SURPRISES!**

9.3   BUILDING UPON THE RELATIONSHIP—THE SUPERSTRUCTURE

Once a solid relationship is present, the engineer can look forward to building upon this foundation to create a lasting relationship. This process works well for client relationships as well as with personal friends and colleagues. The additional elements that follow can help create a mature, lasting client relationship. In engineering terms, this phase of relationship building is like constructing a superstructure upon a firm foundation.

These elements will come into play in a relationship sometime after the foundation of the relationship has developed, and they include maintaining the scope/ schedule/budget relationship, understanding the client's business model and stakeholders, anticipating the client's needs, delivering quality, and remembering the important stuff.

- *Maintaining the scope/schedule/budget relationship*: This relationship of scope, schedule, and budget can be thought of as a connected triangle where each side represents an essential component of the "project" managed by the project manager (PM). The project will be described in a picture and/or text representing the scope of work. The client will have a schedule and budget in mind for ultimate completion and delivery of the project. Many times the client will need assistance defining these terms and the engineer may help (and negotiate) this definition by providing assistance to the client. Once defined, however, the engineer will maintain a positive client relationship by adhering to scope, schedule, and budget constraints.

- *Understanding the client's business model and stakeholders*: Understanding the client's business model will provide a distinct advantage to the engineer when building a strong lasting relationship with the client. This understanding goes beyond the simple grasp of knowledge the client completes his/her task and receives compensation or recognition for it. This understanding includes knowing the competitors, key business terms and concepts, the client's management structure, products, issues, and other related components of the client's business.

- *Anticipating the client's needs*: Anticipating the client's needs before they make a request can help crystallize their thought processes. If the timing is right, this effort can help shape the client's project while the flexibility to define the scope of work still exists. In the public sector, anticipating the public's needs can also help shape the project or allow the engineer to develop a response to any public concerns.

- *Delivering quality*: The quality of the deliverable product will generally be defined in the scope of work. If there is any question about the overall quality of the deliverable, the engineer should clarify the quality issues when discussing the scope of work and schedule. The quality of the deliverable will be related closely to the schedule, the level of effort, and corresponding fee to produce the deliverable. Greater detail about quality is discussed in Chapter 7, Executing a Professional Commission—Project Management.

- *Remembering the important stuff*: The "important stuff" could be anything that is important to the client. If the engineer is a public servant, there may be issues that are politically sensitive. Important things might include quality, schedule, the client's birthday or favorite holiday; the form or package of the deliverable; big issues like security, or little ones like whether the client likes their middle initial on the transmittal letter. In order to enhance the client relationship, the engineer should observe, listen, remember, and employ the important stuff in the relationship. Remembering the important stuff and incorporating this into the deliverables will set the smart engineer apart from others.

9.4   MAINTAINING THE RELATIONSHIP

Maintaining a superior client relationship means going above and beyond the competitors. Clients can acquire engineering services from many sources. High scores in "client service" have become of paramount importance for obtaining and maintaining the "outstanding" rating category. Going above and beyond the competitors most likely will involve personalized service like making deliveries by hand, exhibiting flexibility when scheduling meetings to the client's preferred date or location, providing extra copies, or sending technical articles related to the client's business. Whatever the service is, it should be performed with thought and the intention of benefitting the client (Bachner, J. 1991).

The Value of Keeping Current Clients Happy

"When I was first stockpiling knowledge, I took on Philip Kotler's Kotler on Marketing, a seminal college textbook. Easily the hardest book I've ever slogged through, it practically bored me to tears. I would have vastly preferred falling asleep on the airplane, or going to bed earlier, or just watching the tube.

But I knew the loving thing was to keep trying, to keep reading, to keep working, and that it would pay off. Or, at least, I hoped so.

A few weeks later I finished the book on a plane, with no idea how I would ever use its knowledge. Forty-eight hours later, I was quoting an important section to a thousand people at a marketing conference, telling them that the cost of attracting a new customer is five times the cost of keeping a current customer happy, and that customer profitability tends to grow with the length of customer tenure (long-term customers tend to buy more, they recommend the company more, and they cost the company less). And, the cost to maintain them is lower because they don't complain as much as unhappy customers.

These insights gave me fresh perspective on what marketing is all about: retaining your customers. Until that time, I had always talked about accumulating new customers through marketing."

—Tim Sanders, *Love Is the Killer App*, pp. 198–199

Personalize Your Delivery

Personalized delivery for client deliverables can be an actual delivery made in person by the engineer or submitting the deliverable before the due date. This personal component can set the engineer apart from other competitors.

In addition to exceeding the competitors' performance, several principles contribute to the successful maintenance of the client relationship.

As depicted in Figure 9.2, these are: choosing clients carefully, setting client expectations, maintaining ethical and moral standards, and earning a profit.

- *Choosing clients carefully*: Clients select engineers, but engineers choose to be selected. The chances of maintaining a successful relationship with a client are enhanced if client and engineer are well matched.

  Clients are responsible for administering the process designed to select an engineer. Ideally, the client's selection process starts with the client's clear understanding of the engineer's strengths as a specialist and commitment to a particular market sector. But client selection can begin before the engineer ever meets with the prospective client as a "client." It could start with an informal meeting at a conference or at a convention or by introduction at a local community meeting. The initial meeting may be at a luncheon seminar related to a general subject overview or at a presentation at the client's office.

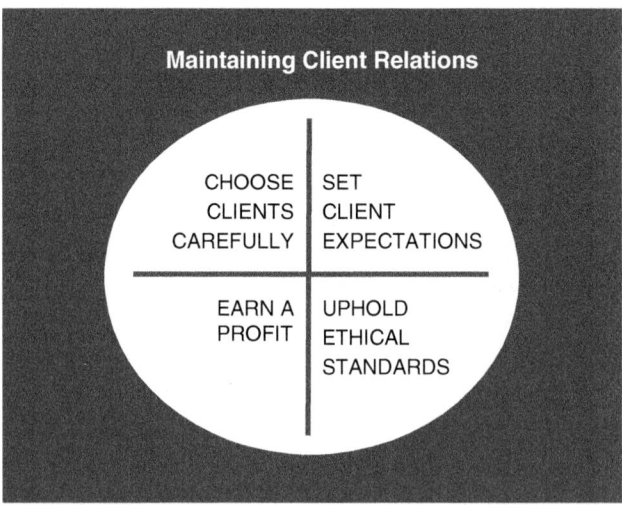

FIGURE 9.2   Maintaining the client relationship.

Obtaining some basic information or doing some research on a prospective client will enhance the engineer's odds of being selected later. Some questions the engineer might ask are:

1. What are the current issues related to this market sector and to this client in particular?
2. Are these issues clearly understood by the engineer and are they in his/her practice area?
3. Has the engineer worked with more than one other client in this market sector on a similar matter?
4. Does the engineer have relevant experience in this area?

Additionally, the engineer should assure that anyone referred to his or her office is a priority because this individual is reaching out and is seeking a valued relationship.

- *Setting client expectations*: Engineers often dive into projects as soon as the client has agreed to work with them. It's common that the engineer gets immersed in a project developing engineering solutions that he/she may tend to lose their clients in the details. Probably the most crucial client expectation to manage is communication. At the outset, client and engineer should establish the frequency, type, and standards of communication desired. Without this understanding, clients may form their own expectations about how the relationship will work. This should be the first expectation to be clarified.

Other details on communications may include:

1. Good times to contact the client
2. When the engineer is available
3. The preferred method of contact, e.g., phone, or e-mail
4. How to handle urgent messages

- *Maintaining ethical and moral standards*: The engineering profession is one of the most highly regarded professions due to the ethical and moral standards held by engineers. Maintaining these standards is the engineer's duty under the business and professional codes, the canons of ASCE, and each state's standards for professional engineers. Ethical treatment of clients is essential for maintaining a positive client relationship.

- *Earning a profit*: In private enterprise maintaining a profit isn't just a principle, it is a requirement. However, there may be short periods of time when a private firm "invests" in a proposal effort for a major project or a key client and actually may not maintain a profit for a month or two or three. But in general, a profit is required for businesses to stay in business. Thus, both clients and engineers need to understand that the engineer's ability to make a profit on a project is a key to maintaining a strong client relationship (Kidde, B. 1992).

9.5 CULTIVATING BUSINESS OPPORTUNITIES

Business can come from many sources. Among these are: participation in professional associations; company press releases; journal and magazine articles; printed brochures; websites; newsletters; awards and competitions; and advertising. These approaches vary in cost and effectiveness. Advertising often is cited as being one of the most expensive and least effective ways of cultivating new business for engineering firms.

Several less expensive approaches are available involving efforts by company employees. These include developing common ground with the client; networking; volunteering; speaking engagements; and asking for referrals:

- *Developing common ground with the client*: Learning if there is some common ground that can be built upon with clients is important. This common ground may be enjoying the same sport, enjoying the same food at a luncheon engagement, hobbies, family, or other areas of common interest. All of us often display pictures of our loved ones or favorite vacation spots in our work spaces. A simple observation can lead to an interesting conversation, if time permits and it seems appropriate.

- *Networking*: Networking is an integral component of business development. Networking can take place through professional organizations such as the ASCE, at business conferences and seminars, through alumni associations and other social clubs, basically anywhere.

- *Volunteering*: Volunteering for public service projects offers an opportunity to share common ground with the public and potential customers. There's a tremendous need for volunteer services at the local shelter for the homeless, creek or beach cleanup activities, local Elks or Lions clubs, scouts, friends of the library or zoo, chamber of commerce, and for those willing to participate in running or walk-a-thons to raise money for good causes. A great deal of personal as well as professional satisfaction can flow from volunteer activities.

- *Speaking engagements*: Another potential area for engineers to gain exposure is by offering to speak on a contemporary issue at a civic organization, school, university, or event. Public speaking offers an opportunity for the engineer to expose local problems that require public attention or to highlight engineering services that are potentially available to solve problems. A public speaking engagement that is done well and timely may create valuable opportunities. Many private firms as well as public agencies will entertain short presentations over lunch time for their staff eager to learn about engineering solutions to contemporary problems.

- *Asking for referrals*: Finally, after completing a successful job or project for a client, the engineer should consider asking for a reference. This request can first be in the form of a feedback questionnaire from upper management in the same company (or agency). Engineers are interested in improvement on their services and feedback from the questionnaire can provide valuable information on how to perform better the next time. Assuming the questionnaire is positive it may allow an opportunity to ask the client if they would mind providing a reference for other potential clients in the future.

Networking Primer

After initially meeting people it's common to inquire about common areas of interest. Establishing these common areas may start with initial questions involving subjects like:

- Where you're originally from
- Whether you have a family or where your children go to school
- Your daily commute route
- Your interests outside of work
- Your latest vacation or weekend outing
- Or other subject areas
- Note: Use caution and common sense when asking personal questions!

Once these areas have been discussed, it's also common to learn how alike we all are and that we may know some of the same people. This initial introductory dialogue can become the foundation for the next potential encounter with this individual or a mutual acquaintance. In these initial encounters you have an opportunity to introduce yourself and your profession and potentially to learn about possible engineering opportunities.

9.6 BUSINESS DEVELOPMENT

Although all employees can be involved in furthering the interests of their companies, most firms formally vest responsibility for business development with certain employees. In smaller firms the principal(s) will be involved in cultivating new business. In larger firms, there may be a Director of Business Development with accompanying staff. Often project managers are expected to bring work into their firms.

An example of a model for business development for private engineering consultants is presented in Figure 4.1 This business development (BD) process begins with identifying an initial problem or a project lead. If the firm determines that the opportunity is worthwhile, the potential project enters the tracking phase, usually managed by a senior engineer in the firm who monitors the pulse of this specific project. This process can take days or more likely months, providing an opportunity for the firm to con-sider initial strategies when entering the positioning phase. It's in this phase where the firm will identify the outstanding accomplishments, tools, or personnel that set this firm apart from competitors.

A "Request for Proposal" (RFP) is an invitation for firms to evaluate the client's problem and to assess whether they can provide a viable, cost-effective solution. The engineering firm must decide whether it should "invest" valuable time, effort, and funds into the proposal efforts. There are books written on creating winning proposals; and needless to say, creating a thorough proposal takes a great deal of technical, administrative, and support effort.

Proposals in the defense industry for sophisticated fighter jet aircraft or rockets can be volumes 5 to 10 feet thick or more to provide a complete answer to the government's request. Typical engineering proposals may be 5 to 10 pages or 1 to 5 volumes, depending on the complexity and project budget. The proposal phase may be a few days or a few months, depending upon the client. However, most engineers feel they never had enough time before the deadline, which accentuates the importance of the tracking and positioning phases preceding the RFP.

These expenses are not related to a specific client or project so they must be charged to the firm's overhead expenses. Firms try to limit their overhead expenses because high overhead drives their services rates up for all customers; and high rates can make the firms uncompetitive. Most firms try to limit these proposal overhead expenses to 1 to 4 percent of the total project fees they anticipate receiving if they win the project. Therefore, firms will often evaluate the client's engineering cost estimate and compare it to their own cost estimate before making a go/no-go decision on the proposal. However, there may be times when a firm goes "all out" for a proposal beyond their own set guidelines to restrict other competitors from establishing a foothold with a business sector or a specific client.

The next phase, client/agency review, may take a few days or a few months again, depending upon the scope of the project and urgency of the need. After the review is complete the interview and selection phases follow. The interview phase is typically referred to as the "shortlist interview," which requires considerable organization and preparation for the private consulting firm. The interview may last from 20 minutes to hours in the client's office. Typically, the client will "drill" the engineer with questions related to their approach, project team or staff, schedule, or budget. These interviews may take anywhere from 2 to 50 hours or more for each of the project participants in the interview, which in turn drives up the engineer's overhead cost.

After the interview, the client/agency will pick the winning team and make announcements to all the participants. One of the most common questions entertained by the client/agency from the losing engineering firms is: "So, how did our firm rank in the process?" Experience has taught the seasoned engineer that the client/agency response is often: "Your firm came in second, but we wish you lots of luck on the next one." Often, in reality the winning firm is number 1 and "all other firms" are ranked as number 2. This secret and odd response to the ranking question often occurs because these competitors don't usually ask one another how they ranked, and the client/agency doesn't want to discourage or restrict future responses from the firms. However, there are clients/agencies that share their ranking process with the competitors but typically not in writing. The engineering firm can gain a great deal of insight into improving their proposal process if they request feedback from the client interview panel.

The final phase before the long-awaited project kick-off meeting is the contract negotiation process. Experienced firms often request a sample contract in the very beginning of the BD process because the terms of the contract may prevent them from proposing in the first place. Some clients/agencies may want to assign "all risk" to the engineering firm for the entire client's staff and other contractors on the project site, or they may request some sort of guarantee for the project before it's even started. There could be other contract terms requiring some sort of insurance protection or payment structure that the engineering firm cannot withstand. This negotiation phase may be an arduous process and may result with the "winning" firm ultimately rejecting the work. This forces the client back to their interview list. Alternatively, the process may be smooth, leading swiftly into the project. Therefore, the negotiation phase is one that requires special attention to detail and experience on project and contract management.

REMEMBER: Work Smarter . . . Not Harder!

9.7   CONFLICT MANAGEMENT

In the professional world, providing thorough and direct communication is an effort to avoid conflict. However, reality sometimes sets in with difficult issues like tight schedules, budget constraints, contract issues, competition for resources, subcontracting issues, and more. Even the most experienced project manager and experienced team can encounter unexpected issues that result in a conflict be-tween the engineer and client or owner. Some basic conflict management skills can help manage these conflicts, repair the damage, and hopefully salvage the relationship (Jones, T. 2008).

Some typical, real-life experiences affecting the engineer and client are listed below. Whether the client or the engineer could have anticipated these events before signing their agreement is doubtful:

- The client PM left the client's firm and the engineer had to temporarily freeze the project until the client could re-assign a new PM.

  The Result: The project was on hold for two months and the engineering PM had a struggle to re-assemble the team and incur re-start costs that could not be billed to the project. Therefore, meeting the client's needs caused the engineering firm to lose revenue and profit on this project.
- The federal government exercised the "stop work clause" in the contract because of priority funding for an international conflict.

  The Result: The engineering firm disbanded the project team, incurred major overhead costs (and lost profits) to keep the employees on the payroll while they were being re-assigned to other projects.
- A subcontractor had a safety incident and one of their employees had a minor laceration that required first aid. The client recognized that this was the second event in six months and temporarily stopped all field work on the West Coast affecting over 50 projects in the entire program to re-examine the Health and Safety Plans for the engineer and subcontractors. The engineer's PM had to redirect the project staff to keep them billable during the three-month evaluation period.

The Result: The engineering firm lost revenue and the client lost confidence in the engineering firm's field safety protocols, even though the firm was not directly responsible for the incident. Finally, the engineering firm did not win the follow-on multi-million-dollar contract one year later.

- The engineering firm's PM left the firm and the firm had two weeks to re-assign another less-experienced PM to the project.

  The Result: The client canceled the contract, and the original PM re-contracted the client with the new engineering firm.

- On another similar project with a different engineering firm, the firm's PM left the firm. The firm hired another more-experienced PM to retain the client and the project.

  The Result: The firm incurred higher overhead costs because the new PM was compensated at a higher rate and was hired through an agency with expensive placement fees. The engineering firm was able to keep the client. However, the firm lost revenue and profit because the project staff (hours and effort) was temporarily reduced by the client until a new engineering PM was hired.

- The engineering firm entered into a contract with a chemical analytical laboratory to perform analytical testing for potentially toxic and hazardous compounds. A large elaborate and expensive data collection field effort was planned and conducted. During a very busy summer period, the analytical laboratory performed the analyses but later realized that the samples were analyzed past their prescribed holding times. The data would not be accepted by the regulating agency and was considered invalid.

  The Result: The analytical laboratory sincerely apologized for the error and offered to re-analyze new samples for a significantly reduced fee. As a result, the client missed the report deadline and was angry with the engineering firm and the laboratory. The engineering PM was able to negotiate with the regulatory agencies for additional time so the client would not be fined. However, the engineering firm suffered a significant cost impact due to the need to re-collect field samples and then have them re-analyzed at a different laboratory for a higher fee. The engineering firm decided to seek restitution from the analytical laboratory.

The point of all these typical, real-life experiences described above is that despite all reasonable care taken by the engineering PM and the client PM, these situations were not anticipated. The engineering PM and the client PM may have common goals on the project, but each PM has a different employer with different stakeholders that can place them in direct conflict with one another.

The engineer should attempt to resolve these conflicts as soon as possible and at the lowest management level possible. Frequent and direct communication on these issues is highly recommended, as described in Chapter 13. The contract or agreement most likely includes language and direction on pursuing solutions to these problems, but often at a greater expense to one party—frequently the engineer. Specific contract issues and terms are discussed in Chapters 4 and 11 for reference.

In addition, applying the 4 Cs of conflict resolution may be beneficial if the atmosphere for resolution presents itself and can be fostered. If either party attempts to integrate their legal representatives into the resolution, the PMs usually lose control of possible alternatives. Often negotiations get very difficult and can get more aggressive at this point, and salvaging the relationship becomes more problematic.

9.7.1   The 4 Cs of Conflict Management

Experienced PMs often implement the 4 Cs of conflict resolution as part of the effort to reach an equitable agreement. Preselecting the targeted strategy of the resolution is important because only two of these strategies offer an equitable solution. A more detailed discussion on this management technique appears in Chapter 13, Communicating as a Professional Engineer.

Collaboration—A collaborative strategy in conflict management is when two parties come up with a completely new idea that pleases them both. Win/Win

Compromise—A compromise in conflict management is when two or more sides agree to accept less than they originally wanted. Draw/Draw

Co-existence—Both parties agree to disagree. Lose/Lose

Capitulation—One party completely gives up resisting the other. Lose/ Win (Maybe)

Some Tips about Building Relationships

- Keep in touch with your clients! In sales jargon, we actually refer to this as "touching," which means staying in contact by phone, e-mail, or meetings on a regular basis. This can be as small as asking about their family, remembering or referring to something they are interested in or their hobbies or a business problem they previously shared with you. It never hurts to take notes. Think about sending them articles that may be of interest that you have come across in your reading. It is also advisable to call sometimes just to "check in" without an agenda.

- Be curious about your clients. People generally like to talk about themselves. A connection opportunity may be the pictures or mementos on their desk. They are there for a reason. You never know what will help you make that "connection" that can lead to further business or referrals.

—Janet Riser, Senior V.P. and Executive Manager, Janney, Montgomery and Scott LLC

9.8 SUMMARY

The fundamental foundation of client relationships rests on trust, respect, commitment, follow-through, and communication. Building upon the foundation of this relationship includes anticipating the client's needs, understanding the client's business model or public's views, excellent scope, schedule, budget control in contractual deliverables and projects, an appreciation for the client's expectations on quality, and remembering the client's important stuff. Guiding principles for maintaining relationships include integrity, ethical standards, and learning to set expectations. The fundamental rule of client relationships is: **NO SURPRISES** (except good ones)!

BIBLIOGRAPHY

Bachner, John. (1991). *Practice Management for Design Professionals: A Practical Guide to Avoiding Liability and Enhancing Profitability.* John Wiley & Sons, Inc, New Jersey.

Jones, Tricia S. and Brinkert, Ross. (2008). *Conflict Coaching: Conflict Management Strategies and Skills for the Individual.* Sage Publications, Inc, Thousand Oaks, CA, ISBN-10: 1-412-95083-X.

Kidde, Barbara and Fetterberg, Fred. (1992). "Succeeding with consultants: Self-assessment for the changing nonprofit." The Foundation Center. (http://foundationcenter.org) ISBN-10: 0-879-54450-3.

Sanders, Tim. (2009). *Love Is the Killer App.* Three Rivers Press, New York.

CHAPTER 10

Leadership

Big Idea

"As engineers, we were going to be in a position to change the world—not just study it."
—Henry Petroski

"Management is doing things right; leadership is doing the right things."
—Peter Drucker

"If you think you can do a thing or think you can't do a thing, you're right."
—Henry Ford

"Effective leadership is putting first things first. Effective management is discipline, carrying it out."
—Stephen Covey

"Outstanding leaders go out of their way to boost the self-esteem of their personnel. If people believe in themselves, it's amazing what they can accomplish."
—Sam Walton

Civil Engineer's Handbook of Professional Practice, Second Edition. Karen Lee Hansen and Kent E. Zenobia.
© 2025 John Wiley & Sons, Inc. Published 2025 by John Wiley & Sons, Inc.
Companion website: www.wiley.com/go/hansen/CivilEngineersHandbook

Key Topics Covered

- Background
- Leadership Styles
- Tools for Leadership and Management
- Four Key Skill Sets Possessed by Effective Civil Engineering Leaders
- Public Service Leadership (for Government Employees) and Marketing Leadership (for Consulting Engineers)
- Secret Recipe for an Effective Leader

Related Chapters in This Book

- Chapter 3: Ethics
- Chapter 4: Professional Engagement
- Chapter 5: The Engineer's Role in Project Development
- Chapter 6: What Engineers Deliver
- Chapter 7: Executing a Professional Commission, Project Management
- Chapter 9: The Client Relationship and Business Development
- Chapter 11: Legal Aspects of Professional Practice
- Chapter 12: Managing the Civil Engineering Enterprise
- Chapter 13: Communicating as a Professional Engineer
- Chapter 14: Balancing Life, Family, and Career
- Chapter 15: Globalization
- Chapter 16: Sustainability
- Chapter 17: Emerging Technologies
- Chapter 18: Human Relations Policies and Employment Practices
- Chapter 19: Construction Management
- Chapter 20: Critical Health and Safety Knowledge for Civil Engineers
- Chapter 21: What Engineers Need to Know

10.1   INTRODUCTION

This chapter will delve into the depths of good leadership models, practices, styles, tools, and skills. By the time Civil Engineers receive their college degrees they have gained a tremendous amount of knowledge in the foundational, technical, and professional skills as described in ASCE's Body of Knowledge (CEBOK). With practical application, this knowledge will be transformed into demonstrated abilities. Continued application of these abilities will become valuable experience.

Leadership is a learned skill, likely enhanced by mentors and role models like parents, siblings, coaches, teachers, good friends, or educational influencers. In civil engineering practice, leaders inspire innovation, original thought, strategy, and invention applied to improvement, profitability, time savings, safety, resilience, and/or efficiency. Leadership skills are applied to design, construction, and maintenance of the physical and naturally built environment, and all components of our "civilization."

Today, in the United States our major companies and government agencies have an increasingly diverse, multigenerational, multilingual workforce. A new kind of leadership is required for success. Civil engineering leaders will have to shift from management hierarchy, deliverable product-focused, profit-driven, homogenous culture to a new paradigm that embraces success stories, continuous personal and company improvement, timely training programs, to a team focused culture that empowers its members across the national and international landscape. This chapter addresses the common elements of effective leadership, tools for leadership and management, and essential skill sets needed for success for the civil engineering leader, including:

- examples of great leaders and ten enduring principles of great leaders
- leadership styles
- tools for leadership and management
- case study examples
- four key skill sets possessed by effective civil engineering leaders
- public service leadership for government employees and marketing leadership for consulting engineers
- a comparison between America's greatest leader in 1776 and contemporary principles of leadership
- a discussion of whether: "Do times really change for leaders?"

10.2   BACKGROUND

Leadership is the ability to influence, direct, or steer another person or persons. In the civil engineering practice, individuals exhibiting leadership qualities often ascend to executive management or specialty technical expert positions. In government, civil engineering leaders frequently rise to department or agency heads.

Let's discuss some tangible examples of great leaders to get a better idea of the leadership qualities and characteristics with names and personalities you likely know. Most contemporary leaders in engineering, society, politics, industry, sports, religious, military, sciences possess positive and negative traits. For example, some great sports leaders in football, bicycle racing, baseball, etc., have record achievements but also some controversy over possible drug enhancement or claims of cheating. Government and political leaders also exhibit positive and negative behavior regarding budget issues, political party affiliations, external activities, etc. Great industry leaders have been exposed to negative claims about unfair labor practices, disproportionate compensation, or human rights issues.

Given that we are all human and no human is perfect, these observations are not surprising. However, in order to initiate a discussion of leadership, examples of virtual, historical, and contemporary leaders are provided in the following textboxes.

Example Leaders—Virtual

Captain America
Source: Schweiz41/Wikimedia Commons/Public domain.

Captain America, as depicted in movies, is the recipient of the Super-Soldier serum in the early days of World War II. This character's name is Steve Rogers, and he fights for American ideals as one of the world's mightiest heroes and the leader of the Avengers. Steve was a physical weakling who was originally rejected from enlisting in the US Army. Displaying determination as a leader, Steve was undeterred and made five more attempts to fight in the dark days of World War II.

Steve's valiant wish was to fight bullies wherever they were existed. He captured the attention of scientist Dr. Abraham Erskine, who handpicked the idealistic young individual for the Army's Super-Soldier program under his supervision. Steve had learned of Erskine's disastrous previous attempts to realize a super-soldier, but he braved the process and emerged from it a new person. With a powerful body and heightened speed and reflexes, Steve faced his first challenge when forced to chase Dr. Erskine's assassin immediately following the experiment.

Steve jumped at the chance to join the real fighting when he received news of the capture of his friends' army division and rushed to rescue the soldiers. Upon the successful completion of the unplanned mission, Steve determined his path as a one-person fighting force, displaying grit, courage, and resolve and had several other action sequences where he showed selflessness, empathy, courage, and intelligence. In one mission, Steve fought onto an enemy plane, battling through numerous opponents and forcing the plane to crash deep in the Artic, where it was lost for many years.

After the plane crash in 1942, Steve, as Captain America, entered into a state of suspended animation for decades. He was recovered in the present day by S.H.I.E.L.D., the successor to the 1942 US's Strategic Scientific Reserve. Awakened, he learned of his current status and began the long journey to acclimating to a brand-new, modern world as a new present-day hero and leader of the Avengers.

The movie allows us to follow the growth and maturity of Captain America. It displays his character and attributes through dialogue, action scenes, and interactions with other relatable characters. Of course, since it's a movie we are not shown any particularly negative traits or actions related to Captain America. He's a superhero, a "good guy" and a leader!

Example Leaders—Historical

George Washington
Source: Daderot/Wikimedia Commons/ Public domain.

An excellent example of one of America's earliest and greatest leaders is George Washington.

After the Declaration of Independence in 1776, young poor American patriots were outnumbered, out-gunned, disorganized, inexperienced, and had very limited funding. Who would even consider organizing and leading an inexperienced Continental Army composed of farmers, trappers, and shopkeepers against a formidable foe – the British Empire? How would this Continental Army be paid for, who would provide weapons and ammunition, feed and clothe the Army across the original Colonies, where a 2-mile trek was considered a good travel day (Steinteinbrueck 2022).

Consider this task, especially since the associated effort would be defined as treason or sedition, punishable by death. George Washington took on the task and it was extremely difficult, for years. For example, in 1777, George Washington witnessed approximately 2,500 American soldiers dying in Valley Forge due to the harsh winter conditions. The soldiers loved freedom; but according to historian David McCollough, they made sacrifices mainly because of their love for Washington.

(Continued)

"They would go anywhere with him and do anything for im. They knew how much Washington cared for them and how he put himself in harm's way. In earlier battles, Washington's two horses were shot out from under him, and four bullets passed through his coat. The American soldiers knew this. And bled for him." (McCullough 2005)

What made Washington such a great leader?

- He believed in his soldiers.
- He had an exemplary character.
- He treated others with the utmost respect.
- He held his men accountable.
- He placed the welfare of his men ahead of his own.
- He had deep faith.
- He did not waver from his guiding principles.
- He took his responsibilities seriously.
- He was personally invested in the cause.

Example Leaders—Contemporary

El Capitan in Yosemite National Park
Source: Anonymous User/PublicDomainPictires.net.

"They did it!" Tommy Caldwell and Kevin Jorgeson climbed a rock that everyone thought was impossible to climb. Much has been written and more will be written about their extraordinary feat. One view of their climbing that has struck me is that there are many lessons for leaders that we can glean from what they did. So as I continue to share vicariously in this ultimate event, I can't help but think about the extraordinary examples of leadership that we have learned from the two individuals.

1. **Focus like a laser.** The precision required for the El Capitan climb in Yosemite National Park leaves me in awe. In dealing with issues of people or tasks, it is essential for a leader to focus with the "eyes of a super creature" (Superman or Catwoman). Regardless of the size of the target, it is essential to be totally

absorbed. Attention must never dip; if it does, the end result may be less than the leader hoped for. The leader must attend with every cell in his or her body, as the watching and listening are the drivers of success. Leaders must then be precise with feedback to their teams.

2. **Share every aspect.** Can you imagine how the mind, heart, and body must have been engaged for each of the climbers? Their success rested upon the constant interplay of these three elements. The experience is total when each player is able to share what is happening—in their minds, hearts, and bodies—as things move along. Leaders can leave nothing to chance when it comes to what followers need to know. Constructing and using a mental or electronic communications matrix will lead to near-perfect knowledge and action on the part of all involved in each situation.

3. **Stick with it**. There was an obvious personal quality that stuck out for me. It was tenacity. On more than one occasion during the climb, Tommy or Kevin could be seen trying to move from one position on the rock to another. There are lots of reasons why that could happen, but the important thing is that a missed move did not dissuade either climber from continuing to move forward. Leaders who get results are never seen giving up. Whatever the driving factors may be (employees, customers, regulators, etc.) to reach a certain place, all must maintain the commitment to reach the outcome desired. (Note: It is better, though, to give up on a result deemed impossible to attain than to waste resources on an ultimate dead end.)

4. **Count on me**. Kevin and Tommy had an astonishingly close relationship during the climb. The loyalty and confidence that grew stronger each hour was something that leaders ought to prize among themselves and their fellow travelers. Loyalty grows when you feel as connected to each other, as these climbers did. This was the result of sharing, as mentioned above, but also a result of using the value that sharing provided. Confidence also grew because each could count on the other for strength and fortitude. Confidence grew in each of the individuals, and it grew in them as a unit. "I can count on you and you can count on me." Imagine the energy to perform when leaders create that kind of trusting synergy.

5. **Touch the surroundings.** The world of "need to know" around Tommy and Kevin was limited. They needed to keep the next place on the rock that would be ascended to in view all the time. Constant monitoring of the weather, especially wind changes, was paramount, too. Plus, they had to understand moisture patterns along the way. All of this understanding allowed them to proceed with

(Continued)

confidence. Leaders who are constantly aware of what is going on around them can be better prepared for what might come next. That preparation must be drenched in flexibility, just as it was for the climbers. Their flexibility had no limits. Does the ability to "turn on a dime" give leaders an opportunity to pause and reflect?

6. **Keep perspective.** Tommy and Kevin knew 1) their climbing skills, 2) how to move cautiously, 3) when to be creative, 4) the distinct nature of their task, 5) their preparation was spot on, and 6) when to reach down for more energy and courage. Leaders who take a self-inventory of these key knowledge factors at any point during a task (individual or team) can go a long way in gaining whatever result is being sought. Their story should inspire all leaders to determine the key success factors in the planning process and stay mindful of them all along the way. The key is not to leave anything out that is important.

7. **Hold yourself accountable.** The climbers held themselves accountable for the results. As leaders, they gained agreement and then executed. As leaders we are responsible for the plan and the execution. Not the other person. Not all those people on the third floor. Just yourself. Keep your eyes on yourself and your execution. Establish contingencies so that everyone in the organization will act on what they are accountable for—after those areas have been thoroughly established via comprehensive training and ongoing communication and feedback. Leaders must insure that a culture of conversation on accountabilities and resources for training are top priorities for the organization.

"OK, leaders. Let's strap on our boots and get climbing!"
—(Robert C. Preziosi, January 2015, SmartBrief –
Leadership lessons from the El Capitan climb | @SmartBrief SmartBlogs)

Figure 10.1 displays leadership in action. Experienced civil engineers likely will recognize these leadership styles. Readers are encouraged to examine the examples provided while developing their own unique leadership styles.

Leadership is both the act of guiding and the process of showing the way to proceed toward a common goal (Northouse). A leader takes risk, exhibits initiative, inspires, and motivates others to follow and perform an activity. In a social order, such as a community or an employer's office, a leader must have the self-confidence to lead others, intelligence and the forethought to plan, and the respect and trust of followers. Engineers can make good leaders because they are generally intelligent and have the ability to plan. If the engineer possesses the additional ingredients of a dynamic personality and strong character, they can gain the respect and trust of their followers. Table 10.1 summarizes these actions and characteristics that good leaders possess and exhibit.

AUTOCRATIC	DEMOCRATIC	DELEGATIVE
Leader Dictates Decisions to Employees	Leader includes Employees, Leader Has Final Decision	Employees are proactive, Leader Has Final Decision

Source: USHMM/Wikimedia Commons/Public domain.	*Source*: Monusco photos/Flickr/CC BY SA 2.0.	*Source*: Obama White House Archived/Flickr.
Employees Hear:	**Employees Hear:**	**Employees Hear:**
"Just Do It – The Way I Asked."	**"Let's do It Together – Have Any Ideas?"**	**"Please Do It, and – Keep Me Informed"** **Have Any Ideas?"**

FIGURE 10.1   Leadership styles.

TABLE 10.1   Actions and characteristics of good leaders.

"Actions" of a Good Leader:	"Characteristics" of a Good Leader:
• Is accountable	• Integrity
• Inspires	• Trustworthy
• Motivates and recognizes	• Intelligent
• Sets a good example	• Courageous
• Demonstrates respect	• Empathy

10.3   LEADERSHIP STYLES

A lot has transpired as leadership over the American-dominating years from the 1950s to the end of the Cold War with the Soviet Union in the late 1980s, from the 1980s to 2001 and the 9/11 Event, from 2001 to 2022, and then likely scenarios for the near future with you, as young civil engineers shape leadership maturity even further. It's worth taking a peek at the historic evolving leadership picture because there are facts and instances where we will see leadership styles evolve to keep pace with ever-increasing demand for efficiency, cost-effectiveness, quality, employee morale, and job satisfaction (Roberts 2018).

10.3.1   Definition of Leadership Styles

Leadership style is the manner and approach of providing direction, implementing plans, and motivating people. Kurt Lewin (1939) led a group of researchers to identify different styles of leadership. This early study has been very influential and established three major leadership styles. The three major styles of leadership identified by Lewin were (Roberts 2018):

- Autocratic
- Democratic
- Delegative or Free Reign

Effective leaders are aware of all three types of leadership styles and adapt these styles to the individual and the situation. It's possible to use all three styles on the same individual in different situations, and it's possible to use all three styles on different individuals for the same situation. Note that the leader is still responsible for the decisions and performance of the team members regardless of the style used (Rogers 2017) Let's look at the definitions of these leadership styles in more detail. Figure 10.1 illustrates these leadership styles.

10.3.2   Autocratic Leadership

An *autocratic leadership style* is one where the leader tells the team members or followers what to do, how to do it, and when to do it without any input from their followers. Typical instances for using an autocratic style might be when there is an emergency situation that requires immediate action like, "Call for help!" or when the leader has all the information to solve a particular problem and requests a specific tool or action. Generally, the leader desires (and requires) little input from the team members (Army Handbook). New or unskilled team members may be accustomed to this leadership style but experienced and skilled team members will not appreciate being dictated to and told what to do and how to do it (Rogers 2017).

Autocratic Style Example

Think about how a military commander might respond when commanding a response to a mortar rocket assault and an overwhelming attacking force on their remote compound.

10.3.3   Democratic Leadership

A *democratic leadership style* occurs when the leader invites the team members or followers to provide input into the decision-making process respecting them and validating their input into the overall process (Army Handbook). This style might be used when the team members have valuable information regarding the process and the leader has other knowledge or information regarding the process (Northouse). The leader cannot be expected to know everything and relies on knowledgeable employees. Generally, the leader maintains the final authority but considers the team members' input. This style displays the leadership strength of the leader and usually generates respect from the team members. There are time requirements for including the team members' input, so this style may not be employable during urgent situations or where the team members have little working knowledge to share on the process (Goffee, Jones 2009).

Democratic Style Example

Think about how a newly trained project manager might react when responsible for an innovative IT software program with undefined concepts, limited budget, and a dedicated team.

10.3.4 Delegative Leadership

A delegative leadership style is one where the leader provides broad guidance to the team members and allows the team members to decide what to do, how and when to do it, and where to do it (Army Handbook). This style is often employed with very senior and skilled members when the team members can correctly analyze the situation, determine what needs to be done, and how to do it (Northouse). The leader understands that they cannot do everything, so they set priorities and delegate certain tasks. However, one of the pitfalls, especially for new leaders, is to simply assume that all the team members are skilled and fully capable. In such situations, the leader needs to "check in" with the team members to verify their level of self-confidence and capability. There's more information on this communication process in Chapter 13, Communicating as a Professional Engineer. This style may be employed when the leader enjoys full trust and confidence in the team members (Goffee, Jones 2009).

Delegative Style Example

Think about how an experienced manager, recently transferred from New York to Texas, might respond when employed by a Fortune 100 Company.

There is no single leadership style employable for all team members at all times especially if the leader experiences a "life-event" forcing maturation on their own natural style. Let us examine a Ted Talk on Tracy Young, the founder of a Start-up called PlanGrid in Silicon Valley.

There is no one leadership style that works for all team members under all circumstances. Conversely, it is unlikely that one leadership style will be the appropriate for each employee all the time.

Tracy Young, cofounder and CEO of PlanGrid

Adapted from Democratic to Delegative Leadership Styles
TEDWomen 2020.

"Tracy Young often worried that employees and investors valued male CEOs more—and that being a woman compromised her position as a leader. In this brave, personal talk, she gives an honest look at the constraints women face when trying to adapt to a male-dominated business culture—and shares how she developed the courage and vulnerability to lead as her complete, raw self."

While Tracy became an expectant mother, her start-up entered a critical phase of product development and near hostile take-over. This unplanned event forced Tracy to focus on raising multi-millions of dollars for a war-chest. Tracy's pregnancy arrived just as a takeover offer nearing $1B surfaced, a stunning contrast to her puny war-chest. When Tracy's baby was born, and she decided the baby needed her more. Talk about managing conflicts! Tracy agonized over the *really big money* at stake, as well as the lives of her fellow employees. In her TED Talk, Tracy discusses the complications and contrasts life decisions for a male CEO in the same situation.

While she assumed all financial responsibilities and liabilities, Tracy's leadership style was forced to move from a democratic style to a delegative style. She simply could not be in two places at once while she struggled to care for her baby.

Dare to lead as your authentic self (Tracy Young | TEDWomen 2020) https://www.ted.com/talks/tracy_young_dare_to_lead_as_your_authentic_self?utm_source=tedcomshare&utm_medium=email&utm_camp *(This talk contains a graphic story. Discretion is advised.)*

There are numerous factors that leaders should consider when choosing a leadership style for a team member and a specific situation. The key is for the leader to be aware of, and consciously choose the most appropriate style—including consideration of their individual "natural" style. Some factors are listed below:

- Urgency of the task
- Competence level and skill level of the team members
- The level of respect and trust between the leader and team member
- The leader's knowledge level and team members' knowledge level of the task
- Task type—meaning the level of structure (existing procedures) associated with the task, complexity of the task, number of components, number of steps, extent

of knowledge required, and technical and cognitive abilities needed to accomplish the task

- Relationship with and knowledge of team members
- The schedule for completion of this task
- The budget for the task
- Other critical factors

A leader's natural style may have components of all three leadership styles described above.

Example Management/Leadership Scenario

Consider the following scenario. As a Senior Manager, you have an urgent, critical task that is influencing you, right now. This task could affect the future of your company's (or agency's) budget and viability.

You have two employees you could assign to this urgent task. They are both competent, dependable, respected, and trusted. One difference, and probably the only major difference, is that *Employee 1* has been on your team for ten years and *Employee 2*, relatively new, has been on your team for one year. Either employee can perform this task, but your leadership style will likely be different.

As Senior Manager, you might feel comfortable assigning this task to *Employee 1* and you might employ your "delegative style" of leadership. In this instance, you might engage *Employee 1* with periodic check-ins, especially if you have other urgent tasks related to this issue that you must address immediately.

Then as Senior Manager, you might want to allow *Employee 2* to gain experience with this task. For *Employee 2*, you might employ your "democratic style" of leadership. In this case, you might likely ask *Employee 2* about their strategy and steps they intend to take since you have the ultimate responsibility for this outcome.

Here is an example of the same scenario with two trusted employees where a different leadership style is required. Additional information is available on the following website: www.nwlink.com/~donclark/leader/leadstl.html.

There's no right or wrong answer. But the message is that leaders should adjust their leadership style to fit the circumstance, timing, and their employees.

Each of us holds the power to inspire confidence in others!

– Anonymous

Critical Thinking for the Successful Achievement of Complex Goals by Dr. Iain A MacLeod, Institution of Engineers in Scotland

There is a story, probably true, that in the 1980s a delegation of American engineers went to Japan to find out the reasons for the success of the latter's manufacturing industry. The Japanese hosts explained how rather than have production targets, everyone in the organization adopted a constant and unremitting drive towards improving the processes that were used to create products and provide services. The visiting party asked about how such a mode of operation arose. Was this part of the Japanese culture? The reply was that it was an American, W. Edwards Deming, who had introduced these ideas to Japanese manufacturers.

Deming (1900–1993) had promoted his ideas in USA with little success, but the Japanese were amenable to adopting them. Manufacturing is one of the most competitive activities in society. It is very difficult to succeed without careful attention to strategies for doing well. Manufacturing firms use a range of methodologies e.g., Lean, 5S, Six Sigma, Total Quality Management, that lead back to the work of Deming. He is the "father" of the concept of continuous process improvement. His principle that one should not set production targets is counter intuitive. How can you produce more widgets unless you have an objective for doing that? Deming was opposed to numerical targets. He is quoted as saying, "Give a manager a numerical target he will make it, even if he has to destroy the company in the process." In meeting the target, a manager may for example, compromise the quality of the product leading to customer dissatisfaction with a corresponding drop in sales. Deming's point was that improvement in production rate should be based on a method of achieving it rather than on exhortation to work harder.

10.3.5 Collaborative Leadership

Table 10.2, Drive in Fear, Drive out Fear, displays hypothetical actions by an employee and the corresponding actions by the "Autocratic" manager and the "Democratic" manager as summarized in Table 10.1. Note that the "Autocratic" manager will threaten the employee with punishment for a non-willful mistake. Conversely, the "Democratic" manager will treat this action as a learning opportunity. So, a Democratic manager will welcome suggestions and use them if appropriate and reward those who make suggestions that are adopted.

W. Edwards Deming explored these still relevant concepts through his work on what is now known as *total quality management*. Deming set out 14 points for good management of manufacturing processes. Point 8 is: "Drive out fear, so that everyone may work effectively for

TABLE 10.2   Drive in fear, drive out fear.

Example Action by an Employee	Action by Manager	
	Drive in Fear, Autocratic Style	Drive Out Fear, Democratic Style
Make a non-willful mistake	Threaten employee with punishment	Treat as a learning opportunity
Point out faults in the organization	Ignore such advice; treat the employee as a whistleblower; seek to have them dismissed	Welcome suggestions and use them if appropriate; reward those who make suggestions that are adopted

the company." The converse, "drive in fear" style of management is common. It is a high-risk strategy because (a) the employee is likely to hide mistakes and (b) working with ideas from all persons directly involved in a project is likely to give better results. For more information on Deming's work, see the textbox, "Critical Thinking for the Successful Achievement of Complex Goals" by Dr. Iain A MacLeod.

Some managers seem to believe that their appointment confers special status in relation to wisdom and that because of this, their ideas must be superior to those from people at lower levels in the organization. This is absurd. Working in safety-critical context tends to result in acceptance that a "drive out fear" ethos is essential. If getting it wrong can be disastrous, then using every strategy for reducing risk becomes a clear choice.

For example, Deming discovered that in some air disasters, crew had information that the captain needed but was not communicated by colleagues because of the rigid command structure in the cockpit. Crew are now trained to share information; captains are expected to consult with their crew members and to welcome information from them. It is called *crew resource management*. Such ethos should be pervasive in organizations. It is whistle-blowing turned on its head. Rather than being vilified for pointing out faults, people are praised for putting forward ideas that might help to improve performance. Teaching people from the earliest practical age that a "drive in fear" style of management is inappropriate benefits all. Those who have experience of working under a "drive in fear" manager will tend to adopt that style when they become managers. Regrettably, in some cases, the abused become the abusers.

A leadership style that focuses everyone in an organization on the goals of the organization seems an obvious option, but this is often not the case. If management creates a "them and us" ethos, the employees will focus on their own goals. Some people believe that competition and the market will solve all problems whereas others believe that co-operation is the answer. It seems clear that both competition and cooperation are necessary in society and a fundamental objective of any management system must be to get the right balance between the two. The Deming approach leans strongly on the co-operation side. The Deming philosophy defines how attitudes should be shaped for success, an approach that can be applied beyond manufacturing.

Additional Thoughts on Leadership

Leadership is a relationship between the leader and the team member(s). A leader can't lead without followers. The leader has to demonstrate continually respect, trust, and courage to maintain leadership and inspire and motivate the followers. The essential ingredients of an effective leader are very similar to those components presented in Chapter 9—The Client Relationship and Business Development, and the foundation elements of a lasting relationship depicted in Figure 9.1. In general, leadership is also *nonhierarchical*. Just because the supervisor has a title, that doesn't make him or her a leader, but it does make them the boss. "Supervision without leadership" presents challenges for those at organizational levels both below and above such individuals. Leaders exist at all levels within an organization (Goffee, Jones 2009).

Quotes from Famous Leaders

- "If your actions inspire others to dream more, learn more, do more and become more, you are a leader." John Quincy Adams
- "Setting an example is not the main means of influencing another, it is the only means." Albert Einstein
- "The ability to learn is the most important quality a leader can have." Sheryl Sandberg
- "Innovation distinguishes between a leader and a follower." Steven Jobs
- "To be a leader, you have to make people follow you, and nobody wants to follow someone who doesn't know where he is going." Joe Namath

—www.thinkexist.com

10.4   TOOLS FOR LEADERSHIP AND MANAGEMENT

Most engineering managers are leaders and they generally have common tools in their tool boxes independent of whether they are involved in engineering consulting or public service. We'll focus on the responsibilities of a "manager" and later discuss how this relates to being a "leader" and how an engineer can be both a manager and a leader. A manager assumes responsibility for a business unit or an element of a government service like an engineering department or an organization. For purposes of this discussion, let's agree to use the term "business unit." The manager's overall responsibility usually includes:

- *Planning* functions of the business unit for short- and long-term operations
- *Organizing and monitoring* the daily operations of the business unit, the organization of the team often including the business unit's marketing and sales functions (if it's an engineering consulting organization), or the public service/public affairs function (if it's a government business unit)
- *Leading t*he strategic planning functions of the company
- *Controlling* the income revenue and cost elements of the business unit, which is referred to as the Profit and Loss (P&L) responsibility (if it's engineering consulting) or the budget responsibility (if it's government)

So, in essence, managers *plan, organize, lead*, and *control* (POLC) to accomplish the mission of the business unit (DuBrin 2000).

10.4.1   Planning

The planning function includes an analysis of where the business unit is going in the short- and long-term periods. The plan may be referred to as a "business plan," a "marketing plan," an "operations plan," a "strategic plan," or "some other plan." But the manager's plan will include a strategy, much like a chess player, that describes where the business unit should be going to respond to the market or to the public's needs. The manager will most likely have a "vision" that will guide and shape the plan (DuBrin 2000).

10.4.2   Organizing

Organizing activities are related to assessing the needs of the business unit with regard to personnel, equipment, services, and resources to operate the unit to accomplish the mission. Organizing also includes the responsibility to accommodate the staff with a safe, comfortable work environment, the required tools and equipment to accomplish staff job responsibilities, training to enhance and increase staff development, output, and efficiency (DuBrin 2000). The effective manager monitors the business unit and staff performance to assess the effectiveness of their organized activities. Managing the organization can be as simple as ordering new computers or software or as complicated as finding a new cost-effective space to house the projected growth (or decline) of the unit (DuBrin 2000).

10.4.3   Leading

The effectiveness of the leadership function separates "managers" from "leaders." A leader will lead the strategic planning function of the business unit in a way that matches the vision of the executive management team (DuBrin 2000). Hopefully, the vision communicated in the strategic plan matches the perceived direction of the market for the consulting engineer leader or the direction of the public's views for the government engineer leader. This is where the leadership ability distinguishes the manager from the leader (DuBrin 2000).

10.4.4 Controlling

Controlling an engineering business unit is a lot like flying a light aircraft from a quiet valley, over the mountains and through a storm. The first thing a new pilot would likely do is:

- Inspect the aircraft (on the ground)
- Study the operator's manual
- Learn about the gauges and navigation system
- Run through some mock scenarios (on the ground)
- Take lessons from a qualified, experienced pilot
- Take more lessons
- Run a solo flight
- Then be ever cautious

Similarly, a new manager coming into an existing business unit would likely:

- Learn about the business unit, the staff, the goals, clients, and deliverables
- Study the existing operating procedures, guidelines, and requirements
- Learn about the revenue and costs for the unit and compare this to the output
- Run some mock scenarios aligned with the business plan or strategic plan
- Counsel a qualified, trusted engineering manager familiar with the unit
- Manage the unit solo while carefully watching the gauges under the guidance of the qualified, trusted engineering manager
- Then be ever cautious

Whether you're a new pilot or new engineering manager, managing a business unit will likely involve some smooth operations, some rocky terrain, and some strong storms. It's important to understand the business unit's plan, the organizational staff, resources, partners and output deliverables, and the appropriate leadership technique for the staff and conditions of the business climate. "Controlling" and managing the unit will likely stretch the engineering manager's skills until he/she gains the appropriate leadership skills (DuBrin, Andrew J. 2000 and Deming, W. Edwards. 1982).

10.5   FOUR KEY SKILL SETS POSSESSED BY EFFECTIVE CIVIL ENGINEERING LEADERS

Remember, the leadership styles, as described earlier in this chapter, are *autocratic, democratic,* or *delegative.* The Manager generally focuses on the direction, management strategy, and deployment of the organization staff and resources. The manager often uses feedback from the past while trying to forge ahead. It's similar to driving a car while focusing on the rear-view mirror but occasionally glancing ahead. The manager will guide the business unit from last month's staff reports, revenue, and cost reports, while trying to stay on track with the business plan.

Leaders often display strong character, strength, fortitude, and courage to lead the way toward the vision in the strategic plan. The leader paves the way described in the strategic plan before the followers can see the vision the leader embraces. The leader is focusing on the path ahead while glancing backward to see the previous destination and gauge progress. Effective leaders do not forget that they need to control the velocity and direction of the business unit (Altier 1999).

Management is needed to run the operation and stay on track with the realities of cost, revenue, and resource usage. Leaders are needed to prevent stagnation, to encourage continuous improvement and innovation to compete in the market. Leaders also create new ways to serve clients and/or the public more effectively (Deming, W. Edwards. 1982).

So, what's the right mix of management and leadership? Are there specific areas in which a leader can excel? Let's look at typical areas of engineering management where engineers excel at leadership. These leadership specialty areas are depicted in Figure 10.2 and are referred to as the four quadrants of leadership. Most effective civil engineering leaders will possess at least two or more of these specialty leadership skill sets (the more acquired skill sets, the better):

- Strategic Leadership
- Financial Leadership
- Technical Leadership
- Marketing Leadership (for consulting engineers) and Public Service Leadership (for government engineers)

These four quadrants, as depicted in Figure 10.2, represent essential developmental skill areas in addition to the other tools previously mentioned in the manager's tool box for the leader to practice honing throughout their career (Altier 1999). Effective leadership requires a delicate balance in all four areas whether the engineer is in private practice or public service. Continuous improvement and development of these skill areas make engineering a career and a practice and not just a job (Deming, W. Edwards. 1982).

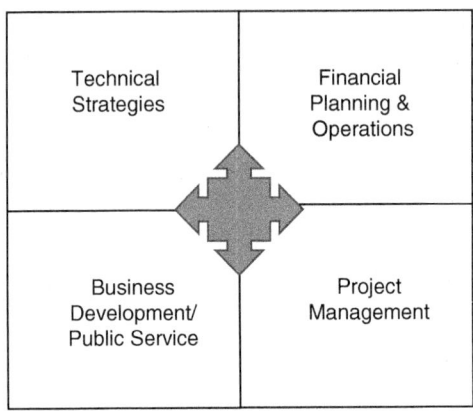

FIGURE 10.2   Four quadrants.

10.5.1   Strategic Leadership

Strategy involves concentrated thought, planning, direction, team organizational skills, and vision. Strategic leadership requires concentrated thought and keen observation that evolves into a vision and action to realize that vision. This means that the consulting engineer leader will observe the surrounding conditions in a particular market and concentrate on where the market is going in order to create a vision of the products or services required. The public service employee will perform a similar function by observing the surrounding conditions within the boundaries of a state/city or political environment and concentrating on the needs and desires of the public in order to create a vision of the products or services required for that entity (Northouse 2010; Roberts 2018).

Strategic leaders understand the balance between managing the business unit based upon past performance measures and carefully investing the unit's time or profit into creating and realizing the vision communicated in the strategic plan. If we could stroll through the strategic leader's mind, we would likely observe the thought patterns on:

Strategy

1. Where do we want to be and where do we need to be?
2. How can I articulate the vision so staff will understand the criticality of realizing that vision?
3. How much time and how many resources can we invest into realizing this vision?
4. Is there any way to accelerate the investment process and just how much can we afford to devote to our future?

Public Service Considerations

1. Do I truly understand the public opinion?
2. What is the view of this business unit to the public?
3. How can I gauge the public opinion and respond to it?
4. Is the current operating procedure meeting or exceeding our mission?

Market Considerations

1. Do I truly understand my client's needs and expectations?
2. Do I understand where my client's market is going, and can the business unit respond to that direction?
3. What can the unit do to increase the client base and diversify?
4. Can I cross-sell my services to clients from other departments within the organization?

Management

1. Can I achieve the operating goals and simultaneously invest in the vision?
2. Are there any major pieces of equipment or other resources that are almost completely depreciated that could break down and significantly impact the unit's overhead or expense cost?
3. What are my operating and investment costs for public affairs (community involvement and donations) and marketing and are they in line with the revenue and other competitors?
4. Are we achieving the unit's profit or service goals and the corporate or agency's goals? Can the output be improved?

These questions and corresponding answers can help to position the business unit strategically and can result in successful profitable operations or public service merit awards.

10.5.2 Financial Leadership

Financial leaders know and comprehend the value of the services and products produced by the business unit in relation to materials, overhead costs, direct costs, and expenses. The leader understands how these costs and expenses are related to one another, how to reduce these cost impacts, and how to increase output. The leader also understands that the unit's exemplary service (or profit) provides the fuel for the unit to pay the salaries, provide training, pay the rent, fund special events like parties and picnics, all of which are necessary to maintain the engine to continue accomplishing the work (Northouse 2010; Roberts 2018).

Financial leaders understand the balance between the cost of business and revenue required for investing in the vision in the strategic plan. If we could stroll through the financial leader's mind, we would observe the thought patterns on:

General Financial Concepts

1. Does staff understand the cost of doing business and how they may help lower overhead expenses?
2. Are our unit's marketing costs and overhead costs in line with the competition and the market?
3. Does the unit have adequate cash flow, reserve capacity, and a relationship with a financial institution to secure quick loans at reasonable costs?
4. Are the unit's direct costs in line with the indirect costs?
5. Am I receiving just compensation for the unit's products and services and are there ways to increase mark-ups for outside vendors or materials?

Risk-Related Financial Concepts

1. Does the unit employ adequate risk and uncertainty analyses to protect the unit's viability?
2. Have the key unit managers received important training on negotiating, interviewing, sensitivity, prevention of sexual harassment, defensive driving, and other mandated training by the human resources department, as discussed in Chapter 18?
3. Does the unit have adequate safety and health practices to protect the employees and is the unit protected/insured from lawsuits on related issues, as discussed in Chapter 20?

Project Management

1. Do the unit's project managers (PMs) understand the budgets we create have a tremendous impact on the unit's overall performance and viability?
2. Do the PMs understand that if a PM loses $50,000 on a particular project that this represents the total profit on a $1 million project if the profit were set at 5 percent?
3. Are the PMs carefully watching for scope creep and prompt invoicing to maintain the unit's revenue stream?
4. Do the PMs understand the cost of money for delayed invoicing and unbilled labor?
5. Does the business unit have adequate reserve funds for unanticipated impacts?
6. Do the PMs understand when an issue is out of the contract's scope and how to create a change order?
7. Do the PMs understand our competition's strengths and weaknesses compared to our own unit?

10.5.3   Technical Leadership

Technical leaders display technical acumen, excellent proficiency in engineering, and ingenious application of solutions to client problems. In addition, the technical leader usually possesses strategic and financial leadership qualities, too. These leaders are often on the cutting edge of new technology and developments and include services like:

- Research and Development
- Process Development
- Technical Services
- Customer Support
- Specific Quality Cells (for example, expertise areas like the Geotechnical Engineering Quality Cell or the Environmental Engineering Quality Cell)
- Product Development, among others

Technical leaders often publish their works in magazines or periodicals to gain professional recognition in their field of expertise. These published leaders can then offer their prospective clients the latest journal articles and may gain technical points for receiving future work (Northouse 2010; Roberts 2018).

Technical leaders understand the balance between the investment in technical innovation and investing in the vision in the strategic plan. If we could stroll through the technical leader's mind, we would observe the thought patterns:

Innovation

1. Are the unit's technical leaders aware of the most recent developments in the areas of expertise locally, nationally, and internationally?
2. Does the organization encourage critical thinking and an atmosphere conducive to process improvement and innovation?
3. Does the organization recognize individuals who have demonstrated innovation or award programs for encouragement?
4. Can the technical leaders identify innovative technologies that fit potential pilot or full-scale applications for projects in the unit's practice area?
5. Are there any grant funds or private funds available for furthering the application of new technologies to the unit's practice area?

Organization

1. Does the organization attract top technical talent and if not, what can be done to improve this condition?
2. Does the organization have a training and/or mentorship program?
3. Does the organization encourage and compensate colleagues for pursuing secondary degrees or other specialized coursework?

4. Does the organization encourage colleagues to communicate potential problems early in the project when the cost to correct a situation is usually lower?

Vision

1. Do technical leaders display critical thinking and offer innovative perspectives to today's challenges?
2. Do technical leaders display strategic thinking for application adopting technology to today's challenges?
3. Do technical leaders comprehend and accept there are realistic limits to the investment of technology for the organization?
4. Are these leaders willing to accept their roles in the daily tasks to run the organization while maintaining their roles as technical leaders?

10.5.4 Marketing Leadership

Marketing Leadership generally applies to consulting engineers performing business development activities. In public service, engineers working for the government also display leadership but generally with a different purpose. Both leaders are discussed below.

For the purposes of this discussion, the information presented for consulting engineers is related to "marketing" the firm and its employees and the information presented for government employees is related to public service. The critical elements for each are very similar and relate to the outreach efforts to communicate the service capabilities of these respective organizations. Marketing and public service leaders understand the value of the services and products produced by the business unit and recognize that if their services do not meet the expectations of the users, the existence of the business unit will be jeopardized. Regardless of whether it's private industry or public service, these leaders are continually challenged with producing more results in less time with improved service. The leadership in this area is very dependent on accurate visual aids, good presentation skills, and clear communications (Northouse 2010; Roberts 2018).

Public service and marketing leaders understand the balance between the cost of presentation and outreach and consistency with the vision in the strategic plan. If we could stroll through the public service and marketing leaders' minds, we would observe the thought patterns on:

Outreach

1. Do the public service and marketing leaders make themselves available for outreach efforts and do they understand the criticality of this effort?
2. Can the leaders prepare and present effective presentations that communicate the message in a concise and deliberate manner?

3. Do the leaders display respectful and tactful responses to questions and comments from the clients or public?
4. Do the leaders participate in open forums, strategy sessions, and workshops in a manner that exhibits genuine concern for the stakeholders?
5. Do these leaders relate well to people from all levels of society?

Marketing and Presenting the Concept

1. For public service, do the leaders understand the costs associated with democratic participation, public awareness of the business unit, and the subject and goals of the presentation?
2. For consulting engineers, do the leaders understand the costs associated with marketing, strategic positioning, sales presentations, proposal preparation, short-list interviews, getting the project, and maintaining the client relationship?
3. Can the leaders express the concepts well and empathize with the impacted listeners' concerns?
4. Do the leaders understand how to plan, prepare, and execute public presentations?
5. Do the leaders understand the value of customer service?

Now that we've covered the Tools for Leadership and Management and the Four Quadrants of Effective Leadership, let's examine the current thoughts of enduring principles of great leaders. Ask yourself this, "How do we identify leadership skills that can stand the test of time?" Robert Glazer, a marketing consultant, suggests there are "10 Enduring Qualities of Great Leaders."

10 ENDURING QUALITIES OF GREAT LEADERS
Robert Glazer, Acceleration Partners

1. **Integrity:** People want leaders they can trust to act for the greater good and tell the truth. We want leaders who act according to their stated principles, are honest with us and keep their word.
2. **Humility:** As we shift away from command and control leadership, we gravitate toward leaders who are approachable and don't hold themselves above others. When leaders show humility and vulnerability, others instinctively want to work with them to achieve their goals.

(Continued)

3. **Empowering Others:** Great leaders trust the people on their team and coach them to make important decisions without micromanagement. They don't do everything themselves—instead, they set clear vision and values, and direct others to work according to those guiding principles.

4. **Great Communication:** Leaders must communicate well, both to move others to action and to ensure their directives are well-understood. It's no surprise that we often celebrate leaders who deliver historic speeches or impactful quotes. Great leaders also give their teams the information they need to excel.

5. **Forward-Thinking:** A great leader sets a compelling vision for the future, attracting and convincing others to want to join their movement. These leaders can share their vision with clarity and specificity, and they are passionate about the execution of those goals.

6. **Empathy:** We want our leaders to demonstrate empathy and an ability to relate to those they lead, especially in moments of crisis. A leader cannot effectively lead someone if they fail to understand their fundamental needs, and if they cannot connect others' fulfillment to their own.

7. **Competence:** Leaders must be capable of doing the job at hand and surround themselves with competent people. Competent leaders don't know how to do everything but are skilled at identifying people whose abilities complement their own, and bringing them into the fold. They also aren't afraid to hire people who are smarter than they are.

8. **Accountable:** Great leaders have a "the buck stops here" mentality. History is filled with leaders who credit their teams for their successes and accept personal responsibility for the team's failures. Poor leaders do the opposite, taking credit for their teams' accomplishments and distancing themselves from accountability.

9. **Gratitude:** One of the core responsibilities of a leader is to consider the needs of the many. A mindset of gratitude pushes leaders to focus less on themselves and more on how they can value and strengthen others.

10. **Self-Awareness:** Leaders must be aware of their own strengths and limitations. They must build a team that magnifies their strengths and limits their weaknesses. Leaders are also open to criticism and willing to do the sometimes-painful work required to improve.

—Aug 24, 2020

While we may never reach a consensus on political leadership, it is crucial to identify an apolitical benchmark by which we evaluate all our leaders. Great leadership should be an objective metric, not a changing threshold viewed through a political or ideological lens; otherwise, the term itself is meaningless. Finally, we should not have to shy away from holding our leaders accountable for results; the great ones do it for themselves.

If one considers Robert Glazer's **Summary of the 10 Enduring Principles of Great Leaders (above_)** and compares this list to the elements that made George Washington such a great leader (presented earlier in this chapter) one will find a tremendous overlap. In some cases, the words vary but the meaning is very similar. The overlapping principles of George Washington's leadership principles in 1776 and those listed by Robert Glazer in 2020 (above) include:

- Belief in his men and empowering others.
- Exemplary character and integrity.
- Treating others with utmost respect and empowering others.
- Holding the Team accountable.
- Placing the welfare of the Team above his own and also humility.
- Guiding principles and forward-thinking.
- Loved his Team and empathy, and also gratitude.
- Take responsibilities seriously and also competence.
- Personally invested and self-aware.

It seems the enduring principles for a great leader in 1776 versus 2020 are truly similar and enduring!

"Step Up"—It's the only way to climb!
Michael Sanchez, P. E. Principal Engineer, Consor Engineering

Success is best measured against:

- Remembering the POINT OF BEGINNING
- Following a path that is PLANNED
- Aimed at the identified GOAL

(Continued)

POINT OF BEGINNING = Remember where you started.

In the professional world, this means document-document-document and requires good record-keeping. Put your scope, schedule, and budget in writing at the beginning of a project to memorialize the plan and promises at the starting line.

PLAN = Set yourself and others on a path.

Staying on the path requires frequent checking in and "course correction." Use your compass and identify north often. Great leaders in engineering continually check their status and direction and exercise excellent project management skills.

GOAL = Be very clear of the end product.

Know your required outcome. It also helps to have all your team members aiming at that same goal. Sometimes the leader needs to shine a bright light to highlight the goal. This could require representing a client or a specific project in front of the public. When placed in this role, the team lead's job is focused on explaining why and how the goal is being pursued. When this is done well, many will join the team in running towards that very goal.

10.6   SECRET RECIPE FOR AN EFFECTIVE LEADER

This chapter has provided a lot of discussion on the tools for effective managers, leadership styles, and the foundational strengths of effective leaders. So, in an effort to describe a virtual effective leader a "perfect recipe" has been devised. When engineering managers carefully shop for the right individual leader(s) to fit their specific environment, they are more likely to achieve success for their organization.

Ingredients of a Good Leader:

- Start with an energetic individual with a thirst for learning. Now add the following:
  - A liter of strong character
  - A liter of humility
  - Two liters of organization
  - A kilogram of intelligence
  - A pinch of empathy
  - A kilogram of courage
  - A liter of confidence
  - Preheated trust and ethics
  - A liter of reliability mixed with follow-through.
  - A genuine smile

- Now, mix the above ingredients together with demonstrated communication skills, add strong listening abilities, and sensitivity training.
- Simmer over even-heated mentorship with years of experience, continued training, advanced technical/business degree, and dedication to common objectives.
- Garnish with politeness, business attire, and a genuine smile (yes, a fresh one).
- Serve warm to the public (or clients) after a brief introduction in a pleasant environment.

"A leader is one who knows the way, goes the way, and shows the way."
—John C. Maxwell

10.7   SUMMARY

The three major styles of leadership are:

- Autocratic—telling or demanding
- Democratic—interactive and consensus building
- Delegative—or free reign

The management style should be adapted to the specific situation and the individuals being managed. Effective managers employ the tools of planning, organizing, leading, and controlling to accomplish the mission of the organization. *A manager/leader carefully balances the use of feedback from previous quarterly results while strategically leading the business unit to the future.*

Dynamic individuals can become leaders in one or more business competence areas including strategic leadership, financial leadership, technical leadership, and public service/marketing leadership. The more areas of technical tools, strategic-thinking, leadership, and financial expertise you possess the greater you will find your fulfillment and career possibilities!

BIBLIOGRAPHY

Altier, William J. (1999). *The Thinking Manager's Toolbox: Effective Processes for Problem Solving and Decision Making*. Oxford University Press, ISBN: 0-195-13196-7.

Deming, W. Edwards. (1982). *Quality, Productivity, and Competitive Position*. Massachusetts Institute of Technology, Center for Advanced Engineering Study.

DuBrin, Andrew J. (January 2000). *The Active Manager: How to Plan, Organize, Lead and Control Your Way to Success*. South-Western College Press, Chula Vista, CA. ISBN-13: 978-0-324-02740-2.

Goffee, Rob and Jones, Gareth. (2009). *Clever: Leading Your Smartest, Most Creative People.* Harvard Business Press, Watertown, MA. ISBN 978-1-422-12296-9.

https://www.inspiringleadershipnow.com/best-ted-talks-on-leadership (Accessed March 22, 2023).

https://www.ted.com/talks/simon_sinek_how_great_leaders_inspire_action (Accessed March 22, 2023).

McCullough, David. (2005). 1776. Simon and Schuster, New York, NY. ISBN-13: 978-0-7432-2671-1.

Northouse, Peter G. (2010). *Leadership: Theory and Practice.* Sage Publications, Thousand Oaks, CA. ISBN: 978-1-412-97488-2.

Preziosi, Robert C. (January 2015). SmartBrief – Leadership lessons from the El Capitan climb. https://corp.smartbrief.com/original/2015/01/leadership-lessons-el-capitan-climb

Roberts, Capt. Ron. (May 25, 2018). *12 Principles of Modern Military Leadership.* Army University Press, NCO Journal.

Rogers, Major Robert. (April, 2017). *Headquarters Department of the Army*, Washington, DC. Training Circular No. 3-21.7.

10 Leadership lessons from George Washington, All Pro Dad, Leadership. https://www.allprodad.com/10-leadership-lessons-from-george-washington.

CHAPTER 11

Legal Aspects of Professional Practice

Big Idea

All civil engineers, whether employed in public or private practice, need to know the legal consequences of their actions.

> Lawsuit: A machine which you go into as a pig and come out as a sausage.
> —Ambrose Bierce, Civil War veteran and author of the
> 19th century *Devil's Dictionary*

Key Topics Covered

- US Legal System
- Statutory Law
- Common Law
- Contracts Law
- Procurement Management
- Risk Management

Civil Engineer's Handbook of Professional Practice, Second Edition. Karen Lee Hansen and Kent E. Zenobia.
© 2025 John Wiley & Sons, Inc. Published 2025 by John Wiley & Sons, Inc.
Companion website: www.wiley.com/go/hansen/CivilEngineersHandbook

- Insurance and Bonds
- Dispute Resolution
- Alternative Dispute Resolution
- Affirmative Action, Equal Opportunity, and Diversity.

Related Chapters in This Book

- Chapter 3: Ethics
- Chapter 4: Professional Engagement
- Chapter 5: The Civil Engineer's Role in Project Development
- Chapter 6: What Engineers Deliver
- Chapter 12: Managing the Civil Engineering Enterprise
- Chapter 13: Communicating as a Professional Engineer

11.1 INTRODUCTION

The purpose of this chapter is to give civil engineers a basic overview of the US legal system and to highlight some potential legal danger zones. The chapter discusses key legal issues that affect design professionals and suggests what can be done to avoid legal pitfalls. Specifically, the chapter addresses tort law, contract formation and clauses, contract structures used in various project delivery methods, litigation, alternative dispute resolution, the role of insurance and indemnification, affirmative action, equal opportunity, and diversity.

The information provided in this chapter does not, and is not intended to, constitute legal advice; instead, all information, content, and materials available in this chapter are for general informational purposes only.

Regardless of scale, constructed facilities—dams, highways, airports, water treatment plants, office buildings, shopping malls, industrial facilities, houses—use the *project* as an organizing device. In the construction industry, projects tend to be arranged in a way that epitomizes the fragmented form of hierarchical network organization, in which contracts are highly specialized. The project team members are assembled temporarily, for just as long as is necessary, from the array of players outlined in Chapter 5, The Engineer's Role in Project Development. The usual approach relies on price-based contracting methods (market transactions). This often results in adversarial relations between participants, in which information is concealed and information flows are disrupted. In other words, the Architectural, Engineering, and Construction (AEC) industry is rife with miscommunications and misunderstandings. Such an environment creates the need to be aware of the legal consequences of one's actions and knowledgeable enough to know when the advice of an attorney should be sought.

11.2   US LEGAL SYSTEM

As depicted in Figure 11.1, the US legal system is divided into two distinct branches: *Statutory Law* and *Common (or Civil) Law*. Statutory law concerns itself with the rules of behavior—statutes—enacted by a legislative body. Common law is divided further into two branches—*Contract Law* and *Tort Law*. Not surprisingly, contract law is based on contracts. Tort law is based on *precedents*, or prior rulings.

Criminal and common law proceedings can arise from the same conduct. One of the more infamous examples involves the famed football player O.J. Simpson, who was found innocent of murdering his wife under an applicable criminal law statute but was held liable for her death in a civil proceeding.

Generally, American legal principles related to contracts and torts are derived from the English Common Law in what can be termed as the "Anglo-American Common Law" or "Common Law system." In contrast, the "Civil Law system" is typically used to signify a legal system based on the Roman Law system. Civil law-based legal systems are found in Continental Europe and in Louisiana in this country. Creating the potential for confusion is the fact that the American legal system is divided into two basic components, criminal proceedings and civil matters. The latter can include both civil actions based on contract obligations and tort (typically negligence) actions. Actions on warranties have elements of both contracts and torts.

Statutory law reflects policies established by a legislative body on the national, state, or local level. Legislative bodies enact statutes making certain conduct criminal. However, statutes are routinely enacted that affect contract obligations and potential tort liability. For example, a state statute may make a pay-if-paid clause (e.g., a prime designer must pay a subconsultant

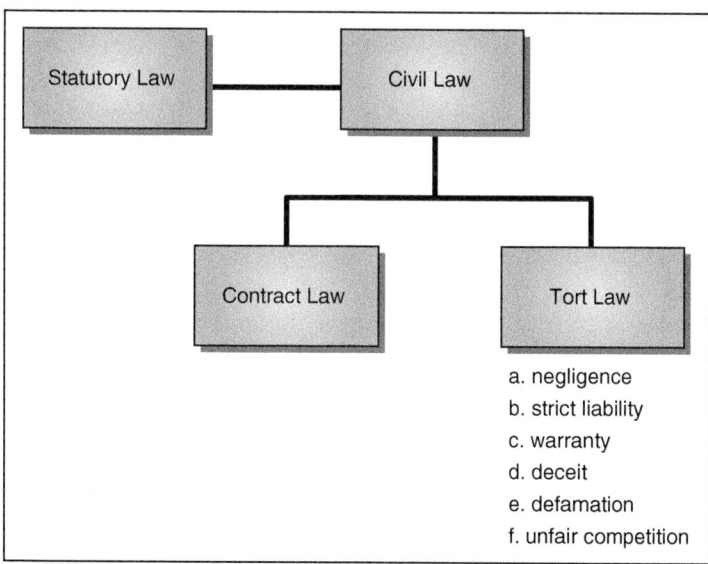

FIGURE 11.1   US legal system.

within a specified period of time after having received payment from the owner) unenforceable on contracts performed in that state. Whether the action is in contract or in tort, court precedents are relied upon to interpret contract language; for example, to determine whether certain conduct (action or inaction) creates or exposes a party to tort liability for damages in a civil (not criminal) action. Even in the area of criminal law, courts rely upon legal precedent in determining the intent of the language in a particular statute.

Contracts are the primary means to define the parties' respective obligations or duties and rights in a commercial transaction. One basic principle of the Anglo-American Common Law system of contracts is the concept of "freedom of contract." That is, the parties to a contract are basically free to agree to the terms of a contract unless those terms violate a public policy, usually set forth in a statute. For example, in some states the parties are free to agree to establish by contract a specific statute of limitations on actions for beach of that contract, for instance, a one-year period to sue on a payment bond. However, many legislatures will limit the parties' freedom to contract away or limit certain rights.

11.3   STATUTORY LAW

Legislative bodies such as the US Congress, state legislatures, county boards of supervisors, and city councils enact statutes or ordinances affecting civil or commercial matters as well as criminal conduct. An example of a statute affecting a potential commercial arrangement is a licensing statute controlling the right to practice engineering in a particular state. That type of law may include provisions affecting civil obligations, for example, the validity of a contract to provide engineering services *and* may also include a criminal sanction, for example, a fine, for failing to comply with the licensure requirements. Another example may be controlling the use of the word "engineer, or engineering" in a company name. For example, a company named "Ace Solar Design and Engineering Company" must have a "professional engineer" on staff if/when engineering services are performed or the company must remove the word "Engineering" from their company name.

While statutory law is sometimes viewed as including only criminal law, that is a too narrow definition. Statutes do define criminal conduct as serious crimes, frequently involving violence; these are called *felonies*, commonly punished through incarceration. Less serious crimes are called *misdemeanors*, usually punished by a relatively small fine, such as parking tickets. Alleged infractions of criminal laws or ordinances are prosecuted by *the people*, for instance federal, state, or county attorneys general. Judgment is rendered by a *trier of fact*—a judge and/or jury—who also determines the penalty.

11.4   COMMON LAW

Though civil engineers can be prosecuted under a criminal statute for criminal negligence, among other crimes, most civil engineering danger zones exist in contract or tort actions based upon Anglo-American Common Law principles, as modified by particular federal, state, or local statutes/ordinances. Actions in contract or tort are brought by companies or individuals, rather than the state. Generally, these actions are termed civil actions, as distinguished from criminal actions. Even though labeled as a civil action, it is based upon the common law rather than the civil law system found on the European continent and in Louisiana in the United States. A civil action is an adversarial system in which an aggrieved party—the *plaintiff*—takes action against the party alleged to be at fault—the *defendant*. One key feature of a civil action in the American legal system is the concept of *discovery*, which allows each party to claim the right to examine the other party's evidence and witnesses prior to court proceedings.

As noted above, the common law has two main divisions: *Contract Law* and *Tort Law*. Contract law's main legal concerns involve whether a contract has been broken, or *breached*. The meaning of certain words and phrases establishes the enforceability of various terms and conditions. If contracts could not be enforced, commerce would come to a standstill.

Tort Law concerns itself with *torts*, civil wrongs for which a court of law will grant a remedy. *Civil wrongs* presuppose *norms of behavior* by which society expects people to abide. The *triers of fact* determine:

- What these norms of behavior are
- Whether or not they have been violated
- What the remedy or damages should be

11.4.1   Tort Law

Tort law also presupposes that individuals have a *duty* to uphold these norms of behavior. Therefore, tort also encompasses the idea of *breach of duty*, for which the law will grant a remedy.

In establishing norms of behavior and related issues, courts look at prior rulings, called *precedents*. Some national design professional associations have legal funds used to appeal unfair or unreasonable rulings to prevent them from becoming standards. Left unchallenged, such decisions could become the equivalent of laws, not only in the jurisdiction where the issue is resolved, but also where the precedent is reviewed. New tort law emanating from California can be adopted by civil courts throughout the nation, including federal courts.

This situation has a significant influence on the way engineering and architecture are practiced. For many years engineers and architects (A/Es) were protected by the concept that they owed a duty of care only to their clients; a third party could not claim negligence. This concept has substantially eroded. Tort law holds that design professionals owe a duty of care to anyone who could be damaged physically or monetarily.

According to John Bachner, author of *Practice Management for Design Professionals: A Practical Guide to Avoiding Liability and Enhancing Profitability*, issues most likely to affect civil engineers in tort law include:

- Negligence
- Strict liability

- Warranty
- Deceit
- Defamation
- Unfair competition

11.4.2   Negligence

Importantly, an error or omission is not necessarily negligence. In order for a plaintiff to win a negligence claim, five conditions must be proven as fact:

1. Defendant was required to abide by standard practice.
2. Defendant owed a duty of care to the plaintiff(s).
3. Defendants breached that duty of care.
4. There was a causal connection between the breach and alleged injury.
5. The injury was real.

Condition 1: Standard of Practice
In claims filed against design professionals, plaintiffs do not usually need to establish that a standard of practice applies. Historically, professionals have been bound to abide by standards of practice. What is important is to establish what the standard of practice was at the time it was allegedly breached.

Standard of practice definition: "The ordinary skill and competence exercised by members of a profession in good standing in the community at the time of the event creating the cause of action."

- Ordinary skill and competence: "ordinary"—professionals are not required to provide the highest level of skill
- Members of a profession in good standing: peers providing same or similar services
- In the community: if very specialized, could be national or global; otherwise local, but it must be established
- At the time of the event creating the cause of action: needs to be evaluated when alleged negligent act was committed

Condition 2: Duty Owed
Absence of a contractual relationship, *privity*, is not a defense. However, the duty must be foreseeable.

Condition 3: Breach of the Standard Practice

Plaintiffs and defendants hire their own expert witnesses, who are supposed to serve the trier of fact (judge and/or jury). Although their role is not one of advocacy, some expert witnesses become "hired guns." Opposing council can defeat or impeach their testimony by pointing out errors, contradictions, weaknesses; however, this can be difficult because most attorneys lack an intimate understanding of civil engineering. Expert witnesses should be required to perform the necessary research, but this does not necessarily happen. The problem can be overcome by working with one's own experts, but there may not be sufficient time. Also, frequently, a jury is more difficult to educate than a judge, so plaintiffs often request jury trials.

Condition 4: Causal Connection

Showing that a civil engineer may have breached the standard of practice is not sufficient to win a suit. The plaintiff must show that the negligent act was the *actual cause* or *proximate cause*. The *actual cause* can be established by the *but for* or *main factor*: but for the defendant's negligence, the damage would not have occurred. The negligent act was the main factor in events resulting in damage. Establishing *proximate cause* requires consideration of three factors: the number of possible causes, the presence of intervening causes, and the foreseeability of the consequences. The plaintiff would need to show that the defendant's actions were closest to the sole cause of the problem.

Condition 5: Real Injury

Finally, plaintiffs must demonstrate the value of the losses in order to establish compensatory damages, in the form of actual losses and/or pain and suffering. If there have been no actual damages, under the principle of *unjust enrichment* no award can be made.

11.4.3   Strict Liability

Strict liability most frequently applies to product manufacturers. Negligence does not need to be proved. The plaintiff simply must show that:

- The product had a defect
- The defect existed when the product was acquired by the purchaser or user
- The defect caused or contributed to an injury
- Under, normal use, the product failed

In some courts, a house can be considered a product; but generally, most constructed facilities are not. A problem can arise when plans, specifications, and reports are viewed as products. Clearly, very few, if any, plans and specifications are flawless. To limit exposure to strict liability laws, in contracts and correspondence between the civil engineer and his or her clients, using the term *instruments of service* is preferable to *plans and specifications*.

11.4.4   Warranty

A warranty promises that things are exactly as they are represented. When civil engineers warrant their work, they are opening the door to the doctrine of strict liability. There are two types of warranties: express and implied. Express warranty exists when language is present such as "I guarantee or warrant that . . ."; when the accuracy of a test is attested to; or when a civil engineer assumes complete responsibility for the accuracy of his or her statements, for instance, makes unequivocal statements. Implied warranty is a problem in jurisdictions where plans, specifications, and reports are treated as products.

11.4.5   Deceit

A typical example of deceit occurs when a salesperson withholds or misrepresents information while a customer is making a purchase decision. The salesperson makes false statements with the intent of altering the customer's position. However, civil engineers may be subjected to charges of deceit when negligence cannot be proved.

11.4.6   Defamation

Civil engineers can be exposed to charges of defamation when they publish derogatory written statements about others that may expose plaintiffs to public hatred and ridicule. Again, the plaintiff must suffer injury and damages. The statement also must be made before witnesses who understand that the statement is disparaging. The spoken version of defamation is *slander*.

11.4.7   Unfair Competition

Charges of unfair competition arise when a plaintiff believes that a civil engineer's actions have interfered with his or her commercial activities. There are several versions of unfair competition:

- Commercial disparagement—a published statement that is false, injurious, made with malice, interferes with the plaintiff's commercial relationships, and results in special damages (e.g., cost of redesign)
- Interference (malicious) with contractual relations—an intent to interfere and actual interference with contractual relations
- Interference with prospective economic advantage—same as above, except that no contract exists

Unfair competition also may be mentioned in some professional organizations' codes of ethics. (For more information, see Chapter 3, Ethics.)

Joint and several liability: Defendants can be held liable both collectively and individually for all damages, regardless of their involvement or degree of fault.

11.4.8   Statutes of Limitation and Repose

Several other concepts are very important when considering the law's influence on civil engineering practice. An effective argument civil engineers can use to fend off claims is that the *statute of limitation* or *statute repose* has expired, that is, that the claim is time-barred. Under a statute of limitation, the clock starts running once a defect has been discovered. If the state's statute of limitation is four years, the aggrieved party has four years to file a suit. Frequently, this can be many years after the design has been completed. A statute of repose begins after construction is substantially complete. Thus, if a defect is found five years after substantial completion and the statute of repose is four years, a claim for construction defects would be banned.

Under some circumstances, designers could have lifelong liability. Professional organizations, such as the AIA and ASCE, monitor developments in various courts around the country and generally mount appeals to decisions that might become dangerous precedents. Conversely, many courts wish to protect statutes of repose to ensure protection to consumers of professional engineering and construction services.

A statute of limitation begins once a defect has been discovered.
A statute of repose begins after construction is substantially complete.

11.5   CONTRACT LAW

A contract is a legally binding agreement that sets forth each party's responsibility to each other. No project should be pursued without a written agreement. To do otherwise would leave the assignment of responsibilities to assumption and key provisions to the vagaries of memory. The contract is a key aspect of client-design professional communication. Violations of responsibility and/or obligation can result in legal action.

There are many terms used to categorize contracts:

- *Bilateral/unilateral*—involving two parties/involving one party
- *Enforceable/unenforceable*—containing all necessary elements/not containing all necessary elements (e.g., statute of frauds requiring certain contracts to be in writing to be enforceable or statute of limitations has passed)

- *Void*—missing one or more elements, perhaps due to an oversight
- *Voidable*—giving a party the right to call the contract void
- *Express/implied*—agreeing explicitly/relying on parties' actions toward one another
- *Written/verbal*—text or orally based

For a Contract to be Binding, Six Elements Must Be Present

1. Agreement—acceptance of an offer
2. Consideration—agreed upon or perceived value
3. Legality—adherence to public policy (enforceability)
4. Authentication—signature or corporate seal (attestation)
5. Capacity—signatories sane and have authority to represent business entity
6. Legal purpose—subject of contract must be legal

11.5.1   Contract Formation

Several issues need to be considered when forming a contract. These include assumption of liability, professional liability insurance, disparate bargaining power, and indemnifications.

11.5.1.1   Assumption of Liability

Liability in tort law revolves around negligence; liability in contract law centers on whether a contract provision has been breached. If a civil engineer *agrees in a contract to act in a nonnegligent manner*, a negligent act would create both tort and contractual liability. A good approach is to avoid provisions that obligate more assumption of responsibility than common law requires. This would include provisions that require civil engineers to perform at the "highest professional level." The standard of practice requires only "ordinary skill and competence."

11.5.1.2   Professional Liability Insurance

Assumption of liability has major implications for professional liability insurance coverage. Most policies exclude coverage of liabilities assumed in the contract, a fact that should be discussed with clients. If well-informed by the civil engineer, clients may be persuaded to drop such problematic clauses, pay the added cost for such coverage, or may become interested in including additional services (e.g., field observation).

11.5.1.3 Disparate Bargaining Power

Disparate bargaining power occurs when one party has an unfair advantage over another during contract negotiation. A *contract adhesion (to be "stuck")* can result when one party to the contract appears to have little power in relationship to the other. This can happen, for example, when a very large client offers the only opportunities for commissions in a small town. A civil engineer, who is forced to accept an onerous contract clause in order to secure a commission, can follow-up with a letter to the client stating that the provision was accepted because the civil engineer needed the work and the principle of disparate bargaining power was at work.

11.5.1.4 Indemnifications

A good definition of *indemnification* is "to secure against loss or damage." Contract clauses can include indemnifications to protect either party. For example, the client could indemnify the civil engineer for any certifications required. Construction contracts between the contractor and the owner often indemnify owners and design consultants to protect them from being sued by workers and/or visitors injured on the jobsite.

The best measures to take when entering into a contract with a client are to:

- Perform professionally
- Create realistic expectations
- Make clients aware of risks
- Minimize one's own (engineer's) risk
- Obtain liability insurance and be sure of coverage
- Consult a knowledgeable attorney

11.5.2 Contract Wording

Words are never more important than when used in contracts, and common words must be selected and used with care. Words such as "all," "every," "none," "whose," "who," "he/she," "his/her" can be significant. Definitions are needed because interpretations of various words and phrases may differ dramatically. For example, "hazardous materials" may mean different things in federal, state, and/or private work.

Attorneys play an important role in reviewing contracts for appropriate language and conformance with applicable clauses and laws. However, civil engineers should not abdicate their responsibility to attorneys. A contract is a communication tool, and reaching a mutual understanding of responsibilities and restrictions with the client can set the tone for all the work that follows.

11.5.3   Typical Contract Formats

Contracts assume various forms; but at a minimum, regardless of the form, all contracts for civil engineering services should include:

1. Scope of services
2. General conditions
3. Performance schedule
4. Fee proposal

Some of the more commonly used contract formats include: conventional proposals, negotiated terms and conditions, multiple contracts, special contracts for major projects, client-developed contracts, purchase orders, and model contracts.

11.5.3.1   Conventional Proposals

As discussed in Chapter 4, Professional Engagement, civil engineers often acquire work by preparing a proposal in response to a Request for Proposal (RFP) or Request for Qualifications (RFQ). In qualifications-based selection (QBS), the work scope included in the proposal becomes part of the contract between the client and civil engineer. If the client is using a fee-based selection method, the client's proposal should have (but not always does have) a well-defined statement (scope) of work (SOW). In either case, the SOW is an important component of conventional proposals. However, if the SOW is not well conceived, the civil engineer can be held responsible for an impossible-to-deliver, unilateral workscope. In QBS, the civil engineer's proposal to the client can clarify the SOW; in fee-based proposals, the civil engineer's cover letter can identify areas needing modification.

These conventional proposals also typically include a section under the heading of *General Conditions*. General conditions are nontechnical understandings between the two parties to the contract. They include the business context and mutual responsibilities, such as when payments are due and any limitations of liability. Most design firms have developed standard general conditions.

Whether the selection method being used is two-envelope QBS or fee-based, conventional proposals will include some mention of fee, generally including staff rates/time, computer use, testing, and a schedule of fees and costs for additional services. A schedule for the services provided also will be incorporated into the contract. The schedule should depict milestones, such as completion dates and dates of major deliverables, noted elsewhere in the contract. It's important that the wording and meaning in the SOW, schedule, and deliverables agree and coincide with the contract. It's also advisable for the CE to include any pertinent assumptions and details like the number of "review meetings", availability and payment of utilities, and security access restrictions, etc.

11.5.3.2   Negotiated Terms and Conditions

Negotiation of contract terms and conditions is not strictly related to monetary issues. Often the discussion centers on scope of services needed. Sometimes the client may not agree to everything in the general conditions, such as indemnification provisions. Clients and the civil engineer may want to shift risk; and contemplating the assignment of risk is best done in the earliest stages in order to minimize its effect. Sometimes general conditions are rewritten, or an *addendum* (*Special Conditions*) may be attached to the contract.

11.5.3.3   Multiple Contracts

Sometimes owners contract separately and simultaneously with the prime (designer or contractor) and the subconsultants or subcontractors. This approach gives the owner more management responsibility and more control. Subconsultants have greater access and may receive more prompt payment. Additionally, the prime designer may reduce "vicarious liability," in other words, exposure to liability stemming from contractual relationships with other design professionals.

11.5.3.4   Special Contracts for Major Projects

For large projects, standard general conditions seldom suffice. Custom contracts with standard contract clauses frequently are developed, requiring attorney involvement.

11.5.3.5   Client-Developed Contracts

Large clients can be powerful, and they sometimes attempt to shift risk (liability) to their designers—engineers and architects. Although the attitudes of large clients may be difficult to change initially, disputes may be even more difficult to win later. Some client-developed contracts attempt to shift more liability to the designer than is required by law or by customs. Typically, these additional risks are not accepted by insurance companies.

In such cases, the civil engineers may become the client's source of "free" insurance. To offset this increased risk, civil engineers can:

- Read RFPs and/or RFQs carefully—responses can become part of the contract
- Have their attorneys review the client-developed contract very carefully
- Charge a fee premium over the usual amount charged for similar services
- List outstanding issues in a cover letter accompanying the proposal and negotiate later
- Attempt to make the client accept risk

Civil engineers need to proceed cautiously with clients who accept risk too willingly. Clients must be able to honor the changes they have agreed to, that is, they need to have sufficient monetary resources to cover increased risk.

11.5.3.6   Purchase Orders

Occasionally, clients use purchase orders to hire civil engineers. This typically happens when a public client, who may have a critical need, does not have time to conduct a formal selection process. Public contract codes limit the amount public agencies can commit through the use of purchase orders, so projects using this type of contract tend to be small. Purchase orders are designed primarily to purchase materials and are not really appropriate for professional engineering services. Civil engineers need to exercise good judgment when signing such agreements.

11.5.3.7   Model (Standard Form) Contracts

Model contracts are developed by professional associations such as:

- American Institute of Architects (AIA)
- ConsensusDOCS LLC (Associated General Contractors (AGC) and 20 other organizations)
- Design Build Institute of America (DBIA)
- The Engineers Joint Contract Documents Committee® (EJCDC®) is a joint venture of three major organizations of professional engineers, the American Council of Engineering Companies (ACEC), the National Society of Professional Engineers (NSPE), and the American Society of Civil Engineers—Construction Institute (ASCE-CI).

The documents produced by these organizations provide a vital function. They offer an economical way for parties to contract with one another without "lawyering up." Each organization has retained attorneys to develop contracts (including general conditions) on behalf of their membership. Most contract clauses have been tested over time, and the documents are revised periodically to reflect changes in law. (See Table 11.1 for a partial list of readily available model contacts and the *Civil Engineer's Handbook of Professional Practice, 2nd ed.* website for examples.)

Internationally, the International Federation of Consulting Engineers (FIDIC) is a leader in standard form contracts. The FIDIC, headquartered in Lausanne, Switzerland, is a coalition of international, independent consulting engineers. Its forms are widely used in developing countries and are recognized by the World Bank. The Joint Contracts Tribunal (JCT) for the Standard Form of Building Contract publishes documents commonly used in the United Kingdom. The Engineering Advancement Association of Japan (ENAA) publishes contracts, also recognized by the World Bank, used on power plant projects constructed on a design-build basis.

Although these organizations attempt to create contracts that are fair and balanced, it should be noted that there may be some bias in favor of their membership.

TABLE 11.1   Readily available model contracts.

Origin	Contract Number	Contract Name
	A101-2017™	Standard Form of Agreement between Owner and Contractor
	A141-2014™	Standard Form of Agreements between Owner and Design/Builder
	A195-2008™	Standard Form of Agreement Between Owner and Contractor for Integrated Project Delivery
	A201-2017™	General Conditions of the Contract for Construction
	A295-2008™	General Conditions of the Contract for Integrated Project Delivery
	A503-2017/2019™	Guide for Supplementary Conditions
	B101-2017™	Standard Form of Agreement Between Owner and Architect
	B102-2017™	Standard Form of Agreement Between Owner and Architect without a Predefined Scope of Architect's Services
	B103-2017™	Standard Form of Agreement Between Owner and Architect for a Large or Complex Project
American Institute of Architects (AIA)	B104-2017™	Standard Form of Agreement Between Owner and Architect for a Project of Limited Scope
	B108-2009™	Standard Form of Agreement Between Owner and Architect for a Federally Funded or Federally Insured Project
	B195-2008™	Standard Form of Agreement Between Owner and Architect for Integrated Project Delivery
	B201-2017™	Standard Form of Architect's Services: Design and Construction Contract Administration
	B202-2020™	Standard Form of Architect's Services: Programming
	B203-2017™	Standard Form of Architect's Services: Site Evaluation and Planning
	B204-2007™	Standard Form of Architect's Services: Value Analysis, for use where the Owner employs a Value Analysis Consultant
	B205-2017™	Standard Form of Architect's Services: Historic Preservation
	B206-2007™	Standard Form of Architect's Services: Security Evaluation and Planning
	B207-2017™	Standard Form of Architect's Services: On-Site Project Representation
	B209-2007™	Standard Form of Architect's Services: Construction Contract Administration, for use where the Owner has retained another Architect for Design Services
	B210-2017™	Standard Form of Architect's Services: Facility Support

(Continued)

TABLE 11.1   (*Continued*)

Origin	Contract Number	Contract Name
	B211-2007™	Standard Form of Architect's Services: Commissioning
	B214-2012™	Standard Form of Architect's Services: LEED® Certification
	B352-2019™	Duties, Responsibilities and Limitations of Authority of the Architect's Project Representative, recommended as a reference document when an Architect's Project Representative is employed
	B503-2017™	Guide for Amendments to AIA Owner-Architect Agreements
	B727-1988™	Standard Form of Agreement Between Owner and Architect for Special Services
	C101-2018™	Joint Venture Agreement for Professional Services
	C401-2017™	Standard Form of Agreement Between Architect and Consultant
	G202-2013™	Project BIM Protocol
DBIA*	DBIA 520	Owner/Design Builder Preliminary Agreement
	DBIA 525	Owner/Design Builder Lump Sum Agreement
	DBIA 530	Owner/Design Builder Cost Plus Fee with Option for GMP Agreement
	DBIA 535	Owner/Design Builder General Conditions
Engineers Joint Contract Documents Committee® (EJCDC)®	E-500 2020	Standard Form of Agreement Between Owner & Engineer for Professional Services
	E-520 2020	Short Form of Agreement Between Owner & Engineer for Professional Services
	E-530 2015	Standard Form of Agreement Between Owner & Geotechnical Engineer for Professional Services
	E-525 2022	Owner-Engineer Agreement for Study and Report
	E-560 2015	Standard Form of Agreement Between Engineer & Land Surveyor for Professional Services
	E-568 2015	Standard Form of Agreement Between Engineer & Architect for Professional Services (2006)
	E-581 2017	Agreement Between Owner, Design Engineer, and Peer Reviewers for Peer Review of Design
	E-570 2020	Standard Form of Agreement Between Engineer and Subconsultant for Professional Services
	E-582 2015	Agreement Between Owner and Program Manager
	E 590 2017	Joint Venture Agreement for Professional Services
	E-990 2022	Owner Engineer Documents, Full Set

TABLE 11.1 *(Continued)*

Origin	Contract Number	Contract Name
	E-991 2015	Engineer Subconsultant Documents, Full Set
	C-700 2018	Standard General Conditions
	C-800 2018	Supplementary Conditions of the Construction Contract
	C-942 2018	Field Order
	D-110 2016	Guide to Request for Qualifications—Design-Build Project
	D-505 2016	Agreement between Design-Builder and Engineer
	D-520 2016	Agreement between Owner and Design-Builder (Stipulated Price)
	CMA 050 2021	Bidding Procedures and Construction Contract Documents—Construction Manager as Advisor
	CMA 200 2021	Instructions to Bidders for Construction Contract—Construction Manager as Advisor
	P3-508 2018	Public-Private Partnership Agreement
ConsensusDOCS**	ConsensusDOCS 240	Owner and Design Professional Agreement (2017)
	ConsensusDOCS 300	Multi-Party Integrated Project Delivery Agreement (Owner, Designer, Construction Manager, and Contractor all sign the same agreement—LEAN construction approach, also known as alliancing or relational contracting) (2017)
	ConsensusDOCS 301	Building Information Modeling (BIM) Addendum (2017)
	ConsensusDOCS 310	Green Building Addendum (2017)
	ConsensusDOCS 400	Owner/Design-Builder Preliminary Agreement—For use in conjunction with ConsensusDOCS 410 or ConsensusDOCS 415 (2017)
	ConsensusDOCS 410	Owner/Design-Builder Agreement (Cost of Work Plus Fee with GMP) (2017)
	ConsensusDOCS 415	Owner/Design-Builder Agreement (Lump Sum) (2017)
	ConsensusDOCS 420	Design-Builder and Design Professional Agreement (2017)
	ConsensusDOCS 421	Design-Builder's Statement of Qualifications (2017)
	ConsensusDOCS 450	Design-Builder/Subcontractor Agreement (2017)
	ConsensusDOCS 470	Design-Builder Performance Bond (Surety Liable for Design Costs) (2020)
	ConsensusDOCS 471	Design-Builder Performance Bond (Surety Not Liable for Design Costs) (2020)
	ConsensusDOCS 472	Design-Builder Payment Bond (Surety Liable for Design Costs) (2020)

(Continued)

TABLE 11.1   *(Continued)*

Origin	Contract Number	Contract Name
	ConsensusDOCS 473	Design-Builder Payment Bond (Surety Not Liable for Design Costs) (2020)
	ConsensusDOCS 481	Design-Build Certificate of Substantial Completion (2017)
	ConsensusDOCS 482	Design-Build Certificate of Final Completion (2017)
	ConsensusDOCS 491	Design-Builder Payment Application (Cost of Work with GMP) (2017)
	ConsensusDOCS 492	Design-Builder Payment Application (Lump Sum) (2017)
	ConsensusDOCS 495	Design-Build Change Order (Cost Plus with GMP) (2017)
	ConsensusDOCS 496	Design-Build Change Order (Lump Sum) (2017)
	ConsensusDOCS 800	Program Management Agreement and General Conditions Between Owner and Program Manager (2017)
	ConsensusDOCS 810	Agreement Between Owner and Owner's Representative (2017)
	ConsensusDOCS 830	Standard Owner and Construction Manager as Agent Agreement (2017)
	ConsensusDOCS 850	Agreement Between Owner and Trade Contractor (Construction Manager is Owner's Agent) (2017)

* Design Build Institute of America.

** Over 40 national endorsing organizations including: Associated General Contractors (AGC), National Association of State Facilities Administrators (NASFA); The Construction Users Roundtable (CURT); Construction Owners Association of America (COAA); Associated Specialty Contractors, Inc. (ASC); Construction Industry Round Table (CIRT); American Subcontractors Association, Inc. (ASA); Associated Builders and Contractors, Inc. (ABC); Lean Construction Institute (LCI); Finishing Contractors Association (FCA); Mechanical Contractors Association of America (MCAA); National Electrical Contractors Association (NECA); National Insulation Association (NIA); National Roofing Contractors Association (NRCA); Painting and Decorating Contractors of America (PDCA); Plumbing Heating Cooling Contractors Association (PHCC); National Subcontractors Alliance (NSA); Sheet Metal and Air Conditioning Contractors' National Association (SMACNA); Association of the Wall and Ceiling Industry (AWCI); National Association of Electrical Distributors (NAED); National Association of Surety Bond Producers (NASBP); The Surety & Fidelity Association of America (SFAA).

11.5.4   Contract Interpretation

Courts interpret contracts as a whole and attempt to give reasonable meaning to all terms. However, specific negotiated provisions are given more weight than general terms. Words and how they are used in a specific industry are very important (*terms of art*). Courts look at the performance of the individuals involved in the contract in question (*course of performance*), as well as how they have performed in previous contracts (*course of conduct*). Unilateral mistakes and/or unexpressed intentions are not considered part of the contract.

When design professionals are asked to utilize BIM (Building Information Modeling), they would be well-advised to seek answers to the following questions:

BIM Checklist

- Determine if BIM is required or desired by the owner. Is the use of BIM an evaluation factor in the award of the contract?
- How does the owner expect the contractor to use BIM on the specific project? Public agencies and private owners may have very different goals and requirements.
- What software is required for the project? Will the current software be fully interoperable with other BIM software used by other contributors on the project?
- Does the owner require the BIM documents to be produced in a certain format? Will the software be fully compatible?
- Are resources (equipment and personnel) needed to implement BIM for a particular project or owner?
- Consider the legal responsibilities for a party's contribution to the model as well as a party's access to the model.
- What is the standard of care for each party's contribution to or use of the model?
- Consider the consequences (cost, time, and responsibility) if the BIM process flags a design conflict.
- What are the procedures and protocols for designating projections derived from a BIM model?
- What legal representations are made as to the dimensional accuracy of the models?
- Who is responsible for any cost, time, and liability related to any design revisions made during a collaborative BIM design process?
- What are the storage and retrieval requirements for electronic files and data?
- What data security issues and various access levels to the BIM models need to be considered?
- What entity will own the design rights (intellectual property rights) for certain data generated during the BIM design process?
- What insurance and bonding risks as BIM is used to facilitate providing preconstruction services?
- Will BIM be used for the RFI process?
- What will be the subcontract/purchase order terms and conditions?

—Adapted from Federal Government Construction Contracts, Second Edition, Kelleher, Abernathy, Bell, and Reed editors (John Wiley & Sons, 2010)

"Think Twice" Contract Clauses

1. Certification
   - Problem—contract may require CIVIL ENGINEER to certify that certain conditions exist before, during, or after construction; but certify can be interpreted as guarantee or warrant
   - Solution—if clause cannot be eliminated from contract, include definition of "certify" acknowledging that CIVIL ENGINEER cannot certify conditions whose existence cannot be known with certainty

2. Consequential Damages
   - Problem—CIVIL ENGINEER may be held responsible for damages as a consequence of an event over which the CIVIL ENGINEER had no control; damages could be completely out of proportion with the CIVIL ENGINEER'S fee
   - Solution—establish a general limit of liability in the contract

3. Construction Cost Estimates
   - Problem—the client may view a construction cost estimate provided by the CIVIL ENGINEER as a "guaranteed maximum"; an inaccurate estimate could trigger claims. The actual construction cost can vary significantly and is definitely related to the project scope of work/schedule/budget components as well as availability of materials and products, seasonal cost variations, possible union work strikes, and more. In addition, the law of supply and demand is at work. For example, the cost of a 4-lane bridge replacement over a two year period will be very different than the same scope of work for a 90-day period. Other factors affecting project costs include specialized labor, union labor requirements, specialized materials, the need for specialized construction equipment, project location and time of year. In addition, depending upon the location, contract type and clauses and specific project there may be a lot of competition or little competition thereby affecting the construction cost estimate greatly.
   - Solution—hire (or have the client hire) a consultant who specializes in preparing construction cost estimates; in the contract, refer to "opinion of probable construction cost" rather than "cost estimate"

4. Construction Monitoring
   - Problem—no set of plans or specifications depicts the project completely; the contractor must complete the design according to "custom," which can be subject to debate in court; regular site visits by prime designers, geotechnical engineers, and structural engineers can be beneficial in recognizing and solving problems in a timely manner
   - Solution—construction monitoring should be an additional service included in the contract and the CIVIL ENGINEER should accept only the responsibility that is spelled out in contract

5. Curing a Breach
   - Problem—difficulties can arise when the method for curing a breach is not addressed in the contract
   - Solution—identify mutual responsibilities and what a breach does, and does not, imply
6. Discovery of Unanticipated Hazardous Materials
   - Problem—injured employee or other party can file claim
   - Solution—when earth work or existing structures are involved, add contract clause acknowledging effects of changed conditions
7. Excluded Services
   - Problem—seeks to eliminate a claim that client was not made aware that certain services were available
   - Solution—identify excluded services
8. Freedom to Report
   - Problem—some contractors file claims against CIVIL ENGINEERS when their reports are critical of the contractor's work
   - Solution—client indemnifies CIVIL ENGINEER or CIVIL ENGINEER reports confidentially to client (client disseminates report rather than CIVIL ENGINEER)
9. Indemnification ("to secure against loss or damage")
   - Problem—some forms of client-proposed clauses are more onerous (distasteful) than others
     - Broad form—CIVIL ENGINEER agrees to hold harmless and indemnify client from any and all liability, including cost of defense, arising out of performance of work; requires CIVIL ENGINEER to cover client's costs, even when problem has been caused solely by the client
     - Intermediate form—CIVIL ENGINEER agrees to hold client harmless from and against liability arising out of CIVIL ENGINEER'S negligence, whether it be sole or in concert with others; CIVIL ENGINEER may be required to pay 100 percent of damages though has caused only 1 percent
     - Limited form—CIVIL ENGINEER agrees to hold harmless and indemnify client against liability arising out of CIVIL ENGINEER'S negligent performance of work; potentially mixes tort law liability with contract obligations making the CIVIL ENGINEER liable both in tort and contract law; liability insurance may only cover tort liability

(Continued)

- Solution—attempt to eliminate such clauses or add clause regarding dispropor-tional payment for liability; have contract examined by legal council and work with professional liability insurer; have contractor's insurance carrier add owner and owner's agents under contractor's liability insurance

10. Jobsite Safety
    - Problem—claims can arise from clauses making the CIVIL ENGINEER respon-sible for acceptance of stop-work authority
    - Solution—avoid clauses that go beyond the CIVIL ENGINEER'S responsibilities normally required by law because these clauses could void liability insurance coverage; suggest language to be used in the client's contract with the general contractor stating that the contractor agrees to waive liability claims against the owner and owner's agents for injury or loss; refuse engagement if you believe safety matters will not be managed effectively

11. Limitation of Liability
    - Problem—some clients may not see the value of limiting CIVIL ENGINEER'S liability
    - Solution—discuss the issue in terms of risk management and add a risk alloca-tion contract clause; a CIVIL ENGINEER always has a liability limit, which is the amount of money available to satisfy claims; the fee should reflect risk—some projects are more risk prone; also include the dollar amount for aggregate liability

12. Maintenance of Service
    - Problem—when a CIVIL ENGINEER is a subconsultant, the prime contract may be cancelled; someone else may do construction monitoring, which can create problems in interpreting plans and specifications
    - Solution—include a contract provision that enables the CIVIL ENGINEER to carry-on work even if owner-prime designer contract is dissolved

13. Ownership of Instruments of Service
    - Problem—client may want to own the plans, specifications, reports, boring logs, field data and notes, laboratory test data, calculations, and estimates, which can result in unauthorized reuse; if the jurisdiction views these as products, any defects (errors and omissions) might be treated as product defects, which could invoke the doctrine of strict liability rather than negligence and could obviate professional liability insurance
    - Solution—include a contract provision that indemnifies CIVIL ENGINEER against unauthorized reuse and compensates CIVIL ENGINEER for the cost of any defense

14. Record Documents
    - Problem—client may want the CIVIL ENGINEER to provide record documents (as-builts) based on information furnished by others; the accuracy of this infor-mation is difficult to verify and the CIVIL ENGINEER may be held liable for losses arising from errors

- Solution—include a contract provision that makes clear the potential for inaccuracies and eliminate terms like "as-built drawings" or "corrected specifications," which imply "without error"; use terms such as "record specifications" and "record drawings" and add a prominent notice on each page of record plans and specs

15. Right to Reject and/or Stop Work

  - Problem—client may want the CIVIL ENGINEER to reject a contractor's work or to stop work if corrections are not made; the CIVIL ENGINEER'S role should be based more on observing and monitoring
  - Solution—add a contract provision that clearly states the CIVIL ENGINEER'S responsibility; advise client to reject work that does not conform with CIVIL ENGINEER'S recommendations, specifications, and design; if the client insists on CIVIL ENGINEER stop work provision, include a contract clause that provides CIVIL ENGINEER'S full waiver from any claim or liability, as well as indemnification

—Adapted from John Philip Bachner (1991). Practice Management for Design Professionals: A Practical Guide to Avoiding Liability and Enhancing Profitability.

11.5.5   Contracts in Project Delivery

Clients are faced with many choices. They must choose the delivery system, procurement method, and contract format most appropriate for each project. (See Table 11.2.) These choices directly affect the civil engineer's role, the type of services provided by the civil engineer, and the civil engineer's means of compensation.

TABLE 11.2   Clients' acquisition strategy.

Project Delivery System	Procurement Method	Contract Format
• Design-Bid-Build (DBB) • Multiple Prime • Construction Management at Risk • Agency Construction Management • Design-Build (DB) • Design Assist	• Sole Source • Limited Competition—Negotiation • Qualifications-Based Selection • Best Value Selection • Fee-Based Selection	• Lump Sum (Fixed Price) • Cost Plus a Fixed Fee (Cost Plus) • Guaranteed Maximum Price • Target • Unit Price

Source: Adapted from Design Build Institute of America (DBIA) *Fundamentals of Project Delivery* Course Materials 2009b.

11.5.5.1   Project Delivery Systems

Clients have several options when selecting a project delivery system, as shown in Figure 11.2. These include: Design-Bid-Build (DBB); Design-Build (DB); Construction Management at Risk; Agency Construction Management; Design-Assist; and Multiple Prime.

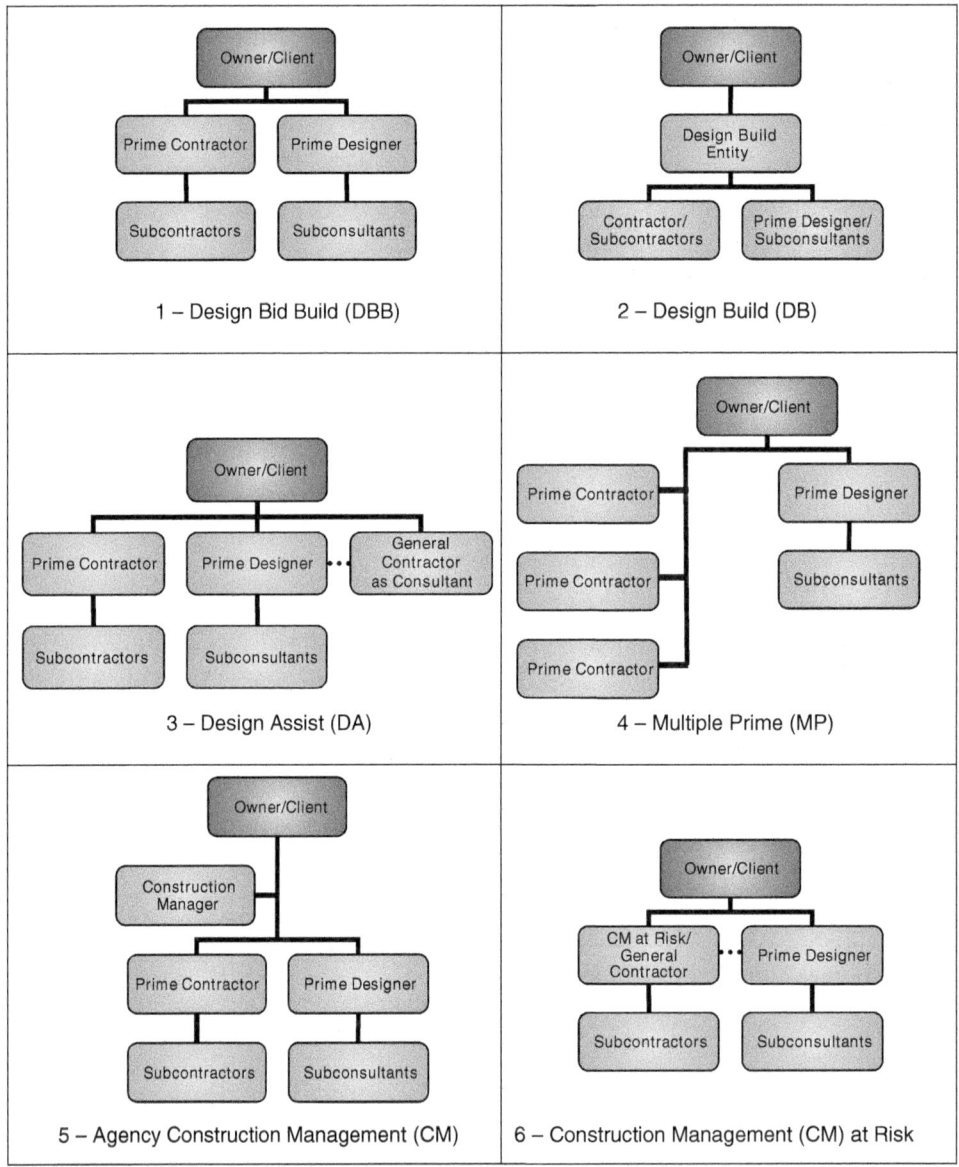

FIGURE 11.2   Project delivery systems.

11.5.5.2   Design-Bid-Build

Design-Bid-Build (DBB) is still the most common method used for designing and constructing projects. As depicted in Figure 5.2—The Flow of Work, DBB is a linear process involving separate stages for design, bid, and construction. In DBB, the prime designer and client enter into a contract for design services. When the design is complete, contractors bid on contract documents prepared by the designers. (See Figure 6.2—Documents Are the Formal Building Blocks of Project Delivery.) After an appropriate period of time, two weeks for small projects and many more for large and/or complex projects, the client enters into a contract with the wining builder. Both Figure 6.1—Contractual Relationships in Design-Bid-Build Project Delivery and Figure 11.3—Project Delivery Systems show the contractual relations involved in DBB, where the prime designer and contractor share no direct, formal contract.

Owners may like DBB because they are able to work directly with the prime designer and are able to exercise control over the entire project, including construction. They also can benefit from having a number of contractors give them prices for the proposed project and being able to select the lowest bid. Also, DBB is allowed in public procurement. The downside of DBB is that the contractor has no input in design, and the process tends not to be collaborative.

11.5.5.3   Design-Build

In Design-Build (DB) project delivery, the client contracts with one entity for both design and construction. This entity can take the form of a limited partnership, joint venture, or fully integrated corporation and can be led by a designer, contractor, or developer. The client is responsible for providing the DB entity with design requirements and performance specifications. Innovation, value engineering, and constructability (the buildability of the design) can be enhanced because these are all in the best interest of the DB entity as well as the client. Schedules can be shortened and costs can be known early.

In DB, the designer and contractor are on the same team and can make unified recommendations to the client. DB takes the owner out of the position between the designer and contractor and consequently results in fewer changes, fewer claims, and less litigation. DB also accommodates complex project phasing and allocates risks to those who can best manage them.

11.5.5.4   Multiple Prime

Sophisticated clients or clients with in-house construction management capability may use a variation of DBB. In multiple prime project delivery, after the design is complete the client hires multiple contractors instead of one general contractor. Thus, a single project may be divided into various contracts such as site development, steel fabrication and erection, mechanical, electrical, and so forth. Or a client may divide a large project into different zones or phases. The advantage of multiple prime contracting is that it eliminates the general

326 Chapter 11 Legal Aspects of Professional Practice

contractor's fees and it is allowed in public procurement. Two primary disadvantages involve the duty to coordinate the multiple prime contractors and the potential for liability for delays caused by one of the prime contractors.

11.5.5.5 Construction Management at Risk

In Construction Management at Risk, the client hires a Prime designer and a construction manager. The construction manager helps to manage the design and then builds the project. The construction manager's involvement enables the client to know construction costs earlier, provides opportunities for value engineering (making changes to maintain or increase value to the client while lowering construction costs), and can reduce schedule time. Also, when budgets are really tight, a guaranteed construction cost can be known earlier.

11.5.5.6 Agency Construction Management

Agency Construction Management is a variation of Construction Management that is marked by a very different role and level of responsibility for the firm performing as a construction manager. In Agency Construction Management, the construction manager (CM) acts as a consultant to the client and manages the design-bid process, and construction; but the CM does not construct the project. Agency CM adds a layer of management and accompanying fees to the project. However, many clients do not have the specialized knowledge required to oversee design and construction and can benefit greatly in terms of managing scope, schedule, and budget from the involvement of a CM. Agency CM is allowed in public procurement.

11.5.5.7 Design-Assist

In Design-Assist, a contractor is hired by the client as a consultant during the design phase of DBB. The contractor can assist the designers with constructability reviews and decisions influencing cost and schedule. It can be argued that the amount of money spent by the client will be exceeded by potential savings, but problems frequently arise when the contractor hired to do the consulting is different from the contractor who is chosen to construct the project. They may have valid but different ways of building the project, resulting in changes during construction. Also, in some jurisdictions the contractor consulting prior to bid is not allowed to bid on the project due to conflict-of-interest laws.

11.6 PROCUREMENT METHOD

Private clients have a range of options open to them for selecting their design consultants and contractors. These include: Sole Source; Limited Competition, for instance, negotiated; Qualifications-Based; Best Value; and Fee-Based selection. Due to public contracting laws, public clients are more limited. Ability to use these procurement methods varies from jurisdiction to jurisdiction, for example, federal agency, state, country, municipality, and so forth.

11.6.1   Sole Source

Sole source selection is as it sounds. The client chooses a designer or builder and enters into a contract. This is a common practice in the private sector, especially when only one firm can provide highly specialized services. Sole source contracts are also awarded to contractors that have previously provided excellent services to the owner and/or where the contractor and owner trust one another. However, it should be noted that most private industry clients take their position in the local community very seriously and these industries try to be fair and attempt to "spread the work" around. Private industries also try to build relationships with multiple suppliers for business reasons. Generally, sole source selection is not permitted in the public sector, unless issues involving national security or emergencies arise.

11.6.2   Limited Competition—Negotiated

Again, negotiated contracts often tend to be the domain of private, rather than public, clients. In this form of procurement, a list of potential service providers is prepared by the client. This list may be based on firms with whom the client has had previous positive experience. In order to gain an adequate competitive basis, a client selection panel interviews each firm. Based on some established selection criteria, the selection panel enters into negotiations with the top one to three firms and makes its selection based on those negotiations.

11.6.3   Qualifications-Based Selection

In qualifications-based selection (QBS), the selection of firms is based on their qualifications and project approach. As discussed in Chapter 4, Professional Engagement, QBS frequently is used to select design firms and is open to many (though not all) public clients, as well as private. QBS is a competitive process involving a two-envelope system. The first envelop contains the engineer's response to the RFP, and the second envelope contains the cost for executing the proposed work. Ideally, the client issues an RFP with a clear definition of project requirements (scope) and well-defined budget and schedule constraints.

Proposing firms respond with the two envelopes and are ranked using the information contained in the first envelop. Those making the short list (from three to ten firms) are interviewed and ranked again. After a firm has been selected based on the merits of their proposal and interview(s), the client opens the second envelop to learn the price. Negotiations with the top-ranked firm ensue. Scope and/or services may be added or deleted until a mutually acceptable price can be reached. If the client and the top-ranked firm cannot reach an agreement, the client may open negotiations with the firm ranked second.

There is another version of QBS where large public entities select several firms based upon their qualifications. Then, when the public entity has a scope of work (SOW) prepared by their in-house engineering department they offer the SOW to these "qualified engineering firms" and ask for cost estimates and/or presentations on the firm's approach to the project.

11.6.4   Best Value Selection

Best Value (BV) selection is a two-phase process that combines technical and cost criteria. In BV selection, the client issues a Request for Qualifications (RFQ), and responding firms are ranked based on their proposals. Firms making the short list are then issued an RFP and perhaps given a stipend or honorarium to cover the costs of phase-two proposal preparation. Their second proposals include both their technical approach and cost, which are scored independently. Consequently, a firm with a higher cost could be selected over a firm with a lower cost, if their technical proposal is superior. See Table 11.3.

11.6.5   Fee-Based Selection

Most engineers do **not** like to compete on the basis of cost alone. Fee-based selection, where the contract award criterion is based on price alone, is not really appropriate for the highly judgmental and creative work required of most engineers. The client's review process is fast and uncomplicated. On the other hand, there are few incentives for innovation or exceptional performance on the part of the engineer.

11.6.6   Contract Format

Clients can choose among several different types of contract formats, including Lump Sum (Fixed Price), Cost Reimbursable—Cost Plus a Fixed Fee (Cost Plus), Guaranteed Maximum Price (GMP), Target—and Unit Price.

TABLE 11.3   Best value selection example (second phase).

Firm	Technical Score (60 pts. max)	Price	Price Score (40 pts. max)	Total Score (100 pts. max)
Acme Civil Engineers	49	$1,700,000	35	84
Benefit Civil Engineers	53	$1,650,000	37	90*
Capital Civil Engineers	47	$1,600,000	39	86

*Winning proposal.

11.6.6.1 Lump Sum (Fixed Price)

Lump sum contracts utilize a single price. For a civil engineer to sign a lump sum contract, the project scope (SOW) should be very clear. In this type of contract, the risk is borne by the engineer; but the engineer also accrues any savings in time/fee. Contracts are easy to administer because the need for cost substantiation is limited to determining progress payments. When all parties are performing ethically, price, schedule, and performance are known at project outset. Consulting engineers prefer this contracting vehicle. When the scope of work is very clear and understood and the consulting engineer has a great deal of experience in providing these services; these contracts can be profitable. An additional bonus for the consulting engineer is that the project invoicing is usually very simple and quickly completed.

11.6.6.2 Cost Reimbursable

Cost reimbursable contracts pass-through the actual cost of the work to the client. These contract types are used typically in negotiated work when the project is technically difficult or unique. Savings are passed on to the client, but price is not necessarily known until the project is complete.

In Cost Plus a Fixed Fee (Cost Plus) contracts, the civil engineer bills the client for the cost of the work performed, plus a fee. The fee is usually a fixed percentage, which is negotiated prior to contract award. Guaranteed Maximum Price (GMP) contracts are a variation Cost Plus; the upper limit of the cost of the project is known initially. As long as project costs remain below the GMP, savings will accrue to the client. There is substantial overhead associated with identifying and reporting costs, and problems can arise regarding wage mark-ups and project staffing levels.

In Target contracts, the client and engineer establish a target price and an incentive fee based on a set of performance criteria at the beginning of the project. The engineer's proposal contains no profit. The budget is reconciled periodically (e.g., quarterly) with the substantiated cost of work. Fee is earned in this period, if the performance and price targets are met. This type of contract format has the potential to lower the client's risk and cost because the engineer has incentives for reducing costs. As in other forms of cost reimbursable contracts, cost reconciliation is a burden.

11.6.6.3 Unit Price

Most Unit Price contracts have to do with units of materials (such as cubic yards of concrete or aggregate), time (such as hours of design services), or completed elements of construction (such as linear feet of roadway paving). They are intended to price variable quantities. Problems can arise when the units are composite (such as linear feet of pipe—where do the excavation, base-course, hangars, rebar, backfill, and so forth fit in?) and, therefore, are not easily defined. Additional confusion is introduced when units are substituted for payment milestones. Activities such as permit acquisition or procurement of special equipment are not measured in quantities. Also, situations may change. A contract may be priced on labor rates for straight time but may require overtime during the course of the project. Assumptions behind unit prices need to be spelled out carefully to the client in the proposal.

In addition to the understanding and, where appropriate, influencing the client's choice in project delivery system, procurement method, and contract format, civil engineers need to understand the underlying risks associated with all projects they undertake.

Dealing with Contract RISK

1. Execute the contract in a professional manner
   - Be quality oriented
2. Thoroughly educate clients
   - Use general conditions to cover costs
3. Identify problems
   - Impose a fee premium or indemnification
4. Identify unavoidable/unacceptable risks
   - For example, modification of existing structure, hazardous materials
5. Use the contract to close loopholes and avoid traps
   - List services the client has declined
6. Establish extralegal conditions
   - Use dispute resolution (mediation/arbitration), reduce statute of limitations/repose, restrict damages
7. Develop clear contract language
   - Make sure that the contract is easily understood by the client and triers of fact
8. If a client is adamant about not taking prudent measures—
   - Walk away

—John Bachner, Practice Management for Design Professionals: A Practical Guide to Avoiding Liability and Enhancing Profitability

11.7   RISK MANAGEMENT

Constructed projects involve the integration of many technical subsystems and components in a context that often involves complex and sensitive economic, social, political, and environmental decisions. An unbroken flow of information from early debates and strategic project definition through design, engineering, procurement, construction, operation, and decommissioning is practically nonexistent. This flow is difficult to achieve because the nature of decisions and types of decision-makers change considerably over project lifecycles.

Some projects experience relatively stable partnerships between firms. This permits long-term interorganizational networks to be established and integrated information systems to be deployed. More commonly there is greater uncertainty. Risk is managed in hierarchical networks of firms. This form of organization is typified by a management and decision-making environment in which problems are highly interdependent while the people, methods, and organizations involved are extraordinarily independent.

"Risk" can be defined as danger, chance, and/or exposure to loss or injury.

11.7.1 Dealing with Risk in General

The most effective "preventive" approach to dealing with risk is to perform professionally. This professional approach to practice concerns not just execution of technical efforts but also a focus on eliminating misunderstandings and unrealistic expectations between civil engineers and their clients and among civil engineers and other professionals. Also very important are:

- A good formal contract
- Good project management
- Knowledge of prevalent risks

Civil engineers can attempt to transfer risk, typically via insurance and/or indemnification. Professional liability insurance is the most common mechanism, though many risk exposures are not covered or are covered only in part. Therefore, professional liability insurance affords only partial transfer of risk.

Indemnification ("to secure against loss or damage") of the civil engineer by the client may be appropriate if risk is created by nature of the project and the civil engineer is powerless to control the risk. Indemnification can be full or partial (risks shared). Partial indemnification can be used to cap the civil engineer's liability, but clients also want to minimize risks. Like insurance, indemnifications are imperfect risk transfer mechanisms.

Transferring *all* risks is impossible. Civil engineers can establish a *loss reserves* account by including risk in their fees; but competition in the marketplace can make this difficult. Also, some problems associated with managing risk are not merely financial. Unsuccessfully managing risk can become an emotional drain and can have a negative effect on professional reputation.

Risk retention is a fact of life. Consequently, civil engineers need to understand and abide by the "rules of the road." They should be thoroughly acquainted with their responsibilities and how to execute them. Being careful with selection of clients and projects is extremely important. Well-managed firms have business plans that identify the types of clients and projects to pursue. Some are far more risk prone than others, for example, hazardous materials remediation. Even before proposing on a project, the civil engineering firm should:

- *Do* background research on the client and the *current* project
- Critically evaluate your firm's capability and experience with the type of work and project
- *Not* rely on past behavior as an indicator of future
- *Not* rely on assumptions

Also necessary before signing a contract, the civil engineering firm should be sure that the contract includes:

- Appropriate clauses
- Explicit general conditions
- A well-defined workscope
- An accurate schedule
- A clear budget
- The CE's list of assumptions

Other actions to take include insisting on an adequate fee or reducing workscope. However, limiting internal plan review, shop drawing review, or construction monitoring is not advisable. Providing additional quality control and applying realistic work assignment and scheduling procedures also help manage risk—less experienced individuals need oversight. Also, civil engineers should not agree to deadlines that cannot be met and may need to hire additional personnel to meet schedule demands.

11.7.2 Establish a Risk Management Program

At least one person in a civil engineering firm should be assigned the task of risk management. If the firm is small, the *Risk Management Program* could be part of quality control, though it should include more than technical quality control. Establishing a risk management program is an attempt to prevent losses that are not really the fault of the firm. Through a risk management program, all staff can be educated on loss prevention. It is important to balance risk and return/reward. (See Figure 11.3.)

Documentation also is another key to managing risk. (See Chapter 13, Communicating as a Professional Engineer.) Proper documentation often can result in claims being dropped.

FIGURE 11.3 Balancing risk and reward.

Notes, memos, e-mails, and meeting minutes can eliminate some of the worst client statements a civil engineer can hear:

"We never said that. . ."
"You said. . ."
"I've never heard [seen] that before. . ."

Finally, civil engineers should react quickly to the symptoms of problems, maintain open lines of communication, and use emotion intelligence—if someone's body language or your "gut" is telling you that there is a problem, there probably is. Deal with issues promptly—unlike most wine, disputes seldom improve with age.

Risk Management

Constructing "Reasonably Believable" Edifices: Lessons from Software, Implications for Construction by Jane E. Millar, PhD and Karen Lee Hansen, PhD
　　Rather than speak of "correct" software in the unqualified sense, one should rather speak of reasonably "believable" software. Rather than say software is error free,

(Continued)

the most one can usually say is that the conditions under which the next errors will be manifested have not yet arisen [1].

Introduction

Software development and construction activities are socially distributed across place and time. As with the many project-based businesses, the principal constituent activities involved in software development and construction work are usually performed within a network, populated by different groups (often within distinct organizations) each responsible for various "stages" of the development process. Relationships between these groups are often strained; historically, conflict, confrontation, and adversarialism have characterized the product development process.

Both software development and construction projects often suffer excessive cost and time overruns and produce poor quality end products. There are several aspects to poor quality products in software and construction, these include products which are unfit for their purpose, inappropriate to end users' needs, unreliable (containing a large number of faults), and those which have poor maintenance and upgrade capabilities. Clearly, construction clients and end users of software alike face serious uncertainties and risks that the product they anticipate will not be delivered.

Risk in Project-Based Industries

Uncertainty becomes risk when the perceived significance of the consequences of an uncertain event becomes critical. Therefore, risk can be defined as any uncertain event which has an impact (usually adverse) on the outcome of the project [2]. Some researchers make a slightly different distinction between risk and uncertainty. In their view, risks are those uncertainties which can be identified and quantified [3]. In other words, risk occurs when the outcome of an event or circumstance can be predicted on the basis of statistical probability. For the purposes of this paper, risk is discussed using the former, broader definition: Risk involves the likelihood that a system will behave in unpredictable ways, the outcome of which may be undesirable.

Risk is endemic to modern society. It is a key feature of economic growth in a period characterized by rapid technological change, the emergence of a pattern of innovation based on technology fusion [4] and the transition to a knowledge-intensive economy [3]. Risk is a precondition for innovation and is intimately connected to learning and change [4]. During innovation, the generative, uncertain, and complex activities involved in generating innovative variety are in conflict with the simultaneous requirements for convergence, control, regulation, and standardization [5]. In order to resolve this conflict, a balance needs to be struck between the demands for risk reduction and the need for variation.

Risk is particularly associated with project-based businesses such as software development and construction. In these industries, product development depends on coordinating the activities of a network of socially distributed (both in terms of their disciplinary base as well as their geographical location) actors, each with their own particular strategic aims and objectives. Concerns over the economic, environmental, and technological hazards connected to such businesses have grown alongside scientific techniques of risk assessment, involving the evaluation of "proneness to failure" [3], and risk management practices (which are often implemented as mere acts of faith).

Networked structures are typical in project-based businesses [8]. In general networks are prejudiced in favor of innovative activities associated with the generation of a variety of new ideas, as distributed groups work independently (often at different locations) on developing aspects of a technological system. This stimulates uncertainty and generates risk in the product/project environment. These take precedence over the generation of ideas which are linked to standardization and control [9].

The risk associated with placing the emphasis on generating a variety of ideas in the product/project environment is compounded by the lack of coordination between members of the network. As a result, in the absence of effective risk management practices, networks can often be considered as a source of risk in themselves. One response to this has been to stimulate cultural change in project-based industries in order to introduce synergy among project teams and end the inherent adversarialism that has previously characterized project-based businesses such as software development [10] and construction [11].

Approaches to Risk in Construction

Within the business of construction, multiple techniques have been adopted to improve risk management in projects. These include PERT (Program Evaluation and Review Technique) used to analyze schedule risk, Monte Carlo simulations, cause and effect diagrams, fault trees, and decision trees, etc. (See Table 11A for a list of these techniques.) Project management experts advocate the use of these methods.

Originally risk management was the domain of insurance companies, where risks are bought and sold. Architectural and engineering firms carry errors and omissions insurance policies. Additionally, many owners require surety bonds of their contractors. A surety bond is a financial instrument issued by a third party (surety company) to a first party (owner) which provides a guarantee that a second party (construction firm) will complete a project according to the terms and conditions of a written contract. In the 1980s the surety bonding business experienced several successive years of heavy losses which has led to the use of innovative approaches such as neural networks as a way of rating new construction bond applicants [16].

(Continued)

However, the approach to risk used by construction firms seldom is based on sophisticated quantitative techniques. In an investigation of strategic decision making in large architectural, engineering, and construction firms, Hansen [17] noted that firms tend to ignore remote possibilities, look at a few outcomes rather than all alternatives, and focus on opportunities rather than dangers. Firms developed several ways of managing risk by using incremental approaches, performing cost justification exercises, limiting expenditures to those which could be funded from existing projects rather than overhead, and passing risk to vendors and subcontractors through the use of performance clauses in contracts.

A study carried out in the UK resulted in a similar finding. In 1992 a survey of expert project management practitioners was undertaken to determine the level of awareness of project risk analysis and management techniques. The sample was drawn from the special interest group concerned with risk of the UK Association of Project Managers (APM). Although the 37 practicing member sample was small, participants were considered to be expert. As indicated by Table 11A, more traditional techniques such as checklists, Monte Carlo simulation, and PERT were favored [18].

TABLE 11A   Project risk analysis.

Technique	Used currently	Used in past but not now	Aware of but not used	Have not heard of
	%	%	%	%
Catastrophe theory	0	0	56	32
Checklists	76	0	8	4
Controlled Interval & Memory Modelling	8	0	48	32
Decision trees	44	0	48	0
Fuzzy set theory	0	0	64	24
Game theory	8	0	48	24
Influence diagrams	28	0	48	12
Monte Carlo simulation	72	4	16	0
Multiple criteria decision making models	24	0	36	28
PERT	64	4	24	0
Sensitivity analysis	60	4	20	8
Utility theory	4	0	48	36

Note: % indicates percentage of positive response from 37 participants.
Source: Adapted from: J.A. Bowers, 1994.

Regardless of the risk analysis technique employed, the focus on remedial action by optimizing within task performance has tended to lead to sub-optimization of the overall project process. This fragmented approach currently is being challenged by sophisticated owners who want more value from their expenditures on constructed products, by project financing vehicles such as the Private Finance Initiative (PFI), and by EC Directives requiring member countries to implement national legislation which demands greater teamwork in the planning and execution of projects (introduced in the UK by the 1994 Construction (Design and Management), or CDM, Regulations) [19].

Therefore, Risk Management, as a distinct activity, is gaining attention in construction. Practitioners and researchers alike advocate a two-phased approach using assessment and action. (2)(20) The figure below depicts the usual com[ponents of the risk management process.

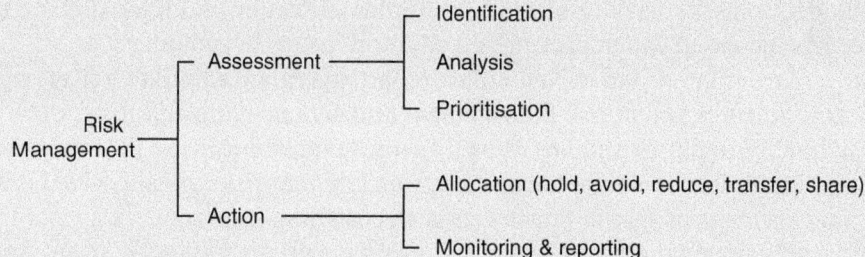

Elements of risk management.

Although techniques for analysis are more robustly developed, the identification of risk is seen as the most critical element of any risk management scheme as risk cannot be managed if it is not identified.

The Changing Nature of Risk Allocation

Recent economic conditions have tended to change the nature of risk allocation in construction. In a recessionary period, the number of business failures generally increases, which leads to the desire to share the risk of financial failure and changing economic conditions [21]. Additionally, due to downsizing, many firms do not have diverse portfolios of projects and operations across which to spread risk. These firms are often interested not in sharing risk but in shifting risk altogether to another party.

A study on perceptions and trends in risk undertaken in the US compares a survey of ENR's (Engineering News and Record) top 100 US contractors with an American Society

(Continued)

of Civil Engineers (ASCE) survey [21]. The results indicate that certain risks are associated consistently with either contractor or owner. Risks allocated to contractors in both surveys include those involving: labor, equipment, and material availability; labor and equipment productivity; quality of work; safety; labor disputes; and competence. Risks allocated to owners in both surveys include: permits and ordinances; site access/right of way; defective design (which normally would be transferred to the architect/engineer); changes in work; and changes in government regulations. Apparently, owners and contractors are willing to accept these risks as they consider themselves able to control these risks or take action to curtail or prevent their occurrence.

The trend, however, appears to be toward sharing contractual/legal risks, which in the earlier survey were borne more by owners [21]. Today many large construction firms either retain lawyers or employ them in their home offices. Contractors may be more comfortable in negotiations and are therefore more willing to share risks associated with change orders, contract delay resolution, and indemnification. Additionally, the use of insurance has increased which provides an addition means of avoiding risk.

Finally, current interest in design build and private finance initiative (PFI) projects also has focused attention on risk management and on achieving objectives in terms of time, cost, quality, and performance as well as future income stream. In these cases the bidding team is under considerable pressure to ensure that risks are allocated correctly.

The management of risk in construction is becoming important as concerns over economic, environmental, and technical hazards have grown alongside scientific techniques of risk analysis. Yet currently available tools for analysis of risk do not seem to have been adopted by the majority of firms. As the next section illustrates, approaches used in software can provide valuable lessons for construction.

Lessons from Software, Implications for Construction

A number of collaborative approaches to software development have capitalized on the awareness that risk can be reduced through effective communication. Among these are the socio-technical systems approach to participative systems design, exemplified by the ETHICS [22] method, and the Scandinavian collective resource approach, see [23]. Common to these many variants is prototyping, whereby continual negotiation among members of the project team iteratively evolves the design, refinement and development of the product under development, for example, Rapid Application Development (RAD) in software.

RAD techniques for software development involve intensive negotiations among heterogeneous and cross-functional teams of specialists. Each specialist in a RAD team is a representative of one of the various functional groups in the product development network. These specialists coalesce on a temporary basis in order to collaboratively and incrementally evolve new software solutions to business problems. Under certain circumstances, this has been shown to lead to software project success through eliminating misunderstandings in communication and enhancing business members' ownership over the software product [24].

In view of these positive results, a revised risk management scheme for use in construction is proposed. (See following figure.) In this enhanced model, risk identification is given a more prominent role and techniques drawn from software, such as RAD, are used to encourage communication and co-operation among members of the project network.

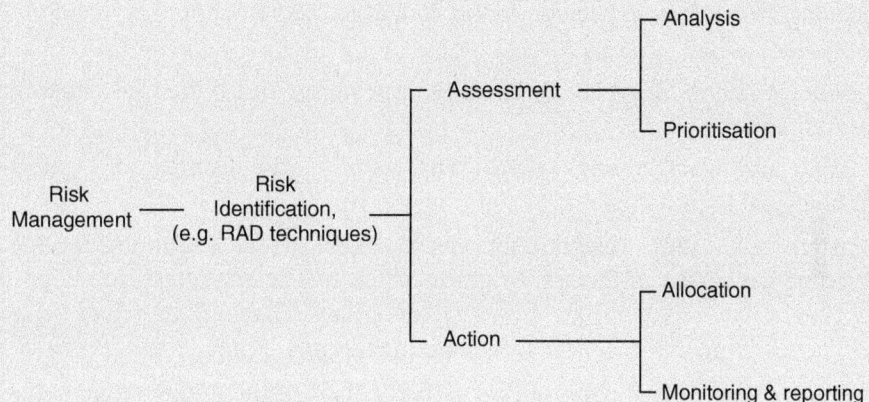

Revised risk management scheme.

Conclusions

Through enhancing communication between the distributed actors involved over the project lifecycle, it is hoped that networked product development processes will become characterized by openness to new ideas and by access to decision making processes. The by-product of this enhanced communication is the early identification of risk associated with the project process.

References

1. A. Macro and J. Buxton. (1987) The Craft of Software Engineering. New York: Addison-Wesley.

2. Crossland, Rose, Jon H. Sims and Chris A. McMahon. (1995) "An Object-Oriented Design Model Incorporating Uncertainty for Early Risk Assessment." 1995 ASME Design Engineering Conference, Boston, USA.

3. R. F. Fellows. (1996) "The Management of Risk," The Chartered Institute of Building, Ascot 65, 1996.

4. F. Kodama. (1991) Emerging Patterns of Innovation. Boston, Mass: Harvard Business School Press.

(Continued)

5. OECD, "Employment and Growth in the Knowledge-based Economy," Paris: OECD, 1996.

6. R. Herbold, "Technologies as Social Experiments. The Construction and Implementation of a High-Tech Waste Disposal Site," in Managing Technology in Society, A. Rip, T. J. Misa, and J. Schot, Eds. London: Pinter, 1995, pp. 185–197.

7. J. Jelsma. (1995) "Learning about Learning in the Development of Biotechnology," in Managing Technology in Society, A. Rip, T. J. Misa, and J. Schot, Eds. London: Pinter, pp. 141–165.

8. M. Hobday. (1996) "Product Complexity, Innovation and Industrial Organisation," Research Policy.

9. D. Foray and M. Gibbons. (1996) "Discovery in the Context of Application," Technological Forecasting and Social Change, vol. 53, pp. 263–277.

10. S. Easterbrook. (1991) "Negotiation and the Role of the Requirements Specification," The University of Sussex, Brighton, Cognitive Science Research Reports 197, July 1991.

11. M. Latham. (1994) "Constructing the Team," HMSO, London, Final Report of the Government/Industry Review of Procurement and Contractual Arrangements in the UK Construction Industry.

12. G. J. Hodgson. (1995) "Design and build—effects of contractor design on highway schemes," Institution of Civil Engineers and Civil Engineering 108, May 1995.

13. Thurston, D.L. (1991) "Design Evaluation of Multiple Attributes under Uncertainty," International Journal of Systems Automation: Research and Applications, vol. 1, pp. 143–159.

14. T. DeMarco. (1982) Software Systems Development. New York: Yourdon Press.

15. I. Sommerville. (1992) Software Engineering. New York: Addison-Wesley.

16. Kangari, Roozbeh and Moataz T. Bekheet. (1995) "Risk assessment of construction bonds underwriting using neural network technique," Integrated Risk Assessment: Current Practice and New Directions, R.E. Melchers and M.G. Stewart, eds. A.A. Balkema, Rotterdam, The Netherlands, pp.139–146.

17. Hansen, Karen Lee. (1993) How Strategies Happen: An Investigation of the Decision to Upgrade Computer Aided Design (CAD) in Architectural, Engineering, and Construction Firms. Doctoral Dissertation, Department of Civil Engineering, Stanford University, Palo Alto, California, USA.

18. Bowers, John A. (1994) "Data for project risk analyses," International Journal of Project Management, vol. 12. no. 1, pp. 9–16.

19. Neale, Brain S. (1994). "Generic health and safety standards—EC Directives and technical policy implications." ISARC Proceedings 1994. Elsevier.

20. Boothroyd, Catherine. (1997) "Managing Risk in Construction," Managing Value and Risk for the Client's Benefit, Members Report 96–21-S for the Construction Productivity Network, CIRIA, London.

21. Kangari, Roozbeh. (1995) "Risk Management Perceptions and Trends of US Construction," Journal of Construction Engineering and Management, American Society of Civil Engineers, Dec. 1995, vol. 121, no. 4, pp. 422–429.

22. E. Mumford. (1995) Effective Systems Design and requirements Analysis: The ETHICS Approach. London: Macmillan.

23. G. Walsham.(1993) Interpreting Information Systems in Organizations. New York: John Wiley and Sons.

24. J. E. Millar. (1996) "Interactive Learning in Situated Software Practice; Factors Mediating the New Production of Knowledge During iCASE Technology Interchange," in SPRU. Brighton: University of Sussex, pp. 248.

—Jane E. Miller, Ph. D. and Karen Lee Hansen, Ph. D.

11.8   INSURANCE AND BONDS

As the frequency of claims increases and balancing revenues and losses becomes more difficult, professional liability insurance has become vital to the practicing civil engineering professional. Professional liability insurance is an essential part of risk management and also may figure in marketing services. Bonds typically are associated with construction; but more design professionals are becoming involved with design-build and integrated project delivery contract structures. Civil engineers should be familiar with the various bonds required by owners. See Chapter 19—Construction Management for more information on bonds.

Liability insurance: A contract under which an insurance company agrees to protect a person or entity against claims arising from real or alleged failure to fulfill an obligation or duty to a third party who is an incidental beneficiary.

Professional liability insurance: Insurance coverage for the insured professional's legal liability for claims arising out of damages sustained by others allegedly as a result of negligent acts, errors, or omissions in the performance of professional services.

Bond: In suretyship, an obligation by which one party (surety or obligator) agrees to guarantee performance by another (principal) of a specified obligation for the benefit of a third person or entity (obligee).

—The Architect's Handbook of Professional Practice, p. 995, p. 999, and p. 985.

11.8.1   Professional Liability Insurance Industry

The principal players in the professional liability insurance industry are:

- *Insureds*—those to whom a policy's coverage is extended
- *Insurance agents*—those who specialize in selling various types of insurance and who obtain commissions based on their sales
- *Insurers*—insurance companies that issue policies and establish what policies do and do not cover, limits and deductibles, and the premium that must be paid
- *Actuaries*—compute the odds that a given risk will materialize and the probable cost of that risk
- *Underwriters*—evaluate each applicant to determine the extent to which various risks may occur and what the premium should be
- *Claims Managers*—advise insureds on what course of action to take when a claim arises
- *Reinsurers*—those who insure insurers, providing both back-up (should a large claim be filed) and stability to the industry

Professional liability insurance is outlined in a policy (contract) between the civil engineer (insured) and insurance company. Most policies are purchased from independent insurance agents (brokers). Protection is provided for the professional in case of negligence. Coverage can be augmented with commercial liability coverage that provides more extended overall protection.

The cost of professional liability insurance is influenced by several factors: risk, demand, and level of coverage, previous claims, among others. Results of the analysis of prevalent risks conducted by insurance company actuaries and underwriters are a prime aspect. In addition, the law of supply and demand is at work. When the supply capacity of insurance is high and demand is low, insurance costs will be relatively low. However, if the demand for insurance increases, insurance costs become more expensive. Level of coverage also affects cost.

11.8.2   Liability Insurance Coverage

Professional liability insurance coverage is designed to create a source of recovery in the event that a civil engineer, or other professional, experiences claims arising out of damages sustained by others allegedly as a result of negligent acts, errors, or omissions in the performance of professional services. Policies do vary in coverage, so the person responsible for purchasing the policy should "read the fine print" and choose an insurance broker with the same care they would exercise in selecting an attorney or accountant. Qualifications, services available, cost, and ability to communicate all matter. Some professional liability insurance companies provide extensive educational and risk management assistance programs; others offer little advice or guidance.

General policy considerations might be:

- What scope of coverage is offered, what endorsements are available to expand the coverage, and what is excluded from the coverage?
- What is the cost of the basic policy and any endorsements?
- Is this coverage part of the firm's overall financial management program?

Specific concerns for professional liability insurance include:

Policy limits: The more protection a policy provides, the more expensive it is. An analysis should be conducted to determine the lowest practical amount of coverage. Another consideration is the amount of coverage required by clients.

Deductible: Usually, raising a policy's limit is relatively less expensive than lowering a deductible. There has to be a balance among the deductible, premium, and level of coverage, considering that a new deductible obligation occurs with each claim.

Cost: The cost of a firm's professional liability insurance is calculated individually by an insurance company's underwriter. The cost of coverage is based on the type of practice (geotechnical, structural, environmental, and so forth), geographical area, project mix, claims history, coverage needs, and resulting risks to the insurer. A firm should ask its insurance agent how its premiums will be calculated—this presents a way to be able to compare policies.

Insurability: Professional liability insurance only covers the professionals for whom it is purchased. Problems arise when owners ask design professionals to indemnify them or require certificates that have the effect of express warrantees or guarantees—none of these is covered by professional liability insurance.

Subconsultants: Prime designers (see Chapter 5, The Engineer's Role in Project Development) routinely retain consultants. Because of this contractual relationship, prime designers are exposed to *vicarious liability* for damages resulting from subconsultants' negligence. Consequently, prime designers need to review the policies of their subconsultants for limits of liability and potential gaps in coverage.

Joint ventures: From a legal point of view, joint ventures are similar to partnerships, even though the joint venture has a more limited purpose. If a professional liability claim is made against a joint venture—or a team or association acting as a joint venture—one or all members can be held liable for any decisions rendered against it. Care needs to be taken regarding the type of professional liability insurance used by the joint venture.

Project professional liability insurance: Project insurance covers the entire project team, even those who practice without insurance. Owners usually pay for project insurance when they desire coverage beyond that normally carried by design firms. When the project scope is greatly increased, this can provide a way to cover consultants who could not otherwise obtain coverage.

Expanded project delivery: Insurance companies have begun to offer coverage for designers involved in design-build or acting as developers. This coverage often is provided through endorsements to existing basic policies. An analysis of this coverage should be conducted to locate and eliminate any potential gaps.

Claims: Claims can be made directly via a demand for money or services based on an allegation of a wrongful act. Claims also can arise from more subjective circumstances, such as the threat of action or a troubling circumstance. Insurance policies should be checked for terms that require timely reporting of claims because not reporting claims on time could jeopardize coverage. A legal term important to know is *barratry*, the fomenting of claims where none exist.

In addition to professional liability insurance, the civil engineering firm may want to carry additional liability insurance, including:

General liability insurance: Civil engineering firms can be exposed to loss from other liability exposures such as slips and falls, libel and slander claims, and property damage to third parties arising from office operations and nonprofessional activities at jobsites. General liability insurance policies usually cover claims involving third-party liability (bodily injury, personal injury, property damage, and so forth). They do not cover professional, automobile, and workers' compensation exposures.

Employment practices liability insurance: In addition to utilizing sound management practices, some firms might choose to purchase employment practices liability insurance to protect against losses arising from employee charges of harassment, discrimination, and wrongful termination.

Insurance may seem deceptively simple; but insurance coverage and cost are influenced by many factors, not the least of which are financial markets. Buying appropriate insurance coverage is a major business decision. For the majority of civil engineering firms, the most devastating professional and business risks stem from litigation based on accusations of negligence in the performance of professional services.

11.8.3   Bonds

Almost all contractors use the services of national surety companies, which provide written bonds guaranteeing the performance of obligations. Like the insurance industry, these companies are subject to public regulation. The most common bonds are bid, payment, and performance bonds, required by the owner. A bid bond guarantees that if a contractor is successful in winning a bid, the contractor will enter into a contract within a specified period of time and furnish any required bonds. A payment bond provides assurance that certain payment obligations associated with a construction project will be satisfied. A performance bond typically is delivered at the time the contract between the owner and contractor is signed. It guarantees that the contractor will perform the work in accordance with the contract documents. The owner usually reserves the right to approve the surety company and the form of the bond. These bond requirements are included in the bidding documents prepared by the prime designer. (See Chapter 6—What Engineers Deliver.)

The class of project (A-1, A, B, and Miscellaneous), the contract amount, and the contract format (lump sum, guaranteed maximum price, unit price, and so forth) combine to determine the cost of bonds. The ability of an entity (contractor, design-build joint venture, partnership, or integrated engineering-procurement-construction firm) to obtain these bonds greatly influences the type and size of projects it can pursue.

Before issuing bonds, surety companies usually undertake thorough investigations. According to Clough, Sears, and Sears, the following are the usual and most important aspects of this type of inquiry:

Essential characteristics of the project under consideration including: size, type, and nature; identity of the owner and the owner's ability to pay; contractor's adequate resources, equipment, expertise, and experience

Total amount of bonded and unbonded uncompleted work in the contractor's current inventory, including work that has yet to be awarded—helps determine if the contractor's working capital, equipment, and organization are becoming overextended

Adequacy of working capital and availability of credit substantiated by financial reports—may prevent the contractor from taking on too much work

Amount of money between the contractor's bid and the next lowest bid, what was "left on the table"—if the spread is more than 5 or 6 percent, there is cause for concern and the surety company may question the soundness of the contractor's bidding practices

Largest contract amount of similar work successfully undertaken and completed to date by the contractor—the surety company is more comfortable if the contractor stays within its realm of expertise; if the contractor wants to enter a new market sector, the surety will advise starting with small projects

Terms of the contract and bonds required, details of how the owner's payments will be made to the contractor, retention (amount of money to be withheld from each of the owner's payments until the project is complete), time allotted for construction, liquidated damages, and required warranties—all affect the contractor's ability to perform the work

Amount of work subcontracted and qualifications of the subcontractors— must possess the necessary organization, financial resources, and experience to carry out the work

After a bonded project is completed, the owner is asked to send a final report to the surety. This report includes a statement regarding the contractor's execution of the contract, changes that were made during the course of the work, and the final total contract amount, which will be used to adjust the amount of the final bond premium.

Professional liability insurance and bonds are both ways of attempting to manage risk. But not even the best risk management plans are guaranteed to yield 100 percent success rates.

11.9   DISPUTE RESOLUTION

Clearly, negotiation is the best way to resolve differences. When that does not work, the parties involved have several courses of action open to them. *Litigation*, the filing of a lawsuit, is the most common form of formal dispute resolution. *Litigation* proceeds in accordance with a complex set of rules. Years may pass between filing a lawsuit, for instance, making a claim, and presenting the case in court. Due to delays and costs, the vast majority of lawsuits are settled before going to court.

Because so much construction litigation involves designers, civil engineers are advised to be familiar with civil litigation procedures and Alternative Dispute Resolution (ADR) practices, such as mediation and arbitration, which are discussed in the following sections.

11.9.1   Civil Litigation

Civil litigation is adversarial and expensive in terms of both time and energy. It can involve considerable emotion because the concepts of battle, winners-losers, and punishment are at play. Civil engineers may find themselves in court for a variety of reasons, such as late or nonpayment by clients or problems arising vicariously from the actions of other design consultants and/or contractors.

There are four stages in civil litigation as illustrated in Figure 11.4: Pleadings, Pretrial, Trial, and Post-trial.

11.9.1.1   Pleadings

The purpose of the pleadings phase is to establish the basis of the claim and to determine if the *plaintiff*, the entity filing the lawsuit, has a case. Figure 11.5 illustrates the typical flow path for pleadings. During this first phase of the lawsuit, the *defendant* must be informed that the plaintiff has initiated action. A representative of the plaintiff (a process server, typically sheriff, marshal, constable, or the like) delivers in person to the defendant:

- Summons and complaint, identifying the defendant(s) and the actions being taken against him/her/them
- Name of the court

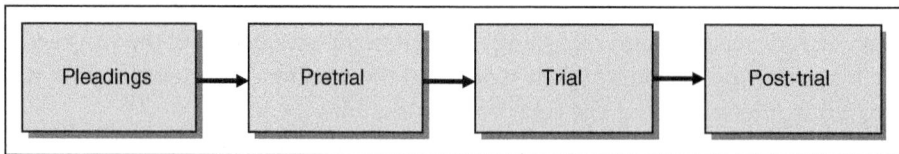

FIGURE 11.4   Stages in civil litigation.

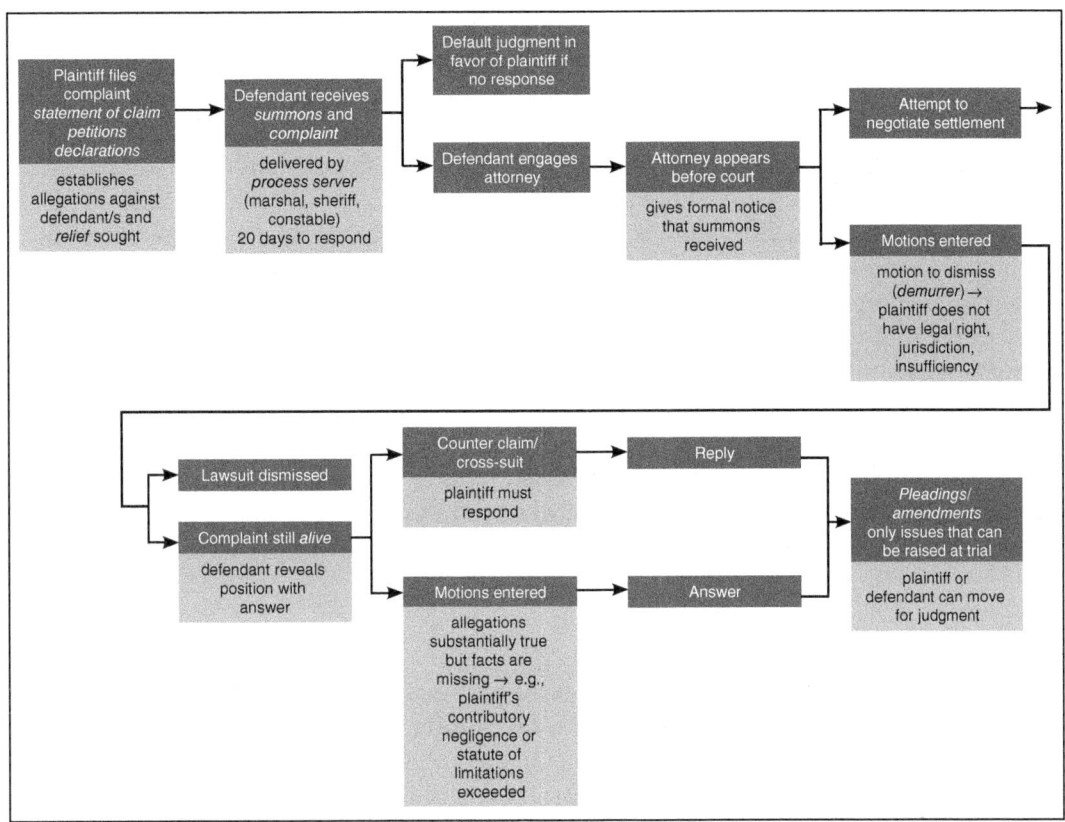

FIGURE 11.5   Pleadings phase of litigation.

- Name(s) of the plaintiff(s)
- Name and address of the plaintiff's legal counsel (attorney)

Defendants have a specific number of days to respond, typically three weeks plus or minus several days. If they do not respond, the court enters a *default judgment* against them. In other words, the judge finds in favor of the plaintiff. The defendant's more prudent course of action is to meet with an attorney, who files a formal notice that the summons has been received. The defendant and attorney can then determine a prudent course of action. Their decision will be based on the legitimacy of the complaint, its potential worth, the plaintiff's ability to sustain an extended legal action financially, the impact of an extended legal action on the defendant and his or her reputation, and the standing of the plaintiff's counsel.

Often both parties attempt to negotiate a settlement and may repeat the process several times if unsuccessful initially. If they are able to reach an agreement, a written document (sometimes called a *stipulation*) is filed with the clerk of the court and the case is recorded as closed. Concurrent with negotiations, the defendant's attorney can file a variety of motions that test the strength of the plaintiff's claims. Among these is the *motion to dismiss*, also known as *demurrer*, which argues that the plaintiff does not have the legal right for a favorable judgment. The defendant's counsel also can file motions arguing that the court does not have jurisdiction over the claim, that the defendant's claim is vague or ambiguous and should be refiled, and/or that portions of the defendant's claim are redundant, immaterial, or scandalous and that they should be removed.

After decisions about the motions have made, the defendant must reveal his or her position with an *answer* to the complaint in the form of a *denial* (*affirmative defense*), a *counterclaim*, or a combination of the two. An affirmative defense alleges the plaintiff's claims essentially are true but that certain explanatory facts have been excluded, such as the plaintiff's contributory negligence or expiration of the statute of limitations. Once filed, a counterclaim becomes a cross-suit to which the plaintiff must answer with a *reply*. The plaintiff must follow the same process that was used initially by the defendant.

Once complete, the complaint, answer, and reply (if the defendant has filed a counter-suit) form the pleadings. Based on preestablished procedures, the pleadings can be amended, but eventually the pleadings identify only those issues that can be raised at trial.

11.9.1.2 Pretrial

Following closure of the pleadings phase of litigation, either the plaintiff or the defendant files a *notice of trial* asking the court to put the suit on its calendar. Figure 11.6 illustrates the typical flow path for the pretrial phase of litigation. Both the plaintiff and defendant can demand a jury trial. If so requested, the court is obligated to provide a jury. The judge may do so even if the parties prefer not to have a jury trial.

Prior to the trial, the judge calls for a mandatory *pretrial hearing* or *conference*. In an effort to save court time and cost, the plaintiff and defendant's attorneys appear before the judge in his or her chambers and may be ordered to:

- Correct defective pleadings
- Eliminate extraneous issues and clarify others
- Agree on the genuineness of various documents
- Limit the number of expert witnesses
- Identify the scope of *discovery*
- Make pretrial admissions, admitting the existence of facts that may help the other side
- Engage in settlement negotiations

Discovery also occurs before the trial. Through the process of discovery, both sides have access to each other's evidence. Discovery saves court time and expense, as well as narrowing the suit's focus and allowing few surprises at trial. The principal methods used in discovery are: *subpoenas duces tecum, interrogatories*, and *depositions*. (See Table 11.4.)

TABLE 11.4 Principal methods used in discovery.

Subpoenas Duces Tecum	Interrogatories	Depositions
Issued by the clerk of the court (or the attorney of record) and delivered by a process server to compel the opposing party to provide certain information	List of fact-based questions sent to the opposing party that must be answered under penalty of perjury	Interrogation of opposing party's witnesses under oath—is transcribed word-for-word and can be used as evidence during the trial

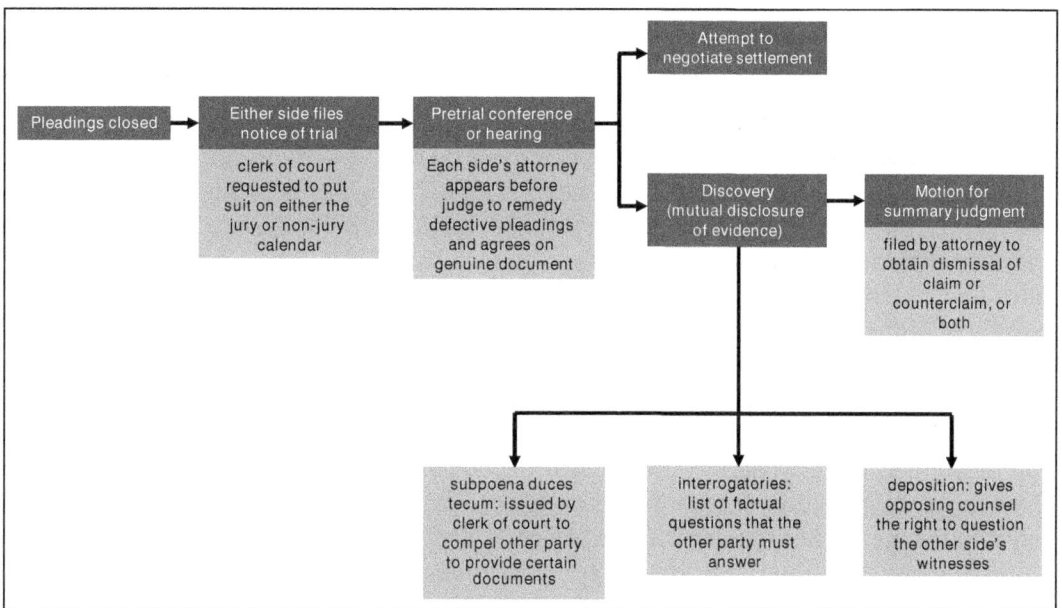

FIGURE 11.6 Pretrial phase of litigation.

Depositions often are taken of expert witnesses, usually with the opposing party's expert witnesses present to assist the attorneys evaluate answers and prepare additional questions.

Before the actual trial phase begins, one of the parties' attorneys may file a motion for *summary judgment*. This motion alleges that the opposing party cannot prove that the facts are true and, therefore, that the case has no merit. If the summary judgment is granted, the claim, counterclaim, and/or both are dismissed. (See Figure 11.6.)

11.9.1.3 Trial

If the suit proceeds, the attorneys answer a *calendar call* and a courtroom eventually is assigned. Figure 11.7 illustrates the typical flow path for the trial phase of litigation. If the trial involves a jury, the judge and attorneys conduct a *voir dire*, where prospective jurors are vetted. Attorneys question prospective jurors. When a juror's answer indicates potential problems with prejudice, financial interest, relationship to an involved party, or the like, the attorneys can challenge the juror stating the reasons. This is called a *challenge for cause*. Attorneys also have a discreet number of *peremptory challenges* and can have a juror dismissed without citing a reason.

After the jury is impaneled and alternate jurors are selected, the trial begins. The plaintiff's attorney begins by summarizing the facts from the plaintiff's perspective. The defendant's attorney follows with the defendant's view of the facts. Then the plaintiff's attorney starts to call witnesses, who are bound by various rules of evidence. One such rule is that lay witnesses may only testify to matters of fact and may not express their personal conclusions. Expert witnesses are allowed to express their opinions and conclusions. Other rules of evidence are included in Table 11.5.

Like jurors, expert witnesses are subjected to *voir dire*. After taking the stand, the expert witness recites his or her credentials. The attorney who has hired the expert witness attempts to have the court recognize the witness as an expert. Though seldom successful, the other party's counsel may challenge the expert's qualifications. Frequently, opposing council and the judge *stipulate* that the expert witness is fit to serve and the witness is not allowed to recite his or her credentials. This can be a disadvantage if these qualifications are more remarkable than those of the other party's expert.

Once a witness—whether factual or expert—takes the stand on behalf of the plaintiff, the plaintiff's attorney begins *direct examination*. When the plaintiff's attorney is finished, the defense attorney initiates *cross examination* of the same witness. On one level cross examination is meant to test the memory and/or knowledge of a witness; on another level, it is used to create doubts about the witness's credibility or to cause the witness to say something that may give the jury a reason to dislike him or her.

The plaintiff's attorney may conduct *redirect examination* of the witness to correct answers given in cross examination that could lead to misunderstandings or to clarify statements that appear to be at odds with evidence developed through discovery. If the plaintiff's attorney redirects, then the defense attorney can pursue *recross examination*. Questions in cross

TABLE 11.5   Various rules of evidence.

Parole Rule	Relevancy Rule	Hearsay Rule	Best Rule
Bars admission of evidence that differs from understandings agreed to formally, as in a contract, if that evidence occurred after the formal agreement(s)	Determines that only evidence outlined in the pleadings may be used; serves to limit *circumstantial evidence*, indirect proof or disproof of a fact in question	Precludes admission of evidence based on what a witness was told by others	Deems that only the best possible form of evidence must be produced at trial, e.g., exhibits, such as documents, should be originals, not copies

examination and recross examination must be limited to the topics introduced in direct examination and redirect examination, respectively. The opposing counsel can file various motions while witnesses are being questioned, including objections to questions and/or answers.

When the plaintiff rests his or her case, the defense can decide to continue with its case or to settle. If the defense elects to continue, the defendant's attorney can file various motions, including: (1) *motion for a directed verdict*—granted when the judge believes that, even if all evidence presented were true, an impartial jury would decide in favor of the defense; and (2) *motion for a voluntary nonsuit*—allows the plaintiff to begin another, potentially stronger, case on the same grounds after the plaintiff has paid court costs.

If these motions are not made or are denied, the defense begins its case. The defense follows the same procedures that were followed by the plaintiff. After all evidence has been presented, the defense can request a directed verdict. If denied, the plaintiff's attorney can offer evidence that rebuts what the defendant's witnesses have said. Then the defendant's attorney can introduce evidence to contradict what was brought up in rebuttal. Once all evidence has been heard, first the plaintiff's attorney and next the defendant's attorney present *summations*. These summations outline the chief issues, the principles of law on which the cases are based.

After both attorneys have completed their statements, the judge *charges the jury*—reviewing the closing statements, outlining the relevant issues of law, going over witness testimony, and giving advice about how the jury should evaluate what they have heard. Usually the jury is advised that the plaintiff (or defendant if there is a counterclaim) has the burden of proof and that proof is based on the *preponderance of evidence*. Both parties' attorneys can file requests to charge, listing special issues to be considered by the judge. They also can object to the charge given by the judge.

After being charged, the jury enters the deliberations phase of the trial. When a decision is known, the jury delivers its verdict. The losing side can request a *judgment notwithstanding the verdict*, essentially asking the judge to overturn the jury's decision. If the *judgment notwithstanding the verdict* is denied, the losing side may make a *motion for a new trial*. When the jury cannot reach agreement, a *hung jury* results and the case must be retried; but if the verdict stands, the judge directs entry of a *final judgment* in favor of the winning side. (See Figure 11.7.)

11.9.1.4 Post-Trial

Either party to the case appeal the court's decision for a variety of reasons—errors were made or perhaps the judgment was too high, for example. Figure 11.8 illustrates the typical flow path for post-trial phase of litigation. The appellant (the party appealing the court's decision) must notify the clerk of the court where the trial was conducted and the appellee that an appeal is being mounted. The appellant prepares a *record of appeal*, citing precedents explaining why the appeal should be granted. The appellant must post a bond to cover the cost of any judgment in the event the appellant's assets are lost during an unsuccessful appeal. The appellee also usually files a record of appeal outlining why the original verdict should stand.

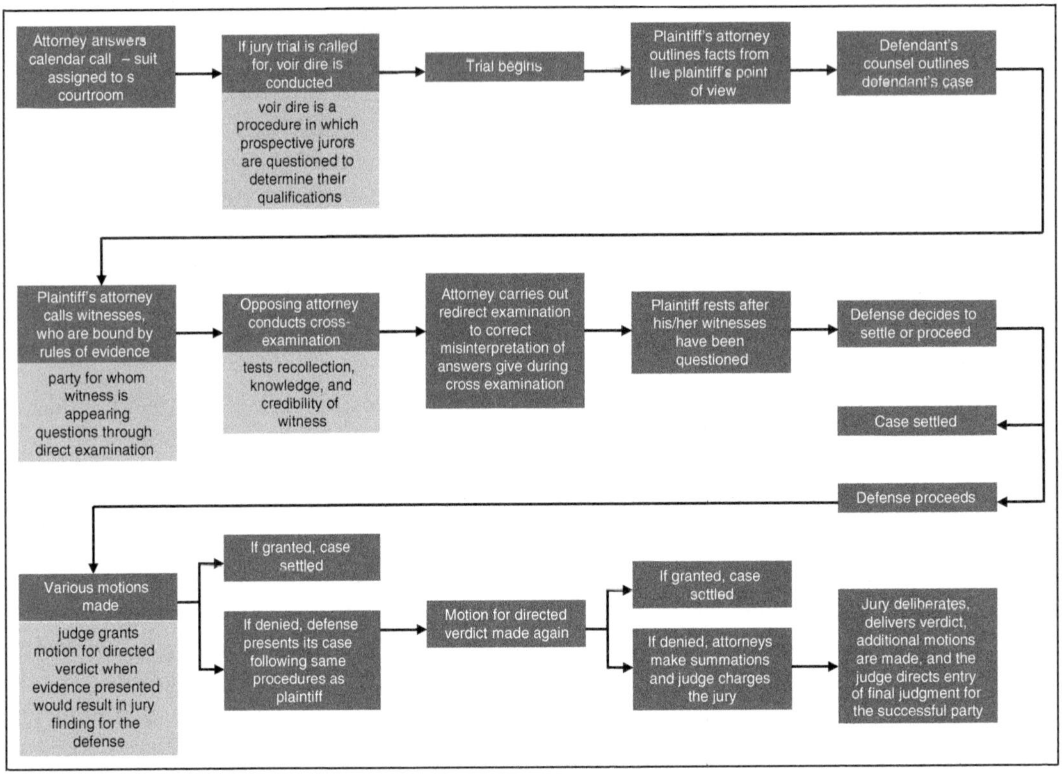

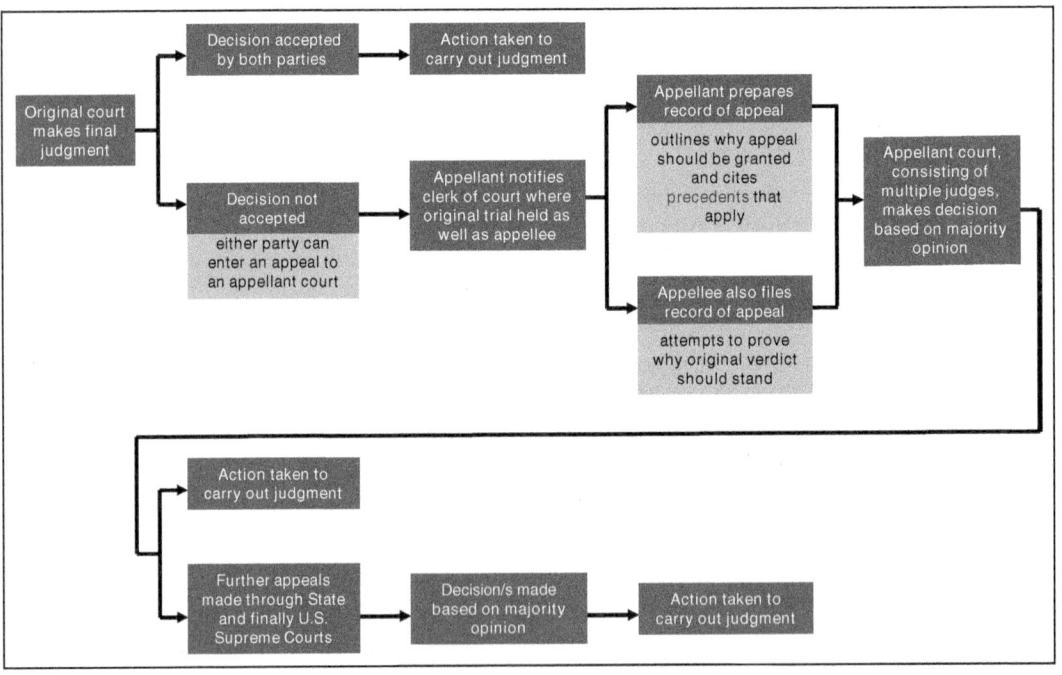

FIGURE 11.7 Trial phase of litigation.

FIGURE 11.8 Post-trial phase of litigation.

Several judges sit on the appeals court. The court's decision is based on the majority opinion. The appeals court may agree with the previous ruling, overturn or modify the judgment, or grant a new trial. If the case presents complex issues of law, the verdict of an appeals court may be appealed to a state supreme or appellate court and, in very rare circumstances, to the US Supreme Court. The aggrieved party may also seek redress in a federal court.

11.10   ALTERNATIVE DISPUTE RESOLUTION

Alternative dispute resolution (ADR) has been used in the US construction industry since the 1870s. ADR is exactly like it sounds—an alternative to the long and draining process of resolving disputes through litigation. ADR consists of several procedures including Mini-Trials, Dispute Review Boards, but mainly variations of Mediation and Arbitration. All of these approaches have advantages and disadvantages, which are explored below.

11.10.1   Mediation

Mediation is the "newest" form of ADR and often precedes other dispute resolution procedures. Mediation is a conciliatory process that can be voluntary or mandatory. Many model contracts require mediation. The Engineers Joint Contract Documents Committee® (EJCDC®), ConsensusDOCS, and the American Institute of Architects (AIA) contracts between owners and design professionals require mediation as a first step in formal dispute resolution. The AIA's most recent contract between an architect and an owner, AIA Document B101™—2007, contains the following contract clause:

§8.2 Mediation

§8.2.1 Any claim, dispute or other matter in question arising out of or related to this Agreement shall be subject to mediation as a condition precedent to binding dispute resolution. . .

In mediation, the disputants meet with a neutral decision-maker, in this case a mediator. The mediator explains the process and reviews the reasons for the parties' participation. The mediator goes over ground rules, decorum, and confidentiality of the process. Then each party states their perception of the matter, the facts, and what is desired. The mediator helps to clarify the facts, identify discrepancies, and assess the relationships among all parties, including the disputants' attorneys. If joint meetings break down, the parties may be put in separate rooms while the mediator shuttles back and forth.

The mediator may be able to find alternatives and creative solutions that the disputants had not yet considered. One of the mediator's objectives is to loosen entrenched positions, which is easier if mediation is pursued sooner rather than later. The mediator can help the parties move toward acceptable adjustments, stress the implications of not arriving at a solution, and put any offers into words. Even if mediation does not result in a formal resolution, important issues are brought to light. Consequently, there may be fewer questions to be resolved in arbitration or at trial. In mediation, the parties, rather than a judge, jury, or arbitrator, control the result.

Joinder: Combining two or more elements into one, as the joinder of parties as coplaintiffs or codependents in litigation or as parties to an arbitration.

11.10.2   Arbitration

Arbitration can be *mandatory* or *voluntary*, and it can be *binding* or *nonbinding*. If arbitration is mandated contractually or court-ordered, it is deemed mandatory. Disputants also can decide to pursue arbitration based on mutual agreement, in which case it is considered voluntary. In binding arbitration, the disputants must adhere to the decision made by the neutral (arbitrator). Possibility for appeal is limited. In nonbinding arbitration the arbitrator provides an opinion and advice, which the disputants decide to follow or not to follow. Based on mutual agreement, nonbinding arbitration can be converted to binding arbitration.

In mandatory binding arbitration the disputants present their cases to one arbitrator or a panel of three arbitrators, depending on the complexity of the problem. Usually, Construction Industry Arbitration Rules of the American Arbitration Association (AAA) are followed. Discovery and the right of appeal are limited; and the arbitrator's decision is binding and enforced by the courts. An AAA-based arbitration procedure begins when one of the involved parties files a letter or AAA form, with an accompanying filing fee, requesting arbitration. Based on information about the claim outlined in this document, the AAA prepares a list of potential arbitrators.

Frequently the disputants and their attorneys convene with a senior arbitrator in a prehearing meeting. At this meeting the parties discuss: undisputed facts, the estimated number of hearings required, schedule, arbitrator compensation, claims and counterclaims, lists of witnesses, and arrangements for site inspection, if necessary. Sometimes this meeting results in a "meeting of the minds," and the parties arrive at a solution to their problem. In a sense, the senior arbitrator has served the purpose of a mediator.

Following the prehearing meeting, disputants select the arbitrator(s) from the AAA list of potential arbitrators. The arbitrator has the power to consider amendments to the claim or counterclaim, control the timing of hearings, and continue without the presence of a party, given appropriate notice. The arbitrator also can subpoena witnesses and documents, when permitted by law.

There are some drawbacks to this process. Discovery is limited, so the potential for surprise at hearings exists, perhaps to the detriment of one of the parties. Also, although arbitrators are familiar with technical issues arising in the construction industry, they may not be well-informed about recent legal precedents guiding the development of contract clauses. Additionally, arbitrators typically write brief decisions and are not required to include their rationale. Courts do not like to interfere with arbitrators' decisions. Seldom is appeal granted, usually only when the arbitrator has been shown to have a conflict of interest or when the rules of arbitration have been breached seriously. See Table 11.6 for a comparison of litigation, arbitration, and mediation.

TABLE 11.6 Litigation, arbitration, and mediation compared.

Factor	Litigation	Arbitration	Mediation
Overall satisfaction	• Generally poor	• Generally high	• Very high
Characterization	• Adversarial	• Adversarial	• Conciliatory
Confidentiality	• Highly public	• Private	• Private
Phase of dispute	• Last resort	• When problems first arise or following mediation	• When problems first arise and when negotiation fails to work
Cost of resolution	• Very expensive	• Administrative fees high, but discovery/trial costs lower than litigation	• Inexpensive
Speed of resolution	• Very slow	• Fast	• Fastest
Sustainability of parties' relationship(s)	• Least possible	• Possible	• Most possible
Discovery	• Extensive	• Limited	• Limited
Possibility of appeal	• Yes	• No	• No
Trier(s) of fact, neutral	• Judge and/or jury	• Arbitrator	• Mediator
Selection of trier(s) of fact, neutral	• Appointed by legal system and drawn from general public	• Informed professional selected by involved parties	• Informed professional selected by involved parties
Atmosphere of surroundings	• Very formal—courtrooms symbolize the power of the law	• More relaxed	• More relaxed
Finality of decision	• None, until appeals exhausted	• Final	• Final, if settlement reached
Fairness of decision	• Varies—courts apply applicable laws; but jurors have little knowledge of issues	• Fair because decision made by experienced arbitrator who is industry professional	• Ultimate in that each party must agree to the resolution

Source: Adapted from Howard Goldburg, Esq., "Dispute Resolution Methods," *The Architect's Handbook of Professional Practice*, pp. 396–404.

11.10.3   Mini-Trial

Mini-trials may be best utilized in complex cases when disputants essentially disagree on the way precedents and the law should be applied to the facts. The term "mini-trial" is somewhat a misnomer because the process is more like solving a business problem than conducting a trial. Because the scope of preparation and presentation is limited, each party must formulate its best case. The parties directly involved in the problem, experts, and attorneys present their case to members of top company management, who have the authority to settle.

The American Arbitration Association has developed a mini-trial process, though the disputants are free to establish their own mutually agreed upon procedures. A neutral party is usually selected to guide the process and to advise the parties about the relative strengths and/ or weaknesses of their cases. Without the well-established rules and limitations imposed by litigation, the parties most knowledgeable about the problem can arrive at a settlement that works for them. The resulting solution is business-oriented and may not have the winner-loser characteristic of litigation.

11.10.4   Dispute Review Board

Dispute Review Boards (DRBs) are an ADR technique used mainly by public agencies. Once a construction contract has been signed and before any disputes arise, a DRB is formed. One member is selected by the owner, subject to contractor approval. Another is selected by the contractor, subject to owner approval. These two members select a third, who serves as board chairperson. Regular DRB meetings are held on-site to review disagreements. The DRB's decisions are not binding and do not preclude later mediation, arbitration, or litigation. However, they do provide an excellent record of real-time events, facts, and opinions of an impartial panel of experts.

11.11   AFFIRMATIVE ACTION, EQUAL OPPORTUNITY, AND DIVERSITY

Various laws affect the practice of civil engineering, including those governing affirmative action (AA), equal opportunity (EO), and diversity in the workplace. Several government organizations are responsible for enforcing federal laws that make discriminating against a job applicant or an employee because of the person's race, color, national origin, sex (including pregnancy), age (40 or older), religion, or disability illegal.

11.11.1   US Anti-Discrimination Laws

The following are specific US anti-discrimination laws:

- *Title VII of the Civil Rights Act of 1964 (Title VII)*
    This law makes discriminating against someone on the bases of race, color, religion, national origin, or sex illegal.

- *The Pregnancy Discrimination Act*

  This law amended Title VII to making discriminating against a woman because of pregnancy, childbirth, or a medical condition related to pregnancy or childbirth illegal.
- *The Equal Pay Act of 1963 (EPA)*

  This law makes paying different wages to men and women if they perform equal work in the same workplace illegal.
- *The Age Discrimination in Employment Act of 1967 (ADEA)*

  This law protects people who are 40 or older from discrimination because of age.
- *Title I of the Americans with Disabilities Act of 1990 (ADA)*

  This law makes discriminating against a qualified person with a disability in the private sector and in state and local governments illegal.
- *Sections 501 and 505 of the Rehabilitation Act of 1973*

  This law makes discriminating against a qualified person with a disability in the federal government illegal. The law also requires that employers reasonably accommodate the known physical or mental limitations of an otherwise qualified individual with a disability who is an applicant or employee, unless doing so would impose an undue hardship on the operation of the employer's business.
- *The Genetic Information Nondiscrimination Act of 2008 (GINA)*

This law makes discriminating against employees or applicants because of genetic information illegal. Genetic information includes information about an individual's genetic tests and the genetic tests of an individual's family members, as well as information about any disease, disorder, or condition of an individual's family members (i.e., an individual's family medical history).

In all of the above laws, retaliating against a person because the person complained about discrimination, filed a charge of discrimination, or participated in an employment discrimination investigation or lawsuit also is illegal.

11.11.2 Enforcement of Anti-Discrimination Laws

The US Equal Employment Opportunity Commission (EEOC) has the authority to investigate charges of discrimination against employers who are covered by the law. Most employers with at least 15 employees are covered by EEOC laws (20 employees in age discrimination cases). Most labor unions and employment agencies are also covered. The laws apply to all types of work situations, including hiring, firing, promotions, harassment, training, wages, and benefits. If discrimination has occurred, EEOC will attempt to settle the charge. If unsuccessful, EEOC has the authority to file a lawsuit to protect the rights of individuals and the interests of the public.

Another organization is the US Department of Labor's Office of Federal Contract Compliance Programs (OFCCP). Since 1965 the OFCCP has ensured that federal contractors comply with the equal employment opportunity and the affirmative action provisions of their contracts. OFCCP administers and enforces Executive Order 11246, as amended, which prohibits federal contractors and federally assisted construction contractors and sub-contractors, who do over $10,000 in government business in one year, from discriminating in employment decisions on the bases of race, color, religion, sex, or national origin. The Executive Order also requires federal contractors to take affirmative action to ensure that equal opportunity is provided in all aspects of their employment.

11.11.3 Affirmative Action Requirements

Each federal contractor with 50 or more employees and $50,000 or more in government contracts is required to develop a written affirmative action program (AAP) for each business entity. The written AAP helps employers identify potential problems in the participation and utilization of women and minorities in their workforce. If there are problems, the employer can specify in its AAP the specific procedures it will follow to provide equal employment opportunity. Companies can include outreach, recruitment, and training as affirmative steps in helping members of protected groups to compete for jobs on a level playing field with other applicants and employees.

For more information on treatment of employees in the workplace, please see Chapter 18—Human Resources.

11.12 SUMMARY

The legal aspects of professional practice are extensive. Knowledge of these important legal aspects of business practice and implementation of defensive and mitigative engineering management practices is imperative for the wise civil engineer. This chapter has covered topics of particular concern to civil engineers: the US legal system; statutory and contract law; contracts used in project delivery; risk management; insurance and bonds; dispute resolution; alternative dispute resolution; and affirmative action, equal opportunity, and diversity. The references listed below provide copious, detailed information and are valuable resources for further study.

The information provided in this chapter does not, and is not intended to, constitute legal advice; instead, all information, content, and materials available in this chapter are for general informational purposes only.

BIBLIOGRAPHY

Bachner, John Philip. (1991). *Practice Management for Design Professionals: A Practical Guide to Avoiding Liability and Enhancing Profitability*. John Wiley & Sons, New York. ISBN 0-471-52205-8.

Beard, Jeffrey L., Loulakis, Michael C., and Wundram, Edward C. (2001). *Design Build: Planning through Development*. McGraw-Hill, Boston, MA. ISBN 0-07-006311-9.

Clough, Richard H., Sears, Glenn A., and Sears, Keoki S. (2005). *Construction Contracting: A Practical Guide to Company Management*, 7th edition. John Wiley & Sons, Inc, Hoboken, NJ. ISBN 0-471-44988-1.

Design Build Institute of America (DBIA). (2009a). *Design-Build Contract & Risk Management.* Course Materials, Design-Build Institute of America, Washington, DC.

Design Build Institute of America (DBIA). (2009b). *Fundamentals of Project Delivery.* Course Materials, Design-Build Institute of America, Washington, DC.

Jackson, Barbara. (2010). *Design-Build Essentials.* Delmar Cengage Learning, Independence, KY. ISBN-10: 1-4283-5303-8/ISBN-13: 978-1-4283-5303-9.

O'Reilly, Michael. (1999). *Civil Engineering Construction Contracts*, 2nd edition. Thomas Telford, London. ISBN: 0 7277 2785 0.

Kelleher, Thomas J., Jr., ed. (2005). *Smith, Currie, & Hancock's Common Sense Construction Law: A Practical Guide for the Construction Professional*, 3rd. edition. John Wiley & Sons, Inc, Hoboken, NJ. ISBN 0-471-66209-7.

The American Institute of Architects. (2003). *The Architect's Guide to Design-Build Services*, Ed. by G. William Quatman II and Dhar Ranjit (Randy). John Wiley & Sons, Inc., Hoboken, NJ. ISBN 0-471-21842-1.

The American Institute of Architects. (2008). *The Architect's Handbook of Professional Practice*, 14th edition. Ed. by J.A. Demkin. John Wiley & Sons, Inc, Hoboken, NJ. ISBN 978-0-470-00957-4.

CHAPTER 12

Managing the Civil Engineering Enterprise

Big Idea

Profit and exceptional client service are essential for the continued existence of the consulting engineering enterprise.

"Our job as leaders—the alpha and the omega and everything in between—is abetting the sustained growth and success and engagement and enthusiasm and commitment to Excellence of those, one at a time, who directly or indirectly serve the ultimate customer."

—Tom Peters

Key Topics Covered

- The Influence of Economics on Project Development
- Financial Reporting
- Professional Human Resources Management

Civil Engineer's Handbook of Professional Practice, Second Edition. Karen Lee Hansen and Kent E. Zenobia.
© 2025 John Wiley & Sons, Inc. Published 2025 by John Wiley & Sons, Inc.
Companion website: www.wiley.com/go/hansen/CivilEngineersHandbook

- Career Planning and Execution
- Specialization
- Certification and Registration
- Professional Services Marketing
- Professional Business Development
- Professional and Trade Organization Activities

Related Chapters in This Book

12.1   INTRODUCTION

Even as technological advances continue to accelerate the technical capabilities of the engineering practice, a fundamental precept that major engineering decisions are based on economic considerations remains unchanged. The basic economics of the engineering practice are essential to understanding their successful application in a larger business administration context.

The discussion of the economic considerations in this chapter is within the convention of the *successful delivery of a project*. Several terms need to be defined to within the following phrase:

A *project* is a temporary endeavor undertaken to create a unique product or service. "Temporary" means that every project has a distinct beginning and definite end. "Unique" means that the product or service provided is different in some way from all other similar products or services (Duncan 1996).

Delivery is the presentation of the final product or service a client asked for at the genesis of the project. It is the reason the project came into existence, and upon its conclusion, the essential lifespan of the project is deemed complete.

Successful is defined in this instance as the submittal of the final product or service to the client that not only meets the agreed-to terms and stipulations of the arrangements made between the engineer and the client, but also provides a mutual sense of accomplishments and satisfaction to both the engineer and the client. In other words, the client feels that the product or service received was well worth the money paid to obtain it, and the engineer considers the fee paid to pro- vide a reasonable profit for the labor and materials expended to produce it. These concepts are also related to the engineer's ethics as presented in greater detail in Chapter 3, Ethics.

The mechanics of executing a successful project were discussed at length in Chapter 7, Executing a Professional Commission. The focus of Chapter 7 is primarily the processes and procedures of project management. This chapter focuses on the business aspects of projects and discusses how to attain a profitable result. This portion of the handbook relates to the application of tools to assist in budgeting a project for a successful business outcome—work performed at a profit for the engineering organization.

12.2   THE INFLUENCE OF ECONOMICS ON PROJECT DEVELOPMENT

During the excitement of planning and initiating the development of a project, the question, "Can we do it?" seldom is followed with the question, "Should we do it?" The "Should we do it?" question addresses whether or not pursuing the project is in the best interest of the engineering firm. Many engineering firms have implemented a process, with varying degrees of formality, commonly referred to as a "Go/No-Go" decision process. In final consideration of a decision to proceed with a project (or not), the engineer should consider the material presented in Chapter 14, Balancing Life, Family, and Career, especially as it relates to Chapter 3, Ethics.

12.2.1   The Go/No-Go Decision Process

While it may seem obvious, frequently decisions are made to pursue the development of a project that are independent of the technical and economic merits of the project outcome. The rationale for these pursuits may be driven by a variety of factors, including relationships with the client, potential expansion into new lines of engineering, political factors, or even the vanity and hubris of a firm's senior leadership. Using a go/no-go process to quantify the economic impacts a project may have on a firm's business should be an essential first step in the project development process. By implementing this process, the engineer may avoid the prospect of performing a project that fails to deliver a desired result within any prescribed schedule, possibly at an enormous cost to the client and/or the firm.

The go/no-go process is essentially a series of questions asked of the potential project manager to consider *prior to* moving into the project initiation stage. While there is no prescribed approach for conducting this exercise, the typical questions that are almost universally asked are:

1. How well is the work understood?
2. How well-defined are the client's expectations?
3. What degree of profit can reasonably be expected?
4. What kind of risk factors are presented by this project? (For instance, is the project in a foreign country, are there unusual terms and conditions, is it a design-build project, is hazardous material/waste present, is there extensive public and/or regulatory agency involvement, and so forth.)

Assuming that the risk and technical factors can be addressed such that the go/no-go process answer to the question "Should we do the work?" is affirmative, the motivational factor of developing a project budget that reflects reasonable and accurate expenditures and commensurate profit becomes essentially the ultimate goal of the project, when viewed from a business economics (as opposed to technical) perspective. One of the first things to understand, then, is how the costs of a project, including indirect and direct overhead, and labor and nonlabor needs affect the creation of an appropriate budget.

12.2.2   Overhead and Direct Labor

Chapter 7 discusses the development of a project budget; key among that budget is the level of effort (hours) needed to perform a given task, and the mix of professional disciplines required to complete that task properly. Each discipline has a different hourly labor rate, but how are these rates determined?

When budgeting for a project, a labor rate that is "loaded" or "fully loaded" is what is typically used to establish budget cost, as opposed to that of a "raw" or "direct" labor rate. In this case, a fully loaded labor rate accounts for the overhead as well as the direct labor costs of a worker on the project. Thus, fully loaded labor rates more accurately reflect the actual costs involved in providing a service. "Overhead," in this instance, is defined as a cost or expense (such as for liability insurance, rent, and utility charges) that relates to an operation of the engineer's firm as a whole, and cannot be applied or traced to any specific unit of output. Overhead is considered an indirect cost.

A subset of the development of a fully loaded labor rate includes the fringe benefits costs associated with the firm's employment of engineering and other professional staff. "Fringe benefits" are compensation paid by the firm in addition to direct wages or salaries. Fringe benefits include items such as medical insurance, vision and dental benefits, paid holidays, 401(k) matches, stock options, and subsidized meals.

Additionally, in the calculation of a fully loaded labor rate, the general and administrative (G&A) costs must be factored into the labor rate calculation. These general and administrative costs reflect those expenditures necessary for the operations of the firm, but are not directly associated with developing a product or providing a service. Examples of G&A-related costs include account invoicing staff, office receptionists, and human resources personnel.

Finally, a calculation of profit may be considered in the development of the fully loaded labor rate, or it may be a calculated value based on the total value of the project. Profit determination varies among contract types, and the description of different contract types is included in Chapter 11. Several examples of the determination of a fully loaded rate, both including and excluding profit calculations, are shown below to better illustrate their development.

Example 1: A newly hired engineer has a base salary that pays $40 per hour. The firm's fringe benefits costs, an indirect cost, average 35 percent for its employees. The overhead costs on an hour of labor for this firm are 65 percent. Lastly, the firm has a G&A rate of 10 percent and an assumed profit of 9 percent for all of the firm's projects. To determine the fully loaded hourly labor rate for this engineer, perform the following calculations:

Multiply raw labor rate by fringe benefits, add to the raw rate: ($40.00 × 0.35) + $40.00 = $54.00

Multiply the result by the overhead rate and add the result: ($54.00 × 0.65) + $54.00 = $89.10

Next, multiply this result by the G&A rate and add the result: ($89.10 × 0.10) + $89.10 = $98.01

Finally, assuming that profit is calculated in the fully loaded rate, multiply the results above by the profit to determine the final fully loaded labor rate: ($98.01 × 0.09) + $98.01 = $106.83

In this example, the firm would realize this projected 9 percent profit (based on this employee's efforts) only if the employee managed to be 100 percent billable to the client's project work, and if the firm actually collected that invoice in full.

Example 2: Perform the same function for a senior project manager with a base salary of $50 per hour, but with profit being considered a separate determination from labor.

Multiply raw labor rate by fringe benefits, add to the raw rate: ($50.00 × 0.35) + $50.00 = $67.50

Multiply the result by the overhead rate and add the result: ($67.50 × 0.65) + $67.50 = $111.38

Next, multiply this result by the G&A rate, add the result: ($111.38 × 0.10) + $111.38 = $122.52

Since profit is a separate calculation in this instance, the result above is the final fully loaded labor rate for this engineer.

Some cautionary words regarding loaded labor rate calculations: the description of the above determination of loaded labor rates is subject to a wide degree of variability. The engineer's firm may calculate their loaded rates in a different fashion, by defining whether certain administrative costs are G&A or true overhead costs instead. Likewise, the ratios shown in the above examples are typical, but by no means standard, for the engineering industry. The fringe benefit costs, overhead, and G&A are determined by the firm through a detailed review of expenditures, and if the firm is performing work for the federal government, the rates may be subject to extensive audits and revision by the auditing agency for which the engineer's firm is proposing to perform contract work. There may also be strategic factors in terms of providing favorable rates to clients in certain circumstances whereby a firm may choose to pro- vide discounted labor rates by lowering their indirect or overhead costs.

12.2.3 Multipliers

In light of the above discussion, the use of multipliers as a means of developing loaded rates and comparing the performance of a project from a business perspective is a very effective tool to determine rapidly if a project is performing in accordance with its budgeted baseline, or if staff usage is consistent with the hours allocated in the original budget. Several distinct multipliers will be discussed: the labor multiplier, the budget multiplier, and the effective multiplier. All are closely related, but can be used to track or budget different aspects of a project.

Labor Multiplier: The labor multiplier is determined by dividing the fully loaded rate in a contract or a budget by the raw or direct labor rate of a given individual. So, for the two examples given above for determining labor rates, the labor multiplier of the new engineer, with profit included, is:

($106.80 ÷ $40.00) = 2.67

For the senior project manager, the labor rate without profit included is:

($122.52 ÷ $50.00) = 2.45

If profit were not included in the loaded labor rates for the new engineer in Example 1 above, the resulting labor multiplier would be 2.45 as well.

Labor multipliers vary for engineering firms, as there are numerous factors that influence their development. Typically, in engineering firms, labor multipliers range from a low of 2.2 to a high of over 4.0. Higher multipliers are often found in higher risk projects, or for extremely technical work where the work product or knowledge required is rare and specialized. Conversely, lower multiplier work is found in projects where there may be a wide range of qualified engineering firms capable of providing the product or services, or the work is considered fairly standard and of fairly low technical complexity. But this is not universal; very complex projects may have low multipliers, and simple projects may have very high multipliers if the terms of the contract are structured in such a manner to track profit or control risks borne by the engineer for their work.

In this example, the manager of this engineering enterprise will rely on the firm's business plan that projects work and the Marketing Department that finds the opportunities and helps procure the work in this firm's business sector. Using three tools the Engineering Manager can effectively manage the business aspects of the firm in a four-step process. The first step calculates the billable versus nonbillable hours. The second step determines the salary of direct labor personnel for billable versus nonbillable/overhead hours. The third step establishes the overhead costs and rates to calculate the "breakeven multiplier." And finally, the Engineering Manager will back-calculate and verify the breakeven multiplier since this manager is business savvy. A detailed calculation of these calculations and labor multipliers is presented in Figure 12.1.

Step 1 - Billable vs. Non-Billable Hours

Direct Labor	Payroll Burden Hours		Indirect Labor Hours			Total Hours	
	1	2	3	4	5	6	7
	Holidays	Paid Leave	Business Development	Administration	Other	Overhead	Billable
Principal	80	160	300	300	100	940	1140
Senior Engineer A	80	160	150	100	100	590	1490
Senior Engineer B	80	120	75	50	100	425	1655
Engineer A	80	120	0	0	100	300	1780
Engineer B	80	120	0	0	100	300	1780
Junior Engineer A	80	120	0	0	100	300	1780
Junior Engineer B	80	120	0	0	100	300	1780

*Total hours per year = 40 hours/week x 52 weeks/year = 2,080 hours/year; total billable hours = total hours/year - total overhead hours/year

****

Step 2 - Billable vs. Non-Billable Hours

Direct Labor	1 Annual Salary	2 Salary per Hour	3 Billable Hours	4 Salary for Billable Hours	5 Overhead Hours	6 Salary for Overhead Hours	7 Total Salary
Principal	$3,00,000	$144	1140	$1,64,423	940	$1,35,577	$3,00,000
Senior Engineer A	$1,30,000	$63	1490	$93,125	590	$36,875	$1,30,000
Senior Engineer B	$1,18,000	$57	1655	$93,889	425	$24,111	$1,18,000
Engineer A	$1,00,000	$48	1780	$85,577	300	$14,423	$1,00,000
Engineer B	$93,000	$45	1780	$79,587	300	$13,413	$93,000
Junior Engineer A	$64,000	$31	1780	$54,769	300	$9,231	$64,000
Junior Engineer B	$57,000	$27	1780	$48,779	300	$8,221	$57,000
Total	$8,62,000			$6,20,149		$2,41,851	$8,62,000

****

FIGURE 12.1 Example calculations of labor multipliers.

Step 3 - Overhead Costs and Rates

Payroll Burden

Holidays	$33,154
Paid Leave	$58,000
Group Insurance	$40,582
Workers' Compensation Insurance	$3,000
Education	$9,000
Professional Due and Licenses	$3,000
Profit-Sharing/Pension	$27,000
Other Benefits	$18,000
Payroll Taxes	$60,000
Total	$2,51,736

Indirect Labor

Accounting	$90,000
Graphics	$66,000
Business Development	$54,000
Administration	$46,000
Other	$40,000
Total	$2,96,000

General and Administrative

Rent	$90,000
Utilities	$15,000
Equipment Leases	$36,000
Maintenance/Repairs	$12,000
Supplies	$11,500
Insurance	$1,35,000
Professional Services	$12,000
Loans (Interest)	$15,000
Other Taxes	$90,000
Other	$75,000
Total	$4,91,500

Total Overhead

Payroll Burden	$2,51,736
Indirect Labor	$2,96,000
General and Administrative	$4,91,500
Total	$10,39,236

Total Overhead	$10,39,236
Billable Salary	$6,20,149
Total	$16,59,385

****

Break-even Multiplier

Total Overhead + Billable Salaries/Billable Salaries   **2.68**

		Salary/Hour	Billed/Hour
Example			
Engineer A	$1,00,000	$48	$129

FIGURE 12.1   *(Continued)*

Budget Multiplier: The budget multiplier is a straightforward calculation. It is the value of the net contract amount of a project divided by the direct (or raw) labor costs. The net contract amount is the total contract amount less any raw sub-contractor costs and less any other direct costs. For example, if a contract were signed for $16,000, with a subcontractor's raw costs being $5,000 of the amount, and with $1,000 of other direct costs such as report reproduction, travel costs, and equipment rental, the net contract amount would be $10,000. If the direct (or raw) labor costs for the contract totaled $4,000, then the budget multiplier for this project is:

$$\$10,000 \div \$4,000 = 2.5.$$

Effective Multiplier: The effective multiplier is a measure of the actual financial performance of the contract when compared with the budget multiplier. It is defined as the actual fully loaded costs for the work, subtracting out all raw nonlabor costs. This number is then divided by the raw direct labor costs to determine the effective multiplier. So, if the contract has had $15,000 of fully loaded costs placed against it, with $5,000 subcontractor raw costs, $500 spent to date for other raw nonlabor costs, and with $4,200 in direct (raw) labor costs being the total expenditures at a certain point in the project's lifecycle, the resulting effective multiplier is:

$$(\$15,000 - \$5,000 - \$500) = \$9,500; \$9,500 \div \$4,200 = 2.26$$

Having an effective multiplier lower than the budget multiplier is an indication that the actual costs for performing the work are exceeding what the budgeted costs were for the same work. This would suggest some type of corrective action may need to be taken, depending on the terms of the contract and the amount of work remaining to be performed to produce the final product or service. While an effective multiplier that is lower than a budget multiplier is an indicator that the profit potential for the project is decreasing, it does not mean necessarily that the project will lose money for the engineering company. That may be dependent on whether the firm has a breakeven multiplier for a project that can demonstrate the lowest effective multiplier threshold a project can reach and not result in a loss of net revenue for the firm.

A Few Words about Multipliers

It's in the best interest of the accounting manager and the engineer to keep comparative multipliers in perspective. Having an effective multiplier lower than a target budget multiplier doesn't mean the project is a failure, but some accountants/managers may make such a black-and-white comparison. A better comparison would be comparing effective multipliers against profit baseline or breakeven multipliers to see if the project is making money or actually costing the firm money to perform. In short, multipliers should not be used as the single "pass/fail" measure of a project's success; there are other factors, most notably client satisfaction and potential for follow-on work that also should be considered when evaluating project performance.

12.3 FINANCIAL REPORTING

12.3.1 Income Statement (Profit and Loss)

An income statement may also be called a profit and loss (P&L) statement. It displays the company's financial position for a specific period of time and shows how the company's gross revenue is transferred to net income. The gross revenue is the total revenue from the sales of products and services and is sometimes referred to as the "top line" on the P&L statement. The company's net income is the amount of revenue remaining after "expenses" is subtracted from the gross revenue and is sometimes referred to as the "bottom line." The income statement shows the company's management and investors how well (or not well) the company is performing during the specific reporting period (typically monthly, quarterly, or annually) (Schroeder, Richard, et al. 2010).

For purposes of this chapter, the income statement is calculated on the basis of the single step method which is the simple approach of totaling revenues and subtracting expenses to find the bottom line. There is a more complex method that involves multiple steps of calculations including inventory, equipment depreciation, and income yield from operations (Makoujy, Jr., Rick J., 2010). Both accounting methods show income before taxes. Net income will show the final income after the applicable taxes are subtracted from the income. However, the income statement does have some limitations.

12.3.2 Example Income Statement

Kando Engineering submits a year-to-date income statement to Capital City Bank to obtain financing for their office expansion for a 12-month period, as of December 31, 2023. Using the figures below, Capital City Bank determines whether to fund **Kando's** planned office expansion.

- Project Revenue = $218,000
- Other Income = $1,000
- Direct Labor = $84,000
- Indirect Labor= $34,000
- Payroll Burden = $25,000
- Rent and Utilities = $12,000
- Equipment Leases =. $5,000
- Maintenance, Repairs = $1,000
- Supplies = $2,000
- Insurance = $15,000
- Professional Services = $2,000
- Interest = $2,000
- Other = $8,000
- Federal Taxes = 30%

This data results in an income statement as follows:

Income Statement—December 31, 2023.

REVENUE		
Project Revenue	$218,000	
Other Income	$1,000	
Total	$219,000	$219,000

EXPENSES		
Direct Labor	$84,000	
Indirect Labor	$34,000	
Payroll Burden	$25,000	
Rent and Utilities	$12,000	
Equipment Leases	$5,000	
Maintenance and Repairs	$1,000	
Supplies	$2,000	
Insurance	$15,000	
Professional Services	$2,000	
Interest	$2,000	
Other	$8,000	
Total	$190,000	$190,000
GROSS PROFIT		$29,000
FEDERAL TAXES	$8,700	–$8,700
NET PROFIT		**$20,300**

This income statement results in the following summary:
1. Kando's *Total Revenue* for 2023: $219,000
2. Kando's Total *Expenses* for 2023: $190,000
3. Kando's *Federal Taxes* for 2023: $8,700
4. Kando's *Gross Profit* for 2023: $29,000
5. Kando's *Net Profit* for 2023: $20,300

Note: There is a difference between *revenue* and *profit*. Revenue is recognized as the monies received (income) during an invoice period. *Profit* is defined as revenue minus expenses.

12.3.3   Limitations of Income Statements

Generally, income statements help company managers, investors, and creditors to assess the past performance of the company. This statement also provides an indication of the future performance and can provide information for generating future cash flows Schroeder, Richard, et al. (2010). However, income statements have limitations:

- Future results are not directly related to past performance.
- Some results on the income statement depend on the specific accounting methods, described above, and cannot measure inventories.
- Some results on the income statement are judgmental, such as final salvage values, useful life, depreciation, or market conditions. For example, the retail industry does more than 50 percent of their business and relies heavily on the fourth-quarter months of September through December. After a detailed evaluation of their income statement in June, a prospective investor may have a significantly different opinion of this company's performance to a similar evaluation of the January results.

12.3.4   Statement of Financial Position (Balance Sheet)

A statement of financial position, also referred to as a balance sheet, is a summary of a company's financial balance. Balance sheets are typically produced at the end of their financial year and reflect a company's assets, liabilities, and owners' equity. The balance sheet is sometimes referred to as a snapshot of a company's financial condition (Williams et al. 2008).

The company balance sheet has three parts: assets, liabilities, and owners' equity. The main categories of assets are usually listed first and typically in order of liquidity (Daniels 1980). Assets are followed by the liabilities. The difference between the assets and the liabilities is known as equity, or the net assets or the net worth or capital of the company and, according to the accounting equation, net worth must equal assets minus liabilities.

In summary, assets equal liabilities plus owners' equity. Looking at the equation in this way shows how assets were financed: either by borrowing money (liability) or by using the owners' money (owners' equity). Balance sheets are usually presented with assets in one section and liabilities and net worth in the other section with the two sections being in "balance" (Williams et al. 2008).

In general, individuals and small businesses tend to have simple balance sheets (Gitman 2005). Larger businesses tend to have more complex balance sheets, which are typically presented in the organization's annual report. It may be useful to compare a balance sheet from one year to the next to look at a company's financial progress in time.

12.3.5   Balance Sheet Example

Kando Engineering also needs to submit a year-to-date statement of financial position to its Board of Directors:

As of December 31, 2023:

- Cash = $20,000
- Short-Term Investments = $7,000
- Work in Progress = $21,000
- Accounts Receivable = $45,000
- Property and Equipment = $29,000
- Accounts Payable and Accrued Expenses = $15,000
- Deferred Income Taxes = $45,000
- Long-Term Debt and Liabilities = $15,000
- Capital Stock = $25,000
- Retained Earnings = $22,000

Statement of financial position—December 31, 2023.

ASSETS		LIABILITIES	Value
Current Assets		**Current Liabilities**	
Cash	$20,000	Accounts payable	$15,000
Short-term investments	$7,000	Deferred income taxes	$45,000
Work in progress	$21,000	Long-term debt and liabilities	$15,000
Accounts receivable	$45,000		
Fixed assets		**Owners' equity**	
Property and equipment	$29,000	Capital stock	$25,000
		Retained earnings	$22,000
TOTAL ASSETS	**$122,000**	**TOTAL LIABILITIES AND OWNERS' EQUITY**	**$122,000**

Kando's total *Current Assets*: $93,000
Kando's total *Fixed Assets*: $29,000
Kando's *Current Liabilities*: $75,000 Kando's
Owners' Equity: $47,000
Kando's *Total Liability and Owners' Equity*: $122,000

12.3.6   Cash Flow

The flow of revenue (cash flow) refers to the movement of money into or out of a project or a company. Cash flow is usually measured over a finite period of time. Evaluating cash flow is used to:

- Assess a project's or company's rate of return on the investment. The amount of money flow into and out of projects is used as inputs and outputs to calculate an internal rate of return, and net present value.
- Assess potential problems with a business's liquidity and net result in the operating checking account. Money is the fuel to keep the employees and vendors paid to continually provide the company's products and services. A profitable company can fail because of a shortage of cash, especially if it cannot pay the employees or vendor partners.
- Evaluate and verify the company's profits when it is believed that accrual accounting concepts aren't accurate. For example, it may appear that a company may be profitable but could be generating cash by issuing shares, selling equipment, or increasing debt.
- Evaluate the actual income generated by accrual accounting. If net income is composed of items other than actual cash, it is considered low quality and potentially suspicious.
- Evaluate risks such as default risk, bonuses, or reinvestment requirements, to mention a few.

Cash flow is considered a generic term and depends on context. Companies that offer professional services commonly wait long periods to receive payment for those services. Cash flow can refer to actual past flows or projected future flows. It can refer to the total of all the flows involved or to only a subset of those flows. Net cash flow is based upon operational, investment, and financing cash flows. To reconcile the ending cash balance, all cash flows should be considered.

12.4   PROFESSIONAL HUMAN RESOURCES MANAGEMENT

Every civil engineering firm has a human resources function; and in large firms there usually is a separate Department of Human Resources. The Human Resources Department typically is responsible for recruiting and training, dealing with performance issues, managing compensation and 401(k) retirement packages, and ensuring that the firm's practices conform to various laws and regulations.

However, within the context of successful project execution, the project manager also plays a critical role in managing human resources. The project manager must balance the limitations imposed by the project budget with the technical requirements and demands of the client for the quality end product. Staffing a project entirely with proven senior professional engineers may be desirable; but if the project budget is based on a preponderance of junior and entry-level labor to produce the work, the likelihood of the project being completed within the original budget is greatly reduced. Even if the senior staff being used are much more efficient than the junior-level staff, the budget likely will be in jeopardy.

Extending this example, if the mix of staff identified to perform the work is appropriate, various technical resource managers will need to be consulted to ensure the availability of staff at key points in the project's life. For example, if a particular junior engineer is the perfect staff member for the job, but is on a remote assignment in a foreign country for the duration of the project, alternative resources need to be allocated. With highly technical or specialized disciplines, staff may be in special demand. Consulting the re-source manager early in the process helps to ensure the project manager is obtaining the right professional staff in a timely manner for a given phase of the project's schedule.

Another aspect in the development of skill mixes is the assignment of categories to specific staff. For a given client or task, is it appropriate to downcharge (charge at a lower billing rate or labor multiplier) an experienced engineer to meet a client's budget, or to upcharge (charge at a higher labor multiplier) a junior staff member to a more senior rate to reflect the actual activities the staff is undertaking? These are situational questions, and some clients, notably the federal government, may explicitly forbid the practice of downcharging or upcharging. Private sector clients may take a less formal approach to these nuances of human resourcing and project budget development, and so may afford the project manager more flexibility in developing profitable project budgets and schedules.

Therefore, in the project planning and budgeting stages, a project manager's key responsibility is to develop an appropriate mix of technical and professional staff (with cost-effective billing rates) to complete the tasks identified for the work. In the course of developing a competitive project budget, the project manager should prepare a (working) cost estimating spreadsheet that shows a representative sample of the labor categories and other direct costs associated with the project. But after the contract award or at the project kick-off meeting, specific names must be assigned to these labor categories, or the project's chances of ultimate success will be hampered.

See Chapter 18—Human Resources for additional information.

12.5   CAREER PLANNING AND EXECUTION

The role of career development is an integral component in any discussion of business administration for projects. The critical role an engineering project manager may play in the development of the staff careers, particularly junior staff, cannot be overstated. The project manager engineer must be mindful of the potential tendency to slot personnel that excel in a given role for a project into that same role repeatedly. The project may benefit, but "pigeon-holing" a given staff member into a particular activity for the sake of project efficiency may

stunt the development of a staff member and frustrate the employee, creating the potential for lowered morale or even the loss of the staff member to another firm.
"Vision without action is a daydream . . .

> Coupling vision and action—in other words, executing strategy—is one of the biggest challenges we face."
>
> —Japanese proverb

In the context of the engineer as project manager serving to advance the financial objectives of the firm, project staffing decisions must relate to the career arcs of the individuals participating in the project. While having an entry-level engineer collect samples in the field as a technician may be appropriate in order to help them understand the nuances of data processing and management, the same engineer should be given new opportunities to expand his or her sphere of experience to gain broader, overall engineering perspectives in professional practice.

The project manager role is not one of supervisor; "mentor" would be a more apt description. The project manager, often even more so than an engineer's supervisors, can gain an accurate feel of the performance strengths, weaknesses, and interests of project staff members. By informally checking in with staff to gauge where their interests lie, and where they seem to excel professionally, the engineer as project manager can assist in guiding staff into opportunities that will further their technical interests in their career, ideally within the structure of the engineer's firm.

12.6   SPECIALIZATION

Discussing the importance of providing a broad range of technical opportunities to engineering staff and then immediately talking about the merits of encouraging technical specialization may appear contradictory. However, if the engineer identifies, either for him/herself or a colleague, a particular area of interest, specialization in a particular technical subarea may lead to extensive opportunities.

> "If you think adventure is dangerous, try routine—it is lethal."
>
> —Paulo Coelho, novelist

Experience by itself is a necessary, but not sufficient, constituent of a successful career. The engineer needs to obtain training to supplement the experience gained through their specialization. There are many professional organizations (ASCE, ACEC, ASFE, NSPE, SAME, to name just a few) that offer training programs to their members. Additionally, many engineering firms have internal and/or external training programs. Training, coupled with specific relevant experience, can lead an engineer to being recognized within the firm as the resident expert on a particular technical matter, and as the "go-to" person to perform that work on a myriad of projects. From a professional development standpoint, there may be commensurate potential salary growth, if the specialized expertise gained is rare and highly sought after.

There's an Old Adage that Relates to "Specialization"

It's "better to be a mile wide and an inch deep than the other way 'round.'" It often seems that the shelf-life of an expert is shrinking as technology accelerates change and improves data accessibility. Technical specialization can be a lucrative career path, but technical or market developments may necessitate a wholesale mid-career change. Before deciding to go down the path of any particular specialization, the engineer is advised to research carefully whether the field is new, established, or undergoing transition. This due diligence may be vital to longevity with this career path.

But, there is a cautionary note to consider with regard to technical specialization. As technology accelerates the forces of change in the engineering profession, an area of expertise that may be in high demand presently, may be supplanted almost entirely by new technology. There are numerous examples; for example, hand-held GPS units now deliver vertical and horizontal data via remote downloads with accuracies that rival formal land surveys. The use of Geographic Information Systems (GIS) for de- sign has transformed the development and presentation of design drawings. Calculations for sizing pipes and pumps that used to rely on careful analysis of multiple complex graphs are completed by computers with minimal input. Information modeling is becoming more and more prevalent in the engineer's workplace.

Therefore, there is a danger in becoming too specialized in a given technical field, as that field's technical advantage may be changed dramatically and quickly with an advance in technology. The expert whose specialized services are in tremendous demand one year may find the next year that his/her body of knowledge is now considered commonplace; suddenly, they must compete with other technical staff to stay gainfully employed.

12.7 CERTIFICATION AND REGISTRATION

As established in Chapters 1, 2, 3, 11, and 21, civil engineering is a recognized profession with a well-established set of standards of professional care. Licensing, or registration, for the engineer is vital in the civil engineering profession; it may be desirable, but is less important, in other more specialized engineering fields. For the civil engineering firm, employing registered engineers is valuable, for it provides the means to perform work in the public sphere may be denied otherwise. Like a law firm without any lawyers who have passed the bar exam, or a medical practice without any licensed physicians, an engineering firm without registered professional engineers will not last long in a competitive business environment.

Nevertheless, in a diverse, multidisciplinary environment, professional registration of every engineer in the firm is not absolutely essential for the firm to be successful, or for individual projects to be profitable. The engineer should not mistake professional registration with financial profitability; one does not guarantee the other. Professional registration offers opportunity, but with it comes professional responsibility, and the two should not be separated.

> "Science is about knowing, engineering is about doing."
> —Henry Petroski, Professor of civil engineering at Duke University and, author
>
> *To Engineer Is Human: The Role of Failure in Successful Design* and *The Essential Engineer: Why Science Alone Will Not Solve Our Global Problems*

In a similar vein, certification of technical staff, such as for industrial hygiene, project management, or other aspects of professional performance, is a more specialized, focused type of recognition of professional competence. The liability issues pertaining to certification are not as great as with registration typically, and the corresponding opportunities and requirements for having certified staff performing tasks on a project not as crucial. Nonetheless, work performed on a project done by certified professionals is likely to make acceptance of the final products or services easier for the firm, because such certifications generally reassure clients.

In the role of project manager, the civil engineer should work with the supervisor of technical staff to support the development of staff in their desire to obtain various professional certificates and/or professional registration. This can be done by advocating to the supervisor on their behalf, expanding staff's work experience to comply with given certification program or registration requirements, and providing the schedule flexibility for staff to pursue these goals without affecting either project performance or their general health and well-being.

> "The higher we are placed, the more humbly we should walk."
> —Cicero, Philosopher

12.8   PROFESSIONAL SERVICES MARKETING

Assuming that the role of marketing and long-range business planning falls outside the purview of the engineer may be tempting. After all, these are administrative functions of the firm, not at all related to technical work, the true domain of the engineering staff. Such a viewpoint is both short-sighted and detrimental to the long-term well-being of the firm. The engineer has vital insights in the workings and relationships of clients and the products or services they demand. Consequently, the engineer absolutely must be included in marketing and the business of planning activities of the firm. It is a quaint, archaic notion for an engineer to assume that he or she can perform their work successfully without interacting and building relationships with their clients. Introversion may come naturally to engineers, but it has no place in developing a successful business model.

Marketing professional services is a formal way of describing and selling the firm and, by extension, the individual engineers employed by the firm. Marketing is not a haphazard, random series of glad-handing events, at least not for those who wish to market their services successfully. Successful firms use a strategic, focused method for selling and marketing. Many Fortune 500 firms use specific marketing processes, and frequently train their professional and technical staff in their application. Why do so many companies use a specific process for selling and marketing? Because they find that using a process works, and as a result, they are able to obtain work. To compete today, successful firms must use focused marketing.

Since the strongest lead for the next project or opportunity for work will always be an existing client, the engineer should continue to develop and ensure strong relationships with current clients. This may seem obvious to most, but it is very easy to take current key clients for granted and not spend the necessary time with them to maintain and continue the relationship. By being client-focused, the engineer will lower business development costs in the long run, while positioning the firm for follow-on business as the incumbent supplier.

Going after every opportunity that uses the word "engineering" or "environmental" or "transportation" (whatever the buzz word is for your group) is like chasing fish or rabbits . . . can you really catch one? Do you really know if the pursuit is worthwhile? Do you know the potential of that client? Does the client know you? The engineer must focus on systematically working to grow the best prospects for the firm.

So how do you gather the information you really need to know about your client? Many civil engineering firms have well-qualified marketing personnel that can help the engineer prepare prior to meeting a potential client. (See Chapter 9 and later in this chapter.) The firm may have identified a process such as the one depicted in Figure 4.1 and may have an established database for tracking leads. Knowing where the firm is in the process is extremely important because all firm representatives need to project a "united front" and a consistent message.

When meeting with the client regarding future opportunities, listening is far more important than selling. Through listening and asking appropriate questions of the client, the engineer may gain the necessary information to help the firm decide to pursue the opportunity (make the go/no-go decision) and develop a proposal.

The process of proposal development is somewhat subjective, and the mechanics of proposal assembly are often the purview of the firm's marketing professionals. However, the content (as opposed to the organization and formatting or appearance) of the proposal often relies extensively, or even exclusively, on the input of the engineer. The next two subsections briefly discuss some of the areas the engineer must be cognizant of when preparing technical and persuasive proposals for winning project work for the firm.

12.8.1   Resume Updates

Just as a civil engineer's personal resume is vital for obtaining employment, a corporate resume is equally vital for successfully obtaining work for the engineer's firm. The value and benefit of the formatting, emphasis, placement, and appearance of engineers' corporate resumes are well-established, but the actual appearance of such resumes varies from firm to firm. A corporate resume should contain the following components for the engineer, or any other technical staff, included in a proposal to a client: the name and title of the individual; their proposed role for the project or contract; their education and certifications/registrations; their general technical experience, including years of service in the field; the *relevant* project experience (defined as that which is pertinent to the scope of services being requested by the client); pertinent publications; and professional citations/memberships. The relative importance of these items is somewhat determined by the nature of a given solicitation for services by the client, but the elements listed are most frequently asked for by prospective clients.

> "All my life, I always wanted to be somebody. Now I see that I should have been more specific."
>
> —Jane Wagner, writer and director

A corporate professional resume should be developed as a comprehensive listing of all technical work performed during the course of the engineer's career, as well as the listing of professional training, certifications, registrations, and publications collected during one's professional tenure. It is not unusual for senior technical engineers with decades of experience to develop corporate resumes that are 60 pages in length or greater; it is their responsibility, however, to work with marketing staff to winnow these resumes down to the page limits specified, and to include only the relevant technical information requested by the client's request for a proposal or qualifications.

Demonstrating that the firm's proposed team members are qualified to meet the client's needs is a critical component of a winning proposal. The client does not want to pay engineers to be educated on how to solve their (the client's) problem; however, clients are willing to pay more for experienced engineers who have learned from their mistakes on someone else's project.

As a rule of thumb, the engineer should update his or her corporate resume every six months and should work with technical staff and supervisors to ensure that other key project personnel are providing these updates to their resumes at the same frequency.

12.8.2   Project Descriptions

Project descriptions, simply put, are the best way of concisely showing a prospective client that the engineer and the firm he/she represents possess all of the experience and skills necessary to perform the project work requested by the client. Project descriptions must be factual, but they also should paint a compelling picture of the work that was performed by the firm on previous, similar or related projects.

As with resumes, the format and organizational presentation of the project description can vary depending on the preference of either the firm or the prospective client, but the content of the project description should be developed by the project manager and pertinent staff most involved with the performance of the project. Project descriptions have essential components that should include: the dollar value of the project; its status (current or completed); the planned and actual duration of the project; the client's information regarding the project (key point of contact for the client and their contact data); a brief description of the scope of services or products provided; any key specialty subcontractors or vendors used to perform the work; where the work was performed; and the role the firm and members of the team proposed, if applicable, played in the performance of the project. The importance of a well-written and timely project description cannot be overemphasized in convincing a prospective client to select the engineer and the firm for the work being solicited.

In a proposal for engineering services, concise information combined with descriptive details, relevant financial information, and sharp photos provide the readers with a good idea of the firm's capabilities!

12.8.3   Business Planning

The role the engineer takes in planning the firm's future year budgets varies by firm. Business plans can vary in length and complexity, but the input of company's operational and marketing staff is essential. In planning the future technical direction and markets the firm wishes to pursue, the accurate and well-thought-out perspectives of the engineer should be sought and incorporated into both short- and long-term planning documents.

The focus and goals of a firm often dictate the components of a business plan. The technical business development components that should incorporate the engineer's input include a description of a given market sector, and the short- and long-term outlook for work or project opportunities in that sector. This information should be derived from the professional relationships developed with clients by the engineer in order to gain as broad (and therefore reasonably accurate) perspective as possible. A discussion of market trends, growth drivers, and short- and long-term growth estimates or goals also should be developed. If possible, an analysis of the competition in the market sector and a comparison of the firm's capabilities and cost effectiveness against this competition should be utilized. Lastly, the goals for growth within the market should be stated clearly, the objectives and plans for achieving these goals clearly laid out, and the responsible parties for implementation identified.

The business plan is less a blueprint and more of a roadmap for growing the size and profitability of a firm. It should be considered a living document, and subject to revision and adjustment throughout the course of its implementation phases. Course corrections to the plan are necessary to ensure its overall success, or else it will be nothing more than a stagnant comparative tool for basing performance data against subsequent business plans.

A Great Business Plan Should Include Five Related Factors Critical to Every New Venture (Adapted from Sahlman 2008)

1. The Specific Venture: The business plan includes a detailed description of the business opportunity, the specific need, the competitors, the potential customers, the product details, the marketing plans, distribution and outlets.

2. The Business Team: The venture team should include specific managers, staff, subcontractors, suppliers, and vendors. Other support staff should be mentioned such as legal staff, human resources, and business support among others.

3. The Work Plan: This section includes specifics like financing options and investment opportunities, potential customer base and demographics, banking details such as interest rates and investment relationship, projected rates of return, schedules, product or service details and delivery plans, inventory control, accounting software, projections and basis of assumptions.

4. The Potential Rewards and Associated Risks: A complete plan will include the details on the total investment funding needed, the likely sources, rates of return, return periods, and product/service delivery schedules. This description will be accompanied by a statement of risks and a mitigation plan that include cash flow projections, sales and marketing, invoicing, cash receivable projections, market risks, and contingencies.

5. A Complete Financial Analysis: This analysis includes key elements of the four components above folded into a summary statement that includes the income statement, balance sheet, and cash flow details.

12.9   PROFESSIONAL BUSINESS DEVELOPMENT

The term "business development" may have a loaded connotation of either engineering staff sitting at a display booth in a large technical conference trying to attract attention to potential clients, or that of connected "players" smoking cigars and making deals while playing a round of golf. The truth surrounding business development for the civil engineering services is that it is simultaneously more prosaic, yet sophisticated. The most basic element of professional business development is contact—contacting potential clients is the essence of any business development program. If the firm wishes to perform work for a client in the future, contact must be made as soon as possible with that client. Staying at the forefront of a client's thoughts—to be "on their radar"—is critical, and to do that requires making frequent contact with that client.

There are numerous publications regarding the art of making sales and the psychology of buying and selling. A recurring theme that seems to be widely accepted is that a client will buy engineering services on emotion. The client will like the work the firm does, will like the staff used to perform the work, and/or will like how they feel about interacting with the engineering staff when project work is being done. While there are many structured and formal approaches to generating the requirements for official requests for proposals, and equally as many logical, objective procedures for evaluating the merits of competitive proposals, ultimately how the client feels about the firm and the staff will affect the selection process the most. A client cannot like a firm they don't know as much as one they do, assuming the established firm has either met or exceeded the client's expectations. The only way for a client to get to know a firm is for the firm's representative to make contact with that client.

> "You can't listen your way out of a sale, but you can sure talk your way out of one."
> —Zig Ziglar

The engineer has to overcome the often-predominant trait of introversion to be successful in business development endeavors. The temptation is to let professional marketers handle these types of contacts, but many clients also have technical or engineering staff who may be introverted by nature. The client often feels intimidated by exuberant marketers but contact with similar technical people can put them more at ease.

There are some fundamental aspects of contacting clients that the engineer should consider before moving down this path. First, keep a positive attitude, even if the initial responses are negative or neutral in tone. Very rarely does new work come from a single business development meeting with a client. The more realistic scenario is that multiple meetings with the same prospective client are required, so following- up is a crucial component. The engineer should be professional in these interactions with the client, so that even if work is not forthcoming in the short run, a favorable impression of the firm and its people is left in the mind of the client. Finally, relationships need to be nurtured; clients need to be thanked for their time, and as the initiator of the meeting, the engineer must be an active listener in these engagements. While the key selling points of the firm should be discussed, the business development meetings that are the most successful are the ones where the prospective client does the majority of the talking.

Every contact made with a prospective client should be designed. There are numerous mechanisms used to contact a client, including phone calls, meetings, formal presentations, e-mails, marketing brochures, or cut sheets. Every one of these mechanisms can be used when designing the contact. Meetings should not be haphazard; would the civil engineer ever construct a high-rise building without a design?

To design a client contact, the engineer should first define the objective of the contact. That is, what is the ultimate outcome desired from this meeting? Once this is done, the strategy to achieve this objective needs to be developed. What is the stated basis of the contact? Then, a game plan to implement this strategy should be scoped out by defining the analyses and methods to be used for the contact. Finally, the game plan should be checked against the objective, or in design terms, a quality assurance check should be performed to ensure that the game plan will get the firm what it wants. More information on forming these relationships is provided in Chapter 9.

Of similar importance is the mindset the engineer must have when performing the work that has been won. A project should be considered an audition for another project in the future. Successfully performing a project for a client, by demonstrating that their needs are understood and by keeping commitments regarding cost and schedule, is arguably the most powerful business development tool of all. It is a cliché, but there is truth in the old saying that "the reward for good work is more work." The converse is true as well—the surest way to lose a client permanently is to produce a poor-quality product or service that is not what the client wanted or expected, and to do so at a cost far higher than originally estimated. While the actions of the engineer performing on a project are necessary to ensure follow-on work, the other key aspects of business development also must be in place to ensure future work with that client (Bachner, J. 1991).

12.10　PROFESSIONAL AND TRADE ORGANIZATION ACTIVITIES

There are numerous benefits to the engineer for engaging in professional and trade organization activities outside of work. From a professional development standpoint, through technical presentations, seminars, and conferences, these organizations offer opportunities for continuing education. Life-long learning is an important aspect of professional development, and many organizations offer formal recognition of this education with continuing education units and/or professional development hours.

Additionally, professional organizations offer a forum for the engineer to interact with colleagues from other firms, to learn about various markets and clients, and to cross-feed technical information. Prospective clients are often members of the same organizations. The engineer should not feel limited only to professional organizations in the technical field; however, there are other organizations, either community- or market-centric where joining would be beneficial for the engineer. Examples include local chambers of commerce, real estate developer associations, and environmental outreach organizations.

12.10.1　Voluntary Activities and Sponsorship

Community involvement generates a number of significant benefits for the civil engineer's firm and its clients. Through volunteer activities, staff from the firm can develop a higher profile in the community and a stronger community connection to the firm, thereby contributing positively to staff recruitment, retention, and public image. Increasingly, clients appreciate firms that are recognized as strong contributors to the community. Such involvement often provides business development opportunities as an additional benefit.

There are numerous opportunities for volunteer activities and sponsorships through involvement in professional and trade organizations within the local community. One theory in marketing is that a potential client has to hear a name, phrase, or company name nine times before it takes root in the mind. When a potential client connects a firm's name and volunteer staff with a community project, it counts as one or possibly more of these connections.

12.11 SUMMARY

Financial management is the life blood of the civil engineering enterprise, whether public service or engineering consulting. "Profit and exceptional client service" are essential for the continued existence of the consulting engineering enterprise. The project manager is at the heart of the exceptional client service aspect and the engineering manager (or financial manager) has to manage the civil engineering enterprise's cash flow to keep the operation viable.

"Genius is one percent inspiration and ninety-nine percent perspiration."

—Thomas Edison

BIBLIOGRAPHY

Bachner, John Philip. (1991). *Practice Management for Design Professionals: A Practical Guide to Avoiding Liability and Enhancing Profitability*. John Wiley & Sons, New York. ISBN 0-471-52205-8.

Daniels, Mortimer. (1980). *Corporation Financial Statements*. Arno Press, New York. ISBN 0-405-13514-9.

Duncan, William R. (1996). *A Guide to the Project Management Body of Knowledge*. PMI Standards Committee, Darby, PA. ISBN: 1-880-41013-3.

Gitman, Lawrence J. (2005). *Principles of Managerial Finance*, 11th edition. Addison Wesley, Boston, MA. Makoujy, RickJ, Jr. (2010). *How to Read a Balance Sheet*. McGraw Hill Companies, New York. ISBN: 978-0-071-70033-7.

Sahlman, William A. (2008). *How to Write a Great Business Plan*. Harvard Business School Publishing Corporation, Harvard Business Review Classic, Watertown, MA. ISBN 978-1-422-12142-9.

Schroeder, Richard. et al. (2010). *Financial Accounting Theory and Analysis—Text and Cases*. John Wiley and Sons, Hoboken, NJ. ISBN: 978-0-470-64628-1.

Williams, Jan R., Haka, Susan F., Bettner, Mark S., and Carcello, Joseph V. (2008). *Financial & Managerial Accounting*. McGraw-Hill Companies, Columbus, OH. ISBN 978-0-072-99650-0.

CHAPTER 13

Communicating as a Professional Engineer

Big Idea

Communication provides the conduit for the engineer to transmit extensive information, knowledge, experience, empathy, and humility for effective use by clients and others on the project team.

"The most important thing in communication is hearing what isn't said."
—Peter F. Drucker, management consultant

Key Topics Covered

- Communication Conduits
- Body Language—Beyond Words—How to Read Unspoken Signals
- Conflict Resolution
- Behavioral Characteristics of Team Members, Friends, or Family
- Typical Report Format
- Useful Forms for the Engineer

Civil Engineer's Handbook of Professional Practice, Second Edition. Karen Lee Hansen and Kent E. Zenobia.
© 2025 John Wiley & Sons, Inc. Published 2025 by John Wiley & Sons, Inc.
Companion website: www.wiley.com/go/hansen/CivilEngineersHandbook

- Useful Letters (or E-Mails) for the Engineer
- Sample PowerPoint Presentation

Related Chapters in This Book

- Chapter 3: Ethics
- Chapter 4: Professional Engagement
- Chapter 5: The Engineer's Role in Project Development
- Chapter 6: What Engineers Deliver
- Chapter 7: Executing a Professional Commission, Project Management
- Chapter 8: Permitting
- Chapter 9: The Client Relationship and Business Development
- Chapter 10: Leadership
- Chapter 11: Legal Aspects of Professional Practice
- Chapter 12: Managing the Civil Engineering Enterprise
- Chapter 14: Balancing Life, Family, and Career
- Chapter 15: Globalization
- Chapter 16: Sustainability
- Chapter 17: Emerging Technologies
- Chapter 18: Human Relations Policies and Employment Practices
- Chapter 19: Construction Management
- Chapter 20: Critical Health and Safety Knowledge for Civil Engineers
- Chapter 21: What Engineers Need to Know

13.1   INTRODUCTION

Communication is a three-dimensional process where information is exchanged to meet a mutual objective in a relevant time period among two or more individuals. This informational exchange typically involved dialogue, body language, and possibly gestures. Effective communication includes:

- A message sent from the sender to the receiving party, signifying the transfer of information
- Acknowledgment of receipt (or clarification, if needed) of this information
- The "time element" for the transfer of this information

Communication is the process of exchanging ideas, information, feelings, or data from one party to another. This information transfer can occur verbally, non-verbally, electronically, physically (by touch), or in writing. Complete and total communications occurs when all modes of information transfer are engaged, understood, and confirmed by the receiving party and agreed upon.

These points make the difference between effective and ineffective communication. For example, just because one party transferred information to another does not mean that the message was received, understood, or acknowledged. The acknowledgment is an essential component of the information transfer and signals to the sender that the message was understood.

Critical communications in the military, police, and security service include requirements for an "acknowledgement response" where the receiving party verifies the message was received and presumably, the request will be delivered. This is a good habit to get into in your personal communication style. Simply take a moment and ask, "Did you receive my message (and exhibits), and do you have any questions or comments"?

With any two (or more) party exchange, a summary statement is an effective means of wrapping up and getting a feel for the level of understanding the other party has regarding the information presented. An engineer should end a meeting with this sort of statement, "So, in summary, we are offering to prepare a work plan that addresses the issues of bridge abutments on the river bank for the fixed fee of $150,000. Does that adequately address your concerns?" This gives the other party, in this case a prospective client, a chance to respond with an affirmation or to come back with their own understanding of the exchange.

This type of communication must take place during the initial conversations on a project or proposal, or both parties may spend countless hours working on something that could wind up being a waste of time. Engineering managers have all heard the excuses from staff and clients, "Sorry, I did not get the message, so I'm not responsible." In order to circumvent this example of failed communication, when an experienced manager does not receive a message confirmation he/she will often send another message to the receiving party and ask what the party perceived as the request and if they would acknowledge or possibly even repeat back what was said. This confirmation loop will demonstrate that all parties received the original transmission and that everyone agrees. A simple acknowledgment shows respect and saves time for all parties.

An example of this informational transfer is shown in Figure 13.1. The figure also illustrates how the sender should consider and vary the level of detail in communication, depending upon the relative comprehension of the subject matter by the listener.

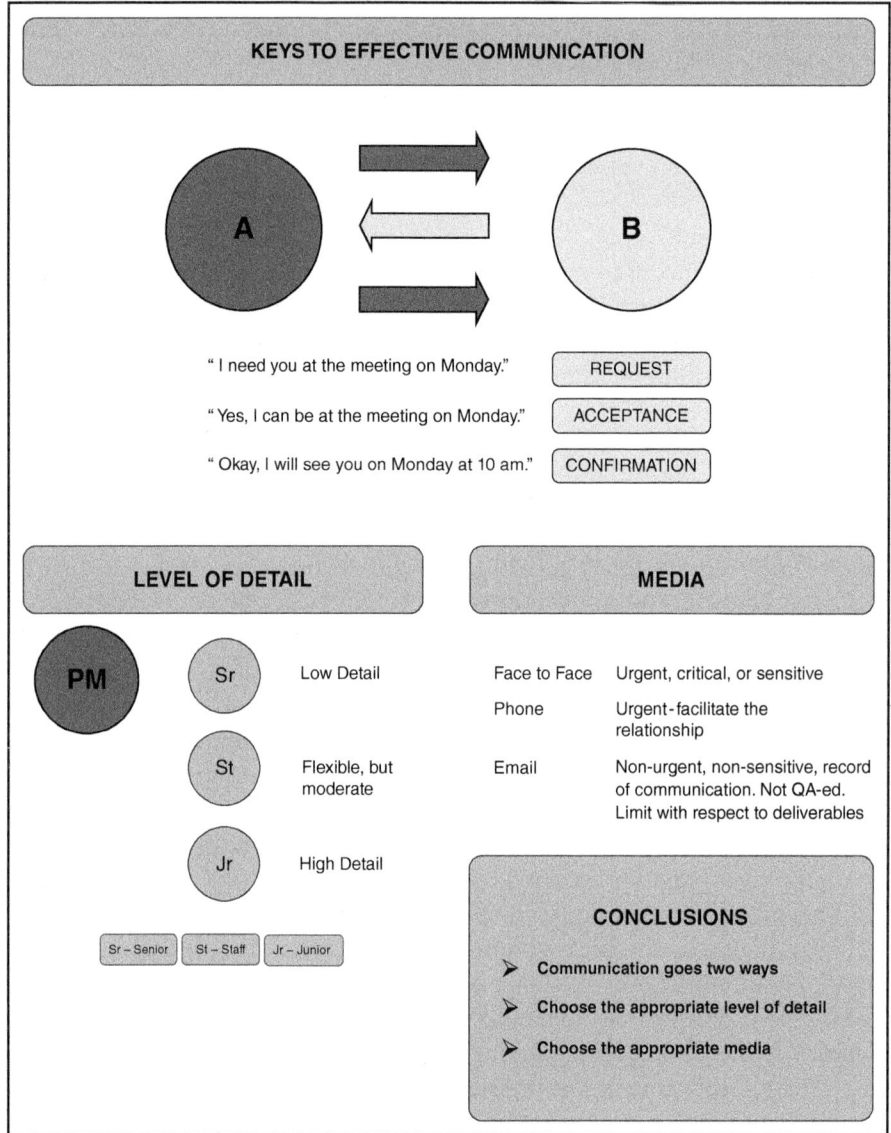

FIGURE 13.1   Keys to effective communication.

Sidebars

Sometimes in the context of an active conversation or e-mail exchange, another urgent but related issue may come up that needs to be immediately addressed. One way to pursue this additional related conversation is referred to as a "sidebar." Engineers have borrowed this informational exchange from the legal profession. During an active trial the judge may call a sidebar where the attorneys approach the judge to have a separate private conversation related to the trial. This process is an acceptable way of calling a timeout from the original active conversation to immediately address a side issue related to the original informational exchange. However, the engineer should exercise judgment and care if this process is used because it may not be appreciated by the other party, and it can be disruptive. However, use of this sidebar tool can sometimes be advantageous if used wisely to break up a tense conversation or to introduce a time delay in the conversation.

Time relevance is also a critical factor for effective communication. Information transfer is related to time. Regardless of whether the information is transmitted via e-mail or face-to-face contact, the receiving party should regard an acknowledgment and/or instruction with a sense of urgency. This urgency may be minutes, hours, or days, but it is professional to respond "as soon as practical" or immediately if the task has become an urgent issue. An outdated acknowledgment or delayed action upon a request that was time critical shows a lack of attention to the job and a lack of consideration regarding those who depend on a timely response.

13.2   COMMUNICATION CONDUITS

Information is exchanged through some type of conduit or medium. The most frequent conduit is the atmosphere, or air, by verbal communication. Another frequent medium is the electronic medium, meaning phone lines or cell towers or electronic messages such as text messages or e-mails. The auditory method, speaking, can be transmitted in air by simple dialogue or electronically by cell tower or land line.

Communication is a process by which we assign and convey meaning in an attempt to create shared understanding. This process requires a vast repertoire of skills in intrapersonal and interpersonal processing, listening, observing, speaking, questioning, analyzing, and evaluating. These processes are developmental and transfer to all areas of life: home, school, community, work, and beyond. It is through communication that collaboration, cooperation, and subsequent action occur.

Tailor communications to the level of knowledge and experience level of the listener(s). Speaking above their level of knowledge may cause your audience to lose interest. Speaking below the level of knowledge of your audience may insult or bore your audience. Know you audience and present data and information appropriately.

Systems of signals, such as voice sounds (intonations or pitch), gestures, or written symbols communicate thoughts or feelings. There are three major parts in human face-to-face communication: body language, voice tonality, and words. According to communications research:

- 55 percent of impact is determined by body language—postures, gestures, and eye contact. Think about this one for a moment. Does your listener (receiving party) truly understand your electronic message with the meaning you intended?

- 38 percent by the tone of voice and volume, and
- 7 percent by the content or the words used in the communication process.

The exact percentage of influence by factors other than spoken words may differ with regard to variables such as the listener and the speaker. There are many scientific studies dealing with "body talk," how to interpret motions and gestures, facial expressions, eye contact, touch, and even foot placement. Kevin Hogan, author of *The Secret Language of Business: How to Read Anyone* in 3 *Seconds or Less*, breaks down body language into eight key elements: eyes, face, gestures, touch, posture, movement, appearance, and voice (Hogan, K. 2008). This example illustrates that body language is a very important component of communication and that "actions may speak louder than words."

Therefore, engineers should be aware of nonverbal communication as a way of interpreting language and also recognize that the importance of picking the proper conduit for transmitting information. For example, experienced engineers will assess the content and nature of a message and choose the most appropriate conduit for delivery of this information. Conveying a convenient meeting time or location is appropriate for e-mail delivery but, conveying an employee injury on the jobsite is an appropriate in-person or phone message delivery.

Communication for Private Sector Consulting Engineers

Communication for private sector consulting engineers is often concise and direct. The project team is very aware of the fixed relationship between scope of work, schedule, and budget. Often in private industry all three elements of this relationship are critical, so efficient communication is essential to stay viable and profitable. Frequently, there is little time for pleasant discussions or conversations.

Communication for public sector engineers is also concise and direct, but often more parties are involved in the communication loop for informational purposes (depending upon the actual agency or department). The public sector team is also very aware of the fixed relationship between scope of work, schedule, and budget. Many public agencies are trying to do more with less, so efficient communication is essential to be responsive to the public and governing political body, e.g., legislature, board of supervisors, city council, and so forth.

13.2.1   E-Mail and Text Message Usage and Limitations

Electronic messaging has its place in communication techniques and styles. However, there are situations where electronic messaging may be totally inappropriate, rude, or possibly dangerous. Delivery of critical and/or sensitive information or data should be conducted by phone or in person. This explains why the military and police employ Chaplains that personally communicate sensitive news about family members to show concern, respect, and empathy.

Most e-mail transmissions are informal and many use verbal shorthand; salutations are minimal or nonexistent and punctuation is often limited to emoticons (angry, frustrated, for example). Here are some of the many advantages of e-mail communication:

- You can communicate quickly with anyone on the Internet.
- You don't have to worry about interrupting someone when you send e-mail.
- You can deal with your e-mail at a convenient time.
- You don't have to be shy about using e-mail to communicate with anyone.
- The cost for e-mail has nothing to do with distance, and in many cases, the cost doesn't depend on the size of the message.

However, although essential, electronic communications do have their drawbacks. Many have great "oops" stories, such as sending messages to the wrong recipient, sending the wrong version of a report or other document, or sending an off-color joke to someone who was offended. There are many limitations of e-mail communication, including the lack of privacy. System administrators can read e-mails, others can bypass security, and still others can save and/or possibly send your message to unknown parties. Additionally, some e-mail systems can send or receive text files only. It's good to remember that:

- It's possible to forge e-mail (think viruses).
- It's difficult to express emotion using e-mail.
- You can receive too much, unwanted e-mail.
- You may not know about the person with whom you are communicating.

Writing (for print) developed over centuries, with an entire industry of writers, editors, proofreaders, publishers, printers, and book stores. E-mail has been in wide use for approximately 30 years and has its own set of unofficial rules. Most of us use our own professional judgment when writing/responding to e-mail communication. Here are some suggestions to keep in mind when using e-mail for business and engineering applications.

13.2.2 Use Clear Subject Lines and Other Recommendations

State your desired response like "Please Review/Comment by a Date/Time," "For Your Records," "Please Approve," and you may receive a faster response. Use a simple, clean format (no fancy fonts, cute symbols, or color). Other suggestions include:

- *Group messaging*: Restrict names in the copy (cc) line to those who need to know. Use the blind copy (bcc) feature when writing to a large list so that the recipient doesn't have to go through a page of names before getting to the message.

- *Discretion*: Don't write anything you wouldn't want your first-grade teacher to see.
- *Monitoring*: Try to check your business e-mail account every hour, if at all possible.
- *Temper*: Don't write or respond in anger. Try waiting for an hour before responding.
- *Proofreading*: Use caution when employing spell check because it's not a replacement for actually reading the message and verifying the attachments are correct. Spell check will not catch all *miss steaks*, as illustrated with this example. Confirm you have included any exhibits or attachments you referenced in your message and they are the correct versions.
- *Efficiency*: Use your electronic address book, which will save time and energy when looking for a contact name.
- *Confidentiality*: If you see a "Confidentiality" statement on the e-mail do not forward it to others.
- *Signature block*: Develop a signature block with your name and contact information. Better yet, set up your e-mail system so that this is attached to all outgoing messages.
- *Subject line(s)*: Often, we respond to e-mail without revising the subject lines. However, if a message was sent to other parties with new responses, possibly over several days, the subject line will likely be inaccurate or outdated. Your receiving party may even dismiss the message because the "old issue" was resolved, or maybe the event is over. *If you update your subject line, you can avoid miscommunication.*

> Do not make the mistake of simply "responding to" OR "forwarding" an electronic message with a "stale" subject line. Sometimes replying to a message that includes an outdated subject line is quick and easy. But doing so can be embarrassing and risk miscommunication. Recipients may quickly review the subject line, think they already have responded, and just delete the message. Seek to be clear with your messaging.

Some servers have limits on the size of attachments, which may restrict the addressee from receiving the intended message and files. It is a good idea to send a test e-mail to the receiving party and/or to verify they received the message and files after it has been sent.

If you would like more information on e-mail communication, visit this site from Microsoft: *https://support.microsoft.com/en-us/office/outlook-best-practices-write-great-email-aa9c7d9e-a3f5-4f8d-bbd3-cf55f11b2df3*

13.3   BODY LANGUAGE—HOW TO READ UNSPOKEN SIGNALS

Note that there are numerous books, articles, and informational courses on body language. The information presented herein is a summary of the importance of knowing body language decoding and encoding in the civil engineering profession. This information is important with potential experiences with clients, the public, your direct supervisor, senior management, and even family and friends. Understanding complete messaging and total communication takes time and experience. Honing these skills will be valuable in your career and personal life.

Total messaging is communication including words, tone of the words we use, eye contact with the person you are talking to, and body and/or hand movements.

Nonverbal cues such as volume and tone of voice, hand and body gestures, and posture all play an important part to the message. In this text box below, body language is briefly defined. In addition, important tips on interpretation are presented so one can understand and communicate with people more effectively (Mind Tools Content Team, 2022).

Nonverbal communication has been shown to carry between 65 percent and 93 percent more impact than actual words spoken, explains Darlene Price, author of *Well Said! Presentations and Conversations That Get Results*.

Darlene Price also explains that body language is a strong contributing factor to how others perceive you, in both a positive and negative light. Eye contact is something that many people need to work on, as it may not necessarily come naturally. The ability to look directly at the person who is speaking with you conveys your confidence and attention (Price, D. 2012).

13.3.1   Body Language Examples

Imagine you are speaking to your colleague, and you present an important conclusion about a problem you are working on together. You make this important statement, and your colleague says, "Really!" You believe your colleague follows your logic and agrees. You're likely correct. Now imagine you make this important statement, and your colleague says, "REALLY?" They tilt their head, furrow their eyebrows, and smirk. You might regard this response as one that questions you, your conclusions, and possibly your analytical skills. And you are likely correct. The words are the same, but the body language tells a very different story. If this message were texted or emailed you would not have the benefit of deciphering the message correctly and likely be misled to the proper conclusion about this interaction.

Here is another example that many friends or couples may have experienced. Together you and your friend or significant other decide to go out to dinner and a movie. You agree on the restaurant and then suggest a movie. Your friend or significant other does not give a clear response indicating a positive yes or no to your movie suggestion. Therefore, you offer another movie suggestion for a similar type of action movie, and you receive a similar non-committal response. Finally, you offer a third documentary movie option, and you hear the response, "Fine." You believe you have both agreed and you may be correct. Now imagine the same scenario as above but after the third movie option is presented you hear, "FINE!" But now thew response is louder and more curt than normal. You might regard the response as one that questions your movie selection, your potential badgering, or whether viewing a movie is even an option at all for the evening activity. And you're likely correct. Again, the words are the same but the body language, specifically the *tone*, tells a very different story.

13.3.2   Body Language

Body Language Primer

Research on *body language* is packed with interesting science. Here, we will discuss the basics of body language, how to read other's body language, and how to improve your own.

The Basics

When we talk about body language, we look at the subtle cues we send and receive to each other nonverbally. Many people want to know how to read body language. To get started, body language can be broken down into a few *different channels*:

Facial Expressions
Researcher Dr. Paul Ekman discovered seven universal microexpressions—or short facial gestures every human makes when they feel an intense emotion. We are very drawn to looking at and observing the face to understand someone's hidden emotions. They are an essential part of body language.

Body Proxemics
Proxemics is a term for how our body moves in space. We are constantly looking at how someone is moving—are they gesturing? Leaning? Moving towards or away from us? Body movements tell us a lot about preferences and nervousness. They are instrumental body language cue.

Ornaments
Clothes, jewelry, sunglasses, hairstyles, are all extensions of our body language. Not only do certain colors and styles send signals to others, how we interact with our ornaments is also telling. Is someone a fidgeter with their watch or ring? Do they constantly self-preen or touch their hair? These are all body language cues.
    There are actually two sides to reading body language in others.
    Decoding is your ability to read people's cues. It is how you interpret hidden emotions, information and personality from someone's nonverbal.
    Encoding is your ability to send cues to other people. This is how you control your personal branding, what first impression you give and how you make people feel when they are with you.
    Read more at: https://www.scienceofpeople.com/body-language

Decoding: Know how to read people and decode body language.

Decoding is one of the most essential people skills. When you think about reading people, you need to understand how to group each body language cue into one of two buckets: a micropositive or a micronegative.

- *Micropositive* signals interest, curiosity, or engagement.
- *Micronegative* signals nervousness, disinterest, or boredom.

In an interaction you want to see more micropositives than micronegatives. Every nonverbal cue you read is about deciding whether it is a micropositive or a micronegative. Here are seven powerful body language cues you should know how to read in people:

1. *Spotting Shame or Embarrassment*
   There is a universal behavior that humans do when they feel ashamed or embarrassed, and it is easy to spot. When people get embarrassed, they often touch the side of their forehead. This is a micronegative. You see this all the time when people are embarrassed.

2. *Blocking*
   Whenever someone feels disengaged, uncomfortable, or closed off, their body shows this feeling with a blocking behavior. This is a micronegative. Blocking is when we cover or block a part of our body as a barrier between us and someone else. We do this subconsciously because we are trying to protect ourselves. Pay attention if someone suddenly crosses their arms, their legs, or frequently holds something— such as or a computer or a notepad or a pillow—in front of themselves.

3. *The Head Tilt*
   Do you hear that? It is a natural human behavior to tilt our head and expose our ear when we want to hear something better. This one is a micropositive.
   If someone's head tilts while the person is speaking with you, it is a great sign. It means they are listening, they are engaged, and they want to hear more.

4. *Mouth Block*
   Have you ever seen a child tell a lie? They frequently tell their lie, and then cover their own mouth. Subconsciously, whenever we are trying to keep something in, we cover our own mouth. As if to say to our brain, "No, don't say it!" The mouth cover is also a common thing to do when we are feeling embarrassed (Van Edwards, V, 2022).

5. *Hands*
   Hands tell us so much about what a person is thinking. In fact, with the raised, open hands gesture, one can imagine the person is saying, "I don't know" or "Huh?," or "What you talking about?"
   Watch someone's hands as they are talking. You should be using more hand gestures. They are a great micropositive because they help keep people engaged.

6. *The Eyebrow Raise*
   When talking about body language, we hear a lot about eye contact—and that's important. But we often forget the eyebrows. Eyebrows are a great little nonverbal secret. This is a great micropositive!

(Continued)

Here's what you want to look out for, the eyebrow raise. Whenever someone is interested, engaged, or curious, they raise their eyebrows. It's almost as if we want our eyebrows to get out of the way so we can see something more clearly.

7. *Facial Expressions*
The last cue is the most important: It's knowing how to decode the face. Facial expressions can be both micropositives and micronegatives, depending on which one you see. Again, Dr. Paul Ekman identified seven universal microexpressions:
Fear, Happiness, Anger, Contempt, Surprise, Sadness, Disgust

Encoding: How to Look and Appear Confident

Let's start with your looks, and no, not your clothes or your hair. This tip is all about your body language.

1. *Your Body Language*
If you want to get dates, win business, and influence people, you have to prep your confidence both inside and out. Often, we're focused only on the words we say in emails, in interviews or in conversation. However, the majority of our communication is nonverbal, the how we say something behind the what we say. Nonverbal communication makes up a minimum of 60 percent of our communication ability. So, if you only focus on your words, you are using only 40 percent of your ability. You have to get into the habit of portraying confidence with both your verbal and nonverbal communication.

2. *How You Carry Yourself, Your Eyes, and Fronting*
When you walk into a networking event or your office or a restaurant, do you look like a winner. Research published in Health Psychology found that participants in a mock interview who sat up straight reported a better mood and higher self-esteem compared to their slouched counterparts.

Next, the eyes have it! Confident people know the power of eye gazing. To increase your confidence, be sure to look people in the eye as you are speaking AND as they are speaking. Too often we look away, check our phone or scope out the rest of the room. This is not only rude, but very low confident.

Finally, engage in fronting. Fronting is when you aim your torso and toes toward the person you are speaking with. Nonverbally, this is a sign of respect. When you do this, you look incredibly focused, confident, put together and charismatic. Be sure to always keep your toes and torso aimed at the person you are speaking with. Show respect, intent, and engagement.

3. *How to Speak Confidently*
One of the biggest mistakes people make with nonverbal confidence is with their voice tone. Confident people never use the question inflection for statements. For example, be careful not to say your name as a question like "My name is Tracy?" You want to make it authoritative by going down at the end of the sentence. "My name is Tracy."

Make sure to use the authoritative tone whenever you answer a question to show you are sure of your words. If you're curious about how to be more confident at work, one of the best ways to feel professionally confident is to have meaningfulness in your job.

Your mood affects your voice. We like hearing happy moods and we don't like hearing irritable moods. Reserve your phone calls for when you're in a quiet place, you're calm and you're settled. Resist the urge to answer when you're stuck in traffic or having a bad day. **Happiness is equated with confidence.**

Read more at: https://www.scienceofpeople.com/how-to-be-confident
Read more at: https://www.scienceofpeople.com/body-language

13.4   CONFLICT RESOLUTION

Even when the proper conduit for delivery of information has been chosen, sometimes communication of a message will cause conflict. The conflict may be a result of miscommunication, the transmittal of unfortunate news (like a delay in the project), or receipt of information in an untimely manner (late transmission).

There are multiple methods to resolve conflict and volumes of publications and many management courses on handling conflict (Weeks, D. 1992) (Erickson, S. et al., 2001). The following discussion of the 4 Cs of conflict resolution is not meant to be a treatise on those publications and courses. Instead, some general tools are offered for handling conflict with colleagues, clients, and stakeholders in engineering business practice.

13.4.1   The 4 Cs of Conflict Resolution

Conflict is inevitable; engineers will experience conflict with clients or stakeholders in their careers and with friends and family in their personal lives. When in a potential or real conflict situation, listening is essential. Empathizing and being cognizant of a client's (or other's) position is crucial, especially when the conflict is related to a project's scope of work or schedule and budget impacts. The engineer should consider alternative approaches that could achieve a winning scenario for all.

The 4 Cs of conflict resolution are four general approaches to dealing with conflict and resolving issues, not always with the best-case scenario result for both parties. As discussed below, the 4 Cs framework includes collaboration, compromise, coexistence, and capitulation.

13.4.1.1   Collaboration

Collaboration is a way of dealing with one another that moves beyond differences to find a "common solution" allowing differing views or desires to be valued, realized, and accomplished. It involves working together to create a win-win solution, or a solution that lets both parties feel validated. One additional tool that can be employed in this situation is referred to as mitigation. Mitigation can be employed when a common solution is available for each party but one party still suffers some damage. It may be possible to offset this damage if the advantaged party considers offering some type of compensation. This compensation may be in the form of additional work, time savings, money, or some other appropriate form of value to be given to the disadvantaged party in order to complete the collaboration. This method of conflict resolution is often the most productive and rewarding for all parties concerned.

> *"Nothing great in the world has been accomplished without passion."*
> —Georg Wilhelm Friedrich Hegel, philosopher

13.4.1.2   Compromise

The art of compromise is the second "C" of dealing with differences. Compromise, similar to co-promise, essentially refers to the effort of negotiating with each other about differences to create a "mutual way" that requires each person to give in to a degree. Ultimately a compromise is a "partial win—partial win" solution for each party. This method of conflict resolution has been employed successfully in the realm of politics and was used often by our founding forefathers when the original colonies agreed to form one nation.

13.4.1.3   Coexistence

This strategy refers to the decision simply to accommodate another person's desire or view and to simply coexist, in essence, agreeing to disagree. For example, this situation exists in many households where one partner in the home is affiliated with a specific political party and the other partner is affiliated with the opposite party. This doesn't mean that they can't live in the same home; it simply means they have agreed to respect one another's position and they agree to disagree on this particular issue. Managers, coworkers, clients, or the public rarely will all share the same opinion on a matter, but that's usually okay as long as there's mutual respect. However, sometimes each party's desires will be so different that the best solution is for each person to go their separate ways.

13.4.1.4   Capitulation

Capitulation involves giving in or acquiescing to the other person's wishes. If you decide to capitulate, you decide to handle a difference by agreeing to the other individual's view or desire. This method of conflict resolution is a "win–lose" solution for the parties involved.

13.5   BEHAVIORAL CHARACTERISTICS OF TEAM MEMBERS, FRIENDS, OR FAMILY

Getting acquainted with general character types and traits may be helpful for the engineer to understand "why" some people do what they do and why they do it a particular way.

Table 13.1 describes general character types and traits that people often display when communicating with one another. Knowing the personality traits of those you work with and for (and those you live with) can make both personal and professional life less stressful. For example, if you know a colleague is a "conflict avoider," when asked whether they will meet a deadline a likely response will be "sure, no problem." The more prudent

TABLE 13.1   Typical behavioral characteristics.

Perfectionist	Controller
Advantage: These individuals are usually detail-oriented and respect quality and accuracy in their work products.	*Advantage:* These individuals have leadership qualities and can provide inspiration to the team members.
Disadvantage: These individuals may need to perform task assignments beyond perfection. There may be an inability to let go of the assignment possibly due to fear of being judged, fear of the next assignment, or other complex reasons.	*Disadvantage:* These individuals may need to perform or be overly involved in many task assignments, which could result in negative impacts on the schedule or budgets for these tasks. This behavior may be due to a lack of self-confidence, the need to have an overabundance of work, the need to be in constant control, or simply the inability to delegate tasks.
People Pleaser	**Conflict Avoider**
Advantage: These individuals want to please people and are generally liked by clients and the public because they appreciate the "can-do" attitude and immediate positive responses from the people pleaser.	*Advantage:* These individuals want to avoid conflict and are generally easy going, very likeable, and generally pleasant. They are team players, and they want to perform well but may be internally conflicted by the need to avoid conflict.
Disadvantage: These individuals have a need to please people and may possibly try to receive immediate gratification from pleasing people with overly positive statements. They may make promises that they cannot keep with a tendency to get the project or assignment deeper in the hole or further behind schedule.	*Disadvantage:* These individuals will avoid direct conflict and sometimes avoid direct questions. They may have a tendency to follow and "go with the flow" and postpone inevitable conflicts that could be avoided earlier in their development. Often when conflicts are recognized early but go unresolved they can become larger and more difficult to resolve. There may also be unstated resentment with "general compliance."

question to this conflict-avoiding colleague may be, "What percentage of the task is currently complete, and may I see the product?" Assuming the colleague complies with the request, the next questions could be, "How much effort is needed to complete the task?" And finally, "Do you believe there is sufficient time between now and the deadline to complete this task?"

This discussion and list of behavioral characteristics is a brief summary of years of management and supervision from a practical perspective. There are numerous published and unpublished references and management training series available on these topics.

By understanding the various types of people and the advantages and disadvantages of each character type and trait, directions and responses can be tailored to get the best performance from each individual.

13.6 TYPICAL REPORT FORMAT

Business, industry, and public service often demand short technical reports. They may be proposals, progress reports, trip reports, completion reports, investigation reports, feasibility studies, or evaluation reports. As the names indicate, these reports are diverse in focus and intent, and differ in structure. However, one goal of all reports is the same: to communicate to an audience. A typical example of an engineering Feasibility Study Report is shown in Appendix C—Example Feasibility Study Report and an example short technical report is shown in Appendix D—Example Short Technical Report.

The following section describes a general format for a short report, which can be adapted to the needs of specific reporting requirements. A format, however helpful, cannot replace clear thinking and strategic writing. Organizing ideas carefully and expressing them coherently is a must. Precision and conciseness also are essential.

13.6.1 Typical Report Sections or Chapters

1. **Title Page**—The essential information here is the company or agency name, the title of the project, authors or contributors, and the date. The title of a report can be a statement of the project and the specific subject of discussion. An effective title is informative but reasonably short. If an engineering report contains conclusions or recommendations, it is generally required that the report be prepared under the direction of a registered professional engineer, signed, dated, and stamped by this engineer. The specific regulations of the state for the project and the location of the engineer's office should be checked.

2. **Table of Contents**—If the report is more than 10 to 20 pages the addition of a Table of Contents will help the audience see the report organization and allow them to quickly navigate to any sections they might want to visit first. The addition of major

subsections, a list of tables, a list of figures, and list of appendices can add clarity and direction to the report.

3. **Executive Summary**—This section contains the salient components of the report in a paragraph for a very short report, or a page or two for a medium length report (10 to 50 pages), or maybe up to 2 to 10 pages for a volume. The whole report is summarized in this one section. Writing one sentence or a paragraph that summarizes each of the traditional report chapters is useful. The problem should be emphasized, and a short narrative on the objective and approach helps set the framework for the reader to understand the scope of work that was derived to reach the conclusion. The data collection, summary analyses, and recommendations also should be addressed. The engineer should not copy a whole paragraph from various sections in the report for placement in the executive summary. The executive summary condenses and emphasizes the most important elements of the whole report and it cannot be written until after the report is complete. This summary should be concise and specific with summary details. A technical report is not a mystery novel and giving the report conclusion right away is recommended.

4. **Introduction**—Whereas the executive summary summarizes the whole report, the introduction of a technical report identifies the subject, the purpose (or objective), and the approach or the plan of development of the report. The subject is the "what," the purpose is the "why," and the plan is the "how." This section acquaints the reader with the problem and the overall approach to solving it. The introduction provides the reader with background information needed before launching into the body of the report. It may be necessary to define the terms used in stating the subject and provide background, such as theory or history of the subject.

5. **Background**—If the introduction requires a large amount of supporting information, such as a literature review or a description of a process, then the background material should form its own section. This section may include the previous history, a review of previous research, and regulations or formulas the reader needs to understand the problem.

6. **Discussion**—This section leads to the most important part of the report. It takes many forms and may have subheadings of its own. The basic components are methods, findings (or results), and evaluation (or analysis). In a progress report, the methods and findings may dominate. A final report should emphasize the evaluation. The discussion should answer the questions: who? when? where? what? why? how?

7. **Conclusion**—The conclusion should be explained in terms of the preceding discussion. It is common to see some repetition of the most important ideas presented in the discussion section, but duplication should be avoided.

8. **Recommendations**—The recommendations should be clearly connected to the results of the rest of the report. Those connections should be made explicit at this

point so the reader does not have to guess at the real meaning. This section also may include a statement as to whether any further investigation or work tasks are required.

9. **References**—References include any sources used to research the report's topic, as well as those directly quoted. A list of references at the end of the report reflects the quality of the research and allows an interested reader to follow up the work.

10. **Attachments**—Typical attachments may include additional references, appendices, or other relevant documents. Research or conclusions derived from other sources referred to in the report must also appear in a list of references at the end of the main report. Appendices may include raw data, calculations, graphs, and other quantitative materials that were part of the research but would be distracting to the report itself. Each appendix should be referred to at the appropriate point (or points) in the text. In industry, a company profile and profile of the professionals involved in a project might also appear as appendices.

An excellent example of a short report titled, "The Benefits of Green Roofs" appears in Appendix D. The author did a good job of incorporating the photos and exhibits into the text to achieve an excellent flow and a thorough, detailed report.

Key Words

To communicate important concepts while reporting information, key words can be used that will relay the information more quickly.

Some words related to *research functions* follow:

Diagnosed	Evaluated	Examined
Extracted	Identified	Inspected
Interpreted	Interviewed	Investigated
Organized	Reviewed	Summarized
Researched	Surveyed	Systematized

Some words related to *detail task functions* follow:

Approved	Arranged	Catalogued
Classified	Collected	Compiled
Dispatched	Executed	Generated
Prepared	Recorded	Completed

"There is nothing more deceptive than an obvious fact."

—Sir Arthur Conan Doyle

13.7   USEFUL FORMS FOR THE ENGINEER

In an effort to do more with less an engineer will often struggle with time constraints and efficiency. Therefore, some useful forms are offered to the reader for carrying out daily or frequently performed tasks.

Much engineering work is conducted through "team efforts" with client interactions. Therefore, some of the most useful forms for documenting pertinent information and decisions are the "telephone record," Figure 13.2, and "meeting notes," Figure 13.3. The forms are designed so the engineer can capture the salient points of the discussion and specifically record the action items, the party responsible for performing these actions, and the targeted completion dates for each action items. These forms offer the basic record information for the engineer. Revisions and modifications to the forms are invited for custom use.

Another form often used by experienced engineers is the "transmittal form," Figure 13.4. This is particularly useful and includes pertinent project information, states the type of deliverable product, delivery information, reason for transmittal (for review or approval), the present form of the deliverable product expressed as draft, final, or 90 percent drawings, and is offered as an alternative to preparing a custom cover letter. Cover letters are a useful tool if time permits custom preparation for each deliverable product. However, the transmittal form serves the same purpose and is easier and quicker to prepare in the event time is limited.

Finally, one additional form offered for consideration is referred to as the "Response to Comment" (RTC) table, Figure 13.5. After transmittal of a deliverable product, such as 60 percent drawings or a draft report to a client, the engineer will inevitably receive comments from one or more other team partners on the project. The RTC table is a valuable tool to record these comments from multiple sources for the record and includes key project data, summary comments, a space for the engineer's response to the comment, and a summary conclusion on how the comment(s) will be addressed. Revisions and modifications to the forms are invited for custom use.

13.8   USEFUL LETTERS (OR E-MAILS) FOR THE ENGINEER

An experienced civil engineer often will find there is a lot more to do or perform in a day than the hours available. Therefore, civil engineers frequently create ways to become more efficient and effective in their careers. One way to become more efficient is to have sample correspondences immediately available for use.

For example, concise messages may be desirable to send when there is a need to:

- Say "thank you" for the support, patience, or information.
- Acknowledge a good job, promotion, thoughtful act, or task.
- Express empathy for a significant loss, or unfortunate event.
- Express an apology for a mistake, a misunderstanding, or a canceled event.

Company or Personal Name Here	Telephone Conversation Record				
Phone: SAMPLE ONLY					
☐ TO: ☐ FROM:		Phone Number			
Name / Title					
Company / Agency		Date:			
Subject:		Time:			
		File / Project #			

Discussion

Item #	Action Item or Task	Responsible Person	Due Date

FIGURE 13.2 Telephone record form.

- Reintroduce yourself to a colleague you haven't spoken to in a while but one you enjoyed working with previously.
- Request a favor, information, or a reference.
- Provide a favor, information, or reference or other types of communications in your area of industry.

Company or Personal Name Here Phone: **SAMPLE ONLY**	**Meeting Notes**	
	Revision #	
Project / File:	**Date**	
Purpose:	**Time**	
Attendees:		

Discussion/Conclusions

Item #	Action Item or Task	Responsible Person	Due Date

FIGURE 13.3   Meeting notes form.

Company or Personal Name Here			Letter of Transmittal		
Phone:					
SAMPLE ONLY		Revision			
To:			Subject		
Company;			Phone Number		
Address:			Date:		
			Time:		
			File / Project #		

Item #	Deliverable	Version
	Remarks	

	Delivered Via		Transmitted
☐	Overnight Delivery	☒	As Requested
☒	Messenger	☐	For Approval
☐	US Mail	☒	For Review and Comment
☐	Email	☐	For Your Reference

FIGURE 13.4   Letter of transmittal form.

			Revision #			
Company or Personal Name Here						
Phone:						
SAMPLE ONLY						

RESPONSE TO COMMENTS TABLE

Agency:		Project Location:		APN		Date:
Project Name:				Plan Check No.:		
Document Being Reviewed/Revision Number:						

COMMENT NO.	SHEET NO./ DWG NO	COMMENT	RESPONDENT	RESPONSE	CONCUR	NON-CUNCUR	FIO
REVIEWER:							
GENERAL AND CIVIL DRAWINGS							

FIGURE 13.5 Response to comments form.

Developing good samples of targeted electronic or written communications can save time. This assumes that electronic messaging or written communication is actually more appropriate that phoning or speaking in person. Filing easily retrievable samples of these types of brief communiques can save time and allow more opportunities for enjoyment of life outside of work.

Valuable Lessons I Learned from My Clients

Bridget Crenshaw Mabunga, Adjunct Professor and Technical Writer

It's amazing how easy it is to forget our audience. Whether through written or oral communication, it is imperative to understand the needs of your audience. When I teach, I consider what my students may or may not be familiar with, and I try to be as transparent as possible. The same transparency is necessary in my work with clients. When in doubt, I make sure to clarify my communication and check in with the client or student to verify that we have a shared understanding of the expectation or goal at hand.

In addition to knowing our audience, we must make sure that we develop what I call multicultural communication awareness. I teach a multitude of students with a variety of cultural backgrounds, and each cultural group has particular communication strategies or comfort levels. For example, I know that often my students of Asian descent (Chinese, Japanese, Filipino, Vietnamese, for example) struggle with asserting their points and opinions because their cultures are largely focused on community and less focused on the individual, as we are in American business culture. Thus, I have to spend a little extra time helping them navigate the rigors of academic writing, which requires asserting points and making arguments.

Another cultural consideration is the relationship with time. After having lived in Costa Rica, I learned that time is not as fixed as it is here in the States. I would make appointments to meet my Tico friends and I would show up on the dot, but they consistently arrived up to an hour late, and were unapologetic. Their relationship with time was more flexible than mine, and I had to adjust to that. In addition, in many Latin cultures, a business lunch will begin with friendly conversation and may take quite some time to get to the business at hand, as it is culturally valuable to develop a friendly relationship before getting into the business aspect of the meeting. Ultimately, when considering multicultural communication, we may have to set our cultural norms aside and do a little more negotiating to make sure our message gets through in the way we intend within the time frame necessary. The best way to navigate these cultural differences is to know your audience; if you are working with a client from another cultural background, take some time to familiarize yourself with the client's cultural norms to stave off communication struggles over the course of your relationship.

13.9   SAMPLE POWERPOINT PRESENTATION

PowerPoint has advantages and disadvantages for presentation of information to groups. However, it is still a popular tool for informational presentations. **An example of a good PowerPoint presentation** prepared by senior civil engineering students is presented on the support website at www.wiley.com/go/cehandbook, Example PowerPoint Presentation. This presentation was prepared in response to the sample "Request for Proposal" discussed in Chapter 4 and included in Appendix A, Example RFP. The presentation is the culmination of the Final Feasibility Study Report prepared by the student group that called themselves CVision Engineering. The Feasibility Study Report is also presented in Appendix C, Example Feasibility Study Report, for consideration as a good example of a report.

13.10   SUMMARY

Excellent communication skills are essential to the civil engineer's successful career. The information in this chapter provides civil engineers the knowledge and tools to improve their communication skills, effectiveness, efficiency, understanding, content, and transmission.

Communication is the process of exchanging ideas, information, feelings, or data from one party to another. The information transfer can occur verbally, nonverbally, electronically, physically (by touch), or in writing. Complete communication occurs when all modes of information transfer are engaged. An acknowledgment of information receipt and a brief reiteration of the message by the listener will provide the sender with an opportunity to confirm or deny the accuracy of the transmission. Examples of efficient communication tools are discussed and included for the engineer's use. Additional examples of typical short engineering reports and medium-length technical reports are also included for reference.

BIBLIOGRAPHY

Erickson, Stephen K. and McKnight, Marilyn S. (2001). *The Practitioner's Guide to Mediation: A Client-Centered Approach*. John Wiley & Sons, Inc., New York. ISBN 0-471-35368-X.

Hogan, Kevin. (2008). *The Secret Language of Business: How to Read Anyone in 3 Seconds or Less*. John Wiley & Sons, Inc., Hoboken, NJ. ISBN 978-0-470-22289-8.

Mind Tools Content Team. (2022). *Body language, beyond words—how to read unspoken signals*. https://www.mindtools.com/pages/article/Body_Language.htm (accessed 12/22/2022).

Price, Darlene. (August 22, 2012). Well said! presentations and conversations that get results. AMACON Publishing.

Van Edwards, Vanessa. (December, 2022). How to be more confident: 11 scientific strategies for more confidence, science of people. December 2022 https://www.scienceofpeople.com/how-to-be-confident.

Weeks, Douglas A. (1992). *The Eight Essential Steps to Conflict Resolution*. Tarcher Putnam, New York. ISBN 0-874-77751-8.

CHAPTER 14

Balancing Life, Family, and Career

Big Idea

Successful engineers must learn how to integrate a busy career with community, family, and self. The balance among these elements is essential for "having a life."

> "Life loves to be taken by the lapel and told: 'I am with you kid. Let's go.'"
>
> —Maya Angelou

Key Topics Covered

- A Concept from Physics and Thermodynamics Called Negentropy and Posentropy
- The Key Components to Your Living Being
- The Mind
- The Body
- The Spirit
- The Effective Combination of Mind, Body, and Spirit
- Laugh and Have Fun

Civil Engineer's Handbook of Professional Practice, Second Edition. Karen Lee Hansen and Kent E. Zenobia.
© 2025 John Wiley & Sons, Inc. Published 2025 by John Wiley & Sons, Inc.
Companion website: www.wiley.com/go/hansen/CivilEngineersHandbook

- Self-Assessment Test – Please Challenge Yourself
- Analysis of Self-Assessment Test
- More Ideas for Creating Life Balance

Related Chapters in This Book

- Balancing Life, Family, and Career is related to Chapters 1 through 21 and Appendices A, B, C, D, E, F, and G

14.1 INTRODUCTION

Ironically, this chapter of the book is the shortest, but it is probably the one of the most important. Some might ask, "Why?" Because ignoring our innermost desires and needs can extinguish a portion of ourselves. This does not mean that we should forgo the pursuit of a meaningful career, love, or family, but we should create time and a mechanism to pursue the things in life that we enjoy and "have fun."

Very experienced engineers have commented on this chapter and provided their input on balancing an engineering career with a personal life. That's just it: life should be balanced. The balance won't look like the scales of justice where the "lady of justice" carries scales that balance the strength of a legal case against the opposition to the case.

Life balancing is more like balancing a pizza on a pencil as depicted on Figure 14.1. The number and size of pizza slices represent the many components of our lives such as career,

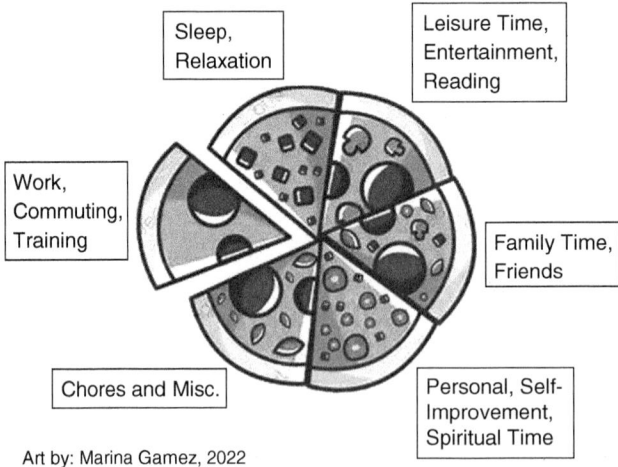

Art by: Marina Gamez, 2022

Creating and maintaining a healthy balance in Life, Family, and Career is Key to success, happiness, and joy! When one segment becomes overloaded OR overlooked, the balance is lost and life can wobble.

FIGURE 14.1 Balancing life, family, and career.

family, personal time, fun, exercise, sleep, volunteer work, and more. One can imagine that when any pizza slice becomes too large and overwhelming, the pizza becomes unstable and out of balance. Once we experience an "out of balance situation" the potential for disruption, conflict, and hurt feelings and can easily occur.

14.2   A CONCEPT FROM PHYSICS AND THERMODYNAMICS CALLED NEGENTROPY AND POSENTROPY

Let us examine life balancing in two separate ways;

- One method at a higher level, intellectually, through physics and thermodynamics referred to as negEntrophy and posEntropy, and
- A second method from a logical common-sense approach by analyzing the key components to your living being

There will not be a detailed treatise of thermodynamics here, but the reader is free to research those details. Based on Carr-Chellman's research into negEntropy, a five-step guide to reverse energy loss (and reduce chaos) in daily life is suggested:

1. **Find the negEntropy—Where am I devoting lots of energy and time?**
   Identify places where energy is lost in the social systems of your daily life. Look at time and energy sinks. Answer the question, "Where do I become bogged down, lose time, and lose energy?" For example, a messy, disorganized office makes it hard to find things, documents, and key items needed for forward progress.
2. **Prioritize the losses—Which tasks and/or activities are the highest drain on me?**
   Identify the largest or most annoying losses and those that draw your attention most often. For example, constantly searching for key documents or your house keys or cell phone may be a priority for improvements to your office organization that would make it more functional.
3. **Come up with a plan—How can I mitigate (fix) this?**
   Identify actions that will reverse the energy losses you noted and plan ways to address the highest priorities first. You could start by dedicating special places for items or maybe setting up a daily and monthly activity file. Analyze the problem and figure out how to mitigate or avoid the problems.
4. **Try it out and pay attention—Implement your priority plan and monitor results.**
   Put the ideas into action, but stay focused on energy gains and losses. As you try to mitigate negentropic traits and behaviors, keep track of what works, how much effort it took and ideas you come up with for future corrective actions.
5. **Go beyond fixing and maintenance—Can I expand this implementation plan into other areas of my life, career, family, and personal time?**
   As you work to reverse energy losses, you may find that at times you are actually maintaining a social system that is not beneficial no matter how smoothly it works.

So, this process can apply to your office and career as well as to your social life. In fact, integrating this process into your career, social life, family, and personal time may actually create an better life balance. The best way to apply corrections of negentropy to social systems is to not only improve the small processes, but also to look at the big picture and see if the status quo itself promotes energy loss.

Carr-Chellman notes: "Seeing things through a negEntropic lens will not solve a bad relationship or help you love a job you hate—those are complicated issues. However, if you begin to notice where energy is lost in your life, it will be easier to prioritize and act in ways that can improve the social systems around you." If you believe these concepts would benefit you in some way, try this for two weeks and monitor your own progress toward balancing life.

Finally, to close the loop completely, let us discuss posEntropy. The concept of *posEntropy* is the process of energy gain in a system. For our discussion here, let's examine how tasks and activities improve your career direction, family interactions, personal time, and life balance. Here is a look at the posEntropy process and this five-step guide to a better-balanced life that complements negEntropy.

1. **Find the posEntropy—Where am I experiencing positive feelings, quality time and accomplishments?**
   Identify places where energy is gained in the social systems of your daily life. Look at time and energy gains. Answer the question, "What actions, tasks, or activities are generating positive results in my career, family life, and personal time?" For example, commuting to work earlier saves time and allows one to leave work earlier in the day to spend quality time with others.

2. **Prioritize the posEntropy—Which tasks and/or activities are providing the highest gains for me?**
   Identify the largest or most productive activities that provide forward direction and momentum. For example, biking with loved ones after work improves our health, promotes relaxation, creates family time, and personal enjoyment.

3. **Come up with a plan—How can I enhance this?**
   Identify actions that promote energy gains. Look for ways to address the highest priorities first. You could start by dedicating time for fun and creative activities at work or home to enhance enjoyment. Search for ways to improve your life and happiness for you and those around you.

4. **Try it out and pay attention—Implement your priority plan and monitor results.**
   Put your posEntropy ideas into action, but stay focused on energy gains and losses. As you try to implement posEntropic concepts and ideas, keep track of what works, how much effort it took and ideas you come up with for future posEntropic actions.

5. **Go beyond maintenance and look for improved happiness and energy gains—Can I expand this implementation plan into other areas of my life, career, family, and personal time?**

As you work to enhance energy gains, you may find time to expand your social system into new areas of engagement. So, this process can apply to your office and career as well as your social life. In fact, integrating this process into your career, social life, family, and personal time will actually create a better life balance that complements the negEntropy process described above.

If you begin to notice where energy is gained and lost in your life, it will be easier to prioritize and act in ways that can improve the social systems around you. Try integrating the negEntropy and posEntropy guide for two weeks, monitor your results, and see if it works for you. This process should enhance your career, family, and personal time to create an improved life balance.

NegEntropy Primer

Based on Alison Carr-Chellman's *A concept from physics called negentropy could help your life run smoother*

Engineers first encounter the concept of entropy in physics and thermodynamics classes. Entropy is discussed in the second law of thermodynamics. The second law states that entropy is a physical property of a thermodynamic system. Yes, your life is considered a thermodynamic system! "Entropy predicts the direction of spontaneous processes, and determines whether they are irreversible or impossible, despite obeying the requirement of conservation of energy. Conservation of energy is established in the first law of thermodynamics" (Carr-Chellman 2021).

"Entropy is a measure of how much energy is lost in a system. If a system loses too much energy, it will disintegrate into chaos. Entropy is also a measure of the molecular disorder, uncertainty, or randomness, of a system. Boiling water is an example of entropy. As water is heated it converts to a gaseous form where the molecules are then free to move independently through space in a random disorderly fashion. Another example more relevant to our daily lives is performing quick minor maintenance on our home which does not require a lot of energy. However, if one does not take care of their home's yard, exterior, interior, drainage, and minor repairs the work adds up to a chaotic home that would take a lot of energy to fix. That chaos will leach away your time and ability to accomplish other things. A concept from physics called negentropy could help your life run smoother" (Carr-Chellman 2021).

In both physics and social systems, energy can be defined as the capacity or ability to do work. Others began to wonder whether applying physics concepts to social systems could help them run better. Hence, the concept theory of negentropy was born. Research "suggests that when people keep the idea of negentropy in mind and take actions that limit or reverse energy loss, social systems are more efficient and effective. This might even make it easier for people to achieve larger goals" (Carr-Chellman 2021). So, yes, it makes sense to apply small amounts of energy to keep your home running smoothly, or perhaps to improve your career and work efforts. Doing so may allow you to see other ways to avoid future energy losses.

"No man is a failure who is enjoying life."

—William Feather, American publisher

14.3   THE KEY COMPONENTS TO YOUR BEING

So, what does *"having a life"* really mean? Some would say, "having time to do what I love to do" or perhaps "being around friends and loved ones more often" maybe "having a meaningful relationship and being part of a family" or possibly "making a contribution to society" or "having children" possibly "being famous" or "being wealthy." Life is more than the tiny electrochemical charges that stimulate our mind and body. Life is laughing, crying, feeling, sharing your time or meals with someone and seeing wonderful things like the colors of nature, hearing the sounds of life in a forest, or feeling the breezes at the beach.

Life is many of these things and *taking the time* to make them happen, celebrating these events, reflecting upon the events, and planning and enjoying new ones. The important element is *taking the time to create our "own personalized balance."* Our lives become cluttered and we almost always have somewhere to go or something to do. So, it will be up to you to take the time to have a life, to feel fulfilled, engaged, happy, and enthusiastic. And, finding a balance among career, family, friends, and personal time in your life will be critical.

The key components to life are:

- Mind—The Command Center
- Body—Our Home
- Spirit—Our "inner self," sometimes referred to as our soul

Establishing a real-life balance with mind, body, and spirit will lead toward achieving a fulfilling life.

14.4   THE MIND

14.4.1   The Command Center of the Body and Our Inner Self

The mind is like the "bridge" on a ship or the "cockpit" on a jet. It's the control center for the body and spirit. The mind is the center for all intellectual thought, memory, emotion, and free will. It's the command center for all the major systems in the body and without it the body cannot function.

"If you maintain the ability to use your memory, concentrate on intellectual thought, and focus critical thinking toward solving a problem, your mind should operate near its highest potential. Studies have shown that frequent use of the brain as you age helps maintain the ability to function effectively. Keeping your mind intellectually stimulated may strengthen brain cell networks and help preserve mental functions" (Katz, Rubin 1999). If you can synchronize the optimum operation of the mind with the body and the spirit, you can operate at the highest level possible for you as an individual.

The most important professional component that our mind controls is the belief in what we do and how we view our job performance. A good self-image shines through to others, especially when combined with a confident, positive attitude. How we handle problems and challenges can be more important than the actual solutions we devise. Self-confidence balanced with professionalism and humility is a tremendous asset to an engineer. When your personality and character are genuine, the art of presenting yourself at your best is simple. Spend some time identifying your best "selling" points and build on those as you enhance your mind.

Believe it or not, we sell ourselves more than we think. Think about it:

- To enter college you had to sell yourself on your application and explain why you should be chosen above other applicants.
- To become employed you had to demonstrate how you could be an asset to your employer.
- To meet people (that are now your friends) you had to relax, be yourself, and show how you might be an interesting person to another.
- We are constantly selling ourselves as responsible citizens, dedicated employees, worthy partners, concerned parents, and good people.

14.4.2   What about Stress?

You just yelled at your life partner. Your children have taken to calling you the Commander. You start daydreaming about drinks after work. You find you're forgetting things—important things. Your family says you're never home. You can't seem to concentrate. What's your problem?

In a word: Stress. (Sternberg, Esther 2001)

Welcome to the club. Almost half of all Americans are concerned about the level of stress in their lives, according to the American Psychological Association's 2009 and 2010 Stress Survey. Chalk it up to "our overscheduled, harried 21st-century lifestyle, which can wreak havoc with our relationships and our work," says Bruce S. McEwen, M.D., a coauthor of *The End of Stress as We Know It.*

Stress also plays an important role in our overall health. Chronic stress weakens the immune system and increases the risk for a range of illnesses, including heart disease and depression. Stress drives people to eat too much, sleep too little, skimp on exercise, and short-change fun. It doesn't have to be toxic; a little stress can sharpen focus, improve memory, and heighten emotions. But sometimes good stress goes bad, and researchers have just begun to figure out how to deal with it. "By understanding what makes it go wrong, we have the power to make it right" (McEwen 2002).

14.4.3   The Stress Response

In its most basic form, the stress response is known as "fight or flight," and it swings into action whenever you're confronted with a novel or threatening situation.

"If you step off the curb in front of an oncoming bus, your body reacts automatically to protect yourself" (Sternberg and Esther 2001). In a matter of seconds and without even thinking, you begin pumping out brain chemicals and hormones, including adrenaline and cortisol. Your heart rate accelerates, oxygen-bearing red blood cells flood the bloodstream, the immune system gears up for the possibility of injury, and energy resources are diverted to your muscles, brain, heart, and lungs and away from functions, such as digestion and hunger, which can wait until the crisis has passed.

Meanwhile, the brain releases a cascade of endorphins, the body's natural opiates, to dull the pain of those potential injuries. You're ready for action, whether it's a full-out battle or a hasty retreat in this case, fleeing back onto the sidewalk to escape the speeding bus.

When the danger has passed, all these systems are restored to their normal resting state. "Your stress response makes you get out of danger. Without it, you'd be dead. Many of the physical changes that energize you to get out of the way of the bus are the same ones at work in more positive situations. Your heart races when you're falling in love or more accurately perhaps, when you decide to love another (McEwen, B, 2002)."

Developing a practice of *mindful relaxation* can be a great help. Physiologically, relaxation is the opposite of stress. When you're relaxed, your breathing and heart rate slow and your mind clears. Mindfulness is a way to achieve this level of relaxation using a variety of techniques, including yoga, meditation, and simple relaxation exercises.

Mindfulness quiets your mind by teaching you how to observe your thoughts and feelings without viewing them as positive or negative. It trains you to use your breathing and an awareness of your body to focus on the here and now. The basic relaxation response was first described in 1975 by Harvard Medical School researcher Herbert Benson (Benson, M.D., H. et.al. 1976).

His approach has two steps: First, close your eyes and focus on your breath (that's the foundation). Second, choose a phrase, a word, or a prayer and repeat it to stay in the moment and be mindful. "I use two phrases," says Bernadette Johnson, director of integrative medicine at Greenwich Hospital, in Connecticut. "I'm breathing in 'relaxation and peace' when I inhale. I'm breathing out 'tension and anxiety' on the exhale."

Ideally, you'd begin and end your day with 10 to 20 minutes of regular relaxation exercise. But should you find your tension rising during the day, "take a deep breath, hold it for a count of four, and exhale for a count of four," Johnson says. "That's what we call a mini, and if it's built on a foundation of regular, longer relaxation exercises, you can tap into it whenever you need it."

If a thought or an emotion intrudes on your mindfulness and threatens to take you out of the moment. Think of it as a leaf floating by on a slow-moving stream. Observe it but don't react to it.

14.5 THE BODY

Consider your body as your home for your mind and inner self. In the engineering profession, we consider the operation and maintenance (O & M) of any structures we design or build. Good O & M practices extend the usefulness, functionality, and viability of these structures. In our career as engineers, these structures may be buildings, roads, bridges, dams, treatment plants, airports, wetlands, landfills, or operational processes. As we carry out our engineering practice, we should recognize that we are operating out of our bodies, our home. And, maintaining our home will extend our own viability and usefulness.

A trip to your medical practitioner will provide detailed advice and instructions for each individual. There is no substitute for this professional counsel. There are literally volumes and millions of pages of references on health and the human body. In general, here are some relatively simple rules from the personal experience of the authors to increase our effectiveness:

- Eat Well
- Sleep Well
- Exercise Well

> "Life is really simple, but men insist on making it complicated."
>
> —Confucius

14.5.1 Eat Well—The Balanced Diet

Food provides the nutrients to run an extremely complicated machine—our bodies. As learned in elementary school and virtually every health magazine and other prestigious references, a balanced diet is required to help maximize performance. Maximum performance will be required to excel in an engineering career.

14.5.2   Sleep Well—Save Space for Dreams

Our bodies require rest to accommodate for daily stresses and to relax the muscular, circulatory, and nervous systems. Most references state that adults should strive for about eight hours of sleep each night. In general, they also state that sleep deprivation often seriously affects memory and cognitive abilities and that deficit hours cannot be made up on the weekends. In an effort to excel at our careers and pack the most into our lives, sleep is often sacrificed. There is a price to be paid for sleep deprivation and it varies from slower response to higher accident rates. Sleep is very important for adults to succeed in their careers. Sleep disturbance can be a sure sign of stress or other impacts on your life, and it should be addressed with your medical practitioner.

If sleep is disturbed with thoughts of "things to do" or the "next day's activities," keeping a pad and pen next to the bed may help to purge these thoughts, to allow sleep to come. Other tips include allowing time for some pleasant reading material (no, not technical journals or homework), soft music, or favorite pleasant sounds like a running stream to induce rest. Some additional advice includes thinking pleasant thoughts as one drifts off to sleep.

14.5.3   Exercise Well—A Healthy Body Will Facilitate a Healthy Mind

Our bodies also require exercise to maintain the muscle groups, the heart, and circulatory system. Most references cite the need for at least 30 minutes of strenuous exercise per day, at least 3 to 4 times per week. The exercise could be as simple as a brisk walk for 30 minutes or mixing exercise and fun like tennis, biking, swimming, or other active sports.

In general, health magazines and references state that the lack of exercise can induce weight gain and early onset of diseases. Like sleep deprivation, the lack of adequate exercise cannot be made up on the weekends. This condition is referred to as a "weekend warrior" and is often accompanied by soft tissue, muscle, or other injuries. Exercise is very important for adults to excel in their careers.

14.6   THE SPIRIT

Spirit is our inner self where we store our feelings and thoughts. Spirit is referred to by some as our soul or inner essence and is therefore metaphysical in nature. In this regard, the spirit is closely related to our mind and body. It can be thought of as the spark of life that actually displays our character, emotion, personality, and consciousness that connects our mind and body to form unique individuals, different from any other in the universe. Many people connect spirit with a religion or their own belief or concept of a god, deity, or higher power. Other people may not believe or accept a god, deity, or higher power but that doesn't mean they don't have spirit.

Our intellect and learned information are housed in our physical brain. Our spirit seems to be primarily in our metaphysical mind. An example might be "instinct" which can be seen as a combination of learned information and feelings.

Instinct can be thought of as our innate behavior. As an example of innate information and pure instinct, one can observe baby sea turtles. After hatching from the sandy beach, they automatically move toward the ocean on their own without any guidance from their parents. These instinctual actions are contrasted with learned information, which is taught to us and stored in our memory (physical brain) for future retrieval and use. Sometimes instincts are hard-wired ready for implementation after a maturation process.

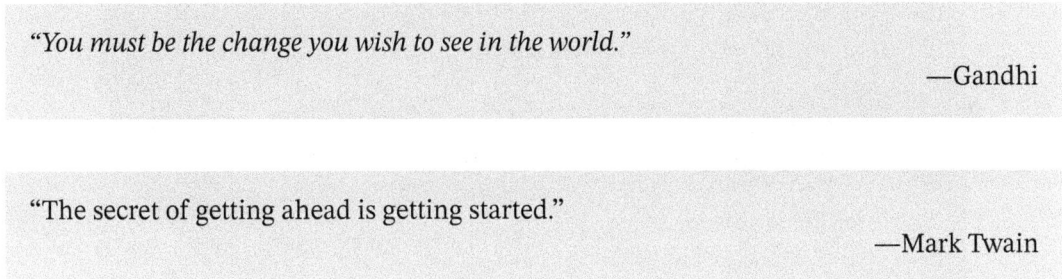

"You must be the change you wish to see in the world."

—Gandhi

"The secret of getting ahead is getting started."

—Mark Twain

14.7   THE EFFECTIVE COMBINATION OF MIND, BODY, AND SPIRIT

Okay, so how does this all relate to *having a life* and being an engineer? It's very difficult to articulate, but it seems that if we can integrate the learned and innate information in our mind while maintaining a well-tuned and healthy body and synergistically combining our individual spirit and personality, we have the ability to solve (or answer) extremely complex and sophisticated problems. In this manner, our conscious mind can draw upon megabytes of learned information from all conscious and subconscious levels (innate thoughts) integrated with our spirit (and personality) operating multi-dimensionally to reveal these answers.

14.8   LAUGH AND HAVE FUN

14.8.1   Laughing—Don't Be Too Serious

What is the value of a laugh? Laughing is fun; it's healthy, and contagious. Humor can make your job enjoyable and rewarding. It can lighten stress and tense moments and be one of your secret weapons. If you have some tense moments, consider whether a lighthearted statement and a brief laugh would lighten the mood, then move on to address the situation.

FIGURE 14.2   Balanced life.
Source: Wikimedia Commons.

Laughter reduces pain and allows us to tolerate discomfort as you might feel when you view Figure 14.2 or Figure 14.3. It reduces blood sugar levels, increasing glucose tolerance in diabetics and nondiabetics alike. It improves your job performance, especially if your work depends on creativity and solving complex problems.

The role of humor in intimate relationships is vastly underestimated and it really is the glue of many good marriages. It synchronizes the brains of speaker and listener so that they are emotionally attuned. Laughter establishes—or restores—a positive emotional climate and a sense of connection between two people. In fact, some researchers believe that the major function of laughter is to bring people together. And all the health benefits of laughter may simply result from the social support that laughter stimulates.

Recently, there has been hard evidence that laughter helps your blood vessels function better. It acts on the inner lining of blood vessels, called the endothelium, causing vessels to relax and expand, increasing blood flow. In other words, it's good for your heart and brain, two organs that require the steady flow of oxygen carried in the blood.

"The research doesn't say for sure exactly how laughter delivers its heart benefit. Thirty minutes of exercise three times a week, and 15 minutes of laughter on a daily basis is probably good for the vascular system" (McGhee 1999).

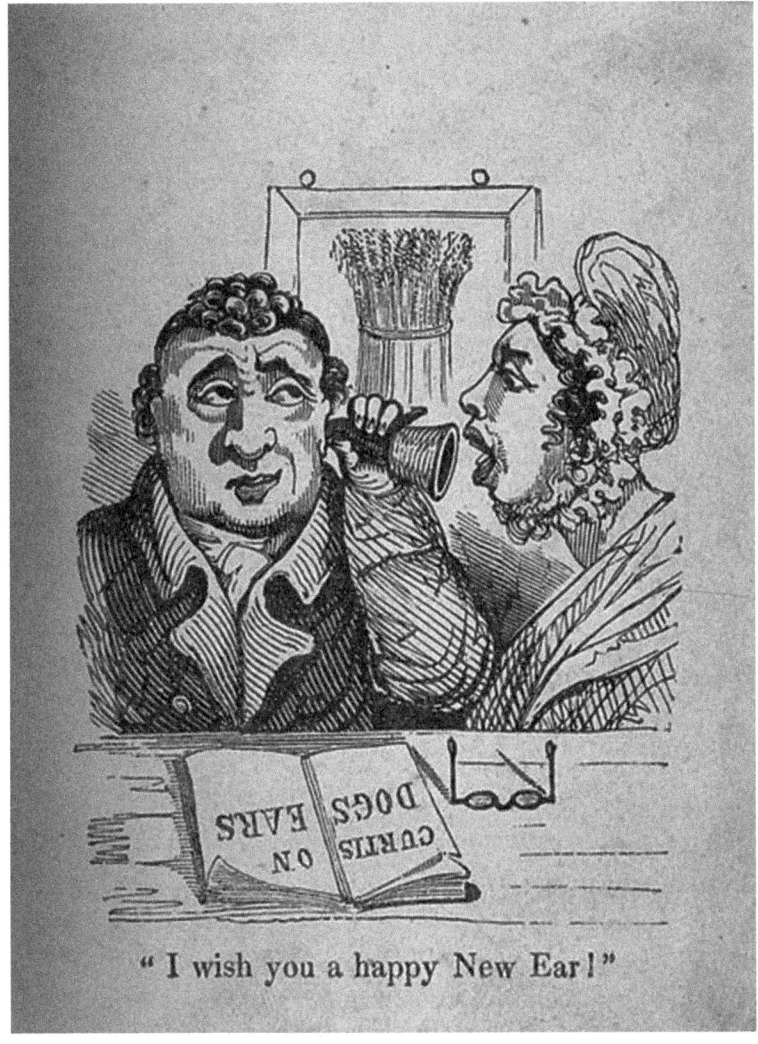

FIGURE 14.3 Humor is important.
Source: Wikimedia Commons.

14.8.2 Personalize Your Fun Time

Personalized fun time can be a real treat. Think of something you're good at and something you really like to do. Maybe this activity will be swimming, fishing, running, biking, volunteering, snowboarding, or skiing. Now, think about how you might personalize this activity, with your own style and grace. Use something unique about you that applies your personal attributes to the activity. Then, embrace it, increase your passion for it, learn from others, and share your knowledge and experiences.

14.9 SELF-ASSESSMENT—PLEASE CHALLENGE YOURSELF

14.9.1 Mind: What Do You Think, Really?

It's often a challenge for analytical people like engineers to assess their own thoughts and feelings and express them, even to ourselves. It's important to get in touch with your inner thoughts and to capture these feelings for yourself. It may help to go to a very quiet place where you feel safe and very comfortable. Try this activity for 10 minutes weekly. Then compare what is going right with any items that require your attention. Congratulate yourself for items going well. Prioritize any items that need attention, create solutions, seek help if needed, and act upon tasks for correction.

1. Concentrate on your life components and "having a life."
2. Think about how you feel emotionally.
3. Focus and come up with some things that are going right for you.
4. Now consider some things that may need attention.
5. Develop a plan to address any issues such as the ones described earlier in PosEntropy and NegEntropy.

14.9.2 Body: How Does Your Body Feel, Really?

Many of us don't like to look too closely in the mirror. Make an appointment with your closest glossy surface and take a close look. Check for signs of stress, your range of motion, try touching your toes, or just stand up straight. When was the last time you ran or walked or even noticed the weather? If the weather question stumped you, walk to a window in your office at least once a day for a month. You'll probably feel more like going outside if you see daylight on a regular basis. Give yourself permission to take a 15-minute walk during your lunch break. Try this exercise, take brief notes, and focus on your body. Does anything need attention?

1. What is your honest assessment after the mirror exercise?
2. Think about how you feel physically.
3. Focus and come up with some things that are going right for you.
4. Now consider some things that may need attention.
5. Develop a plan to address any issues such as the ones described earlier in PosEntropy and NegEntropy.

14.9.3   Spirit: How Do You Feel, Really?

Check in with yourself regularly. Ask yourself how things are going, where they are going, where you would like them to go. Include all "the slices of the pizza" such as your career, your hobbies or favorite sport, your family and friends. Now, try this activity for ten minutes weekly and consider the positive feedback and any action items from the "mind and body" assessments you performed above. Congratulate yourself for things going well. Stay focused.

1. What is your honest assessment of your feelings?
2. Think about your relationships with family, friends, coworkers, and community.
3. Focus and come up with some things that are going right for you.
4. Now consider some things that may need attention.
5. Develop a plan to address any issues such as the ones described earlier in Posentropy and Negentropy.

14.10   ANALYSIS OF THE ASSESSMENT

14.10.1   On a Scale of 1 to 10...

How happy are you today? How happy are you compared to last week, last month, or last year? What could you do to be even happier? Is it time to make a plan?

- So, how do you feel—mad, sad, glad, or scared? Consider your mission or direction in life and in your career. Do you need to make any adjustments or revisions? Are any life components out of balance? Is it time to make a plan? It may help to take some brief notes to capture your thoughts. Now, place these feelings or notes in a little box in your mind or in a drawer, or someplace safe if it's your physical notes. Temporarily put them away for the night or for a day or two.
- Periodically, take out the box and re-examine the thoughts you had last time you put them away. How do you feel now? How does the current score compare with your previous score? Are any revisions necessary? Do you need to take any action at all?

Life is about balance among mind, body, and spirit. When things are given equal importance and one thing does not overshadow the other, then you have the perfect "pizza on a pencil" balance. If you are out of balance, perhaps you could identify the area of weakness and plan to fortify this area. For example, if you need to sharpen your mind, take some college courses, learn a foreign language, or take a cooking class. If your body needs attention, find a personalized sport or activity where you can combine fun with exercise. If your spirit is waning, reconnect with an old friend, make new connections with colleagues or neighbors, or spend a day with your child and see the world through their eyes.

> "Vision without action is a daydream. . .Coupling vision and action—in other words, executing strategy—is one of the biggest challenges we face."
>
> —Japanese proverb

Take Time . . .

Take time to enjoy the little things. Make a conscious effort to notice things you enjoy, focus on them during the moments you have, and remember them as part of your day. For example, on your commute to work one might notice something interesting like flowers in someone's yard, school children laughing while waiting for a bus, a smile on a stranger's face, a creek or a river, a really large old tree or an interesting building? Can you remember this interesting mind-picture, reframe it and compare it to the previous day or save it for a brief discussion with a colleague? Savor the treat and consider it as an asset to your day.

Take time to enjoy your family and friends. When visiting family or friends you haven't seen for a while, have you ever said, "Wow, the kids have grown," or "he/she looks older or different than I remember" or something similar? People are busy, we're all busy, but time marches on and we can't get it back . . . ever. The question to ask yourself is, "Are your family and friends a priority?"

Take time during your career to learn something unique from your colleagues. Make an effort to use this learned skill to become a more qualified engineer, a smarter and better person.

More Ideas for Creating Life Balance

Five Reasons Why Helping You Helps Me by Kristin Hendrix

Recently, I shared an older post about helping each other through the chaos on LinkedIn. Someone in my community commented on how much our engagement has helped her ride the waves of change during COVID.

My reaction was immediate. Helping her helped me.

I participate periodically in office hours with my colleague and friend, Michael Santarcangelo at Security Catalyst. In one of our sessions, we talked about the work I've done as a coach and mentor, and as the executive sponsor for our technical women employee resource group.

Michael asked me "Who gets more from these engagements, them or you?" It's me, hands down.

It may sound counter-intuitive, but it's true. Helping others gives me so much. I don't want to presume it's more or less than what others are getting from the conversation, but there's no question about the multiplying factor of giving.

The world is pulling us in many directions. We may not feel like we have much energy left to give more. Why should we help when we are already giving (or it seems like the world is taking) so much?

Because helping others helps us.

Perspective

When I'm having conversations with friends, colleagues, or clients, I'm often asked for my views on a particular challenge. Frequently, there is similarity to something I have experienced or am going through.

During these discussions, I can pull from my experience, research I've done, etc. However, hearing someone else's challenges gives me new ideas I might not have considered previously. Often, it not only helps that person, but gives me a different perspective or idea on my own situation.

It's easier to see a problem from the outside in than the inside out. When we see our challenges manifest with someone else, it removes us from the center. This gives us more objectivity, which we can then apply to ourselves.

Acceptance

Years ago, I struggled with feedback I received on an annual review. Eventually, I was able to see it as a gift. The gift of empathy for anyone else who received unexpected feedback. I was in a better position to help prevent it from happening on my teams, and help guide someone else through the experience.

It took me about 9 months to get there. During that time, I was angry and resentful. Now? I remember those feelings, but I'm not carrying them with me. Instead, I carry the awareness that the experience makes me a better coach and leader.

Each time I pull from that experience to help someone else, it comes from a place of acceptance. Acceptance of my struggle, and blessing for the insight and empathy it provided.

(Continued)

When we help others, it allows us to transform our challenges and pain from burdens into gifts. Ones that can be shared, understood, and valued. What might have been a dark period can turn into a light for others, and ourselves.

Belonging

Recently, a co-worker introduced me to someone new and we got to talking. We each tend to be the helpers/givers in relationships, and wanted to figure out how we could help the other. What we didn't expect is how much we'd each get in return.

We are both single working mothers, struggling at times with home-based schooling, work demands, and household stuff. We are also on journeys of re-learning about ourselves and the world around us.

During our conversations, we frequently end up laughing or getting choked up. It's moments where we realize we are not alone. That someone else is going through this journey too.

Sometimes we laugh in joy that someone else is with us. Other times, we are emotionally overwhelmed because it was a scary thing we thought we were experiencing in isolation. Throughout, it's a sense of belonging and being seen.

Helping others may start as offering the hand we wish someone had extended us. We are rarely the only ones who have experienced something hard or painful or scary. When we share and realize we aren't the only ones, it can help lighten our load.

Energy & Wellness

When we see someone else getting value in our advice or experience, it can be fuel to energize us.

I call it "the juice." That good feeling that comes with helping others. But there's science behind it too.

While engaging with others too much could be draining (calling all my fellow introverts here), relating to others can be a reward. Looking at NLI's SCARF model, relatedness can either be a threat (when we feel disconnected from others) or a reward (when we find connection).

In the book "Burnout," the authors share science that demonstrates emotional exhaustion is the primary driver for burnout in women. One of the ways to prevent burnout is through positive, meaningful social interaction.

What is more positive and meaningful than helping someone else?

Helping others can refill our cup and give us the energy for all those other demands life is throwing at us.

Humility

That this is last is an irony that isn't lost on me. When I was gathering my thoughts and energy for this article, it's the last one that popped into my head. And yet one of the most important.

None of us has all the answers. Not even a fraction of them, relative to all the possible things that could work in a given situation. No friend, colleague, coach, or "expert" does. What we do have is the experience of poor outcomes that puts in a position to help others navigate similar situations. Hopefully better than we did.

Humility—acknowledgement of being a perfectly flawed human—is a gift in itself. When I share my stories, such as through my writing or coaching conversations, I often realize I have to take my own advice. There are also times when listening to someone else recount their challenges and what they want to try, helps me add a new tool to my toolkit.

When we help others, we often open ourselves to a wider world of possibility. It can feed a growth mindset, if we are as open to receiving as we are to giving.

Be a giver and a taker

It's been a number of years since I read the book, but I loved reading "Give and Take" by Adam Grant.

I like to see myself as a giver. As a helper. I've struggled, however, with perceiving myself as a taker when I open myself up to help from others.

The how's and why's of that mindset will wait for another day. However, what I have realized is helping others can help me. . .but only if I'm open to it.

I have to be willing to receive the gift of others' help. Instead of hording the good feelings of giving, allowing someone else to give can fuel them as well. While whatever advice I think I'm offering may be a gift, allowing them to give in return could be an even greater one.

When someone I'm coaching or mentoring offers thanks for my help, I thank them in return. I'll share something I got from the conversation. If an interaction inspires an article, I share it and let them know their contribution. Allow them to see that they weren't just taking in our interaction. . .that they were givers too.

All of these benefits go both ways. When we let someone see the gifts we are receiving from an interaction—even when we are the ones "helping"—it can be a source of belonging, acceptance, and energy for them.

Back to the original question "who gets more?" Is it me or those I help. Instead, it's exponential, giving all of us more. More relatedness, energy, and belonging that we can then gift to others.

It's the way we'll change the world, one person at a time.

14.11   SUMMARY

There are many competing factors relating to your engineering career including actual time at work, commuting, family time, community service and participation, sleep, personal time, fun, leisure activities, and more. Balancing these components of our lives is a lot like trying to balance a pizza on a pencil, with each life component represented by a slice. Being "out of balance" can lead to a "wobble" in our lives. These disruptions may manifest themselves as struggles in your personal relationships, lack of concentration, lack of patience, self-medicating with alcohol or drugs, or actual physical health problems and more.

We should ask ourselves, "Do we want to keep these healthy components in our lives, do we enjoy having a mix of interesting components of our life, and do we want to be well-balanced?"

Integrating mind, body, and spirit is a means to a balanced life. Creating time and space to do the things we truly enjoy and spend time with the people we love requires the ability to "balance" the components in our life. The methods to create an improved life balance are presented and discussed.

A relatively new term, *negEntropy*, is described. NegEntropy is a reduction in entropy inducing a corresponding increase in order. More simply, negentropy is a process that will increase efficiency and order and reduce chaos. So, the negEntropy process presented herein may enhance your well-being and life balance. A five-step guide for the negEntropic process to reverse energy loss (and reduce chaos) in daily life was presented for possibly enhancing the reader's life balance.

The key components to human life are:

- Mind—The Command Center
- Body—Our Home
- Spirit—Our "inner self," sometimes referred to as our soul

Establishing a real-life balance with mind, body, and spirit will lead toward achieving a fulfilling life.

When we learn to integrate the learned and innate information in our mind while maintaining a well-tuned and healthy body and synergistically combining our individual spirit and personality, we have the ability to solve (or answer) extremely complex and sophisticated problems. Then, we will have achieved the effective combination of Mind, Body, and Spirit.

As an engineer, we learn to look at the *big picture*. The big picture here is your life, your happiness, your family, and your career. A well-balanced life will likely reflect in a well-balanced individual with more joy, happiness, creativity, humility, empathy, humor, and love.

Apply these life-balancing skills to your own life; take the time to perform an assessment of happiness and life balance as presented herein. Ask yourself:

- What's working well?
- Does anything need attention or modification?
- Do I need to develop a plan for improvement?
- Am I truly as happy as I can be?

- Can I personalize my happiness?
- Is there anything that's fun that I want to try, to increase my happiness and life-balance?

Feeling Out of Balance, Overwhelmed, Thinking of Hurting Yourself?

IF you've performed the exercises above and tried to establish a real-life balance with mind, body, and spirit to achieve a fulfilling life balance the components in your life and you feel desperate and/or thinking of hurting yourself THEN, please refer to the important text boxes below for guidance and help:

Remember that "if" you feel completely overwhelmed and have any potential thoughts of hurting yourself (or others), help is available.

Your local crisis center usually serves your entire community, often 24/7 and free of charge. These centers connect callers to providers in their community that can support their needs. You can call the National Suicide Prevention Hotline at the time of this printing in 2023 and 2024: 988 (OR 911)

You can contact: *https://988lifeline.org/our-crisis-centers*

for information or to chat on-line.

In addition, many employers may have help available through their Health Plans and their employee assistance contractors.

The National Suicide Prevention Lifeline is now: 988 Suicide and Crisis Lifeline

IF the reader is not in the United States and urgent help is needed, then contact your region's "emergency services" phone number and/or web site.

988 has been designated as the new three-digit dialing code that will route callers to the National Suicide Prevention Lifeline. While some areas may be currently able to connect to the Lifeline by dialing 988, this dialing code will be available to everyone across the United States starting on July 16, 2022.

LEARN MORE ABOUT THE LIFELINE & 988

BIBLIOGRAPHY

Benson, Herbert, MD. et al. (1976). *The Relaxation Response*, HarperCollins Publisher, New York, NY. ISBN 0-380-81595-8.

Carr-Chellman, Alison. (March 15, 2021). A concept from physics called negentropy could help your life run smoother. The Conversation.

Hendrix, Kristin. (May 6, 2021). *Five Reasons Why Helping You Helps Me.* Leadership Vitae.

https://phys.org/news/2021-03-concept-physics-negentropy-life-smoother.html. (Accessed March 26, 2023).

Katz, Lawrence and Rubin, Manning. (1999). *Keep Your Brain Alive.* Workman Publishing Company, New York. ISBN 13: 978-0-761-11052-1.

McEwen, Bruce S., MD (2002). *The End of Stress as We Know It.* Dana Press, New York, NY. ISBN-13: -978-1932594553.

McGhee, Paul. (1999). *Health, Healing, and the Amuse System: Humor as Survival Training.* Kendall/Hunt Publishers. ISBN-13: 978-0-787-25797-2.

National Suicide Prevention Lifeline web site. https://suicidepreventionlifeline.org/our-crisis-centers (Accessed March 26, 2023).

Sternberg, M.D. and Esther, M. (2001). *The Balance Within: The Science Connecting Health and Emotions.* W. H. Freeman and Company, New York. ISBN-13: 978-0-716-74445-0.

CHAPTER 15

Globalization

Big Idea

"Globalization is a term used to describe how trade and technology have made the world into a more connected and interdependent place. Globalization also captures in its scope the economic and social changes that have come about as a result. It may be pictured as the threads of an immense spider web formed over millennia, with the number and reach of these threads increasing over time. People, money, material goods, ideas, and even disease and devastation have traveled these silken strands, and have done so in greater numbers and with greater speed than ever in the present age."

—National Geographic Society

Key Topics Covered

- The Globalization Process
- Global Climate Change—From a World View and a State Perspective
- Outcomes of Globalization and Climate Change

Civil Engineer's Handbook of Professional Practice, Second Edition. Karen Lee Hansen and Kent E. Zenobia.
© 2025 John Wiley & Sons, Inc. Published 2025 by John Wiley & Sons, Inc.
Companion website: www.wiley.com/go/hansen/CivilEngineersHandbook

- Learning to Project Manage a Mega-Project—The Case of BAA and Heathrow Terminal 5
- Civil Engineering Practice—A Wider Community Viewpoint

Related Chapters in This Book

- Chapter 3: Ethics
- Chapter 4: Professional Engagement
- Chapter 5: The Engineer's Role in Project Development
- Chapter 6: What Engineers Deliver
- Chapter 7: Executing a Professional Commission, Project Management
- Chapter 8: Permitting
- Chapter 9: The Client Relationship and Business Development
- Chapter 10: Leadership
- Chapter 11: Legal Aspects of Professional Practice
- Chapter 12: Managing the Civil Engineering Enterprise
- Chapter 13: Communicating as a Professional Engineer
- Chapter 14: Balancing Life, Family, and Career
- Chapter 16: Sustainability
- Chapter 17: Emerging Technologies
- Chapter 18: Human Relations Policies and Employment Practices
- Chapter 19: Construction Management
- Chapter 20: Critical Health and Safety Knowledge for Civil Engineers
- Chapter 21: What Engineers Need to Know

"Where there is no vision, there is no hope."
—George Washington Carver, American scientist, inventor

15.1   INTRODUCTION

This chapter gives civil engineers a basic understanding of how globalization affects competition, utility of engineering services, design, cost-effectiveness, construction, and decision-making. The effects of globalization are being felt now and just are beginning to be understood. Civil engineers can expect additional, needed attention on infrastructure, increased requirements for ethical standards as the global community gets smaller, more complex problem solving, and increased competition for financial resources.

The term became widely used and accepted by economists and other social scientists in the 1960s. Its use was widespread in the world news in the late 1980s. Since its inception, the concept of globalization has inspired numerous competing definitions and interpretations. For the purpose of this handbook, globalization is the continuous evolution of separate economies, nations, and cultures toward homogenization and integration through a worldwide network of communication, travel, and trade Steger, Manfred B. (2009).

In an economic context, globalization refers to the reduction and removal of barriers between national borders in order to facilitate the flow of goods, capital, services, and labor. Although considerable barriers remain on the flow of labor, globalization is not a new phenomenon. The Silk Road could be considered an early example of globalization. However, the Industrial Revolution, with increased speed of travel and communication, has caused exponential growth. It began in the late 19th century, but its spread slowed during the period from the start of World War I until the early 1970s. This slowdown can be attributed to the inward-looking policies pursued by a number of countries in order to protect their respective industries. However, the pace of globalization picked up rapidly during the last 25 years of the 20th century Steger, Manfred B. (2009).

15.2   THE GLOBALIZATION PROCESS

According to Rosebeth Moss Kanter of Harvard University, globalization is a process of change stemming from a combination of increasing cross-border activity and information technology (IT) enabling virtually simultaneous communication worldwide Kanter, Rosebeth Moss. (1995). Its promise is to make the world's best accessible to everyone. Four broad processes (shown in Figure 15.1) put more choices in the hands of individual consumers and organizational customers, generating a "globalization cascade" with reinforcing feedback loops that strengthen and accelerate globalizing forces. More information on these processes follows.

> **Process 1: Mobility—Capital, People, Ideas.** The key business ingredients of capital, people, and ideas are increasingly mobile. Capital mobility is noted often, but migrant professionals and managers are now joining more traditional migrant workers in an international labor force. Ideas move around the world through academic research and through global media, such as CNN, the Internet, and popular websites. High-speed information transfer generally makes location irrelevant: American Airlines' data entry point for tickets is in Barbados (at the time of this writing, anyway); British credit bureaus, European patent offices, and switching networks are located in the Philippines; and a Swedish fire department reaches its databases of street routes through a computer in Ohio (https://www.youmatter.world), (Ritzer, George, 2010).
>
> **Process 2: Simultaneity—Everywhere at Once.** Globalization means that goods and services are increasingly available in many places at the same time. The time lag between

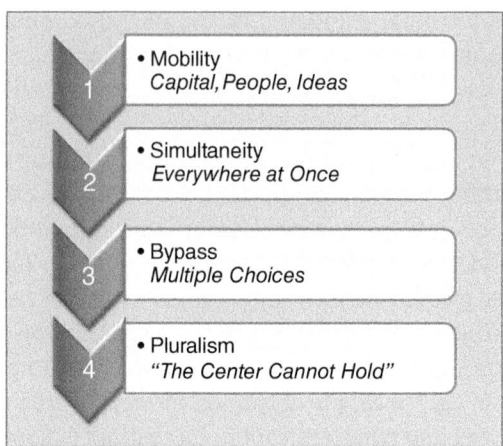

- Mobility
  Capital, People, Ideas

- Simultaneity
  Everywhere at Once

- Bypass
  Multiple Choices

- Pluralism
  "The Center Cannot Hold"

FIGURE 15.1   Four processes contribute to globalization.

the introduction of a product or service in one place and its adoption in other places is decreasing radically (unless we incur a global crisis like COVID-19). The slow rollout from local test to home country launch to adjacent country availability is becoming less common. The newer the technology or application, the more likely it is to be designed with the whole world in mind. "This" is another reason why globalization, sustainability, and emerging technologies are "all" so closely related to one another as presented in the beginning on all three of these key chapters in this book. And this explains why this concept is key to the CE's career and hopefully one's personal mission in life (https://www.youmatter.world).

Process 3: Bypass—Multiple Choices. Cross-border competition supported by easier international travel, deregulation, and privatization of government monopolies aids globalization and increases alternatives. Innovators can use alternative channels and new technology to go around established players rather than competing with them head-to-head. "Bypass" first referred to the rise of private switching networks that went around American regional telephone operating companies' wires; now wireless networks such as cellular and satellite systems bypass even more easily. Bypass creates numerous alternative routes to reach and serve customers. Japanese mail-order companies save 20 to 30 percent of postal costs by sending catalogues to Hong Kong for mailing back to Japan, thereby bypassing Japan's expensive postal service monopoly.

Process 4: Pluralism—"The Center Cannot Hold." There is a relative decline of monopolist centers of expertise and influence; activities concentrated in a few places are being decentralized. Traditional centers often still direct action and are main beneficiaries, but their automatic dominance or power to shape events declines when expertise and influence spread. "National champions," especially government-owned enterprises, are being reorganized and opened up to competition. Some corporate headquarter functions are being dispersed with local services or regional "centers (Ritzer, George. 2010).

An idea left over from the industrial economy now being discredited is that power comes from control over the means of *production*. In the global information economy, power comes from influence over *consumption*. Globalization of markets increases customers' choices, requiring producers to think more like customers. For example:

- Producers think they are making products; customers think they are buying services.
- Producers want to maximize return on the resources they own; customers care about whether resources are applied for their benefit, not who owns them.
- Producers worry about visible mistakes; customers are lost because of invisible mistakes.
- Producers think their technologies create products; customers think their needs create products.
- Producers organize for internal managerial convenience; customers want their convenience to come first (Ritzer, George. 2010).

Companies positioned to be successful in global markets put an emphasis on in- novation, learning, and collaboration. They:

- Organize around customer logic
- Set high goals
- Select people who are broad, creative thinkers
- Encourage enterprise (on the part of employees)
- Support constant learning
- Collaborate with partners

An Official Definition of Globalization by the World Health Organization (WHO)

According to WHO, globalization can be defined as "the increased interconnectedness and interdependence of peoples and countries. It is generally understood to include two inter-related elements: the opening of international borders to increasingly fast flows of goods, services, finance, people and ideas; and the changes in institutions and policies at national and international levels that facilitate or promote such flows."

—youmatter.world

15.3   GLOBAL CLIMATE CHANGE—A WORLD VIEW AND A STATE PERSPECTIVE

Though there are those who will debate its causes, global climate change appears to be real. The implications for civil engineers regarding global climate change are significant, perhaps more so than for other engineering disciplines. This section reports on the findings of three organizations: 1) the US Environmental Protection Agency (EPA); 2) the Intergovernmental Panel on Climate Change (IPCC), a scientific body established by the United Nations Environment Program (UNEP) and the World Meteorological Organization (WMO); and 3) the State of California Natural Resources Agency, in conjunction with numerous other state agencies.

15.3.1   A World View

In this chapter we define "Globalization," discuss the globalization process, interactions, outcomes, international projects, impacts on civil engineering practice and your career. And, you will discover the symbiotic relationships globalization has with sustainability and emerging technologies and the secondary interactive relationships with ethics, what engineers deliver, the client relationship, leadership, communicating as a professional engineer, balancing life, family, and career. Of course, armed with this understanding and enlightenment, civil engineers will realize, if they do not do so already, the unique position they occupy and the distinctive ability they possess to help realize positive change.

But first, it's important to note a critical fact one needs to recognize related to globalization…perspective. Take a close look at picture of "earth from space" below and imagine this is the view from the International Space Station (ISS) where you can readily see the continents, oceans, storms, clouds, and more. Actually, this is a view from the International Space Station (ISS). Hopefully, you may have an opportunity to see this view in your lifetime. But notice, we cannot see international boundaries, property lines, nations, families, or individuals. ALL those details are blurred, and our new perspective is that it's all one big, connected, thing, Our Home.

We ALL live here together!

Here's a quick question for you: Have you taken anything for granted today? Anything…at all?

Anything…like our atmosphere, the air you breathe, the water in your morning drink or shower, your dinner plans, the power to heat or cool your home, the raw materials to build your home, office, and buildings in your city, the technologies to run our cities, hospitals, universities, our infrastructure, transit system, supply chains, our world?

Source: CC0 Community/publicdomainpictures.net.

There is a unique international organization we can turn toward to gain a global perspective and this organization is the "United Nations." What better organization to learn from than the UN, a collection of the world's major countries and governments working together and toward important, collaborative, peaceful goals and objectives. According to United Nations Resolution 70/1, Transforming our world: the 2030 Agenda for Sustainable Development:

> *The United Nations is an intergovernmental organization aiming to maintain international peace and security, develop friendly relations among nations, achieve international cooperation, and be a center for harmonizing the actions of nations." It is the world's*

largest and most familiar international organization founded in 1945 after the Second World War by 51 countries committed to maintaining international peace and security, developing friendly relations among nations and promoting social progress, better living standards and human rights (United Nations Resolution 70/1).

Today the UN includes 193 Member States, the UN and its work are guided by the purposes and principles contained in its founding Charter. The UN has evolved over the years to keep pace with a rapidly changing world. But one thing has stayed the same: it remains the one place on Earth where all the world's nations can gather together, discuss common problems, and find shared solutions that benefit all of humanity.

The UN has the mission and reach to solve global problems at a time when collective action is needed. The UN, all Nations, recognize many issues and goals to work together on. But there is an overreaching goal that "globally" we need to embrace, develop a master plan, and implement corrective actions. All Nations must accept the "Sustainable Development Goals." These goals are depicted in the figure below and include the 17 major goals that can be classified in one of three general categories:

- Social
- Environmental
- Economic

Source: UNITED NATIONS/https://www.un.org/sustainabledevelopment/blog/2015/12/sustainable-development-goals-kick-off-with-start-of-new-year/ / last accessed April 08, 2023.

Again, according to United Nations Resolution 70/1:

> *The 17 goals fall into these three categories because people and governments* (social)
> *hold the power to provide and implement the solutions. These solutions will all likely*
> *require economic power, meaning money, collaboration, prioritization, some sacrifice,*
> *and hard work* (economic). *Finally, the UN and its 193-member nations recognize we*
> *must work within our environmental factors to maintain healthy and biodiversity such*
> *as clean air/water, land emissions reductions, zero waste, and environmental justice*
> (environment).

OK, one might ask "why" these particular goals and how do they impact the United
States, my career, me? What can civil engineers do to work toward these goals and "when"
will these lofty goals be planned, and "what" is the implementation plan? "How" are these
"Global Goals for Sustainable Development" related to this chapter on "Globalization." These are
all good questions; please be patient and consider this:

1. In September 2015, The UN General Assembly adopted "Resolution 70/1: Transforming our world—the 2030 Agenda for Sustainable Development which is a plan of
   action for people, planet and prosperity. It also seeks to strengthen universal peace
   in larger freedom. We recognize that eradicating poverty in all its forms and dimensions, including extreme poverty, is the greatest global challenge and an indispensable
   requirement for sustainable development."
   - So, civil engineers are in a very unique position as leaders, designers, construction
     managers, members of society to work with our government leaders, the public,
     and clients to implement and apply our knowledge and tools for sustainable
     development.
2. The 2030 Agenda and the achievement of the Sustainable Development Goals are
   underpinned by a healthy and productive environment.
   - The General Assembly (193 Nations) agreed that these 17 Goals will help
     achieve a healthy and productive environment for all the world's people. So,
     this is why these goals are important and when we hope to achieve these goals,
     which is 2030.
3. SDGs 6, 13, 14, and 15 and the Strategic Plan for Biodiversity 2011–2020 are at the
   heart of the UN's work. The focus is to foster ecosystem management that delivers
   on all Sustainable Development Goals, including poverty reduction, food/water and
   energy security, employment, and gender equality.

The Implementation Plan is briefly outlined here in the UN's "5P Plan":

- People—We are determined to end poverty and hunger, in all their forms and dimensions, and to ensure that all human beings can fulfil their potential in dignity and equality and in a healthy environment. (IPCC 2007; Friedman 2009), (UN Resolution 70/1)
- Planet—We are determined to protect the planet from degradation, including through sustainable consumption and production, sustainably managing its natural resources and taking urgent action on climate change, so that it can support the needs of the present and future generations.
- Prosperity—We are determined to ensure that all human beings can enjoy prosperous and fulfilling lives and that economic, social, and technological progress occurs in harmony with nature.
- Peace—We are determined to foster peaceful, just, and inclusive societies which are free from fear and violence. There can be no sustainable development without peace and no peace without sustainable development.
- Partnership—We are determined to mobilize the means required to implement this Agenda through a revitalized Global Partnership for Sustainable Development, based on a spirit of strengthened global solidarity, focused in particular on the needs of the poorest and most vulnerable and with the participation of all countries, all stakeholders, and all people.
- The interlinkages and integrated nature of the Sustainable Development Goals are of crucial importance in ensuring that the purpose of the new UN Agenda is realized. If we realize our ambitions across the full extent of the Agenda, the lives of all will be profoundly improved and our world will be transformed for the better.

4. So, the inherent question above is, "How" are these "Global Goals for Sustainable Development" related to this chapter on "Globalization"?

- The answer is: "Global Goals for Sustainable Development" are related to globalization because every nation on the globe, the entire world, is needed to accomplish these goals in an expeditious manner. The figure below displays how the UN's Global Goals for sustainability promote and fit into our global economy, society, and biosphere. These goals are discussed here to illuminate the civil engineer's path and provide a road map for future direction and impacts on the civil engineering careers and development. This action plan has begun in part but requires more collaboration, leadership support, and economic stimulus. The leadership, experience, and support of all civil engineers are needed to make this a reality!

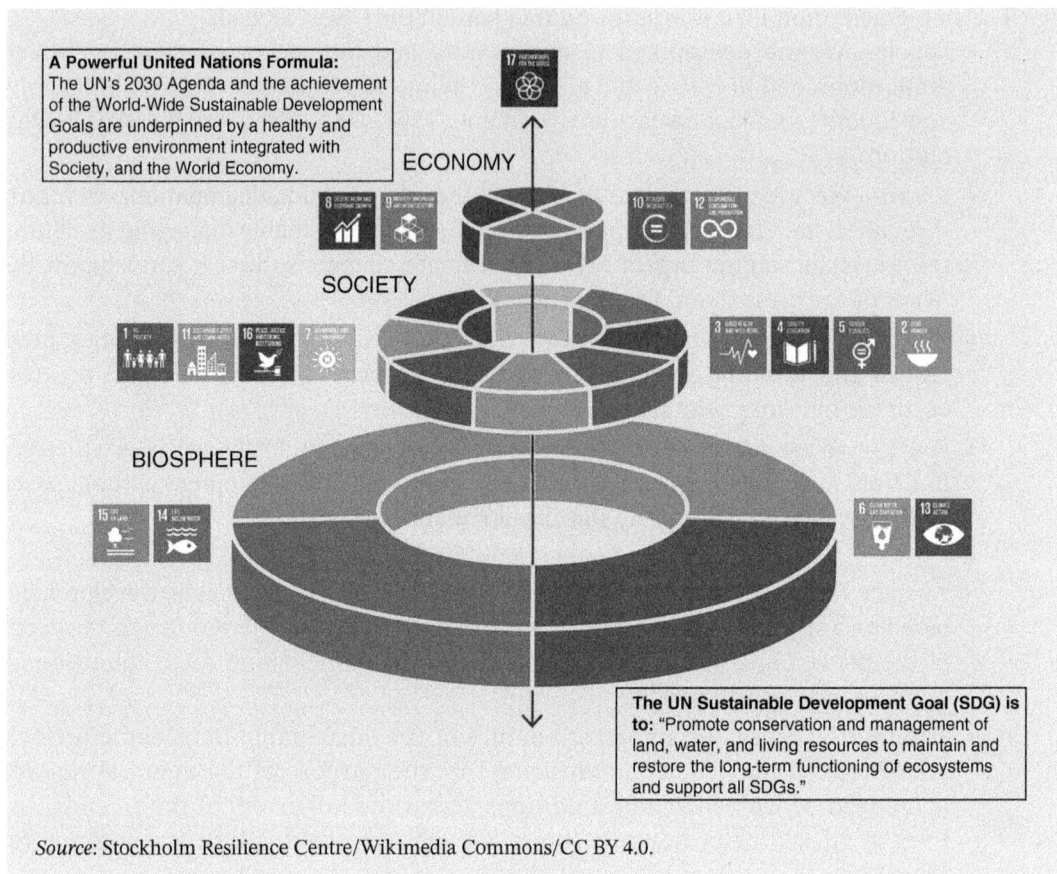

A Powerful United Nations Formula:
The UN's 2030 Agenda and the achievement of the World-Wide Sustainable Development Goals are underpinned by a healthy and productive environment integrated with Society, and the World Economy.

ECONOMY

SOCIETY

BIOSPHERE

The UN Sustainable Development Goal (SDG) is to: "Promote conservation and management of land, water, and living resources to maintain and restore the long-term functioning of ecosystems and support all SDGs."

Source: Stockholm Resilience Centre/Wikimedia Commons/CC BY 4.0.

15.3.2   Potential Global Impacts

According to the US EPA:

> *"Many elements of human society and the environment are sensitive to climate variability and change. Human health, agriculture, natural ecosystems, coastal areas, and heating and cooling requirements are examples of climate-sensitive systems.*
>
> *Rising average temperatures are already affecting the environment. Some observed changes include shrinking of glaciers, thawing of permafrost, later freezing and earlier break-up of ice on rivers and lakes, lengthening of growing seasons, shifts in plant and animal ranges, and earlier flowering of trees." (www.epa.gov/climatechange/effects/index.html)*

There are inter-relationships of climate impacts to conditions affecting public health that will have direct and indirect impacts on civil engineering design requirements, the earth, and civilization (see Figure 15.2).

The US EPA has additional relevant information on their website and also has directed readers to the Intergovernmental Panel on Climate Change (IPCC) website. In 2007, IPCC shared the Nobel Peace Prize with former Vice-President Al Gore for their work on climate change, which the Nobel Prize Committee recognized as having a connection with peace and war. The IPCC website contains much useful information, including an (IPCC) report titled, Climate Change 2007: Impacts, Adaptation and Vulnerability and more recent updates from NOAA Fisheries in March 2022 titled, "Climate Change 2022: Impacts, Adaptation, and Vulnerability". In addition to these reports, further information is available from recent IPCC and NOAA meetings and publications. (The Intergovernmental Panel on Climate Change (IPCC) 2022). The reader is encouraged to visit the website to access these materials to draw your own conclusions.

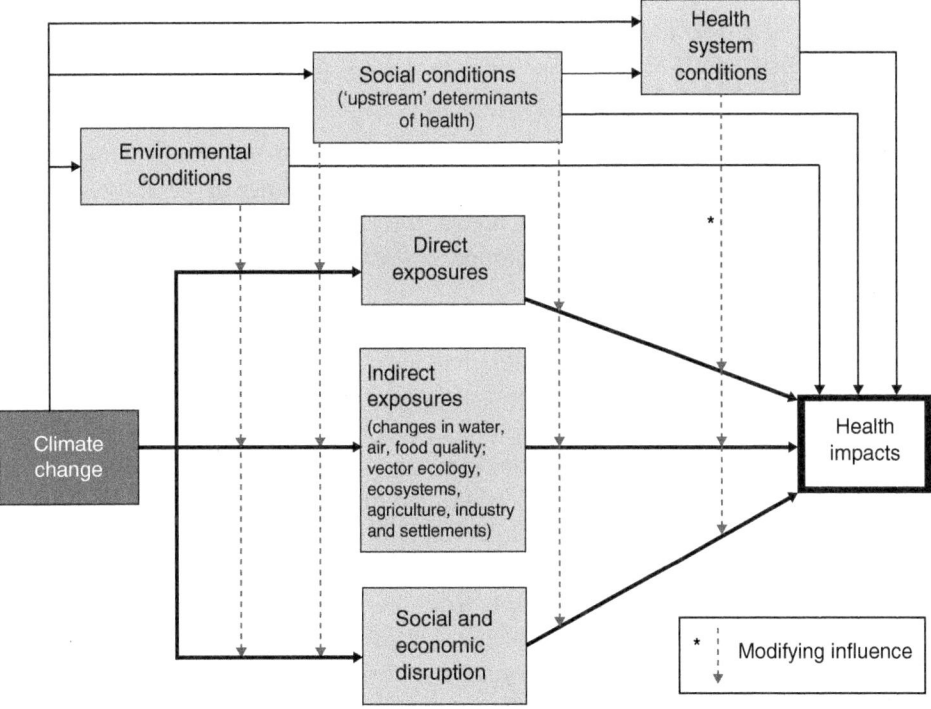

FIGURE 15.2   Flow diagram of effect of climate change on health.
Source: IPCC Fourth Assessment Report, 2007.

IPCCs 2007 report is rather detailed, so a brief summary is presented here:

"Observational evidence from all continents and most oceans shows that many natural systems are being affected by regional climate changes, particularly temperature increases (very high confidence). A global assessment of data since 1970 has shown it is likely that anthropogenic warming has had a discernible influence on many physical and biological systems . . ."

"For physical systems, climate change is affecting natural and human systems in regions of snow, ice and frozen ground, and there is now evidence of effects on hydrology and water resources, coastal zones and oceans . . ."

"There is more evidence, from a wider range of species and communities in terrestrial ecosystems than reported in the Third Assessment, that recent warming is already strongly affecting natural biological systems. There is substantial new evidence relating changes in marine and freshwater systems to warming. The evidence suggests that both terrestrial and marine biological systems are now being strongly influenced by observed recent warming . . ."

"The number of people living in severely stressed river basins is projected to increase significantly from 1.4-1.6 billion in 1995 to 4.3-6.9 billion in 2050 . . . (medium confidence)."

"The resilience of many ecosystems (their ability to adapt naturally) is likely to be exceeded by 2100 by an unprecedented combination of change in climate, associated disturbances (e.g., flooding, drought, wildfire, insects, ocean acidification), and other global change drivers (e.g., land-use change, pollution, over-exploitation of resources) (high confidence) . . ."

This information is accompanied by detailed descriptions, graphics, and references from around the globe. Some of the more interesting graphics are presented in Figure 15.3 to illustrate the potential severity and gravity of climate change.

Figure 15.3 also illustrates the potential impacts to systems/resources due to an average temperature rise of from 1 to 5 degrees Celsius. The impacts on freshwater systems, ecosystems, and food production are reflected in stress on public health, pressure on water resources, increased species extinction, a greater number of wildfires, decreased production in some crops, and changing coastlines. The impacts from 0 to 2 degrees Celsius are significant but become very dramatic and extremely problematic from 3 to 5 degrees Celsius.

Figure 15.4 illustrates the potential impacts to the major continents, polar regions, and small islands related to rises in average temperatures from 1 to 5 degrees Celsius. The specific systems and resources impacted in these land masses are highlighted. The potential impacts to North America will be an increased need for cooling systems within buildings, increased

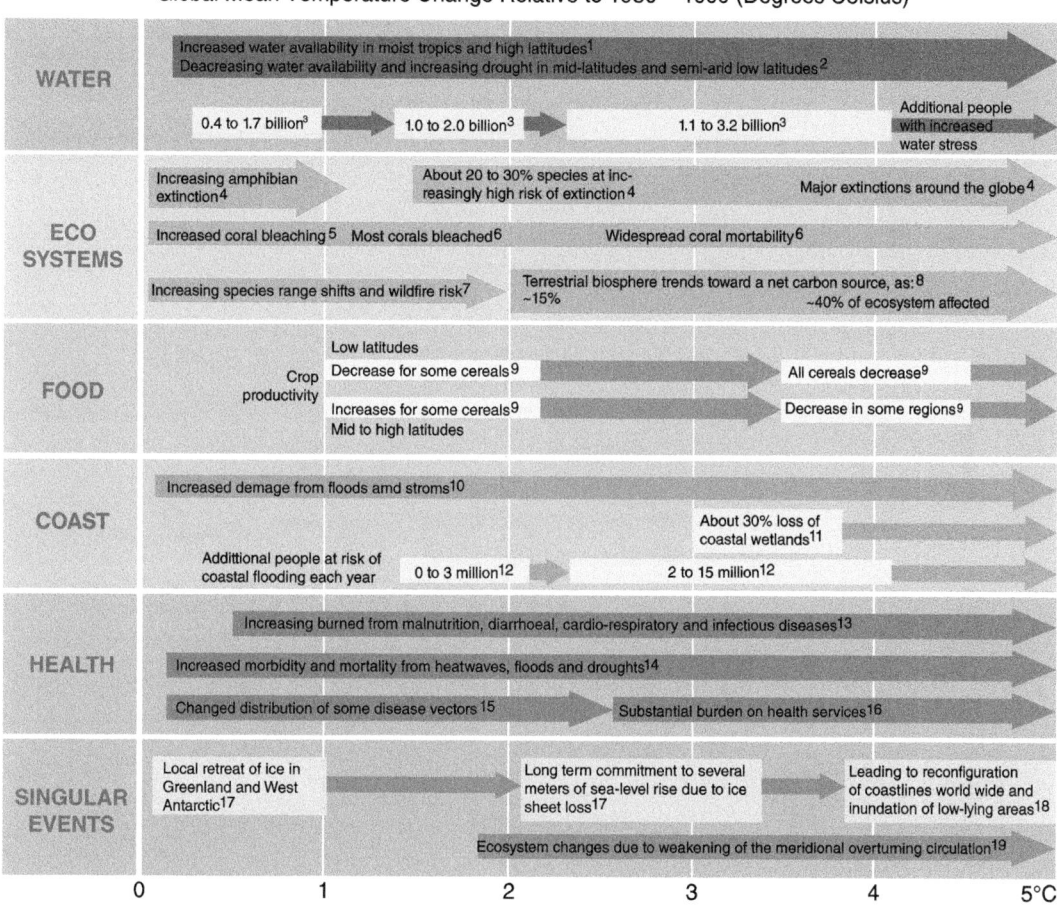

FIGURE 15.3 Example of impacts to systems and resources related to temperature rise.
Source: Adopted from Intergovernmental Panel on Climate Change (IPCC) – Honored With the 2007 Nobel Peace Prize; http://www.ipcc.ch/publications

frequency of high ozone pollution days, an increase in crop yield where reliable water supplies are available, and increased wildfires. The impact for Europe will be an increase of water resources in northern Europe with an accompanied decrease of water resources in southern Europe, and variable increase in crop yield where reliable water resources exist. Other continents have varied impacts over the range of temperature increases.

Some conclusions that can be drawn from Figures 15.3 and 15.4, the 2007 IPCC report, and the 2022 NOAA Fisheries report are included in the textbox, Summary of Main Findings.

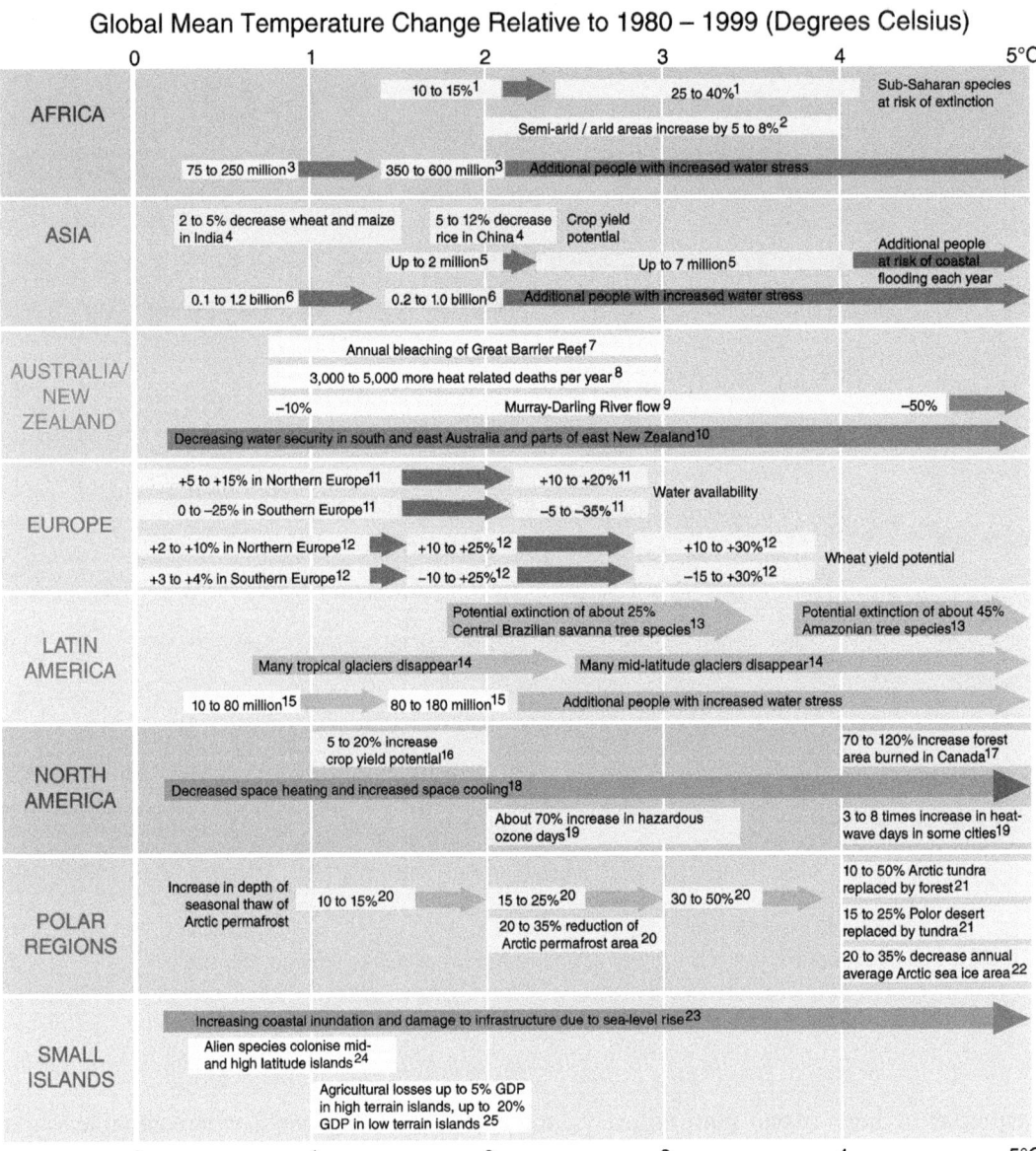

FIGURE 15.4   Projected global temperature rise and impacts throughout the world.
Source: Adopted from Intergovernmental Panel on Climate Change (IPCC) – Honored With the 2007 Nobel Peace Prize; http://www.ipcc.ch/publications

Summary of Main Findings

- Observational evidence from all continents and most oceans shows that many natural systems are being affected by regional climate changes, particularly temperature increases.

- A global assessment of data since 1970 has shown it is likely that anthropogenic warming has had a discernible influence on many physical and biological systems.

- Other effects of regional climate changes on natural and human environments are emerging, although many are difficult to discern due to adaptation and non-climatic drivers.

- More specific information is now available across a wide range of systems and sectors concerning the nature of future impacts, including for some fields not covered in previous assessments.

- More specific information is now available across the regions of the world concerning the nature of future impacts, including for some places not covered in previous assessments.

- Magnitudes of impact can now be estimated more systematically for a range of possible increases in global average temperature.

- Impacts due to altered frequencies and intensities of extreme weather, climate and sea-level events are very likely to change.

- Some large-scale climate events have the potential to cause very large impacts, especially after the 21st century.

- Impacts of climate change will vary regionally but, aggregated and discounted to the present, they are very likely to impose net annual costs which will increase over time as global temperatures increase.

- Some adaptation is occurring now, to observed and projected future climate change, but on a limited basis.

- Adaptation will be necessary to address impacts resulting from the warming which is already unavoidable due to past emissions.

- A wide array of adaptation options is available, but more extensive adaptation than is currently occurring is required to reduce vulnerability to future climate change. There are barriers, limits and costs, but these are not fully understood.

- Vulnerability to climate change can be exacerbated by the presence of other stresses.

- Future vulnerability depends not only on climate change but also on development pathway.

(Continued)

- Sustainable development can reduce vulnerability to climate change, and climate change could impede nations' abilities to achieve sustainable development pathways.
- Many impacts can be avoided, reduced or delayed by mitigation.
- A portfolio of adaptation and mitigation measures can diminish the risks associated with climate (Climate Change 2022: Impacts, Adaptation, and Vulnerability. NOAA Fisheries, March 2022)

Source: IPCC Technical Report Titled: Climate Change 2007: Impacts, Adaptation and Vulnerability: www.ipcc.ch/publications_and_data/publications_ipcc_fourth_assessment_report_wg2_report_impacts_adaptation_and_vulnerability.htm, (Climate Change 2022: Impacts, Adaptation, and Vulnerability), (NOAA Fisheries, March 2022), (Science Basis Contribution of Working Group, 2007)

15.3.3   Potential Impacts on California and the Western United States

Two similar reports to the US Environmental Protection Agency (EPA), the Intergovernmental Panel on Climate Change (IPCC), and the NOAA Fisheries reports on the world's impact from climate change is also available from the State of California on the impact to the western US. Two reports, titled *2009 and 2021 California Climate Adaptation Strategy*, were chosen because they present some global climate impacts that will affect the State, as well as a strategy for how California will handle climate change in the 2021 report. Specifically, the 2009 report summarizes the best-known science on climate change impacts in seven specific sectors. Here are some excerpts from the report to the Governor of the State of California in response to the Governor's Executive Order S-13–2008:

1. "Climate change is already affecting California. Sea levels have risen by as much as seven inches along the California coast over the last century, increasing erosion and pressure on the state's infrastructure, water supplies, and natural resources. The state has also seen increased average temperatures, more extreme hot days, fewer cold nights, a lengthening of the growing season, shifts in the water cycle with less winter precipitation falling as snow, and both snowmelt and rainwater running off sooner in the year."
2. "These climate driven changes affect resources critical to the health and prosperity of California. For example, forest wildland fires are becoming more frequent and intense due to dry seasons that start earlier and end later. The state's water supply, already stressed under current demands and expected population growth, will shrink under even the most conservative climate change scenario."

3. "If the state were to take no action to reduce or minimize expected impacts from future climate change, the costs could be severe. A 2008 report by the University of California, Berkeley and the non-profit organization 'Next 10' estimates that if no such action is taken in California, damages across sectors would result in 'tens of billions of dollars per year in direct costs' and 'expose trillions of dollars of assets to collateral risk.' More specifically, the report suggests that of the state's $4 trillion in real estate assets '$2.5 trillion is at risk from extreme weather events, sea level rise, and wildfires' with a projected annual price tag of up to $3.9 billion over this century depending on climate scenarios."

4. "California's ability to manage its climate risks through adaptation depends on a number of critical factors including its baseline and projected economic resources, technologies, infrastructure, institutional support and effective governance, public awareness, access to the best available scientific information, sustainably managed natural resources, and equity in access to these resources."

5. "To effectively address the challenges that a changing climate will bring, **climate adaptation and mitigation (i.e., reducing state greenhouse gas (GHG) emissions) policies must complement each other** and efforts within and across sectors must be coordinated. For years, the two approaches have been viewed as alternatives, rather than as complementary and equally necessary approaches. Adaptation is a relatively new concept in California policy. The term generally refers to efforts that respond to the impacts of climate change—adjustments in natural or human systems to actual or expected climate changes to minimize harm or take advantage of beneficial opportunities." California Climate Adaptation Strategy Report, (2021 June), California Climate Adaptation Strategy Report, (2009 June)

15.3.4   A Collaborative Approach to the Adaptation Strategy

"The development of the adaptation strategies was spearheaded by the state's resource management agencies, California Natural Resources Agency (CNRA) staff who worked with seven sector-based Climate Adaptation Working Groups (CAWGs) focused on the following areas:

- Public health
- Ocean and coastal resources
- Water supply and flood protection
- Agriculture
- Forestry
- Biodiversity and habitat
- Transportation and energy infrastructure

This adaptation strategy could not have been developed without the involvement of numerous stakeholders. Converging missions, common interests, inherent needs for cooperation, and the fact that climate change impacts cut across jurisdictional boundaries will require governments, businesses, non-governmental organizations, and individuals to minimize risks and take advantage of potential planning opportunities." (www.climatechange.ca.gov/adaptation)

The 2011 California Climate Adaptation Strategy report went on to conclude potential impacts to public health and biodiversity. These conclusions are highlighted below. The 2021 California Climate Adaptation Strategy: Priorities, Goals, and Actions appear after the stated impacts.

Public Health Impacts Due to Sea-Level Rise

- Wastewater issues with flooding of septic systems near coastline
- Saltwater intrusion—risks to drinking water quality and quantity
- Threats of injury and even death during coastal storms
- Emotional and mental health impacts related to more coastal flooding and erosion
- Emotional and mental health impacts related to internal displacement and migration of coastal residents

Public Health and Environmental Impacts Due to Warming

- Higher rates of mortality and morbidity
- Increased air pollution
- Seasonal changes and increases in allergens
- Changes in prevalence and spread of disease vectors
- Possible decrease in food quality and security
- Reduction in water availability
- Increased pesticide use

Biodiversity and Habitat Impacts Due to Warming

- Higher Barriers to species migration and movement
- Temperature Rise—lakes, streams, and oceans

- Increase in invasive species potential
- Changes in natural community structure
- Threats to rare, threatened, or endangered species
- Altered timing of phenological events
- Timing disruptions between predators and prey and pollinators and plants
- Loss of ecosystem goods and services

Biodiversity and Habitat Impacts Due to Sea-Level Rise

- Inundation of permanent coastal habitat
- Alteration of dune habitat and coastal wetlands
- Coastal habitat loss of migratory birds, shellfish and endangered plants
- Reduction of fresh water resources due to salt water intrusion for coastal land areas
- Sedimentation increases may increase pollution and run off
- Degradation of aquatic ecosystem
- Increase in invasive species
- Competition shifts in urban growth and development
- Agricultural relocation
- Alterations of ecological reserves, wildlife areas, undesignated lands, mitigations sites and easements
- Groundwater recharge & over drafting
- Water management and water transfer conflicts
- Reduction in wetland habitat on commercial and sport fisheries

The *2021 California Climate Adaptation Strategy: Priorities, Goals, and Actions* report makes a series of recommendations. The California Natural Resources Agency staff, other California Agencies, public comments, stakeholder comments covered many topics and contributed to these recommendations, which are listed below. Detailed content may be viewed at: https://climateresilience.ca.gov/overview/docs/20220404-CAS_Priorities_Goals_Actions.pdf.

2021 California Climate Adaptation Strategy: Priorities, Goals, and Actions

A resilient California can help all communities weather and adjust to climate change, and can help all communities thrive. A climate resilient "California for All" requires focused support for those communities most vulnerable to climate threats and experiencing compounding inequalities. Therefore, one of California's climate adaptation priorities is to strengthen climate resilience in communities most vulnerable to the climate crisis.

2021 California Climate Adaptation Strategy: Priorities, Goals, and Actions

GOAL A: Engage with and build capacity in climate vulnerable communities

- Action 1: Support California Native American tribes' development of climate change and health equity resilience planning tools and capacity.
- Action 2: Support state resources and promote partnerships to expand the capacity of under-resourced communities, including California Native American tribes, to lead and implement climate change mitigation, adaptation, and resilience plans, programs, and projects.
- Action 3: Prioritize social equity, tribal nations, and disadvantaged communities in climate adaptation planning and strategies.
- Action 4: Increase community participation in planning transportation projects to build climate resilience.
- Action 5: Partner with underserved communities, including tribal communities in California, to build coastal and ocean climate resilience.
- Action 6: Support Technical Assistance, through providers such as the University of California Division of Agriculture and Natural Resources, to make climate smart agricultural knowledge and incentives available to Socially Disadvantaged Farmers and Ranchers and disadvantaged communities.
- Action 7: Enhance adaptive capacity of communities through increased climate-related service and volunteerism.
- Action 8: Through meaningful multilingual and culturally-relevant outreach and engagement, seek and facilitate community partnership and input into health impact studies of climate-related hazards, including air pollution, wildfire smoke, and heat, while highlighting disparities and ways to minimize exposures in vulnerable communities.
- Action 9: Engage with communities most likely to be affected by state climate

2021 California Climate Adaptation Strategy: Priorities, Goals, and Actions

GOAL B: Improve understanding of climate impacts on California's communities, including what drives vulnerability

- Action 1: Identify the most climate vulnerable communities in California to direct and implement actions across sectors and regions that reduce risk and build resilience.
- Action 2: Improve and refine quantitative wildfire risk assessments across California to identify the most wildfire vulnerable communities and populations for inclusion in California's Vulnerable Communities Platform.
- Action 3: Increase the collection, analysis, and reporting of data on climate related health impacts.
- Action 4: Support actionable, community-driven, and equitable research partnerships to inform and accelerate climate adaptation action based on best available climate science.
- Action 5: Assist emergency managers and planners in identifying, locating, and deploying resources to populations at greater risk from climate impacts.
- Action 6: Document the impacts of climate change on tribal nations, their communities, traditional foods, cultural resources, and ecosystems by collaborating with tribes to document, as part of the Indicators of Climate Change in California Report, their perspectives, expertise, and traditional knowledges on climate change-related stressors.
- Action 7: Improve the understanding of climate change and its impacts on California's cultural heritage, including tribal nations.
- Action 8: Hold a Public Health Workgroup of the California Climate Action Team on the mental and behavioral health impacts of climate change, including climate grief, and strategies to bolster personal and community resilience as the climate change health crisis accelerates.

2021 California Climate Adaptation Strategy: Priorities, Goals, and Actions

GOAL C: Build resilience in climate vulnerable communities through state programs

- Action 1: Prioritize actions that reduce wildfire risks to tribal nations and climate vulnerable communities.
- Action 2: Support wildfire prone communities by expanding the Regional Forest and Fire Capacity Program beyond high-risk areas throughout the state and increase

(Continued)

local and regional governments' capacity to build and maintain a pipeline of forest health and fire prevention projects.

- Action 3: Update transportation infrastructure competitive programs' guidelines to incentivize climate adaptation and climate risk assessments/strategies.

- Action 4: Review local government plans to meet housing needs (Housing Elements) with a lens toward climate resilience, adaptation, and protection of vulnerable communities to relevant climate impacts.

- Action 5: Strengthen protections for people who are experiencing homelessness and those people extremely vulnerable to climate risks through funding programs for permanent and interim housing.

- Action 6: Invest Community Development Block Grant funds in long-term disaster recovery and resilience building that targets the unmet housing recovery needs of low and moderate-income households in in a way that mitigates disaster risk and reduces future losses among vulnerable communities.

- Action 7: Promote sustainable land use planning and transportation investments that support walkable and bikeable communities and infill development to build resilience of climate vulnerable communities.

- Action 8: Increase access to locally and traditionally grown food and produce in low-income communities to build climate resilient food systems and increase agricultural economic sustainability.

- Action 9: Consider and integrate environmental justice principles in permit decisions and planning documents that drive on California's climate adaptation priorities.

- Action 10: Promote equity, community engagement, and culturally competent emergency response program design.

- Action 11: Prioritize climate resilience and health equity resources for people and places experiencing the most need and risks due to historical and continuing disinvestment and inequities.

- Action 12: Ensure the Affordable Housing and Sustainable Communities Program advances California's climate resilience priorities.

- Action 13: Ensure projects funded through the Transformative Climate Communities Program advance California's climate resilience priorities.

- Action 14: Enact policies for California High-Speed Rail that establish sustainability, climate change adaptation, and social equity goals.

- Action 15: Enact policies in the California State Rail Plan that support climate change adaptation for rail infrastructure and support emergency maintenance and repairs.

- Action 16: Increase awareness and understanding, and reduce climate impacts on children, and pregnant people, by providing public health expertise and collaborating with state, tribal and local agencies, community organizations and other entities taking actions to build climate resilience.
- Action 17: Incorporate cultural heritage in California's climate adaptation actions.

—California Climate Adaptation Strategy report, (June 2021),
California Natural Resources Agency

The 2021 Climate Adaptation Strategy shows how the state's efforts fit together to deliver on six priorities for climate resilience action in California:

- Strengthen Protections for Climate Vulnerable Communities
- Bolster Public Health and Safety to Protect Against Increasing Climate Risks
- Make Decisions Based on the Best Available Climate Science
- Build a Climate Resilient Economy
- Accelerate Nature-Based Climate Solutions and Strengthen Climate Resilience of Natural Systems
- Partner and Collaborate to Leverage Resources

15.4   OUTCOMES OF GLOBALIZATION AND CLIMATE CHANGE

Analysis of the information contained in these climate impact reports summarized above and background literature leads to the conclusion that globalization and climate change will influence civil engineering applications, education, and business on a very large scale. Several possible outcomes of globalization and climate change include:

Outcome 1: The competition for resources and financial capital will become greater.

There appears to be a growing backlog of deferred maintenance in the United States and other large economies that has caused or may result in failures. The "American Society of Civil Engineers Infrastructure Report Card" documents this problem and gives the nation's infrastructure a barely passing grade. (https://infrastructurereportcard.org) (Odeh, David, 2022)

Searching the Report Card yields the most recent "engineering failures." The root cause of many of these failures is disregard for the engineer's recommendations, including often deferred periodic maintenance. Civil engineers recognize that capital budgeting is necessary and critical to the maintenance of civil structure and/or improvements. However, once the improvement is placed in service, the maintenance budgets are often drawn from general or operating funds. Finance and budgeting staff often ignore Operation and Maintenance (O & M) requirements for infrastructure and remove capital budgeting from the hands of engineers.

In the past, frequently the O & M budgets were lumped into the general, local, or municipal funds to compete with all other general expenditures. There are many competing segments of the public exercising their influence over the budgeting of these general funds, and maintenance often falls to the bottom to satisfy the voting constituency.

The *Infrastructure Investment and Jobs Act (IIJA)*, also known as the *Bipartisan Infrastructure Bill* (H.R. 3684), was passed by the 117th Congress in 2021 and became Public Law No: 117–58. "This *Bipartisan Infrastructure Law* will rebuild America's roads, bridges and rails, expand access to clean drinking water, ensure every American has access to high-speed internet, tackle the climate crisis, advance environmental justice, and invest in communities that have too often been left behind. The legislation will help ease inflationary pressures and strengthen supply chains by making long overdue improvements for our nation's ports, airports, rail, and roads. It will drive the creation of good-paying union jobs and grow the economy sustainably and equitably so that everyone gets ahead for decades to come. Combined with the President's Build Back Framework, it will add on average 1.5 million jobs per year for the next 10 years" (https://www.whitehouse.gov/briefing-room/statements).

However, we still need to see how State and local governments will pay for operation and maintenance (O&M) once the new investments are complete.

Outcome 2. Civil engineers will see increased emphasis on cost and low maintenance requirements for project goals.

One way to combat maintenance is in the selection, preparation, or original assembly of the civil structure. However, low maintenance options in civil design are often initially more expensive than other options. This competing scenario (higher initial cost versus lower maintenance and overall costs) will require the engineer to skillfully communicate "and record" the advantages of various alternatives presented to the client or owner (Shigley, Paul, 2008; Odeh, David, 2022). If the Owner should request the least cost alternative potentially requiring more O & M costs, then the CE should impress upon the Owner that higher O & M future costs are required to protect this capital improvement and the public. It's recommended the CE should record the Owner's decision to protect themselves and their firm in case there are potential future failures due to lack of adequate O & M that would have protected the capital improvement and the public.

The Rise of China and India—Tectonic Economics

"The two countries have one thing in common: their transformations—and the way they will transform the globe—are as stunning as any the world has seen since America itself emerged onto the world economic stage. The impact can be seen from the falling prices on Walmart's shelves, the rising prices at local gas stations, the shrinking size of many

American paychecks, even in the air we breathe. It can be heard in the voices on the end of tech-support phone calls. It is noticeable from the way freighters float low in the waters of the South China Sea because they are so heavily loaded with goods flowing out of new Chinese factories. Most plainly, it can be seen in the raw numbers: India and China have become the fastest- growing big economies on the planet. They look set to stay that way for decades and are on their way to becoming economic giants within a generation.

Yet the rise of India and China is about much more than jobs moving overseas: it is about a major shift in post-Cold War geopolitics, about quenching a growing thirst for oil, and about massive environmental change. This is tectonic economics: the rise of India and China has caused the entire earth's economic and political landscape to shift before our eyes."

—Robyn Meredith (2008). *The Elephant and the Dragon*, pp. 11–13, Norton & Company, New York. ISBN 978–0-393-33193–6 pbk.

Outcome 3. American civil engineers will need to consider potential economic impacts from India and China when serving global clients and pricing competitively.

But India and China not only consume engineering services, but they also educate young engineers. Both countries' cultures value education and view engineering as a well-respected profession. The sheer size of the populations of these countries and the growing supply of engineers and other labor will be factor in the global economy. The key question is whether these countries will graduate more engineers and other skilled labor in greater numbers that their own countries require (Meredith, Robyn, 2008).

Outcome 4. Civil engineers will need to consider global climate change and more severe weather in their civil designs.

The US Army Corps of Engineers has a publication that predicts the height of rising oceans in the next 30 years. Many parts of the nation and the world seem to be experiencing greater fluctuations in temperature and rainfall. Civil engineers would be prudent to clearly identify project design criteria that consider these trends in the "basis of design." Having the client or owner approve (or reject with the engineer's confirmation) these climate-related design criteria, assumptions, and details as part of the project records also would be prudent (ASCE, 2006, 2021, 2022).

Outcome 5. Civil engineering and many other technical disciplines will continue to move toward a higher degree of specialization.

A higher degree of complexity in projects seems to be a prevalent condition. This complexity is leading to increased specialization within engineering disciplines. ASCE recognizes this trend, which is evidenced in the discussion in Chapter 1. An engineer equipped with a B.S. degree in civil engineering may not be sufficiently prepared for the projected complexity of engineering tasks on many projects in the future (ASCE, 2006, 2021, 2022).

Accompanying the trend toward increased specialization will be a need for civil engineers to be conversant in advanced computer tools involving simulation and data modeling. Clients, such as the federal government, already are requiring designers to supply building information models (BIM) as part of a project's "as-built plan documents" referred to as "as-builts". Contracting methods, like design build and integrated project delivery, and new techniques, like lean construction, are converging with 3D computer-aided design (CAD), 4D CAD (3D plus project scheduling), and 5D CAD (4D plus budget/cash flow). (See Chapter 17, Emerging Technologies.)

Outcome 6. Project Management Excellence—Excelling at project management will be a great asset in the 21st century.

If projects seem to be more complex and the competition for resources and financial capital is increasing, then it seems logical that the need to remain on schedule and within budget will become even more critical. As discussed in Chapter 4, Professional Engagement, many clients believe in the "open window" concept where the longer the window (their project) remains open and active the more dollars (their money/budget) fly out. And, in some ways, these clients are correct because most schedule slippages are accompanied by budget increases. Many experienced civil engineers will agree that, depending upon the specific project, labor costs are a very significant component of the overall budget. Each hour the project is extended generally is accompanied by additional labor hours to manage, direct, and construct the project. Therefore, the demand for effective project managers will likely increase and we may also see an increase in innovative project delivery systems such as design build (DB) and integrated project development (IPD), as clients struggle to keep their projects on schedule and within budget.

Outcome 7. Civil engineering must respond proactively to increasingly complex challenges related to public health, safety, and welfare (McNeil Engineering 2021).

The 2001 American Society of Civil Engineers (ASCE) published the report titled Engineering the Future of Civil Engineering, and in 2006 ASCE prepared The Vision for Civil Engineering in 2025. Both publications acknowledged that civil engineers will need to respond proactively to increasingly complex challenges related to public health, safety, and welfare. If we review the six outcomes presented above in addition to the data in the previous section titled, Global Climate Change—From a World View and a State Perspective, it is apparent that civil engineers will have a challenging future responding to increased needs for infrastructure; water resources; building design and construction; response to disasters from high winds, floods, and wildfires, among other public health and safety demands (ASCE, 2006, 2021, 2022).

15.5   LEARNING TO PROJECT MANAGE A MEGA-PROJECT—THE CASE OF BAA AND HEATHROW TERMINAL 5

The following case study looks at ways that a project team responded to scarce resources and financial capital accompanied by an increased emphasis on low cost and very low maintenance. The example considered is the mega-project for construction of Terminal

5 at London's Heathrow airport. BAA Airports Ltd. The BAA (British Airport Authority), formerly the owner and operator of seven UK airports, was one of the largest transport companies in the world. BAA was purchased and currently is managed by Heathrow Airport Holdings Limited, in turn owned by FGP Topco Limited, a consortium owned and led by the Spanish infrastructure specialist Ferrovial S.A. (25.00 percent), Qatar Investment Authority (20.00 percent), Caisse de dépôt et placement du Québec (CDPQ) (12.62 percent), GIC of Singapore (11.20 percent), Alinda Capital Partners of the United States (11.18 percent), China Investment Corporation (10.00 percent), and Universities Superannuation Scheme (USS) of the UK (10.00 percent).

Learning to Project Manage a Mega-Project—The Case of BAA Airports Ltd. And Heathrow Terminal 5

By Dr. Tim Brady, Principal Research Fellow, CENTRIM, University of Brighton, UK

Abstract

This case study examines how BAA implemented a strategic program of capability building to improve the management of its construction projects, ranging from routine capital projects to a one-off mega-project—Heathrow's Terminal 5 (T5). It focuses on the learning gained from previous projects, individuals and organizations that contributed to the innovative approach used to manage T5—Europe's largest and most complex project. The project is an example of a "mega-project" because of its scale, complexity and high cost, and its potential to transform the project management practices of the UK construction industry.

BAA Airports Ltd., (BAA) owner and operator of seven UK airports is one of the largest transport companies in the world. BAA stems originally from British Airports Authority and was purchased by Airport Development and Investment Limited, a company formed and owned by a consortium led by Grupo Ferrovial, a Spanish firm specializing in infrastructure.

Introduction

This chapter examines how BAA—a major construction client—implemented a strategic program of capability building to improve the management of projects at its airports. These range from routine capital projects to a one-off "mega-project"—Heathrow's Terminal 5 (T5).

(Continued)

The focus is on the learning gained from previous projects, individuals and organizations that contributed to the innovative approach used to manage the T5 project. The T5 project used "integrated team working" to ensure that safety, time, budget and quality constraints were met. It was completed in March 2008, on time, within budget and with a high safety record. BAA developed an innovative form of cost-reimbursable contract—the "T5 Agreement"—under which BAA held all the risks associated with the project, rather than transferring them to external suppliers, and guaranteed a level of profit for suppliers.

Background

BAA is the world's largest commercial operator of airports, responsible for seven UK airports and managing a range of activities in various other airports around the world. It undertakes hundreds of capital projects as part of its ongoing operations, as well one-off mega-projects, such as the design and construction of Stansted Airport and Heathrow's T5. In 1994 BAA's capital projects program was running at £500 million but it was also planning a series of major projects, including the development of T5 (the largest project BAA has undertaken). This ambitious capital projects programme had to be undertaken against a background of steadily rising costs of building, while the charges they could levy on their building occupants were rising at a lower rate since they were subject to regulation. The T5 project was one of Europe's largest and most complex projects. It is an example of a mega-project (Flyvbjerg et al., 2003[i]) because of its scale, complexity and high cost and its potential to transform the project management practices of the UK construction industry. It was broken down into 16 major projects and 147 sub-projects. At any one time the project employed up to 6000 workers, and as many as 60,000 people were involved in the project over its lifetime. The goal of the project was to increase the airport's capacity of 67 million passengers a year to 95 million (https://www.heathrow.com/company/about, Accessed 3/27/23).

Towards Developing Repeatability and Predictability in Construction

Based on his previous experience in the car industry, Sir John Egan, BAA's Chairman during the early 1990s recognized that BAA could make radical improvements to the way it traditionally delivered projects. Egan wanted to emulate the continuous improvements in performance achieved in car manufacturing, by creating an orderly, predictable, and replicable approach to project delivery. Egan brought in experienced people from outside BAA to spearhead this strategy—people who had worked on massive projects for demanding and sophisticated clients in other sectors.

This new thinking was embodied in a UK government-sponsored report called Rethinking Construction (1998) authored by Sir John Egan, which became a manifesto for the transformation of the UK construction industry. The report argued that the client could play a role in this transformation by abandoning competitive tendering and embracing long-term partnerships with suppliers, based on clear measures of performance. Partnerships would provide a continuous stream of work over an extended period and the stable environment that contractors needed to achieve systematic improvements in the quality and efficiency of their processes. The report recommended that firms focus more strongly on customer needs, integrating processes and teams, and on quality rather than cost.

Under Egan's guidance, BAA's senior managers began applying the principles laid out in the Egan Report to improve BAA's project processes and relationships with suppliers, and by the late 1990s BAA had made significant improvements to its project execution capabilities, reflected in a greater degree of predictability in terms of time, cost and quality.

BAA's first phase of moving towards being able to deliver T5 centered around trying to get more predictability in terms of time and cost in its projects. There were four key strands to this:

- Developing new and improved project processes
- Managing the supply chain
- Standardization and prefabrication
- Integrated team working

Project Process Improvement

BAA's desire to improve its project processes was led by its CIPP (continuous improvement of the project process) program to develop a new process for organizing projects in its capital investment program—which was introduced in 1995. It aimed to establish a consistent best practice process applied to all projects with a value of over £250,000. The process was designed around a typical £15 million building project, but it was expected that it could be used regardless of the size of the project right across BAA's business. A taskforce was set up, with project representatives from all parts of the group, with the aim of capturing all the best parts of existing practice and creating a single system. More than 300 people were consulted over a period of 18 months, both from within BAA and from other companies and industries. The issuing of the CIPP handbook was accompanied by a long-term training program to provide a firm basis of understanding for implementation.

The CIPP handbook laid down a set of key policies or principles, safe projects, a consistent process, design standards, standard components, framework agreements (see below), concurrent engineering and pre-planning which all capital projects had to adhere to. It provided a template for the organization of BAA projects and outlined a

(Continued)

seven-stage process covering the project life cycle, from inception through to operation and maintenance. Each stage included a series of checkpoints which had to be completed and a series of evaluation gateways, where the project was assessed by an evaluation team before going to approval gateways for sign-off from local and/or group capital project committees. To successfully pass through a gateway, eight key sub-processes needed to be managed and the outputs from each co-ordinate: development management, evaluation and approval, design management, cost management, procurement management, health and safety, implementation and control, and commission and handover. BAA developed a process map showing each stage along the top and the sub-processes within each stage, outputs and gateways.

Managing the Supply Chain: The Framework Program

At the same time as it was developing the CIPP process, BAA started the development of what it called the Framework Program to work with a number of preferred suppliers on an ongoing basis. Up to that point, every time BAA embarked on a capital project it went to tender and through a whole process of qualifying, with the result that everyone had to go up the learning curve every time. The five-year framework agreement provided suppliers with an opportunity to learn and to make continuous improvements year by year that would benefit both BAA and the supplier.

This was first attempted in 1993–1994 and subsequently became widely used. Framework agreements were not restricted to first-tier suppliers—they encompassed a wide range of services, including specialist services, consultancy services (design and engineering), construction services, etc. Each agreement was structured slightly differently, depending on the nature of the service, but the concept was applied consistently. Standardization of components helps to reduce unit costs, but the framework agreements also incentivize the suppliers to improve the products and performance. For example, in the case of lifts and escalators, the installation contractors went to BAA and would analyze the process, saying that if they did it like this they could get the lift installed quicker and this would save time and money.

In 1998 BAA still had as many as 23,000 suppliers working throughout the company, and each of its seven UK airports had developed its own unique approach to supply chain management. BAA recruited Tony Douglas as Group Supply Chain Director and tasked him with profiling and reducing BAA's number of suppliers to a much more manageable level (Douglas, 2002[ii]). He was able to draw upon his own experience in the car industry and many studies of electronics, aerospace and other industries that had already developed sophisticated approaches to supply chain management. Strong internal capability had to be developed so that BAA could better manage its external suppliers.

By 2002 BAA had developed a second generation of framework agreements to achieve more accurate project costs, to implement best practice and to work with suppliers in longer-term partnerships. The key difference between first- and second-generation framework agreements is that the latter were devised to source the best-in-class capability and were valid for a period of 10 years rather than five. BAA developed strict criteria to select the best partner for each project. All aspects of its business are examined, including all its systems and processes covering quality, people, its supply chain, finance, R&D, and business development.

Under the second-generation agreements, suppliers worked with BAA in integrated project teams to cultivate close co-operation, to leverage the right expertise needed for specific projects and to reduce costs. BAA injected commercial rigor into its second-generation framework agreements by creating an annual review to measure the supplier's performance against projects that it delivers. If the supplier failed to perform, as measured against an agreed set of performance criteria, BAA set an improvement plan. A supplier would remain in BAA's family of framework suppliers if it achieved the targets set in the improvement plan. If it failed, it was deselected and replaced by another firm from the list of framework suppliers.

Standardization and Prefabrication

Several projects carried out at BAA prior to the T5 project were experiments in predictability and repeatability. CIPP was applied to four different clusters of BAA products: car parks, pavements, baggage handling, and buildings. For example, an early attempt to create predictability in BAA's project delivery was the design of standardized BAA office products. Standardized design was first developed for three offices at BAA's World Business Centers in Heathrow, which were built in succession. Previously each office would have been dealt with as a separate project, starting from scratch each time. Under the new approach BAA designed and built the first one and, rather than starting from scratch again with a new design, decided to replicate the first on a site next door but with a target of reducing the cost by 10% and the time by 15%. BAA then built a third to the same design with similar targets for time and cost reduction over the second one.

The design was then used at Gatwick Airport and reused at lower cost for Stansted Airport's second satellite office buildings. For the Stansted project 90% of the team from Gatwick was retained. The frame that had been used for the Gatwick project had had some problems related to component fit and watertightness, but these were overcome at Stansted, where the frame utilized pre-cast columns with cast-in slab/column connectors (patented by Laing O'Rourke plc). Prefabrication of the external stairs in entire floor height modules was another feature. BAA reaped significant learning-curve cost advantages from this approach to product standardization. The above-ground construction time was reduced by five weeks, saving around £250,000.[iii]

(Continued)

Involvement in framework agreements on a long-term basis has enabled other suppliers to develop repeatable processes and improvements in productivity over time, as lessons have been learned and shared. For example, AMEC—which has framework agreements covering building services, airfield pavements, aircraft stand services and general consultancy to support the development and upgrade of airport infrastructure across the UK—claims that in excess of 100 projects have been delivered on time with savings of 30% achieved overall.[iv]

Integrated Team Working

One of the central features of the BAA approach was integrated team working. The Heathrow Express tunnel collapse in 1994 played a key role in the development of this form of partnering. BAA had contracted Balfour Beatty as the main contractor for the construction of a fast passenger link between Heathrow and London, the Heathrow Express. Balfour Beatty had been chosen because they had just completed the Channel Tunnel and a crucial engineering concern was the tunnels needed within the airport.

The collapse of the Heathrow Express tunnel put the entire £440 million project in jeopardy. The normal construction industry response would have been to go down the litigation route and place the blame on the main contractor, Balfour Beatty. Instead BAA chose to work together with Balfour Beatty and the other main parties as partners to resolve the situation. A new contract was drawn up between BAA and Balfour Beatty, which established what became known as "the single team," in which all the parties worked as one team instead of a collection of separate project groups. This integrated team, which also included Mott MacDonald, succeeded in delivering the new project in June 1998—only six months behind the original schedule.

BAA appointed a new Construction Director for the recovery project, who demonstrated and promoted a culture of co-operation instead of rivalry. He employed:

> "a team of specialist change facilitators and behavioral coaches to work with the team on changing the tradition of adversarial working in order to make the single team "statement of intent" a reality. This was demonstrated in the everyday behavior of the construction team."[v]

The success of the Heathrow Express recovery project showed how partnering and trust could be made to work.[vi]

At one point Heathrow Express was 24 months behind program, but eventually became operational only nine months after the original projected date. It was a huge success story within BAA. Once finished, the Construction Director recruited to work for BAA on that Heathrow Express recovery project was transferred to the central BAA capital projects team as Group Construction Director. Much of the learning was channeled into developing the CIPP and applying the integrated team working concept to different clusters: the pavement team (for runways, taxiways, links and resurfacing); the baggage handling team (clusters of suppliers); and a cluster of suppliers for buildings (shell and core, fit-out, etc.). Teamwork was mentioned as a major success factor in the Terminal 1 International Arrivals concourse refurbishment project, which we studied in 1999. There it was claimed that teamwork had been excellent, both at the Heathrow Airport Limited level and through to construction activities where the co-location of the team provided huge benefits. It was also noted that the team members "left their companies at the door" when they came to work on the project.

Preparing for T5: Creating the T5 Approach

The steps that BAA had taken up to this point—the development of the CIPP, the first and second-generation framework agreements, and integrated team working had improved project predictability and repeatability. However, a more radical approach was needed to deliver T5, where there was a much higher level of uncertainty involved. T5 had to be constructed while causing minimal disruption to the operations of Heathrow Airport.

Between 2000 and 2002 BAA carried out an in-depth analysis of every major UK construction project in the last 10 years (valued at over £1 billion) and every international airport that had been opened over the last 15 years. This analysis showed that:

- No single UK construction project of that size had been delivered on time, on budget, safely and to the quality standards that had originally been determined, and
- Not a single international airport had worked properly on day one.

Based on this analysis BAA predicted that T5 would be 18 to 24 months late, over budget by a £1 billion, and six people would be killed during its construction.[vii] The airport case studies showed that it would take three years to build up to moving 30 million passengers a year through the new terminal. BAA expects T5 to do that in its first year of operation, since the passengers are already there in Terminals 1 and 4.

Following the public inquiry, severe restrictions were placed on traffic volumes and routes outside the site and there was only one viable entrance to the site. According to Tony Douglas, Managing Director of the T5 project:

> . . . if we were to build it conventionally, not only would we require about 7000 additional workers on site, but there would be a 40 ft vehicle load passing through the gate every 40 seconds or so for the next four years. . .

The only possible solution is to develop standard solutions that include pre-assembled products, manufactured off site, then assembling at the airport with the minimum of human intervention. This also brings environmental benefits by reducing the impact of

(Continued)

thousands of construction workers on local infrastructures and, as everybody recognizes, safety performance is infinitely better in factory conditions than on a construction site.[viii]

There was recognition in BAA that the only way to deliver T5 was to change "the rules of the game" by creating a set of behaviors that allow people to be constructive and come up with innovative solutions to problems. The success of the Heathrow Express recovery project demonstrated that such an approach was viable. The T5 approach combined two main principles: the client always bears the risk; and partners are worth more than suppliers. These principles were embodied in the T5 Agreement—a new form of contract developed by BAA which guaranteed suppliers' profits while the client retains the risk.

The T5 Agreement provided an appropriate environment for integrated team working. Rather than transfer risks to its suppliers, BAA assumed responsibility for all project risks all the time and worked with its integrated team members to solve problems encountered during the project. The agreement included an incentive payment if a supplier achieved exceptional performance. This was designed to enable suppliers to work effectively as a part of an integrated team and focus on meeting the project's objectives not only in relation to the traditional time, budget, and quality measures, but also in relation to safety and environmental targets.

BAA decided to adopt this approach since traditional liabilities—such as negligence, defective workmanship, and the like are extremely difficult to prove in an integrated team environment. BAA recognized that if suppliers were made jointly responsible for running significantly over budget, then this would probably put them out of business. It decided to reimburse the costs of delivery and to reward exceptional performance and penalize inferior performance only in terms of profitability.

BAA's approach to the delivery of T5 has resulted in a number of innovations and examples of best practice in a number of areas (NAO, 2005)[ix] including:

- The contract form—the T5 Agreement is a cost-reimbursable contract, in contrast to the fixed-price contracts generally seen on large projects.
- Risk management—the client takes on the risk so that project management becomes a tool of risk management rather than vice versa.
- Sponsorship and leadership—the T5 project managing director is an executive main board member and receives regular and overt support from other board members.
- Logistics management—the constraints on access to the site have led to:
  - a focus on pre-assembly and off-site fabrication and testing wherever possible
  - the creation of two consolidation centers: the Colnbrook Logistics Centre—consisting of a railhead for the supply of all bulk materials, a factory for the prefabrication and assembly of rebar, and a laydown area—and the Heathrow South Logistics Centre for the preassembly of pile cages and later as an area for assembly of materials into work packages ready to deliver on site
  - the use of "demand fulfilment software" designed to pull materials on site on a just-in-time basis.

- Insurance—BAA took out insurance for the whole project, lowering costs and avoiding unnecessary duplication.
- The approval process—a five-stage approval process based on the changing level of risk, which enables the project to move forward to the next stage without completing production design.
- Teamwork—the T5 Agreement incorporates integrated teams working to a common set of objectives and based on a capability approach, so that the team for each sub-project is assembled with the best possible expertise within the partner firms.

T5 opened on 27 March 2008, it was delivered on schedule and within budget. It was a major breakthrough in project management as it avoided the trend of all similar mega-projects being delivered late, over budget, and often below quality expectations.

References

i. Flyvbjerg, B., Bruzelius, N. and Rothengatter, W. (2003), *Megaprojects and Risk: An Anatomy of Ambition*, Cambridge: Cambridge University Press.

ii. Douglas, T. (2002), "Talking about supply chains," *Solutions: Projects and News*, WSP Group plc, Spring Issue, 5.

iii. Source: "BAA Lynton roll out Mark 3," May 2000, www.m4i.org.uk.

iv. Source: "Upgrading airports in partnership with BAA," www.amec.com.

v. Lownds, S. (1998), "Management of change: building the Heathrow Express. Leveraging team skills to get a railway business rolling—the story of the change of culture on the construction of the Heathrow Express Railway," presented to Transport Economists' Group, University of Westminster, 25 November.

vi. "Case Study 2 in Project Recovery: breaking the cycle of failure," Major Projects Association seminar, London, June 2000.

vii. National Audit Office (2005), "Case studies: improving public services through better construction," 15 March.

viii. Douglas, T. (2002), "Talking about supply chains," Solutions: Projects and News, WSP Group plc, Spring Issue, 5.

ix. National Audit Office (2005), "Case studies: improving public services through better construction," 15 March.

15.6   CIVIL ENGINEERING PRACTICE—A WIDER COMMUNITY VIEWPOINT

Additional insights on practicing civil engineering on an international basis are offered in the article titled, "Civil Engineering Practice—A Wider Community View." Several approaches are presented that may spark the imagination of those civil engineers wishing to become involved with civil engineering beyond the borders of their own discipline or nation.

Civil Engineering Practice—A Wider Community Viewpoint

By Brian S. Neale, Chartered Engineer, United Kingdom

Introduction

When qualified with their degree, civil engineers tend to concentrate on one particular branch of the many branches within the discipline (at a time, perhaps), learning more about it and then practicing within that arena. The wider vision can be exciting, however, as the practicing civil engineer develops and becomes more involved in the wider built environment community as well as the wider community as a whole. This can include clients and the public who use that infrastructure, for example, as that essential broader scope develops.

These growing interactions and experiences can include many enriching aspects in a developing professional life. Some thoughts on examples of international and pan-discipline working can be considered in groupings such as those listed below:

- Forensic Engineering—A Worldwide Community
- Engineering Disciplines—The Similarity of Focus
- Working with a Range of Other Disciplines
- Risk Management—How Safe?
- Professional Communications—Bridging Cultural and National Practices
- Qualifying to Practice—Some Different Approaches
- Codes of Practice—*Sans Frontière*
- Conclusion—Continuing Professional Development

Forensic Engineering—A Worldwide Community

There is plenty of evidence to support the saying that travel broadens the mind and this is confirmed by the personal experience of many people. As an example, a visit to the American Society of Civil Engineers[1] (ASCE) Convention in Minneapolis in 1997 proved to be an inspiration because, among other things, the first congress on Forensic

Engineering[2] was being held in parallel. Although not known at the time, this event stimulated the growth of a wider community of people interested in promulgating the understanding of both the fundamental reasons and also the wider reasons (or context) for our built environment sometimes not behaving as it was thought it would—or indeed as it was thought that it should. In other words, why did performance sometimes fall short of expectations to varying degrees, including those of the client and also of the designer? Occasionally that poor performance has proved to be catastrophic and caused loss of life.

The following year, a conference on Forensic Engineering[3] was held in London. This was organized by the UK-based Institution of Civil Engineers (ICE) in association with other relevant professional disciplines. It attracted wide interest, including from the BBC's World Service where it was featured in a radio program. An international series of conferences was thus born where, with the support of ASCE, it helped to widen the worldwide family of civil engineers and other disciplines. In the more recent conferences, papers have been received from all five continents of the world. More Forensic Engineering conferences had been held elsewhere in the world, for example, in Taiwan, India, Italy, and Spain. Conference proceedings[4,5,6] capture this knowledge base and thus the performance of facilities in ways that were as not intended. Building on the 1997 congress, a second one[7] was held in 2000 which established them as series, also, which had continued on a three yearly basis, since. Moreover, ASCE produces a publication, "'The Journal of Performance of Constructed Facilities"—that builds on the success of that venture by their Technical Council for Forensic Engineering between congresses to continue the flow and exchange of information with regular publication dates.

The two series of Forensic Engineering conferences mentioned above had enjoyed continued support both in the USA and in the UK, where ASCE and ICE, respectively, continued to organize them with significant work and support from their members and many others. However, ICE's series has now stopped. There is an international culture to "give something back" to the profession and thus the community as a whole.

Engineering Disciplines—The Similarity of Focus

When one examines catastrophic events that receive worldwide publicity and thus attention, original perceptions may change. Those events can include both "natural" events and "man made" ones. Here it is possible to think of examples of earthquakes and perhaps incidents in chemical works, respectively, where each type of event can (and has) led to significant loss of life. The international community pulls together to help where they feel they can—and learn for the future and thus future generations. Many engineering disciplines come into play including a number of specialist elements where civil and structural engineering are to the fore in respect of the all-important residual structural stability during and after rescue operations, for example.

(Continued)

Investigators in this context tend to be professional engineers, usually civil or structural, although chemical, electrical, fire, control systems and mechanical engineers share similar interests, for example, as well as professions such as metallurgists and materials specialists. The fascinating behavioral study of people and why we humans do things or omit to do things in a particular way, or in a particular order or time, becomes more understandable when listening to the input from ergonomists and psychologists, for example, who can and do contribute.

We thus have community which can be considered rich in diversity in ways such as:

- Many engineering disciplines
- Disciplines other than engineering
- National and international variations in culture—and thus possible approaches

Working with a Range of Other Disciplines

Working in Forensic Engineering, for example, often means interacting with a range of other disciplines. Hence, the experiences of investigators, clients, law enforcers, lawyers, code writers, loss adjusters, researchers, and teachers—to name just some—have been shared to help disseminate these global experiences of performance and "knock-on" activities. The rich mix of disciplines can help inspire by cross-fertilization of ideas with pan-discipline cultures. This compliments the, albeit, expected inspiration from experiencing the variety of national cultures with, perhaps, some different approaches, as mentioned above.

A further example of this nexus can be seen in an organization based in Europe that was set up following major incidents worldwide with the objective of acting as a forum, primarily for engineers, to exchange views and discuss significant issues at a senior level to help contribute to prevention of future occurrences of such incidents. A major strength of this body, the Hazards Forum,[8] is its rich mix of members and contributors. The focus is not Forensic Engineering, as such, as it focuses on future developments and management, including the avoidance of—or appropriate mitigation or amelioration of hazards—and thus potential risks, particularly pan-discipline. To look at the future, however, there is a need to examine the catalogue and pathology of past events and these aspects are included as appropriate.

For example, the Hazards Forum has looked at the next generation of new nuclear power station design issues; risks with some alternate fuels; and carbon capture in a particular series. To show the variety of topics presented and discussed, however, the event immediately preceding that Energy series' was from the medical sector and from an ergonomics point of view. Whereas this might have been perceived as near the outer bounds of the remit of the forum and of little interest to engineers, discussions centered on the ease with which prescriptions could be (and are) sometimes incorrectly administered through people error. A solution offered that significantly helped was better labelling and packaging in more than one way—perhaps not surprising. One speaker presented the way forward in the form of a hospital direction sign to the various internal departments where the strap line for a better way forward could simply be—Safety by better design. Does this sound familiar to the construction industry and to civil engineers? If not, perhaps every civil engineer could consider this as a fundamental maxim to use as appropriate.

The sign, for further thought, is reproduced in the following figure:

A Wider Community Viewpoint
(Acknowledgment: Professor Peter Buckle, Robens Centre for Public Health, University of Surrey from Hazards Forum Newsletter No. 64, Report of Seminar How ergonomics improves patient safety. [Peter Buckle, p.buckle@surrey.ac.uk and www.hazardsforum.org.uk/publications/publications_newsletters.asp]

Risk Management—How Safe?

This brings us into the realms of a discussion on hazards, risk, and how to approach this and which approaches for the assessment of hazards and risk would be better for which situations. For example, international variations include:

- The principles of the risk hierarchy[9]
- The precautionary principle
- As low as reasonably practical (ALARP)

(Continued)

- So far as is reasonably practical (SOFARP)
- The traffic signals system (red, amber, green)
- Others.

There will be various tools that can be used within assessments, some of which will include tools for analysis, a word sometimes used where "assessment" would be more appropriate and therefore better advised. Where analysis takes place the users need to be very aware of the accuracy and significance of any data used—and of course of the significance of their outputs.

National enforcers can, of course, contribute to the discussion on optimum approaches to ameliorate hazards and risks, although international influences are often incorporated in various ways. National regulations may also affect the approach, such as Building Regulations which may have requirements to resist disproportionate collapse, such as in the UK. Other legislations, such as for occupational health and safety and the safety of the public as a result of those activities, can affect the construction sector, including designers where they may become subject to criminal law—again, such as in the UK.[10] The risk creators have much guidance available to them, as do other duty holders.

Professional Communications—Bridging Cultural and National Practices

We can learn so much from others, therefore, if we open up our minds to other cultures and ways of doing things. This can include, for example, the international civil engineering community and still further in the wider engineering professional community—and others. When doing this, however, we need to ensure that our communications are effective, and thus accurate. Do we know how successful we are at this?

Communication is an essential part of a professional's job and one that sometimes achieves less than 100 percent success. We know from studies outside the engineering community the verbal communication can be less than 10 percent of that which happens in face-to-face encounters, partly depending on whether it is a social interaction or a business one—apparently. There are, of course, many tools that engineers use for communication, but how often do we know that those with whom we communicate have exactly the idea of what we expect to happen, that is, the "correct" idea.

Many examples of poor performance and other failures have been shown to be the result of "misunderstandings."

In one's professional practice, how often do we ensure that ideas have been communicated accurately, bearing in mind that there are two (or maybe more) parties to any communication? We may be satisfied with what is transmitted, but what assurance do we have that the receiver has it correctly and if not, what margin of error would be acceptable? Perhaps none! Well documented fatalities have occurred around the world because of inaccurate communications—or misunderstandings. Part of communication relies on particular competencies and perhaps assumptions about competencies of people and organizations, although it has been said that nothing should be assumed. Within parts of occupational health and safety legislation in Europe there are requirements for the competence of those involved, both organizations and individuals, which are in criminal legislation. This is generally seen as not unreasonable. How does one define competence, however, and even more difficult, how does one measure appropriate competence for the numerous tasks that are undertaken in creating and maintaining the built environment and not forgetting the removing, reusing, and recycling stages also?

Defining competence has been seen described as problematic but tends to include a combination of relevant and appropriate education, training, and experience as a start—although there are variants. Each of those elements then needs to be expanded for effective application by both those attempting to demonstrate their competence and also by those seeking to be convinced of those relevant competencies, of course.

Mentioning Europe above brings to mind another communications thought—how many continents are there in the world? When next with a group of people, perhaps a group that you may not normally be with, it can be interesting to ask them—and to note the variations in the answer. We may be used to America being divided into two—north and south, with the Arctic and Antarctic adding a further two. These would give more than the basic five. We may be used to sub-divisions also, such as the Indian sub-continent. One example that might take some people by surprise is the combining of two continents rather than splitting down into smaller sets as mentioned above, such as the word "Eurasia," which is sometimes used. This is another communications example of one person's knowledge being outside the scope of another's—or is that competence? How careful we need to be!

Qualifying to Practice—Some Different Approaches for Competency

Approaches to professional practice vary across the globe where, for example, in the USA engineers are licensed to practice on a state-by-state basis and on defined or particular aspects. This can be seen as a qualification of competence—for the period of the license which will need to be renewed and perhaps altered to suit developments, new requirements (perhaps in legislation), or interests of the applicant, for example. The need for continuing professional development is seen as an essential component for practicing engineers. This is seen as applicable for professional engineers across the world, although the basis and way in which the state gets involved varies. As an example there is not a renewable type of PE qualification in the UK, but a CEng award. CEng is the recognized shortened form for Chartered Engineer which is a qualification awarded

(Continued)

by the Engineering Council UK[11] under a Royal Charter which recognizes a level of competence gained—at a particular time. That level is tested by a professional institution such as the UK—based Institution of Civil Engineers for corporate membership of that body.

To achieve success a candidate must have appropriate academic qualifications from an accredited university together with appropriate post- graduate experience, where the latter is the particular focus of the examination for corporate membership. This can result in some people with extensive experience being tempted to apply to a number of institutions which can result in a multitude of letters after a person's name—a system which some cultures find bemusing! To return to CEng, however, that is a one-off award and is usually retained for the whole of a career. To put this into context, however, these engineers are expected to keep up with current relevant developments appropriate to their personal work area through recognized Continuing Professional Development (CPD). Other routes to professional membership of UK[12] engineering institutions are available although the above is the most usual.

Hence, although there are many similarities between, for example the ASCE and the ICE, one fundamental difference is that ICE is a qualifying body for its members, with most engineers who qualify choosing to apply for the CEng award that corporate membership allows them also.

Codes of Practice—Sans Frontière

Codes of practice, or standards, can be an interesting area for example of sharing experiences and approaches (especially across national boundaries) where many options will be explored before deciding on a common, or agreed path to follow and adopt. They are therefore a uniting force across cultures for both users and drafters. As national standards organizations look more and more to unified codes on a global or world region basis, others look more to having their own codes adopted more widely among their membership communities. The existence of ISO[13] (International Organization for Standardization) codes tends to be well known, with many being adopted as both world region and also as national standards. As an example, in Europe and the UK, there are now standards with the nomenclature BS EN ISO, which can be seen as a British Standard Euro Norm International Standard! There may also be another version, for example.

How are these achieved? An example close to civil engineers is the Structural Eurocodes produced for CEN[14] (European Committee for Standardization), the European standards-making body. The number of states in volved is twenty plus, with almost as many languages and of course, many diverse cultures with a very wide variation in weather conditions. How does one move forward to establish agreement on codes when often a member state will have thought that it had produced the ideal code for its own purposes, for example. Perhaps above all, how does one move forward with the plethora of languages in use?

There are just three official languages—German, French, and English. The official versions of each code are produced in those three languages with other national versions produced by member states in other languages as required from that base. To prepare these codes, however, they are drafted, discussed, developed, and finalized in one language. That working language was English and was used even outside the meeting caucus where knowledge of further languages may still not offer a compatible mode of communication. An interesting side effect of being a natural English speaker in a community where English is the working language but not the home language of most, there was, on returning to the UK, a short period of readjustment that was required to help to re-order words as usually spoken! Another positive spin-off was that the English used in those CEN meeting tends to be more precise than used at home and thus aided better communications, it is thought. The following example is given from a "Loads" code[15]:

- Climatic—instead of environment for weather conditions
- Actions—instead of loadings
- Execution—instead of construction which can include demolition, for example

The wider context for this code is the head code[16] for the actions series. There are codes, however, that do not as yet come under the ambit of CEN and whereas member states are required to withdraw codes that are covered by equivalent CEN codes, they are free to develop or maintain those national codes that are not. An example[17] in the UK is one that includes sustainability issues with respect to environmental consideration and the broad issue of demolition, partial demolition, decommissioning, and structural refurbishment where a number of options can be considered on a broad risk management approach. A performance-based code for demolition activities has proved advantageous and is one used in other countries too. The code also includes planning ahead for ways of dealing with waste streams where considerations such as re-use and re-cycling are helpful in establishing the materials that are to be removed before starting the works. A demolition protocol can help. Minimizing waste during construction is also helped by planning ahead, of course.

Hazard assessments and risk assessments are part of a way of civil engineering life with assessments for accidental actions becoming more relevant. There is a Structural Eurocode[18] on the topic that may be of interest to those considering accidental actions.

(Continued)

Conclusion

Are we alone as civil engineers? The answer is a resounding No!, of course. It is good to pause occasionally and think of the wider (worldwide) community in which we live, practice, and develop. It could be said that we need to keep developing and expanding our menu of experiences for the greater good of our profession and the global community—as well as ourselves, perhaps. This may lead us to a performance-based approach to engineering solutions that can both stimulate thought and inspire innovation—when considered with an open mind and with forward thinking which anticipates future demands and needs.

Hence, if continuing professional development (or continuing education and training) becomes a way of thinking for personal development there are many optimistic avenues to pursue, some of which are mentioned above.

Brian Neale CEng, FICE, FIStructE, Hon FIDE
Cardiff, UK

References

1. www.asce.org

2. Rens, Kevin L., Editor. *Forensic Engineering: Proceedings of the First Congress*. ASCE. Reston, USA, 1997

3. Neale, B.S., Editor. *Forensic Engineering—a professional approach to investigation*, Proceedings of the Institution of Civil Engineers First International Conference, 28–29 September 1998. Thomas Telford Ltd, London. 1999.

4. Neale, Brian S., Editor. Forensic Engineering—from failure to understanding. Proceedings of the Institution of Civil Engineers Fourth International Conference, 2–4 December 2008. Thomas Telford Ltd, London. 2009.

5. Neale, B.S., Editor. Forensic engineering—diagnosing failures and solving problems. Proceedings of the Institution of Civil Engineers Third International Conference, 10–11 November 2005. Taylor and Francis, London. 2005.

6. Neale, B.S., Editor. Forensic engineering—learning from failures, Proceedings of the Institution of Civil Engineers Second International Conference, 12–13 November 2001. Thomas Telford Ltd, London. 2001.

7. Rens, Kevin L.; Rendon-Herrero, Oswald; Bosela, Paul A.; Editors. Forensic Engineering: Proceedings of the Second Congress. ASCE, Reston, USA. 2000.

8. www.hazardsforum.org.

9. www.hse.gov.uk

10. www.hse.gov.uk

11. www.engc.org.uk

12. Kardon, Joshua B. The responsible engineer: the concept of the "Engineer of record." Proceedings of the Institution of Civil Engineers Fourth International Conference, 2–4 December 2008; Neale, Brian S., Editor. Forensic Engineering—from failure to understanding. Thomas Telford Ltd, London. 2009.

13. www.iso.org

14. www.cen.eu/cenorm/homepage.htm

15. BS EN 1991–1-6:2005 Actions on structures—Actions during execution (A part of Eurocode 1). BSI, London. 2005. To use standard the appropriate National Annexe is required, such as for the UK—BS EN 1991–1-6:2008 National Annex for Actions on structures—Actions during execution.

16. BS EN 1990:2002 Eurocode—Basis of structural design. Eurocode 1. BSI, London. 2002.

17. BS 6187:2011 Code of practice for full and partial demolition. BSI, London. 2000.

18. BS EN 1991–1-7:2008 Actions on structures—Accidental actions during execution (A part of Eurocode 1). BSI, London. 2008. To use standard the appropriate National Annex is required, such as for the UK—BS EN 1991–1-7:2008 National Annex for Actions on structures—Accidental actions.

15.7 SUMMARY

There has been a "global explosion" of technical achievements in the last 225 years since the initial age of industrialization, and civil engineers have been involved in many of these achievements. The growth and maturity of the profession has been astounding and complementary.

Judging from the potential impacts of the four broad processes (depicted in Figure 15.1) that place more choices in the hands of individual consumers and organizational customers, we can expect a "globalization cascade" with reinforcing feedback loops that strengthen and accelerate globalizing forces. These forces, including global climate change, will have dramatic effects on the civil engineering profession with anticipated demand for engineers, innovation, client service, and decentralized centers of expertise and influence.

Hot, Flat, and Crowded

"We are the first generation of Americans in the Energy-Climate Era. And what we do about the challenges of energy and climate, conservation, and preservation, will tell our kids who we really are. After all is said and done, I am still an optimist that we will rise to this challenge. I am certain that my children and grandchildren will live in a cleaner world and a safer world and a more sustainable world. Why? Because technology today is allowing us to connect and leverage more and more brainpower than ever before. Whole swaths of the world that really could not collaborate in solving problems are being brought into the discussion. That is hugely important and the reason that I believe we will work this out—we will learn as nations and individuals that we cannot afford to grow the old-fashioned way—by just mining the global commons and by thinking that the universe and nature revolve around us, and not the other way around."

—Thomas L. Friedman (2009). *Hot, Flat, and Crowded*, p. 474.

In addition, seven possible outcomes of globalization and climate change are presented for the reader's consideration, including:

Outcome 1: The competition for resources and financial capital will become greater.

Outcome 2: Civil engineers will see increased emphasis on cost and low maintenance requirements as project goals.

Outcome 3: American civil engineers will need to consider potential economic impacts of India and China when serving global clients and pricing competitively.

Outcome 4: Civil engineers will need to consider global climate change and more severe weather in their civil designs.

Outcome 5: Engineering and many other technical disciplines will continue to move toward a higher degree of specialization.

Outcome 6: Project management excellence—Excelling at project management will be a great asset in the 21st century.

Outcome 7: Civil engineering must respond proactively to increasingly complex challenges related to public health, safety, and welfare (Odeh, David, 2022; ASCE, 2006, 2021, 2022).

Civil engineers will play an important role in globalization and sustainability for the implementation of the UN's Sustainability Development Goals (SDG) themes. These themes will have a significant impact on all international and US civil engineering programs and projects. Civil engineering political, corporate, and organizational leaders will have to take a "lead role" in the creation of a truly sustainable society. Furthermore, civil engineers will need to exercise their knowledge, leadership, and communication skills to assist world leaders in resolving social, economic, and sustainability issues through national and international business practices that leverage their strengths in technology and collaboration.

> GLOBALIZATION is more than networking business to maximize profits, locating the lowest cost source of labor, securing the cheapest raw material despite the environmental impacts—because that is the definition of exploitation.
>
> At the time of this book's publication, according to estimates, more than 7 billion people live on our planet. Each day, more than 200,000 new babies add to this figure, which works out to about 140 additional people per minute, every minute, every day, 24/7/365! This fact alone should explain how globalization has a symbiotic relationship with sustainability.
>
> WE need to manage the world differently, with everyone working together in a "global sustainable way." The Better Concept: WE need to think of this globe as One World, all connected.

BIBLIOGRAPHY

ASCE. (2006). "The vision for civil engineering in 2025." American Society of Civil Engineers, ASCE Steering Committee. (Accessed June 21, 2006).

ASCE. (2021). "American Society of Civil Engineers Infrastructure Report Card ASCE." Washington D. C. https://www.infrastructurereportcard.org (Accessed March 3, 2021).

ASCE. (2022). "Step into the future: Civil engineers release visualization tool for engineers to imagine a Mega City in 2070." American Society of Civil Engineers—Civil Engineering Source Magazine. (February 2, 2022).

California Climate Adaptation Strategy 2021. (2021). "California climate adaptation strategy: Priorities, goals, and actions." https://climateresilience.ca.gov/overview/docs/20220404-CAS_Priorities_Goals_Actions.pdf. (Accessed June 12, 2021).

California Climate Adaptation Strategy Report. (2009 June). California Natural Resources Agency, Sacramento, California.

California Climate Adaptation Strategy Report. (2021 June). "California Natural Resources Agency." https://resources.ca.gov/Initiatives/Building-Climate-Resilience/2021-State-Adaptation-Strategy-Update. (Accessed June 12, 2021).

Climate Change. (2022 March). 2022: Impacts, Adaptation, and Vulnerability. NOAA Fisheries. https://www.fisheries.noaa.gov/video/climate-change-2022-impacts-adaptation-and-vulnerability.

Friedman, Thomas L. (2009). "Hot, flat, and crowded: Why we need a green revolution—And how it can renew America". Picador/ Farrar, Straus And Giroux, New York. ISBN-13: 978-0-312-42892-1, ISBN-10: 978-0-312-42892-2.

https://education.nationalgeographic.org/resource/globalization (Accessed 3/27/23).

https://resources.ca.gov/CNRALegacyFiles/docs/climate/Statewide_Adaptation_Strategy.pdf. (Accessed May 5, 2022).

https://www.asce.org/publications-and-news/civil-engineering-source/society-news/article/2021/03/03/asces-infrastructure-report-card-gives-us-c- (Accessed March 27, 2023).

https://www.heathrow.com/company/about (Accessed 3/27/23).

https://www.whitehouse.gov/bipartisan-infrastructure-law (Accessed 3/27/23).

https://www.whitehouse.gov/briefing-room/statements-releases/2021/11/06/fact-sheet-the-bipartisan-infrastructure-deal (Accessed 3/27/23).

https://www.youmatter.world/en/definition/definitions-globalization-definition-benefits-effects-examples (Accessed 3/27/23).

Investment in Our Nation's Infrastructure. (2021, July) "The bipartisan infrastructure bill," The White House.

Kanter, Rosebeth Moss. (1995). *World Class: Thriving Locally in the Global Economy*. Simon & Schuster, New York.

McNeil Engineering. (2021). "The future of civil engineering." https://www.mcneilengineering.com/the-future-of-civil-engineering (Accessed May 5, 2021).

Meredith, Robyn. (2008). *The Elephant and the Dragon*. Norton & Company, New York. ISBN: 978-0-393-33193-6.

Odeh, David. (2022). "What is future world vision?." American Society of Civil Engineers—Civil Engineering Source Magazine. (March 1, 2022).

Ritzer, George. (2010). *Globalization: A Basic Text*. Wiley-Blackwell, West Sussex, United Kingdom. ISBN: 978-1-4051-3271-8.

(2007). Science Basis Contribution of Working Group I to the Fourth Assessment Report of the IPCC. (ISBN 978 0521 88009-1 Hardback; 978 0521 70596-7 Paperback).

Shigley, Paul. (2008). Prepare for Inevitable Climate Change. November 25, 2008. https://www.cp-dr.com/articles/node-2194.

Steger, Manfred B. (2009). *Globalization—A Very Short Introduction*. Oxford University Press, New York. ISBN: 978-0-199-55226-9.

The Intergovernmental Panel on Climate Change (IPCC) – set up jointly by the World Meteorological Organization and the United Nations Environment Programme, Climate Change 2007 – The Physical.

The Intergovernmental Panel on Climate Change (IPCC). (2022). IPCC–58. https://www.ipcc.ch (Accessed May 5, 2022).

United Nations Resolution 70/1. (October 21, 2015a). Transforming our world: The 2030 Agenda for Sustainable Development.

United Nations Resolution 70/1. (2015b). "Transforming our world: The 2030 agenda for sustainable development." https://sdgs.un.org/2030agenda. (Accessed May 5, 2022).

CHAPTER 16

Sustainability

Big Idea

As the global population grows and standards of living improve, there will be increasing stress on the world's limited resources. Thus, engineers of the future will be asked to use the Earth's resources more efficiently and produce less waste, while at the same time satisfying an ever-increasing demand for goods and services

"Knowledge of the principles of sustainability, and their expression in engineering practice, is required of all civil engineers."
  —*ASCE'S Civil Engineering Body of Knowledge for the 21st Century, BOK2*, p. 128

"Challenging issues such as climate change, urbanization, and the rapid pace of technological advancement create opportunities. These issues also require serious re-evaluation of current professional practice and standards."
  —ASCE's Roadmap to Sustainable Development: Four Priorities for Change
(January 2023)

Civil Engineer's Handbook of Professional Practice, Second Edition. Karen Lee Hansen and Kent E. Zenobia.
© 2025 John Wiley & Sons, Inc. Published 2025 by John Wiley & Sons, Inc.
Companion website: www.wiley.com/go/hansen/CivilEngineersHandbook

Key Topics Covered

- Sustainability Defined
- Sustainable Engineering
- Systems Thinking
- Ecodesign
- Toward New Values and Processes
- Expanded Project Delivery Process
- Integrative Approaches
- Sustainable Design and Materials Strategies
- Lifecycle Cost Analysis
- Leadership in Energy and Environmental Design
- Future Directions
- Recognizing the Need and Criticality for Sustainability—Here's What We Can Do About It
- Four Ways We Can Become More Sustainable as Professional Engineers

Related Chapters in This Book

- Chapter 5: The Engineer's Role in Project Development
- Chapter 6: What Engineers Deliver
- Chapter 15: Globalization
- Chapter 17: Emerging Technologies
- Chapter 19: Construction Management
- Chapter 21: What Engineers Need to Know

16.1   INTRODUCTION

For the past 200 years, humanity gradually has been polluting and modifying the natural environment at an increasing rate. Added to the pollution is the quantum depletion of non-renewable energy in fossil fuels. Why should civil engineers, architects, and builders care? As Kenneth Yeang, author of *Ecodesign*, states:

> *Our health as human beings, and as one of the millions of species in nature, depends upon the air that we breathe and the water that we drink, as well as on the uncontaminated quality of the soil from which our food is produced.*

The built environment is constructed from renewable and nonrenewable energy and material resources. Built systems are dependent upon the earth as a supplier of energy and material resources. Understanding the basis of sustainability enables us to comprehend the interconnections and processes that make up the environment and can help us to recognize the causes of degradation due to humankind's development and progress.

This chapter explores why ASCE has called for renewed professional commitment to stewardship of our natural resources and the environment and has stated that the knowledge of the principles of sustainability, and their expression in engineering practice, is required of all civil engineers. The chapter also describes what civil engineers can do to promote sustainability and how sustainability can be incorporated into the contemporary design process.

16.1.1 Background

A *sustainable humankind living plan* is a lifestyle plan that conserves earth's natural resources. This plan is essential for all nations and humans to adopt because our lives depend upon it. Our earth's ecosystems are communities of living flora and fauna that *coexist* while interacting with their physical environment. Natural resources need to be managed like a checking account. We cannot write checks that our account cannot afford, and we cannot continue with irresponsible consumption that places our valuable resources into a *threatened* or *extinction* status without future humankind paying a very high price.

The United Nations (UN) has recognized the importance and criticality of sustainable lifestyles and has adopted the Sustainable Development Goal titled, UN Goal 12: Sustainable Consumption and Production. Here is what the UN states about sustainable development and why it's important:

> *One of the greatest global challenges is to integrate environmental sustainability with economic growth and welfare by decoupling environmental degradation from economic growth and doing more with less. Resource decoupling and impact decoupling are needed to promote sustainable consumption and production patterns and to make the transition towards a greener and more socially inclusive global economy.*

Responsible sustainable growth and development involve more than one person, one community, one state, one nation. Civil engineers are an important part of this solution. A prudent CE might ask, "How can I do my part?" "What exactly can I do as a civil engineer and how can my company/organization help in this critical endeavor?" The UN Nation's committees have thought a lot about these questions. Figure 16.1 depicts specific actions that we ALL can implement to become more sustainable.

Do we have any option other than adopting *sustainable lifestyles*? The journey to arrive at our present position took a long time, with mostly unsustainable practices pursued

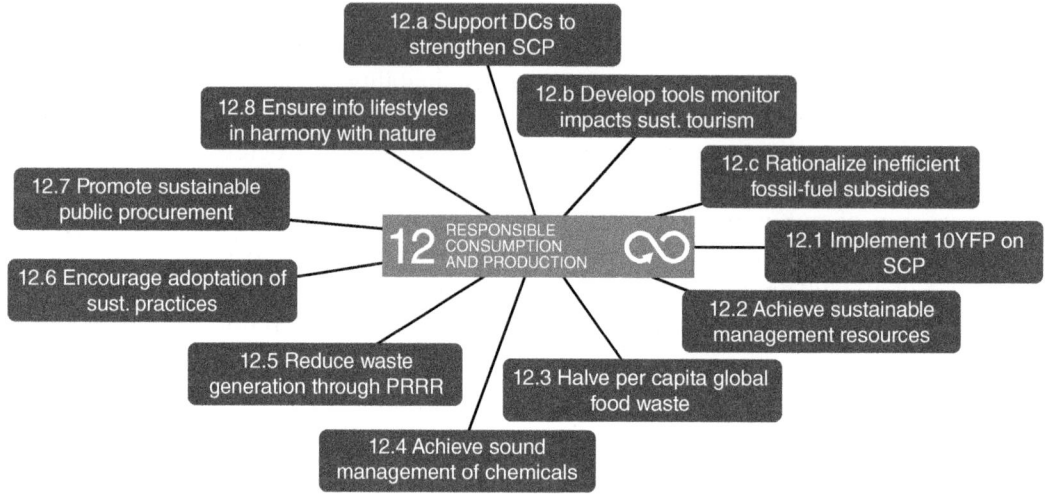

United Nations Sustainable Development Goal 12 (SDG 12)– encourages
more sustainable consumption and production (SCP) patterns through various
measures in order to promote increased human well-being while decoupling
economic growth from resource use and environmental degradation.

KEY

DCs: Developing Countries

SCP: Sustainable Consumption and Production

10YFP: 10 Year Framework Program

PRRR: Prevention, recycling, reduction, and reuse

FIGURE 16.1   UN Goal 12: Sustainable consumption and production."
Source: Adapted from UNEP, *https://www.unep.org/explore-topics/sustainable-development-goals/why-do-sustainable-development-goals-matter/goal-12*

globally over several centuries. To achieve a course correction toward sustainable lifestyles will take time and dedicated effort. Civil engineers need to be aware of plans, such as UN Goal 12, as well as impending mandates and practices within their own local jurisdictions.

> "We cannot solve our problems with the same thinking we used when we created them."
> —Albert Einstein

16.2   SUSTAINABILITY DEFINED

"Sustainability" has numerous meanings in the English language. The most widely quoted definition is from the UN's 1987 Brundtland Commission Report, *Our Common Future*: "meeting the needs of the present without compromising the ability of future generations to meet their own needs." The Brundtland Report also pointed out the importance of evaluating actions in terms of what has been called the Three E's:

- Environment
- Economy
- Equity

The "Three E's" force us to examine the cause and effects of our actions in relationship to the systems of the Earth and also to examine issues of justice—both economically and socially—for our fellow humans. This is a very challenging concept!

Our Common Future laid the foundation for the "Earth Summit" at Rio de Janeiro, Brazil, in 1992, which marked the real beginning of international environmental protection.

Sustainability Definitions

In 1983, Gro Harlem Brundtland, a female physician who later became Prime Minister of Norway, was invited by then United Nations Secretary-General Javier Pe'rez de Cue'llar to establish and chair the World Commission on Environment and Development (WCED), also known as the Brundtland Commission. Through extensive public hearings, the Commission published a report in April 1987 called Our Common Future.

The complete quote from the Brundtland Commission Report is:

"Sustainable development is development that meets the needs of the present without compromising the ability of future generations to meet their own needs. It contains within it two key concepts:

1. the concept of 'needs,' in particular the essential needs of the world's poor, to which overriding priority should be given; and
2. the idea of limitations imposed by the state of technology and social organization on the environment's ability to meet present and future needs."

Additional definitions of sustainability abound, and they are many and varied. However, as does the Brundtland definition, most definitions of sustainability include three basic components: 1) environmental protection; 2) economic development; and 3) equity (social).

Civil engineers have tremendous influence on new legislation, development trends, design concepts, innovation, building methods and codes, and much more. Now is the time to display this leadership and influence, the earth as we know it and society needs civil engineers.

16.3   SUSTAINABLE ENGINEERING

As the global population grows and standards of living improve, there will be increasing stress on the world's limited resources. Thus, engineers of the future will be asked to use the Earth's resources more efficiently and produce less waste, while at the same time satisfying an ever-increasing demand for goods and services. To prepare for such challenges, engineers will need to understand the impact of their decisions on built and natural systems, and must be adept at working closely with planners, decision-makers, and the general public (Birkland, Janis 2002). Figure 16.2 illustrates how engineers should seek opportunities in planning and designing to improve:

- Biodiversity
- Ecological connectivity
- Biointegration with local habitants

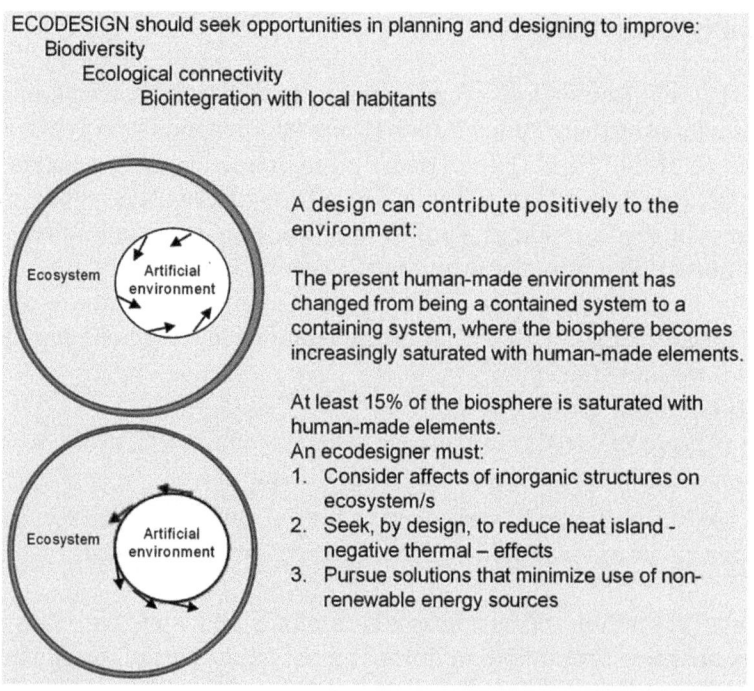

FIGURE 16.2   Ecodesign.
Source: Adapted from Kenneth Yeang, *Ecodesign*.

In 2005, Carnegie Mellon University, the University of Texas at Austin, and Arizona State University established the *Center for Sustainable Engineering*, supported by the National Science Foundation (NSF) and the Environmental Protection Agency (EPA). The Center for Sustainable Engineering uses the Brundtland definition of sustainable development and cites the following examples of sustainable engineering:

- Using methods that minimize environmental damage to provide sufficient food, water, shelter, and mobility for a growing world population
- Designing products and processes so that wastes from one are used as inputs to another
- Incorporating environmental and social constraints as well as economic considerations into engineering decisions

The American Society of Civil Engineers has taken a strong position in support of sustainability:

ASCE embraced sustainability as an ethical obligation in 1996, and policy statements 418 and 517 point to the leadership role that civil engineers must play in sustainable development. The 2006 ASCE Summit on the Future of Civil Engineering called for renewed professional commitment to stewardship of natural resources and the environment. Knowledge of the principles of sustainability, and their expression in engineering practice, is required of all civil engineers.
Civil Engineering Body of Knowledge for the 21st Century, 2008, p. 128

Overcoming Obstacles to Sustainability

At a 2009 workshop in Sacramento, California, members of ASCE, ASCE's Environmental & Water Resources Institute (EWRI), and the Floodplain Management Association (FMA) came together to discuss sustainability. Participants observed the following:

Although many California communities are eager to incorporate sustainability into their planning efforts, a number of challenges make this difficult.

- In many cases, communities lack a sustainability vision, state leadership, tools, incentives, and indicators to effectively ensure their own sustainability.
- Community sustainability needs financing, political will, and appropriate protective regulations.

(Continued)

- Communities are not linked with State actions and policies on sustainability.
- Addressing the connection between land-use planning, flood management, water supply, energy consumption, and natural resource protection is still in the early stages of implementation.
- Water policy does not sufficiently incorporate the economics of water, or the significant connection between water use and greenhouse gas production.
- Planning efforts at local, regional, and State levels lack sufficient coordination and integration.

These challenges are not unique to California. What is encouraging is that civil engineers are developing plans to overcome these obstacles.

More recently, ASCE's position is as follows:

ASCE and its members are dedicated to ensuring a sustainable future in which human society has the capacity and opportunity to maintain and improve its quality of life indefinitely, without degrading the quantity, quality, or the availability of natural, economic, and social resources.

ASCE encourages global leaders to stand behind the reduction of carbon emissions ensuring for a more resilient and sustainable future through its Conference of Parties (COP) 26 statement.

ASCE has long considered sustainability a strategic issue confronting practicing civil engineers. Its integration into professional practice is required to address changing environmental, social, and economic conditions ethically and responsibly. ASCE has outlined a roadmap to transform the profession.

Source: ASCE, (Jan 2023), Sustainability at ASCE, *https://www.asce.org/communities/institutes-and-technical-groups/sustainability*

16.4 SYSTEMS THINKING

Sustainability encompasses a set of diverse concerns such as global climate change, environmental degradation, pressures on food and water supplies, loss of species, consumerism, and pollution. In essence, sustainability includes many systems and subsystems. These systems include very complex issues, such as climate, demographics, and global economics. The reductionist analytical approach, which breaks a whole into its parts to be studied independently, fails to enable understanding of systems-level questions. As the system gets larger, more complex, and more dynamic, the analytical approach becomes even less effective in dealing with the complexities of sustainability. Thus, sustainable engineering begins with study and analysis that requires a systems perspective.

Systems thinking provides a framework based on the theory that the component parts of a system can be understood best in the context of relationships with other components and with other systems, rather than in isolation. Understanding why a problem occurs and persists requires understanding the part in relation to the whole.

Since built systems can be composites containing both natural and artificial components, establishing the boundaries between these components is crucial. *System theory* defines a system boundary as:

> *A physical or conceptual boundary that contains all the system's essentials and effectively isolates it from its external environment, except inputs and outputs that can move across the system's boundary. Establish the designed system's boundaries in relationship to the ecosystem.*

16.5 ECODESIGN

Much can be learned about the systems approach from exploring the work of noted architect, Kenneth Yeang. He has developed the concept of *Ecodesign*, which he uses in both theory and practice. Ecodesign integrates artificial systems with natural systems. *Integrate* is the key word. In Ecodesign, design of the built environment should reflect the relationship between our human-made environment and the natural environment. However, Ecodesign is not an assembly of ecological-technological systems and gadgets in a single project or product. *The ultimate objective is environmental integration by design.*

Since Ecodesign involves the integration of artificial systems with natural systems, determining the level of environmental integration that can be achieved is essential:

- Theoretically, there are limitless ways of making a design ecologically responsive.
- The critical issue is where to stop biointegration.
- There is a dynamic balance between the standard of living and environmental consequences.

Ecodesign incorporates the entire ecosystem concept where maintaining the web of species within an ecosystem and the web connecting all four systems—the biosphere, the ecosystem, the community, and the population—is essential. When the web is damaged, then the overall health of the ecosystem becomes more fragile. Figure 16.3 illustrates this concept.

Ecodesign planning principles provide guidance to the design engineer and facilitate the integration of ecosystem concepts into our civil structures and our communities. Figure 16.4 describes the key principles to accomplish this goal.

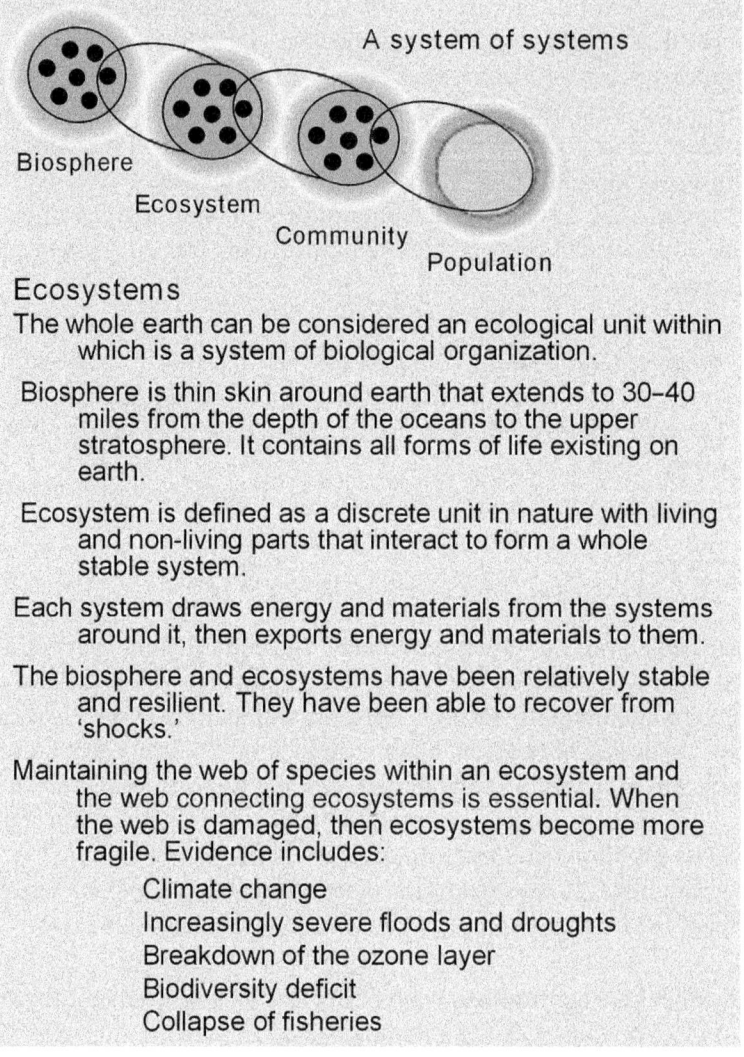

FIGURE 16.3   Ecosystems.
Source: Adapted from Kenneth Yeang, *Ecodesign*.

Effective planning begins with an initial assessment in the early design phase of the input and output of materials and their environmental consequences, followed by integrating efficient and eco-friendly transportation into the community, preserving landforms for ecological corridors, encouraging human inhabitants' perception of the natural beauty of the environment by enhancing distinctive site features, limiting waste output, enhancing resource conservation, controlling noise, and incorporating energy efficiency.

An award-winning project for sustainability is depicted in Figure 16.5. This project followed the Ecodesign planning principles and won accolades for this accomplishment.

Key planning principles can include:

➤ Plan, manage, and integrate human vehicular and urban transportation system:

- Minimize routes traversing ecologically sensitive sites
- Avoid fragmentation of existing ecosystems that would create isolated island habitats
- Shape routes and landscape to assure continuity of vegetation patterns
- Encourage patterns that minimize the use of fossil fuels and emissions that add to greenhouse gases

➤ Preserve what exists or use opportunities in landforms to create new ecological corridors

➤ Encourage human inhabitants' perception of the natural environment by enhancing distinctive site features like unusual rock formations, topographic configurations, vistas, etc.

➤ Limit the amount of sewage emissions and other outputs, such as waste generated for disposal off-site

➤ Increase water conservation

- Filter run-off from impervious surfaces and return back to the ground
- Harvest rainwater

➤ Control noise emissions

➤ Assess the designed energy system in terms of its use and management

➤ At beginning of design, assess the input and output of materials and their environmental consequences

➤ Manage all demolition and construction activities to minimize their effects on ecological systems

FIGURE 16.4   Ecodesign planning principles.
Source: Adapted from Kenneth Yeang, *Ecodesign*.

16.6   TOWARD NEW VALUES AND PROCESSES

In the traditional business environment within the architectural/engineering/construction (AEC) industry, projects typically are defined and delivered as a *linear* process that begins with response to a problem, a need, an opportunity, or a desire and ends with the delivery of a completed facility or civil infrastructure system. (See Chapter 5, The Engineer's Role in Project Delivery.) The focus of this linear definition and delivery process is on the efficiency and productivity of: (1) the management of the planning, the design, the procurement and construction, and the commissioning and start-up processes; and (2) the management and use of the resource base, which can include economic and financial resources; physical resources such as materials, equipment, and tools; human resources such as technical,

FIGURE 16.5   Singapore national library.
Source: Terence Ong/Wikimedia Commons/CC BY 2.5.

nontechnical, and administrative personnel; technological resources such as computing, communication, collaboration, and management of information technologies; and miscellaneous other resources such as data/information, knowledge/experience, abilities/skills, and technological proficiency.

Fortunately, there are alternatives available to these potentially environmentally wasteful processes. They begin with considering the design, construction, and operations over the entire life time of the building. An *integrated design* process helps to establish goals for the design, foresee the impacts of construction, and plan the operations and maintenance of a building.

It's worth remembering that:
Human welfare is related to the
    material standard of living, which depends on the
        provision of manufactured goods, which requires the
            consumption of natural capital, which means the
                extraction of natural resources, which involves
                    discharges of waste, which result in
                        impacts on human welfare.

Ecomimicry

Properties of ecosystems that can be applied to design and construction, including:

- Optimization of energy and materials consumption
  - Minimization of waste generation
    - Use of effluents (discharge/waste) of one process serve as raw materials for another

Specific human ecomimicry objectives:

1. Reduce dependency on nonrenewable energy in a system's lifecycle
2. Change from nonrenewable to renewable sources of energy
3. Increase efficiency in energy use
4. Reduce wasteful use of nonrenewable energy resources
5. Recycle materials and outputs
6. Balance producers, manufacturers, and services with consumers
7. Increase diversity to stabilize system
8. Increase compact spatial efficiency
9. Have high community organization (many networks to provide information)
10. Provide global protection from environmental perturbations
11. Conserve resources, use sustainably in order to buffer and cope with changes
12. Adopt self-correcting systems for environmental stability

Sources: Kenneth Yeang, *Ecodesign*, and Janis Birkland, *Design for Sustainability: A Sourcebook of Integrated Ecological Solutions*

Sustainability Principles

Specific education and research opportunities exist within the broad body of general knowledge on sustainability. For example, the International Institute for Sustainable Development provides an extensive compilation of sustainable development principles from numerous sources that address three major aspects: environment, economy, and community (Yeang, Kenneth 2006).

(Continued)

There is also an extensive body of specific knowledge on built environment sustainability. The key is to investigate how to adapt and customize this knowledge to the specific reality of an architectural/engineering/construction (AEC) project. For example, some selected examples of principles that can be used to implement and achieve best practice include:

- The Precautionary Principle, which guides human activities to prevent harm to the environment and to human health.
- The Earth Charter Principles, which promote respect and care for the community of life, ecological integrity, social and economic justice, and democracy, nonviolence, and peace.
- The Natural Step System Conditions, which define basic principles for maintaining essential ecological processes, and recognizing the importance of meeting human needs worldwide, as integral and essential elements of sustainability.
- The Daly Principles, which address the regenerative and assimilative capacities of natural capital, and the rate of depletion of nonrenewable resources.
- The Ceres Principles, which provide a code of environmental conduct for environmental, investor, and advocacy groups working together for a sustainable future.
- The Bellagio Principles, which serve as guidelines for starting and improving the sustainability assessment process and activities of community groups, nongovernment organizations, corporations, national governments, and international institutions, including the choice and design of indicators, their interpretation, and communication of the results.
- The Ahwahnee Principles, which guide the planning and development of urban and suburban communities in a way that they will more successfully serve the needs of those who live and work within them.
- The Interface Steps to Sustainability, which were created to guide the interface company in addressing the needs of society and the environment by developing a system of industrial production that decreases their costs and dramatically reduces the burdens placed upon living systems.
- The Hannover Principles, which assist planners, government officials, designers, and all involved in setting priorities for the built environment, and promoting an approach to design which may meet the needs and aspirations of the present without compromising the ability of the planet to sustain an equally supportive future.
- Design through the 12 Principles of Green Engineering, which provide a framework for scientists and engineers to engage in when designing new materials, products, processes, and systems that are benign to human health and the environment.

Source: Karen Lee Hansen and Jorge A. Vanegas. (2006). "A Guiding Road Map, Principles, and Vision for Researching and Teaching Sustainable Design and Construction." *American Society for Engineering Education (ASEE), Chicago, IL, Conference Proceedings.*

16.7   EXPANDED PROJECT DELIVERY PROCESS

Construction costs can represent 5 to 15 percent of a facility's lifecycle cost; design costs are typically less than 1 percent. Operations and renovations constitute most of the remaining costs. For the least expense, design can have the greatest impact on long-term sustainability. Therefore, planning and designing facilities and civil infrastructure systems sustainably is critical.

For built environment sustainability, the project delivery process needs to:

- Address architectural/engineering/construction (AEC) projects from their complete lifecycle perspective, including operations and maintenance, and also the end-of-service life decision
- Emphasize the use of sustainable resources
- Monitor and document the outcomes resulting from the use of the facility or civil infrastructure system delivered
- Provide a post-occupancy evaluation and feedback to the project originator that, depending on what the project driver was, answers the question: Did the delivered project solve the problem? Satisfy the need? Capitalize on the opportunity? Realize the desire?

The fundamental approach for enabling and achieving sustainable facilities or civil infrastructure systems at a *strategic level* is shown in Figure 16.6. The basis for implementation is to frame an AEC project within a contextual envelope, which (1) is defined by the *requirements and characteristics* of the specific facility or civil infrastructure system, the processes *followed in its delivery and use*, and the *resources consumed in its delivery and use*; and (2) uses sustainability as a fundamental criterion in making decisions, choosing among various options, or taking actions.

The contextual envelope for the project is represented as an (x, y, z) diagram defined by the specific requirements and characteristics of the specific facility or civil infrastructure system x, its processes y, and its resources z, expressed as relative points on an axis, with a scale that spans from what is unsustainable to what is sustainable in each one, and with the thresholds that separate the two extremes within each axis as point (0, 0, 0). The strategy then, is to maintain project decisions, choices, and actions within the *octant* where all three (x, y, z) points are sustainable. While this may be conceptually simple, in reality it is quite complex, and much research is yet to be done to provide clear and absolute definitions of what is sustainable, what is unsustainable, and what is the threshold that separates them (Vanegas, J. 2004).

16.8   INTEGRATIVE APPROACHES

If the model of sustainable engineering uses a systems approach that integrates diverse natural and man-made elements into an ecological system, then the project delivery process also should be integrated and balanced. The design, construction, and operation of facilities and civil infrastructure systems should be part of a restorative, integrated process, rather than a wasteful, linear one.

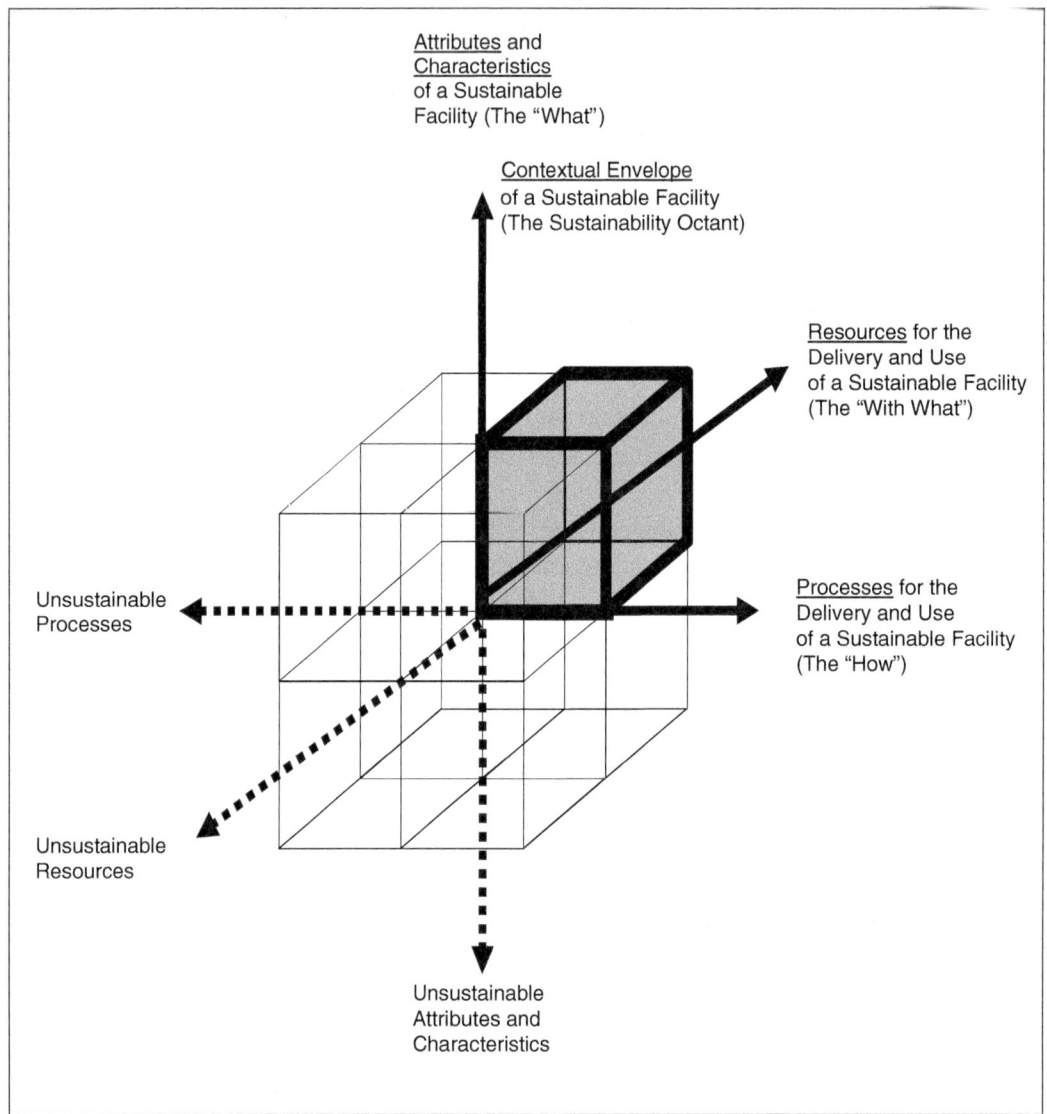

FIGURE 16.6   Strategic level of implementation of built environment sustainability.
Source: Adapted from Dr. Jorge A. Vanegas, Dean Emeritus, College of Architecture, Texas A&M University.

Collaborative project delivery processes facilitate integration of design strategies among all disciplines and players. Integrating design requires an inclusive project team that involves users, contractors and subcontractors, and future maintenance staff. Using this approach, the design team has more opportunities to discover creative solutions, to save time, and to minimize costs. For example, contractors included in the design phase understand project goals and can work toward suitable solutions with the designers. Integrated design also can lead to better facility performance, thereby saving costs. For example, overall energy consumption can be optimized if site design, facility "envelope" and electrical and mechanical systems are thought of together.

These synergies and savings require collaboration among engineers, architects, and others. Synergies can provide multiple benefits, much like the collaboration of civil engineers, mechanical engineers, traffic engineers, and maintenance staff can lead to low-impact development, preserving open space, decreasing stormwater treatment, and reducing radiant heat and the heat island effect from the landscape.

Impact of the Construction and Operation of Facilities on the Environment

The construction and operation of facilities have both direct and indirect impacts on the environment. Direct impacts include the energy used for electricity and heating, ventilating, and air conditioning (HVAC) systems, materials and resources required for construction, the water used in facilities/buildings, and the stormwater and open space impacts of displacing greenfield sites.

In the United States, buildings consume 39 percent of primary energy use (including fuel input for production) and contribute 38 percent of CO_2 emissions. Buildings represent 72 percent of U.S electricity consumption and use 13.6 percent of all potable water, or 15 trillion gallons per year. The EPA estimates that 170 million tons of building-related construction and demolition debris was generated in the United States in 2003, with 61 percent coming from nonresidential and 39 percent from residential sources.

Indirect impacts include the inputs to building material manufacturing processes, the fossil fuels used to transport the materials, and the CO_2 emissions from both. Additionally, the emissions of volatile organic compounds (VOCs) from many building materials adversely impact people's well-being.

Source: U. S. Green Building Council Website (www.usgbc.org)

Solving multiple environmental challenges requires an integrated, multidisciplinary design team. Using integrated design practices to maximize sustainability, rather than focusing on minimum requirements, opens up synergies and solutions that otherwise might remain undiscovered. The project team needs to set sustainability goals early, design to achieve those goals, and verify that performance meets those goals. Through this process, sustainability becomes integral to design and construction, not an "add-on."

16.9   SUSTAINABLE DESIGN AND MATERIALS STRATEGIES

Realization of a sustainable facility or civil infrastructure system starts long before the actual design process begins. The owner, perhaps in conjunction with a professional consultant, needs to establish sustainable design goals early, embed a multidisciplinary effort into the design process, and budget for the necessary meetings and analyses critical to ensuring sustainability from design through construction through post occupancy.

16.9.1   Sustainable Design Strategy

Sustainable design and materials strategies are defined by *requirements and characteristics* of the specific facility or civil infrastructure system, the processes followed in its delivery and use, and the *resources* consumed in its delivery and use. These strategies include, but are not limited to, the following:

- Siting and design considerations that optimize local geographic features to improve sustainability, such as proximity to public transportation and maximizing use of vistas, microclimate, and prevailing winds
- Durable systems and finishes with long lifecycles that minimize maintenance and replacement needs
- Layouts and designed spaces that can be reconfigured for adaptive reuse (versus demolition).
- Systems designed for optimization of energy, water, and other natural resources
- Optimization of indoor environmental quality
- Utilization of environmentally preferable products and processes, such as recycled-content and recyclable materials
- Procedures that monitor, trend, and report operational performance as compared to the optimal design and operating parameters
- Durable systems and finishes with long lifecycles that minimize maintenance and replacement
- In order to operationalize these strategies, the owner must allocate adequate budget and schedule. Enovity and HOK, two firms with expertise in sustainability and alternative energy solutions, suggest that the following meetings, analyses, and actions be incorporated into the project:
- Sustainable design goals meeting
- Design and strategy charrette with entire multidisciplinary project team
- Energy and daylighting analysis
- Lifecycle cost analysis
- Scope and extent of commissioning
- Comprehensive maintenance plan
- Post occupancy evaluation, including measurement and verification
- Regular design sessions to develop and implement integrated design

See Table 16.1 for more information regarding actions that support a sustainable design.

TABLE 16.1 Useful meetings, analyses, and actions that support sustainable design.

	Meeting, Analysis, or Action	Description
1	Sustainable design goals meeting	During predesign phases (Scoping, Feasibility, Programming) conduct a sustainable design goals meeting. Owner personnel and the project team should commit to a sustainable design vision, measurable goals, and methods of measuring and verifying the achievement of the goals. Prioritize according to project-specific opportunities and constraints. The results of the sustainable design goals meeting should inform the equivalent of the Owner's Project Requirements (OPR) document, as required for Leadership in Energy and Environmental Design (LEED) certification. The OPR should explain the owner's goals and expectations of the facility's program, operation, energy efficiency, and sustainability.
2	Design a strategy charrette with entire multidisciplinary project team (*Charrette: intense effort to complete a project within a limited period of time*)	Hold workshop intended to establish specific ways to achieve the goals and intentions identified in the sustainable design goals meeting. Define a design methodology based on analysis, integrated design, and optimized sustainability, rather than simply meeting minimum standards. Discuss the cost implications of sustainability strategies, such as a set-aside budget for technologies with a good lifecycle value and cost-effective energy-efficiency upgrades that may help supplement construction budget. Consider design strategies' costs and benefits. Discuss the schedule implications of sustainability measures, such as commissioning and testing.
3	Energy and daylighting analysis	Optimize designs with the assistance of energy and daylighting analysis tools and software that can help evaluate performance. Computer programs to assess energy efficiency include: DOE-2, Energy-10™, and eQuest 1. Daylighting evaluation programs include Radience, Lumen Micro, and the Lightscape visualization system. A Helidon, a device that uses a light source and physical model, can also simulate the effects of sunlight.
4	Lifecycle cost analysis	Require designers to incorporate lifecycle cost analysis (LCCA) into the design to predict long-term building costs and minimize energy and replacement costs. LCCA compares the total costs and benefits over the entire lifecycle of a system, component, or material. It allows for future costs and benefits to be incorporated into present-day decision-making.

(Continued)

TABLE 16.1 *(Continued)*

	Meeting, Analysis, or Action	Description
5	Scope and extent of commissioning	Select commissioning authority, determine scope and extent of commissioning. This information is needed early so that it can be incorporated into the owner's budget and the project schedule.
6	Comprehensive maintenance plan	After the design and strategy charrette, develop a maintenance plan that helps foresee operations challenges and ensures that original design intentions are met for the lifetime of the facility. Newer facilities will have advanced technologies which may require training or different operations practices. For these reasons, those responsible for building operations and maintenance should be included in early goals and strategy meetings.
7	Post occupancy evaluation, including measurement and verification	Plan and schedule a post occupancy evaluation from the outset. Set regular reviews of energy and water end use, air quality and heating, ventilation, and air-conditioning (HVAC) performance, and waste metrics. Implement the measurement and verification plan to ensure that energy and HVAC design intentions are being met. Use the measurement and compliance strategies appropriate to the project.
8	Regular design sessions to develop and implement integrated design	Hold regular project meetings, including designers, stakeholders, and—if possible—key contractor(s) and subcontractors. Collaborative, multidisciplinary, integrated design teams need to communicate well and regularly in order to be effective. Some team members may be required to perform tasks outside their typical roles. For example, civil engineers may be asked to design biologically based stormwater treatment systems.
9	Miscellaneous actions	Include sustainability in RFPs and RFQs.
		Build a design team with members who are experienced and committed to sustainable design and working collaboratively.
		Appoint a Sustainability Champion. Design for flexibility and adaptability.
		Give priority to building materials and systems of durable/repairable assemblies (replacement or repair of isolated areas when possible, without the need to replace the entire system).

Source: Adapted from Enovity—HOK.

Sustainable Materials Strategies

- **Limit the Use of Materials with Hazardous Content**

Materials with hazardous content can increase health risks of facility users, generate hazardous construction waste, or require a toxic manufacturing process. Avoiding these materials whenever possible is recommended.

- **Look for Salvaged Materials**

Used bricks and timbers, asphalt paving, and concrete all present opportunities for reuse or refurbishment. Alternatively, upon disassembly or tear-down of old facilities, owners can retain reusable materials for future use or sale to a salvage yard.

- **Seek Products with Recycled and Renewable Content**

The EPA promotes the recovery of materials from solid waste streams through its Comprehensive Procurement Guidelines. Consult the guidelines at www.epa.gov/cpg/products.htm.

- **Identify Local Manufacturers, Regionally Appropriate and Locally Available Materials**

Locally produced goods reduce the costs and environmental impact of transportation. Local goods are also more likely to be adapted to the regional climate and building requirements. At the same time, they support the local economy. Regional and local materials should account for at least 20 percent of material costs in new construction. A list of local manufacturers and materials should be maintained and incorporated into standard designs and future projects.

- **Use Low-Emitting Materials**

Adhesives, sealants, carpeting, paint, coatings, and composite wood products that emit low amounts of volatile organic compounds (VOCs) are all available on the market and should be considered for use. As demand for sustainable buildings increase, more options are becoming available.

- **Create a Policy of Diverting 95 Percent of Construction Waste**

Instituting the 95 percent goal as a policy will provide design teams with an official and identifiable goal to strive for, facilitating design optimization and creative reuse.

Source: Enovity—HOK

16.10 LIFECYCLE COST ANALYSIS

Lifecycle Cost Analysis (LCCA) can inform decisions about a multitude of factors, thereby helping to control the initial and the future costs of facility ownership and maintenance. LCCA is a measure of "cradle to grave" costs that can be performed on a full range of projects, from entire site complexes to specific system components, both large and small. For example, civil engineers can use LCCA to choose between concrete or asphalt paving and between a reinforced concrete or steel structures. As contrasted with many decisions made in conventional project development processes, initial cost is a factor, but not the only factor.

Lifecycle Cost (LCC) is the total discounted (present value) dollar cost of owning, operating, maintaining, and disposing of a facility or system over a period of time. The LCC equation contains three variables: the pertinent *costs* of ownership, the period of *time* over which these costs are incurred, and the *discount rate* that is applied to future costs to equate them with present-day costs.

The Basic Lifecycle Cost Analysis Process

- Estimate costs and benefits
- Estimate timing
- Discount future costs and benefits
- Compare net present values

Following are the key concepts used in LCCA.

16.10.1   Costs

The two major cost categories by which projects are evaluated in an LCCA are initial expenses and future expenses. *Initial expenses* are all costs incurred prior to occupation of the facility. *Future expenses* are all costs incurred after occupation of the facility. Defining exact costs of each expense category tends to be difficult since few costs are known at the time of the LCCA. Careful, well-documented assumptions are necessary for preparation of a realistic LCCA.

16.10.2   Residual Value

One future expense that warrants further explanation is residual value. *Residual value* is the net worth of a facility or system at the end of the LCCA study period. Unlike other future expenses, an alternative's residual value can be positive or negative. LCC is a summation of costs, so a negative residual value indicates that there is value associated with the facility at the end of the study period. The value could be in a component that was replaced recently or in the facility's superstructure that could function for another 30 years; or the costs could be related to abatement of hazardous material or demolition of the structure. A positive residual value indicates disposal costs associated with the facility at the end of the study period. The residual value is a tangible asset that should be included in the LCCA.

Zero residual value indicates that there is no value or cost associated with the facility at the end of the study period. This situation is rare and occurs, for example, if the intended use of the facility terminates concurrently with the end of the study period, the owner is unable to sell the facility, or the owner is able to abandon the facility at no expense.

16.10.3   Study Period

Time is the second component of the LCC equation. The *study period* is the time over which ownership and operations expenses are to be evaluated. A typical study period ranges from 20 to 40 years. The length of the study period depends on the owner's preferences, the stability of the user's program, and the designed overall life of the facility, though the study period usually is shorter than the intended life of the facility.

Some LCCA approaches, such as that defined by the National Institute of Standards and Technology (NIST), break the study period into two phases: the planning/construction period and the service period. The planning/construction period is the time period from the start of the study to the date the building becomes operational (the service date). The service period is the time period from the date the building becomes operational to the end of the study.

16.10.4   Discount Rate

The third component in the LCC equation is the discount rate. The *discount rate* is the rate of interest reflecting the investor's time value of money—either the minimum acceptable rate of return for investment purposes (frequently used by owners in private industry) or the current rate of interest for borrowing (frequently used by public owners). As world economics change, so does the discount rate.

16.10.5   Constant versus Current Dollars

Constant dollars can be defined as dollars of uniform purchasing power tied to a reference year, exclusive of general price inflation or deflation. *Current dollars* can be defined as dollars of nonuniform purchasing power that include general price inflation or deflation.

The use of constant dollars simplifies LCCA. For example, suppose one wants to evaluate a product over a 30-year period. However, one product must be replaced after 20 years. How much will the replacement of the product cost in 20 years? By using constant dollars, estimating the escalation of labor and material costs is eliminated. The future constant dollar cost (excluding demolition) to install a new product in 20 years is the same as the initial cost to install it. Any change in the value of money over time will be accounted for by the discount rate.

Escalation must be considered when future costs differ from inflation. For example, energy costs may rise more quickly, perhaps 1 to 2 percent, than inflation.

16.10.6   Present Value

To combine initial expenses with future expenses accurately, the present value of all expenses must be determined. *Present value* can be defined as the time equivalent value of past, present, or future cash flows as of the beginning of the base year. The present value calculation uses the discount rate and the time a cost was or will be incurred to establish the present value of the cost in the base year of the study period.

Initial expenses are considered to occur during the base year of the study period since most of these expenses occur at approximately the same time. Consequently, the present value of these initial expenses does not need to be calculated because their present value is equal to their actual cost. The present value of future costs is time dependent. The time period is the difference between the time of initial costs and the time of future costs. Initial costs are incurred at the beginning of the study period at Year 0, the base year. Future costs can be incurred anytime between Year 1 and Year n. The present value calculation is the summation of initial and future costs. Along with time, the discount rate also dictates the present value of future costs. Because the current discount rate is a positive value, future expenses will have a present value less than their cost at the time they are incurred.

16.10.7   Future Costs

Future costs can be broken down into two categories: one-time costs and recurring costs. *Recurring costs* are costs that occur every year over the span of the study period. Most operating and maintenance costs are recurring costs. *One-time costs* are costs that do not occur every year over the span of the study period. Most replacement costs are one-time costs. To simplify the LCCA, all recurring costs are expressed as annual expenses incurred at the end of each year and one-time costs are incurred at the end of the year in which they occur. If costs in a particular cost category are equal in all project alternatives, they can be documented as such and removed from consideration in the LCC.

Present value calculations can be made using the formulas included in Table 16.2.

16.10.8   Alternatives

Prior to beginning an LCCA, several (usually three) distinctly different and viable solutions—*alternatives*—should be established. The LCCA describes each project alternative as well as the rationale for its inclusion.

Some possible options that can be considered while selecting the most viable, reasonable, and cost-effective alternatives include:

- Renovation and addition to an existing facility.
- Rental and remodeling of an existing facility.
- Purchase and remodeling of an existing facility.

TABLE 16.2 Present value formulas.

Type of Cost	Application	Constant Dollars	Current Dollars
First	Initial capital investments	$P = P$ (no conversion)	$P = P$ (no conversion)
Future	Replacements or alterations	$P = F \times \dfrac{1}{(1+i)^n}$	$P = F \times \dfrac{(1+e)^n}{(1+i)^n}$
Future Series	Energy or maintenance	$P = A \times \dfrac{\left(\dfrac{(1+e)^n}{(1+i)^n}\right) - 1}{1 - \left(\dfrac{(1+e)^n}{(1+i)^n}\right)}$	$P = A \times \dfrac{\left(\dfrac{(1+e)^n}{(1+i)^n}\right) - 1}{1 - \left(\dfrac{(1+e)^n}{(1+i)^n}\right)}$

P = present value
F = future value
i = real interest rate
i = total interest rate (real interest rate plus inflation)
e = escalation
n = time (expressed as number of years)
A = annual amount

- Demolition of existing facility and construction of a new facility on the same site.
- Sale of existing facility and construction of a new facility on a new site.
- The use of double labor shifts or some other solution not involving construction that increases facility capacity.
- Alteration of the facility's function in some way, also a non-construction approach. A "No Action" alternative also frequently must be considered.

Lifecycle Cost Analysis Terminology

Lifecycle Cost: the total discounted (present value) dollar cost of owning, operating, maintaining, and disposing of a facility or system over a period of time
Alternative: distinctly different and viable solution
Constant dollars: dollars of uniform purchasing power tied to a reference year, exclusive of general price inflation or deflation
Current dollars: dollars of nonuniform purchasing power that include general price inflation or deflation

(Continued)

Discount rate: rate of interest reflecting the investor's time value of money
Future expenses: costs incurred after occupation of the facility
Initial expenses: costs incurred prior to occupation of the facility
One-time costs: costs that do not occur every year over the span of the study period
Present value: time equivalent value of past, present, or future cash flows as of the beginning of the base year
Recurring costs: costs that occur every year over the span of the study period
Residual value: net worth of a building at the end of the LCCA study period

16.10.9   Limitations of LCCA

Defining accurate costs and discount rates necessary for realistic LCCA is difficult at best. Where to establish cost boundaries often is not obvious. For example, fuel costs are considered in transportation input, but what about the costs associated with manufacturing vehicles and maintaining highways? Naturally, manufacturers have incentives for limiting boundaries in their studies. Uninformed users may employ LCCA data without realizing its limitations.

Lifecycle Assessment is a next step beyond LCCA. LCA translates the flow of resources and waste into overall environmental impact. The lifecycle of any given product may result in many emissions, each affecting the environment to varying degrees. Because the impact of each emission varies, the overall effect of each emission on greenhouse gases is expressed in units of carbon dioxide (CO_2) equivalents. The Environmental Protection Agency (EPA) uses a similar process to estimate environmental impacts in other categories. EPA's Tool for the Reduction and Assessment of Environmental Impacts (TRACI) includes variables such as ozone depletion, global warming, acidification, photochemical smog, human carcinogenic effects, ecotoxity, fossil fuel use, land use, and water use, among others. (See Figure 16.7.)

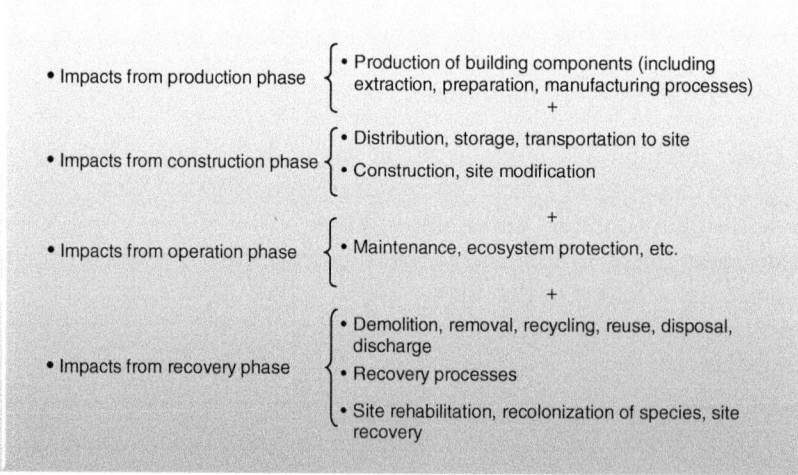

FIGURE 16.7   Lifecycle impacts of a designed system.
Source: Adapted from Kenneth Yeang, Ecodesign.

Cradle to Grave

Cradle to Cradle (2002) by architect William McDonough and chemist Michael Braungart proposes that an industrial system that "takes, makes and wastes" can transform itself into a creator of goods and services that generates ecological, social, and economic value. The authors maintain that products and manufacturing practices stemming from the Industrial Revolution have resulted in a series of unintended, disastrous consequences. But with today's greater understanding of Earth as a living system, the authors imagine that nature and commerce can coexist in new products, industrial systems, buildings, and regional plans. The book itself is an example of "cradle to cradle" principles: It is printed on a synthetic, treeless "paper," made from plastic resins and inorganic fillers, designed to be recycled and used again (McDonough, W. 2002).

16.11   LEADERSHIP IN ENERGY AND ENVIRONMENTAL DESIGN

Leadership in Energy and Environmental Design (LEED), developed by the US Green Building Council (USGBC), is a voluntary certification program that measures how well a project or community compares to others in achieving predetermined sustainability goals. The LEED framework provides metrics involving energy savings, water efficiency, CO_2 emissions reduction, improved indoor environmental quality, and stewardship of resources and sensitivity to their impacts. The framework can be applied throughout the facility lifecycle—design and construction, operations and maintenance, tenant improvement, and significant retrofit. A category called Neighborhood Development and Cities and Communities extend the LEED framework beyond the building footprint into the surrounding neighborhoods and cities.

As shown in Figure 16.8, LEED evaluates sustainability performance in several areas:

- Building Design and Construction
- Interior Design and Construction
- Building Operations and Maintenance
- Neighborhood Development
- Homes
- Cities and Communities
- LEED Recertification
- LEED Zero

Other certification programs exist—Green Globes is used in Canada and BREEAM (Building Research Establishment Environment Assessment Method) in the United Kingdom. However, LEED currently is the most widely adopted third-party sustainability certification system, perhaps because many public owners have put the requirement for LEED certification in their requests for proposals (RFPs) and requests for qualifications (RFQs). Figure 16.9 depicts a typical LEED rating scorecard.

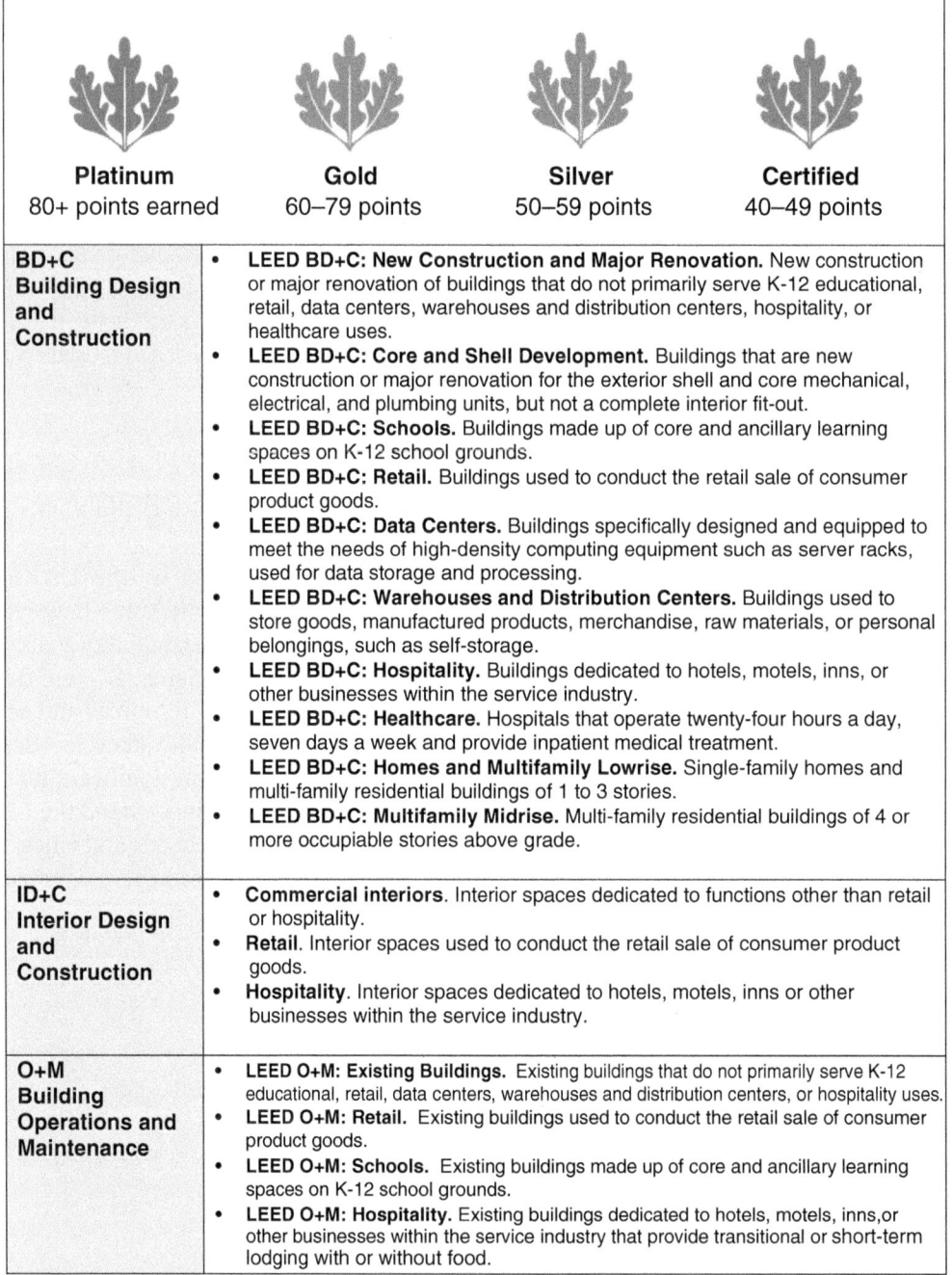

Platinum	Gold	Silver	Certified
80+ points earned	60–79 points	50–59 points	40–49 points

BD+C **Building Design and Construction**	• **LEED BD+C: New Construction and Major Renovation.** New construction or major renovation of buildings that do not primarily serve K-12 educational, retail, data centers, warehouses and distribution centers, hospitality, or healthcare uses. • **LEED BD+C: Core and Shell Development.** Buildings that are new construction or major renovation for the exterior shell and core mechanical, electrical, and plumbing units, but not a complete interior fit-out. • **LEED BD+C: Schools.** Buildings made up of core and ancillary learning spaces on K-12 school grounds. • **LEED BD+C: Retail.** Buildings used to conduct the retail sale of consumer product goods. • **LEED BD+C: Data Centers.** Buildings specifically designed and equipped to meet the needs of high-density computing equipment such as server racks, used for data storage and processing. • **LEED BD+C: Warehouses and Distribution Centers.** Buildings used to store goods, manufactured products, merchandise, raw materials, or personal belongings, such as self-storage. • **LEED BD+C: Hospitality.** Buildings dedicated to hotels, motels, inns, or other businesses within the service industry. • **LEED BD+C: Healthcare.** Hospitals that operate twenty-four hours a day, seven days a week and provide inpatient medical treatment. • **LEED BD+C: Homes and Multifamily Lowrise.** Single-family homes and multi-family residential buildings of 1 to 3 stories. • **LEED BD+C: Multifamily Midrise.** Multi-family residential buildings of 4 or more occupiable stories above grade.
ID+C **Interior Design and Construction**	• **Commercial interiors.** Interior spaces dedicated to functions other than retail or hospitality. • **Retail.** Interior spaces used to conduct the retail sale of consumer product goods. • **Hospitality.** Interior spaces dedicated to hotels, motels, inns or other businesses within the service industry.
O+M **Building Operations and Maintenance**	• **LEED O+M: Existing Buildings.** Existing buildings that do not primarily serve K-12 educational, retail, data centers, warehouses and distribution centers, or hospitality uses. • **LEED O+M: Retail.** Existing buildings used to conduct the retail sale of consumer product goods. • **LEED O+M: Schools.** Existing buildings made up of core and ancillary learning spaces on K-12 school grounds. • **LEED O+M: Hospitality.** Existing buildings dedicated to hotels, motels, inns,or other businesses within the service industry that provide transitional or short-term lodging with or without food.

FIGURE 16.8 Leadership in energy and environmental design (LEED) categories.

Source: Adapted from U.S. Green Building Council, https://www.usgbc.org/leed.

	• **LEED O+M: Data Centers.** Existing buildings specifically designed and equipped to meet the needs of high density computing equipment such as server racks, used for data storage and processing. • **LEED O+M: Warehouses & Distribution Centers.** Existing buildings used to store goods, manufactured products, merchandise, raw materials, or personal belongings (such as self-storage). • **LEED O+M: Multifamily.** Existing multifamily projects with 20 or more units, either a single building with 20 units or multiple buildings within the same complex (e.g., 20 townhomes).
ND Neighborhood Development	• **LEED ND: Plan.** Projects in conceptual planning or master planning phases, or under construction. • **LEED ND: Built Project.** Completed development projects.
Homes	• **Single Family Homes**. New single-family homes that are attached or detached, and multifamily buildings with up to four units. • **Multifamily Homes**. Applicable to any predominantly multifamily building with two or more units and any number of stories. • **Multifamily Homes Core and Shell**. Multifamily buildings that are new construction or major renovation for the exterior shell and core mechanical, electrical and plumbing units, but not a complete interior fit-out.
Cities and Communities	• **Encompasses** social, economic and environmental performance indicators and strategies with a clear, data-driven means of benchmarking and communicating progress. • **Aligned** with the United Nations Sustainable Development Goals.
LEED Recertification	• **Available** to all occupied and in-use projects that have previously achieved certification under LEED — including BD+C and ID+C, regardless of their initial rating system or version.
LEED Zero	• **Complement** to LEED that verifies the achievement of net zero goals in existing buildings. • **Carbon** recognizes net zero carbon emissions from energy consumption through carbon emissions avoided or offset over a period of 12 months. • **Energy** recognizes a source energy use balance of zero over a period of 12 months. • **Water** recognizes a potable water use balance of zero over a period of 12 months. • **Zero Waste** recognizes buildings that achieve GBCI's TRUE certification at the Platinum level.

FIGURE 16.8 *(Continued)*

LEED also offers a credential program for design and construction professionals. The LEED Green Associate and LEED AP (Accredited Professional) with Specialty designations signify current proficiency in sustainable design, construction, and operations standards:

- **LEED Green Associate**: A foundational professional credential signifying core competency in green building principles. The exam measures general knowledge of green building practices and how to support others working on LEED projects and is ideal for those new to green building.

LEED v4.1 BD+C: Healthcare
Project Checklist

Project Name:
Date:

Y	?	N			
Y			Prereq	Integrative Project Planning and Design	Required
			Credit	Integrative Process	1

			Location and Transportation		**9**
			Credit	LEED for Neighborhood Development Location	9
			Credit	Sensitive Land Protection	1
			Credit	High Priority Site and Equitable Development	2
			Credit	Surrounding Density and Diverse Uses	1
			Credit	Access to Quality Transit	2
			Credit	Bicycle Facilities	1
			Credit	Reduced Parking Footprint	1
			Credit	Electric Vehicles	1

			Sustainable Sites		**9**
Y			Prereq	Construction Activity Pollution Prevention	Required
Y			Prereq	Environmental Site Assessment	Required
			Credit	Site Assessment	1
			Credit	Protect or Restore Habitat	1
			Credit	Open Space	1
			Credit	Rainwater Management	2
			Credit	Heat Island Reduction	1
			Credit	Light Pollution Reduction	1
			Credit	Places of Respite	1
			Credit	Direct Exterior Access	1

			Water Efficiency		**11**
Y			Prereq	Outdoor Water Use Reduction	Required
Y			Prereq	Indoor Water Use Reduction	Required
Y			Prereq	Building-Level Water Metering	Required
			Credit	Outdoor Water Use Reduction	1
			Credit	Indoor Water Use Reduction	7
			Credit	Optimize Process Water Use	2
			Credit	Water Metering	1

			Energy and Atmosphere		**35**
Y			Prereq	Fundamental Commissioning and Verification	Required
Y			Prereq	Minimum Energy Performance	Required
Y			Prereq	Building-Level Energy Metering	Required
Y			Prereq	Fundamental Refrigerant Management	Required
			Credit	Enhanced Commissioning	6
			Credit	Optimize Energy Performance	20

0	0	0	**Materials and Resources**		**19**
Y			Prereq	Storage and Collection of Recyclables	Required
Y			Prereq	PBT Source Reduction–Mercury	Required
			Credit	Building Life-Cycle Impact Reduction	5
			Credit	Environmental Product Declarations	2
			Credit	Sourcing of Raw Materials	2
			Credit	Material Ingredients	1
			Credit	PBT Source Reduction–Mercury	1
			Credit	PBT Source Reduction–Lead, Cadmium, and Copper	2
			Credit	Furniture and Medical Furnishings	2
			Credit	Design for Flexibility	1
			Credit	Construction and Demolition Waste Management	2

0	0	0	**Indoor Environmental Quality**		**16**
Y			Prereq	Minimum Indoor Air Quality Performance	Required
Y			Prereq	Environmental Tobacco Smoke Control	Required
			Credit	Enhanced Indoor Air Quality Strategies	2
			Credit	Low-Emitting Materials	3
			Credit	Construction Indoor Air Quality Management Plan	1
			Credit	Indoor Air Quality Assessment	2
			Credit	Thermal Comfort	1
			Credit	Interior Lighting	1
			Credit	Daylight	2
			Credit	Quality Views	2
			Credit	Acoustic Performance	2

0	0	0	**Innovation**		**6**
			Credit	Innovation	5
			Credit	LEED Accredited Professional	1

0	0	0	**Regional Priority**		**4**
			Credit	Regional Priority: Specific Credit	1
			Credit	Regional Priority: Specific Credit	1
			Credit	Regional Priority: Specific Credit	1
			Credit	Regional Priority: Specific Credit	1

0	0	0	**TOTALS**	Possible Points:	**110**

Certified: 40 to 49 points, Silver: 50 to 59 points, Gold: 60 to 79 points, Platinum: 80 to 110

FIGURE 16.9 Leadership in energy and environmental design (LEED) professional credentials.
Source: Adapted from U.S. Green Building Council/https://www.usgbc.org/resources/bdc-v41-credit-overview / last accessed 16 March 2023.

- **LEED AP with specialty**: An advanced professional credential signifying expertise in green building and a LEED rating system. To earn a LEED AP with specialty, candidates must first pass the LEED Green Associate exam. The exams measure knowledge about green building, a specific LEED rating system and the certification process and are ideal for individuals who are actively working on green building and LEED projects. (https://www.usgbc.org/credentials/leed-green-associate)

See Figure 16.10 for more information regarding LEED's credential programs.

At this point, LEED is oriented more toward buildings than civil infrastructure. However, several rating categories involve civil engineers, such as stormwater provisions in New Construction and some of the land use and transportation categories in Neighborhood

---

LEED Green Associate
foundational professional credential signifying core competency in green building principles

LEED AP with specialty
advanced professional credential signifying expertise in green building and a LEED rating system.

- **LEED AP Building Design + Construction (LEED AP BD+C)**

  Suits professionals with expertise in the design and construction phases of green buildings, serving the commercial, residential, education and healthcare sectors.

- **LEED AP Interior Design + Construction (LEED AP ID+C)**

  Serves participants in the design, construction and improvement of commercial interiors and tenant spaces that offer a healthy, sustainable and productive work environment.

- **LEED AP Operations + Maintenance (LEED AP O+M)**

  Distinguishes professionals implementing sustainable practices, improving performance, heightening efficiency and reducing environmental impact in existing buildings through enhanced operations and maintenance.

- **LEED AP Neighborhood Development (LEED AP ND)**

  Applies to individuals participating in the planning, design and development of walkable, neighborhoods and communities.

- **LEED AP Homes**

  Suited for those involved in the design and construction of healthy, durable homes that use fewer resources and produce less waste.

FIGURE 16.10 Leadership in Energy and Environmental Design (LEED) Professional Credentials
Source: Adapted from U.S. Green Building Council https://www.usgbc.org/resources/bdc-v41-credit-overview.

Development. Even if LEED is not a part of project requirements, the LEED approach is useful in helping to identify areas where sustainable principles can be applied.

Importance of Evaluation, Measurement, and Verification

"Studies have shown that simply by operating most existing buildings as they were originally designed, a 20 percent energy savings could be achieved. Unfortunately, this does not happen often enough because those operating the building were not part of the original design, or designers do not stay in touch with the operations of the building. Ensuring buildings perform as well as they are designed is critical to sustainability . . . Design and LEED certification are therefore only the beginning of sustainability. And many LEED credits acknowledge the need for follow-up evaluation, measurement and verification."

Source: Enovity—HOK

Other organizations have noted the importance of sustainability and have developed rating systems appropriate for infrastructure. Danny Nguyen, a civil engineering graduate student at California State University, Sacramento, has compiled the list outlined in the *Sustainable Transportation Industry Systems textbox.*

Sustainable Transportation Industry Systems

Source: https://www.spur.org/news/2022-04-19/how-we-got-parkway-people

Transportation is responsible for a significant portion of environmental impacts – according to the US Environmental Protection Agency, the transportation sector was responsible for 27% of greenhouse gas emissions in 2020. To promote sustainability in transportation, various ranking systems have been developed to account for transportation-specific project needs that do not exist in traditional building scope.

Greenroads, developed through the combined efforts of the University of Washington and CH2M HILL, Inc. with assistance from various organizations – Washington State Transportation Improvement Board, American Public Works Association, and Institute of Transportation Engineers – is one system that seeks to provide a basic template for US projects but can be extended to work for international projects as well. Credits are awarded for aligning with green best practices that then determine the certification level of a project. The Presidio Main Post Tunnels and Parkway project in San Francisco, California, achieved a Bronze-tier certification for its tree conservation and replacement program among other enhanced considerations.

However, *Greenroads* is hardly the only transportation sustainability scoring system – other systems also aim to certify projects internationally, while other regional systems account for extra considerations unique to the area, such as:

- Envision (Institute for Sustainable Infrastructure)
- Illinois-Livable and Sustainable Transportation Rating System and Guide (I-LAST)
- Green Leadership in Transportation and Environmental Sustainability (GreenLITES)
- Infrastructure Voluntary Evaluation Sustainability Tool (INVEST)
- Civil Engineering Environmental Quality Assessment and Award Scheme (CEEQUAL)
- Sustainable Transportation Analysis Rating System (STARS)
- GreenPave (Transportation Association of Canada – TAC)
- Sustainability National Road Administrations (SUNRA)
- Building for Environmental and Economic Sustainability (BE²ST)

Adoption of transportation ranking systems like *Greenroads* is imperative for curbing the effects the sector has on climate change via increased adoption of greener best practices. Governments should provide financial incentives to speed system adoption, while in academia students should be more widely exposed to these systems for increased understanding of how the systems positively affect projects.

16.12   FUTURE DIRECTIONS

A serious movement began in 1969 with the US National Environmental Policy Act (NEPA). NEPA declared as its goal a national policy to "create and maintain conditions under which [humans] and nature can exist in productive harmony, and fulfill the social, economic and other requirements of present and future generations of Americans." Most Americans recognize that a balance between nature and humans must be achieved but progress has been slow. In the short run, ignoring the concepts expressed by Ecodesign is less expensive than dealing with the impacts of non-sustainable development on public health and the environment. Ultimately the cost of those impacts has to be paid, usually at a substantially higher price.

Five hundred years ago, ignorance of the environment led to mass deaths from polluted water and dirty air. Even in the late 20th century civil engineers struggled to quantify and control waste discharges. In December of 2009, on the eve of the United Nations Climate Change Conference Copenhagen, the EPA Administrator signed two distinct findings regarding greenhouse gases under Section 202(a) of the Clean Air Act:

Endangerment Finding: The Administrator finds that the current and projected concentrations of the six key well-mixed greenhouse gases—carbon dioxide (CO_2), methane (CH_4), nitrous oxide (N_2O), hydrofluorocarbons (HFCs), perfluorocarbons (PFCs), and sulfur hexafluoride (SF_6)—in the atmosphere threaten the public health and welfare of current and future generations.

Cause or Contribute Finding: The Administrator finds that the combined emissions of these well-mixed greenhouse gases from new motor vehicles and new motor vehicle engines contribute to the greenhouse gas pollution, which threatens public health and welfare.

This action means that the EPA is now authorized and obligated to take reasonable efforts to reduce greenhouse pollutants under the Clean Air Act.

Another organization, Architecture 2030, was established by architect Edward Mazria in 2002. Edward Mazria, author of *The Passive Solar Energy Book* in 1979, has been a long-term supporter of sustainability principles. Architecture 2030's mission is to transform the US and global building sector from the major contributor of greenhouse gas emissions to a central part of the solution to the global warming crisis. The organization has established specific targets for the AEC industry.

These examples are indicative of a movement from pollution control to pollution prevention and now to sustainability (United Nations Resolution 70/1, 2015 a,b; Table 16.3).

The 2030 Challenge Targets

All new buildings, developments, and major renovations shall be designed to meet a fossil fuel, greenhouse gas (GHG)-emitting, energy consumption performance standard of 50 percent of the regional (or country) average for that building type.

At a minimum, an equal amount of existing building area shall be renovated annually to meet a fossil fuel, GHG-emitting, energy consumption performance standard of 50 percent of the regional (or country) average for that building type (Mazria, E. 2002).

The fossil fuel reduction standard for all new buildings and major renovations shall be increased to 60 percent in 2010, 70 percent in 2015, 80 percent in 2020, and 90 percent in 2025.

And, finally, to meet the 2030 challenge targets new buildings shall be Carbon-neutral in 2030, meaning that they will use no fossil fuel GHG-emitting energy to operate.

Source: 2030 Challenge, https://architecture2030.org/2030_challenges/

The actions and influence of civil engineering professionals can have a positive impact on sustainability for clients, communities, nations, and individuals. We recognize the criticality for implementing sustainable actions worldwide and now!

16.13   FOUR WAYS CIVIL ENGINEERS CAN BECOME MORE SUSTAINABLE

1. CAREER ACTIONS: As a CE Professional—Become more sustainable in your career. Consider sustainable options with your clients; show them how sustainability can save money in construction and operator and maintenance. Look for opportunities to demonstrate sustainable design and implementation of innovative ideas. Consider LEED—Where these Professionals believe "putting our ideals into practice through measurable steps and strategies, and compiling the data to compare them with others, brings us closer to our targets for carbon reduction, human health and cost reduction" (https://www.usgbc.org/leed).

   LEED Professionals believe green buildings are the foundation of something bigger: helping people, and the communities and cities they reside in—safely, healthily, and sustainably thrive. The heart of our green building community's efforts must go well beyond construction and efficiency, and the materials that make up our buildings. We must dig deeper and focus on what matters most within those buildings: human beings (https://www.usgbc.org/leed).

   Please review this award-winning concept to help save humans and wild animals where both people and animals will have a safe crossing over a six-lane highway in a Texas City.

The Largest Wildlife Bridge in the US, San Antonio, Texas

(Continued)

"In 2021, San Antonio opened the country's largest natural bridge designed for both pedestrian and animal use to safely get across a busy stretch of highway. The Robert L.B. Tobin Land Bridge stretches over a six-lane highway, connecting two sections of the Phil Hardberger Park, 330 acres of land named for a former San Antonio mayor. Hardberger served one term as mayor from 2005 to 2009, preceding Julian Castro, the former presidential candidate and Secretary of Housing and Urban Development under President Obama.

Wildlife bridges that cross highways or other obstacles that pose dangers for animals originated in France in the 1950s and gained popularity throughout Europe in the past few decades. Recently, the trend has caught on in the United States. There are similar on Interstate 80 in eastern Nevada to protect wild hers of Elk and also to protect the drivers that use the same migration route.

It's working too—Cameras captured moose, deer, bears, and coyotes using the crossing—which surprised state officials who thought wildlife would need more time to adjust after construction was completed in 2018.

Several studies have shown that wildlife bridges or underpasses make highways significantly safer for both animals and drivers. In Wyoming, a 2012 project that constructed two wildlife bridges and six underpasses produced an 81% drop in vehicular collisions with animals over the next three years. Before construction, the Wyoming Department of Transportation estimated that collisions at that particular stretch of the state highway were costing over $500,000 each year. The state concluded that the cost savings from wildlife crossings will pay for the $11 million project in about 17 years."

Coleman, Emma. (December 14, 2020). The Largest Wildlife Bridge in the U.S. Opens in San Antonio, Route Fifty.
Source: Stimson Studios/https://www.route-fifty.com/management/2020/12/san-antonio-wildlife-bridge/170750/ / last accessed 31 May 2024.

2. PERSONAL DEDICATION: Let's Put Our Monies and Investments Where Our Intention Is, Through Environmental, Social, and Governance (ESG) Investing. As professionals with extensive mathematical training and detailed knowledge of engineering economics hopefully wise CEs will make smart investments in ESG companies for growth and eventual retirement. Therefore, ESG investing is worth a look for possible investment opportunities. At least it makes sense to become aware of Companies dedicated to sustainable, responsible, and socially responsible investing. It seems these Companies perform very well so evaluating their corporate culture, best practices, and performance is prudent. For more information and details on rewarding ESG investing the authors suggest the reader contact their own stockbroker, investment counselor, or other financial experts.

3. CEs CAN INFLUENCE CHANGE and LEAD US TO A SUSTAINABLE FUTURE: The "Winds Are Shifting"—Some cities' building codes are eliminating fossil fuels for most space and water heating. CEs need to be aware of impending changes regarding energy use, operation, and maintenance cost impacts, building efficiencies, utility cost trends, building code revisions, and more. Here's an example of significant trends on building code revisions and utility use.

4. BE THE EXAMPLE—AWARD-WINNING CORPORATE HEADQUARTERS. The Wilo Group is one of the world's leading premium suppliers of pumps and pump systems for building services, water management, and the industrial sector. Wilo invested in their own Corporate Headquarters in the Germany to show their clients their capabilities to "be the premium supplier of pumps and pump systems for building services." Wilo doesn't just talk about their premium capabilities they do it!

Major Building Code Changes

"Seattle's energy code is not only among the strongest in the nation, it prioritizes protecting the health of our most impacted populations and is a critical mechanism to support our City's transition to a clean energy future," Seattle Mayor Durkan wrote in a statement.

The code, which applies to new commercial buildings and large multi-family buildings, eliminates fossil fuels from most space and water heating, increases energy efficiency and increases access to onsite renewables. The city is receiving support from the Bloomberg Philanthropies American Cities Climate Challenge as it adopts the updates.

The code requires all commercial and multi-family buildings taller than three stories to wire for future electrification of appliances; increase on-site solar photovoltaics; reduce envelope heat loss and air leakage; and reduce interior lighting power allowances, among other measures. The code also restricts fossil fuel space heating and most fossil fuel water heating systems to reduce dependence on gas and oil. Implementation is pending.

Electrification of the building sector has become a trend along the West Coast, particularly as jurisdictions struggle with poor air quality due to out-of-control wildfires. Berkeley made history in July 2019 as the first city to ban natural gas infrastructure in new low-rise residential buildings, sending a ripple effect across California where more than 40 cities, including San Jose, San Francisco and Oakland, have now implemented measures to push fossil fuels out of new construction.

These efforts even led California to the top of the American Council for an Energy-Efficient Economy's (ACEEE) annual State Efficiency Scorecard in December, for the first time since 2016."

TABLE 16.3 Examples of innovative approaches to increasing sustainability.

Example	Description
  California Academy of Sciences Building—San Francisco, California Source: Flickr/Doug Letterman. https://calacademy.org/our-strategic-plan	• The Academy's home is a forward-looking example of sustainable design that minimizes its footprint. The Academy's mission is to regenerate the natural world through science, learning, and collaboration. Architect Renzo Piano, working with the Academy, designed the world's first building to receive a double-platinum LEED (Leadership in Energy and Environmental Design) rating. The building employs a wide range of innovative elements and approaches to operations including: 1) a living green roof; 2) eco-friendly materials that maximize water and energy efficiency; and 3) building operations that continue to advance sustainable practices.

Example	Description
 WiloPark—Dortmund, Germany Source: Macina GmbH for WILO SE.	• WiloPark includes an ultra-modern digital smart factory, office building, customer service center, and product development facilities of almost 200,000 m2 (2,152,782 ft2). The Wilo Group is an international supplier of pumps and pump systems for building services and water management in the industrial sector. Over a decade, the company transformed itself into a sustainability champion. The facility has reduced energy consumption by nearly 40% and has lowered CO_2 emissions per year by an anticipated 3,500 tons, equivalent to the same amount of CO_2 that 280,000 tress can absorb.

(Continued)

TABLE 16.3 (Continued)

Example	Description
 *Cityscape with Mt. Rainier—Seattle, Washington* *Source:* Wikimedia Commons.	• Seattle and many other state and municipal governments have adopted new building codes that: 1) eliminate fossil fuels from most space and water heating; 2) increase energy efficiency and; 3) boost access to onsite renewable energy sources. The city has received support from the Bloomberg Philanthropies American Cities Climate Challenge as it adopts the updates. Similar efforts among forty Californian cities resulted in California topping the American Council for an Energy-Efficient Economy's (ACEEE) annual State Efficiency Scorecard in 2016.

16.14   SUMMARY

The Architecture, Engineering, and Construction (AEC) industry plays a critical role in delivering a diverse range of facilities and civil infrastructure systems, including residential, building, industrial facilities, and transportation, energy, water supply, waste management, and communications systems. It also plays a critical role in maintaining their quality, integrity, and longevity. At the same time, the AEC industry contributes to natural resource depletion, waste generation and accumulation, and environmental impact and degradation. As a result, a range of constituencies have been attempting to define the attributes and characteristics, the processes for the delivery and use, and the resources consumed in the delivery and use of facilities and civil infrastructure systems as possible mechanisms to slow, reduce, and eliminate these impacts.

> "Your present circumstances don't determine where you can go; they merely determine where you start."
>
> —Nido Qubein, businessman and motivational speaker

Traditional approaches of environmental regulatory compliance or reactive corrective actions have proven to be consistently costly, inefficient, and often ineffective. In a sustainable approach to design and construction, decision-makers integrate sustainability at all stages of the project lifecycle, particularly the early funding allocation, planning, and conceptual design phases.

Our actions and influence as CEs and professionals can have a positive impact on sustainability to our clients, our community, our nation, and ourselves. Intelligent, critical-thinking CEs, and Professionals recognize the criticality for implementing sustainable actions worldwide and now! Really, what is the choice? Do we believe we have infinite fisheries in the oceans, unlimited land for development and agriculture, unlimited fuel for our lifestyles, unlimited clean water? We've detailed reals ways above on "How we can promote sustainability" to the public, clients, even our peers? Summarizing, here's how some CE's promote sustainability:

- As a CE Professional—Become more sustainable in your career. Consider sustainable options with your clients, show them how sustainability can save money in construction and operator and maintenance. Look for opportunities to demonstrate sustainable design and implementation of innovative ideas, like LEED concepts. These concepts and sustainability provide "Value Engineering"!
- Let's Put Our Monies and Investments Where Our Intention Is—Through Environmental, Social, and Governance (ESG) Investing. As professionals with extensive mathematical training and detailed knowledge of engineering economics hopefully wise CEs will make smart investments in ESG companies for growth and eventual retirement. ESG investing promotes sustainability and pays off. It's a Win/Win scenario for All!

- CEs CAN INFLUENCE CHANGE and LEAD US TO A SUSTAINABLE FUTURE: The "Winds Are Shifting"—Some cities' building codes are eliminating fossil fuels for most space and water heating. CEs need to be aware of impending changes regarding energy use, operation, and maintenance cost impacts, building efficiencies, utility cost trends, building code revisions, and more. CEs should encourage these building code revisions that promote sustainability and also provide "Value Engineering" for clients, owners, and citizens!
- BE THE EXAMPLE—Strive to Provide Civil Engineering Award-Winning Projects and Concepts that *promote sustainability and value engineering.* Look at the example of "The Wilo Group" as one of the world's leading premium suppliers of pumps and pump systems for building services, water management, and the industrial sector. Wilo invested their profits into their own Corporate Headquarters in Germany to demonstrate their capabilities to their clients on how Wilo is "the premium supplier of pumps and pump systems for building services." Wilo incorporated their profits into a capital investment for their World Headquarters that displays their commitment to sustainability and demonstrated expertise used as "hands-on" marketing for their company.

Existing standards, such as LEED (Leadership in Energy and Environmental Design) and BREEAM (Building Research Establishment Environment Assessment Method), are helpful but may give designers the perception that meeting these prescribed targets will result in satisfactory environmental performance. *Sustainable engineering* must address issues that go beyond checklists.

REFERENCES

The American Institute of Architects. (2007). *The Architects Handbook of Professional Practice*, J.A. Demkin, ed. John Wiley & Sons, Inc, New York.

American Society of Civil Engineers. (2008). *Civil Engineering Body of Knowledge for the 21st Century*, 2nd edition. American Society of Civil Engineers (ASCE) Report, Reston, VA.

ASCE's Roadmap to Sustainable Development: Four Priorities for Change. (January 2023). ASCE, Washington D. C.

Birkland, Janis. (2002). *Design for Sustainability: A Sourcebook of Integrated Eco-logical Solutions.* Earthscan, London.

Enovity—HOK. (2008). *Sustainable and Strategies Report for California State University.* Sacramento. July 2008.

Hansen, Karen Lee and Vanegas, Jorge A. (2006). "A guiding road map, principles, and vision for researching and teaching sustainable design and construction." In *American Society for Engineering Education (ASEE)* Chicago, IL. Conference Proceedings.

Mazria, Edward. (2002). *2030 challenge.* http://www.architecture2030.org/2030_challenge/index.html.

McDonough, William and Braungart, Michael. (2002). *Cradle to Cradle.* North Point Press, New York. ISBN 0-865- 47587-3.

Sustainability at ASCE. (January 2023). ASCE, Washington D. C. https://www.asce.org/communities/institutes-and-technical-groups/sustainability.

United Nations Resolution 70/1, Transforming our world: the 2030 Agenda for Sustainable Development. (21 October) 2015a.

United Nations Resolution 70/1, Transforming our world: the 2030 Agenda for Sustainable Development. (2015b). and https://sdgs.un.org/2030agenda.

Vanegas, Jorge A., ed. (2004). *Sustainable Engineering Practice: An Introduction.* ASCE Press, Reston, Virginia. ISBN 10: 0-784-40750-9, ISBN 13: 978-0-784- 40750-9.

Yeang, Kenneth. (2006). *Ecodesign: A Manual for Ecological Design.* Wiley Academy, London.

https://www.asce.org/communities/institutes-and-technical-groups/sustainability/sustainability-roadmap.

A FINAL WORD ON SUSTAINABILITY AND THOUGHTS FOR ALL CEs

Robert Redford: What I See When I Think about the Future. November 1, 2021

Robert Redford is a director, actor, producer, and environmental activist. He co-founded The Redford Center with his late son James Redford in 2005a nonprofit organization that uses the power of storytelling to galvanize environmental justice and regeneration.

Here are excerpts from his speech:

"There is so much I can envision for this planet. I can see a world where human and planetary health and justice are fundamental values. I can see a world where everyone has a safe place to live. Where environmental and economic injustices are addressed and repaired."

"What do you see when you think about the future of our communities, our country and our world?"

"For me, this is the first moment in my life in which there seems to be both a will and a way to address the environmental crisis at the scale that science and justice demand. But the clock is ticking, and we need to take immediate action to achieve a cleaner, healthier and more sustainable future."

"I see crisis after crisis if the current inaction persists. I see no community safe from extreme weather events—fires, hurricanes, floods or drought. I see people becoming ill. I see countless lives being lost."

"I've spoken out about the environment for decades. It's comforting to have more voices in the mainstream pushing for change, especially now that the situation is increasingly dire. How many more voices are needed for our elected leaders to listen, move beyond their own interests, and act in the interest of the greater good—and generations to come?"

CHAPTER 17

Emerging Technologies

Big Idea

Changing technologies, as well as change in general, are facts of life. To use change to their advantage, civil engineers must develop ways of maximizing the application of new technologies.

"Time is a sort of river of passing events, and strong is its current; no sooner is a thing brought to sight than it is swept by and another takes its place, and this too will be swept away."

—Marcus Aurelius

Key Topics Covered

- Current Emerging and Innovative Technologies
- The Nature of Change
- Information Technology—Enabled Process Change
- Engineering Thinking

Civil Engineer's Handbook of Professional Practice, Second Edition. Karen Lee Hansen and Kent E. Zenobia.
© 2025 John Wiley & Sons, Inc. Published 2025 by John Wiley & Sons, Inc.
Companion website: www.wiley.com/go/hansen/CivilEngineersHandbook

Related Chapters in This Book

17.1 INTRODUCTION

Technologies have helped form both our physical environment and world socioeconomic systems. In his book *Guns, Germs, and Steel: The Fates of Human Societies*, Jared Diamond argues that advances in military technologies have been a key factor in shaping world history and development. By enabling the use of armored cavalry, something as simple as the adoption of the stirrup has determined who were the vanquished and the victors in more than one historic battle.

Today, civil engineers are exposed to a heady array of new technologies. With such variety, decisions about which technology to adopt can be baffling. Civil engineers may ask:

- How can this technology create value for our clients/customers?
- What problems can this technology solve?
- What processes can this technology improve?
- How have some organizations enhanced their effectiveness/profitability with this technology?

This chapter addresses the use of emerging information technologies and processes. It highlights several new developments in engineering materials and methods and outlines some of the world's key engineering challenges. The chapter concludes with suggestions on how the development of *engineering thinking* can assist civil engineers to make the most of change.

tech•nol•o•gy: the practical application of knowledge, especially in a particular area
emer•ging: newly formed or prominent
—Merriam Webster OnLine/www.merriam-webster.com

The Globalization, Sustainability, and Emerging Technologies chapters of this handbook are all closely related and interrelated. Civil engineers are involved in defining, creating, and implementing sustainable solutions on a global scale while sharing and using emerging technologies for the betterment of humankind. We ask the reader to integrate the information and concepts presented within these related chapters into their problem-identification and problem-solving approaches.

17.2 CURRENT EMERGING AND INNOVATIVE TECHNOLOGIES

Technology has impacted the appearance and function of our civil structures for thousands of years. Figure 17.1 displays the maturation of some of these structures.

The term *emerging technologies* includes:

- Innovative ideas and practices that show promise for civil engineering and other applications.
- New technologies that have been pilot-tested and are in the demonstration phase.
- New demonstrated technologies that may be applicable and expanded to other areas of CE practice and/or construction.

Within the rapidly changing environment of disruptive technology development, the following technologies stand out (Vanegas 2023):

1. **AV & UAV—Autonomous and/or Unmanned Vehicles (Land, Air, and Water):** Autonomous vehicles, also known as driverless, self-driving, or robotic cars, are capable of fulfilling the transportation capabilities of a traditional car, sensing their environment, and navigating without human input. Unmanned aerial vehicles, commonly known as drones, unpiloted aerial vehicles, and remotely piloted aircraft, are aircraft that fly without a human pilot aboard. Autonomous boats, and underwater vehicles navigate without a human pilot aboard.

2. **UC—Ubiquitous Computing:** Ubiquitous computing makes many computers available to a user throughout the physical environment, while making them effectively invisible to the user. UC enables the user to remotely interact with people and the natural, built, and virtual environments; to remotely monitor, collect, and access data, information, knowledge, experience, and wisdom; and to remotely control devices.

3. **UPT—Ubiquitous Positioning Technologies:** UPT enables the location of people, objects, or both, anytime, whether they are indoors or outdoors or moving between the two, at predefined location accuracies, with the support of one or more location-sensing devices and associated infrastructure.

EARLY DEVELOPMENT	CONTEMPORARY

Bridges

Lierganes Bridge

Source:
Jose Ignacio Soto/Adobestock.

Helix Bridge, Singapore

Source:
Sirikunkrittaphuk/Shutterstock.

Roads

Ancient Roman roads, Leptis Magna, Libya

Source:
Wikimedia Commons.

'Bridge' The Gap: Animal Crossing, India

Source:
Jeroen/Adobe Stock.

Water Resources

Les Ferreres Aqueduct, Tarragona Spain

Source:
Roberto Al/flickr/CC BY-NC-ND 2.0.

California Aqueduct, CA State Water Project

Source:
NEW.

FIGURE 17.1 Example of civil structures maturing through advancements in technologies.

4. **GIS—Geographic Information Systems:**
   Systems designed to capture, integrate, store, edit, manipulate, analyze, manage, share, and display spatial or geographic data, with a spatial data infrastructure that has no restrictive boundaries. GIS applications are tools that enable user-created searches and interactive queries, analyze spatial information, edit data in maps, and present the results of all these operations to support decision-making and change. GIS also refers to the academic discipline that studies geographic information systems as the science underlying geographic concepts, applications, and systems.

5. **CC—Cloud Computing:**
   A style of computing in which capabilities related to Information Technologies (IT) are provided to users "as a service" allowing them to access technology-enabled services from the Internet ("in the cloud") without requiring knowledge of, expertise with, or control over the technology infrastructure that supports the services.

6. **AR—Augmented Reality:**
   A term for a live direct or indirect view of a physical real-world environment whose elements are *merged* with, or *augmented* by, virtual computer-generated imagery.

7. **A&R—Automation and Robotics:**
   The application of **science**, **engineering**, and **technology** (particularly **electronics**, **mechanics**, **control systems**, **computer-aided technologies**, **hardware** & **software**, and **artificial intelligence**), in the design, manufacture, and application of **autonomous devices and robots for industrial, consumer,** or **entertainment use**, which reduce the need for human sensory and mental requirements, and which perform tasks that are too dirty, dangerous, repetitive, or dull for humans.

8. **3DP—3D Printing:**
   A form of additive manufacturing technology where a three-dimensional object is created by laying down successive layers of material. 3D printers are generally faster, more affordable, easier to use than other additive manufacturing technologies, and offer product developers the ability to print parts and assemblies made of several materials with different mechanical and physical properties in a single build process. Innovators have used this technology to create actual buildings.

9. **CI—Collective Intelligence:**
   A shared or group intelligence that emerges through collaboration, innovation, and competition, from the capacity of human communities to evolve towards higher order complexity and integration, which (a) appears in a wide variety of forms of consensus decision making in bacteria, animals, humans, and computer networks; and (b) is studied as a subfield of sociology, business, computer science, mass communications, and mass behavior—from the level of quarks to the level of bacterial, plant, animal, and human societies.

10. **BD&A—Big Data and Analytics:**

BD&A refers to data sets that are so large or complex that traditional data processing application software is inadequate to deal with them, including data capture, storage, analysis, curation, search, sharing, transfer, visualization, querying, updating, and information privacy. The term "big data" often refers to the use of large-scale modeling and simulation, and of predictive analytics, user behavior analytics, or other advanced data analytics methods that extract value from data.

11. **AI—Artificial Intelligence (AI):**

AI refers to the intelligence of machines and the branch of computer science that aims to create it. AI textbooks define the field as "the study and design of intelligent agents," where an intelligent agent is a system that perceives its environment and takes actions that maximize its chances of success.

12. **NBIC—Nano-Bio-Info-Cogno Convergence:**

The synergistic combination of four major provinces of science and technology, each of which is currently progressing at a rapid rate: (a) nanoscience and nanotechnology; (b) biotechnology and biomedicine, including genetic engineering; (c) information technology, including advanced computing and communications; and (d) cognitive science, including cognitive neuroscience.

Table 17.1 highlights several innovative technologies that are particularly applicable to today's civil engineering consulting practice, site preparation, infrastructure, and building construction. Figure 17.2 depicts actual 3D printed homes in Williamsburg, Virginia and also in Beckum, western Germany.

TABLE 17.1   Innovative uses of currently available technologies.

Technology	Example
Unmanned Aircraft Systems (UAS's)–drone technology–used for: • Design details on steep dangerous slopes, • Health and Safety Plan details for inaccessible areas on the river, banks, and slopes, • Construction Procedures, and • Post-Construction Monitoring and Performance 	*ASCE Sacramento Section–2017 Project of the Year–Storm Damage Emergency Rehabilitation* Client: California State Department of Water Resources (DWR) Engineer: AECOM Scope of project: Repair of 500 levee sites, part of 1,000 miles of levees in California's State Plan of Flood Control, which were damaged significantly by intense rainfall combined with snow melt and subsequent flooding. These distressed levee sites were rated as a "critical emergency," where one more storm could cause tremendous damage to State infrastructure.

TABLE 17.1   *(Continued)*

Technology	Example
For more information see: Zenobia, Kent, David Wheeldon, and Rob Nixon. (2020): *Aircraft System Surveys in Levee Repairs*. Presentation and Publication, Floodplain Management Virtual Conference, California, USA, September 2020	DWR and AECOM: • Developed remedial designs • Expedited permitting strategies • Created emergency construction and remediation design plans and innovative construction techniques for damaged sites on the deep and fast flowing Sacramento River and other nearby tributaries
Combination of Geographic Information System (GIS) and Augmented Reality (AG)–used for: • Survey of terrain and spatial information • Creation of virtual computer-generated imagery • Communication of design concepts to the public and other decision makers 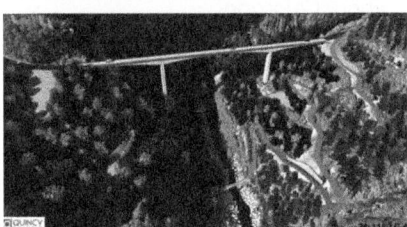 For more information see: Source: County of El Dorado. Mountain Democrat / https://www.youtube.com/watch?v=7INFrAUscjw / last accessed 16, March 2023.	*Mosquito Bridge Replacement Project* Client: El Dorado County Department of Transportation, State of California Engineer: Quincy Engineering in partnership with SYSTRA IBT Contractor: Shimmick Construction *Scope of project:* Replacement of one-lane wooden suspension bridge, originally constructed in 1867 and rebuilt on the original foundations in 1939. New bridge is approximately 1,180 foot-long, three span, post-tensioned concrete box girder, 400 feet above the South Fork of the American River. The estimated construction cost is approximately $93M. Quincy Engineering: • Created GIS model of the site and key features • Analyzed spatial information • Digitally merged natural elements with proposed designed • Used the integrated model for crucial design decisions and for communicating with the client

(Continued)

TABLE 17.1   *(Continued)*

Technology	Example
Three-dimensional (3D) Printing–used for: • Long-term reduced construction costs and shortened construction schedules • Decreased reliance on skilled construction labor • Ultimately, lower construction costs resulting in more affordable homes • With concrete, improved resilience in dealing with hurricanes, heavy storms, fire, and other severe weather  For more information see: Figure 17.2–3D Printed Homes. Source: Jarett Gross/https://www.youtube.com/watch?v=qWBA-6NglJg / last accessed 16 March 2023.	*Case Study–Large Two-Story House, Houston, Texas* Architect: Hannah Builder: PERI 3D Construction (Germany) and CIVE (Houston), which provided engineering support *Scope of project:* Print three-bedroom, 4,000 square foot multi-level family home using concrete. Incorporates traditional steel Design and construction team: • Used design-build approach • Pushed 3D printing technology capacity • Learned the limits of the material (concrete) when employed this way • Optimized the speed of printing • Explored various design possibilities
Collective Intelligence–used for: • Solution of societal issues • Sharing knowledge, data, and skills • Mobilization of a wide range of information, ideas, and insight • Reaching consensus  For more information see: • thegovlab.org/collective-intelligence.org • MIT Center for Collective Intelligence (CCI) • YouTube: Lakewood's Sustainable Neighborhoods Program	*Sustainable Neighborhoods Program, Lakewood, Colorado* Owner: City of Lakewood Program: Sustainable Neighborhoods *Scope of project:* Program gives residents the opportunity to make their city sustainable by participating in a certification program where they learn how to reduce their ecological footprints and make their neighborhoods more livable. These residents are trained to organize workshops in their neighborhoods, extending the reach of the program. Lakewood Sustainability Division: • Engaged more than 20,000 residents by running 500 sustainability events and projects • Enhanced environmental and social strength of the community

Habitat for Humanity, first 3-D printed house, Williamsburg, Virginia, USA

Source: ALQUIST.

3D-printed house is in Beckum, in western Germany.PERI

Source: PERI SE.

FIGURE 17.2 "3D" printed homes.

17.3   THE NATURE OF CHANGE

For several decades, most organizations have found themselves in the midst of rapid business, technological, and process change. Now environmental change has been thrown into the mix. (See Chapter 15, Globalization and Chapter 16, Sustainability.)

Some reasons for rapid change in the business environment include:

- Globalization of competition
- Strengthened role of powerful clients
- Increased regulation
- Internationalization of technologies and tools
- Growing distance between world-class firms and local "backbone"

These changes have led to a fundamentally different way of considering business and economics. Table 17.2 contrasts the old and new views of economics.

Intertwined with these changes are rapid, continuous changes in information technology hardware and software. Some of these developments include:

- Ever increasing processor speeds
- Miniaturization
- Exponentially expanded storage capacity
- "Hardened" hardware, i.e., hardware that functions on dirty construction sites
- Global positioning systems (GPS) and geographic information systems (GIS)
- Radio frequency (RF) tracking

TABLE 17.2   A newer perspective on complex issues.

Old Economics	New Economics
Based on 19th-century physics—equilibrium, stability, deterministic dynamics	Based on biology—structure, pattern, self-organization, lifecycle
People identical	Focus on individual life; people separate and different
If only there were no externalities and all had equal abilities, we'd reach Nirvana	Externalities and differences become driving force; no Nirvana, system constantly unfolding
Elements are quantities and prices	Elements are patterns and possibilities
No real dynamics in the sense that everything is at equilibrium	Economy rushes forward—structures constantly coalescing, decaying, changing
See subject as structurally simple	See subject as inherently complex
Economics as soft physics	Economics as high-complexity science

Source: Adapted from Waldrop, *Complexity: The Emerging Science at the Edge of Order and Chaos.*

- Integration of systems
- Interoperability standards
- The Internet
- Social networking
- Mobilization

Responses to these changes have been varied. Many involve efforts to improve processes with the help of new technologies and tools. Process improvement efforts include:

- Business Process Reengineering (BPR)
- Lean Value Chains (e.g., Lean Construction)
- Concurrent Engineering
- Continuous Improvement
- Total Quality Management (TQM)
- Total Enterprise Management

These approaches have ranged from the Business Process Reengineering (BPR) perspective to highly quantitative methods. BPR essentially is an integration mechanism for holding together organizational forms and linking the operations of an organization to the requirements of customers. One dramatic result of BPR has been the pervasive outsourcing of all but the most essential functions in many US corporations. Toward the other end of the process improvement spectrum are methods used in manufacturing. In this approach, a list of activities is coupled with the way in which these interact with resources. The resulting data can then be used to generate a mathematical model that can be solved to arrive at a production plan or distribution plan or plant design. See Figure 17.3 for the principles of BPR and Figure 17.4 for some process definition models used in process improvement.

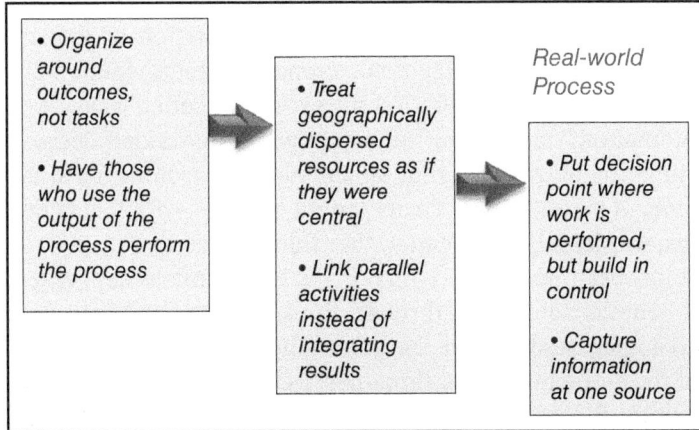

FIGURE 17.3   Fundamental principles of business process reengineering.

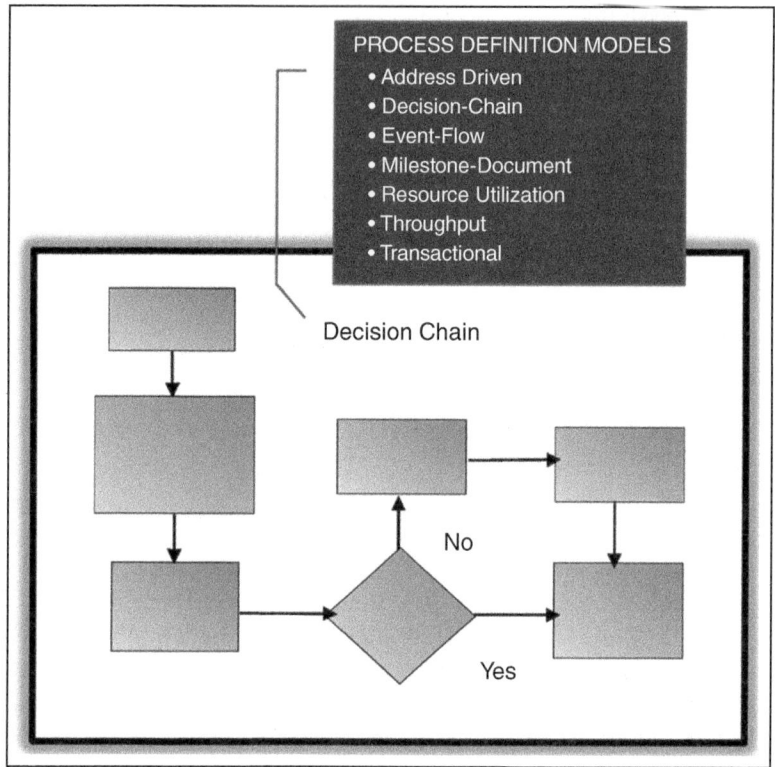

FIGURE 17.4   Process definition models, decision chain example.

17.4   INFORMATION TECHNOLOGY—ENABLED PROCESS CHANGE

For many years, architectural, engineering, and construction industry (AEC) practitioners and academics have believed that appropriate implementation of information technology (IT) design, engineering, and management support systems could help significantly to improve performance in project delivery. Yet the experience often was one of implementation in "islands of automation," such as in the use of computer- aided design (CAD) by design firms, or cost-estimating and scheduling systems by contractors. This, at best, resulted in small improvements in performance. The potential benefits, which may flow from the use of integrated systems, was largely imagined, rather than realized.

An approach has been needed in which firms learn how to deploy technology to move from the limited benefits achievable through the substitution of information technologies for existing technologies, resulting in the automation of individual work steps, to a wider transformation of the entire process. A three-part model of the information technology adoption process illustrates the possibilities for firms to improve performance incrementally by moving through each phase, and more radically by pursuing strategies to transform processes with the assistance of information technology systems. (See Figure 17.5.)

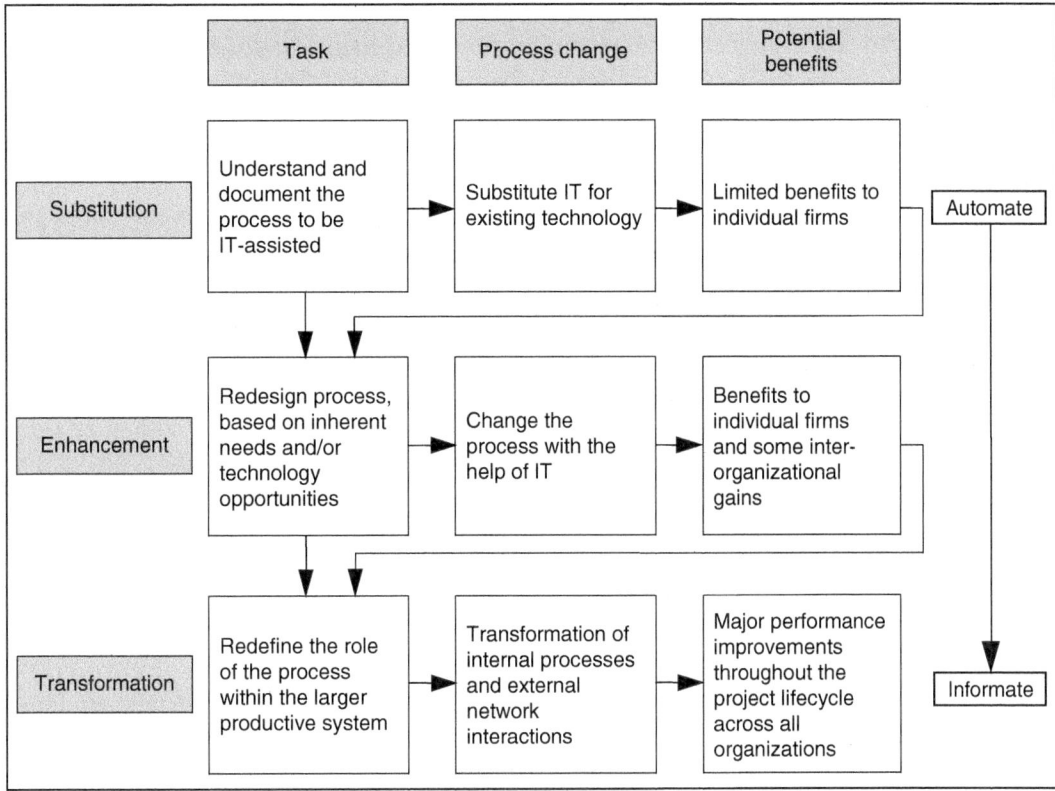

FIGURE 17.5 IT adoption and business process change.
Source: Adapted from Gann & Hansen, IT Decision Support and Business Press Change in the United States.

17.4.1 Early Developments

The seeds for this transformation were sown many years ago. AEC firms have responded to change through the adoption of new tools such as:

- 3D CAD
- Simulation
- VR (Virtual Reality)
- Rapid Prototyping
- Modeling and Workflow
- Automated Process Improvement

Perhaps for historical reasons, many of these tools have focused on design. Over the last 40 years or so, there have been numerous attempts to devise explicit models of the design process, including approaches which assist in finding ways of construing a rational model for architecture, for example, exploring the total industrial process of construction, sometimes linked to an even larger project process including land development; or developing a generic approach to design, including product design, manufacturing design, chemical engineering, and so forth.

The most obvious thrust of design method work was toward CAD drawing systems and attempts to link these either to prospective component or building systems or to form-finding methods that exploited classic theories of typology, proportion, and composition. There also were attempts to link such programs to databases of components used as part of a mainte-nance strategy for buildings. In the 1970s and 1980s, architectural researchers were unable to devise the types of coherent computer-aided design computer-aided manufacturing (CAD CAM) systems developed in some manufacturing engineering industries, neither were they able to develop a total computerized architecture in the vision of Negroponte's 1970 book, *The Architecture Machine*. (Nicolas Negroponte was a pioneer in the field of computer-aided design and has been a member of the MIT faculty since 1966.)

One of the difficulties was that, by contrast, until relatively recently there have been poor methods for representing construction processes. Consequently, designers have relied extensively on visual representation as a means of maintaining a holistic account of the project. However, in the mid-1990s some firms sought to introduce single project databases and new communica-tions media into the design process. These technologies brought with them a new dimension to the integration of design and construction activities; they changed the type of involvement of each participant and altered the ways in which decisions were made. They also provided design, engineering, and construction organizations with opportunities to carry out new types of work, offering customers new services, and thus developing better client relationships.

Examples of these efforts include:

1. Architect Frank Gehry's design studio took a hands-on approach to the entire process of design and construction. Gehry's sculpted physical models were digitized using 3D scanning technologies (transferred from the field of medicine). These were used to drive a 3D model developed with the CAD application, CATIA (used by Chrysler and Boeing to design with curved surfaces). CATIA was then used to provide a rapid prototyping capability which removed a number of intermediate (paper-based) steps in the process, allowing new physical models to be validated. The final digital data was transferred from the architects and structural engineers to the general contractor, steel fabricator, and steel erector. Thus, Gehry and his colleagues radically altered the design-construct process.

2. Stone & Webster's Advanced Systems Development Services (ASDS) was an early user and developer of 3D modeling (instead of the prevailing 2D systems). It was special-ized in project-specific systems integration, developing and customizing applications software to link CAD with databases and knowledge-based systems. The result was an "as needed" approach to integrating systems, creating a "bricolage model." ASDS solu-tions have been used effectively in mechanical engineering (by firms such as Chrysler

and Mercedes-Benz). This was a particular approach to middleware development in which the project appears to induce the selection of data, knowledge, and applications, which are then transformed into practical tools.

3. Parsons adopted an approach to sharing project information aimed specifically at extending the market for their services from engineering, procurement, and construction, forward into early project decision-making and downstream into facility management. To achieve this, they developed their existing Computer-Integrated-Engineering IT support systems to form a new Computer-Integrated-Project system. This was supported by a variety of technologies such as GIS in early project decision-making. The adoption of this approach resulted in the need for internal business process changes and new relationships with suppliers and other design and construction organizations. It created opportunities for the company to provide its clients with new value-added services, extending Parsons' markets.

4. Bechtel used Virtual Reality systems to share information with their clients in order to reduce risk and uncertainty and improve predictability in design decisions. The technology provided the client with a decision-making tool, in which transnational customer links facilitated visualizations of prospective facilities enabling clients to modify design decisions at little expense. The system also helped to reduce overall project times and saved on travel costs. Other developments at Bechtel included simulations of heavy lifting processes and digital data collection tools for site work.

As the new millennium has come into focus and then moved on, these earlier approaches have become less leading-edge and more the norm.

17.5 BUILDING INFORMATION MODELING

Advances in *interoperability*, the ability to exchange data between computer programs, have hastened the development of new technologies. Among these is building information modeling (BIM). One of the promises that BIM holds out is the potential to bridge the gap between conceptual and technical problem solving. (See Figure 17.6.)

Design, engineering, and construction work involves integrating and assembling different subsystems and components. Many of the challenges facing the AEC industry relate to problems encountered at the interfaces between the work of different professional disciplines. It is here that BIM may be of particular assistance in helping to improve information flows between different experts, professionals, technicians, and trades, and across building lifecycles.

No single computer program has yet been able to support all tasks associated with design and construction; but interoperability allows data to flow from one application to another. Thus, many experts and applications can be included in the design-construct process. Still, two applications can export or import different information for describing the same object. In the United States, there is an effort to standardize the data required for particular workflow

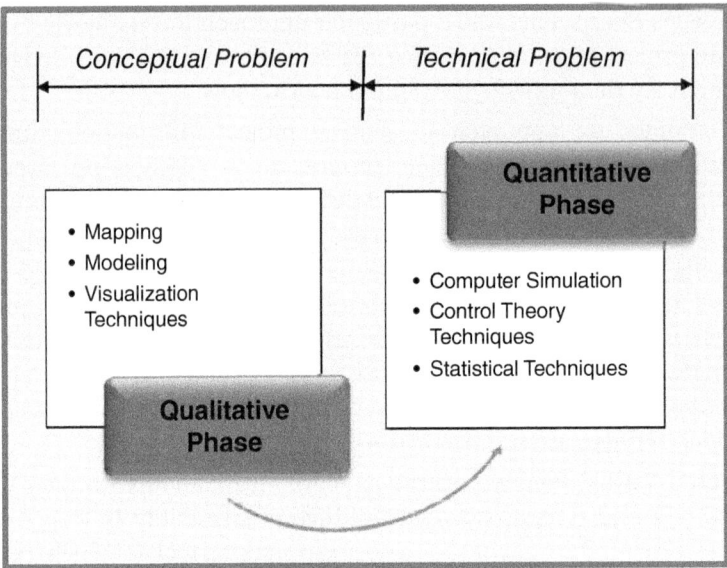

FIGURE 17.6   Bridging the gap between conceptual and technical problems.

exchanges. The main endeavor is called the National BIM Standards (NBIMS) and is being conducted by the National Institute of Building Sciences. Table 17.3 summarizes the primary methods used in information exchange.

As specialized programs have emerged, an accompanying plethora of exchange formats has developed. Common exchange formats used in the AEC industry are shown in Table 17.4.

agcXML

The agcXML project is a top priority of the Associated General Contractors (AGC) Electronic Information Systems Committee. It will result in a set of XML schemas for the transactional data that is now commonly exchanged in paper documents. Examples of such documents include owner/contractor agreements, schedules of values, requests for information (RFIs), requests for proposals (RFPs), architect/engineer supplemental instructions, change orders, change directives, submittals, applications for payment, addenda, and the like. To ensure compatibility with related efforts, the agcXML Project is being executed as part of the building SMART Initiative. For more information, see www.agcXML.org.

Communication among project participants occurs formally and informally, and on a variety of levels. Eastman, Teicholz, Sacks, and Liston have identified four different types of communication exchange (as opposed to data transfer) that transpire in a BIM process. These are shown in Table 17.5.

TABLE 17.3 Methods of exchanging data.

	Medium	Description
1	Direct, proprietary link between specific BIM tools	Provides an integrated connection between two applications. Relies on middleware software interfacing capabilities like ODBC or COM or proprietary interfacing capabilities like ArchiCad's GDL or Bentley's MDL, all of which use C, Cþþ, or C# languages.
2	Proprietary file exchange format, primarily dealing with geometry	Interfaces within specific company's own applications. In the AEC industry, well-known formats include: DXF (Data eXchange Format), developed by Autodesk; SAT by Spatial Technology; STL for stereo lithography; and 3DS for 3D-Studio.
3	Public product data exchange format	Involves an open-standard building model. IFC (Industry Foundation Class) and CIS/2 for steel are the principal options. Carries information regarding object and material properties and relations between objects as well as geometry.
4	XML-based exchange format	Supports exchange of many types of data between applications. XML is eXtensible Markup Language, an extension of HTML, the base language of the Web. Especially good at exchanging small amounts of business data between two applications set up to do so.

Source: Adapted from Eastman, Teicholz, Sacks, and Liston, *BIM Handbook: A Guide to Building Information Modeling for Owners, Managers, Designers, Engineers, and Contractors.* p. 67.

TABLE 17.4 Common exchange formats used in the AEC industry.

Format Type	Variability	File Extensions
Image (raster)	Compactness, number of possible colors per pixel, data loss with compression	JPG, GIF, TIF, BMP, PIC, PNG, RAW, TGA, RLE
2D Vector	Compactness, line widths and pattern control, color, layering, and types of curves supported	DXF, DWG, AI, CGM, EMF, IGS, WMF, DGN
3D Surface and Shape	Types of surfaces and edges represented, whether surfaces and/or solids are represented, material properties of shapes (color, bitmap, texture map), and viewpoint information	3DS, WRL, STL, IGS, SAT, DXF, DWG, OBJ, DGN, PDF(3D), XGL, DWF, U3D, IPT, PTS
3D Object Exchange	Geometry according to the 2D or 3D types represented, object properties, and relations between objects	STP, EXP, CIS/2
Game	Types of surfaces, whether they carry hierarchical structure, types of material properties, texture and bump map parameters, animation, and skinning	RWQ, X, GOF, FACT
GIS	Geographical information system	SHP, SHX, DBF, DEM, NED
XML	Information exchanged and workflows supported	AecXML, Obix, AEX, bcXML, AGCxml

Source: Adapted from Eastman, Teicholz, Sacks, and Liston, *BIM Handbook: A Guide to Building Information Modeling for Owners, Managers, Designers, Engineers, and Contractors.* p. 69.

TABLE 17.5   Types of communication in BIM processes.

	Communication	Description
1	Published snapshots	One-directional, static views that provide the receiving party with access only to visual or filtered meta-data, such as bitmap images.
2	Published BIM views and meta-data	Viewing access to the model with limited ability to edit or modify data, such as PDF or DWF. Receiving party can perform query functions on the model, comment, mark-up, and change certain view parameters.
3	Published BIM files	Access to the native data through proprietary and standard file formats such as DWG, RVT, and IFC.
4	Direct database access	Access to the project database through a dedicated or distributed project server. Model data controlled through access privileges or more sophisticated edit and change capabilities.

Source: Adapted from Eastman, Teicholz, Sacks, and Liston, *BIM Handbook: A Guide to Building Information Modeling for Owners, Managers, Designers, Engineers, and Contractors* pp. 123–124.

Based on interviews with hundreds of owners; architects; civil, structural, and MEP (mechanical, electrical, plumbing) engineers; construction managers; and general contractors and subcontractors currently using BIM, a McGraw-Hill report (*Smart Market Report: Building Information Modeling*, p. 27) found that the most valuable aspects of BIM are:

- Easier coordination of different software and project personnel
- Improved efficiency, production, and time savings
- Lifecycle analysis, including modeling energy usage
- Better communication
- Improved quality control/accuracy
- Visualization (ability to keep owners informed and to clarify construction tasks to workers)
- 3D modeling and coordination, including interference checking/clash detection
- Keeping pace with advances by competition and others in the marketplace

The report also found that the US Army Corps of Engineers is requiring BIM- based deliverables as part of its Centers for Standardization program, an effort involving 43 standard facility types.

Hurdles on the Path to BIM Adoption

Adequate Training

Training is often the biggest challenge with the adoption of any new technology. Because few users have expert backgrounds, there is a shortage of training resources. As more expertise develops in universities, within firms, and from consultants/trainers, the challenge of training should be reduced. Among architects, engineers, contractors, and owners, engineers are most concerned about training.

Costs of Software and Hardware Upgrades

Issues related to cost also are common with the adoption of new technologies. Increased costs of software and hardware upgrades are significant concerns in BIM adoption. These costs are of greater importance to architects and engineers than contractors and owners.

Senior Management Buy-In

Higher-level management is less likely than any group to embrace BIM adoption. This could be because they must justify the costs of adoption (in terms of training, software, and hardware) or because they are more comfortable with "tried-and-true" methods. Junior-level staff buy-in is considered least challenging, possibly because they may have been exposed to BIM as part of their education, are more open to change, and/or are less aware of associated risks.

Other Factors

Among architects, engineers, contractors, and owners, engineers are most likely to see a lack of external incentives or directives moving them to use BIM.

Both architects and engineers are challenged by the potential loss of intellectual property and increased liability associated with BIM.

—McGraw-Hill Construction, *Smart Market Report: Building Information Modeling*, p. 9

More recently, developments in computer information technologies have converged with new project delivery methods and contract structures, such as Design-Build, Design-Assist, and Integrated Project Delivery. (See Chapter 11, Legal Aspects of Professional Practice.)

17.6 INTEGRATED PROJECT DELIVERY

The American Institute of Architects (AIA) has become active in promoting integrated project delivery (IPD). From the AIA's perspective [AIA National and AIA California Council, *Integrated Project Delivery: a Guide*]:

> *Integrated Project Delivery (IPD) is a project delivery approach that integrates people, systems, business structures and practices into a process that collaboratively harnesses the talents and insights of all participants to optimize project results, increase value to the owner, reduce waste, and maximize efficiency through all phases of the design, fabrication and construction.*
>
> *IPD principles can be applied to variety of contractual arrangements and IPD teams can include members well beyond the basic triad of owner, architect, and contractor. In all cases, integrated projects are uniquely distinguished by highly effective collaboration among the owner, the prime designer, and the prime constructor, commencing at early design and continuing through to project handover.*

IPD changes contract structures, the way project teams are formed, the manner in which the teams interact, and the technologies used in project delivery. See Table 17.6 and Figure 17.7.

IPD is a new process enabled by software, but it is not the software itself. BIM and its accompanying interoperability aid the project team in accomplishing the primary goal: satisfying (or more than satisfying) a client's need within a specific time period and for a given budget.

The basic principles of IPD include:

1. *Mutual respect and trust*: All members of the integrated team—owner, designers, consultants, contractor, subcontractors, and suppliers—value collaboration and are committed to working as a team in the best interests of the project.
2. *Mutual benefit and reward*: All participants and/or team members benefit from IPD. Compensation structures recognize the need for and reward early involvement. Compensation is based on value added, such as incentives tied to achieving project objectives.
3. *Collaborative innovation and decision-making*: Freely exchanged ideas among all participants stimulate innovation. Ideas are judged on merits, not on their author's role or status. To the greatest extent possible, decisions are made unanimously.
4. *Early involvement of key participants*: Participants are involved at the earliest practical moment, thereby improving decision-making through the influx of knowledge and expertise of all key participants. Decisions that are made early have the greatest effect. (See Chapter 4, Engineer's Role in Project Development.)

TABLE 17.6 Traditional versus integrated project delivery.

Element	Traditional Project Delivery	Integrated Project Delivery
Teams	Fragmented group of prime designers and subconsultant representatives, assembled on "just-as-needed" or "minimum necessary" basis; strongly hierarchical, controlled	Integrated team entity composed of key project stakeholders (owner, architect, engineers, contractor, subcontractors, others) assembled early in the process; open, collaborative
Process	Linear, distinct, segregated; knowledge gathered "just-as-needed"; information hoarded; silos of knowledge and expertise	Concurrent and multilevel; early contributions of knowledge and expertise; information openly shared; stakeholder trust and respect
Risk	Individually managed by each entity; transferred to the greatest extent possible	Collectively managed; appropriately shared
Compensation/ reward	Individually pursued by each entity; minimum effort for maximum return; (usually) first cost basis	Team success tied to project success; value based
Communications/ technology	Paper-based; analog; 2 dimensional	Digitally based; Building Information Modeling (3-, 4-, 5-dimensional BIM)
Agreements	Unilateral effort encouraged; risk allocated and transferred; no risk sharing	Multilateral open sharing and collaboration encouraged, fostered, promoted, and supported; risk sharing

Source: Adapted from AIA National and AIA California Council, *Integrated Project Delivery: A Guide*, p. 1.

5. *Early goal definition*: Project goals are developed early, agreed upon, and respected by all participants. Project outcomes are held at the center of a framework of individual objectives and values.

6. *Intensified planning*: Increased efforts in planning result in increased efficiency and savings during execution. The thrust of IPD is not to reduce the design effort but to improve the design results and thereby streamline and shorten the construction period.

7. *Open communication*: Team performance is based on open, direct, and honest communication. A no-blame culture leads to early identification and resolution of problems. Disputes are recognized as they occur and are resolved promptly.

8. *Appropriate technology*: Proper technology is specified at project initiation to maximize functionality, generality, and interoperability. Technology that complies with open standards is used whenever possible because it best enables communications among all project participants.

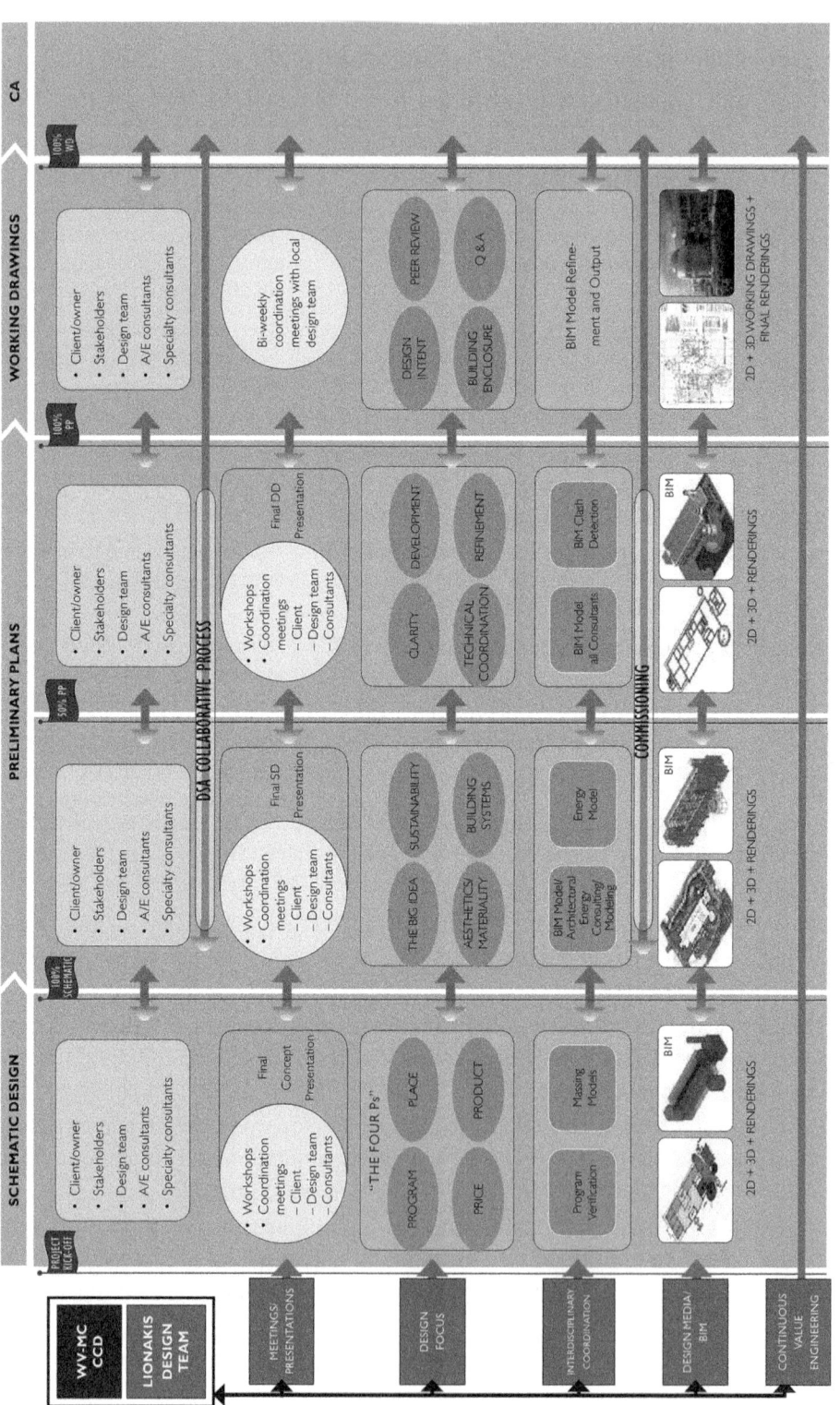

FIGURE 17.7 Integrated project delivery.

9. *Organization and leadership*: Project team members are committed to the team's goals and values. Leadership is taken by the team member most capable with regard to specific work and services—often design professionals and contractors lead in their areas of traditional expertise. Roles are defined clearly but do not create artificial barriers.

Dimensions Defined

2D—2-dimensional project representation; x and y coordinates only; often paper-based

3D—3 dimensions; x, y, and z coordinates included in a geometric digital model, sometimes with additional "intelligence" attached to drawing objects

4D—dimension of time incorporated into the 3D digital model so that the construction schedule can be visualized

5D—dimension of cost incorporated into the 3D digital model in order to automate quantity take-offs. When used with the 4D feature, 5D also can predict cash flow.

As the basic principles of IPD indicate, the level of effort in design phases changes from design-bid-build (DBB). (See Chapter 5, Engineer's Role in Project Development and Chapter 6, What Engineers Deliver.) Because of the collaborative environment, many decisions are brought forward in the design process. Consequently, two very important aspects of IPD are: (1) having the right contract for the specific professional services needed; and (2) selecting project team members who have the right attitudes, aptitudes, and knowledge.

Lionakis, a California-based architectural practice known for innovative design that incorporates sustainability and technology, has developed an approach to workflow that recognizes the changes required by the IPD process. Integrated Design Workflow (IDW) integrates the design phase, design focus, meetings, the design team, the design media/software, and a unique concept referred as clash detection for the building modeling and energy. This concept is all integrated across the key project phases of pre-design and concept, schematic design, and the design development. Figure 17.8 displays the integrated design workflow.

Model contracts do exist for IPD, and an attorney familiar with IPD also should be consulted. (See Chapter 11, Legal Aspects of Professional Practice.) Building the IPD teams involves selecting people who can work together effectively. According to the AIA, in addition to being committed to the collaborative process, IPD team members also should:

1. Identify participant roles as soon as possible
2. Prequalify firms and individuals who will be on the team
3. Seek involvement of additional, interested parties (building officials, utility companies, insurers, sureties, and other stakeholders)

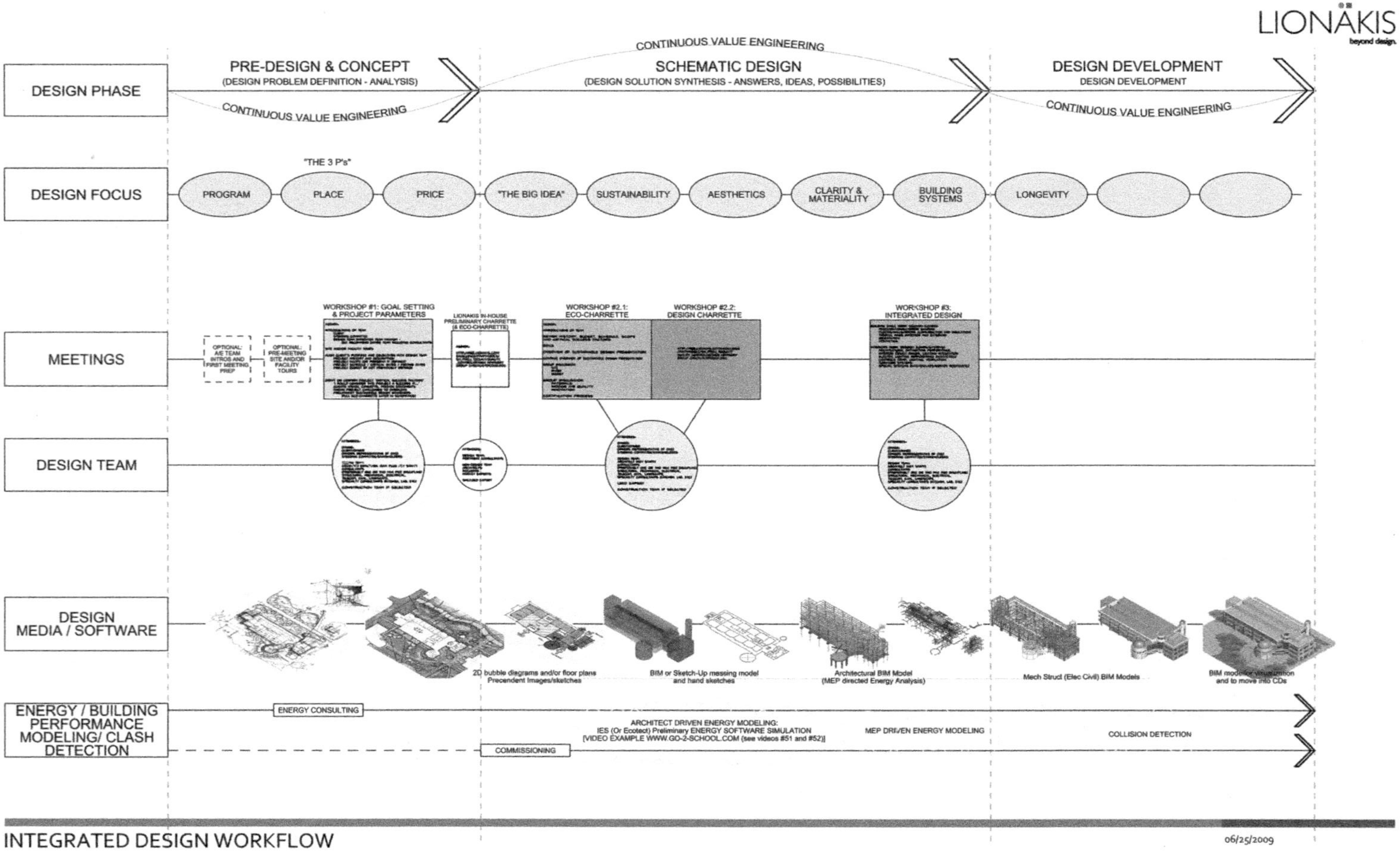

FIGURE 17.8 Integrated design workflow.

4. Define mutually understood values, goals, interests, and objectives of team members and participating stakeholders
5. Identify the IPD organizational and business (contract) structure that is best suited to the participants' needs and constraints
6. Develop project agreements to define the roles and accountability of participants, including key provisions regarding compensation, obligation, and risk allocation

17.7   FIATECH ROADMAP—AN ORGANIZING PRINCIPLE

FIATECH is a nonprofit organization that grew out of the Construction Industry Institute, an independent research center at the University of Texas at Austin. Formed in 2000, FIATECH is a consortium of leading capital project industry owners, engineering construction contractors, and technology suppliers who advocate development and deployment of fully integrated and automated technologies. FIATECH members are grounded in business and their motivation is to deliver the highest business value throughout the lifecycle of capital projects.

FIATECH's *Capital Projects Technology Roadmap* (see Figure 17.9) presents a vision for the capital projects industry (i.e., the industry that executes the planning, engineering, procurement, construction, and operation of predominantly large-scale buildings, plants, facilities, and infrastructure). According to FIATECH and many others, the capital projects industry greatly lags other sectors in exploiting technological advances:

It is characterized by vast disparities in business practices and levels of technology application. It is fragmented, with great divergence in tools and technologies from company to company and across its supply chains. New pressures, such as Homeland Security in the US, have moved infrastructure security to the forefront of our national consciousness . . . The capital projects industry generally is not well prepared for this far-reaching response, which extends beyond the boundaries of control for any one organization.

All these issues can and should be addressed in a collaborative environment for shared success. FIATECH was formed to provide that integrating entity in partnership with invested stakeholders across the industry. The Capital Projects Technology Roadmap is open to all companies, consortia, associations, and research institutions interested in addressing these critical issues to the industry. Presently, there is no concerted effort to define common goals, leverage available resources, and cooperate to deliver dramatic improvements in capability and cost-effectiveness. This initiative fills that void. [https://www.irbnet.de/daten/iconda/CIB16491.pdf]

Certainly, with increased interoperability, a willingness of the AEC industry to embrace technologies such as BIM that are beginning to deliver on the promise of integration, and new forms of project organization such as IPD, the FIATECH Roadmap is closer to being reality than it was ten years ago.

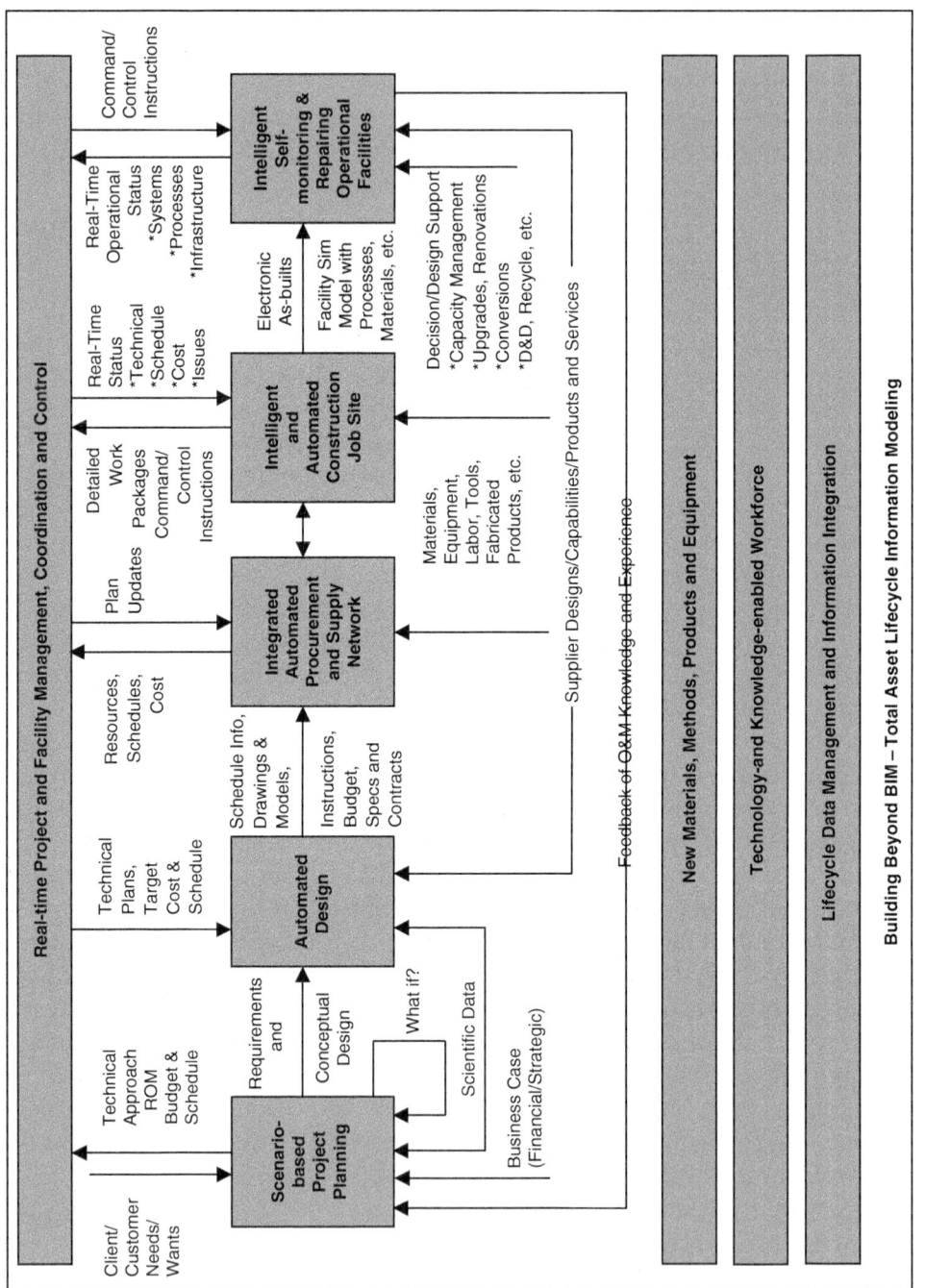

Real-time Project and Facility Management, Coordination and Control

Client/
Customer
Needs/
Wants

**Scenario-
based
Project
Planning**

Business Case
(Financial/Strategic)

Scientific Data

What if?

Requirements
and
Conceptual
Design

Technical
Approach
ROM
Budget &
Schedule

**Automated
Design**

Instructions,
Budget,
Specs and
Contracts

Schedule Info,
Drawings &
Models,

Technical
Plans,
Target
Cost &
Schedule

**Integrated
Automated
Procurement
and Supply
Network**

Materials,
Equipment,
Labor, Tools,
Fabricated
Products, etc.

Resources,
Schedules,
Cost

Plan
Updates

**Intelligent
and
Automated
Construction
Job Site**

Detailed
Work
Packages
Command/
Control
Instructions

Real-Time
Status
*Technical
*Schedule
*Cost
*Issues

Electronic
As-builts

Facility Sim
Model with
Processes,
Materials, etc.

**Intelligent Self-
monitoring &
Repairing
Operational
Facilities**

Real-Time
Operational
Status
*Systems
*Processes
*Infrastructure

Command/
Control
Instructions

Decision/Design Support
*Capacity Management
*Upgrades, Renovations
*Conversions
*D&D, Recycle, etc.

Supplier Designs/Capabilities/Products and Services

Feedback of O&M Knowledge and Experience

New Materials, Methods, Products and Equipment

Technology-and Knowledge-enabled Workforce

Lifecycle Data Management and Information Integration

Building Beyond BIM – Total Asset Lifecycle Information Modeling

FIGURE 17.9 FIATECH's capital projects technology roadmap.

Emerging Technology Resources

The American Institute of Architects

Integrated Practice Information
www.aia.org/ip_default

The American Institute of Architects, California Council

Resources related to IPD including Frequently Asked Questions
www.ipd-ca.net

Associated General Contractors of America

BIM Guide for Contractors
http://agc.org/

Center for Integrated Facility Engineering (CIFE)

Research center for virtual design and construction AEC industry projects
www.cife.stanford.edu

Construction Specifications Institute

MasterFormat
www.csinet.org/s_csi/docs/9400/9361.pdf

Construction Users Roundtable (CURT)

Owners' views on the need for integrated project delivery
www.curt.org

Design Build Institute of America (DBIA)

Library of information and case studies related to design-build
www.dbia.org

FIATECH

Consortium of leading capital project industry owners, engineering construction contractors, and technology suppliers that provides global leadership in development and deployment of fully integrated and automated technologies
http://fiatech.org

(Continued)

International Alliance for Interoperability (IAI)/buildingSMART Alliance

International organization working to facilitate software interoperability and information exchange in the AEC/FM industry
www.iai-na.org

LEAN Construction Institute

Nonprofit corporation dedicated to conducting research to develop knowledge regarding project-based production management in the design, engineering, and construction of capital facilities
www.leanconstruction.org

McGraw-Hill Construction

Source for design and construction industry information regarding IPD
www.construction.com/NewsCenter/TechnologyCenter/Headlines/archive/2006/ENR_1009.asp

National Institute of Building Sciences, National BIM Standards (NBIMS) Committee

Many related articles on IPD and BIM
www.facilityinformationcouncil.org/bim/publications.php
Cost Analysis of Inadequate Interoperability in the US Capital Facilities Industry
www.bfrl.nist.gov/oae/publications/gcrs/04867.pdf

UNIFORMAT II Elemental Classification for Building Specifications, Cost Estimating, and Cost Analysis

www.bfrl.nist.gov/oae/publications/nistirs/6389.pdf

OmniClass

Classification structure for electronic databases
www.omniclass.org

Open Geospatial Consortium

International, voluntary consensus standards organization that is leading the development of standards for geospatial and location-based services
www.opengeospatial.org

Open Standards Consortium for Real Estate

Standards related to information sharing—BIM
http://oscre.org/

US General Services Administration

Nation's largest facility owner and manager's program to use innovative 3D, 4D, and BIM technologies to complement, leverage, and improve existing technologies to achieve major quality and productivity improvements
www.gsa.gov/bim

Source: AIA National and AIA California Council, *Integrated Project Delivery: A Guide*

Some Technologies on the Horizon for Civil Engineering Projects

1. Transdisciplinary, Transinstitutional, and Transnational: Eliminating the artificial boundaries among disciplines and knowledge domains, institutions (public and private), and nations, in the pursuit of solutions

2. Ubiquitous Computing: Making many computers available to a user throughout the physical environment, while making them effectively invisible to the user, enabling the user to remotely interact with people and the natural, built, and virtual environments; monitor, collect, and access data, information, knowledge, experience, and wisdom; and remotely control devices.

3. Ubiquitous Positioning Technologies: Enabling the location of people, objects, or both, anytime, whether they are indoors or outdoors or moving between the two, at predefined location accuracies, with the support of one or more location-sensing devices and associated infrastructure.

4. Cloud Computing: A style of computing in which capabilities related to Information Technologies (IT) are provided to users "as a service" allowing them to access technology-enabled services from the Internet ("in the cloud") without requiring knowledge of, expertise with, or control over the technology infrastructure that supports the services.

5. Augmented Reality: A term for a live direct or indirect view of a physical real-world environment whose elements are merged with, or augmented by virtual computer-generated imagery, creating a mixed reality.

(Continued)

6. Collective Intelligence: A shared or group intelligence that emerges through collaboration, innovation, and competition, from the capacity of human communities to evolve toward higher-order complexity and integration, which (1) appears in a wide variety of forms of consensus decision-making in bacteria, animals, humans, and computer networks; and (2) is studied as a subfield of sociology, of business, of computer science, of mass communications, and of mass behavior—from the level of quarks to the level of bacterial, plant, animal, and human societies.

7. Automation and Robotics: The application of science, engineering, and technology (particularly electronics, mechanics, control systems, computer-aided technologies, hardware and software, and artificial intelligence), in the design, manufacture, and application of autonomous devices and robots for industrial, consumer, or entertainment use, which reduce the need for human sensory and mental requirements, and which perform tasks that are too dirty, dangerous, repetitive, or dull for humans.

8. Nano-Bio-Info-Cogno Convergence: The synergistic combination of four major provinces of science and technology, each of which is currently progressing at a rapid rate: (1) nanoscience and nanotechnology; (2) biotechnology and biomedicine, including genetic engineering; (3) information technology, including advanced computing and communications; and (4) cognitive science, including cognitive neuroscience.

17.8   ENGINEERING THINKING

With so many changes unfolding simultaneously, how can civil engineers be prepared to answer the questions posed in the introduction to this chapter?

- How can this technology create value for our clients/customers?
- What problems can this technology solve?
- What processes can this technology improve?
- How have some organizations enhanced their effectiveness/profitability with this technology?

In truth, ancient, Renaissance, 19th-century engineers, basically all engineers who have preceded today's civil engineers, have pushed the limits of the technologies known to them. One way to thrive in an environment that involves continuous and rapid change is to develop *engineering thinking*.

Engineering Thinking

A desirable attribute of a professional engineer is to be a clear-thinking, innovative problem solver. The following section explores how such competence may be developed.

1 Knowledge

For the purposes of the discussion which follows, the following definitions are inferred:

Knowledge— that which is contained in the brain. What a person knows.
Information—a representation of knowledge outside the brain in the form of text, speech, graphics, mathematical models, and so forth.

Two types of knowledge are:

Explicit knowledge—can be represented as information
Tacit knowledge—cannot or has not yet been represented as information

1.1 Features of Tacit Knowledge

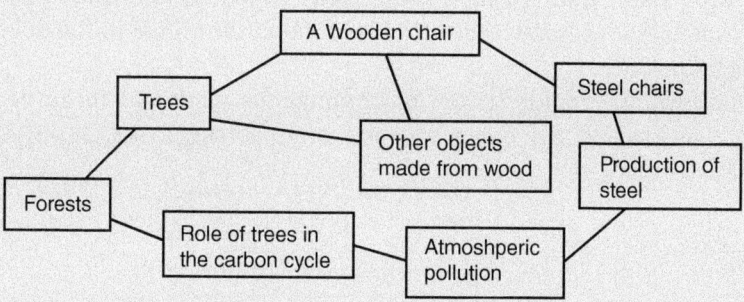

FIGURE 1 Example of knowledge associations.

Since tacit knowledge, by definition, cannot be identified directly, it can only be inferred from outcomes which required its use. Features of tacit knowledge include:

Associativity—Items of knowledge can be deeply interconnected in the brain. For example, consider a simple object such as a specific wooden chair. The accompanying organizational chart shows a small set of the issues and entities that can be

(Continued)

linked to a wooden chair. This does not form a simple hierarchy for which there are limits to the directions of the interconnections but is a distributed network for which there can be links between any node. While we can present as information small sets of such knowledge, the totality of the associations among items of knowledge in the brain is very, very large. Computers, as yet, do not come near to simulating the structure of such interconnectivity in the human brain. Therefore, the associations among the items of knowledge are mainly tacit. Such associativity is a major feature of the power of the human brain.

Intuition—We often know things without knowing why we know them. The brain is a phenomenally complex engine which can process knowledge subconsciously. Some believe that we do not take adequate advantage of our intuition (Gigerenzer, 2004).

Judgment—An important feature in the use of nondeterminate processes (Section 2, Process Thinking within this text box) is that decisions can seldom be based on logic alone. Use of the word "judgment" tends to relate to such contexts. Computers, probably because of their low level of associativity among items of information, cannot match the power of the brain in making judgments.

Understanding—This may be defined as the structuring of knowledge in the brain such that it can be used. It depends on associativity and, as discussed in what follows, is improved by working the brain.

Some people assert that the tacit component is greater and more important than the explicit component of knowledge. This is an important issue in the development of engineering competence.

Figure 2 shows an engine model of competence. One can think of knowledge as the "fuel" needed to drive professional engineering tasks. Input information

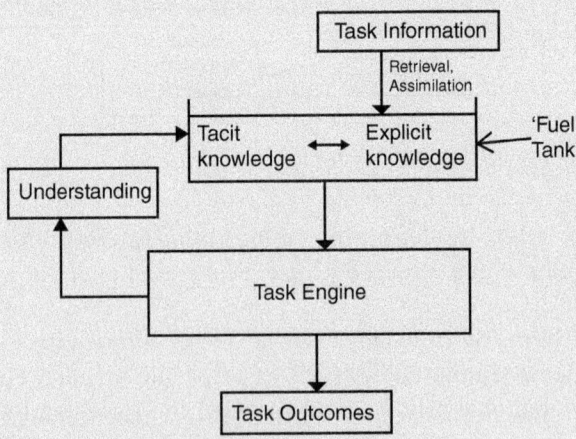

FIGURE 2   An engine model for information tasks.

becomes explicit knowledge in the fuel tank. In the task process, explicit and tacit knowledge are mixed to achieve task outcomes. The processes of the task engine have another important outcome: They develop understanding which feeds back into the tank as new tacit knowledge. Therefore, the mental task engines, unlike a combustion engine, enhance the quality of the fuel rather than consume it. The quality of the combination of explicit and tacit knowledge in the brain is fundamental to competence. Education tends to focus on explicit knowledge—one cannot transmit tacit knowledge. It develops by thinking.

It is clear that one cannot have too much explicit knowledge but its value is significantly less if it has not been used in tasks so as to develop corresponding tacit components. Therefore, to be a good engineering thinker one has to get the brain working hard.

2   Process Thinking

A determinate process has a unique outcome whereas a nondeterminate process may have more than one valid outcome. For example, when using a mathematical model the decision about which model to use will be nondeterminate but the calculation process will be determinate. With determinate processes the difference between the outcome and the correct result is error whereas with nondeterminate processes, acceptance of outcomes is in the realm of uncertainty. Most professional engineering processes are nondeterminate, often with determinate subprocesses.

All processes can be viewed as having three basic components:

- Inception—the requirements are established and information is gathered.
- Conception—the process is defined.
- Production—the process is implemented.

These components need to be controlled by asking relevant questions such as:

Inception: Are the requirements complete and clear? Is the input. information adequate? (The assessment question)

Conception: Is the process capable of satisfying the requirements? Is the process the best in the context? (The validation question)

Production: Has the process been correctly implemented? (The verification question)

Table 1 shows review/control activities relevant to the three process components.

(Continued)

TABLE 1   Basic Process Model.

Stage	Activity	Review/Control
Inception	Define the requirements, acquire information, investigate	Assess requirements, assess other input information
Conception	Identify options, evaluate, decide	Validate (ensure that the process can satisfy the requirements), optimize (seek to identify the best process)
Production	Implement	Verify (ensure that the process has been correctly implemented), interpret outcomes, revalidate

While successful engineers might not be explicit in expressing that they use the control strategies listed in Table 1, simple logic shows that they must do this. Under what circumstances can you achieve good outcomes if you do not have a **clear idea of your objectives (requirements assessment) or if your process is not capable of satisfying the requirements (validation) or if the process has not** been correctly implemented (verification)? The answer to this question is: "Only **by luck!**"

2.1   Process Model for Design

The process model of Table 1 applied to engineering design is shown in Table 2.

TABLE 2   Basic Process Model for Engineering Design.

Stage	Activity	Review/Control
Inception	Define the requirements, acquire information, investigate the context, equip (in terms of staff competence, software, hardware, etc.)	Assess requirements, assess input information, assess equipment
Conception	Identify design options, evaluate, decide on the design solution	Validate the options, optimize the solution
Production	Technical design, produce drawings and specifications	Verify the outcomes against the requirements

2.2   Process Model for Analysis Modelling

The process model of Table 1 applied to analysis modelling (i.e., use of mathematical models to predict behavior of engineering entities) is shown in Table 3.

TABLE 3   Basic Process Model for Analysis Modeling.

Stage	Activity	Review/Control
Inception	Define the requirements, equip	Assess requirements
Conception	Establish the model	Validate the model (ensure that it can satisfy the requirements), optimize the model, validate, and verify the software
Production	Prepare data and carry out calculations	Verify the results (ensure that the model has been correctly implemented); interpret the results to identify behavior of the system being modeled. Carry out sensitivity analysis to gain understanding of behavior of the system

2.3   How Should the Process Model Be Used?

It is important to be constantly asking control questions and challenging outcomes. This is a main component of good engineering thinking.

The process model is used recursively, that is, it is used for the overall context and for detailed issues. For example, it is used for an overall design and for a detailed part of the system being designed. It can even be used on itself. For example, if you have to produce requirements, it may be worthwhile to establish requirements and a process for doing that.

2.4   Example of Challenge to Outcomes

This is an example from structural analysis. Figure 3 shows the deflected shape from a plane frame analysis model of a bridge truss. It is supported vertically at nodes 1 and 4 and has a single vertical point load at the central node 11. Even people who are not engineers are likely to suggest that there is something wrong with this shape. They would expect the deflection to be in the form of a smooth curve rather than being more of a "V" shape, as in the diagram.

If you make such an observation and go on to rationalize the situation you can be in a win-win situation. If there is an error then you can identify the reason and you have discovered something very important by making the challenge. If the challenge is unfounded and you explain why this is so, then you learn about the behavior of the system. It is the latter situation that pertains in relation to Figure 3. (Both bending and shear mode deformation contribute to the deflection of the truss. In the case of the truss in Figure 3 the shear mode component, which gives linear displacement with constant shear force, dominates the behavior.)

(Continued)

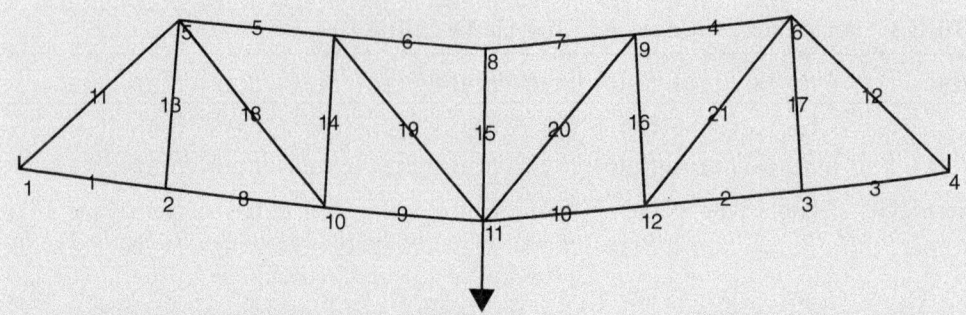

FIGURE 3   Deflected shape of a bridge truss.

3   Thinking for Innovation

One can define **creativity** as the production of new ideas/concepts whereas **innovation** is the development of creative ideas into useful outcomes. (MacLeod et al., 1996)

3.1   Producing Creative Ideas—Free Thinking

Albert Einstein said that "combinatory play seems to be the essential feature in productive thought" (Einstein 1954). That is, new ideas tend to result from the bringing together of concepts that were previously unrelated.

The combining of ideas may be affected by the following factors (MacLeod et al. 1998):

- **Innate Ability**—This is the ability that is not dependent on practice and received knowledge.
- **Knowledge**—Two types of knowledge are important. First, knowledge within the domain being considered is essential. Second, general knowledge is important because it may be advantageous to combine ideas from diverse areas. Being a specialist and a generalist is therefore useful for innovative engineers.
- **Knowledge Distance**—At high levels of creativity one is making connections between items of knowledge which were previously unconnected and the further these ideas are "apart" then the more difficult it will be to combine them. The "distance" between items is a measure of how close they appeared to be before any combining action takes place.
- **Practice**—The degree to which the person has practiced making combinations is likely to be an important factor for success in creative work.
- **Effort**—The amount and intensity of "effort" spent on the situation for which combination is required may be of significant importance in creative situations. Good creative ideas may need very intense mental effort.

There are techniques for developing ideas in groups such a brainstorming. When developing ideas with others it is a good strategy to arrange that one does not feel bad about being wrong. It can be worthwhile to agree that a session is for "free thinking" where ideas can pour out without being necessarily well thought out. Sometimes ideas that seem crazy at first turn out to have substance.

3.2 Making Creative Ideas Work—Focused Thinking

Producing creative ideas may be the easiest part of the innovative process. Converting them into useful outcomes can be the more difficult task. Since risk when innovating is normally increased, one has to pay closer attention to the process control strategies discussed in Section 2. Challenging outcomes is especially important. "Focused thinking" is needed where it is important not to be wrong.

3.3 Subconscious Thinking

A strategy used by people in innovative situations is to hold back from making decisions for as long as is practical. The subconscious mind can shape ideas. To get the subconscious to work requires hard conscious thinking followed by incubation periods where you move your thinking elsewhere—then come back to the problem. In the conscious thinking there should be a focus back to the requirements. Decision time looms up. It is important to leave enough time for implementation but not to rush to an early decision.

3.4 Knowing When to Innovate

It is important to know when innovation is needed. If you are innovating when there are standard ways of achieving a better result, then you may be considered to be incompetent.

4 Conclusion

We know that ability is a combination of what we were born with and lifetime experience. While nothing can be done to change the former attribute, the structuring of the latter is of fundamental importance. We are strongly influenced by principles that evolved in the culture in which we are raised. Some of these principles may not stand up to rational analysis but people still cling to them. The engineering approach is to analyze the way ideas are approached and to cut away the components of thinking which can be shown to be based on false logic, or no logic.

The way that we think, and, hence, behave also is deeply dependent on our interaction with others—our parents, siblings, friends, colleagues, managers. A very useful strategy is to seek to identify those whose mode of thinking is good, to identify

(Continued)

principles that contribute to this competence, and to try to use these principles. Such principles can come from all walks of life. For example, the Duke of Wellington, one of the most successful battle commanders of all time, always did his own reconnaissance of a battleground rather than rely on reports from his subordinates. A good engineer also must satisfy herself or himself of the reliability of information that has been given.

Muscles need to be worked on to develop strength especially when one is young. It seems likely that the human brain has the same attribute. It needs to be worked hard in order to achieve optimum performance. The engine model of Figure 2 reflects this principle as does the discussion in Section 3.1.

Finally, engineering thinking involves ethical thinking. A main feature of a successful society is that resources are shared by the populace to an acceptable level. To achieve this, the society must be, in general, free from corruption. The behavior of the professionals in general, and of professional engineers in particular, has a very important role in setting the ethical standards for a country. For example, giving or taking a bribe is totally negative to good professional behavior. The work of professional engineers often affects the safety of the public. This issue should be at the forefront of their thoughts and actions.

Reference

MacLeod, I.A., B. Kumar, and J. McCullough. (1998). "Innovative Design in the Construction Industry." *Proc Inst of Civil Engrs*, Vol 126, No1, February, pp. 31–36.

—Iain A. McLeod, Ph.D., Chartered Engineer Professor Emeritus, Department of Civil Engineering, Strathclyde University, Glasgow, Scotland

"It is said that the present is pregnant with the future."

—Voltaire

"It is not the strongest of the species that survives, nor the most intelligent, but the one most responsive to change."

—Charles Darwin

"If you don't like change, you're going to like irrelevance even less."

—General Eric Shinseki, Chief of Staff, US Army

17.9   SUMMARY

Recent developments in emerging information technologies will have a major impact on the way civil engineers work. As always, critical thinking—*engineering thinking*—is a key component in developing a successful and productive career in civil engineering.

Today civil engineers are exposed to a heady array of new technologies. With such variety, decisions about which technology to adopt can be baffling. With so many innovations to consider, civil engineers should review the new tech and ask themselves, "So What"? How can this new tech help my clients, help me, or help society?" "Next, ask, where does this new tech stand in the development stage? Has this new tech been demonstrated, applied, accepted/approved, permitted, safe or safer to use, or cost-effective?"

Specifically, Civil engineers should ask:

- How can this technology create value for our clients/customers?
- What problems can this technology solve?
- What processes can this technology improve?
- How have some organizations enhanced their effectiveness/profitability with this technology?
- How and why are emerging technology applications related to globalization and sustainability?

These considerations are not meant to hamper implementation of new technology but simply to aid in evaluating the total picture.

REFERENCES

AIA National and AIA California Council. (2007). *Integrated Project Delivery: A Guide*. American Institute of Architects, Washington, D.C.

Eastman, Chuck, Teicholz, Paul, Sacks, Rafael, and Liston, Kathleen. (2008). *BIM Handbook: A Guide to Building Information Modeling for Owners, Managers, De- Signers, Engineers, and Contractors*. John Wiley & Sons, Hoboken, New Jersey. IBSN: 978-0-470-18528-5.

Gann, David M. and Hansen, Karen L. (1996). *IT Decision Support and Business Pro Cess Change in the U.S.* Final Report to the U.K, Department of Trade and Indus- try (DTI) for Overseas Science and Technology Expert Mission.

http://fiatech.org/capital-projects-technology-roadmap.htm (accessed December 15, 2009).

McGraw-Hill Construction. (2008). *Smart Market Report: Building Information Modeling (Bim)—transforming Design and Construction to Achieve Greater Industry Productivity*. The McGraw-Hill Companies, New York. ISBN: 978-1- 934-92625-3.

Reid, Robert L. (March 29, 2021). Would you believe what wood can achieve? ASCE, Civil Engineering Source Magazine.

Vanegas, Jorge A. (2009). *Is the capital projects industry observant? is it prepared?* Invited Speaker within the Breakout Forum on a Futurist View: What's on the Horizon? at the 41st Annual ECC Conference: The Perfect Storm: Navigating through the Turbulence of Risk and Change, Engineering & Construction Contracting Association (ECC), Bastrop, Texas. September 2009.

Vanegas, Jorge A. (2023). *Compiled List of Emerging Disruptive Technologies.* College of Architecture, Texas A&M University. College Station, TX.

Zenobia, Kent, Wheeldon, David, and Nixon, Rob. (2020): *Aircraft system surveys in levee repairs.* Presentation and Publication, Floodplain Management Virtual Conference, California, USA. September 2020.

Human Relations Policies and Employment Practices

Big Idea

"Always treat your employees exactly as you want them to treat your best customers."

— Stephen R. Covey

"Clients do not come first. Employees come first. If you take care of your employees, they will take care of the clients."

— Richard Branson

Key Topics Covered

- Rules of Engagement
- Compliance with Employment Laws and Typical Human Resources Policies
- Forecasting
- Hiring
- Training

Civil Engineer's Handbook of Professional Practice, Second Edition. Karen Lee Hansen and Kent E. Zenobia.
© 2025 John Wiley & Sons, Inc. Published 2025 by John Wiley & Sons, Inc.
Companion website: www.wiley.com/go/hansen/CivilEngineersHandbook

- Compensation and Benefits Plans
- Health and Safety
- Employee Retention

Related Chapters in This Book

- Chapters 1 through 21 and Appendices A, B, C, D, E, F,G

18.1   INTRODUCTION

This chapter provides Civil Engineers (CEs) with important information on typical governmental or private engineering organizations' human resources policies and practices. Human resources involve the hiring (recruiting/vetting/selecting), administration (paying, promoting, firing), and onboarding/training of personnel. An older term for "Human Resources" is "Personnel." In some sectors, the term "Human Resources" has been replaced with "People Operations," "Employee Experience," "Employee Success," and "Partner Resources," for example.

Awareness of organizations' standard policies and practices is essential for civil engineers to perform optimally, professionally, and legally in their careers. This chapter introduces an array of topics related to human relations and employment laws and is not intended to be an "Employee Handbook." CEs within any organization should be aware of these general HR policies and employment laws to guide them through their career effectively. At a minimum, every civil engineer should be aware of:

- Rules of Engagement
- Compliance with Employment Laws and Typical Human Resources Policies
- Forecasting
- Hiring
- Onboarding and Training
- Compensation and Benefits Plans
- Worker Health and Safety
- Employee Retention

Employees must follow their company or agency's human resource (HR) policies and employment practices; failure to do so could result in disciplinary action, termination, and/or civil legal action (California Department of Human Resources, Human Resources Manual, 2023).

Employers need employees, and vice versa, to succeed and achieve their primary mission goals. Usually, a company or public agency first will welcome a new employee into the organization. This "onboarding" involves integrating a new employee into the organization and familiarizing the new employee with the organization's clients, constituents, and services. The organization provides the new employee with a welcome statement, overview, mission statement, and duty statement and then introduces key managers, supervisors, and colleagues. Early on, the HR Supervisor likely will direct the new employee to the "Employee Handbook of Policies and Procedures." (More detailed information and legal requirements for the CE follow.)

One typical question the CE has in their initial interviews or later with the HR staff is a typical pathway for their career at the company or organization. Please see Figure 18.1 titled, "Engineering Career Pathways and Typical Relative Compensation Packages," for an example career pathway.

KEY:
A: Engineering Specialist, Lead Designer, Senior Engineer
B: Engineering Management, Chief Engineer, Department Manager
C: Executive Level, Chief Executive, President, Financial Officer
D: Engineering Marketing, Business Development, Client Managers
E: Program Manager, Senior Project Manager

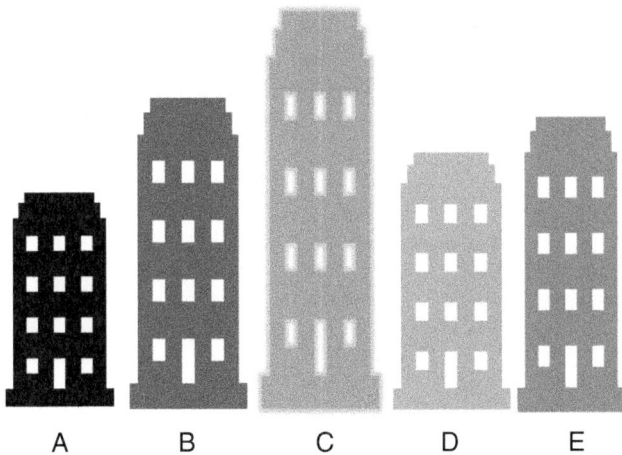

A   B   C   D   E

FIGURE 18.1   Engineering career pathways and typical relative compensation packages. These Civil Engineering pathways depict "relative compensation" packages. An individual can excel well beyond these illustrations. For example, a CE on the "A" pathway working as an "expert witness" may have a compensation package greater than the "C" pathway as an executive.

18.2   RULES OF ENGAGEMENT

Once the new CE employee dives into the "Employee Handbook of Policies and Procedures," they typically will find an overview statement about the policy of working together cooperatively in a *team* atmosphere to achieve the organization's mission. Note that the policies summarized below *illustrate the typical overall content and intent.* These typical policies are provided as an example for discussion to illuminate an organization's expectations from their employees. Actual company and agency policies are usually much more involved and detailed. These statements and policies provide valuable insight into the company's or agency's values and furnish the CE with typical policies and legal requirements about behavior in the workplace and in public. Significantly, employee rules for conduct in the office also pertain to conduct in the field, at other business offices, and in public. Please see Figure 18.2 titled Typical corporate human relations (HR) manual versus the actual human relations.

Employers expect employee integrity, and an employer policy statement might read:

- ***Integrity***: Our core business, personnel, and client relationships are conducted in accordance with ethical and honorable beliefs. We respect human rights and treat all people fairly. We are an ethical business entity and base our actions on honesty and equality for all employees.

In addition, most Organizations will offer recognition to their place or role in the community. Naturally, the organization will likely employ staff from the local community. Likewise, the community usually has some say about the overall operations of the organization possibly through the town council or local government representatives. Being a respected member of the community makes good business sense. A typical statement about "community interaction" might be:

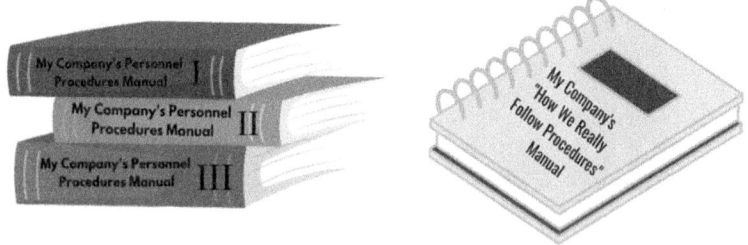

Hopefully, civil engineers will find that the corporate HR Manual is the same as the HR field manual!

FIGURE 18.2   Typical corporate human relations (HR) manual versus the actual humen relations.

- **Community Commitment**: As a responsible corporate member of this community, we strive to provide a positive economic and social influence in our community. We encourage employee interaction with the community and provide employees some paid volunteer time to foster active participation with the local schools, recreational sports teams, food banks, and senior programs.

    NOTE: Many Fortune 500 Companies, including those in the engineering consulting, information technology (IT), construction, financial and banking, chemical and industrial processing sectors employ this type of community commitment and even offer internships, scholarships, and sponsor their employee's time for active volunteering, such as team coaching.

Another area a company or public agency typically will include in the early phase of "new employee introductions" is a type of personal commitment to the individual. This policy could be one that is mutually beneficial to the individual and the organization. It might read:

- **Innovation**: Innovation is the creativity that allows our organization to excel beyond our competitors. Innovation is encouraged and recognized by management and colleagues. Innovation includes "thinking outside of the box," as well as raising diverse ideas and opinions for discussion and possible implementation. Innovation is coupled with possibility thinking.
  or
- **Continuous Improvement**: Continuous improvement is encouraged. It allows our organization to excel beyond our competitors. Continuous Improvement is encouraged and recognized by management and colleagues. It encourages us to become better, leaner, more effective, and efficient. Continuous improvement means we invest our time and diligence today to make a better tomorrow.

Employee rules for conduct in the office also pertain to conduct in the field, at other business offices, and in public (Dawson, 2019)

18.3 COMPLIANCE WITH EMPLOYMENT LAWS AND TYPICAL HUMAN RESOURCES POLICIES

Following is a discussion of the laws governing US employment practices, as well as an examination of typical organizational human resources policies.

18.3.1   Equal Employment Opportunity

In addition to organizational policies and rules of conduct, US (federal) employment laws apply to all organizations operating in the United States. Specific state laws may vary from one state to another, but they apply in addition to the federal laws. An employer policy statement might read:

- ***Equal Employment Opportunity***: Our organization interviews and employs qualified individuals for our open positions and we provide the opportunity for advancement. We do not discriminate because of race, color, religion, sex or gender, age, national origin, ancestry, marital status, pregnancy, disability, sexual orientation, military service, veteran status, or other grounds protected under applicable state and federal laws regulation, and/or executive orders (Dawson, 2019; CA Department of Human Resources, 2023)

    Note the "discrimination categories" listed above (race, color, religion . . .) are referred to as "protected status categories." Some States may have additional protected categories.
- ***Harassment***: Any form of harassment of any individual because of protected status noted above is a very serious offense and is strictly prohibited (Dawson, 2019). Harassment is illegal, discriminatory conduct due to that person's legally-protected characteristics. Sexual harassment is a form of sex discrimination, including sex stereotyping or discrimination. Sexual harassment involves unwelcome conduct of a sexual nature. Sex stereotyping involves preconceived ideas about how a person of a specific gender should behave.
- ***Retaliation***: Any form of retaliation against an employee who reports discrimination or harassment to the Company in accordance with this policy, files a charge of discrimination or harassment, or who cooperates with the investigation of a charge, is prohibited (CA Department of Human Resources, 2023)

Any incident of discrimination or harassment must be reported immediately, in confidence, to the Human Resources Manager or any other supervisor. Every effort is made to investigate promptly all allegations of discrimination and/or harassment in as confidential a manner as possible, and to take the appropriate corrective action. Any employee who is determined, after an investigation, to have engaged in discrimination and/or harassment in violation of this policy will be subject to disciplinary action, up to and including discharge (CA Department of Human Resources, 2023).

Affected employees may report any harassment incidents to any organization or agency manager or supervisor. No manager, supervisor, or employee may retaliate in any manner against any applicant/employee because they report discrimination or harassment in accordance with the organization's policy, file a discrimination charge, or cooperate or provide testimony during an investigation.

So, in some States (and likely more States as time passes) an Organization may be liable for harassment even if the Senior management had no knowledge about the behavior, intentions, was unaware, or even tried to prevent the misconduct. This is a powerful incentive for organizations to provide training, monitoring, and constant reminders about these employment laws and staff interactions. Many organizations have even announced a "zero tolerance" to their employees meaning, the perpetrator and immediate Manager can be held personally liable for monetary damages for their misconduct. These are powerful laws with very serious consequences and can impact ones' career, family, and finances.

History of Sexual Harassment Laws

As early as the 1920s, some companies provided women with literature on how to ward off "lecherous" unwanted advances by coworkers. For years thereafter, federal law did not prohibit any form of harassment.

The 1960s saw change regarding sexual harassment. In 1964, the US Congress passed Title VII of the Civil Rights Act. Title VII prohibited sex discrimination in the workplace under federal law. At the time, this meant employers could not give men preferential treatment simply because of their gender.

Then in the late 1960s and into the 1990s, California broadened prohibitions on workplace discrimination and harassment. The changes were brought about through case law, legislative updates, and state constitutional interpretation. In the early 1970s, California law's definition of sexual harassment was expanded. California continued broadening its laws in the late 1970s and into the 1990s.

In the late 1990s, the US Supreme Court ruled that a company must take steps to prevent sexual harassment in the workplace in two prominent cases. The Court also ruled that companies can be liable for incidents of workplace sexual harassment. This ruling was significant and garnered attention with senior management and chief executive officers (CEOs).

In the 2000s to the 2020s, as awareness of sexual harassment increases, more and more states enact laws clarifying the definition of sexual harassment and raising requirements on organizations to prevent this type of misconduct. In 2007, California became the third state to require certain employees to take training on sexual harassment. In 2018, California expanded its training requirement to require all employees (not just managers) to receive training on the prevention of harassment (Time Magazine, Cohen, Sascha, Apr 2016).

Liability for Sexual Harassment

Under California Law, and some other States, an organization is liable for *quid pro quo* harassment by its managers. *Quid pro quo* meaning "a trade" or "this for that." This liability extends to the Employer whether or not the organization knew about the behavior, intended for it to occur, or knew nothing about, and even tried to prevent the

(Continued)

misconduct. Managers can be held personally liable for monetary damages for their conduct. Therefore, Employers have an added incentive to prevent this harassment.

Quid pro quo harassment, a frequent type of sexual harassment, occurs when job benefits are offered for sexual favors. (CA Code of Regulations, Title 2 Administration, 2023).

Another type of sexual harassment, and possibly the more common form, is *hostile environment* sexual harassment. This harassment occurs when someone's unwelcome and offensive conduct of a sexual nature interferes with an employee's performance or work environment.

Both types of harassment are very serious; organizations and employees guilty of this type of harassment face significant liability (California Code of Regulations, Title 2 Administration, 2023).

What Is Harassment?

Harassment generally involves illegal, discriminatory conduct directed at someone due to that person's legally-protected characteristics, such as sex, age, race, or others. Harassment is illegal under US federal and State laws. It is hurtful both to the person(s) being harassed and to the coworkers who witness or learn of it.

Harassment includes offensive behavior that can be motivated by stereotypes about a protected characteristic or simply that makes the targeted person uncomfortable. The targeted person may be a member of a protected group or simply associated with such a group. For instance, a man can be offended by harassment directed at women.

Harassment is disrespectful to all of an organization's employees, undermines a fair and non-discriminatory workplace, is bad business practice, and demeans a culture of mutual respect. In summary, it is in everyone's best interest to prevent, identify, and stop sexual harassment in the workplace.

What Is Sexual Harassment?

Sexual harassment is a particular type of discrimination. Sex discrimination in general includes:

- Any type of bias on the basis of a person's sex
- Sex stereotyping
- Discrimination on the basis of sexual orientation or gender identity
- Discrimination on the basis of pregnancy

What is Unwelcome Conduct?

- Sexually harassing behavior is unwelcome conduct of a sexual nature meaning it is uninvited, uninitiated, and/or unwanted.
- Even conduct that is tolerated, or even participated in (like laughing at jokes), may still be considered unwelcome.

What is Conduct of a Sexual Nature?

- Conduct of a sexual nature includes sexual advances, requests for sexual favors, and other sexually-oriented verbal or physical conduct. This can include real or electronic postings such as pictures or calendars photos.

What is Conduct of a Sex Stereotyping?

- Sex stereotyping includes preconceived ideas about how a person of a specific gender might behave.

A manager's role, under law, is to act as an extension of the organization. The organization is liable for a manager's actions. Managers are held to a higher standard of behavior since they are in the position of authority for the Organization. Their actions must not be abused and they are expected to act as leaders and display professional behavior and be aware of questionable behaviors around them both on and off the job.

Managers have a "reporting requirement." They are required to report harassment behaviors they observe or behaviors reported to them to senior HR management. Managers are required to report these harassment behaviors even if staff does not object. Even trivial harassment behaviors should be reported as they may be part of a pattern reported by others. The Manager is required to report these behaviors even if the impacted party asks that it be kept confidential or not be reported!

Managers will/should be disciplined for allowing harassment to continue, failing to report harassment incidents, and/or engaging in retaliation on the impacted person(s) who report the information (CA Department of Human Resources, 2023).

So, what can a CE do to avoid these situations and maintain a positive working environment and career path?

1. **Examine**: First examine your own behavior and do not engage in conduct that could be perceived to be humiliating, threatening, intimidating, or generally unprofessional. Ask yourself, would I act this way if my supervisor, spouse, partner, mother, or father were here? This is, in effect, filtering your content before expressing it.

2. **Confront**: Confront an abusive person in a calm, non-threatening, professional, and objective manner. For example, one might say something like:
   a. "Please don't put anyone down in front of peers. It is bully-like and disrespectful."
   b. "Please lower your voice. Shouting is unproductive and demeaning."
   c. "It is demeaning to say words like that to someone."
   d. "Your words can hurt people. I suggest you apologize."

3. **Deflect**: Deflect a verbal attack by simply excusing yourself with a sample response like these listed below and then leaving the area:
   a. "I'm on my way to the restroom. I'll talk to you later."
   b. "I'm not sure I agree with you." Then leave the area.
   c. "I'll see you later. I have a call to make."

4. **Document**: Document these incidents and include details, such as: people in attendance, witnesses, time, date, and a summary of the event(s). Continue documenting and determine if there is a pattern.

5. **Seek Help**: Seek help if you witness or unfortunately experience abusive conduct in the workplace or in public from peers at your workplace. If you are uncomfortable confronting the abuser or if your efforts fail, you could:
   a. Seek help from a Company Manager, HR representative, or other resource.
   b. Document meetings or discussions with those you seek help from because retaliation is also unlawful. (CA Code of Regulations, Title 2 Administration, 2023).

18.3.2 Reasonable Accommodations

Typically, a company or agency will provide reasonable accommodations to qualified individuals with disabilities. Reasonable accommodations are provided so the employee may perform essential job functions and participate and/or enjoy any employment practice, condition or benefit of employment. Human Resources can provide more details regarding reasonable accommodation for a disability.

Ethical Business Policies Apply Beyond the Workplace

Many organizations hold managers and supervisors responsible for upholding the "policies and laws" noted in this chapter *both on and off the job*. This means if a manager were to go to lunch (off company or agency property) and saw or heard a fellow employee violating the "Equal Opportunity, Harassment, or Retaliation" policies noted above, this manager should step in, put a stop to it, and report it to the senior HR manager immediately.

This responsibility includes after hours, such as meals in restaurants, company or agency sponsored events, and weekend team building activities. Some companies or agencies believe that if a manager is aware of or should have been aware that organizational policy violations or possible illegal civil rights violations have occurred, the manager should act to stop these violations. If the manager fails to act to stop this behavior immediately, the Manager may be considered complicit in the act and may also be subject to punishment or adverse action.

18.3.3 Business Ethic Policy

The public confidence and reputation of a company or agency is based upon fair business practices and ethical conduct of management and employees. The company or agency's reputation for honesty and integrity is based on employees following all applicable Federal and State laws and the company or agency's most recent employee handbook.

Whistleblowers' Protection Act

If you have any concerns about the business ethics of a company or agency and its employees, you are encouraged to bring these concerns to management's attention without fear of retaliation. In compliance with the Whistleblowers' Protection Act, an organization should protect employees from discharge, threats, or discrimination because an employee or a person acting on behalf of an employee: (1) reports a violation, or (2) participates in an investigation or hearing. Compliance with this policy of business ethics and conduct is the responsibility of every company or agency employee.

18.3.4 Drug-Free Workplace

Companies and agencies must provide a safe and productive environment for their employees, sub-contractors, visitors, and customers. Most organizations implement a Drug and Alcohol Program that requires strict adherence to the policy, and this policy is enforced. The Drug and Alcohol Program may also apply to the use of recreational marijuana, even in States where marijuana possession and use is legal. For example, employees cannot be "under the influence" at work. The substance abuse policy is intended to support existing safety programs and to avoid alcohol and illegal drug-related work performance problems by striving for an alcohol and drug-free working environment.

18.3.5   Medical Examination/Drug Screening

Some companies or agencies may offer employment contingent upon a prospective employee's satisfactory completion of a medical examination and/or drug test by a health professional of the organization's choice. The company or agency likely will pay the costs associated with the medical examination and drug tests. In addition, employment may include periodic medical examination and drug testing to verify an employee's mental and physical ability to perform the essential job functions. Information gathered from medical and drug screening is required to be kept confidential and separate from other employee information. Access to this information should be limited to those who have a legitimate need to know.

18.3.6   Honesty

Most companies or agencies include a clause about honesty. Typical honesty clauses state that: "Honesty, trustworthiness, and integrity are absolute requirements." Organizations may include the additional requirement that honesty and trustworthiness include both "on and off" the job activities. The policy may state that the company or agency will prosecute anyone, whether customer or employee, who steals money, merchandise, or other property or who falsifies time records. Failure to cooperate fully regarding investigations concerning dishonesty is extremely serious and may result in disciplinary action, up to and including discharge.

18.3.7   Timekeeping

In accordance with company or agency policies, all personnel must record hours worked in the official timekeeping program. Employees' hours should be recorded in this timekeeping program, and employees are responsible for the record's accuracy. Typically, employees are required to record their time at least daily and are required to submit timesheets to supervisor/s for approval weekly. Employees must record only their own time and may not record time for another employee. Any changes must be made and approved by the supervisor.

18.3.8   Federal and State Employment Regulatory Review Office

Major companies or agencies may have a different names and acronyms for this office, but the duties are typically the same. The duty of the Federal and State Employment Regulatory Review Office is to monitor and record the compliance, implementation, and coordination of the company's or agency's Affirmative Action and Equal Employment Opportunity Programs (AA and EEO), and to verify compliance with state and federal laws requiring equal employment opportunities for all current and prospective employees.

18.3.9   Safety Policy

This policy usually states that the company or agency will comply with all applicable federal, state, and local governmental safety laws and regulations. The work tasks performed by the staff will comply with industry safe work practices to maintain a safe work environment, a high level of efficiency, and acceptable safety record among employees.

Usually, the company or agency policy also states that all employees, sub-contractors, and others have a right to a safe work environment and that incidents are preventable. Safety is a core value for the company or agency, and the company or agency is committed to a safe work environment for employees, contractors, and visitors to prevent occupational injuries and illnesses. The Authors want to stress the importance of a Safety Policy and have included a comprehensive chapter on this subject. Please see Chapter 20 for this information.

18.3.10   Prevention of Workplace Violence

The organization will provide a safe and secure work environment for its employees and visitors. Threats or acts of violence in the workplace are strictly prohibited. Threats of violence include statements, intimidation, bullying, coercion, or any act threatening violence. A verbal threat is a violation of the policy. Possession of firearms and/or dangerous weapons in the workplace is a serious offense.

18.3.11   Disciplinary/Corrective Action

The purpose of this policy is to state the organization's policies toward administering equitable and consistent discipline for unsatisfactory conduct in the workplace.

Employment with the company or agency is based on mutual consent. Both the employee and the organization have the right to terminate employment at will, with or without reason or notice. The company or agency may use progressive discipline at its discretion.

Disciplinary/corrective action may be applied in a progressive four-step manner depending upon the severity of the misconduct, whether there have been previous verbal warnings or written warnings, combinations of employment violations, or the number of occurrences. There may be circumstances when one or more steps are bypassed due to the severity of the violation(s). Typically, "progressive discipline" may include the following:

1. A verbal warning (first offense)
2. A written warning (second offense)
3. A suspension with or without pay (third offense)
4. Termination of employment (still another offense)

The company or agency subscribes to progressive discipline policy to allow employees to correct their infractions, learn from their mistakes, and still be a productive member of the company.

The above typical employment policies and procedures have been presented to set expectations and familiarize the new CE employee with behavioral requirements, employment laws, ethical, and honesty standards. The detailed contents of an organization's "Employee Handbook of Policies and Procedures" will depend on the specific company or agency, state in which the organization conducts business, management team, service sector, whether the organization does business with their State or Federal Government, and other factors. There are likely many other policy and procedural requirements on subjects such as:

- Smoking in/around the organization's facilities
- Drug or substance abuse,
- Disability,
- Employment contracts
- Benefit plans
- Retirement plans
- Other company regulations

18.4   FORECASTING

Civil Engineering projects and general construction are often project-based, potentially seasonal;and dependent upon the economy, interest rates, local needs, and funding. Forecasting includes the prediction of new work or projects and projecting the needs for personnel to perform this new work. Personnel requirements can vary a lot and can change dramatically in a few days. Personnel forecasting is dependent upon several key factors including:

- Company or agency's Mission Statement
- Most recent Business Plan and intent
- Marketing and Projections for new work, projects, or assignments
- Funding and banking relationships
- Private engineering firm's profitability and public agency's secure budgeting and funding
- Other factors

18.4.1   Mission Statement and Business Plan

The organization's Mission Statement is likely the foundation for the organization and future forecasting that predicts new work, new projects, and new personnel positions. The latest Business Plan is much more detailed and will reflect the immediate future objectives and methods to achieve these goals. It serves as the compass for the organizations' operations. The organizations' operations division must synchronize with the marketing plans, growth projections, and *hit rate* for winning new work and projects in private industry and engineering consulting. In the public sector, the budgeting and capital appropriations must synchronize with the parent agencies' growth plans and legislative funding.

18.4.2   Marketing Projections and Winning New Work or Projects

The marketing efforts and projections for new work must coordinate directly with actually securing (winning) this new work and forecasting the needs for additional personnel. Projections are just another form of forecasting. Winning new work and projects provides the fuel to run the organization. Forecasting and hiring new positions within the organization will likely lag behind the start-up of the new work, unless the completion and closeout of previous project work matches the initiation of new work, thereby allowing existing staff to move into these positions. This is challenging for the operations staff since they most likely need to work extra-long or hard to complete their old projects while starting up new project work. In addition, there is the challenge of providing the capital to fund new positions while still paying for the existing staff and project invoices. Often private firms find "growing" a real challenge because the capital is siphoned off the hard-earned profits, which are carefully guarded by the principal owners. Often, this capital funding is the restrictive element in forecasting and achieving new work and growth in the organization.

18.4.3   Funding and Banking Relationships

Funding and excellent banking relationships are essential in private industry and engineering consulting. This is especially true because there is so much slack time between the private firms' delivery of their products and the corresponding creation of the current invoice, inclusion of sub-contracting and material costs, preparation and delivery of the invoices to the client, and the time for the clients to review, approve, and pay the invoice. This lag time or slack time between delivery of private industry's interim products and receiving the client's payment may take 60–120 days; consequently, private industry carries these costs for a considerable time.

18.4.4   Private Engineering and Construction Firm's Profitability and Public Agency's Secure Budgeting and Funding

Secure funding and a good banking relationship are essential for private engineering and construction firms. There is a very long-time gap between engineering and construction work and capital costs applied to jobs/projects and the actual deposit of clients' payments for these accomplishments. Banks will monitor the private firms claimed accomplishments and "booked work" so the banks can safeguard their investments. Several years ago, there were engineering firms that tended to over-claim their "percent complete" statements resulting in over-inflated invoices to their clients. Then, the job/project stalled significantly in the (90+ percent) completion range as the staff struggled to make up the difference between the "claimed and invoiced" percent complete and the "actual" percent complete. The construction company lost money, the banks that guaranteed the project funding lost money, the Clients overpaid and suffered losses, and the construction company's stock value suffered. These losses due to *front end loading* impacted the value of the public's invested monies into these companies on the stock exchange, so the Security and Exchange Commission (SEC) became involved to set parameters to avoid this issue in the future.

Now, publicly-traded construction and engineering firms must comply with a different method of stating job/project completion. The company's accounting department and the job Project Manager (PM) must report an invoiced item typically referred to as "Effort to Complete" (ETC). The invoice can include the current personnel charge rates and materials but must also state the estimated ETC in effort and materials as compared to the original cost estimate for the completed job. This accounting method presents a more accurate depiction of the actual job status and prevents the company, PM, or managers from overstating their accomplishments and also safeguards public investors' stock values. Any "out of scope work" is reported and tracked separately.

In the public sector, cost tracking in relation to product delivery is also essential so that the agency can track, monitor, and report on accomplishments. A good relationship is essential with the parent agency, the budgeting office, the finance department, and public officials. And, just as in the private engineering and construction sectors, cost tracking and product delivery are monitored and tracked. Product and service delivery products resulting from engineering and construction contributions are typically costly, and time sensitive. All scopes of work need to be established with careful engineering estimates, material and supply costs, assumptions, and detailed descriptions of what the product or service includes.

Forecasting provides a "look-ahead" for the organizations' upcoming needs and must be carefully coordinated with the organizations' previous accomplishments to refine future predictions for accuracy. The duties and responsibilities of the HR manager include collaborating with construction managers, project managers, contractors, and others to identify and document each project's roles and responsibilities, as well as detail the end-to-end processes required on a project (or series of projects) to determine labor needs.

18.5   HIRING

Organizations (private engineering and construction firms and public entities, such as agencies) usually are very cautious about hiring additional staff above and beyond their current employee teams. During a period of growth, or maybe a result of an organization's business plan, sometimes staff are overloaded with work and are asked to put in extra time beyond the typical 40-hour work week. Alternatively, work is simply backordered and placed in a queue awaiting staff availability. Hiring is a major function of the HR Department, and civil engineering applicants should realize that the hiring process is highly competitive.

18.5.1   Growth Situations Resulting in the Need for Hiring

Backordering of engineering work is typically a rare event since it is usually engineering work that is needed for repair, replacement, expansion, development, or company profit. Organizations frequently try very hard to respond to the demand for their services. Sometimes during an interim period, there is more work that one CE unit ("unit" is expressed as one full-time 40-hour per week position) can accomplish but two CE units are not needed. In this case, organizations may request that staff put in overtime and work harder and longer, with the promise that help is coming. Additionally, during this interim period, some managers may claim they are organizing work very efficiently and are being extra profitable, potentially in a bid for higher recognition and/or bonuses.

Often there is a serious price to pay after long periods of overworking staff. Staff may resign, seeking to find a better "balance" in life and career, as discussed in Chapter 14 of this handbook. However, if the very high workload is the result of an unusual condition like a regional emergency or if management genuinely is addressing the staffing situation, then most professional staff will be empathetic, understanding, and patient until the situation resolves itself. During these periods, tenured staff may have opportunities for upward mobility and promotion.

Another situation resulting in the need for hiring occurs when staff leave an organization. This situation can provide an opportunity for CE applicants, depending upon the real reason the previous staff has left their positions. During the interview process, potentially near the end of the interview, the CE applicant might inquire about the reason for the new civil engineering position. The applicant simply could ask: "Is this position a vacancy or is it a result of growth?" Much can be learned from the answer.

18.5.2   The Application Process—Hiring Is a Competitive Atmosphere

The recent civil engineering graduate enters the full-time working world in a competitive atmosphere and should prepare to act accordingly. Organizations carefully craft their position descriptions in to define the work tasks accurately. The civil engineering applicant should address their resumes and applications toward their direct experience that matches the position description. The applicants should fully explain their experience on the application and refrain from statements such as "refer to resume."

Social Media

The civil engineering applicant's goal is to obtain an interview so they can display their experience, character, and qualifications. The applicant should be aware that prior to interviewing (and/or hiring), many organizations monitor applicant's social media for additional insight into their backgrounds. This publicly available information is potentially valuable in hiring decisions.

Civil engineers should exercise great caution in what they say and display on social media.

If the civil engineering applicant provides concise, accurate, and honest responses that fit the position description, the applicant may expect to be invited to interview. Much happens between application submission and a positive or negative response. However, many organizations do not respond to applicants due to an over-worked staff, simple neglect, and/or rudeness. Periodically monitoring the sponsoring organization for position announcements and to stay involved in the hiring status is prudent.

During the hiring process and even after a civil engineer is hired, the organization's human resources department may monitor the state's Department of Motor Vehicles (DMV) records about their staff's motor vehicle violations. Organizations feel justified performing this monitoring because they believe employees are representing the organization and may subject the organization to additional liability if they drive or act irresponsibly. If you, as an applicant, do not receive a response about your application for an organization's open position, you might want to check your social media or DMV records.

18.5.3   Interviewing

Civil engineering applicants should be knowledgeable about the organization prior to the interview. Questions like, "What does this organization do?" or "What products or services do you provide?" will not impress potential employers. If applicants do ask these types of questions, they are wasting their time and that of the interviewers. There might be an opportunity for discussions and questions during or after the interview in order for the applicant to display a genuine interest in the organization. Key sources of information that aid in preparation for the interview include the organization's:

- Mission statement and key divisions
- Products or services and record reviews on these products or services
- Website, links, public announcements, stock values if applicable
- Posted business plan or any information about growth and expansion

Other sources include:

- Any relevant or applicable legislation that might impact the organization
- Applicant's own knowledge and/or interest

The interviewers will seek to learn more about the applicant as an individual in order to gauge the applicant's fit with the position and the organization. The interview goes both ways—the applicant also is interviewing the organization for details on whether they fit with the organization. The savvy civil engineering applicant will be prepared with additional information on their own character. Hopefully, they can create an opportunity to present and discuss topics like:

- Key awards earned
- Publications and/or presentations made
- Recent examples of their work, especially examples that relate directly to the interviewing organization's products or services (but nothing that would violate previous organizations' rules of confidentiality or sharing information)
- Membership in related organizations such as the American Society of Civil Engineers (ASCE), National Society of Professional Engineers (NSPE), American Council of Engineering Companies (ACEC), or others
- Community service and contributions of their personal time

Finally, the civil engineering applicant should dress for success, practice looking into the eyes of the interviewer/s, be confident but humble, be honest, provide complete but concise answers to questions, and practice interviewing with a friend.

18.6 TRAINING

Training is an essential element of a successful and fruitful career as a practicing civil engineer. Most states have training requirements in the form of "continuing education units" (CEUs) required to maintain a Professional Engineer's license. This makes sense to most civil engineers because of the need to stay abreast of new technology, updated legislation related to engineering and/or construction, new personnel policies and laws, quality management, innovative software and updates, changing requirements for sustainable and resilient design, new energy applications, and many other areas.

Training usually invigorates the civil engineering students, and training courses provide interaction opportunities with peers from other organizations who may have alternate viewpoints and opinions. One of the most important elements of training is the opportunity for civil engineers to enjoy "continuous improvement" for their own careers and for their organizations. A college degree does not complete a civil engineer's learning. Lifelong learning and an effective training program are essential for professional growth and continuous improvement.

An organization's training program provides opportunities for civil engineers and other staff to gain upward mobility. Group, branch, department, and regional manager job descriptions have differing requirements. Many of these requirements can be met through training and subsequent experience. Civil engineers and most staff view training programs as a job benefit so they can achieve more job satisfaction. Many organizations promote college training and advanced degrees as long as these courses are related to employees' current job descriptions or other possible jobs in the organization. Many organizations reimburse (or partially reimburse) their staff if they have achieved a grade A or B in their classes.

Two tools are described here for civil engineers to assess their current experience and qualification and then to identify gaps in qualifications needed to be promoted to their next position or another higher position within the Organization.

1. Using the first training tool, here's an example for Sarah Smith, P. E., Senior Civil Engineer and Section Manager specializing in transportation engineering with four years of experience at a private engineering consulting firm. Sarah is a Section manager managing a staff of three other civil engineers. She wants to prepare herself and apply for a Department Manager position in the firm. Please refer to Table 18.1 titled, *Personal Detailed Evaluation of Skills, Abilities, Experience Matrix—Leading to Areas for Self-Improvement.*

TABLE 18.1   Personal Detailed Evaluation of Skills/Abilities/Experience Matrix Leading to Areas for Self-Improvement.

EXAMPLE MATRIX Assessment of Sarah, Smith, PE, Sr. Eng Skills, Abilities, and Experience Job or Project Task Activity	Personal Assessment of Experience, Skills as of December 1 = Little Experience 3 = Possess Knowledge 5 = Highest Level Expert 1 . . . 2 . . . 3 . . . 4 . . . 5	Personal Goal for Higher Level of Expertise by December, Next Yr 3 . . . 4 . . . 5	Applicable Senior Eng or Project Mgr (SE/ PM)	Comments and Demonstrated Progress
Surveying (Performing Land Surveying)	2.5	*	SE	*Opportunity with Sub
Surveying (Handling DTM/Survey Data)	4	4.5	SE	Main St, Frenchtown
Drainage Analysis and Design	4	4.5	SE	Spring Road
AASHTO Standards	3.5	4.5	SE	Ongoing
Local Standards	3.5	4.5	SE	City of Baker, Jones City
Horizontal Alignment	5	5	SE	All projects ongoing
Vertical Alignment	5	5	SE	2nd Dorado Street
Superelevation Design	4	4.5	SE	All projects ongoing
Typical Sections (Structural Section)	4	4.5	SE	2nd Street
Roadside Features (MBGR/Dike/Etc.)	4	4.5	SE	All project ongoing
Design Tools (Microstation)	4	4.5	SE	Broadway
Land Development Desktop (AutoCAD)	5		SE	3rd Street
Pavement Delineation (Signs & Striping)	4	4.5	SE	All projects ongoing
Structure Design	1	3	SE	—

TABLE 18.1   *(Continued)*

EXAMPLE MATRIX Assessment of Sarah, Smith, PE, Sr. Eng Skills, Abilities, and Experience Job or Project Task Activity	Personal Assessment of Experience, Skills as of December 1 = Little Experience 3 = Possess Knowledge 5 = Highest Level Expert 1 . . . 2 . . . 3 . . . 4 . . . 5	Personal Goal for Higher Level of Expertise by December, Next Yr 3 . . . 4 . . . 5	Applicable Senior Eng or Project Mgr (SE/PM)	Comments and Demonstrated Progress
Quantities	5	5	SE	All projects ongoing
Right-of-Way (Evaluation of Impacts)	3.5	4	SE	Opportunity on Rt 120
Redlines—Review prepare redlines	3	4	SE	Spring Rd, other projects
Establishing Project Design Criteria	3	4	SE/PM	Opportunity on Rt 120
Environmental (Process and Coord.)	3	4	SE/PM	Opportunity on Rt 120
Geotech & Materials	3	4.5	SE/PM	Opportunity on Rt 120
Coordination of Roadway, Bridge Plans	3.5	4.5	SE/PM	Opportunity on Rt 56
GAD / 30% Plans	4	4.5	SE/PM	Opportunity on Rt 2
100% Plans, Specifications, and Estimates (PS&E)	3.5	4.5	SE/PM	Opportunity on Rt 2
Prep Engineer's Estimate	2.5	4.5	SE/PM	Opportunity on Rt 2

Senior Engineer: (Highest Level Project Engineer—responsible for complex design and project oversight roles): My personal evaluation confirms that I am an overall strong engineer with better than average experience on technical aspects of roadway project design, development of plans, generation of quantities, and preparation of engineer's estimate. My greatest weakness is lack of recent experience with preparation of detailed technical specs, and engineer's cost estimate for large projects. Base diagram contents adapted from Michael Sanchez, PE.

A civil engineer preparing for their annual performance review or simply documenting their overall qualifications can use this tabular format to perform a specific personal assessment of their performance, skills, and qualifications. A brief summary of Sarah's overall qualifications and her personal assessment is presented.

Sarah has completed her honest assessment of her overall qualifications, which is listed on the Table 18.1. Now these qualifications can be compared to the known skills described on the organization's position description for a Department Manager, her next targeted position.

Using the information in the table, Sarah's Supervisor can help develop a list of in-house learning opportunities on other jobs/projects to spot potential opportunities where the civil engineer might gain this valuable experience as "on-the-job-training."

The comparing acquired skills with needed skills is relatively easy for Sarah but gaining the needed experience may pose a challenge. Sarah's Supervisor will speak to the other Project Managers for several jobs on Sarah's list showing where she might gain valuable experience. The Supervisor will request that the Project Managers assist Sarah by including her on their projects. Sarah's Supervisor might be able to justify this additional job/project work for Sarah as a "stretch assignment."

The information gained from Table 18.1 will illuminate gaps in experience and overall qualifications.

2. These gaps can be now transferred onto Table 18.2 to create a personal development training plan. Table 18.2 is titled, *"Critical Knowledge/Skills—Sarah Smith, P. E.— Personal Training and Development Plan for Mutually Agreed-Upon Goal.* In our case, Sarah met with her Supervisor and mutually agreed upon her goal to apply for an upcoming position in the firm for a Department Manager. So, Sarah's goal is a "Promotion from Senior Engineer to Department Manager." Let's examine Sarah's development plan for the next year.

From the results of Table 18.1, Sarah concluded her focused learning objectives and listed them in Column 1. She learned that she had some experience and in other instances she needed much more experience for some of these learning objectives.

Sarah then listed her action items to fill the focused areas of knowledge where she possessed limited knowledge or no knowledge and created action items like, "Take the Organizations on-line 'Team-Building' course." These action items appear under the second column titled, Required Assistance, Resources, & Opportunities from My Company/Agency.

Finally, she created a schedule divided into quarters over the next year. She decided when and how she would gain this valuable focused learning and targeted the time period for the action.

This information contained in Tables 18.1 and 18.2 now form Sarah's Training and Development Plan for the next year so Sarah can reach her goal for promotion to Department manager.

TABLE 18.2 Critical Knowledge/Skills-Sarah Smith, P.E.- Personal Training and Development Plan for Mutually Agreed-Upon Goal: Promotion from Senior Engineer to Department Manager.

Sarah Smith, P. E, Senior Engineer Personal Assessment and Focused Learning	Required Assistance, Resources, & Opportunities from My Company / Agency	Planned Development Ratings (1–5) by Quarters 1 = Little Progress / 3 = Average Progress 5 = Significant Progress				
		Planned Activity	Q1	Q2	Q3	Q4
Proposal Prep for Design/Build Project and Project Team	1. Take the Organizations on-line "Team-Building" course	Take on-line "Team-Building" course	3	5	–	–
• Set up Project Team and Experts for Design/Build	2. Read new book on Advanced Proposal Writing	Read new book, Advanced Proposal Writing, Class "A" Contracting Licensing	3	5	–	–
• Write the "Big Job" Proposal	3. Learn details of State's Class "A" construction contracting licenses.					
• Focus on "what sets us apart" with our Class "A" Contractor	4. Become very comfortable with public speaking	Apply for Class "A" Contracting Licensing	1	1	3	5
• Prepare Previous Experience Summaries, Awards, and Client Recommendations		Join Toastmasters	1	3	5	5
		Planned Activity	Q1	Q2	Q3	Q4
Refine Our Innovative Approach and Cost-Saving Methods	1. Identify new innovative construction techniques for this "Big Job."	Innovative construction techniques	3	5	5	–
• Develop Basis of Design and Innovations Construction Plans, Review with Construction Firm	2. Take the Organizations on-line "Advanced Negotiations" course	Take on-line advanced negotiation courses	1	3	5	–
• Finalize Innovations and Our Demonstrated Experience	3. Work with my Mentor on learning the details on contracting terms, conditions, and liability clauses.	Learn details on contracting terms, liability	1	3	3	5
		Planned Activity	Q1	Q2	Q3	Q4
Management Skills	1. Attend one-week seminar on Personnel Management. Ask Supervisor to approve my Training	Personnel Management seminar	3	5	–	–
• Improve my Personnel Management skills. Request $5K Travel and Training Funding	2. Take private course on "Know Body Language"	Body language courses	–	3	5	5
• Learn the basics on Body Language	3. Take community college night course on "Leadership"	College night course on "Leadership"	–	3	3	5
• Improve my Leadership skills						

Development Plan for Promotion from Senior Engineer to Department Manager: My personal assessment allowed me to identify several areas in which I will need significant growth and experience to successfully create a winning project proposal, win the Client interview, manage the project, and deliver the project. A detailed development plan for achieving the skills and experience to receive a promotion from Senior Engineer to Department Manager is summarized above. My personal training and development Plan is generated from the "Focused Learning" goals and the "Required Assistance and Resources" list to achieve these goals.

18.7   COMPENSATION AND BENEFITS PLANS

Civil engineers are often very fortunate to have access to generous compensation and benefit plans, depending on their organization. Private engineering and construction firms usually have more freedom with varied benefits than federal, state, and local government organizations.

18.7.1   Compensation

The civil engineer's largest compensation typically is an annual salary paid periodically through bi-weekly or monthly paychecks. There can be additional options related to the annual salary, such as options for partial matching contributions to retirement plans. For example, the Organization may offer to match the employee's retirement contribution up to a fixed percentage, like 2 percent. Taking advantage of these retirement plans even in the early stages of a career makes sense because of the long period of time for small contributions to grow to a reasonable sum at retirement.

There may be other options provided by the organization, such as having the payroll deduction of union funds (for state or local government employees, if applicable) and maybe even an option for referral to network of attorneys for a small fee.

18.7.2   Health Care

The most common benefit plan that people seek is health care insurance. The majority of organizations offer a selection of health care plans. These plans can be organized as a *preferred provider organization* (PPO) like Blue Cross or a *health maintenance organization* (HMO), which requires members to visit only their in-network professionals. Organizations often provide a fixed health care contribution depending on the employee's management level within the organization. The employee can use the organization's funds (per employee) to choose a budget plan or a more expensive health plan. More expensive health care plans may have lower deductibles or offer a greater choice of doctors or specialty services. These plans have options, such as:

- Inclusion of family members, spouses, or possibly domestic partners for an extra cost depending how many additional members are included.
- Dental care
- Eye care

These plans are usually available at an extra cost for the employee.

18.7.3   Other Benefits

Organizations may offer other benefits that typically increase as one is promoted to higher ranks in the organization. Examples include:

- *Vacation days*—Most organizations offer paid time off for vacation days. The vacation days (time periods) generally begin with two weeks paid time off for new employees and can range to four weeks per year or more depending upon seniority and/or time with the organization.

- *Personal Time*—Some organizations simply offer "personal time," which can include time dedicated for sick days or just personal time. These time periods generally begin with three weeks paid time off for new employees and can range to five weeks per year or more depending upon seniority and/or time with the organization. Both private engineering and construction firms and government engineering organizations recognize the value of this benefit and offer similar packages, though government agencies may offer more time off with pay.

- *Sick Days*—Most organizations have a set number for paid time-off for sick days, ranging from 5 to 20 sick days annually. As stated above, some organization combine paid time off for sick days with "personal time," which can combine time dedicated for sick days and personal time.

- *Personal Leave*—Most organizations offer other paid time off for events like doctor appointments, legal, or other appointments usually ranging from 2 to 5 personal days annually. As stated above some organizations include these types of paid time off in one package including vacation, sick, and personal time referred to as Personal Leave.

- *Holidays*—Most organizations recognize major holidays and offer paid time off for these events. Typically, holidays can range from a minimum of 8 major holidays up to 14 holidays, depending upon the organization. Many civil engineers combine personal time with paid holidays to extend their mini-vacations.

- *Educational Funds*—Organizations may pay for university/college degree programs or specialty courses, if related to their position description. For example, organizations may offer full or partial reimbursement toward each course assuming the CE achieves a minimum grade of an "A" or "B."

- *Publications*—Many organizations offer bonus programs for civil engineers (possibly other staff) if they publish a *peer-reviewed* paper or make a presentation as a representative of their organization at a seminar or conference. Typically, the civil engineer asks their supervisor about sharing the organization's innovative information and obtaining permission to attend the conference with costs reimbursed. Next the civil engineer submits an *abstract* to the conference sponsor. If the sponsor accepts the abstract, the civil engineer completes the publication and presents the information as an organization representative. Later the civil engineer submits paperwork for reimbursement of conference expenses and/or the bonus for preparing and presenting a publication at a conference.

- **Bonuses**—Many private engineering and/or construction firms have bonus structures for their key staff. Usually, bonuses are distributed annually or upon completion of a significant job or project. Some firms announce their bonus structure during the hiring process. They may mention that their salary structure may be slightly less than competitors' but that bonuses encourage top performance that offer rewards; or they may mention that their bonuses can result in salaries that are significantly higher than competitors'. These bonuses can range from a few hundred dollars to 5 percent to 25 percent, or more, of annual salaries. Except in rare circumstances, government engineering departments do not offer bonuses.

- **Summer or Winter Party Events**—Many organizations recognize the value of high morale and offer company time for sponsored events such as summer BBQs, outdoor party, or winter party events. These events can include brief statements from senior management (possibly to justify the reason for the event) and are usually very pleasant and fun.

- **Free parking or Contributions Toward Mass Transit**—Some organizations offer free parking or encourage use of mass transit by contributing to the cost of bus or rail passes.

- **Funeral Time**—Most organizations allow paid time off for the death of immediate family members. These days are not recognized or budgeted on an annual basis and may be provided "as needed" to recognize the needs of the employee and offer the organizations' condolences and respect.

18.8   HEALTH AND SAFETY

An organization's respect and regard for a vigorous and robust Health and Safety (H&S) Program is essential for long-term success, profitability, and employee safeguard. Organizations appreciate the need for the "Health" portion of this Program and offer health benefits and paid sick days as part of their key benefits. Civil engineers should clearly recognize and appreciate these characteristics in an organization.

The "Safety" portion of the Program is a little more variable among organizations. The risk of lax safety protocols is recognized; unsafe practices trigger higher accident rates and safety violations, sometimes resulting in death. Organizations carry liability insurance, and high accidents rates can impact these premiums and their states workers compensation insurance ratings.

Workers' compensation insurance makes payments to employees if they are injured at work or become sick due to their work. Workers' compensation includes payments to employees to cover their:

- Wages while workers are not capable of performing their work
- Medical expenses
- Rehabilitation back to a healthy condition

Workers' compensation insurance protects employees and provides benefits to those who get injured or sick from a work-related cause. This insurance includes disability benefits, lost wage recovery, and death benefits. Workers' compensation insurance reduces an organization's liability for work-related injuries and illnesses unless the organization can be shown to have been grossly negligent.

Organizations (employers) pay for workers' compensation insurance, regardless of state. An organization's cost for workers' compensation is a percentage of payroll costs. Employees do not have any payroll deductions for workers' compensation insurance. Insurance carriers will raise organizations' insurance premiums significantly if they have a high number of accident and workman's compensation claims. (The Hartford Insurance Company, Worker's Compensatsion Definition & FAQs, 2023).

Health and Safety Programs are very important to organizations, civil engineers, and staff. Much more detailed information on Health and Safety is included in Chapter 20 of this Handbook. **Please refer to this important information—it does impact all employees' personal well-being.**

18.9   EMPLOYEE RETENTION

Employee retention is an important factor in profitable, responsible, and respected organizations. Organizations expend considerable time, funds, and energy to create position descriptions, screen applicants, select applicants for interviews, conduct interviews, prepare offer letters and packages, initiate the "on-boarding" process for hiring, and train and bring employees to full potential. Therefore, retaining employees makes sense. Losing an employee can result in reduced production or revenue for months until a new person fills that position and comes up to speed.

Organizations may attempt to demonstrate their intent toward employee retention through compensation and benefit plans. But many civil engineers and employees recognize there is more to retention other than money and a health plan. The sections above illustrate the importance of excellent training programs to maintain competence, career growth opportunities, continuous improvement goals, and eligibility to move into positions with more responsibility. Experienced civil engineering managers acknowledge the importance of recognizing their employees' accomplishments and providing positive and productive feedback on their efforts. Retention often comes down to a personal relationship with a supervisor or manager, knowing that they care about the civil engineer's career accomplishments, growth, and direction.

Another very important element related to retention is flexibility and time off as discussed in *Chapter 14—Balancing Life, family, and Career*. Flexibility is needed for employees to respond to family/friend obligations, attend events, and be available when required.

> Some organizations may periodically remind employees that "We are a family, and you are part of it!" Realistic civil engineers recognize that organizational managers ultimately put the organization's mission and profits first. The experienced, successful civil engineer strives to remain employable by developing marketable skills, superior experience, satisfied clients and supervisors, innovative thinking, and a reliable character

18.10   SUMMARY

All employees should read and understand their organization's "Employee Handbook of Policies and Procedures." Being aware of employment laws and policies is imperative, as is keeping up with the training an organization offers. Civil engineers should practice politeness, humility, and professional performance on and off the job, being careful with language and behavior. They should know that some people do not like to be touched in any way, even with a pat on the back for a job well-done. Most organizations hold managers and supervisors responsible to uphold their HR policies and procedures as noted above. Many employment law violations are punishable by the employer and even may become civil liabilities. Knowing organization policies, as well as state and local jurisdiction employment laws, contributes to a civil engineer's successful career, health, compensation, happiness, and avoidance of liability and possible lawsuits.

BIBLIOGRAPHY

CA Code of Regulations, Title 2 Administration. (2023). https://govt.westlaw.com/calregs/Browse/Home/California/CaliforniaCodeofRegulations?guid=IFAACB1F05A0911EC8227000D3A7C4BC3&originationContext=documenttoc&transitionType=Default&contextData=(sc.Default)&bhcp=1.

CA Department of Human Resources. (2023). Personnel Policies. https://www.calhr.ca.gov/state-hr-professionals/Pages/Personnel-Policies.aspx.

Dawson. (2019). A Subsidiary of the Hawaiian Native Corporation. Employee Handbook, January.

The Hartford Insurance Company, Worker's Compensatsion Definition & FAQs. (2023). https://www.thehartford.com/workers-compensation/definition.

Time Magazine, Cohen, Sascha. (2016). Before Anita Hill: History of Sexual Harassment in the U.S. https://time.com/4286575/sexual-harassment-before-anita-hill. April.

CHAPTER 19

Construction Management for Engineers

Big Idea

"By failing to prepare, you are preparing to fail."
— Generally attributed to Benjamin Franklin

Key Topics Covered

- Project Planning
- Parties to a Project
- Work Breakdown Structure
- Project Scheduling
- Project Estimating
- Project Close-out

Related Chapters in This Book

Chapters 3, 4, 5, 6, 7, 8, 9, 10, 11, 12, 13, 14, 18, 20 and Appendices A, B, C, D, E, F, G

Civil Engineer's Handbook of Professional Practice, Second Edition. Karen Lee Hansen and Kent E. Zenobia.
© 2025 John Wiley & Sons, Inc. Published 2025 by John Wiley & Sons, Inc.
Companion website: www.wiley.com/go/hansen/CivilEngineersHandbook

19.1   INTRODUCTION

According to Ken Simonson, Chief Economist of the Associated Contractors of America:

> *Construction is a major contributor to the US economy. The industry has more than 680,000 employers with over 7 million employees and creates nearly $1.3 trillion worth of structures each year. Construction is one of the largest customers for manufacturing, mining and a variety of services.*

The Construction Industry, or more generally the Architecture/Engineering/Construction (AEC) Industry, is rife with challenges. Figure 19.1 highlights some of these challenges.

This chapter addresses how AEC professionals manage the complex assortment of activities inherent in the design and construction process. Specifically, it examines the nature of projects, project planning, project teams, project management, and construction management, including pre-design, design, and construction.

Other topics related to construction management included in this handbook are:

- Maintaining ethical conduct (Chapter 3)
- Conducting problem recognition and solving (Chapter 4)
- Maintaining a high-quality product (Chapter 4)
- Maintaining the engineer's role in project development (Chapter 5)
- Producing engineering deliverables (Chapter 6)
- Following sound project management principles (Chapter 7)
- Managing the permitting requirements (Chapter 8)
- Maintaining the client relationship (Chapter 9)
- Conducting oneself as a leader (Chapter 10)
- Conducting the project in accordance with the contractual and legal aspects (Chapter 11)
- Preparing and reviewing the invoices (Chapter 12)
- Managing the civil engineering enterprise (Chapter 12)
- Communicating as a professional (Chapter 13)
- Having a life (Chapter 14)
- Following human relations (HR) policies and employment laws (Chapter 18)
- Recognizing the need for robust health and safety practices (Chapter 20)

19.1.1   The Project

Most of these complex construction challenges are resolved through activities organized within projects. Construction projects involve the integration of many technical subsystems and components in a context that often involves complex and sensitive economic,

Project Planning			
Scope	**Budget**	**Schedule**	**Quality**
Description of project	Labor costs	Design schedule	Design criteria
Rationale	Materials and supplies	Construction schedule	Finance
Justification	Capital equipment	Materials/equip delivery	Supervision
Assumptions	Productivity	Government review and requirements	Worker skill level
Constraints	Equipment		Material availability
Exclusions	Time delays	Permits	Inspection
Acceptance criteria	Legal disputes	Productivity/delays	
Deliverables	Inflation	Changes	

Budget-Schedule-Quality Trade-offs
Union agreements
Training
Value engineering
Analysis of high cost areas

Construction Management		
Pre-design	**Design**	**Construction**
Owner	*Engineer/architect*	*Constructor/Builder*
Feasibility	Scope confirmation	Plans/specifications review
Budget/finance	Team selection	Contract type
Scope definition	Work breakdown structure	Bonds and insurance
Operating requirements	Design schedule	Personnel selection
Project schedule	Cost and value control	Labor relations
Quality	Procedures	Construction means and methods
Resources	Constructability/buildability	Value engineering
Project delivery system	Review plans, specs, and cost estimates	Construction schedule
Procurement method		
Contract type and terms	Constructability review	Cost analysis/cash flow
Project controls	Value engineering	
Bidder selection		

FIGURE 19.1 Construction challenges.

Source: Adapted from Barrie and Paulson, Professional Construction Management.

social, political, and environmental decisions. The form of project organization depicted in Figure 19.1 is typified by a management and decision-making climate in which problems are highly interdependent while the people, methods, and organizations involved are extraordinarily independent.

As opposed to manufacturing, civil engineering projects tend to be one-off undertakings with beginnings, middles, and ends. The ends of projects may have various phases, eventually leading to owner and/or user occupancy. Construction projects involve many players, seldom co-located, and often having different points of view and objectives. Project participants typically include: owners/clients; investors; designers (architects, engineers, interior designers); builders; various specialists (environmental, fire protection, etc.); government agencies; special interest groups; and the general public.

Project:

". . .an endeavor that is undertaken to produce the results that are expected from the requesting party."

Planning:

". . .the formulation of a course of action to guide a project to completion."

—Garold D.Oberlender

19.2   PROJECT PLANNING

Delivering facilities and civil infrastructure systems involves a staggering number of players. Often with limited local knowledge, Architecture, Engineering, and Construction (AEC) professionals must produce one-of-a-kind products with stringent cost, schedule, and quality standards.

The success of any project depends on sound planning. Planning helps turn an intangible idea into something real. Thus, planning facilitates communication among stakeholders, establishes what is required, identifies who will be responsible for specific tasks, facilitates monitoring and control of the project, and examines inherent risks.

Planning for most construction projects consists of four major components:

- Scope – work to be accomplished
- Budget – costs measured in dollars or hours
- Schedule – task identification, duration, logical sequencing, and timing of work to be performed, as well as key milestones
- Quality – intersection of scope, budget, and schedule

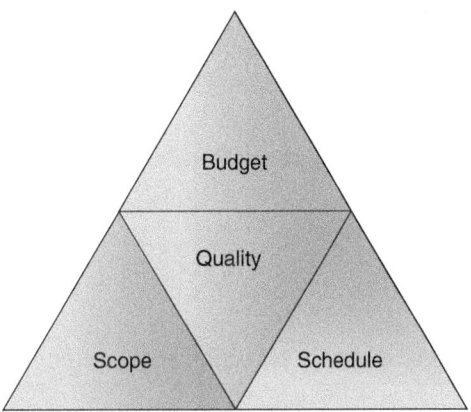

FIGURE 19.2 Construction project components.

Thorough planning assures that issues of scope, budget, schedule and quality are addressed adequately (See Figure 19.2).

19.2.1 Scope

A project scope identifies the work to be accomplished. Creating a clear scope definition during the initial stages of any project is critical to meeting budget, schedule, and quality goals. With superior understanding of clients' needs, designers can shorten the product delivery cycle and provide an end product with greater customer satisfaction.

Typically, the client/owner is responsible for developing the scope. This makes sense because the client/owner is the party who has a need or desire that requires attention. Frequently, clients hire consultants (architects and/or engineers) to assist with scope development through a process known as programming.

In addition to the project's functional requirements, the scope should incorporate the project's rationale and justification, assumptions, constraints, and exclusions. The scope also should delineate expected deliverables and the stakeholders' acceptance criteria.

Chapter 5 – Engineer's Role in Project Development deals extensively with scope development and methods for establishing the project's features and functional requirements, that is, the program.

19.2.2 Budget

Most clients/owners cannot achieve project approval from private boards of directors, city councils, boards of supervisors, or public agencies without an estimation of probable project costs, measured in dollars or hours. A project budget relies on a cost estimate (or possibly a series of cost estimates) to allocate costs over the life of a project. A budget recognizes initial capital costs, as well as recurring costs, such as operations and maintenance. Cost estimates assess the costs for each work package or activity that comprise the project.

Early cost estimates, when scant scope definition exits, incorporate large contingencies, perhaps as much as ±50–100%. Table 19.1 lists some of the types of costs estimates used in in planning and design.

TABLE 19.1 Types of cost estimates in planning and design.

Project Phase	Accuracy	Construction Documents % Complete	Purpose	Who Creates It
Pre-design (No design work)	L: −20% to −50% H: +30% to +100%	0% to 2%	• identifies order of magnitude costs • screens concepts	• usually generated by client with possible assistance from consultant • frequently developed with the help of a designer
Pre-design (Conceptual design)	L: −15% to −30% H: +20% to +50%	1% to 15%	• establishes project feasibility or impracticality • enables client/owner to gain internal approval for project	• usually generated by client with possible assistance from consultant • frequently developed with the help of a designer
Preliminary design	L: −10% to −20% H: +10% to +30%	10% to 40%	• assigns costs to project parameters used by client to instruct design team • assists client select among alternatives • indicates whether design is adhering to budget	• frequently developed with the help of a consultant • validated by project team, including owner, to assure that client requirements have been understood
Design development	L: −5% to −15% H: +5% to +20%	30% to 75%	• helps fix functional relationships and major elements of the design • indicates whether design is adhering to budget	• frequently developed with the help of a consultant • validated by project team, including owner, to assure that client requirements have been met

TABLE 19.1 *(Continued)*

Project Phase	Accuracy	Construction Documents % Complete	Purpose	Who Creates It
Construction documents	L: −3% to −10% H: +3% to +15%	65% to 100%	• provides a basis of comparison to contactor's bid • indicates whether design is adhering to budget	• compiled by the design team and their consultants • carefully reviewed by client to confirm conformance to requirements

Source: Adapted from AACE International (Association for the Advancement of Cost Engineering) Recommended Practice No. 18R-97.

19.2.3 Schedule

Project scheduling transforms the scope and budget into an operational timeline. A project schedule can provide a clear picture of requirements and an opportunity to detect potential problems early. It also helps manage project teams and provides accountability. If accurate, project scheduling makes the entire project run more smoothly by:

- Making the project more tangible to participants, that is, recognizing important activities and dates
- Helping to prioritize tasks
- Identifying clear milestones and decision points
- Establishing reporting guidelines
- Obtaining project participants' buy-in
- Giving all players the ability to allocate resources efficiently and to anticipate when resources will be available and/or required
- Reducing rework and overall project time and costs
- Preparing for the unexpected
- Minimizing conflicts and potential legal action
- Diminishing stress

Establishing a preliminary schedule as soon as possible in the project planning process is advisable since the schedule has direct impacts on budget and quality. Later sections of this chapter specifically address scheduling techniques such as Gantt charts, Critical Path Method (CPM), Program Evaluation and Review Technique (PERT), 4D and 5D Computer Aided Design (CAD), lean construction scheduling, and Artificial Intelligence (AI).

19.2.4   Quality

". . .the most important decisions regarding the quality of a completed facility are made during the design and planning stages rather than during construction. It is during these preliminary stages that component configurations, material specifications and functional performance are decided. Quality control during construction consists largely of insuring conformance to these original design and planning decisions." (cmu.edu)

The level of quality required by the owner/client affects scope, budget, and schedule. A reduction or tightening of scope, budget, or schedule typically results in lowered quality. Or said another way:

$$\textbf{\textit{Scope + Budget + Schedule = Quality}}$$

A change in any of the four project planning components affects one or more of the others. This is not to say that projects delivered ahead of schedule or under budget are of an inferior quality. The point is that scope, budget and schedule should explicitly address desired project quality in the project planning and definition phase.

Ford vs. Ferrari? (COLOGNE, GERMANY/The Ford Motor Company)

The Oxford Dictionary defines *automobile* as: "a road vehicle, typically with four wheels, powered by an internal combustion engine or electric motor and able to carry a small number of people." Suppose you needed to help a client write a scope for an automobile. Their desires regarding price, driving range, handling, speed, interior finishes, life-expectancy, availability, safety, and residual value would need to be considered. Depending on the client's desires, budget, and the amount of time they are willing to wait for their car to be delivered, a Ford Fiesta or a Ferrari 488 could satisfy the brief. However, a Ford Fiesta can be purchased for approximately 10% of the cost of a Ferrari 488. (carsguide.com). Clearly, scope, budget, schedule, and quality all would factor into the client's decision.

Source: Equatore/Adobe Stock.

Figure 19.3 – Project Framework identifies the primary elements of project planning as well as the work that follows the project planning effort. Success of the project plan depends on the care with which it is developed, as well as the involvement and buy-in of key parties to the project.

19.3   PARTIES TO A PROJECT

AEC projects involve a myriad of participants, including:

- *Client/Owner*
  Initiates project, has the need and the capital, sometimes occupies the finished project as the end user

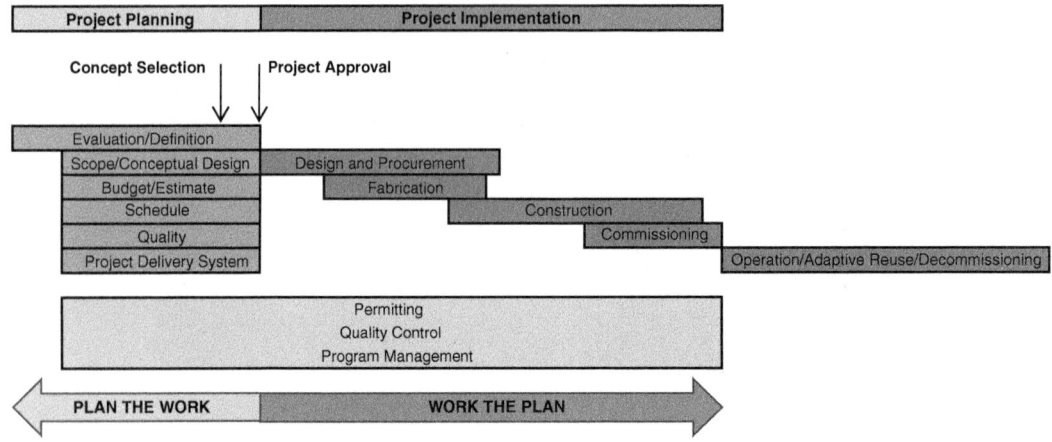

FIGURE 19.3   Project framework.
Source: Adapted from Plummer, Project Engineering: The Essential Toolbox for Young Engineers.

- *Designer (Engineer/Architect)*
  Originates the contract documents, including programming, studies, calculations, reports, drawings and specifications (Project Manual), addenda, and advises the owner
- *Contractor/Builder*
  Organizes resources (personnel, capital, materials, equipment) and actualizes the design
- *Regulatory and Permitting Agencies, Public and Special Interest Groups, Legislative and Statutory Bodies*
  Provide guidance regarding public policy and special requirements

Cooperation and a team approach among these project participants yield positive results. If problems arise among the players within these groups or between players in different groups, relationships can become adversarial.

Table 19.2 outlines the usual involvement of the owner, designer, contractor, government agencies, and the public.

Management

- Management can be organized using several different structures:
  - Functional/discipline management–grouping of personnel by areas of specialty, sometimes creating "silos"
  - Matrix management–multi-discipline approach with personnel reporting to both functional and project managers
  - Project management–system focused on achieving discrete goals within accepted criteria on a single endeavor

- Functions of project management are:
  - *Planning*–development of a course of action to guide a project to completion
  - *Organizing*–arrangement of resources (labor, money, equipment) in a logical manner to fit the project plan
  - *Staffing*–selection of personnel who have the ability to complete the work
  - *Directing*–administration of the work required to complete the project
  - *Controlling*–establishment of a system to measure, report and forecast progress toward project completion

(Adapted from Oberlender)

TABLE 19.2 Usual Involvement of A/E/C project participants.

Owner/ Client	Engineer/ Architect	Contractor/ Builder	Government Agencies	Public
• Provide project rationale and justification	• Create and help owner/ client evaluate design alternatives	• Furnish all labor, equipment, material, and know-how necessary to construct project	• Assure adherence to public policy	• Provide input regarding special interests/ needs
• Establish assumptions, constraints, and exclusions	• Provide visualization to support timely client/owner decision-making	• Build project per designers' plans and specifications	• Develop standards regarding structural safety, energy use, the environment, etc.	• Review and comment on reports, plans, and specifications
• Provide information regarding project's expected lifecycle	• Generate deliverables – plans, specifications, reports, calculations, • cost estimates	• Develop detailed budget and schedule and maintain accurate project controls	• Provide inspection criteria to establish compliance with standards	• Attend public meetings

(Continued)

TABLE 19.2   *(Continued)*

Owner/ Client	Engineer/ Architect	Contractor/ Builder	Government Agencies	Public
• Set operational criteria	• Assist client in meeting government requirements	• Provide timely cost and schedule information regarding changes (client initiated and otherwise required)	• Review and comment on Federal, State, and local building codes • Enforce compliance with any relevant Health and Safety Requirements and labor laws	• Occasionally participate in designer/ builder selection panels
• Fix parameters on total cost, payment of costs, major milestones, and completion dates	• Prepare and present project information to client decision makers and at public meetings	• Obtain client/ designer approval regarding various materials, equipment, and processes (submittals and shop drawings)	• Provide periodic inspections and oversight • Potentially collect fees and/or taxes	• Frequently provide labor for A/E/C projects
• Identify their level of involvement with review and approval process	• Design to budget and time constraints; act on behalf of the owners; advise owners during construction	• Work with various required inspectors	• Strive for a continuing, effective, and trusted partnership with the owner, A/E/C professionals, and the public	• Strive for a continuing, effective, and trusted partnership with the owner, A/E/C professionals, and government agencies

19.3.1   Project Teams

All project team members are vital to the success of a project. Design teams include architects, a variety of engineers, modeling/visualization and other specialists. Construction teams include the general contractor, key subcontractors, material and equipment suppliers, and inspectors. Frequently an owner's representative joins these project teams. The design and construction teams come from different cultures and their perspectives may vary.

As team members frequently are drawn from different departments and organizations, team building requires effective communication. All members must realize that they have a common customer. Continuity of the project team helps assure that information given is understood. The client's requirements (scope, budget, schedule, and quality) need to be well-defined. A clear understanding of the resources available to the team, as well as the process for decision-making, is crucial. Relationships play a minimal role in low bid public projects, so a specific challenge is to develop one singular focus; contract structures that address the self-interests of all parties are advantageous.

Each project team needs to be an integral part of a larger organization; and to function properly, project teams need unambiguous goals and objectives. The team leader (project manager, project executive) must guide the overall effort and develop a leadership style that is respected and accepted by the project team. (For more information, see Chapter 10 – Leadership.)

19.3.2   Project Manager

Part of the project manager's responsibilities include acting as a coach, answering questions, motivating, and making the way clear for the team, while assuring the desired outcome is understood. The project manager needs to influence a diverse set of individuals with some-times competing goals, needs, and perspectives. Due to the complexity of most projects, the project manager must lead a team effort so that all technical and managerial (scope, budget, schedule, quality) aspects of the project are addressed adequately.

Project managers focus on the success of the project. They have to be able to work under pressure and know how to exert pressure when required. Key words associated with project management include *responsibility, honesty, humility, respect, effectiveness*, and *communication*. (For more information, see Chapter 7 – Project Management.)

Project management:

". . .the art and science of coordinating people, equipment, materials, money, and schedules to complete a specified project on time and within approved cost."
—Garold D. Oberlender

Ideally, the project manager will have been involved in their company's proposal/bidding process. When assigned a new project, the project manager's initial review includes:

- Gathering all background material prepared by the client and/or designer
- Becoming familiar with owner's objectives and overall project needs
- Identifying additional information required
- Determining the client's orientation, their level of construction expertise, and expected level of involvement

- Ascertaining who are the authorized representatives of each key project participant
- Organizing a team kick-off meeting or retreat

Since the owner controls the flow of money and the final project sign-off, the owner's orientation must be known. Owner's representatives on project teams can provide information and clarify project requirements. They also can review and approve team decisions. Establishing the owner's representative's level of authority can assure that requests for cost increases and/or schedule extensions are dealt with efficiently. Effective planning, an accurate cost estimate, and a well-conceived schedule are keys to successful project management.

19.4   WORK BREAKDOWN STRUCTURE

As discussed in Chapter 7 – Project Management, a Work Breakdown Structure (WBS) is the cornerstone of a project work plan. The WBS creates a framework for organizing and sequencing the activities that make up a project. The WBS:

- Defines work to be performed
- Identifies the needed expertise
- Assists in selection of the project participants
- Establishes a base for project scheduling and control
- Provides a graphical display of the project

The process of creating a WBS is continuous, and the effort starts when the project is assigned to the project manager and the project team. The project manager may identify the major areas of the project and team members help define the work to be performed in more detail.

The preliminary WBS identifies major tasks and work to be accomplished. A detailed list of tasks should be grouped into phases that show sequencing. Then resources and technical expertise required can be identified. The smallest unit of detail in the WBS is the work package, defined in sufficient detail so the work can be measured, budgeted, scheduled, and controlled. The initial WBS should be followed up with a review and the creation of a work plan.

A typical WBS may have five levels:

- Level 1 – Project, as defined by the owner–contractor agreement
- Level 2 – Subproject, either individual components of the project or project stages
- Level 3 – Subnetwork, activities dealing with discrete components of the subprojects
- Level 4 – Activity, lowest level of WBS addressed in project schedule
- Level 5 – Subactivity/Work package, used for cost accumulation and control

Level 5 can be divided further into labor, materials, equipment, subcontractors, and any other useful category. It also can be used for resource planning and scheduling. (See Figure 19.4 – Work Breakdown Structure.)

Level 4 of the WBS identifies the activities that will be included in the project schedule. Each project can be subdivided into individual activities that must be carried out in order to complete the project.

19.4.1   Activities Defined and Identified

During the construction phase of a project, drawings and specifications form the basis of scope. The identification and organization of activities (Level 4 in the WBS) enable achievement of specific time, cost, and quality objectives.

An activity can be defined as an operation or process that can be managed by an individual or work team (crew). It is a measurable element of the project plan. Activities have the following characteristics (Glavinich). They:

- Consume time
- Usually expend resources
- Have a definable start and finish
- Are assignable
- Are measurable

The first step in scheduling a construction project is identifying these individual activities. In order to identify activities, a construction project must be understood thoroughly. A complete review of contract documents will help determine what has to be done. As noted in Chapter 6, contract documents (used by the client/owner and contractor) include:

- Agreement/contract forms (contract between client/owner and contractor)
- Conditions of Contract (general and supplementary conditions)
- Drawings (design information in graphical format)
- Technical specifications and any required calculations (design information in text format)
- Addendum/a (changes made during the bidding process)
- Modifications to the contract (changes made after the client/owner and contractor have signed the contract)

Owner-supplied labor, materials, and equipment are important to single out, as are project milestones. Project milestones indicate points in time representing specific project events, such as owner approval of various phases, e.g., inspections, substantial completion.

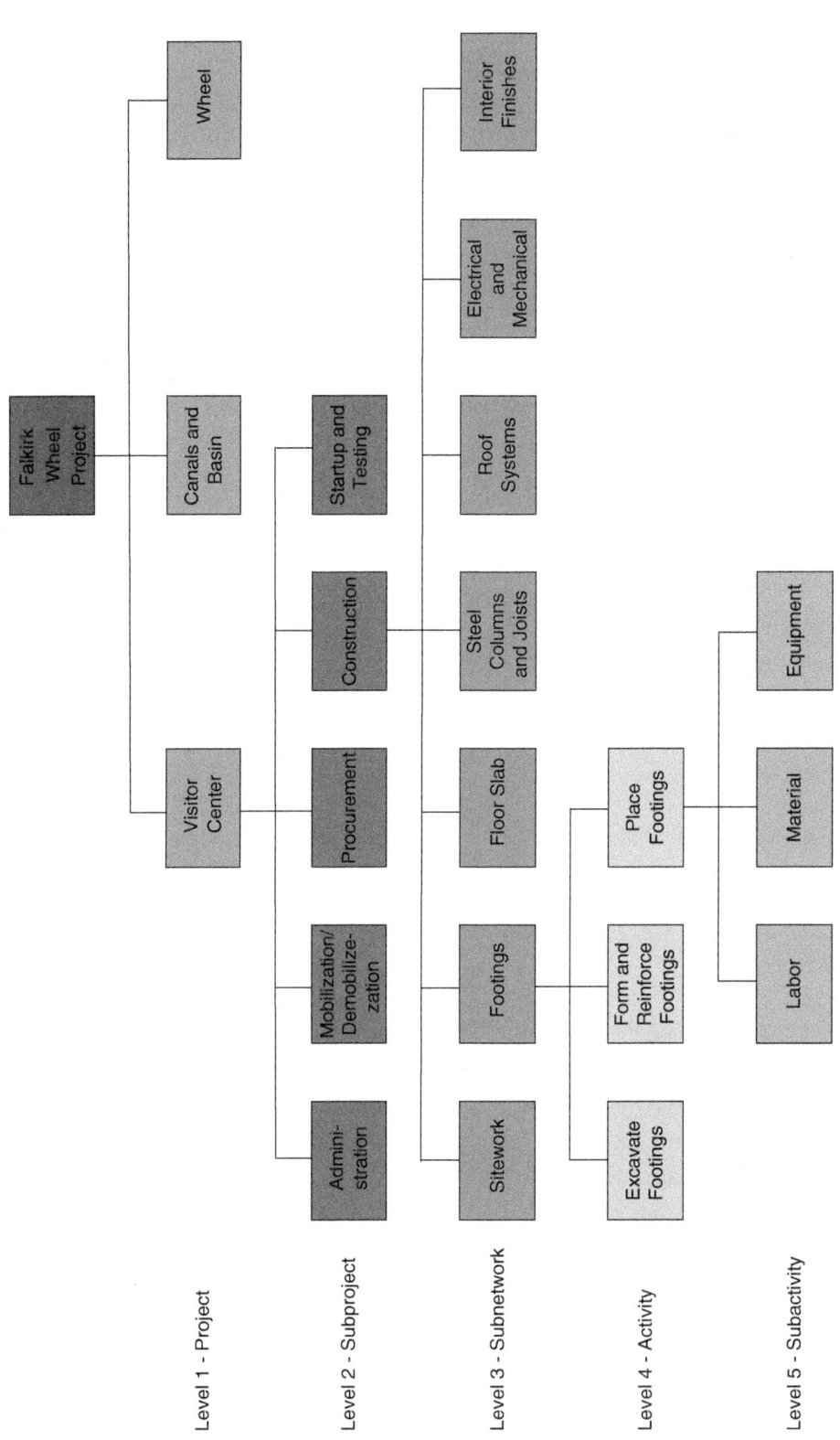

Level 1 - Project

Level 2 - Subproject

Level 3 - Subnetwork

Level 4 - Activity

Level 5 - Subactivity

FIGURE 19.4 Work breakdown structure (WBS).

19.4.2 Activity Categories

Identification of activities forms the foundation of the construction planning and scheduling process. Activities should be defined with clear scopes and responsibilities for performance. The value of the schedule is influenced greatly by quality of activity identification.

Broad categories of construction activities include:

- Administration
- Mobilization/demobilization
- Procurement
- Construction
- Start-up and testing

Table 19.3 lists examples of activities that fall into these broad categories.

TABLE 19.3 Broad categories of construction activities.

Broad Categories	Examples
Administration	• Request and receive *Notice to Proceed* • Apply for and receive permits and licenses • Request and receive technical information • Request and receive contract changes • Request and receive approval for means and methods (contractor, not architect/engineer, activity) • Review and receive acceptance of part or all of work Closeout • Receive acceptance of corrective work for substantial and final completion • Present final accounting of quantities and/or costs • Inspect/certify equipment and systems (performed by third party) • Submit operation and maintenance manuals • Provide warranties and guarantees
Mobilization/ demobilization	Mobilization • Set up field office, storage facilities, on-site equipment (e.g., cranes) • Perform site surveys and testing • Construct temporary access roads, laydown areas, and perimeter fences • Attain the necessary utilities for construction • Secure, train, and certify specialized labor • Acquire, set up, test, and certify construction equipment Demobilization • Remove onsite services and equipment at end of project (field office, storage facilities, equipment, temporary utilities, etc.) • Dismantle personnel lift when facilities elevators are operational

TABLE 19.3 *(Continued)*

Broad Categories	Examples
Procurement	• Identify type and quantity of materials and equipment needed • Arrange for shipping and deliveries to project site • Procure long lead items • Prepare shop drawings, samples, submittals for designer/owner review • Submit shop drawings, samples, etc. to designer/owner for review • Construct sample or mock-up installations for designer/owner review • Design/layout customized systems (e.g., fire suppression system) • Test materials, equipment, and systems for performance requirements • Schedule designer/owner review/approval of materials, equipment, and systems • Obtain raw materials for assembly of materials, equipment, and systems • Unload, inspect, inventory, clean materials and equipment for use at site • Fabricate materials, equipment, and systems
Construction	• Clear, excavate, and grade project site • Build foundations, including any footings, piers, piles • Form, reinforce, and place floor slabs • Build structure (e.g., layup masonry walls, erect steel columns and beams) • Install roofing system • Make exterior finish weather tight (e.g., install cladding) • Rough in electrical, mechanical, and plumbing • Finish electrical, mechanical, and plumbing • Apply interiors finishes
Start-up and testing (Commissioning)	• Test/run major equipment such as chillers, switch gear, communications, etc. • Test/balance heating, ventilation, and air conditioning system (HVAC) • Flush and clean domestic water system • Test safety and security systems (emergency power, smoke detection, etc.)

Source: Adapted from Glavinich, Construction Planning and Scheduling.

19.4.3 Activity Durations

Activity durations can be represented by a variety of units: years, months, weeks, days, or hours. For very large infrastructure projects, years may be appropriate. (See Table 8.1 – How Permitting is Integrated with Project Phases.) In the early stages of a construction project, months may be suitable. A construction schedule expressed in weeks can provide a framework for developing a later, more detailed schedule. For most construction projects, days are the most common unit of time used, enabling activity start and finish dates to be scheduled on specific calendar days, making incorporation of specific contract milestones easier. For short-duration, complex schedules, such as shutting down a process or system, hours may be more appropriate.

Short interval schedules can be developed in using durations of days or hours from the overall construction schedule for managing day-to-day on-site construction operations. Day to day activities often are depicted in a three week *look ahead* schedule that is coordinated with the main schedule. Look ahead schedules are the products of weekly coordination meetings involving the construction superintendent and relevant subcontractors. These meetings help create buy-in and coordinate actions and equipment use. Any units of time can be used effectively in planning and scheduling. However, units of time should be consistent within each schedule (See Figure 19.5).

The duration of an activity is a function of the amount of work required and the rate at which the work is completed. In other words, activity duration is the quantity of work divided by the rate of production:

$$\textbf{Activity Duration} = \frac{\textbf{Quantity of Work}}{\textbf{Production Rate}}$$

Quantity of work is expressed in some measurable amount of materials and/or equipment, such as square feet or cubic yards. The quantity of work serves as a basis for measuring progress. Production rate is the planned daily production and must incorporate required set-up and downtime associated with the activity.

Ways of determining activity durations include:

- Analyze historical records
- Refer to commercial manuals that provide costs and production rates, e.g., *R.S. Means*
- Talk with the person who is responsible for performing the work and who has previous experience and good judgment
- Coordinate with subcontractors to determine optimal activity duration/sequencing
- Refer to the construction estimate

Factors influencing activity durations depend on:

- Quantity of the work
- Quality/complexity of the work
- Number of people assigned to the activity
- Delivery and availability of materials
- Amount of equipment assigned and its availability
- Worker skills
- Work environment/effectiveness of supervision
- Site conditions, access to work, space to work
- Weather

Kando Construction Company

Project Name:

Project Location:

Superintendent/Crew Leader:

Project Manager:

Week ending date:

Short interval scheduling:

1. Helps assess progress toward short term (typically weekly) goals

2. Identifies problems

3. Creates opportunities for early corrective actions

4. Assist in optimization of labor hours, equipment, and materials use

Act No.	Activity/Equipment	Budget/Hours Remaining	Mon	Tue	Wed	Thr	Fri	Sat	Sun	Total Planned	Total Actual	Delta	Reasons/Comments

FIGURE 19.5 Short interval schedule.

Source: Gareth Figgess.

- Quality of the design (drawings and technical specifications) and participation of architects/engineers
- Unexpected delays
- Major changes in the scope of work

Activity durations determine the critical path (the minimum time necessary to complete a project) in Critical Path Method (CPM) scheduling. Activity durations also govern timing of activities and distribution of costs. Additionally, they influence the utilization of resources: people (labor); equipment (such as cranes); and money.

19.5 PROJECT SCHEDULING

Table 19.4 lists the types of project scheduling techniques widely used in the AEC Industry: Critical Path Method (CPM); Program Evaluation Review Technique (PERT); Gantt Chart; Line of Balance (LOB); Quantitative (Q) Scheduling; and Last Planner System® of Production Control (LPS).

TABLE 19.4 Types of construction schedules.

Name	Description and Use
Critical Path Method (CPM)	• Used to calculate minimum project completion time as well as start and finish times for project activities • Links activities in a network and identifies most critical activities needing attention in order to maintain or shorten overall schedule • An increase in duration of an activity on the critical path increases the entire project duration by the same amount
Program Evaluation Review Technique (PERT)	• Used when there is a high level of uncertainty • Three durations can be used to measure degree of uncertainty of each activity • Major difference between CPM and PERT is estimation of durations
Gantt Chart	• Used to illustrate schedule in bar chart format • Each activity is represented by a bar – the bar's position reflects the start date, duration, and end date • Relatively easy to create and use

(Continued)

TABLE 19.4   (*Continued*)

Name	Description and Use
Line of Balance (LOB)/Linear Scheduling Method (LSM)/ Location Based Management System (LBMS)	• Used when the schedule includes repetitive processes • Activities planned according to a rate of production • Can simplify the schedule's graphical representation • In LSM, activities described by one line instead two lines used in LOB • In LBMS, location as well as duration of different activities included
Quantitative (Q) Scheduling	• Used to reveal relation of activity sequence and cost incurred • Is suitable for varying volume of repetitive activities at different locations • Derived from LOB with provisions for non-repetitive activities
Resource Oriented Scheduling	• Used to assign resources in an orderly manner throughout the project's life • Considers all who will need resources ahead of time • Limits bottlenecks while waiting for a particular resource – can increase productivity and reduce delays
Last Planner System® of Production Control (LPS)	• Used to guide project through design and construction phases • *Last Planner* refers to people on the design or construction team responsible for making final assignment of work to those performing the work
Critical Chain Project Management (CCPM) (See Figure 19.8)	• Similar to CPM – both methods link activities using networks • Resources (people, equipment, and material) needed for the project considered in CCPM, like CPM • Based on theory of constraints (TOC) • Three types of buffers (project, feeding, and resource) utilized to absorb project delays
Discrete Event Simulation (DES) Scheduling	• Used to model construction projects decomposed into separate processes • Deals with uncertainties of durations and productivity rates for activities • Uses various software, such as Stroboscope, Cyclone, AnyLogic, Simphony.NET, etc.
Integrated Scheduling Systems	• Combines many scheduling functions into one software package, such as Vico Office • Incorporates Gantt charts, linear scheduling method, 3D, 4D, and 5D capabilities in one integrated system

19.5.1   Critical Path Method (CPM)

The purpose of Critical Path Method (CPM) scheduling is to help plan work, guide progress, and provide a baseline for project control. Many clients require that the contractor provide a CPM schedule as part of the contractual requirements. Deviation, and the reason for this deviation, from these schedules form the basis of legal disputes regarding delayed completion. If the contractor fails to complete the project on time, a *liquidated damages* contract clause would require that the contractor be charged daily until the project has been finished. The CPM schedule is instrumental in establishing the legal complaint.

To create a CPM Sequence (Precedence) Diagram (Network), three things need to be known:

- Activities
- Activity durations
- Activity sequences

The previous Work Breakdown section contains information on how to determine activities and their durations. Following is a discussion of activity sequences used in CPM scheduling.

19.5.1.1   Activity Sequence

Activity sequence refers to successor/predecessor relationships in the CPM network of activities (precedence diagram). See Figure 15.5 – Critical Path Method (CPM) Precedence Diagram for an example of a CPM network.

There are several types of successor/predecessor relationships of activities used in CPM scheduling:

- F-S = finish to start (*pure*)
- S-S = start to start
- S-F = start to finish
- F-F = finish to finish
- Lag = delay in relationship of one activity to another

There is no industry software standard for handling lags, and caution needs to be used in inserting lags between activities. Lags can cause unnecessary confusion and can create problems in determining the critical path. An activity can be added vs. a lag, for example, cure time of concrete.

19.5.1.2   Procedure

Creating a CPM network requires understanding several relevant concepts and terms. Please see Table 19.5 and Figure 19.6.

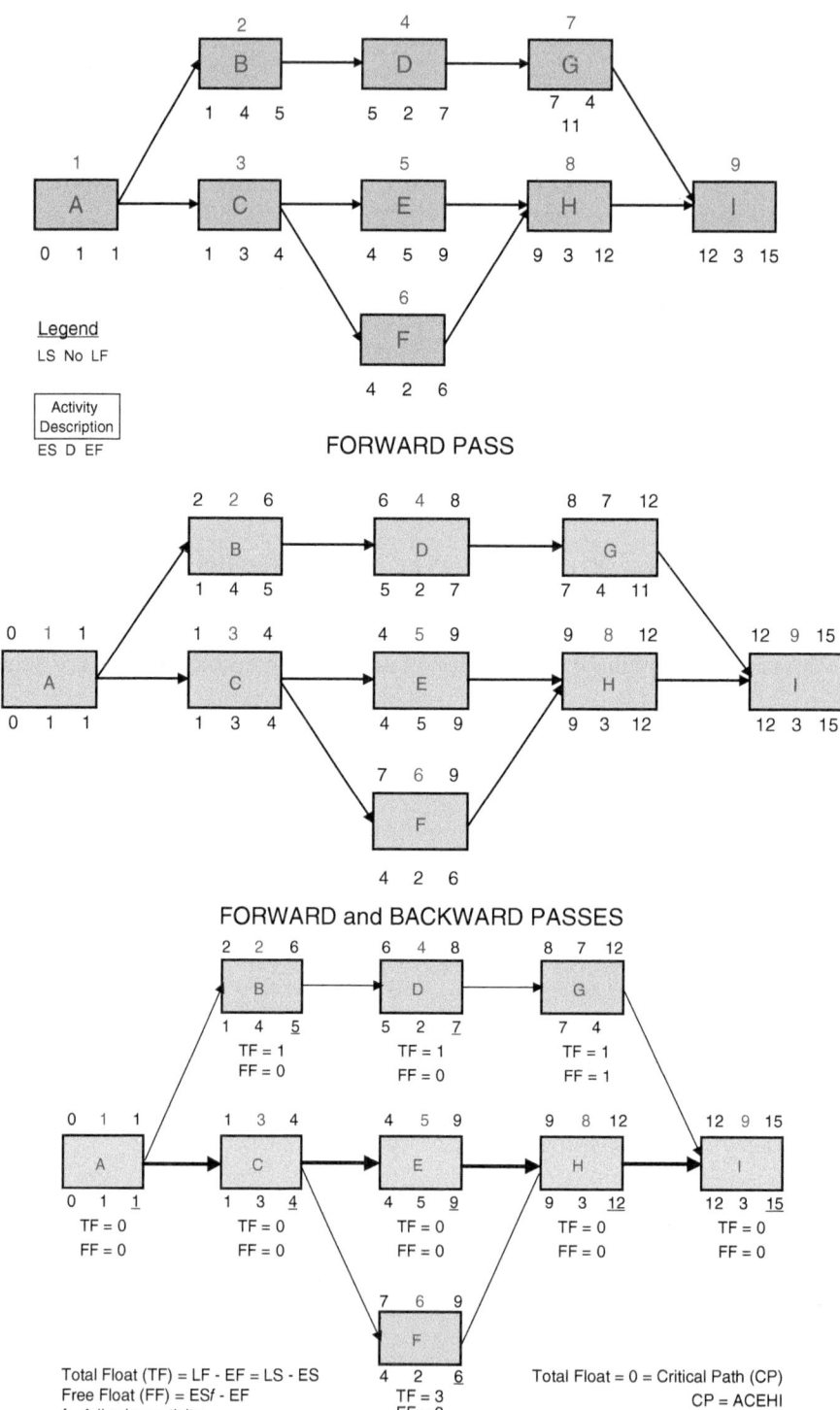

FIGURE 19.6 CPM.

TABLE 19.5 Useful CPM scheduling concepts and terms (Figure 19.7).

Term/concept	Meaning
Network	Diagram showing the sequence and duration of all project activities
Precedence Diagram	Network diagram that uses boxes, representing activities, and arrows, representing sequences
Arrow Diagram	Network diagram that uses arrows only to represent both activities and sequences
Critical Path	Sequence of activities that must be finished for the project to be considered complete—if the duration of any activity on the critical path is lengthened, the project finish date will lengthened by the same amount
Activity	Operation or process that can be managed by an individual or work team (crew) requiring time, cost, or time and cost
Dummy Activity	Activity used to indicate that an activity following the dummy activity cannot start until the activity preceding the dummy activity is completed; does not require time
Duration (D)	Estimated time required to perform an activity, including start-up and down times
Early Start (ES)	Earliest time an activity can start
Early Finish (EF)	Earliest time an activity can finish $EF = ES + D$
Late Finish (LF)	Latest time an activity can finish
Late Start (LS)	Latest time an activity can start without delaying the project completion date $LS = LF - D$
Free Float	Amount of time an activity can be delayed without delaying the early start time of the immediately following activity $FF = ES_f - EF_p$, where f represents the following activity and p represents the preceding activity
Total Float	Amount of time an activity can be delayed without delaying the project completion date $TF = LF - EF = LS - ES$

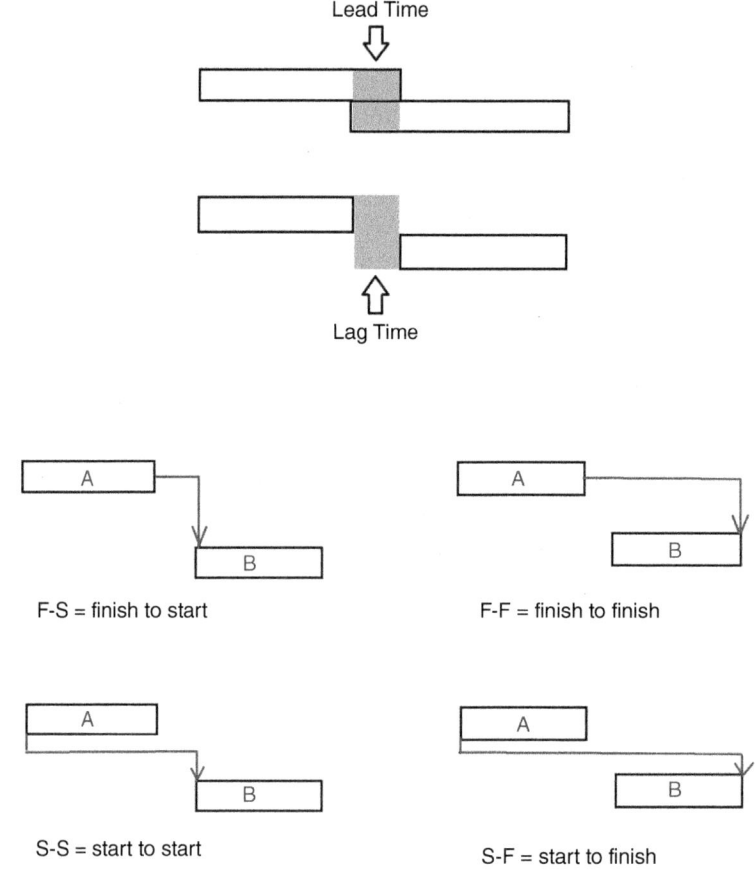

FIGURE 19.7   CPM concepts and terms.
Source: Dr Tarek Salama.

Float/Slack Time

A useful component of CPM scheduling is float or slack time. *Float*–or *slack*–is the time that an activity in a CPM network can be delayed without causing a delay to:

- Subsequent activities – *Free Float* (associated with an activity)
- Project completion date – *Total Float* (associated with a path)

Said another way, *Free Float* is the amount of time that an activity can be delayed without delaying the early start date of any successor activity. *Total Float* is the amount of time that an activity can be delayed from its early start date without delaying the project finish date.

Most contractors are advised against showing float. Questions arise, such as: who owns the float, the contractor or owner?

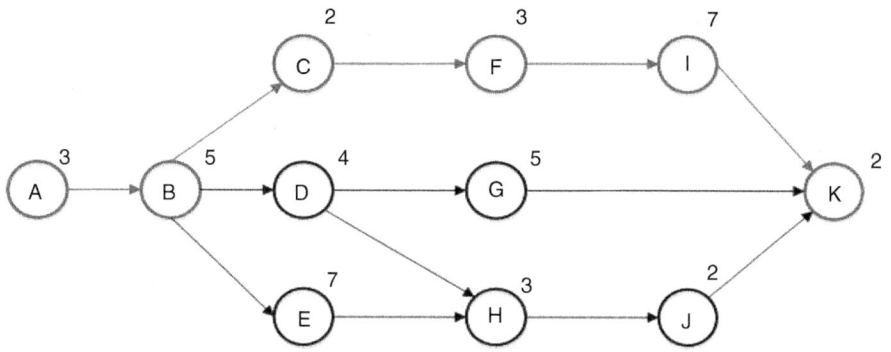

Critical path for a network before resolving resources conflict

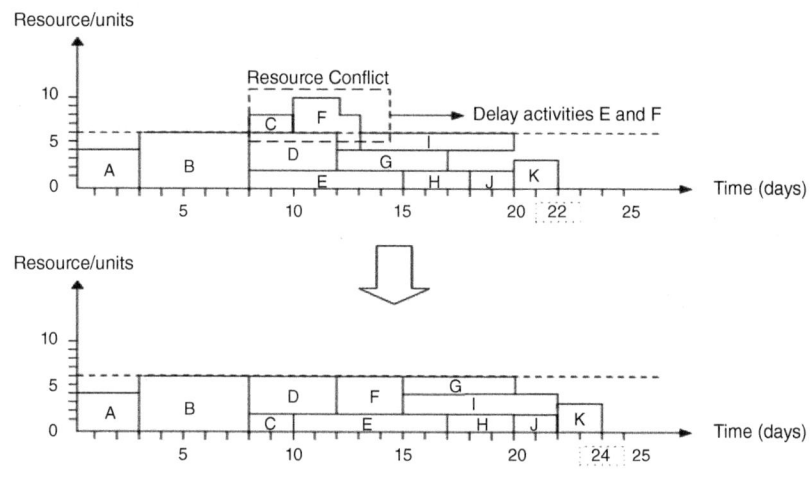

Resolving resources conflict

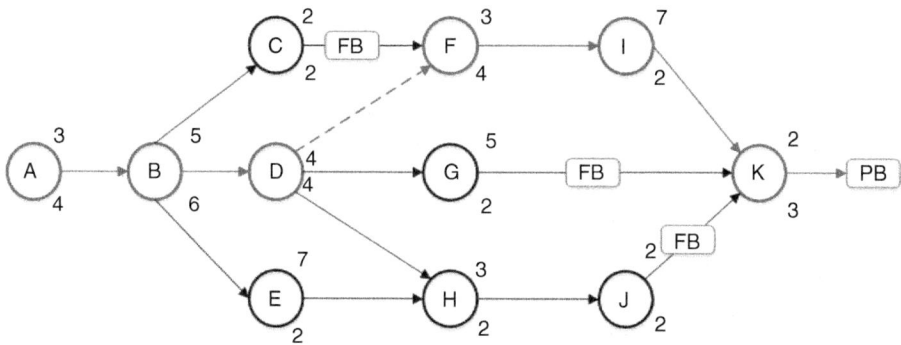

Critical Supply Chain after Resolving resources conflict

FIGURE 19.8 CCPM.
Source: Dr Tarek Salama.

The procedure for creating a CPM network follows:

1. Understand the meaning of the numbers above and below the Activity Description boxes. (See Figure 19.6.)
2. Make a Forward pass by adding the Activity Duration to the Early Start number. Place this number in the Early Finish location.
3. Transfer the Early Finish number to the Early Start position of the following Activity. If an Activity has more than one precedent, choose the larger number, e.g., Act G. Perform the Forward Pass from Act. A through Act. I.
4. At the end of Network/Precedence Diagram, transfer the Early Finish number to the Late Finish location, e.g., Act. I.
5. Now perform a Backward Pass starting with Act. I and working backwards to Act. A. Subtract the Duration from the Late Finish number, and place this number in the Late Start location.
6. Transfer the Late Start number to the Late Finish location of the preceding Activity. If an Activity is linked to more than one Activity, choose the smaller number, e.g., Act C.
7. Activities whose Early Finish = Late Start on a following Activity are on the *Critical Path*.
8. Also observe that most Precedence Diagrams start with a single Activity (e.g., Mobilization) and end with a single activity (e.g., Project Completion).

The CPM schedule must be in alignment with the owner's desired completion date. Consequently, establishing a CPM schedule involves considering external realities as well as construction activities and durations.

19.5.2 PERT (Program Evaluation Review Techniques)

Program Evaluation and Review Technique (PERT) is used when there is a high level of uncertainty. Forward and backward passes are performed as in CPM.

The major difference between CPM and PERT is estimation of activity durations. PERT uses a weighted average of three times to determine activity duration:

- a = optimistic time = shortest possible time in which activity can be performed
- b = pessimistic time = longest time that an activity might require considering the worst case
- m = most likely time = time a manager would assign if only one duration were used
- t = expected time of an activity

Note: a and b are not symmetrical around m

The simple formula used to find the weighted average of the three times to determine the expected duration of an activity (time $= t$) is:

$$t = \frac{a + 4m + b}{6}$$

The three durations can be used to measure the degree of uncertainty of each activity as well as the overall optimistic, pessimistic, and most likely project durations. The probability that the project will be completed within these times also can be determined.

19.5.3   Gantt (Bar) Charts

Henry Gantt, an American mechanical engineer, designed the Gantt chart in the early 20th century. A Gantt chart is commonly used to depict a project schedule graphically. It uses horizontal bars to show the start and finish dates (hence, durations) of a project's activities. Project activities are represented on the vertical axis and time is represented on the horizontal axis.

Originally, the bars on Gantt charts were just shapes with no connection to other bars. Current project management tools, such as Microsoft Visio, Project, SharePoint, and Excel, or specialized software, such as Gantto or Matchware, can help in creating Gantt charts. These digitally based Gantt charts can help create relationships among activities, e.g., if an activity's duration extends, the start dates for succeeding activities similarly extend.

19.5.4   4D and 5D Scheduling

Four and five dimensional (4D and 5D) scheduling approaches also exist. In 4D scheduling an additional dimension–time–is added to a three dimensional (3D) building information model (BIM). (Note: See Chapter 17 – Emerging Technologies for more information on BIM.) Both building professionals and clients can visualize and understand the process of construction events more easily when an 3D model includes the dimension of time. Construction activities are simulated through the 4D model, enabling users to evaluate and communicate activities' interdependencies. Each activity of a Gantt chart (generated in project management software such as Primavera or Microsoft Project) can be linked to a project's BIM. Programs such as SYNCHRO or Navisworks can be used to create 4D models.

The overall aim of a 4D model, or simulation, is to understand time-based processes, foresee conflicts, and study potential solutions. A 4D simulation can expose issues such as out-of-sequence work, scheduling conflicts between multiple trades, "what if" scenarios, and program-level construction phasing strategies to optimize the construction schedule. A simple example of a problem averted is having too many construction workers in the same location at the same time; the sequence of overlapping activities may be logical but difficult, or impossible, to achieve in reality.

In 5D scheduling, the standard estimating technique of relying on two-dimensional drawing measurements is replaced with a three-dimensional project model that brings costs into the virtual building life cycle. A BIM is linked to a construction cost estimate through material quantities that are generated from the data within the model. 5D estimation can assist with budgeting, provide cost-loaded schedules (cash flow), and display interactive forecasts to make comparison among various scenarios. When the design is changed, cost adjustments can be calculated in real time. 5D scheduling can enable the general contractor, designer, and owner improve design with value-engineering decisions.

Useful Project Scheduling Software

Primavera (Oracle®) – scheduling, risk analysis, opportunity management, resource management, collaboration and control capabilities; widely used by contractors

Microsoft Project (Microsoft) – scheduling, assigning resources to tasks, tracking progress, managing the budget, and analyzing workloads; widely used by civil engineering firms

SYNCHRO (Bentley®) – visual planning, automated quantity take-off and "what if" scenarios; enables reviews with AEC professionals, clients, and stakeholders

Navisworks (Autodesk®) – opening, combining, and reviewing 3D models; plug-ins add interference detection, 4D time simulation, and photorealistic rendering

ALICE Technologies – using artificial intelligence (AI) to explore multiple construction scenarios in order to optimize project bidding, project duration, and labor and equipment costs

---

19.5.5   Artificial Intelligence (AI)

Artificial intelligence (AI) techniques offer a new approach to scheduling. These tools can assist in the planning process by making informed suggestions as to what durations, sequence, and cost of activities should be. Even developing one or two schedules for large construction projects is a time-consuming but necessary challenge. By using AI, multiple fully resource-loaded schedules with a clear cost and time impact of each iteration can be developed within hours.

19.6   PROJECT ESTIMATING

There is a difference between project *budgeting* and project *estimating*. Typically, budgeting begins at the beginning of a project's life, in the pre-design phase, and must be in alignment with the owner's funding limits. An accurate project budget helps ensure that funding is available and has been appropriated. (See Chapter 7 – Executing a Professional Commission–Project Management for information regarding early cost estimates.) The budget is built using cost estimates and the project schedule. The budget provides a view of how much the project is estimated to cost and provides a basis of the cost performance baseline. This baseline is a critical ingredient in performing earned value analysis and other cost tracking techniques.

Costs Associated with Constructed Facilities

The owner's costs of a constructed facility include both the initial capital cost and the subsequent operation and maintenance costs.

The *capital cost* for a construction project includes the costs related to initial land acquisition, design, and construction:

- Land acquisition, including rights-of-way and improvements (demolition, utilities)
- Planning and feasibility studies
- Architectural and engineering design
- Construction, including materials, equipment, and labor
- Field supervision of construction
- Construction financing
- Insurance and taxes during construction
- Owner's general office overhead
- Equipment and furnishings not included in construction
- Inspection and testing

The *operation and maintenance cost* in subsequent years over the project life cycle includes the following expenses:

- Land rent, if applicable
- Operating staff
- Labor and material for maintenance and repairs
- Periodic renovations
- Insurance and taxes
- Financing costs
- Utilities
- Owner's other expenses

—Adapted from Carnegie Melon University, Project Management

From the contractor's perspective, estimating involves determining costs associated with work packages or activities within the project schedule. After the scope has been defined, cost estimation can be used to forecast financial and other resources needed to complete a project. Cost estimation accounts for each element required for the project—materials, labor, equipment, subcontracts, etc.—and calculates the total amount, determining a project's cost.

19.6.1   Types of Construction Cost Estimates

As shown in Table 19.1 – *Types of Cost Estimates in Planning and Design*, cost estimates vary depending on the information available and level of accuracy sought. Availability of knowledgeable personnel is a major factor in creating and analyzing construction cost estimates. Estimators use a combination of techniques producing varying levels of accuracy. The most accurate estimates rely on definitive, rather than conceptual, methods. Table 19.6 highlights several construction cost estimating methods.

Construction cost estimates can be either *substantive* or *definitive*. A substantive estimate uses acceptably finalized drawings and specifications (approximately 90% complete) to create a moderately accurate cost estimate. A substantive estimate is accurate enough to generate a bid or to control project costs. A definitive estimate is created when drawings and specifications are almost fully defined (nominally 100% complete). The most accurate and reliable estimates are definitive (bottom-up estimating), and these estimates are used to produce bids, tenders (formal offers), and cost baselines.

Capturing the entire scope is the most critical aspect of construction cost estimates; everything else develops from the scope. When the scope is well-defined, then the cost estimator can quantify the scope and establish production rates. If an aspect or item of the scope is missed, the estimate will be inaccurate. This can be the difference between a profitable or loss-making job.

Estimating becomes more straightforward when complete drawings and specifications are available and the site can be investigated. Site inspections should include information regarding:

- Site transportation access
- Utilities
- Water supply
- Meteorological data
- Geological evidence/surface exposures of rock or other materials (in addition to the geotechnical report)
- Type and availability of natural materials (e.g., aggregate for fill and/or concrete mix)
- Local conditions (e.g., terrain, protection of private property or natural features)
- Local labor conditions (e.g., availability, housing, union vs. non-union labor)
- Local sources of supply (e.g., equipment, equipment repairs, vendors)
- Potential site layout for storing materials and equipment

TABLE 19.6 Construction cost estimating methods.

Name	Description and Use
Analogous	<ul><li>Top-down estimating leveraging historical data from previous projects</li><li>Costs on previous projects adapted to current project</li><li>Useful for producing estimates when information is limited</li><li>Can be created quickly</li><li>Used only when the projects are similar in fact, not solely in appearance</li><li>Relies heavily on estimators' experience and expertise</li><li>Special consideration given for inflation, location, current/future market conditions, availability of labor and equipment</li></ul>
Parametric	<ul><li>Large project elements, e.g., demolition, cut and fill, highway pavement, structural streel, expressed in various parameters (cost per linear foot, square feet, cubic yards, tons, etc.)</li><li>Costs aggregated to arrive at an overall project cost</li><li>Parameters must be quantifiable</li><li>Not as accurate as a bottom-up estimate</li></ul>
Bottom up	<ul><li>Determines cost for each resource needed to create project components and sums costs of all resources</li><li>Includes a contingency reserve</li><li>Aggregates the total into a bottom-up estimate</li><li>Used when project is well defined and project details are known</li><li>Time consuming</li><li>Typically, prime designers/general contractors compile costs from information provided by subconsultants/subcontractors</li></ul>
Three-point	<ul><li>Used when uncertainty, risk, and/or unknowns are high</li><li>Provides a range and an expected cost</li><li>Three estimates are made based on the best case (optimistic), most likely, and worst-case (pessimistic) scenarios</li><li>Most common way of calculating a weighted average (cost expected) is $c_e = (c_o + 4c_m + c_p) / 6$</li><li>Very time consuming</li><li>Provides a justification for contingency (cost reserve)</li></ul>

However, not even definitive estimates remain static through project execution. Estimators make numerous assumptions based on risks of all magnitudes. If these base assumptions change significantly, or if additional risks are recognized, cost estimates are updated. Should this happen, the estimator will adjust the project cost baseline accordingly so that project performance can be tracked accurately.

19.6.2   Sources for Construction Cost Information

Most engineers need consultants to perform estimates. Construction estimators draw on various types of information. Following are some sources for construction cost information:

- **Historical Data:** Many construction cost estimation methods utilize historical data, such as cost per square foot to construct past projects, the average labor costs per hour for trade specialties (such as carpenters or electricians), and the units of work, such as the cost per pile cast. Data can come from a company's previous projects or from external industry references. Historical production rates largely remain constant for the same type of work; however, costs must capture a specific location and moment in time.

- **Cost Research:** The most current data can be obtained from contemporaneous cost checks by contacting suppliers to get price quote for each individual cost element (all materials, labor, and equipment) This kind of information increases accuracy, but the process is exceedingly labor intensive.

- **Cost References:** Manuals including *Walker's Building Estimator's Reference Book* and industry databases, such as RSMeans data, provide cost data as well as other information on productivity rates, city cost indexes, crew composition, and contractors' overhead and profit rates.

- **Cost Indexes:** Using cost indexes, which reflect trends in prices for various construction inputs, as well as productivity and inflation, offer an additional way to estimate current costs. Estimators can factor construction industry reports with government data together. Private indexes, like the CBRE (Coldwell Banker Richard Ellis) Group and Turner Building Index, the Engineering News-Record, US Census Bureau, and US Federal Reserve Bank produce cost index information.

- **Expert Judgment:** Experienced construction personnel and subcontractors (steel erectors, concrete providers, etc.) can be a source of highly accurate information. By gathering costs for each substructure or from each subcontractor, the estimator can assemble an overall estimate. Estimators can impress accuracy by averaging the numbers obtained for each element of the project from multiple subcontractors.

If the cost data obtained has not been modified to reflect factors such as inflation, regional variances, or site conditions, the estimator or estimating team *must* revise their numbers accordingly.

Cost Estimating Resources

- **AACE** International (American Association of Cost Engineers)
- **ASPE** (American Society of Professional Estimators)
- **ASTM E1557-09(2020)e1**: Standard Classification for Building Elements and Related Sitework—UNIFORMAT II
- **ASTM E2083-05(2016)**: Standard Classification for Building Construction Field Requirements, and Office Overhead & Profit
- **GAO-20-195G** (US Government Accountability Office): Cost Estimating and Assessment

19.6.3   Factors Considered in Estimates

Many factors need to be considered in preparing accurate cost estimates. The primary factors in developing estimates include:

- Standardization of the process
- Alignment of objectives between owner's bid documents and contractor
- Selection of appropriate methodology
- Collection of project data and historical costs
- Organization of estimate into desire format
- Documentation of basis, accuracy
- Review and checking
- Complete accounting of entire scope

The construction cost estimator specifically needs to consider:

19.6.3.1   Level of Accuracy Needed

Before any estimate can begin, both the quality of information available (How complete are the drawings and specifications? What is known about the site?) and the level of estimate desired need to be determined. The time required/available to complete the estimate also needs to be established. The level of accuracy required bears directly on the time required to perform the estimate.

19.6.3.2   Organization

Cost estimators assemble their work in ways that can be understood by various project participants. Additionally, as time is frequently of essence, estimates are arranged so that a team can work of the estimate simultaneously. Many estimates are organized around the project's specification system, such as the Construction Specification Institute's (CSI) MasterFormat®. (For a discussion of MasterFormat® see *Chapter 6 – What Engineers Deliver.*) Another commonly used specification system that can be used to format estimates is UNIFORMAT II, which organizes major building elements into discreet classifications (see Figure 19.9).

19.6.3.3   Quantity Take-Off Procedures

In the pre-construction phase, cost estimators review the project drawings and "take off" measurements to forecast construction costs. "Taking" the information "off" drawings enables estimators to list items to be valued with measurable quantities. As a crucial step in a project's early stage, accurate quantity take-offs can make or break project success.

ASTM UNIFORMAT Classification of Building Elements (E1557–97)		
Level 1 Major Group Elements	Level 2 Group Elements	Level 3 Individual Elements
A. SUBSTRUCTURE	A10 Foundations	A1010 Standard Foundations A1020 Special Foundations A1030 Slab on Grade
	A20 Basement Construction	A2010 Basement Excavation A2020 Basement Walls
B. SHELL	B10 Superstructure	B1010 Floor Construction B1020 Roof Construction
	B20 Exterior Closure	B2010 Exterior Walls B2020 Exterior Windows Exterior Doors
	B30 Roofing	B3010 Roof Coverings B3020 Roof Openings
C. INTERIORS	C10 Interior Construction	C1010 Partitions C1020 Interior Doors C1030 Specialties
	C20 Staircases	C2010 Stair Construction C2020 Stair Finishes
	C30 Interior Finishes	C3010 Wall Finishes C3020 Floor Finishes C3030 Ceiling Finishes
D. SERVICES	D10 Conveying Systems	D1010 Elevators D1020 Escalators & Moving Walks D1030 Material Handling Systems
	D20 Plumbing	D2010 Plumbing Fixtures D2020 Domestic Water Distribution D2030 Sanitary Waste D2040 Rain Water Drainage D2050 Special Plumbing Systems
	D30 HVAC	D3010 Energy Supply D3020 Heat Generating Systems D3030 Cooling Generating Systems D3040 Distribution Systems D3050 Terminal & Package Units D3080 Controls & Instrumentation D3070 Special HVAC Systems & Equipment D3080 Systems Testing & Balancing
	D40 Fire Protection	D4010 Fire Protection Sprinkler Systems D4020 Stand-Pipe & Hose Systems D4030 Fire Protection Specialties D4040 Special Electrical Systems
	D50 Electrical	D5010 Electrical Service & Distribution D5020 Lighting & Branch Wiring D5030 Communication & Security Systems D5040 Special Electrical Systems
E. EQUIPMENT & FURNISHINGS	E10 Equipment	E1010 Commercial Equipment E1020 Institutional Equipment E1030 Vehicular Equipment E1040 Other Equipment
	E20 Furnishings	E2010 Fixed Furnishings E2020 Movable Furnishings
F. SPECIAL CONSTRUCTION & DEMOLITION	F10 Special Construction	F1010 Special Structures F1020 Integrated Construction F1030 Special Construction Systems F1040 Special Facilities F1050 Special Controls & Instrumentation
	F20 Selective Building Demolition	F2010 Building Elements Demolition F2020 Hazardous Components Abatement

FIGURE 19.9 UNIFORMAT.
Source: https://www.uniformat.com/

Take-offs can involve both digital and manual estimating. Many quantity take-off software packages exist. The contractor may adopt a standard computer take-off system, or the owner may require that a specific system be used. Determining the approach to be taken early is essential, along with:

- Documenting assumptions
- Determining adjustments for the estimate of risk, inflation, and geographic location
- Establishing the use of cost indexes and where/how contingencies will be used, e.g., will contingencies be applied to each item or as one line-item at the end of the estimate.

19.6.3.4   Site

For contractors operating locally, many site factors will be well-known. However, no two sites are exactly alike and various conditions can change over time. Consequently, all site information should be current and complete.

19.6.3.5   General Concept of Job Operations

Before starting a cost estimate, a cost estimator must review the contract documents and develop a concept of:

- Main features of the project
- How the contractor is going to build the project
- Major phases of construction
- General site accessibility and layout
- Major on-site facilities that might be used, e.g., concrete batch plant
- Production rates required to construct the project within the agreed schedule
- Types of equipment to be used

19.6.3.6   Materials

Waste factors and pricing methods need to be established. A waste factor is normally expressed as a percentage (%) of the required amount. Materials costs may escalate on long term projects.

19.6.3.7   Labor

The cost estimator needs to develop production rates and to adjust the estimate for a wide variety of factors that affect production rates, such as:

- Weather
- Site and job working conditions

- Labor skill, training, and motivation
- Quality of drawings and specifications
- Expertise of superintendent's jobsite management, project management, and company management

In addition to establishing site information and labor rates, the estimator also needs to determine:

- Basic wages and whether prevailing, Davis-Bacon, wages will be used
- Fringe benefits
- Tax, insurance, and social benefits not included in indirect labor costs
- Special occupational safety and health provisions
- Equal employment opportunity regulations

19.6.3.8   Equipment

The estimator needs to determine the types of equipment to be used and the related production rates. The equipment may be owned by the construction company, leased, or purchased specifically for the project; it may incur freight costs if being shipping to the site. Availability of skilled labor is an important factor in determining equipment hourly rates. In heavy civil construction, equipment operating costs can form a very large percentage of the overall estimate.

19.6.3.9   Overhead and Profit Markups

The estimator must establish indirect labor costs, such as:

- Insurance – general liability
- Safety and personal protective equipment (PPE)
- Job site trailers, bathrooms, fencing
- Office equipment and supplies
- Computer costs
- Office salaries
- Project managers
- Superintendents and support staff
- Equipment and vehicle costs, including maintenance, insurance, licenses, depreciation, and fuel
- Payroll details — including health insurance and employer-paid taxes, worker's compensation
- Taxes

Additionally, the contractor's profit margin needs to be established. The profit plus overhead, site direct costs, and indirect costs form the contract price.

19.6.3.10 Unit Prices

The estimator must know the terms of the contract for construction. Many large civil infrastructure projects are costed using unit pricing. In unit price contracts, the contractor is paid for the actual quantity of each element of work performed per field measurements. The unit price of each element of work includes all labor, material, equipment, overhead, and profit attributable to it. Appendix X has an example of a bid for a bridge that utilizes unit pricing.

19.6.3.11 Other Factors

Other factors that need to be considered in estimate preparation include:

- Project financing (owner's and contractor's cash flow needs)
- Bonding capacity
- Availability of skilled estimating personnel
- Availability of construction crews to perform the work competently
- Geographical location in relationship to site
- Severity of the competition

See Appendix G for an example of a construction cost estimate.

Davis-Bacon and Related Acts

The Davis-Bacon and Related Acts apply to contractors and subcontractors performing on federally funded or assisted contracts in excess of $2,000 for the construction, alteration, or repair (including painting and decorating) of public buildings or public works. Davis-Bacon Act and Related Act contractors and subcontractors must pay their laborers and mechanics employed under the contract no less than the locally prevailing wages and fringe benefits for corresponding work on similar projects in the area.

The Davis-Bacon Act directs the Department of Labor to determine such locally prevailing wage rates. The Davis-Bacon Act applies to contractors and subcontractors performing work on federal or District of Columbia contracts. The Davis-Bacon Act prevailing wage provisions apply to the "Related Acts," under which federal agencies assist construction projects through grants, loans, loan guarantees, and insurance.

For prime contracts in excess of $100,000, contractors and subcontractors must also, under the provisions of the Contract Work Hours and Safety Standards Act, as amended, pay laborers and mechanics, including guards and watchmen, at least one and one-half times their regular rate of pay for all hours worked over 40 in a workweek. The overtime provisions of the Fair Labor Standards Act may also apply to DBA [Defense Base Act]-covered contracts.

US Department of Labor
https://www.dol.gov/agencies/whd/government-contracts/construction

Other Construction Related Information Contained In This Text

- Tracking costs – see Chapter 4
- Change orders and modifications to the contract – see Chapter 6
- Contract documents – see Chapter 6
- Contract types (lump sum, cost plus a fixed fee, etc.) – see Chapter 11
- Project delivery systems (design-bid-build, design-build, etc.) – see Chapter 11
- Insurance and Bonds – see Chapter 11
- Model contracts (American Institute of Architects-AIA, Design Build Institute of America-DBIA, Engineers Joint Contract Documents Committee®-EJCDC®, Consensus Docs-Associated General Contractors et al.) – see Chapter 11

19.7   PROJECT CLOSE-OUT

Project closeout, the final stage of a construction project, frequently absorbs more time than typically reflected in initial project schedules and can result in delaying completion dates. The key components in project close-out include:

- System testing and start-up
- Final inspection
- Guarantee and warranties
- Lien releases
- Record drawings
- Disposition of project files
- Post project critique and owner feedback

19.7.1   System Testing and Start-Up

Though particularly complex for industrial plant projects, system testing and start-up is an essential component of the construction process. Testing of individual components takes place throughout construction, and, as the project nears completion, testing of entire systems and major components occurs. Developing a plan early for system start-up that includes owner representatives/construction managers, designers, the general contractor, and key subcontractors is essential.

19.7.2   Final Inspection

According to Garold D. Oberlender:

The start of project close out begins near the end of the project, when the contractor requests a final inspection of the work. Prior to the request, a punch list is prepared listing all items still requiring completion or correction. To develop this punch list, the field inspection personnel must carefully review their daily inspector's log to note all work items which have been entered that require corrective actions. It is sometimes necessary to recycle through the punch list process several times before the work is satisfactory for acceptance. The final walk-through inspection should include representatives of the owner, contractor, and the key design professionals (the architect, as well as, civil, electrical, mechanical engineers, etc.) who have worked on the project. The project manager should schedule and conduct the final walk-through inspection.

Acceptance of the work and final payment to the contractor must be done in accordance with the specification in the contract documents. Substantial completion of the project is the date when construction is sufficiently complete in accordance with the contract documents so the project can be used for the purposes it was intended. This means that only minor items remain to be finished and that the project is complete enough to be put in use. The contractor may issue a Certificate of Substantial Completion with an attached list of all the work remaining to be done to complete the project.

19.7.3   Guarantees and Warranties

Before final payment, the contractor must submit guarantees and that all material, equipment, and work is of good quality and complies with the requirements of the contract documents, usually for one year after construction completion. This period can be extended contractually. Warranties, operating instructions, manuals, and spare parts for individual pieces of equipment and components also must be submitted to the owner.

The warranty periods on individual pieces of equipment typically vary from one to five years, and the contractor cannot receive final payment until all warranties have been submitted to the owner.

19.7.4 Lien Releases

Subcontractors, material suppliers, and workers who have worked on a construction project but have not been paid can file liens against the owner's property. These (mechanic) liens can be filed even if the owner has paid the general contractor fully. To encourage the general contractor to pay subcontractors, material suppliers, and workers, the owner typically withholds payment from the general contractor. Typically, ten percent is retained from each invoice submitted. This *retainage* is returned to the general contractor on the final invoice when all tiers of subcontractors, suppliers, and workers have provided lien releases. The release of liens is provided for in the General Conditions (see Chapter 6) of the construction contract between the owner and contractor.

19.7.5 Record Drawings

Most General Conditions of the construction contract between the owner and contractor also include provisions for record drawings. Record drawings include all changes made throughout the construction project and enable owners to have drawings that closely reflect the finished project. These documents are useful for maintenance and later modifications.

The contractor records deviations from the drawings that do not require change orders. Toward the end of the project, the contractor transmits these modifications to the designer/s for incorporation into the record drawings. Though referred to as "as-builts" in everyday language, the term "record drawings" is used for legal reasons. Essentially, designers are relying on the accuracy of the contractor's notes regarding these changes and cannot really verify that the changes accurately reflect what was built (see Chapter 11).

19.7.6 Disposition of Project Files

During construction, project managers and superintendents often maintain record files and working files. The record file contains copies of information related to contract documents (see Chapter 6) and other legal material. The working file is used for day-to-day management and usually includes copies of the documents in the record file, correspondence, meeting minutes, logs, reports, etc. These files contain a large amount of relevant information. Most organizations have defined policies regarding the disposition of these files, as they form the basis for any future legal action regarding the project.

19.7.7 Post Project Review

There are lessons learned from every project and a post project review is beneficial. These lessons can be used to improve performance of future projects. The American Society of Civil Engineers (ASCE) has a policy regarding peer review. Unfortunately, due to the fragmented nature of most construction projects, post projects reviews are all too rare. Ideally, a post project review should include the owner, end users, designers, contractor, and key subcontractors. The review might address the following questions:

- What did the owner and end user want?
- How well did the design team translate the owner's and end user's needs?
- Does the constructed project meet these aspirations?
- Does the project provide the optimal environment for users?
- Have sustainability targets been met?
- Is any additional or remedial work required?
- What can be done better next time?
- How can the current building and project delivery processes be improved?
- What specific lessons or opportunities can be learned from this project?
- How could communication among project participants be improved?

A post project review is most likely to occur when the project participants genuinely want to improve their performance and when the owner is willing to pay for the activity.

Table 19.7 summarizes the duties during the close-out phase of the owner, construction manager, designers, and contractor.

TABLE 19.7 Checklist of duties for project close-out adapted from oberlender.

Close-out item	Owner	CM	Designer	Contractor
Certificate of Substantial Completion	Approve	Review, approve, file	Review, approve	Originate
Clean-up	Observe and comment	Coordinate, enforce	Observe	Respond
Punch list	Approve as required	Coordinate, expedite work	Prepare, evaluate work	Respond
Call backs (after construction)	Request	Coordinate, arrange	Review, approve work	Respond

Source: Adapted from Oberlender.

American Society of Civil Engineers (ASCE) Policy Statement 351 – Peer Review

Approved by the Engineering Practice Policy Committee on February 8, 2022

Approved by the Policy Review Committee on March 4, 2022

Adopted by the Board of Direction on July 22, 2022

Policy

The American Society of Civil Engineers (ASCE) promotes and supports the use of peer reviews for projects. Peer review is the practice of obtaining an independent, unbiased evaluation of the adequacy and application of engineering principles, standards, and judgment from an independent group of professionals having substantial experience in the same field of expertise. Peer reviews are in addition to the normal quality control and checking procedures required on any engineering assignment.

Issue

Projects are expected to meet industry standards of performance. Peer review is a separate, important step in the design process for selected projects to provide an evaluation of design concepts and management to meet performance objectives. Peer reviews are recommended on projects:

- That are critical or could endanger public health, safety, welfare, the environment, or national defense or security.
- Where performance under emergency conditions is paramount.
- When using innovative or untested materials, techniques, or design methods.
- That could result in significant financial or operational losses.

Peer reviews should be initiated or requested by the owner/client or regulatory agency with full cooperation from the designer or consultant. Reviews address a defined scope as set forth by the initiating party. The number and timing of peer reviews vary by project. Reviews should normally occur before critical decisions or criteria are adopted as well as prior to the completion or release of contract documents. Peer reviews should be performed by independent or external professionals not associated with or accountable to the original design team.

19.8   SUMMARY

This chapter has addressed construction management and scheduling for civil engineers. Additional information on these topics can be found in other chapters of this book:

- Tracking costs – see Chapter 4
- Change orders and modifications to the contract – see Chapter 6
- Contract documents – see Chapter 6
- Contract types (lump sum, cost plus a fixed fee, etc.) – see Chapter 11 Project delivery systems (design-bid-build, design-build, etc.) – see Chapter 11
- Insurance and Bonds
- Model contracts (American Institute of Architects-AIA, Design Build Institute of America-DBIA, Engineers Joint Contract Documents Committee®-EJCDC®, Consensus Docs-Associated General Contractors et al.) – see Chapter 11

As with all topics covered in the *Civil Engineer's Handbook of Professional Practice*, the authors hope that developing a fuller understanding of the A/E/C industry will lead to more profitable projects and successful careers.

BIBLIOGRAPHY

Barrie, Donald S. and Paulson, Boyd C. (1991). *Professional Construction Management*, 3rd. edition. McGraw-Hill, Inc. New York, New York.

Frein, Joseph P. ed. (1980). *Handbook of Construction Management and Organization*, 2nd edition.. Van Nostrand Rheinhold, New York City, New York, ISBN-0-442-22475-3.

Glavinich, Thomas E. (2004). *Construction Planning and Scheduling*, 2nd edition. Associated General Contractors, Arlington, VA, ASIN: B00DS92A0I.

Oberlender, Garold D. (2014). *Project Management for Engineering and Construction*, 3rd edition. McGraw Hill, New York City, New York, ISBN-10: 0071822313; ISBN-13: 978-0071822312.

https://www.agc.org/learn/construction-data (accessed November 16, 2021).

https://www.asce.org/advocacy/policy-statements/ps351—peer-review (accessed August 12, 2022).

https://www.cmu.edu/cee/projects/PMbook/05_Cost_Estimation.html (accessed August 11, 2022).

https://www.cmu.edu/cee/projects/PMbook/13_Quality_Control_and_Safety_During_Construction.html (accessed November 21, 2021).

https://www.dol.gov/agencies/whd/government-contracts/construction.html (accessed August 11, 2022).

https://www.imperial.ac.uk/estates-projects/project-procedures/stages/.html (accessed August 12, 2022).

Health and Safety Knowledge for Civil Engineers

Big Idea

"An ounce of prevention is worth a pound of cure."

—Benjamin Franklin

"Safety has to be everyone's responsibility . . . everyone needs to know that they are empowered to speak up if there's an issue."

—Captain Scott Kelly

Key Topics Covered

- The Occupational Safety and Health Administration (OSHA)
- Health and Safety Programs and Project Plans
- Civil Engineers Health and Safety Tool Box
- OSHA Quick Reference Card for Top 4 Construction Site Hazards and Construction Personal Protective Equipment (PPE)
- Organization/Company/Employee Responsibilities and Employee Rights

Civil Engineer's Handbook of Professional Practice, Second Edition. Karen Lee Hansen and Kent E. Zenobia.
© 2025 John Wiley & Sons, Inc. Published 2025 by John Wiley & Sons, Inc.
Companion website: www.wiley.com/go/hansen/CivilEngineersHandbook

Related Chapters in This Book

- Chapter 3: Ethics
- Chapter 4: Professional Engagement
- Chapter 5: The Engineer's Role in Project Development
- Chapter 6: What Engineers Deliver
- Chapter 7: Executing a Professional Commission, Project Management
- Chapter 9: The Client Relationship and Business Development
- Chapter 10: Leadership
- Chapter 11: Legal Aspects of Professional Practice
- Chapter 12: Managing the Civil Engineering Enterprise
- Chapter 18: Human Relations Policies and Employment Practices
- Chapter 19: Construction Management
- Chapter 21: What Engineers Need to Know

20.1 INTRODUCTION

Civil engineers are viewed as "individual and/or team leaders" and must be aware of critical Health and Safety (H&S) Laws and methods to safeguard employees, the general public, their firms or organizations, and themselves. Civil engineering staff and mangers need to possess a passion and knowledge about H&S Laws, organizational requirements, hazards and methods to mitigate hazards, required documentation, and training requirements. A civil engineer's career and success depends upon this knowledge and implementation of these requirements. H&S legal, ethical, and utility requirements are not usually taught in university; and yet they are one of the many hands-on subjects that are critical elements in a successful civil engineering career. H&S requirements are extremely important, legally required, complex, and should be of the highest priority for civil engineers and all professionals on/near jobsites. Figure 20.1 summarizes an employee's rights and responsibilities and the employer's responsibilities.

Civil engineers need to be aware of the details and requirements codified in, "The Occupational Safety and Health Act of 1970, Executive Order 12196 and 29 CFR 1960" (OSHA). Practicing "job safety and health" can safeguard one's life and others in and around the job site. Civil engineers often work in or around the construction industry which is in the top four industries as most the dangerous.

Job Safety and Health
IT'S THE LAW!

All workers have the right to:

- A safe workplace.
- Raise a safety or health concern with your employer or OSHA, or report a work-related injury or illness, without being retaliated against.
- Receive information and training on job hazards, including all hazardous substances in your workplace.
- Request a confidential OSHA inspection of your workplace if you believe there are unsafe or unhealthy conditions. You have the right to have a representative contact OSHA on your behalf.
- Participate (or have your representative participate) in an OSHA inspection and speak in private to the inspector.
- File a complaint with OSHA within 30 days (by phone, online or by mail) if you have been retaliated against for using your rights.
- See any OSHA citations issued to your employer.
- Request copies of your medical records, tests that measure hazards in the workplace, and the workplace injury and illness log.

This poster is available free from OSHA.

Contact OSHA. We can help.

Employers must:

- Provide employees a workplace free from recognized hazards. It is illegal to retaliate against an employee for using any of their rights under the law, including raising a health and safety concern with you or with OSHA, or reporting a work-related injury or illness.
- Comply with all applicable OSHA standards.
- Notify OSHA within 8 hours of a workplace fatality or within 24 hours of any work-related inpatient hospitalization, amputation, or loss of an eye.
- Provide required training to all workers in a language and vocabulary they can understand.
- Prominently display this poster in the workplace.
- Post OSHA citations at or near the place of the alleged violations.

On-Site Consultation services are available to small and medium-sized employers, without citation or penalty, through OSHA-supported consultation programs in every state.

1-800-321-OSHA (6742) • TTY 1-877-889-5627 • www.osha.gov

FIGURE 20.1 OSHA job safety and health—it's the law.

According to the most recent data available in 2020, depending on the measure used:
    The construction industry is ranked number one in the top four industries as most danger-
ous, and construction experienced the most workplace deaths. (MacLaury 1981; National
Safety Council (NSC)).

OSHA covers most private sector employers and their workers in all 50 states, the
District of Columbia, and other US jurisdictions either directly through Federal OSHA
or through an OSHA-approved state program. In addition, this law requires the heads of
Federal agencies to furnish to employees with workplaces and conditions of employment
that are free from job safety and health hazards. See Table 20.1 for additional details on
OSHA requirements.

20.2   THE OCCUPATIONAL SAFETY AND HEALTH ADMINISTRATION (OSHA) AND ITS IMPORTANCE TO CIVIL ENGINEERS

Every working American worker is affected by OSHA standards. All working Americans have
a right to work in a safe working place. OSHA (both the law and the agency) was created
"to assure so far as possible every working man and woman in the nation safe and healthful working conditions and to preserve our human resources." OSHA provides rules and
requirements for employees and employers for the creation of a safe working environment
for workers that also benefits employers. OSHA protects over 130 million employees in the
US, and increases employee productivity, keeps employees at work, increases organizations'
overall profitability and effectiveness, and protects workers and organizations from lawsuits
or legal ramifications.

"To achieve this, federal and state governments work together with more than 100 million working men and women and eight million employers. Some of the things OSHA does
to carry out its mission are:

- Providing training programs to increase knowledge about occupational safety
  and health,
- Developing job safety and health standards and enforcing them through worksite
  inspections,
- Maintaining a reporting and recordkeeping system to keep track of job-related injuries
  and illnesses.

OSHA also assists States in their efforts to assure safe and healthful working conditions,
through OSHA-approved job safety and health programs operated by individual states.
State plans are OSHA-approved job safety and health programs operated by individual

TABLE 20.1 OSHA safety and health protection summary—responsibilities for employers, employees, and rights of employees.

Occupational Safety and Health Protection for Employees of the (Insert Your Agency Name Here)*

The Occupational Safety and Health Act of 1970, Executive Order 12196 and 29 CFR 1960 require the heads of Federal agencies to furnish to employees places and conditions of employment that are free from job safety and health hazards.

Responsibilities of Your Agency

1. General Requirements

The head of your agency will furnish Your agency employees places and conditions of employment that are free from on-the-job safety and health hazards.

2. OSHA Regulations

Your agency will comply with applicable regulations of the Occupational Safety and Health Administration.

3. Reporting Hazards

Your agency will respond to employee reports of hazards in the workplace.

4. Workplace Inspections

Your agency will insure that each workplace is inspected annually for hazardous conditions. Your agency will post Notices of Unsafe or Unhealthful Working Conditions found during the inspections for a minimum of three working days, or until the hazard is corrected, whichever is later.

8. Reporting Accidents, Injuries and Occupational Illnesses

Supervisors must submit a supervisor's report of accidental injury/illness for all work-related accidents, injuries, or occupational illnesses experienced by employees under their supervision.

9. Safety and Health Committees

Your agency will support any safety and health committees that are formed from management and employee representatives.

Employee Responsibilities

1. Compliance with Standards

Employees shall comply with all OSHA and approved Your agency occupational safety and health standards, policies and directives.

3. Reporting Hazards

Employees and their representatives shall have the right to report unsafe or unhealthful working conditions to appropriate officials and to request an inspection of the workplace. The name of the employee making the report will be kept confidential if requested.

4. Freedom from Fear of Reprisal

Employees and their representatives are protected from restraint, interference, coercion, discrimination, or reprisal for exercising any of their rights under the Your agency Safety and Health Program.

(Continued)

TABLE 20.1 *(Continued)*

5. Correction of Unsafe Conditions Your agency will take prompt action to assure that hazardous conditions are eliminated. Imminent danger conditions will be corrected immediately. **6. Safety and Protective Equipment** Your agency will acquire, maintain and require use of appropriate protective and safety equipment. **7. Safety and Health Training** Your agency will provide occupational safety and health training for employees.	**2. Safety and Protective Equipment** Employees shall use appropriate protective and safety equipment provided by Your agency. **Rights of Employees and Their Representatives** **1. Participation in Safety and Health Program** Employees and their representatives shall have the right to participate in the Your agency Safety and Health Program. Employees shall be authorized official time for these activities. **2. Access to Records and Documents** Employees and their representatives shall have access to copies of applicable OSHA and other recognized standards and regulations; Your agency safety and health policies and directives; accident, injury and illness statistics of the Your agency.	**Responsible Officials** The Designated Agency Safety and Health Official (DASHO) for Your agency is Name, Title. The Safety and Health Designee for this workplace is: and may be contacted at (Telephone and location) **Further Information** This notice highlights the Your agency employee job safety and health program. More information about the Your agency program or its standards and procedures may be obtained from the workplace Safety and Health Designee.

*Section 5(a)(1) of the Occupational Safety and Health Act is referred to as the "General Duty Clause" which requires an employer to furnish to its employees: "employment and a place of employment which are free from recognized hazards that are causing or are likely to cause death or serious physical harm to his employees..." There3frore the OSHA requirements apply to employers and not just the Federal Government. Employers must comply with all applicable OSHA standards.

Source: Occupational Safety and Health Act (1970) /U.S Department of Labour/Public Domain.

states instead of federal OSHA. States with approved plans cover most private sector employees as well as state and local government workers in the state" (Miller 2020).

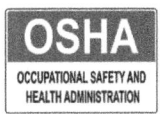

20.2.1   How Does OSHA Increase Employee Productivity and Keep Employees at Work?

OSHA focuses on:

- Safe and healthy work conditions,
- Providing training programs,
- Providing and enforcing standards for varying workplace sectors,
- Fostering general principles of safety,
- Providing specific workplace standards for fields like research and medical practice, and
- Overseeing research about new rules.

Simply put, OSHA's mission is to provide safe and healthful working conditions. This mission should therefore be the adopted mission for every employer and organization and thus inherited by every civil engineer, employee, and manager (MacLaury 1981).

Employers can increase happiness among employees by showing they care and by providing them with a safe and healthy working environment. If employees feel unsafe or uncared for it is likely to have a direct impact manifested by lower productivity. Benefits of increased productivity for employers and employees include:

- Providing excellent customer service that boosts loyalty
- Building a safety culture among employees, caring for the employee's health and welfare
- Boosting employee morale
- Providing (possible) better incentives to your employees as the company prospers
- Spending less money on employee accidents, workman's compensation claims, and resulting healthcare
- Maintaining a Company's or Organization's reputation

OSHA regulations help reduce future incidents by identifying potential hazards, reviewing safety procedures with employees to make sure they are well-known, and recordkeeping information about events. A safer environment keeps employees at work by reducing the chances of accidents or health problems (Zak 2018).

Which groups DO NOT come under OSHA's coverage?

1. The self-employed.
2. Immediate members of farming families not employing outside workers.
3. Mine workers, certain truckers and transportation workers, and atomic energy-workers who are covered by other federal agencies.
4. Public employees in state and local governments, although some states havetheir own plans that cover these workers (Miller 2020).

In the study conducted by the University of Warwick, a researcher, Dr. Sgroi, stated: "The driving force seems to be that happier workers use the time they have more effectively, increasing the pace at which they can work without sacrificing quality." A happy employee doesn't count the minutes until his or her shift is over, but instead actively participates in the workplace.

The Importance of H&S Training

Even though OSHA has had an impact on worker safety and health, significant hazards and unsafe conditions still exist in US workplaces (Miller 2020).
Each year:

- An average of 15 workers die every day from job injuries;
- Over 5,600 Americans die from workplace injuries annually;
- Over 4 million non-fatal workplace injuries and illnesses were reported; and
- The estimated cost of occupational injuries and illnesses are from $145 billion to $290 billion a year for direct and indirect costs (MacLaury 1981, Miller 2020).

20.2.2   Protection from Lawsuits and/or Legal Ramifications

OSHA can send inspectors to business locations for standards compliance and assessments. When a company passes inspections, documentation is provided. Then, if any personal injury lawsuits are filed by workers, these lawsuits are more difficult to defend when there is a recent inspection on file. If lawsuits are not a company's or organization's biggest concern, accidents on the job also can cost organizations a great deal of money from the missing personnel to medical costs.

Death or Multiple Injury Reporting Requirements

Contractors working for a Federal Agency under a US Army Corps of Engineers (USACE) contract are responsible for notifying OSHA in accordance with 29 CFR 1904.39 within 8-hours when their employee(s) is fatally injured or 1 or more persons are hospitalized as inpatients as a result of a single occurrence.

Most State H&S Agencies, Private Companies, and other Organizations have similar reporting requirements since Federal OSHA requirements dictate this requirement.

20.2.3 Civil Engineers Are Encouraged to Continue Learning about OSHA

Civil engineers should learn about the importance of OSHA in the workplace and consider requesting management to invest in OSHA compliance training for employees through outside consultants or contractors. There are many, many training classes, from back injury prevention to ergonomics training.

Civil engineers should also remember the importance of OSHA standards applicable to themselves, their colleagues, any contractors/stakeholders or people at or near the job site, their workspaces, and their industry. Safety should be extended and practiced at home to maintain a safe and healthy lifestyle. Many industry and public clients ask to assess the engineering firms' safety records as part of their assessment for hiring qualified firms. H&S training requirements, good practices, and other training mandated by the Human Resources Department should be kept up to date.

The Importance of Reporting/Discussing Mishaps and Lessons Learned

The purpose of investigating a mishap from the accident is to share information in hopes that it will not happen again, to mitigate the real cause(s) of the incident, and to track accident trends for regulation updates. This data is sent to OSHA, State, and many local H&S Agencies, as well as the company/organization insurance carriers.

20.2.4 How OSHA Affects Civil Engineers in the Workplace

The civil engineer's project, job, or construction site is typically a busy and important place with a lot of potential human health hazards. The project site may include multiple pieces of moving and heavy equipment, raw material deliveries and storage, overhead and underground utilities, electrical motors and switchgear, excavations, enclosed spaces, pressurized gases and cylinders, slip/trip and fall hazards, all of which are typically compounded by tight timeframes and schedule requirements with big-dollar price tags. These conditions create a situation ripe for injuries or mistakes. In addition, typically many parties are involved in the project as part of the project team including any mix of the following: civil engineering designers, civil engineer construction management team, the owner, stakeholders, permitting agencies, contractors and sub-contractors, raw material providers, utility company operators, H&S staff, possibly local citizens, skilled and unskilled laborers. In general, staff come from different companies, with differing cultures and backgrounds, differing and sometimes competing missions must all work together safely in and around job sites (MacLaury 1981, Zak 2018).

Construction site employees are typically in one of these categories:

- Management such as the owners and stakeholders,
- Technical and specialty staff such as civil engineers, permitting staff, H&S staff, construction management staff, project manager, and support staff, and
- Laborer's work force including highly skilled workers like crane operators, skilled workers such as plumbers, and unskilled staff.

The project/construction staff includes an eclectic mix of knowledge, experience, and training sometimes with opposing objectives. The management staff has vision and direction. They may or may not be aware of design requirements/limitations, construction procedures, permitting requirements, or division of labor tasking. There are personnel with high educational qualifications, some highly experienced, some with limited experience, and still others with detailed knowledge of schedule and budget requirements. And, many skilled laborers with valuable experience, some highly educated or with limited education, may or may not be aware of innovative solutions or detailed knowledge of schedule and budget requirements. This eclectic mix of project personnel are required to follow the job site H&S requirements.

Typical construction site safety requirements just to visit the site:

- Hard Hat and High Visibility clothing to be worn by all visitors and workers
- Eye protection and hearing protection
- Written Record of Visitors—Sign in at the Job Trailer
- Review and sign off on Construction company's daily hazard assessment listed on Job Hazard Analysis (JHA). The typical term for a JHA in private industry is a Job Safety Analysis (JSA) or an Activity Hazard Analysis (AHA).

Generally, all staff, management, technical, skilled, and semi-skilled workers are at riskof being injured, death, or various illnesses originating at the project/construction site. The level of risk varies with personnel classification, their job tasks and activities, or whether they are in the field or office or both. The point is that all staff and site visitors potentially can be subjected to potential risks or harm. They "all" need to be aware of these known and potential unknown risks, how to eliminate or mitigate these risks, and how to stay safe and healthy. A hazard is a potential source of harm that can inflict an adverse health effect on that person and/or others. Often, a potential hazard and a potential risk are synonymous. Figure 20.2 illustrates two excellent references from OSHA, the Top 4 Construction Site Hazards and Construction Site Personal Protective Equipment (PPE) recommendations.

Engineering firms, construction firms, public agencies, and organizations are responsible for providing the overall programmatic guidance for developing, managing, and implementing a safety and occupational health (SOH) Program. SOH and H&S are used interchangeably here. Each project or Job should have an approved site-specific H&S Plan that fulfills the requirement of the Organization's H&S Program which in turn, follows the Federal, State and local requirements. In turn each and every project phase, element, task, and sub-task that is part of the job/project should be included in the H&S Plan.

Applicable Health and Safety Regulations for different construction sites:

- OSHA—Federal site/Federal Job and all private sector employers and their workers in the 50 states and all territories and jurisdictions under federal authority.
- EM385—US Army Corps of Engineers (USACE) Site.
- Cal/OSHA OR "Your State's OSHA"—California Site or "Your State"/State lands/ State work.

Many OSHA standards apply to residential construction for the prevention of injuries, illnesses, and fatalities (MacLaury 1981, Zak 2018, US Army Corps of Engineers (USACE), Manual No. 385-1-1 2014).

There are volumes of reference material on H&S Plans, their content, requirements, and applicability which are well beyond the content that will be covered here. The point is, civil engineers need to be aware of the H&S requirements, the Organizations' H&S Program, the legal requirements and need for site-specific H&S Plans, typical Plan contents, complexity, H&S specialty staff, individual, and management responsibilities. These key components and details will be briefly covered and discussed in the following sections.

Top Four
Construction Hazards

The top four causes of construction fatalities are: Falls, Struck-By, Caught-In/Between and Electrocutions.

Prevent Falls
- Wear and use personal fall arrest equipment.
- Install and maintain perimeter protection.
- Cover and secure floor openings and label floor opening covers.
- Use ladders and scaffolds safely.

Prevent Struck-By
- Never position yourself between moving and fixed objects.
- Wear high-visibility clothes near equipment/vehicles.

Prevent Caught-In/Between
- Never enter an unprotected trench or excavation 5 feet or deeper without an adequate protective system in place; some trenches under 5 feet deep may also need such a system.
- Make sure the trench or excavation is protected either by sloping, shoring, benching or trench shield systems.

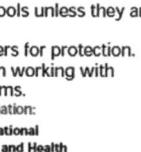

Prevent Electrocutions
- Locate and identify utilities before starting work.
- Look for overhead power lines when operating any equipment.
- Maintain a safe distance away from power lines; learn the safe distance requirements.
- Do not operate portable electric tools unless they are grounded or double insulated.
- Use ground-fault circuit interrupters for protection.
- Be alert to electrical hazards when working with ladders, scaffolds or other platforms.

For more complete information:

OSHA Occupational Safety and Health Administration
U.S. Department of Labor
www.osha.gov   (800) 321-OSHA

OSHA 3216-8N-06

Protect Yourself
Construction Personal Protective Equipment (PPE)

Eye and Face Protection
- Safety glasses or face shields are worn any time work operations can cause foreign objects to get in the eye. For example, during welding, cutting, grinding, nailing (or when working with concrete and/or harmful chemicals or when exposed to flying particles). Wear when exposed to any electrical hazards, including working on energized electrical systems.
- Eye and face protectors – select based on anticipated hazards.

Foot Protection
- Construction workers should wear work shoes or boots with slip-resistant and puncture-resistant soles.
- Safety-toed footwear is worn to prevent crushed toes when working around heavy equipment or falling objects.

Hand Protection
- Gloves should fit snugly.
- Workers should wear the right gloves for the job (examples: heavy-duty rubber gloves for concrete work; welding gloves for welding; insulated gloves and sleeves when exposed to electrical hazards).

Head Protection
- Wear hard hats where there is a potential for objects falling from above, bumps to the head from fixed objects, or of accidental head contact with electrical hazards.
- Hard hats – routinely inspect them for dents, cracks or deterioration; replace after a heavy blow or electrical shock; maintain in good condition.

Hearing Protection
- Use earplugs/earmuffs in high noise work areas where chainsaws or heavy equipment are used; clean or replace earplugs regularly.

For more complete information:

OSHA Occupational Safety and Health Administration
U.S. Department of Labor
www.osha.gov   (800) 321-OSHA

OSHA 3260-09N-05

FIGURE 20.2   OSHA quick card reference for top 4 construction site hazards and construction personal protective equipment (PPE).
Source: OSHA / Public Domain.

20.3   HEALTH AND SAFETY PROGRAMS AND PROJECT PLANS

Contractors such as engineering firms and construction firms working for the Federal Government, Department of Defense, and other federal agencies working under a US Army Corps of Engineers (USACE) contract must comply with the USACE's Safety and Health Requirements published in Manual No. 385–1-1, (including interim changes). So, if the USACE is managing a contract for any of these Federal Agencies then Manual 385–1-1 must be complied with unless a Command like, the US Navy, has other similar documents that Agency/Command prefers for required H&S compliance. But, the contents of a complete H&S Manual will be very similar (US Army Corps of Engineers (USACE), Manual No. 385–1-1 2014).

20.3.1   USACE Manual 385–1-1

USACE Manual 385–1-1 is approximately 930 pages long and its corresponding Table of Contents is 11 pages. The civil engineer is strongly encouraged to review the depth and breadth of Manual 385–1-1 Table of Contents. Refer to this information below, *to understand the complexity and seriousness and legal requirements of the engineering and construction industry and compliance with H&S Laws.* Again, the term "safety and occupational health (SOH)" and "Health and Safety (H&S)" are used interchangeably in this chapter.

The Manual 385–1-1 refers to some *basic references* the reader and the CE should become familiar with include:

- The 29 Code of Federal Regulation (CFR) 1910, Occupational Safety and Health Standards for General Industry,
- The 29 CFR 1926, Occupational Safety and Health Standards for Construction,
- The 29 CFR 1960, Basic Program Elements for Federal Employees, OSHA,
- The Executive Order (EO) 12196, Occupational Safety and Health Programs for Federal Employees, 26 Feb 1980,
- The Federal Acquisition Regulation (FAR) Clause 52.236–13, Accident Prevention, Nov 1991, and
- The Department of Defense Instruction (DODI) 6055.1, DOD Safety and Occupational Health Program, 14 Oct 2014, among others (US Army Corps of Engineers (USACE), Manual No. 385–1-1 2014).

20.3.2   USACE Manual 385–1-1 Required Plan Elements

Here are some required elements that all civil engineers should be aware of that *apply to the office and field settings*. These requirements are summarized here and referenced in more detail in the UASCE Manual 385-1-1:

General

This Section provides the overall programmatic guidance for developing, managing, and implementing a safety and occupational health (SOH) program.

- No person shall be required, instructed, or allowed to work in surroundings or under conditions that are unsafe or dangerous to his or her health.
- The employer is responsible for initiating and maintaining a H&S program that complies with the US Army Corps of Engineers (USACE) SOH requirements.
- Each employee is responsible for complying with applicable SOH requirements, wearing prescribed SOH equipment, reporting unsafe conditions or activities, preventing avoidable mishaps, and working in a safe manner.
- Supervisors shall remove employees from exposure to work hazards, or the work site when they are observed acting in an unsafe manner, or otherwise pose a potential SOH threat to themselves or others. Employees may return to the work environment after appropriate supervisory action has occurred (i.e., re-training on proper safe procedures, etc.).
- H&S programs, documents, signs, and tags shall be communicated to employees in a language that they understand.
- Worksites with non-English speaking workers shall have a person(s), fluent in the language(s) spoken as well as English, on-site when work or training is being performed, to interpret and translate as needed.
- H&S Bulletin Board. The Contractor or USACE Project shall erect and maintain a SOH bulletin board in a commonly accessed area in clear view of the on-site workers. The bulletin board shall be continually maintained and updated and placed in a location that is protected against the elements and unauthorized removal. It shall contain, at minimum, the following SOH information: (Reference EM 385-1-1)
  a. A map denoting the route to the nearest emergency care facility.
  b. Emergency phone numbers.
  c. A copy of the most current Accident Prevention Plan (APP) or Project Safety and Occupational Health (SOH) Plan, mounted on/adjacent to the bulletin board, or a notice on the bulletin board stating the location of the Plan. The location of the Plan shall be accessible on the site by all workers.
  d. The Occupational Safety and Health Administration (OSHA) Form 300A, Summary of Work-Related Injuries and Illnesses, posted in accordance with OSHA requirements (from February 1 to April 30 of the year following the issuance of this

form). It shall be mounted on/adjacent to the bulletin board, accessible on the site by all workers.

    e. A copy of the SOH deficiency tracking log mounted on/adjacent to the bulletin board or a notice on the bulletin board shall state the location where it may be accessed by all workers upon request.

    f. SOH promotional posters.

    g. Date of last lost workday injury and date of last OSHA recordable injury.

    h. OSHA Safety and Health Poster.

    i. A copy of the hazardous material inventory, identification of use, approximate quantities and site map detailing location.

USACE Business Process

- USACE Project Managers (PMs), in accordance with the SOH Reference Document (Ref Doc 8016G) contained in the USACE Business Manual, shall ensure that a SOH plan is developed for funded projects and incorporated into each Project Management Plan (PMP)/Program Management Plan (PrgMP).

    a. The PM shall collaborate with the customer and the local SOH office (SOHO) on project safety goals and objectives and communicate these through the PMP/PrgMP and Project Delivery Team (PDT) meetings.

    b. Coordination between local SOHOs of the design district and the construction district shall occur during the development of the PMP.

USACE Project Management Plan

- USACE PMs and the PDT shall develop the SOH program requirements to be incorporated in the PMP and are responsible for assuring that SOH requirements are properly addressed and executed throughout the life cycle of each project.

    a. The PM shall ensure that identified hazards, control mechanisms, and risk acceptance are formally communicated to all project stakeholders. EM 385-1-1

    b. The current Unified Facilities Guide Specification (UFGS) for Safety and Health in effect on the date of solicitation shall be used in all USACE contract work administered on behalf of the USACE under the provisions of Federal Acquisition Register (FAR) Clause 52.236–13 and on other contracts as deemed appropriate based on the risk assessment.

    c. Military Construction (MILCON) Transformation contracts will include the Federal Acquisition Regulation (FAR) Clause 52.236–13 as well as the Model Request for Proposal (RFP).

    d. Locally developed SOH requirements will not be included in contract requirements without the concurrence of the Contracting Officer (KO) and local SOHO.

    e. *When an employee is deemed to be in imminent danger, a designated representative shall immediately stop the unsafe work being performed.* See Federal Acquisition Regulation (FAR) Clause 52.236–13(d).

USACE Project SOH Plan

- For USACE activities where USACE employees are engaged in functions other than routine office or administrative duties, a Project SOH Plan shall be developed, implemented, and updated as necessary.

  a. Such activities include operations and maintenance; recreational resource management; in-house conducted environmental restoration (investigation, design, and remediation); surveying, inspection, and testing; construction management; warehousing; transportation; research and development; and other activities when the Government Designated Authority (GDA) and the command's local SOHO agree on the benefit of such a program for accident prevention.

  b. The Project SOH Plan shall address applicable items listed in Appendix A, and in addition, any local SOPs or requirements identified in the USACE Command's SOH Program.

  c. For Hazardous Waste Operations and Emergency Response (HAZWOPER) sites, refer to Section 33 for Site Safety and Health Plan (SSHP) guidance.

Position Hazard Analyses (PHA)

- A PHA shall be prepared, updated as necessary, documented by the supervisor, and reviewed by the command's SOHO for each USACE position according to the hazards associated with the position's tasks.

  For example, a PHA could be developed if needed or required for a specific classification of employees like surveyors, or engineering aids according to the hazards associated with the position's tasks. Another possible example might be a generic PHA may be used for groups of employees performing repetitive office/administrative tasks where the primary hazards result from ergonomic challenges, lighting conditions, light lifting and carrying tasks.

Accident Prevention Plans (APP) for Contract Work

- Before initiation of work at the job site, an APP shall be reviewed and approved by the Contract Manager.

  APPs shall be developed and submitted by the Contractor. The Contractor shall address each of the elements/sub-elements in the outline contained in Appendix A of EM 385-1-1 document. If an item is not applicable because of the nature of the work to be performed, the Contractor shall state this exception and provide a justification.

  a. The Contractor shall identify each major phase of work that will be performedon this contract. Within each major phase, all activities, tasks or Definable Features of Work (DFOWs) shall be identified that will require an Activity Hazard Analysis (AHA). See Section 01.A.14 and Appendix A, paragraph 3.J.

  b. The APP shall also address any unusual or unique aspects of the project or activity (US Army Corps of Engineers (USACE), Manual No. 385-1-1 2014).

These important elements from the UASCE Manual 385-1-1 have been summarized here so that the civil engineer is at least familiar with the overall scope and content of the requirements for all employees, owners, and visitors to jobsites, and even in the home organization's office. A Health and Safety Tool Box for the civil engineer is presented below as a valuable resource to civil engineers.

20.4   CIVIL ENGINEER'S HEALTH AND SAFETY TOOLBOX

Civil engineers should employ leadership qualities to share this knowledge with staff, managers, stakeholders, contractors/sub-contractors, and nearby impacted parties to the project or construction job. This can be accomplished by providing an accessible library, think of it as an H&S Toolbox. Every "craftsperson and engineering professional" should possess their own H&S toolbox to respond to their organizations', clients', company staff, or even their own personal needs. This toolbox should contain key references, training courses, records, personal protective equipment (PPE), and safety equipment.

This Toolbox contains details, outlines, general forms, typical example and real forms, OSHA reference information and user-friendly OSHA quick reference cards. Simply having a toolbox is not sufficient to assure a safe work site—*these are just tools, and they must be used correctly and consistently*. Every person on-site needs to heed to the requirements in the H&S Plan and be reminded. This suggested toolbox contents include:

- Typical Organization's H&S Program Contents
- **Typical Example of a Health and Safety Plan and Accident Prevention Plan (HSP/APP) for a Large Project**
- The Activity Hazard Analysis (AHA) Form, Content, and Use and Example above
- OSHA Quick Card Reference for Top 4 Construction Site Hazards and Construction Personal Protective Equipment (PPE)

More specifically, the suggested toolbox contents appear below.

20.4.1   Typical Organization's H&S Program Contents

Become familiar with your organization's H&S program or adopt these practices, plans, policies, training program, and recordkeeping requirements. The civil engineer can assume a leadership role working with their supervisor or manager and facilitate these key practices:

Very Important Organizational Safety and Health Plans and Practices

- ✓ Initiate General Office Safety Practices
- ✓ Initiate General Field Safety Practices

- ✓ Implement Safety Training and Recordkeeping
- ✓ Establish a Safety Document Information Location
- ✓ Establish a Safety Filing System

Other Very Important Plans and Procedures – The contents and details of these Plans and Procedures may be found on the OSHA web site or the Cal/OSHA Guidance for Construction Employers web site.

- ✓ Prepare and implement your customized Emergency Action Plan (EAP)
- ✓ Pre-work Safety Plans (PWSP)
- ✓ Injury and Illness Prevention Plan (IIPP)
- ✓ Heat Illness Prevention Plan (Including Site Specific Procedures)
- ✓ Fire Prevention Plan
- ✓ Applicable Code of Safe Work Practices
- ✓ Activity Hazard Analysis (AHA) or Job Hazard Analyses (JHA'S)
- ✓ Organization's Safety Policy and Other Specific Plans Listed Below

The details and content of several of these top 5 important practices and activities (listed above) are explained below with references and exhibits. In addition, it is recommended that CEs create a Health and Safety Library of key documents, implement employee safety training that corresponds with each person's job duties, and establish a training record filing system. Or, alternatively, your Company or Organization may have dedicated H&S Personnel or Officers that have these documents for use.

Please note these documents and **the list are key documents and are not intended to be comprehensive documents for every type of job or project that civil engineers manage or participate in**. These Plans and Practices should be "living documents" that are kept up to date, edited, reviewed, and practiced on a routine basis. It's important to note that **the CE should create their own customized Tool Box and these documents provide a good start to the Tool Box contents.** Please see below for the list of required safety documents, filing system contents and requirements. Actual photos of the documents are provided for your use.

20.4.1.1   Initiate General Office Safety Practices

1. Discuss your Organization's, Safety Program and employee safety expectations during new employee orientation.
2. Begin all job or project meetings with a safety moment and document that safety moment for specific conditions at the time of the meeting.
3. Monthly, discuss results of your Organizations H&S meetings at your office, branch, project/job, or section meetings.
4. Insure employees read and understand the Emergency Action Plan (EAP).

5. Identify a central location and readily available EAPs at the field offices for Staff.

6. Prepare a Staff Evacuation Plan so Emergency Response Personnel can easily identify if employees (staff) are inside a potentially impacted building.

7. Keep your section, branch, office Staff Evacuation Checklist up to date.

8. Establish a responsibility matrix for completing Staff Evacuation Checklist during an emergency if you or your delegate is not in the office during an emergency evacuation.

9. Employ "periodic practice sessions" to familiarize staff with everyone's responsibility and the EAP.

20.4.1.2   Initiate General Field Safety Practices

1. Ensure all field staff is trained on preparation of Pre-work Safety Plans (PWSP), Activity Hazard Analysis (AHA) forms, or Job Hazard Analyses (JHA), more on this later.

2. Prior to initiating any field work, prepare a Pre-work Safety Plan (PWSP), Activity Hazard analysis form, Job Hazard Analysis (JHA), or have a tailgate/field safety meeting, based the hazards of the work. Discuss the potential job hazards and how they will be eliminated or mitigated. For example, a simple site reconnaissance may include subjects such as fencing or barriers, poisonous insects or reptiles, etc.

3. Hold tailgate/field safety meetings at least weekly and more often if work, crew, job location, or conditions change.

4. Document tailgate/field safety meetings using standard forms such as, Record of Tailgate or Field Safety Meeting, and have employees sign form.

5. When the temperature is expected to be 95°F or greater, hold daily tailgate/field safety meeting specific to Heat Illness Prevention and document on the H&S meeting form.

6. Report all near misses and accidents to your organization or Safety Engineers within 1 hour. Typically, Safety Engineers or the H&S Manager will determine if OSHA notification applies. If OSHA notification is required, then OSHA must be notified immediately, or typically within 8 hours of the occurrence of accident/near miss.

7. Fill out a Supervisor Initial Incident Report Form within first 24 hours.

20.4.1.3   Implement Employee Safety Training and Recordkeeping

1. Identify job specific training requirements for each employee.

2. Develop a prioritized safety training implementation schedule.

3. File employee training records in secure location.

20.4.1.4   Establish a Safety Document Information Location

1. Identify a location that is accessible to all staff and designate a bookcase or specific shelves for the location of general safety documents. At minimum, make the following documents available:
   a. Emergency Action Plan (EAP)
   b. Injury and Illness Prevention Plan (IIPP)
   c. Heat Illness Prevention Plan (HIPP) that Includes Site Specific Procedures
   d. Fire Prevention Plan
   e. Applicable Code of Safe Work Practices (SWPs)
   f. Organization's Safety Policy
   g. Organization's Hazard Communication Plan
   h. Organization's Energy Isolation Plan, if needed
   i. Organization's Code of Safe Work Practice (CSWP)
2. Meet with your Organization's Safety Engineer or Senior manager or Principal to determine what additional plans, procedures, or code-of-safe-work practices that may be needed.

20.4.1.5   Establish a Safety Filing System

1. Dedicate one lockable file cabinet for filing and keeping safety records for a minimum of 5 years. Verify details and requirements with your State and local requirements.
2. Safety records to be filed include but are not limited to:
   a. Activity Hazard Analysis (AHA) or Job Hazard Analyses (JHA)
   b. Safety Tailgate/Field Safety Meetings—Daily at Every Field Job/Project
   c. Safety Audits/Inspections
   d. Safety Training Records—Actual Signed Records are Very Important
   e. Pre-Work Safety Plans (PWSP'S)
   f. Vehicle/Equipment/Boat pre-operational inspection records
   g. Confined Space Permits - These Permit and Plans are very detailed and demand specialists, management approvals, and potential integration with emergency services such as the local fire department.
   h. Crane Operation Pre-Lift Plans
   i. Hazard Correction Records
   j. Safety Moments—Brief Discussions or statements applicable for the day's specific activities and tasks
   k. Other safety documentation as needed for the work process
   l. Original Employee Training Rosters, if possible, with Agenda and/or Training materials attached

3. Confidential safety records must be made available within hours upon State's H&S Agency or OSHA. These typical documents that could be requested include:

   a. Incident Investigation Reports

   b. Progressive Discipline memos for safety rule violations

   c. Employee safety training records

   d. OSHA 300 LOG (maintained by Organization's support staff)

   e. Any other confidential employee safety documentation State's H&S Agency or OSHA may request.

Table 20.2 depicts the typical contents for an Organizations H & S Program that includes the components listed above.

TABLE 20.2   Typical organization's H&S program contents.

Example	Depicted
	• Some Key Health and Safety Equipment—hardhat, safety glasses, high visibility vest • First Aid Kit • Vehicle and Equipment Reservation Book
	• Lockable Employee Training Files
	• Routine Inspection Record Books

20.4.2   Typical Example of a Health and Safety Plan and Accident Prevention Plan (HSP/APP) for a Large Project

This Health and Safety Plan and Accident Prevention Plan (HSP/APP) Outline depicted in Table 20.3 is a typical example for a large Department of Defense project with multiple tasks for multiple sites with hazardous materials. This project total cost effort was on the order of about $2 M in 2020. The project description is presented below to provide the reader with a good idea of the field effort and corresponding details in the HSP/APP document.

The project consisted of providing all necessary labor, materials, equipment, subcontracts, and services in connection with the abatement of environmental hazardous materials, demolition of buildings, and restoration of the site within the footprint for the buildings specified.

The purpose of this project is to perform abatement of asbestos containing material (ACM), lead containing material (LCM), mercury containing material (MCM), and polychlorinated biphenols (PCB) prior to the demolition of the twenty-nine (29) residential housing units and eight (8) buildings, one (1) two story office building, one (1) single residential unit to include detached enclosed garage structure and stand-alone shed, and three (3) stand-alone metal open carport structures. The project site consists of eight (8) abandoned two-story residential family units that are grouped together within an area that is accessed by two (2) cul-de-sac parking areas. Within the cul-de-sac parking areas are stand-alone open parking structures. An abandoned single residential unit with two (2) adjacent structures are located across from the Family Housing Units.

> H&S ACCIDENTS in the FIELD: One of the most frequent field accidents is a "slip, trip, or fall." It's important to be aware of uneven surfaces in the field and to wear sturdy shoes with good soles and ankle support. Watch for loose rocks, sand, or gravel on hard angled surfaces that create a slip plane surface, and trip hazards like sticks or weeds.

20.4.3   The Activity Hazard Analysis (AHA) Form, Content, and Use and Example

A "Definable Features of Work" (DFOW)' is a specific task required to accomplish an element of the overall scope of work. For example, for an initial site reconnaissance for a development project on a one-acre property a civil engineer might want to walk to property boundary and stake the corners. So, this small task would likely be titled, "initial site-walk and property staking." Another DFOW (or task) might be soil sampling and then testing. The point is that each DFOW or field task is typically included in the "scope of work for the project" and the overall schedule.

There is a valuable tool to use for DFOWs' in the H&S Plan. These details include:

- Contract details like name and number, dates, project location
- Task numbers corresponding to the scope of services and schedule
- Prime Contractor name and signatures, other notes

TABLE 20.3 Typical table of contents, large project H&S, and accident prevention plan.

Table of Contents

(Continued)

TABLE 20.3   *(Continued)*

List of Tables

List of Figures

List of Appendices

APPENDIX A	MAPS
APPENDIX B	ZERO ACCIDENT PLAN
APPENDIX C	ACTIVITY HAZARD ANALYSES
APPENDIX D	COMPETENT PERSONS
APPENDIX E	FORMS
APPENDIX F	RESUMES & CERTIFICATES
APPENDIX G	INSPECTION DEFICENCY TRACKING LOG
APPENDIX H	EXPOSURE CONTROL PLAN

List of Acronyms and Abbreviations

ACM	Asbestos Containing Material
AED	Automated External Defibrillator
AHA	Activity Hazard Analysis
ANSI	American National Standards Institute
APP	Accident Prevention Plan
ASSE	American Society of Safety Engineers
ASTM	American Society for Testing and Materials
BBP	Bloodborne Pathogens
CFR	United States Code of Federal Regulations
COR	Contracting Officer's Representative
CPR	Cardiopulmonary Resuscitation
CQC	Contractor Quality Control
CSCP	Confined Space Competent Person
CSIR	Contractor Significant Incident Report
DHSS	Department of Health and Human Services
dB	Decibel
DFOW	Definable Features of Work
DOD	Department of Defense
EAP	Employee Assistance Plan
EM	Engineer Manual
EMR	Experience Modification Rate
HCP	Hearing Conservation Program
HEC	Hazardous Energy Control
HEPA	High-Efficiency Particulate Air

TABLE 20.3 *(Continued)*

KO	Contracting Officer
IEEE	Institute of Electrical and Electronics Engineers
LCM	Lead Containing Material
MPH	Miles per Hour
NFPA	National Fire Protection Association
NOAA	National Oceanic and Atmospheric Administration
NPRCS	Non-Permit Required Confined Space
NRC	Nuclear Regulatory Commission
OSHA	Occupational Safety and Health Administration
PCB	Polychlorinated Biphenols
PPE	Personal Protective Equipment
PRCS	Permit Required Confined Space
RAC	Risk Assessment Codes
RSO	Radiation Safety Officer
SDS	Safety Data Sheet
SSHO	Site Safety and Health Officer
SS	Site Supervisor
USACE	United States Army Corps of Engineers

- Risk Assessment Code (RAC) Matrix, displaying the severity of the risk for a specific task compared to the probability for occurrence
- RAC Chart description and definitions
- Detailed table displaying each work step, specific anticipated hazards for performing that task, mitigative controls required to safeguard employee's safety
- Overall RAC rating for the task including mitigative safeguards
- Detailed Equipment Table displaying equipment type, training requirements for competent operating personnel, and inspection requirements

A blank AHA form including these details is depicted on Table 20.4. This form is included here for information and future use.

The Contractor shall identify each major phase of work that will be performed on this contract. Within each major phase, all activities, tasks, or DFOWs shall be identified that will require an Activity Hazard Analysis (AHA) form. This blank form presented above and an example presented below are for reference. In addition, the AHA shall also address any unusual or unique aspects of the project or activity.

The best way to protect personnel is to avoid a hazard and eliminate it with an alternate approach if possible. For exbtxtample, hazardous ladder work could be avoided with scaffolding or a construction lift or boom lift.

An example AHA form used on an Army base for backfilling, soil import, and site restoration is included here for information and future use. (See Table 20.5.)

TABLE 20.4 Blank AHA form.

Contractor Name (Performing the Work):

Foreman Signature (Performing the Work):

Activity/Work Task:

AHA Signature Log #

Contract Number:

Date Prepared:

Project Location:

PRIME CONTRACTOR SECTION: REVIEWED BY: SIGNATURE REQUIRED

SSHO Signature:

QC Manager Signature:

Prime Superintendent Signature:

Notes:

Overall Risk Assessment Code (RAC) (Use highest code)

Risk Assessment Code (RAC) Matrix

Severity	Probability				
	Frequent	Likely	Occasional	Seldom	Unlikely
Catastrophic	E	E	H	H	M
Critical	E	H	H	M	L
Marginal	H	M	M	L	L
Negligible	M	L	L	L	L

Step 1: Review each Hazard with identified safety "Controls." Determine RAC (see above).

Probability: Likelihood the activity will cause a Mishap (Near Miss, Incident, or Accident). Identify as Frequent, Likely, Occasional, Seldom or Unlikely

Identify as Catastrophic, Critical, Marginal, or Negligible

Step 2: Identify the RAC (probability vs. severity) as E, H, M, or L for each "Hazard" on AHA.

Annotate the overall highest RAC at the top of the AHA

RAC CHART

E = Extremely High Risk

H = High Risk

M = Moderate Risk

L = Low Risk

Job Steps (Work Sequences)	Specific Anticipated Hazards	Controls	RAC

TABLE 20.5 REAL EXAMPLE AHA form for backfilling, importing, restoration.

Contractor Name (Performing the Work): American Contractor Services, Inc.	Overall Risk Assessment Code (RAC) (Use highest code)

Foreman Signature (Performing the Work):

Activity/Work Task: Contract Task 2.2 Backfilling, Importing, Restoration

AHA Signature Log #

Contract Number: W XXX -20-C-00900

Date Prepared: December 3, 2025

Project Location: EXAMPLE: Army Base Residential Building Demolition

PRIME CONTRACTOR SECTION: REVIEWED BY: SIGNATURE REQUIRED

SSHO Signature:

QC Manager Signature:

Prime Superintendent Signature:

Notes:

Risk Assessment Code (RAC) Matrix

Severity	Probability				
	Frequent	Likely	Occasional	Seldom	Unlikely
Catastrophic	E	E	H	H	M
Critical	E	H	H	M	L
Marginal	H	M	M	L	L
Negligible	M	L	L	L	L

Step 1: Review each Hazard with identified safety "Controls". Determine RAC (see above).

Probability: Likelihood the activity will cause a Mishap (Near Miss, Incident, or Accident).

Identify as Frequent, Likely, Occasional, Seldom or Unlikely

Identify as Catastrophic, Critical, Marginal, or Negligible

Step 2: Identify the RAC (probability vs. severity) as E, H, M, or L for each "Hazard" on AHA.

Annotate the overall highest RAC at the top of the AHA

RAC CHART

E = Extremely High Risk

H = High Risk

M = Moderate Risk

L = Low Risk

(Continued)

TABLE 20.5 *(Continued)*

Job Steps (Work Sequences)	Specific Anticipated Hazards	Controls	RAC
Backfilling Operations and Site Restoration using Heavy Equipment	• Pedestrians or unauthorized vehicles entering the work area • Striking overhead lines or objects with the equipment bucket or the dump bed of the truck • Injury to hearing from equipment noise • Inhalation hazards from dust from backfilling activities • Struck by/crushing hazards from moving equipment or trucks • Physical hazards from power tools or compacting equipment, hand or foot injuries to personnel from use of walk behind or jump jack type compaction equipment, struck by/crushing hazards from heavy equipment or trucks, back strain • Physical hazards with use of hand tools such as back strain, hand or foot injuries, repetitive motion, or struck by/crushing hazards from equipment or trucks • Heat stress	• Set up an exclusion zone. Use spotters to watch for unauthorized pedestrians or vehicles entering the exclusion zone, stop work if an unauthorized person enters the exclusion zone • Before beginning work, walk the exclusion zone looking for overhead obstructions or power lines • Wear approved hearing protection when working close enough to equipment or trucks that you have to speak louder than your normal voice to someone standing next to you • Wear appropriate PPE to protect from dust such as an N-95 dust mask at a minimum or a respirator with P100 cartridges, control dust with water (Refer to the "Dust/Odor Control" AHA) • Wear proper PPE including a reflective vest, stay out of the swing or travel radius of the equipment and trucks, good communication/hand signals and eye contact with the operator/truck driver, do not approach equipment or trucks until they have acknowledged you, equipment bucket is to be placed on the ground, parking brake set and engine shut down during conversations, same is to occur when approaching trucks, stay out of blind spots, all equipment and trucks must have functioning back-up alarms, spot equipment or trucks working in tight locations and while backing	M

Job Steps (Work Sequences)	Specific Anticipated Hazards	Controls	RAC
		• Wear leather gloves, hearing protection and steel toed boots when using power tools/equipment, proper training in the use of power tools/equipment, use compacting equipment on level or slightly sloping surfaces, keep hands between yourself and the equipment to avoid hand injuries on adjacent walls, stay visible, good communication and eye contact with other equipment operators and truck drivers, don't get in their blind spots, get help when needing to move compacting equipment to other locations	
		• Use the proper tool for the job, wear leather or level 2 gloves, keep your back straight, bend at the knees and don't twist at the waist while shoveling, pushing, pulling or carrying objects, get help with awkward or loads over 50 lbs, watch hand and foot placement for pinch points or crushing hazards, take rest breaks as needed, stay visible, good communication and eye contact with other equipment operators and truck drivers, don't get in their blind spots	
		• Stay hydrated drinking water, take additional rest breaks in the shade when sweating excessively, watch out for others working around you for signs of heat stress overtaking them	
Backfilling Operations and Site Restoration using Hand/ Power Tools	• Hand/body injuries–cuts, scrapes, crushing • Struck by/crushing hazards from other equipment or trucks working • Excessive noise • Back strain • Heat stress or repetitive motion • Dust inhalation	• Proper PPE including leather gloves, use the correct tool for the job, inspect tools for signs of wear or defect before use • Stay visible, proper PPE including a reflective vest, good communication and eye contact with other equipment operators and truck drivers, don't get in their blind spots	M

(Continued)

TABLE 20.5 *(Continued)*

Job Steps (Work Sequences)	Specific Anticipated Hazards	Controls	RAC
		• Proper hearing protection including ear muffs or ear plugs • Use proper posture while working, keep your back straight, bend your knees when lifting or pushing, don't twist at the waist, get help with loads over 50 lbs or with awkward loads • Take rest breaks as needed in a shaded area, stay hydrated, take additional rest breaks with excessive sweating, watch for signs of heat stress in others working around you • Use of an N-95 dust mask for dust or use of a respirator with P100 cartridges, perform control dust with water as needed (See Dust Control AHA)	M
Backfilling Operations and Site Restoration using Compaction Equipment	• Hand injuries-cuts scrapes crushing • Struck by/crushing hazards from other equipment or trucks working • Excessive noise causing hearing loss • Back strain from moving the vibratory plate/wacker etc. to another location • Dust inhalation • Heat stress	• Proper PPE including leather gloves, proper training in the use of the vibratory plate/wacker etc., keep hands between yourself and the equipment to avoid hand injuries from adjacent walls • Stay visible, proper PPE including a reflective vest, good communication and eye contact with other equipment operators and truck drivers, don't get in their blind spots • Proper hearing protection including ear muffs or ear plugs • Get help when needing to move the vibratory plate/ wacker etc. to other locations, use the skip loader bucket to transport, lift with your legs in sync with help from others, keep your back straight, don't twist at the waist • Use of an N-95 dust mask for dust or use of a respirator with P100 cartridges, perform control dust with water as needed (See Dust Control AHA) • Take rest breaks as needed in a shaded area and stay hydrated, take additional rest breaks with excessive sweating, watch for signs of heat stress in others working around you	

Job Steps (Work Sequences)	Specific Anticipated Hazards	Controls	RAC
Trucks Importing to the Site	• Trucks entering the exclusion zone to dump their load that are not familiar with the site entering/ leaving in the wrong direction • Wrong material brought to the site • Vehicle accident/Backing over something or someone • Truck overturning while dumping the load • Injury to hearing from equipment noise. • Inhalation hazards from dust from site restoration activities. • Slips, trips and falls. • Physical hazards to personnel from heavy equipment or trucks.	• Spotter stopping the truck when entering the site, give directions for entering and leaving the exclusion zone and PPE requirements if the driver has to get out of the truck • Verify the material the driver is delivering before allowing the truck into the exclusion zone • Spotters as needed to direct truck to the proper dump location and spotting the truck when backing up staying visible to the driver in his mirrors, equipment operators remain alert to trucks entering the exclusion zone maintaining eye contact of their location • Spotter to place truck on level solid ground before raising the bed of the truck, ask an operator to smooth out the dump area as needed • Wear approved hearing protection when working close enough to equipment or trucks that you have to speak louder than your normal voice to someone standing next to you • Wear appropriate PPE to protect from dust such as a N-95 dust mask at a minimum or a respirator with P100 cartridges, control dust with water (Refer to the "Dust/ Odor Control" AHA) • Watch your footing while walking around the exclusion zone, ask an operator to smooth out walking areas • Wear proper PPE including a reflective vest, stay out of the swing or travel radius of the equipment and trucks, good communication/hand signals and eye contact with the operator/truck driver, do not approach equipment or trucks until they have acknowledged you, equipment bucket is to be placed on the ground, parking brake set and engine shut down during conversations, same is to occur when approaching trucks, stay out of blind spots, all equipment and trucks have functioning back-up alarms, spot equipment or trucks working in tight locations and while backing	M

(Continued)

TABLE 20.5 *(Continued)*

Equipment to be used	Training Requirements and Competent or Qualified Personnel Name(s)	Inspection Requirements
• Hand tools • Vibratory Plate or Jumping Jack Wacker • Loader • Excavator • Compactor • Dump truck • Water truck • Water hose • Temporary power box (spider) with GFCI installed.	• 40-Hour OSHA HAZWOPER Training • 8-Hour OSHA HAZWOPER Refresher • OSHA 10-Hour Construction Safety Training • Site-Specific Safety Training and AHA Training • Equipment Training • First Aid/CPR/AED Training	• Daily equipment inspection • Inspect tools/equipment before use • Inspect PPE before use

UFGS 013526 11/15 1.9 Government reserves the right to require the Contractor to revise and resubmit the AHA if it fails to effectively identify the work sequences; specific anticipated hazards, site conditions, equipment, materials, personnel and the control measures to be implemented.

UFGS 013526 1.9.1 Review the AHA list periodically (at least monthly) at supervisory safety meetings, update when procedures, scheduling or hazards change.

UFGS 013526 1.9.2 Each employee performing work . . . must review the AHA and sign a signature log for that AHA prior to starting work. The SSHO must maintain a signature log on site for every AHA

20.5   OSHA Quick Reference Card for Top 4 Construction Site Hazards and Construction Personal Protective Equipment (PPE)

Civil engineers working in the field or at construction sites will find these OSHA Quick Card references very helpful to remind their staff, contractors, visitors, and themselves of the four most frequent dangers on construction sites. Please refer to Figure 20.2—OSHA Quick Card Reference for top 4 construction site hazards and construction "Personal Protective Equipment" (PPE). This second OSHA Quick Card depicted on Figure 20.2 is also a valuable reference for Personal Protective Equipment on/near construction sites or in the field that is helpful.

The four top causes of construction fatalities are:

- Falls,
- Struck-by,
- Caught-In/Between, and
- Electrocution

Typical construction personal protective equipment includes:

- Eye and Face Protection
- Foot Protection
- Hand Protection
- Head Protection
- Hearing Protection

As previously mentioned the most efficient method for creating a safe work environment is by *eliminating hazards*. For example, consider a work task that includes cutting steel beams for on-site building erection. This dangerous task could be eliminated if you instead had these beams cut at the steel fabricator's yard. Employing this alternative eliminates danger on the job site and uses a safer method of steel preparation.

However, "if" eliminating a safety hazard is not possible, the next best method for creating a safe work environment is by *mitigating the hazard*. For example, consider the work task where you must perform a field recon (reconnaissance) in an area where poisonous snakes may exist. You could mitigate potential snake bites by using high protective boots or snake-bite proof footwear.

20.6   ORGANIZATION/COMPANY/EMPLOYEE RESPONSIBILITIES AND EMPLOYEE RIGHTS

A civil engineering organization or company and every employee have important responsibilities to foster a positive attitude toward H&S compliance and practice safe work. The responsibilities for organizations/employees and employee rights are listed and summarized below.

20.6.1   Organization/Company Responsibilities

1. **General Requirements**—Your Company or Agency will furnish employees at their places and conditions of employment that are free from on-the-job safety and health hazards.

2. **OSHA Regulations**—Your Company or Agency will comply with applicable regulations of the Occupational Safety and Health Administration OR your applicable regulations for your State and Local area.

3. **Reporting Hazards**—Your Company or Agency will respond to employee reports of hazards in the workplace.

4. **Workplace Inspections**—Your Company or Agency will ensure that each workplace is inspected annually for hazardous conditions. Your Company or Agency will post Notices of Unsafe or Unhealthful Working Conditions found during the inspections for a minimum of three working days, or until the hazard is corrected, whichever is later.

5. **Correction of Unsafe Conditions**—Your Company or Agency will take prompt action to assure that hazardous conditions are eliminated. Imminent danger conditions will be corrected immediately.

6. **Safety and Protective Equipment**—Your Company or Agency will acquire, maintain, and require use of appropriate protective and safety equipment.

7. **Safety and Health Training**—Your Company or Agency will provide occupational safety and health training for employees.

8. **Reporting Accidents, Injuries, and Occupational Illnesses**—Supervisors must submit a supervisor's report of accidental injury/illness for all work-related accidents, injuries, or occupational illnesses experienced by employees under their supervision.

9. **Safety and Health Committees**—Your Company or Agency will support any safety and health committees that are formed from management and employee representatives.

10. **Designated Agency Safety and Health Official (DASHO)**—Your company, organization, or agency needs to post the Designated Agency Safety and Health Official (DASHO) as depicted below.

The Safety and Health Designee for this workplace is: ————————————————
and may be contacted at, (Telephone and Location): ————————————————-
Further Information—This notice highlights that Your Company or Agency practice an employee job safety and health program. More information about the Your Company or Agency program or its standards and procedures may be obtained from the workplace Safety and Health Designee.

20.6.2 Employee Responsibilities

1. **Compliance with Standards**—Employees shall comply with all OSHA or applicable regulations for your State and Local area occupational safety and health standards, policies, and directives.
2. **Safety and Protective Equipment**—Employees shall use appropriate protective and safety equipment provided by Your Company or *Agency*.

20.6.3 Rights of Employees and Their Representatives

1. **Participation in Safety and Health Program**—Employees and their representatives shall have the right to participate in the Your agency Safety and Health Program. Employees shall be authorized official time for these activities.
2. **Access to Records and Documents**—Employees and their representatives shall have access to copies of applicable OSHA and other recognized standards and regulations; your agency safety and health policies and directives; accident, injury, and illness statistics of your agency.
3. **Reporting Hazards**—Employees and their representatives shall have the right to report unsafe or unhealthful working conditions to appropriate officials and to request an inspection of the workplace. The name of the employee making the report will be kept confidential if requested.
4. **Freedom from Fear of Reprisal**—Employees and their representatives are protected from restraint, interference, coercion, discrimination, or reprisal for exercising any of their rights under the Your agency Safety and Health Program.

Civil engineering staff and mangers need to possess a working knowledge about H&S laws, organizational requirements, hazards and methods to mitigate hazards, required documentation, record keeping, and training requirements. Safeguarding human health and safety should be of the highest priority for civil engineers and all engineering professionals.

20.7 SUMMARY

Health and Safety (H&S) legal, ethical, and detailed requirements are not usually taught in university but are one of the many hands-on subjects that are critical elements in a successful civil engineering career. The information presented in this chapter aims to provide civil engineers important information on mandatory Health and Safety (H&S) laws and methods to safeguard employees, the general public, their firms/organizations, and themselves. H&S requirements are extremely important, legally required, and complex. A civil engineer's career and success depends upon this knowledge and implementation of these requirements. Stay safe and healthy!

BIBLIOGRAPHY

MacLaury, Judson. (March, 1981). "The job safety law of 1970: its passage was perilous." U. S. Department of Labor, *Monthly Labor Review.*

Miller, Danica. (October 7, 2020). "What is OSHA - topic 1 - why is OSHA important to you?." EHS Software - EHS Insider Blog, *https://blog.ehssoftware.io/safetyinsiderblog/what-is-osha.*

National Safety Council (NSC). "Injury facts, industry incidence and rates—most dangerous industries." https://injuryfacts.nsc.org.

US Army Corps of Engineers (USACE), Manual No. 385-1-1. (November 30, 2014). *https:// www.usace.army.mil/Missions/Safety-and-Occupational-Health/Safety-and-Health-Requirements-Manual.*

Zak, Bob. (January 16, 2018). The importance of OSHA in the workplace, Zota Pro.

What Civil Engineers Need to Know

Big Idea

"Like other professions, for instance, law and medicine, civil engineers are granted authority based on their extensive education and are afforded community sanction, in the form of licensure or registration. Civil engineers are held to well-documented ethical codes and are expected to be their clients' trusted advisors. Civil engineers also have a culture of their own that involves providing valuable services to society, a Bachelor's degree in civil engineering represents the first step toward professional registration or licensure, and the degree program itself is accredited by a professional body, such as Accreditation Board for Engineering and Technology (ABET). After completing an accredited degree program, the civil engineer must satisfy a range of requirements (including work experience and exam requirements) before becoming registered or licensed on their way to a licensed civil engineer."

—ASCE, Civil Engineering Body of Knowledge, Third Edition

Civil Engineer's Handbook of Professional Practice, Second Edition. Karen Lee Hansen and Kent E. Zenobia.
© 2025 John Wiley & Sons, Inc. Published 2025 by John Wiley & Sons, Inc.
Companion website: www.wiley.com/go/hansen/CivilEngineersHandbook

Key Topics Covered

- Civil Engineering as a Profession
- Civil Engineering Education
- Civil Engineering Careers
- What Civil Engineers Need to Know

Related Chapters in This Book

- Chapters 1 through 20 and Appendices A, B, C, D, E, F

21.1   BACKGROUND

Chapter 21 examines civil engineering as a profession. The chapter provides background on various career specializations, as well as typical educational and licensure requirements for achievement of professional status. The chapter also provides insights into what knowledge civil engineers need to internalize or "know in their heads."

21.2   CIVIL ENGINEERING AS A PROFESSION

Until modern times there was no clear distinction between civil engineering and architecture, and the term *engineer* or architect referred to the same person. In the western world, the origins of civil engineering as a profession can be found in the years immediately preceding and including the Industrial Revolution, the late 18th and early 19th centuries. The scientific discoveries of the Age of Enlightenment and the new commercial needs of the Industrial Revolution converged to create an ideal environment for innovation. During this period, certain *military* engineers began to work on nonmilitary, or *civil*, projects. The term *civil engineer* was adopted to emphasize this difference. In response to the growth of these new civil projects, the British Institution of Civil Engineers (ICE) was chartered in 1818 and the American Society of Civil Engineers (ASCE) was founded in 1852. Other professional civil engineering organizations followed: Institution of Civil Engineers India (ICEI) in 1860; Spanish Asociación de Ingenieros de Caminos, Canales, y Puertos (AICCP) in 1903; South African Institution of Civil Engineers (SAICE) in 1903; Japan Society of Civil Engineers (JSCE) in 1914; and Chinese Institute of Civil Engineering (CICE) in 1936, among others.

These organizations, as well as those in other countries, helped to formalize civil engineering as a profession. The geotechnical engineer and author, John Philip Bachner, lists five characteristics of a profession. These are:

- Systematic body of theory
- Authority
- Community sanction
- Ethical codes
- A culture

These characteristics help define today's professional civil engineer, who must be adequately prepared with a *systematic body of theory* that incorporates a spirit of rationality. This theory is based on mathematics and natural sciences, such as physics and chemistry. Like other professions, for instance, law and medicine, civil engineers are granted *authority* based on their extensive education and are afforded *community sanction*, in the form of licensure or registration. Civil engineers are held to well-documented *ethical codes* and are expected to be their clients' trusted advisors. Civil engineers also have *a culture* of their own that involves providing valuable services to society, behaving appropriately, and sharing a rich history and folklore.

Attributes of a Profession

1. Systematic body of theory
   - Skills flow from an internally consistent system
   - Spirit of rationality; expansion of theory
2. Authority
   - Extensive education in systematic theory highlights the layperson's comparative ignorance
   - Functional specificity
3. Community sanction
   - State-sponsored boards
   - License or registration
4. Ethical codes
   - Ethical
     - professional

- Client-professional
  - impulse to perform maximally
- Colleague to colleague
- Cooperative
  - egalitarian
  - supportive

5. A culture
- Social values
- Services valuable to the community
- Various modes of "appropriate" behavior
  - sounding like a professional
  - saying "no" gracefully
  - making presentations and conducting meetings
- Symbols
  - argot, jargon
  - insignia, emblems
  - history and folklore

Adapted from John Bachner, Practice Management for Design Professionals: A Practical Guide to Avoiding Liability and Enhancing Profitability.

21.3   CIVIL ENGINEERING EDUCATION

Increases in the civil engineering body of knowledge have resulted in a formalized approach to civil engineering education. The École Polytechnique was founded in Paris in 1794, and the Bauakademie was started in Berlin in 1799, but no such schools existed in Great Britain or the United States until several decades later. The University of Glasgow, Scotland, was the first university school of engineering in the United Kingdom to establish a chair in civil engineering. The first degree in Civil Engineering in the United States was awarded by Rensselaer Polytechnic Institute, New York, in 1835.

Today's civil engineering is linked to advances in understanding of physics, mathematics, and the social and political forces of its time. Civil engineers typically earn a Bachelor of Science (B.S.) degree with a major in civil engineering, though some universities award a Bachelor of Engineering. Students usually pursue their studies for four to six years. Typical civil engineering programs initially cover most, if not all, of the subdisciplines of civil engineering. Students then choose to specialize in one or more subdisciplines toward the end of their degrees.

As discussed in Chapter 1, ideally the degree should include units covering topics in four major categories:

- *Foundational*—mathematics, natural sciences, social sciences, and humanities
- *Engineering fundamentals*—materials science, engineering mechanics, experimental methods and data analysis, and critical thinking and problem solving
- *Technical*—project management, engineering economics, risk and uncertainty, breadth in civil engineering areas, design, depth in civil engineering areas, sustainability
- *Professional*—communication, teamwork and leadership, lifelong learning, professional attitudes, professional responsibilities, and ethical responsibilities

According to the ASCE's Body of Knowledge 2:

Engineering does not occur in a vacuum, and engineers must be able both to explain the impact of historical and contemporary issues on engineering and to explain the impact of engineering on the world. (American Society of Civil Engineers. 2019).

In most countries, a Bachelor's degree in civil engineering represents the first step toward professional registration or licensure, and the degree program itself is accredited by a professional body, such as ABET. After completing an accredited degree program, the civil engineer must satisfy a range of requirements (including work experience and exam requirements) before becoming registered or licensed.

The National Council of Examiners for Engineering and Surveying (NCEES) administers the civil engineering professional engineer (Civil PE) exam. After passing the EIT (Engineer in Training) exam, the prospective engineer is tested with a:

- **Breadth exam** (morning session): This exam contains questions from all five areas of civil engineering: Construction, Geotechnical, Structural, Transportation, and Water Resources and Environmental
- **Depth exams** (afternoon session): These exams focus more closely on a single area of practice in civil engineering. Examinees must choose one of the following areas: Construction, Geotechnical, Structural, Transportation, and Water Resources and Environmental. [NCEES]

Once licensed, the civil engineer is designated the title of Professional Engineer (in the United States, Canada, and South Africa), Chartered Engineer (in most British Commonwealth countries), Chartered Professional Engineer (in Australia and New Zealand), European Engineer (in much of the European Union), and *Professional Engineer* in many Asia countries. There are international engineering agreements between relevant professional bodies that are designed to allow engineers to practice across international borders.

Civil Engineering Associations

American Society of Civil Engineers
Canadian Society for Civil Engineering
Chi Epsilon, Civil Engineering Honor Society
Earthquake Engineering Research Institute
Engineers Australia
Institution of Civil Engineers (UK)
Institute of Structural Engineers (UK)
Institute of Transportation Engineers
Royal Academy of Engineering (UK)
Transportation Research Board
The Institution of Civil Engineering Surveyors
A list of international societies with which ASCE has cooperative agreements is available at: https://www.asce.org/communities/find-a-group/regions/region-10/agreements-cooperation

The advantages of registration or licensure vary depending upon location. For example, in the United States and Canada most licensing organizations use something like the following quote: "only a licensed engineer may prepare, sign and seal, and submit engineering plans and drawings to a public authority for approval, or seal engineering work for public and private clients." This requirement is enforced by state and provincial legislation. In other countries, no such legislation exists. Most professional associations of civil engineers, such as the American Society of Civil Engineers, the British Institution of Civil Engineers (ICE), and the British Institute of Structural Engineers (ISE) maintain a code of ethics by which members are expected to abide or risk expulsion. In this way, these organizations play an important role in maintaining ethical standards for the profession. (See Chapter 3, Ethics, of this text.)

Even in countries where licensure has little or no legal bearing on work, engineers are subject to contract law. In cases where an engineer's work fails he or she may be subject to the tort of negligence and, in extreme cases, the charge of criminal negligence. An engineer's work must also comply with numerous other rules and regulations, such as building codes and legislation pertaining to environmental law. (In this book, see Chapter 8, Permitting, and Chapter 11, Legal Aspects of Professional Practice.)

21.4   CIVIL ENGINEERING CAREERS

There is no one typical career path for civil engineers. Most engineering graduates start with entry-level positions, and as they prove their competence, they gain more and more significant tasks. In some fields and firms, entry-level engineers are put to work primarily monitoring construction in the field, serving as the "eyes and ears" of more senior design engineers. In

TABLE 21.1   U.S. Bureau of Labor Statistics Occupational Outlook.

Quick Facts: Civil Engineers	
2021 Median Pay (May)*	$88,050 per year $42.33 per hour
Federal government, excluding postal service	$100,730
Local government, excluding education and hospitals	$99,330
Engineering services	$93,520
State government, excluding education and hospitals	83,390
Nonresidential building construction	$77,450
Typical Entry-Level Education	Bachelor's degree
Number of Jobs, 2020	309,800
Job Outlook, 2020–30	8% (As fast as average)
Employment Change, 2020–30	25,300

*The median wage is the wage at which half the workers in an occupation earned more than that amount and half earned less. The lowest 10 percent earned less than $60,550, and the highest 10 percent earned more than $133,320.

Source: Adapted from US Bureau of Labor Statistics *Occupational Outlook Handbook*.

other areas, entry-level engineers perform routine tasks of analysis or design and interpretation. Senior engineers can execute complex analysis or design work. They also can work in project management of design projects, or management of other engineers, or specialized consulting. Civil engineers are in high demand at financial institutions and management consultancies because of their analytical skills. They can find many career opportunities in high technology for the same reason.

Civil engineers' salaries vary with education, training, experience, and geographic location. Table 21.1 lists some quick facts provided by the U.S. Bureau of Labor Statistics regarding civil engineers' pay.

Areas of civil engineering specialization have changed over time due to society's needs and the complexities of projects and technologies. Currently, the ASCE incorporates the following Institutes:

- Architectural Engineering (AEI)
- Coasts, Oceans, Ports, and Rivers (COPRI)
- Construction (CI)
- Engineering Mechanics (EMI)
- Environmental and Water Resources (EWRI)
- Geo (G–I)
- Structural Engineering (SEI)
- Transportation & Development (T&DI)
- Utility Engineering and Surveying Institute (UESI)

The ASCE also recognizes the following technical groups:

- Aerospace
- Changing Climate
- Codes & Standards
- Cold Regions
- Computing
- Energy Engineering
- Forensic Engineering
- Infrastructure Resilience
- Sustainability
- The activities and responsibilities of civil engineers working in these various areas are included in Table 21.2.

21.5   WHAT SUCCESSFUL CIVIL ENGINEERS NEED TO KNOW

To have successful careers, civil engineers acquire a unique set of skills. As noted earlier, the profession embraces a systematic body of theory. Some competencies are based in science and technology; other expertise derives from social sciences, theories on innovation and management, the law, and even history. This knowledge is acquired over time through education and experience.

Table 21.3 contains a wealth of information regarding what successful civil engineers need to know. By the time their careers reach maturity, civil engineers are knowledgeable about the contents of Table 21.3 "off the top of their heads," in other words, this knowledge has been internalized. This list is a summary of some of the typical knowledge items and how and why it's important to a civil engineer. These knowledge areas are based on discussions with ten experienced licensed civil engineers.

TABLE 21.2 Civil engineering areas of concentration.

Area	Activities and Responsibilities
General Civil	Focuses on the overall interface of projects with their environments
	Applies the principles of geotechnical engineering, structural engineering, environmental engineering, transportation engineering, and construction engineering to residential, commercial, industrial, and public works projects of all sizes and levels of construction
	Works closely with surveyors and specialized civil engineers
	Designs grading plans, drainage, pavement, water supply, sewer service, electric and communications supply, and land divisions
	Visits project sites, develops community consensus, and prepares construction plans and specifications
Coastal	Helps manage coastal areas
	Defends against flooding and erosion
	Designs ports
	Also works to reclaim land
Construction	Plans and executes the designs from transportation, site development, hydraulic, environmental, structural and geotechnical engineers
	Writes and/or reviews contracts
	Evaluates logistical operations
	Controls prices of necessary materials, operations, and equipment
Environmental	Deals with the treatment of chemical, biological, and/or thermal waste, the purification of water and air, and the remediation of contaminated sites
	Works with pollution reduction, green engineering, and industrial ecology
	Reports information on the environmental consequences of proposed actions and the assessment of effects of proposed actions for the purpose of assisting society and policymakers in the decision-making process, i.e., writes environmental impact reports (EIRs)

(Continued)

TABLE 21.2 *(Continued)*

Area	Activities and Responsibilities
Geotechnical	Deals with complex nature of rock and soil, subsurface investigation and testing, foundations and earth structures (dams, levees, engineered fills, etc.)
	Depends on knowledge from the fields of geology, material science and testing, mechanics, and hydraulics to design foundations, retaining structures, landfills and similar structures
	Can specialize further to use biology and chemistry to devise ways of disposing of hazardous materials and groundwater contamination (called geoenvironmental engineering)
	Contrasts with the relatively well-defined material properties of steel and concrete used in other areas of civil engineering
Land Surveying (considered a distinct profession in the United States, Canada, the United Kingdom, and most Commonwealth countries)	Establishes the boundaries of a parcel of land using its legal description and subdivision plans
	Lays out the routes of railways, tramway tracks, highways, roads, pipelines, and streets as well as positions other infrastructures, such as harbors, before construction
	Employs surveying equipment, such as levels and theodolites, for accurate measurement of angular deviation, horizontal, vertical, and slope distances
	Makes use of electronic distance measurement (EDM), total stations, global position system (GPS) surveying, and laser scanning with computerization, have supplemented (and to a large extent supplanted) the traditional optical instruments
Municipal or Urban Engineering	Involves specifying, designing, constructing, and maintaining municipal infrastructure, such as streets, sidewalks, water supply networks, sewers, street lighting, municipal solid waste management and disposal, storage depots for various bulk materials used for maintenance and public works (salt, sand, etc.), public parks, and bicycle paths
	Includes the civil portion (conduits and access chambers) of the local distribution networks of electrical and telecommunications services
	Focuses on the coordination of infrastructure networks and services, as they are often built and managed by the same municipal authority

TABLE 21.2 *(Continued)*

Area	Activities and Responsibilities
Structural	Analyses and designs the structures of buildings, bridges, towers, overpasses, tunnels, offshore structures like oil and gas fields in the sea, and other structures
	Identifies the loads which act upon a structure and the forces and stresses that arise within that structure due to those loads
	Considers strength, stiffness, and stability of the structure when it is subjected to its own self weight, other dead loads, live loads, including furniture, wind, seismic, crowd or vehicle loads, or transitory, such as temporary construction loads Also takes into account aesthetics, cost, constructability, safety, and sustainability wind engineering and earthquake engineering
	Can specialize further (wind and earthquake engineering)
Transportation	Deals with moving people and goods efficiently, safely, and in a manner conducive to a vital community
	Plans this movement using queuing theory, Intelligent Transportation Systems (ITS), and infrastructure management
	Designs, constructs, and maintains transportation infrastructure, including streets, canals, highways, rail systems, airports, ports, and mass transit
	Investigates and specifies paving materials
	Involves transportation design, transportation planning, traffic engineering, some aspects of municipal/urban engineering
Water Resources	Combines hydrology, environmental science, meteorology, geology, conservation, and resource management in the collection and management of water as a natural resource
	Relates to the prediction and management of both the quality and the quantity of water in underground resources (aquifers) and above ground resources (lakes, rivers, and streams)
	Analyzes and models very small to very large areas to predict the amount and content of water as it flows into, through, or out of a facility such as pipelines, water distribution systems, drainage facilities (including bridges, dams, channels, culverts, levees, storm sewers), and canals

TABLE 21.3   What successful civil engineers need to know.

General category and topic	Specific knowledge item and why it's important	Comments and how it's important
Engineering Related		
How to perform a basic site reconnaissance (recon)	Understand the basic components of a basic site recon. Strive to understand the importance of site details, take notes, collect photo records, and list unknowns.	Site details are critically important to the CE. List what you know, what you don't know, list details, and show dates.
Physics	$F = ma$	The Force equation used in Statics and Dynamics is a key tool in many CE applications and solutions.
Basic Health and Safety Rules	Understand, employ, and protect yourself and your staff by understanding federal "Occupational Safety and Health" (OSHA) laws and local requirements when interviewing/hiring new staff and interacting with staff and others.	"You" are in charge of your safety and health in the office, field, and at home. Be your own safety advocate.
Basic Human Resources/ Personnel Laws	Understand, employ, and protect yourself and your staff by understanding basic human resources and personnel rules, laws, and local requirements when interviewing/hiring new staff and interacting with staff and others.	"Basic Human Resources and Personnel Laws" define the rules of engagement in the workplace and even off-site. Knowledge of these rules is essential to a successful career to avoid personnel actions and liability for the CE.
Dimensional Analyses	Know how to mathematically convert a parameter or measurement to another measurement.	An example is how you could calculate the number of gallons in "one acre-foot". This knowledge allows the CE to avoid memorizing conversion factors and simply using math and dimensional analysis to convert from one measurement to another.
Slope Stability	Slope design is a common skill and knowledge in most civil work projects.	Presenting slope stability solutions is essential to many infrastructure designs, such as levees, dams, highways, foundations, landfills, development projects, etc.

TABLE 21.3 *(Continued)*

General category and topic	Specific knowledge item and why it's important	Comments and how it's important
Conceptual Design and Cost Estimates	Conceptual designs present initial concepts and represent a partially completed design for a project. The concept design can be used to build upon and refine future design considerations. As the design matures and proceeds toward a final design, other important features like construction method, materials of construction, sustainability, high first cost, and long-term maintenance cost should be considered.	Conceptual designs present initial concepts and generally mature through the design process toward a final design. Typical design completions may be expressed as a "per-cent completion". For example, an initial design or concept design may be expressed as a 10% or a 30% design. Cost estimates for a specific design, like a 30% design, should be referenced with caution because there are so many design components that have not been discussed or included.
Water always flows downhill	Impact of a solution may not be limited to the solution or the issue itself. Be aware of the side-effect, impact, or the related consequence to the upstream and downstream communities.	A solution could be perfect for the current condition or current circumstance. Think of the impact for the community at large and the long-term impact.
Earthwork/Site Analysis	The process of project site grading, from design to construction, is one that is full of uncertainty. Predicting the earthwork balance of a project site is not exact science. Engineers should expect during the design of a project there could be as much as a 10% swing in the actual amount of earthwork. Geotechnical soils reports may provide some details on how to best predict these fluctuations.	Design engineers should have a good grasp of how earthwork calculations work and understand the concepts surrounding excavation and fill quantities. Additionally, engineers should understand how the following can impact a project site earthwork balance and ultimately the overall project costs: Compaction (shrinkage), Consolidation, Site Conditions, Over-Excavation, Stripping, Trench Spoils, and Accuracy of Aerial Topography. Earthwork has the potential for great conflict and is inherent in most projects. Understanding these concepts will prepare engineers for some of the pitfalls associated with projects containing a significant grading component.

(Continued)

TABLE 21.3   *(Continued)*

General category and topic	Specific knowledge item and why it's important	Comments and how it's important
Cost Estimation	The ability to breakdown a project into the various components for bidding/ construction and applying and presenting historical cost data in a clear and concise manner is an essential skill for all engineers.	Engineers need to be able to perform quantity take-offs on their projects and organize these quantities into clear and concise estimates that can be used for bidding and estimation of overall project costs. Proficiency requires experience and the ability to stay connected to changes in material and construction costs. Consult and understand building cost index tables to compare historical trends in the construction industry. When providing bids to clients for determining required funding or for gathering bids from contractors, engineers should always remember to not provide the lowest estimate (unless specifically asked to do so by a client).
Construction Plans	The ability to prepare, read, and understand construction plans is extremely valuable.	Understanding scale (true and/or exaggerated), construction stationing, specifications, and the organization of construction plans is essential. As a design engineer, one should understand that the money is in the detail$ (you either make money or lose money in the details of your plans, it a balancing act). Well prepared and detailed plans get constructed with a minimal amount of issues and/or change orders. As a project manager, knowing how to efficiently read and understand how a project will be constructed from the proposed plans will assist in the oversight of the project scope, schedule, and budget.
Visit Job Sites	Every job/project is different and has their own specific challenges. Gain as much experience/knowledge of a sites specific field conditions.	Site visits should be conducted in the beginning of a project, during construction, and before final approval. Do not make the mistake of visiting the site/project for the first time to solve a problem.
Business, Marketing, Personal Relationships		
Humility	Be confident but be humble.	Humility is so important because it tells people you're "human" and an equal to them. Keep your ego in check.

TABLE 21.3   *(Continued)*

General category and topic	Specific knowledge item and why it's important	Comments and how it's important
Listening Skills	Listen, hear what's stated/unstated, comprehend	Learn when/how to be quiet and listen, when to clarify and confirm, when to display empathy and concurrence.
Guard Your Reputation and Social Media Content	Use caution on written (and sometimes and someplace even verbal) content on social media	Many employers, agencies, possibly Clients may check out your public statements and content of social media sites and more.
Know the value of politeness	Politeness invites interaction and forms the foundation of a relationship and represents positive professional behaviors.	Politeness is important because it tells people you're "human" and an equal to them.
Know and understand "Normalcy Bias" sometimes referred to as "gamblers fallacy."	The "Normalcy Bias" is a fallacy some people believe, practice, and often unexpectedly become sadly aware of. This fallacy is, "this has never happened in the past so, we don't expect anything to change in the future." Or, from a gambler's point of view—they might believe that since a number "7" hasn't come up in the last 100 rounds that it will surely come up next. So, bet the house on this?	An example of this fallacy might sound like, "This person has always met the deadline in the past, it will be just fine." Or, "I always run that stop sign on that country road and it's never been a problem." Be prepared, don't skip required steps, follow your Plan, expect the unexpected.
Know and understand the "Pygmalion Effect" where applicable in your business and personal life.	The "Pygmalion Effect" is that positive expectations and beliefs in your students, colleagues, teams, or children leads them to behave in ways (consciously or not) that promotes confidence, effectiveness, and positive results.	Employ the "Pygmalion Effect" to promote confidence and results and be certain your Team has the tools, understanding, and capability to accomplish the desired result. This means, do not just "wish it to be true." It's important to note that the Pygmalion Effect works both ways meaning that if you express lower expectations in your Team, you will likely experience lower performance. This is referred to as the "Golem Effect".

(Continued)

TABLE 21.3 *(Continued)*

General category and topic	Specific knowledge item and why it's important	Comments and how it's important
Professional Courtesy	A good engineer will act with professional courtesy to others, including peers, staff, management, client, and everyone.	Professional Courtesy presents positive professional behaviors, such returning email and phone call promptly. It carries positive vibe in the professional relationship.
Respect Different Ideas	A different idea is not a wrong idea. Different engineer would look at the same subject with different angle and from different background.	Always listen to others' ideas fully and follow up with question if not fully understandable. Different ideas can lead to different dimensions of solutions that may not be expected.
In engineering, the best salesman is not always the one that can talk.	Base your recommendation on your best knowledge, experience, data, and analysis.	Client and staff need your sincere and honest opinion.
Display "Respect" when working on and with a Team. Teams generally work effectively in Civil Engineering.	Teams provide effective solutions. Remember in Teamwork: $1 + 1 > 2$	A "Team" carries different skillsets, knowledge, and expertise. These skillsets are not always obvious during normal operation. However, when there is an emergency or a special need, one with a special expertise may save the team or project.
"Soft Skills"	Soft skills are non-technical skills that relate to how you work and interact with others. Soft skills include, but are not limited to the following: Interpersonal "People" skills, communication, listening, time management, empathy, organization, etc.	Soft skills are so important that most employers typically design interviews to give them an idea of your behavior and interpersonal skills rather than focusing purely on your technical knowledge. These are transferable skills and can be used regardless of the position you hold. These skills more than anything will provide more opportunities for success because they highlight the value you can bring to an organization. Soft skills can be learned and honed with knowledge and practice. Simple things like making eye contact, shaking hands properly, remembering names, listening, taking notes, and developing verbal and written communication proficiency will assist you in becoming an adaptable and respected employee/colleague.

TABLE 21.3   *(Continued)*

General category and topic	Specific knowledge item and why it's important	Comments and how it's important
Personal Presentation, Outlook		
Professional Dress Code	Your personal presentation and outlook to others is important	Often, right or wrong, people may evaluate you based on your outward presentation, confidence, dress code, speech, and body language. Take pride in your appearance. Not only do you feel good about yourself, but it shows others that they matter to you and that you respect them professionally.
Code(s) of ethics and mission	Know your "personal" code of ethics, relate your code to the defined. Create and commit to "your mission in life."	Ethics and your organizations' mission statement and your personal mission statement provide direction, like a ship's rudder, for your career and daily interactions. If you are called to be a street sweeper, sweep streets even as Michelangelo painted, or Beethoven composed music, or Shakespeare wrote poetry. Sweep streets so well that all the hosts of heaven and earth will pause to say, "Here lived a great street sweeper who did his/her job well."
Continuous Improvement	It's critical for CEs to practice "Continuous Improvement" in your life and career	Avoid stagnation, strive to improve yourself, your life, and career. Never compare yourself to someone else, strive to be the best you can be. Just think if you could work on being 1% better every day how much growth you would realize in one month, one year, one decade.
Winner's attitude	The right attitude could be the difference between a solution or a dead-end.	Dr. Henry Lee, Legendary Investigator: • The winner sees a green near every sand trap; the loser sees two or three sand traps near every green. • The winner says, "It may be difficult, but it is possible;" the loser says, "It may be possible, but it's too difficult."
Personal Value/ Self Esteem	Be humble and grateful for everything you have in life and don't fixate on results. Focus on the process of being better rather than immediate results.	Results have a way of slanting our viewpoint. Results are not always positive, but they always provide an opportunity to learn and grow. Remember your worth is never predicated on any one individual performance.

TABLE 21.3   *(Continued)*

General category and topic	Specific knowledge item and why it's important	Comments and how it's important
Attitude	You have two choices every morning—to be in a bad mood and let things get you down or to be in a good mood and enjoy the day.	Words of wisdom: • Mistakes are learning opportunities, not failures • Look for the good in all situations, motives, and people • Your attitude toward others will determine their attitude toward you • If you make another person feel important and appreciated, they will usually return this to you • Don't be embarrassed to share visions, desires and goals • Deal with people in an honest, ethical, and moral way • Think Positively. Positive thoughts result in positive attitudes and actions and vice versa
Engineering, Science Formulas		
Water, Flow, Volumes, Mass	$Q = VA$	Flow, Q = Volume × Area
Pythagoras Theorem	$a^2 + b^2 = c^2$	Handy tool for use in many civil engineering applications
The 1st Law of Thermodynamics	The 1st law of thermodynamics states that energy cannot be created or destroyed.	Handy tool for use in many civil engineering applications
The 2nd Laws of Thermodynamics	The 2nd law of thermodynamics states that energy changes from one form to another, or matter moves freely, disorder (entropy) in a closed system increases.	Handy tool for use in many civil engineering applications
Understand the relationships of proportion, symmetry, and scale.	Understand three key constants, "phi", "pi", and "e". All three constants are imbedded in natural form and essence	Think about objectivity versus subjectivity. Understand the essence, true meaning of these constants.

TABLE 21.3 *(Continued)*

General category and topic	Specific knowledge item and why it's important	Comments and how it's important
Average End Area Method	Basic method for quickly estimating earthwork volumes	Analyzing the end areas of excavation and embankment cross-sections along a given length to determine the quantities of cut and fill.
Earthwork Volumes	Typically presented in cubic yards	Cubic Yard = 27 Cubic Feet
Pond Capacity	Basic formula for estimating pond capacity	Pond Capacity (CF) = ((A1+A2) + ((SQRT(A1*A2))*h))/3
		A1 = Area of the Bottom (CF)
		A2 = Area of the Top (CF) at water surface
		H = Pond depth from bottom to design water surface elevation
Determining Projected Elevations	Using known elevations to find other needed elevations	Elevation 2 = Elevation 1 ± (Slope * Distance)
		Slope is typically given in % (i.e., 2%)
		Distance is the distance between Elevation 1 and Elevation 2
		Be able to visualize slopes and grading field conditions in your head.

Communication in Civil Engineering and Project Management

Clear, concise communications in CE, collaborative team discussions, and project related discussions is essential	Learn how to listen and reiterate a sender's message to confirm accuracy, take detailed discussion notes in team meetings with action items and responsible parties noted for the deliverables.	CE projects typically involve significant investments in time, funds, and energy. Clear, concise communications help keep the project, the Team (and the CE), and many details on the path for success. Mistakes can be costly in many ways. Some issues with writing clearly and effectively are samples that don't always lend themselves to a quick screen edit. A good tool to review is "The Clarity Index." This index is based on the words used and sentence length. Basically, it allows the writer an opportunity to eliminate long words and long sentences without changing the meaning, making the writing sample clearer.

(Continued)

TABLE 21.3 *(Continued)*

General category and topic	Specific knowledge item and why it's important	Comments and how it's important
Effective Team Leaders and Project Managers require excellent communication skills.	Organize interdisciplinary teams and clearly communicate the project's goals and objectives, and how each team member contributes to those. Understand & convey the purposes of a project to non-technical clients and stakeholders.	Presentation skills are an essential skill to the successful CE. This is especially important when explaining the benefits of civil works but also their constraints, risks, limitations, and costs.
IF you are not clear about an instruction or detail, ask for clarification	It's been said that building a new home involves 10,000 decisions—Imagine how many decisions there are in a complex CE project. If you as the CE are unclear about a directive, think about it for a while, and if still unclear ask for clarification.	Complex CE projects tend to move swiftly and often you may receive a request or directive in a work order or email. However, most CEs have received requests from their Supervisors or Clients over lunch, in the break room, and sometimes in the washroom. Track these requests and clarify them if needed.
Keep a pad and pen by your bedside	Sometimes CEs get ideas when their minds are at ease, like just before we drift off to sleep. Rather than trying to remember these ideas, jot them down so you fully relax.	An experienced CE will be relied upon for project delivery. Show you are ready and available at all times to assist your Supervisor or Client. You will be appreciated and likely rewarded.
When your Supervisor or Client calls you into an unplanned meeting bring a notepad or device with you to record notes and action items.	The CE should be prepared to be available to his/her Supervisor or Client at a moment's notice. They are not likely calling you to discuss last night's game so be prepared to address their requests.	Handy tool for use in many civil engineering applications.

TABLE 21.3 *(Continued)*

General category and topic	Specific knowledge item and why it's important	Comments and how it's important
Managers only have two responsibilities; deliver results and retain your staff.	Delivering results is essential to being considered an effective manager.	Managers are expected to deliver results or projects within a defined scope, schedule, and budget. Without an effective team that is willing and happy to work for you, this will be an uphill battle. The single most important thing a manager can do is to implement regular, and rarely missed/canceled, "One-On-One" meetings with their staff. These meetings should typically occur weekly and last no more than 15–20 minutes. These meetings will allow you to build relationships with your team, discuss expectations, and effectively assign tasks to the appropriate individuals based on their skill set. Managers who properly and regularly implement One-On-One meetings with their staff will be amazed at how much easier it is to deliver the project results required by their organization.
Project/Task Management (PM)		
Preparing a scope of work, with a budget, and schedule	The Scope, Schedule, and Budget are all linked	A good understanding of your industry and what is required to deliver successful projects is essential to preparing an effective Scope of Work, Schedule, and Budget. Be able to clearly define the items in your schedule that are "Critical Path" items. These items will affect the entire project delivery if not carefully planned and monitored. Communication with the Project team is key.
Basic Accounting	Understand debits, credits, invoicing and payments, jurisdiction, severability	Funding gives us the ability to deliver projects and provide professional engineering services. It will be essential that Project Managers have a solid grasp of estimating work, tracking expenses, and processing payments to be effective.

(Continued)

TABLE 21.3   *(Continued)*

General category and topic	Specific knowledge item and why it's important	Comments and how it's important
Proposal Preparation	When preparing a proposed scope of work be sure to include your deliverables, specify details, and state any key assumptions	Often the CE's scope of work is incorporated directly into a legal contract. Details and assumptions are important and can limit the CE's risk.
Learn how to balance "speed, quality, and price" for Client deliverables	This is a continuous challenge for task and project managers.	Handy tool for use in many civil engineering applications
Expect the "unexpected"	Unanticipated turns and events will come up in your projects, career, and life.	Strive to have a "back-up" plan available if possible
Stay organized	A clear mind is the best mind. If one does not know a solution to an issue, always know the way how and where to find one.	As one grows with his profession, staying organized is one of the best keys to success.
Time Management	Do not procrastinate. Prioritize the tasks and find what is needed the most.	There is always very little time, but there are so much to do. Prioritize and focus will be the key to find the solution in a timely manner.
Contracts and Legal Related		
Know basic contract clauses	Understand/comprehend contract terms like force majeure, liability, payment terms, warranties, and more.	Contracts are the legal instrument for professional services. A CE could be sued even "if" they just provide free advice.
Know your limits in legal matters	Know "when" you might need to seek legal advice.	Practice smart. Understand that your Professional Engineering License may allow you to provide services outside your area of expertise. Be careful; just because your license allows you to design a two-story building doesn't mean you should be providing these services without first acquiring the knowledge and years of experience necessary to do so.

TABLE 21.3 *(Continued)*

General category and topic	Specific knowledge item and why it's important	Comments and how it's important
Keep legal opinions confidential	Legal opinions from a lawyer carry a client privilege.	Do not forward your lawyer's email, voice message, or other opinions to anyone outside your organization.
The "Final signature, date, and location" on a contract are very important.	Many Clients send their pending contracts back to the Corporate or Main Office for review and signature. The "location" of the "final dated signature" often determines the location of the court where this contract will be judged if necessary.	The final approval signatures establish the court location if the contract is challenged. Engineering companies may think twice about challenging a contract for a relatively insignificant item if the challenge occurs in another State thereby incurring travel and additional attorney fees.
Stamping or Certifying Plans/Reports	Using your PE stamp to certify designs or reports is rewarding but comes with some professional responsibility and potential liability.	Always be sure to understand that when you stamp plans or reports you're stating that said documents were completed under your direction. You should never stamp documents that you weren't involved in the development of or direction of the work. This gets many engineers into trouble.

21.6 SUMMARY

In the broad sense, civil engineering has been a necessary component of life since the beginning of human existence. As time marched on, new materials, methods, and societal demands evolved. The need for specialization and regulation also grew. Professional organizations were launched and laws were created requiring the licensing of professional civil engineers.

As a "formal" profession, civil engineering dates from the Industrial Revolution. Over time, the profession has developed several subdisciplines including coastal engineering, construction engineering, environmental engineering, geotechnical engineering, municipal or urban engineering, structural engineering, surveying, transportation engineering, water resources engineering and utility engineering and surveying.

"Design is not just what it looks like and feels like. Design is how it works."
—Steve Jobs, Apple co-founder

Civil engineers now have specialized educations involving diverse topics that enable them to recognize and solve problems. In addition to mathematics and science, practicing civil engineers need to know about design, sustainability and other contemporary issues, risks and uncertainties, project management, communication, public policy, contract law, business, leadership, teamwork, and professional and ethical responsibility.

University education can provide a Bachelor's degree in civil engineering that "partially" develops an engineer's ability to identify, explain, and apply essential concepts and principles in order to perform critical thinking and problem solving (BOK3 Outcome). But this level of knowledge and training only represents the first step toward civil engineering practice. Seasoned civil engineers recognize engineering as a "profession of practice." The preparation of a civil engineer requires both formal education and mentored experience. With the assistance of a mentor, the graduate engineer's experience should progress with increasing complexity, quality, and responsibility.

Although the average citizen may not recognize the role of civil engineers in society, civil engineers continue to shape the quality of our lives. As the ASCE puts it: Civil engineers build the world's infrastructure. In doing so, they shape the history of nations around the world.

While every effort has been made to ensure the accuracy and completeness of the information provided in this handbook, readers are advised to consult with a qualified professional before applying any of the practices or techniques discussed to specific projects or situations.

ASCE History and Heritage Website

http://content.asce.org/history/index.html

Know your own personal code of ethics, practice continuous improvement in your life and career, be humble and respectful. Become a licensed civil engineer!

BIBLIOGRAPHY

American Society of Civil Engineers. (2019). *Civil Engineering Body of Knowledge for the 21st Century*, 3rd edition. ASCE report, ASCE, Reston, VA. ISBN (print): 9780784415221ISBN (PDF): 9780784481974.

Example Request for Proposal (RFP)

REQUEST FOR PROPOSALS
FOR
ENGINEERING SERVICES
FOR A

Pipeline Routing Study

FOR THE

Applegate to North Auburn Wastewater Conveyance Pipeline

CSUS
CE 190 SENIOR PROJECT

Proposals Due: 9:00 a.m.
Dr. Ed Dammel
Department of Civil Engineering
Room RVR 4030

I INTRODUCTION

Placer County (County) owns and operates a Wastewater Treatment Plant (WWTP) and small collection system that serves the unincorporated community of Applegate, California. The Applegate Sewer County Service Area (CSA 28, Zone 24) consists of a wastewater collection system and pond treatment facility serving 27 connections (36 equivalent dwelling units). The WWTP was constructed in 1975 and consists of three unlined storage ponds and a chlorination system (Assessor's Parcel Number 073-120-013). The ponds are located over rocky soil in an area with high groundwater and spring activity. Increasingly stringent waste discharge requirements and a small customer base make improving the WWTP financially infeasible.

Furthermore, during wet weather, the ponds are prone to filling with rainwater, and on occasion have discharged to an adjacent creek to keep from overtopping. These discharges are violations of the WWTP's permit issued by the Central Valley Regional Water Quality Control Board (CVRWQCB). The CVRWQCB has issued several enforcement actions in response to these discharges; the most recent of which was Administrative Civil Liability Complaint R5-2005-0510 (ACLC). On October 10, 2006, the County Board of Supervisors signed an agreement with the CVRWQCB that established the terms and conditions for settlement of the ACLC. These terms included the construction of a pipeline to convey flows from Applegate to the North Auburn (Sewer Maintenance District 1, SMD 1) collection system. This project is considered a separable element of a larger effort to regionalize several, small collection systems in the Auburn area. The County is soliciting proposals from engineering firms to provide professional services for the preparation of a Pipeline Routing Study.

II PROJECT DESCRIPTIONS

The Consultant shall prepare a Pipeline Routing Study that considers the connection of a pipeline from the existing WWTP to one of four different possible connection points in the SMD1 collection system, one Consultant team per connection point:

A. Dry Creek Road at Windsong Place/Blue Grass Drive
B. Winchester Club Drive at Sugar Pine Road
C. Ridgemore Drive at Meadow Vista Road
D. Christian Valley Road at Williams Drive / Williams Court

The major tasks to be included in the Scope of Services are as follows:

Task 1 – Project Management
Task 2 – Project Research
Task 3 – Develop Alternatives
Task 4 – Prepare Cost Estimates

Task 5 – Evaluate Alternatives

Task 6 – Prepare Pipeline Routing Study Report

Task 7 – Oral Presentation

Task 1 – Project Management

The Consultant shall assign a Project Manager (PM) who shall coordinate the activities of the team. The Consultant shall perform a project kickoff meeting. The agenda for the kickoff meeting shall include project team member introductions, a description of the roles and responsibilities, delineation of the methods and lines of communication, any project safety issues, a review of the scope and schedule of services, and a review of the relevant information the Consultant seeks from the Client. Thereafter, the PM, or his designee, shall prepare regular billings indicating the hours expended to date and budget remaining for each of the tasks performed under the Scope of Services, and maintain the project schedule.

Task 2 – Project Research

The Consultant shall perform site visit(s) as needed to determine and/or confirm the physical constraints of the project and opportunities to develop unique solutions and design alternatives. The Consultant shall also compile and review all available data regarding the project site including prior studies, mapping and aerial photos. Existing mapping should be used to prepare the design layouts included in Task 3.

Task 3 – Develop Alternatives

Each connection point may include several possible routes, and may contain a mix of gravity flow and force main technology. The Consultant shall prepare three or more unique and feasible solutions (alternatives) for the project that seek to achieve these goals:

- Address the needs of diverse project stakeholders
- Meet or exceed acceptable design standards
- Avoid or minimize impacts to physical, environmental, or other constraints in the project area
- Maximize project benefits at the lowest initial and operating cost

The alternatives shall be developed to the extent necessary to establish the basic scope, feasibility, and cost of each. Design elements shall include a pipeline profiles for representative sections, and horizontal and vertical geometrics for the entire pipeline. The footprints of any required pump stations should be laid out with accurate horizontal and vertical geometrics along with a typical pump selection (horsepower, type, number) and associated power requirements. The proposed designs should be consistent with all appropriate standards as determined by the Consultant including Federal (e.g., Clean Water Act), State (e.g., California Department of Transportation, Department of Health Services, and California Public Utility Commission), Local (Placer County), and any other appropriate standards. The proposed alternatives shall include detail regarding construction materials and component technology, but further detailed structural or geotechnical design is excluded from this scope of services.

Task 4 – Prepare Cost Estimates

The Consultant shall prepare planning-level estimates of probable costs for each alternative. These estimates should address at a minimum:

1. Probable cost for construction
2. Probable cost for operations and maintenance
3. Probable cost for land acquisition
4. Estimated engineering fees

The estimates shall be provided in current dollars for the purpose of comparing the alternatives and in future dollars (escalated to the construction year) to determined probable cost for budget programming purposes.

Task 5 – Evaluate Alternatives

The Consultant will develop a set of ranking criteria to compare the benefits, costs, and potential effects of the alternatives. Criteria should be based on input from the Client and major stakeholders. The criteria should be specific enough to differentiate the alternatives. Each alternative shall be systematically evaluated and ranked using the weighted criteria, giving the Client a basis for selecting a recommended alternative for construction.

Task 6 – Prepare Pipeline Routing Study Report

The Consultant shall prepare an engineering feasibility report that addresses the specifics of each project. The report should address the main features and benefits, rankings, and any concerns or risks associated with each alternative. In addition, the estimated cost of each alternative should be presented. The Consultant shall recommend a course of action to the County (i.e., adopting a recommended alternative). The Consultant shall submit both a 90% Draft Report and a Final Report. Dates and formats for the Draft and Final reports are specified below.

Draft Report due on or before November 14 at 9:00 a.m.

Three printed copies of the Draft Report shall be submitted in a three ring binders with all pages easily accessible.

Final Report due at the Oral Presentation (Task 7)

Four spiral-bound printed copies of the Final Report shall be submitted.

Four copies of a diskette, compact disk, or DVD disk which contains electronic copies of pertinent files for all reports that the Consultant has prepared on a computer shall be submitted. The electronic copies shall meet the following criteria:

a. Be inclusive of all graphics (e.g., page orientation, photographs or other images, charts, and tables) and be suitable for printing in final form.
b. Be optimized for use by Adobe Acrobat Reader 5.0.

Task 7 – Oral Presentation

The oral presentation should utilize Microsoft PowerPoint graphics and be a maximum of 30 minutes in length. The presentation should summarize the findings and recommendations of the final report. The date for the oral presentation is specified below.

Date and Location on the CSUS campus provided to students separately.

III   PROPOSAL SUBMITTAL REQUIREMENTS

The Consultant shall submit a Proposal for Engineering Services (Proposal) in conformance with this Request for Proposal (RFP) and in particular, Attachment A – Proposal Submittal Requirements. Proposals should clearly demonstrate the Consultant's possession the following:

• Adequate qualifications and experience,
• Understanding of the project, and
• A detailed work plan including scope, schedule, and budget.

It is highly recommended that the Consultant perform a site visit prior to submittal of the Proposal. Late proposals will not be accepted or considered. The County shall not be responsible for proposals delivered to a person or location other than that specified in this RFP.

EXHIBIT 1

Guidelines for Preparing A Proposal for Professional Services

A1 INTRODUCTION

These guidelines were developed to standardize the preparation of proposals by Consultants for engineering services on a project. The purpose of these guidelines is to help assure consistency in format and content of proposals that are prepared by Consultants and submitted to the Client.

The Consultant shall submit four copies of the proposal.

The proposal shall contain the following information in the order listed:

1. Introductory Letter
2. Title Sheet
3. Table of Contents
4. Project Description
5. Work Plan
6. Schedule of Work
7. Consultant Assets
8. Qualifications and Capability
9. Supportive Information

Civil Engineer's Handbook of Professional Practice, Second Edition. Karen Lee Hansen and Kent E. Zenobia.
© 2025 John Wiley & Sons, Inc. Published 2025 by John Wiley & Sons, Inc.
Companion website: www.wiley.com/go/hansen/CivilEngineersHandbook

A2 RECOMMENDED DETAIL

Introductory Letter

The introductory letter should be addressed to:

See Professor Dammel for Client names and addresses.

The firm submitting the proposal shall give its name, mailing address, telephone number, FAX number, and the name of an individual to contact if further information is desired. This letter should contain a statement of the Consultant's basic understanding of the project. This should be based on existing information available in the Request for Proposal, from a site visit, and from applicable regulations or requirements. This letter should also contain an expression of the Consultant's interest in the work, a statement regarding the qualifications of the Consultant to do the work, and any summary information on the project team or the Consultant that may be useful or informative to the Client.

Title Page

Include the title of the project, who it was written for, who wrote it, and the date.

Table of Contents

The Table of Contents should contain page numbers and descriptions for each section. It should enable the reviewer to find proposal sections quickly.

Project Description

Describe the problem and the goals of the project. This does not have to be in great detail (after all, the Client knows the situation). Sufficient information should be provided, however, to assure the Client that you understand the project.

Work Plan

The work plan ultimately becomes part of the contract by reference to the proposal. It should describe in a specific and straightforward manner the proposed approach to achieving the objectives and accomplishing the tasks described in this Request for Proposal. It should be concise, yet include sufficient detail to completely describe the planned approach. Descriptions of how the objectives will be achieved shall be presented through a logical, rational, and innovative plan. The plan should describe each phase or task of the work to be undertaken including the man-hour level of effort for each class of personnel and for each sub consultant. The plan should detail the prosecution of the work including the submission of plans,

Exhibit 1 711

documents, reports, etc. Results shall be presented in terms of the language and working tools of the practicing engineer or administrator so as to be immediately useful.

Schedule of Work

The prospective consultant shall prepare a comprehensive schedule to reflect the time, in terms of working days required to complete each of the activities listed in the Scope of Services. A schedule should be included showing each activity, when that activity will begin and how long it will continue. Provide the completion date for each activity and identify activities that are interdependent. The schedule shall clearly differentiate between those functions carried out by the Consultant, the Client, and other interested parties. Ideally, the schedule will be presented in graphical format such as a GANTT chart or bar diagram.

Consultant Assets

Identify assets the Consultant shall use to accomplish the work. This may include office space, methods of transportation to project site and meetings, subcontractors and computer tools.

Qualifications and Capability

Identify the key individuals who are proposed to be part of the team, along with their qualifications and experience as related to the project. Experience on similar or related projects should be included. The information should include the expected amount of involvement and time commitment for each of these individuals. Proposal should contain a listing of current work commitments to other projects or activities in sufficient detail to indicate that the organization and all of the individuals assigned to the proposed project will be able to meet the schedule outlined in the Proposal. Consultant shall clearly identify the project team to the extent that individual staff members are clearly defined at each stage of the design. Changes in key personnel after the award of a project must be approved in writing, by the Client before the change is made. Describe the Consultant's capability for actually undertaking and performing the work. Types and locations of similar work performed in the last three years that best characterize the quality and cost control of the Consultant should be included.

Names and phone numbers of individuals that can provide information related to work quality and cost control should be provided. Other resources, including management and organization capabilities, should be addressed.

Supportive Information

Supportive information may include graphs, charts, photographs, resumes, and references. Content is to the prospective Consultant's complete discretion.

Format

The proposal shall be prepared in a professional manner (i.e., typed with computer-generated graphics) and shall be bound in a single volume. The proposal shall be limited to no more than fifteen (15) single-spaced pages (inclusive of references). Include appropriate headings and subheadings throughout the document to assist reviewers in following the proposal narrative. Use font size of at least 11 point with at least one inch margins all around. Number all pages.

An Example CSUS Student Proposal for This RFP is Presented Next. The Example Proposal Is Then Followed by a Different Student Group Feasibility Study for the Same Project:

Example Proposal

October 3, 20XX

Mr. Bob Jones, P.E.
Mr. John Smith, P.E.

**RE: Request for Proposal to Provide Professional Engineering Services for the
Applegate to North Auburn Wastewater Conveyance Pipeline Routing Study**

Dear Mr. Jones and Mr. Smith,

Global Hydraulic Engineers, Inc. is pleased to submit our proposal for the Applegate to North Auburn Wastewater Conveyance Pipeline Routing Study. Our team's experience in planning, designing, and managing public works improvement projects and our thorough understanding of Placer County's requirements for this project will allow our team to develop a pipeline routing study that is fully accepted by Placer County and other major project stakeholders.

Global Hydraulic Engineers, Inc. (GHEI) understands the need for a new pipeline to connect the community of Applegate's wastewater collection system to the larger Placer County Sewer Maintenance District 1 (SMD1) system. The new connection will allow for decommission of the obsolete Applegate Wastewater Treatment Plant and will provide the opportunity for more of Placer County's residents to connect to the SMD1 collection system.

GHEI is aware of the challenges that this project will encounter. Crossing of both Interstate-80 and the Union Pacific Rail Road will play major roles in the development of this project. However, GHEI sees this as an opportunity for Placer County to commission a project that serves as a model for the type of coordination that is required between local and government agencies to provide the public with a well thought out and useful project. Our team's project manager has over three years experience working with Caltrans on various joint and oversight projects including the Sheldon Rd./SR 99 Interchange Reconstruction Project. His experience will prove to be a unique and valuable asset in the development of well thought out and feasible designs ideas.

GHEI is committed to designing projects that are both economical and environmentally sound. We are a well known and respected Sacramento-based firm that takes pride in employing engineers who see possibilities in the challenges we face. We are especially excited and look forward to working with you and your staff on this important project. If you have any questions or require additional information, please do not hesitate to contact our project manager at (555) 555-1212 or email@globohydro.com

Sincerely,
Global Hydraulic Engineers, Inc.

_____	_____	_____
Project Manager	Deputy Manager	Project Engineer

_____	_____
Project Engineer	Project Engineer

6000 J Street, Sacramento, CA 95819

Proposal to Provide Professional Engineering Services for the Applegate to North Auburn Wastewater Conveyance Pipeline Routing Study

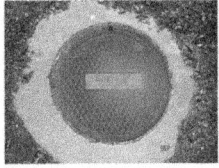

Presented to:

Placer County
California

Bob Jones, P.E.
John Smith, P.E.

Presented by:

Global HYDRAULIC
ENGINEERS INC.

6000 J Street, Sacramento, CA 95819
(555) 555-1212
email@globohydro.com

Placer County
California

Proposal for Applegate to North Auburn Wastewater Conveyance Pipeline Routing Study

TABLE OF CONTENTS

6000 J Street, Sacramento, CA 95819

Placer County
California

Proposal for Applegate to North Auburn Wastewater Conveyance Pipeline Routing Study

Section 1 - Project Description

1.1 Background

Placer County has initialized an effort to regionalize several small wastewater collection systems in the Auburn area. As a separable element of this large project, construction of a pipeline is needed to convey flows from the existing, obsolete wastewater treatment plant (WWTP) that serves the small unincorporated community of Applegate. This new pipeline would then need to connect to the existing North Auburn Sewer Maintenance District 1 (SMD1) collection system, allowing for decommission of the Applegate WWTP. The connection point that has been identified is near the intersection of Winchester Club Drive and Sugar Pine Road. Wastewater captured by the new pipeline will ultimately be conveyed to the SMD1 WWTP on Joeger Road in North Auburn. For this project to materialize, a pipeline routing study is needed. The study will identify and analyze possible alternatives to accomplish the task of rerouting Applegate's wastewater to the SMD1 collection system.

1.2 Project Details

GHEI's preliminary investigation of the project has revealed that routing of a pipeline from the Applegate WWTP to the connection point at Winchester Club Drive and Sugar Pine Road will cross both Interstate-80 (I-80) and the Union Pacific Rail Road (See Figure 1 – Project Area Map). Crossing of these facilities will require close involvement with representatives from both Caltrans and the Union Pacific Rail Road (UPRR) in the acquisition of encroachment permits, adherence to design standards, and discussion of design alternatives. Furthermore, the regional topography presents a unique challenge to the routing of the pipeline and may present the need for force main technology and pump stations in order for the wastewater to reach its destination (See Figure 2 – Project Area Topography and Figure 3 – Elevation Profile). Each of these major issues will be considered in the development of pipeline routing alternatives and study.

1.3 Site Description

The project area includes the Applegate WWTP located approximately 8 miles northeast of Auburn and the route of the pipeline to the connection point at Winchester Club Drive and Sugar Pine Road (approximately 3.2 miles southeast of the Applegate WWTP). The area surrounding the WWTP consists of the rail road tracks directly to the west (which run east parallel to Interstate-80), and several rural, single family homes. Also present is a natural gas pipeline that runs parallel to the UPRR tracks and directly adjacent to the WWTP.

Placer County
California

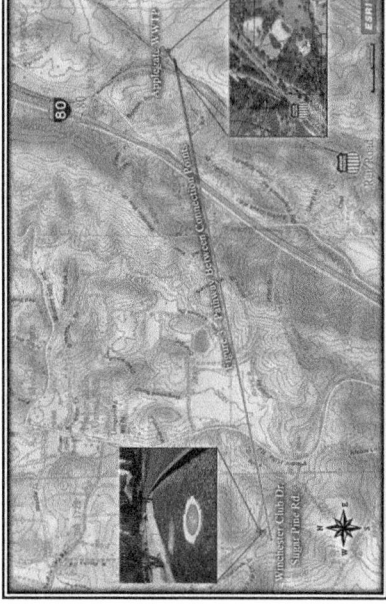

Figure 2 – Project Area Topography

Figure 1 – Project Area Map

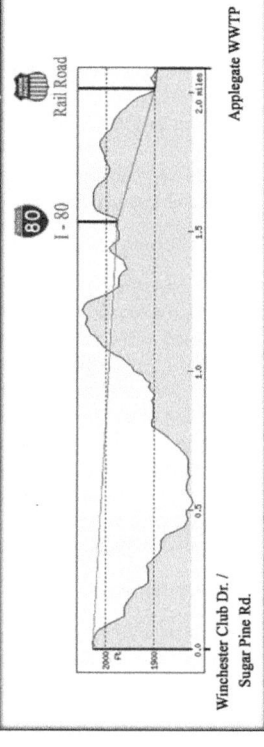

Figure 3 – Elevation Profile: Applegate WWTP to Connection Point

Global Hydraulic
Engineers, Inc.

Placer County
California
Proposal for Applegate to North Auburn Wastewater Conveyance Pipeline Routing Study

Section 2 - Work Plan

2.1   Task 1 – Project Management

Our team has selected Jeff Werner as Project Manager. He will provide ongoing project management throughout the life of the project. In addition to coordinating the team's activities, Mr. Werner will serve as a readily available point of contact for the client and project team.

Task 1 Subtasks:

2.1.1   *Kick Off & PDT Meetings*

GHEI will organize and conduct a project kickoff meeting. The kickoff meeting will provide an opportunity for the client to be introduced to GHEI's project development team (PDT) members and to get an idea of the roles and responsibilities each will play. At the kickoff meeting GHEI will determine the preferred lines of communication for the client and will allow for input on streamlining of the project scope, schedule, safety issues, and relevant information. GHEI will hold additional PDT meetings at least weekly depending on the nature of the project development and the needs of the client.

2.1.2   *Project Schedule*

GHEI will utilize Microsoft Project to establish and maintain a critical path project schedule to meet the client's project schedule. The schedule will be distributed at the kickoff meeting to allow the client to review, provide feedback, and endorse the project's schedule. Updates to the schedule will be provided to the client monthly to show the progress of each task and any revisions due to continued project development.

2.1.3   *Client Billings*

GHEI will prepare regular billings for the client. The billings will show the number of hours expended by project team members for each task and a breakdown of the hours remaining for those tasks. The billings will allow the client to further track the progress of the project and ensure that GHEI has allocated sufficient time to complete the tasks.

Task 1 Client Deliverables:

- ➤ Kickoff Meeting Agenda and Meeting Minutes
- ➤ Billing Statements
- ➤ Updates to Project Schedule

Section 2 - Work Plan

2.2   Task 2 – Project Research

Project research will consist of the following subtasks:

Task 2 Subtasks:

2.2.1 Preliminary Research

GHEI will gather pertinent existing information for the project locations to aid in choosing routing alternatives. This information includes, but is not limited to:

- Placer County Design Standards
- Aerial Site Mapping
- Available Roadway and Sewer As-Builts
- Caltrans Design Standards
- Available Utility Information/Maps
- Topographic Site Mapping
- Available Prior Studies
- Required Key Permits

2.2.2 Project Stakeholders

Through its ongoing project research GHEI will develop and maintain a list of stakeholders for the project. Some of the major project stakeholders have been identified as follows:

- Placer County
- UPRR
- Caltrans

2.2.3 Site Investigation

GHEI will perform up to three site visits to verify existing information obtained from preliminary research in addition to determining the physical constraints to the preliminary routing alternatives. Onsite exploration will help establish potential constraints and possible innovative ways to solve the design issues related to the project.

2.2.4 Constraints

GHEI has identified the following constraints during our initial investigation:

- Crossing of UPRR
- Crossing of I-80
- Regional Topography

GHEI will continue to determine project constraints which will be identified in a Project Constraints Table that includes information on how the constraint might impact the routing alternatives and possible mitigation for the each constraint.

Placer County
California

Proposal for Applegate to North Auburn Wastewater Conveyance Pipeline Routing Study

Section 2 - Work Plan

Task 2 Client Deliverables:

- Stakeholders List
- Project Constraints Table

Task 2 Assumptions:

- Pipeline is only for wastewater only; storm water drainage is not included.

2.3   Task 3 – Development of Alternatives

GHEI will develop a minimum of three (3) design alternatives for the Applegate/North Auburn Wastewater Conveyance Pipeline Routing Study to the extent necessary to establish the basic scope, feasibility, and cost for each. Each developed alternative will at a minimum address the costs, benefits, constructability, stakeholder interests, and impacts to the environment.

Task 3 Subtasks:

2.3.1   *Drawings and Specifications*

GHEI will compose conceptual plan and profile drawings for each alternative in accordance with applicable design standards. Along with the pipeline layout drawings there will be detailed drawings and specifications of required pump stations including horizontal and vertical geometries gravity and force main technology and specific pump selection specifications and sizing. Other necessary specification for each alternative will be developed to the extent required for the feasibility study.

2.3.2   *Material Lists*

GHEI will provide a materials list for each alternative developed. The lists will later be used for estimating the cost of each alternative.

2.3.3   *Available Construction Technology*

GHEI will research available construction technology and or techniques that may help avoid or minimize impacts to physical, environmental, or other constraints in the project area.

Task 3 Client Deliverables:

- Pipeline Layout Drawings
- Specifications
- Pump Drawings and Specifications
- Materials Lists

Placer County
California

Proposal for Applegate to North Auburn Wastewater Conveyance Pipeline Routing Study

Section 2 - Work Plan

2.4   Task 4 – Prepare Cost Estimates

GHEI will prepare planning-level cost estimates for each developed alternative. Each cost estimate will include:

> Construction cost
> Costs for operations and maintenance
> Land acquisition and Right of Way (R/W) services costs
> Cost for engineering fees

Cost estimates prepared by GHEI will be provided in both current year dollars and future year dollars. Current year cost estimates will allow for direct comparison of alternatives. Future year cost estimates are provided to allow for client budget programming. Engineering costs will be calculated as a percentage of the total construction cost.

Task 4 Client Deliverables:

> Cost Estimates for Each Alternative

2.5   Task 5 – Evaluate Alternatives

GHEI, with input from (client) and major project stakeholders, will develop a set of ranking criteria for the purpose of evaluating each alternative. The developed criteria will potentially be divided into the following 5 major categories:

> Benefits   > Environmental Impacts
> Costs   > Stakeholder Concerns/Benefits
> Constructability

Each alternative will be evaluated and ranked according to the final developed ranking weights and criteria.

Task 5 Subtasks:

2.5.1   Summary Table for Recommendation of Alternatives

GHEI will provide a summary table highlighting the evaluated alternatives and describing the positives and negatives for each. A recommendation for implementation of the highest ranked alternative will also be included. Viable evaluation information will be presented in the summary table in order to assure the client that a thorough and systematic evaluation process was followed.

Global HYDRAULIC
ENGINEERS INC.

Placer County
California
Proposal for Applegate to North Auburn Wastewater Conveyance Pipeline Routing Study

Section 2 - Work Plan

Task 5 Client Deliverables:

- Detailed List of Ranking Criteria
- Meeting with Client & Meeting Minutes
- Evaluation Summary Table

2.6 Task 6 – Prepare Pipeline Routing Study Report

GHEI will prepare an engineering feasibility report that should be used as a basis for the client to select one of the design alternatives. The report will summarize the specifics of each alternative and will highlight their main features including the benefits, risks, ranking, and cost. Furthermore, the feasibility report will contain GHEI's recommended alternative and other recommended courses of action for Placer County. Submission dates for the report are as follows:

- 90% Draft Report due on or before November 14, 2008
- Final Report due on or before December 12, 2008

Task 6 Client Deliverables:

- 90% Draft Report
- Response to Comments Table
- Final Report
- Copies of Pertinent Files

2.7 Task 7 – Oral Presentation

GHEI will provide an oral presentation of our findings for the routing study on December 12, 2008; at a minimum length of 30 minutes. The presentation will be open to the Client, major project stakeholders, and general public. Within this presentation GHEI will utilize Microsoft PowerPoint to introduce our company, the project description, and the evaluated alternatives. At the conclusion of this presentation GHEI will provide time for questions and any comments.

Task 7 Client Deliverables:

- Oral Presentation

Placer County
California

2.8 Project Schedule

ID	Task Name	Start	Finish
1	Placer County/Applegate WWTP Pipeline Routing Study	Mon 10/6/08	Fri 12/12/08
2	**Task 1 – Project Management (115 hrs.)**	Mon 10/6/08	Fri 12/12/08
3	Team Meeting (80 hrs.)	Wed 10/8/08	Wed 12/10/08
14	Discuss Proposal with Client (10 hrs.)	Fri 10/10/08	Tue 10/14/08
15	Kickoff Meeting (2-3 hrs.)	Fri 10/17/08	Fri 10/17/08
16	Outline Submission	Fri 10/17/08	Fri 10/17/08
17	Annotated Outline Submission	Fri 10/31/08	Fri 10/31/08
18	Buyoff Meeting (2-3 hrs.)	Mon 11/3/08	Mon 11/3/08
19	90% Draft Report Due	Fri 11/14/08	Fri 11/14/08
20	Discuss 90% Report with Clients (10 hrs.)	Mon 11/17/08	Fri 11/21/08
21	Quality Control (30 hrs.)	Mon 10/6/08	Fri 12/12/08
22	Billing Statements (10 hrs.)	Mon 10/13/08	Mon 11/24/08
26	Final Billing Statement	Mon 12/1/08	Mon 12/1/08
27			
28	**Task 2 – Project Research (115 hrs.)**	Mon 10/6/08	Fri 10/24/08
29	Preliminary Research (70 hrs.)	Mon 10/6/08	Fri 10/17/08
30	Site Investigation (20 hrs.)	Mon 10/13/08	Fri 10/24/08
31	Develop Constrains Table (15 hrs.)	Wed 10/15/08	Wed 10/22/08
32	Develop Stakeholder List (10 hrs.)	Wed 10/15/08	Fri 10/17/08
33			
34	**Task 3 – Development of Alternatives (160 hrs.)**	Mon 10/13/08	Fri 10/31/08
35	Brainstorming Alternative Ideas (10 hrs.)	Mon 10/13/08	Thu 10/16/08
36	Develop Minimum 3 Alternatives (30 hrs.)	Wed 10/15/08	Fri 10/24/08
37	Identify Constraints (20 hrs.)	Mon 10/20/08	Fri 10/24/08
38	Develop Drawings for Each Alternative (80 hrs.)	Mon 10/20/08	Fri 10/31/08
39	Specifications for Each Alternative (20 hrs.)	Mon 10/27/08	Fri 10/31/08
40	Materials Lists (10 hrs.)	Mon 10/27/08	Fri 10/31/08
41			
42	**Task 4 – Preparation of Cost Estimates (25 hrs.)**	Mon 10/27/08	Wed 10/29/08
43	Cost Estimates for Each Alternative (25 hrs.)	Mon 10/27/08	Wed 10/29/08
44			
45	**Task 5 – Evaluation of Alternatives (40 hrs.)**	Wed 10/29/08	Fri 11/7/08
46	Develop Ranking Criteria (20 hrs.)	Wed 10/29/08	Tue 11/4/08
47	Develop Summary of Alternative Evaluation (20 hrs.)	Wed 11/5/08	Fri 11/7/08
48			
49	**Task 6 – Prepare Routing Study Report (185 hrs.)**	Mon 10/27/08	Fri 12/12/08
50	Prepare 90% Report (110 hrs.)	Mon 10/27/08	Fri 11/7/08
51	Review 90% Report (40 hrs.)	Sat 11/8/08	Tue 11/11/08
52	Print 90% Report	Tue 11/11/08	Tue 11/11/08
53	Submit 90% Report	Fri 11/14/08	Fri 11/14/08
54	Response to Comments (5 hrs.)	Mon 11/17/08	Mon 11/17/08
55	Prepare Final Report (30 hrs.)	Tue 11/18/08	Tue 12/9/08
56	Print Final Report	Tue 12/9/08	Tue 12/9/08
57	Deliver Final Report	Fri 12/12/08	Fri 12/12/08
58			
59	**Task 7 – Oral Presentation to Clients (100 hrs.)**	Mon 11/17/08	Fri 12/12/08
60	Prepare Oral Presentation (80 hrs.)	Mon 11/17/08	Fri 12/5/08
61	Rehearse Oral Presentation (20 hrs.)	Mon 12/8/08	Thu 12/11/08
62	Oral Presentation	Fri 12/12/08	Fri 12/12/08

Legend: Task — Split — Progress — Milestone — Summary — Project Summary — External Tasks — External Milestone — Deadline

Global HYDRAULIC ENGINEERS INC.

Proposal for Applegate to North Auburn Wastewater Conveyance Pipeline Routing Study

Section 3 - Consultant Assets and Qualifications

3.1   Assets

GHEI has offices conveniently located within 30 miles of the Applegate WWTP. GHEI is equipped with the necessary tools and software to design and analyze pipeline routing options; including: AutoCAD, Microsoft Office, Google Earth Plus, SewerCAD, ArcGIS and more.

For site visits GHEI will furnish its team with cell phones, vehicles, laptop computers, digital cameras, video recorders and all other equipment required for performing site visits. GHEI assures that each employee can effectively use these tools and is competent to operate them. GHEI provides each employee with extensive training to ensure safe and effective field operation procedures are followed.

In addition to the tools and capabilities mentioned, GHEI has working relationships with many agencies in the Sacramento and Placer regions including Caltrans and UPRR.

3.2   Team Organization & Level of Effort Summary

Figure 4 – Project Organization Chart

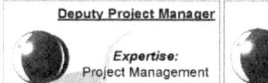
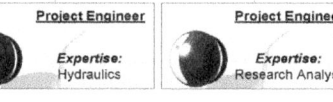
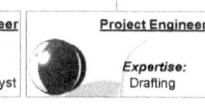

Table 1 – Project Availability

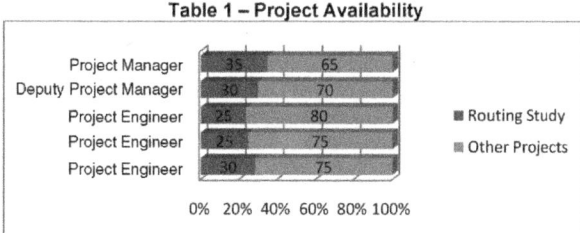

Placer County
California

Proposal for Applegate to North Auburn Wastewater Conveyance Pipeline Routing Study

Table 2 - Level of Effort

Project Tasks	Global Hydraulic Engineers, Inc.						
	Project Manager	Deputy Project Manager	Project Engineer	Project Engineer	Project Engineer	Budget Hours	Percentage of Budget Hours
Project Management	35	35	15	15	15	115	16%
Project Research	15	15	35	15	35	115	16%
Development of Alternatives	25	25	25	60	25	160	22%
Prepare Cost Estimates	5	5	5	5	5	25	3%
Evaluate Alternatives	8	8	8	8	8	40	5%
Prepare Routing Study	40	40	40	25	40	185	25%
Oral Presentation	20	20	20	20	20	100	14%
Total Hours	148	148	148	148	148	740	100%

Section 4 – Supportive Information

Project Manager
Project Availability   35%
Budget Hours   148

Education	BS, Civil Engineering, California State University, Sacramento, December 2008
Experience Summary	He has more than five years experience working in the civil engineering industry. His experience comes from his work as a student intern for Sacramento County's CSD1 and more recently with Interwest Consulting Group, Inc at the City of Elk Grove. He excels as a project manager, ensuring projects stay on track and under budget.
Project Experience	**2008 – Sheldon Road/State Route 99 Interchange Reconstruction Project:** Assistant Project Manager. Prepared and secured the "Request to Proceed With Construction" (Caltrans: E-76) documents for the authorization and release of federal funding. **2007– Sheldon Interchange Demolition Project:** Project Manager. Prepared PS&E and monitored construction operations through completion.

Placer County
California

Proposal for Applegate to North Auburn Wastewater Conveyance Pipeline Routing Study

Section 4 – Supportive Information

Deputy Project Manager
Project Availability   30%
Budget Hours   148

Education	BS, Civil Engineering, California State University, Sacramento, December 2008 Certification, Web Publishing, American River College, December 2008
Experience Summary	He has five years experience as a construction project manager for Ureta Construction. He managed and supervised all construction activities, coordinated client and company meetings, and communicated extensively with client to ensure customer satisfaction.
Project Experience	**2007 – National Cooperative Highway Research Program 12-74:** Played a key role in the development, fabrication, testing, and analysis of the experimental pre cast bent-cap system.

Project Engineer
Project Availability   25%
Budget Hours   148

Education	BS, Civil Engineering, California State University, Sacramento, December 2008
Experience Summary	He has worked in the civil engineering industry for over five years working for Caltrans and Nolte Engineering. His experience has come in the transportation, water, and wastewater engineering fields. He is adept in designing projects using AutoCAD.
Project Experience	**2006 to 2008 – Esparto Community Services District (CSD) Water/Wastewater System Improvements:** Assisted in the re-design and construction of major repairs to the Esparto, CA CSD Domestic Water and Wastewater Distribution System. **2006 to 2008 – City of Wheatland Wastewater System Improvements:** Assisted in the design and construction of major repairs to the City of Wheatland wastewater collection system.

Placer County
California
Proposal for Applegate to North Auburn Wastewater Conveyance Pipeline Routing Study

Section 4 – Supportive Information

Project Engineer
Project Availability   30%
Budget Hours   148

Education	BS, Civil Engineering, California State University, Sacramento, December 2008
Experience Summary	She has three years experience in the civil engineering field working for David Evans and Associates. She has experience in all aspects of land development and transportation engineering. Tasks performed include, designing subdivision improvement plans, preparing cost estimations, assisting in interchange analysis, highway design, and assisting in project coordination.
Project Experience	**2008 - Hwy 65-70 Interchanges:** Assisted in the development of the feasibility studies as required by Caltrans. **2005 to 2007 Bickford Ranch:** Assisted in the design of trunk sewer and land development for 2000 plus homes.

Project Engineer
Project Availability   25%
Budget Hours   148

Education	BS, Civil Engineering, California State University, Sacramento, December 2008
Experience Summary	He has two years civil engineering experience working for Sacramento County in the fields of water resources and land development. His work involved collecting meter data, designing grading and drainage plans, and preparation of cost estimates.
Project Experience	**2006 - PFE Road Water Transmission Pipeline Project:** Performed design calculations and site visits **2006 – Lincoln Bypass Project:** Performed design calculations

APPENDIX C

Example Feasibility Study Report

December 12, 20XX

Mrs. Amanda, P.E.

Mr. Ryan, P.E.

Subject: Pipeline Routing Study for the Applegate to North Auburn Wastewater Conveyance Pipeline

Dear Clients,

CVision Engineering (CVE) is pleased to submit this Pipeline Routing Study for the connection of the existing wastewater collection system of Applegate to Sewer Maintenance District 1 (SMD-1) in Auburn at the intersection of Dry Creek Road and Blue Grass Drive. The study presents a summary of the findings of CVE's project research, based on which three feasible alternatives were identified and developed. The report also discusses in detail the various physical, environmental and design constraints for the project, as well as the multiplicity of stakeholders and their potential impact on the project. Included for each of the three alternatives are preliminary design drawings, a pump selection, a cost estimate and a detailed evaluation against a ranking criterion. Lastly, based on the ranking of the proposed feasible alternatives, CVE provides a recommended course of action for Placer County.

Should you have any questions, please feel free to contact CVE's Project Manager, at (555) 555-1212 or email@cvisionengineering.com Thank you.

Sincerely,

CVision Engineering

Project Manager	*Assistant Project Manager*	

Project Engineer	*Project Engineer*	*Project Engineer*

CVision Engineering · CSU Sacramento · 6000 J Street, Sacramento, CA 95819

Civil Engineer's Handbook of Professional Practice, Second Edition. Karen Lee Hansen and Kent E. Zenobia.
© 2025 John Wiley & Sons, Inc. Published 2025 by John Wiley & Sons, Inc.
Companion website: www.wiley.com/go/hansen/CivilEngineersHandbook

December 12, 20XX

Applegate to North Auburn
Wastewater Conveyance Pipeline
Pipeline Routing Study

December 12, 20XX

PREPARED FOR:

COUNTY OF PLACER, CALIFORNIA

PREPARED BY:

CVISION ENGINEERING

California State University of Sacramento
6000 J Street, Sacramento, CA 95819

TABLE OF CONTENTS

December 12, 20XX

LIST OF TABLES

December 12, 20XX

LIST OF FIGURES

December 12, 20XX

1.0   INTRODUCTION

1.1   Project Overview

The Applegate to North Auburn sewer pipeline extension project is a part of Placer County's effort to regionalize the sanitary sewer system in the Auburn area. The need for the regionalization had been triggered by high costs of maintaining detention pond treatment facilities, such as the existing Wastewater Treatment Plant (WWTP) in Applegate, and lack of reliable performance as documented by the Central Valley Regional Water Quality Control Board (CVRWQCB). CVRWQCB has issued a number of Notice of Violations due to the consistent violations of the WWTP's permit. Furthermore, the current water treatment standards and effluent limitations are expected to become more stringent in the near future. To satisfy the terms and conditions of settlement as established by CVRWQCB, Placer County has agreed to decommission the Applegate WWTP and convey the existing wastewater flows to Sewer Maintenance District 1 (SMD-1) in North Auburn.

1.1.1 Project Location

Applegate is located in the foothills of Placer County, approximately 10 miles north-east of Auburn. The project area includes the proposed pipeline and the Applegate's WWTP with its treatment ponds and dechlorination facilities (Figure 1.1). The WWTP is located on the south side of Interstate-80 (I-80) and immediately east of the Union Pacific Railroad tracks at the location where Merry Lane ends. The north part of the project, from the WWTP to Placer Hills Road, is parallel to I-80. The south part of the project, from Placer Hills Road to the connection point with SMD-1 at Blue Grass Drive, follows Lake Arthur Road, which becomes Dry Creek Road at the intersection with Christian Valley Road. The entire project falls within the American River Watershed (Appendix A).

1.1.2   Existing Wastewater Collection and Treatment System

The existing wastewater collection system of Applegate was constructed in 1975 (Reference [1]) and consists of a combination of force main and gravity sewer leading to detention ponds and on-site dechlorination facilities (Figure 1.2). The capacity of the ponds is inadequate for the wastewater flows both because of inflow of rainwater during wet weather and also inflow of groundwater under artesian conditions. Per CVRWQB (Reference [2]), the groundwater inflow during the winter months is sufficient to overflow one of the detention ponds even if no wastewater is discharged into it. To avoid these violations, the ponds are temporarily decommissioned between the months of September and March and the wastewater is pumped into storage tanks and subsequently hauled away to SMD-1 by large tanker trucks.

1.1.2.1 Location and Service Area

The existing wastewater collection system is mostly within the Placer County right-of-way. Throughout much of the community, the pipe network is a gravity sanitary sewer conveyance system consisting of a 6-inch-diameter trunk with approximately 28 service connections. Within a section of Applegate Road, from Brick Road to Apple Court, the system is a 4-inch force main for approximately 2,000 feet (Figure 1.2). An existing lift station at the intersection of

---

Applegate to North Auburn Wastewater Conveyance Pipeline

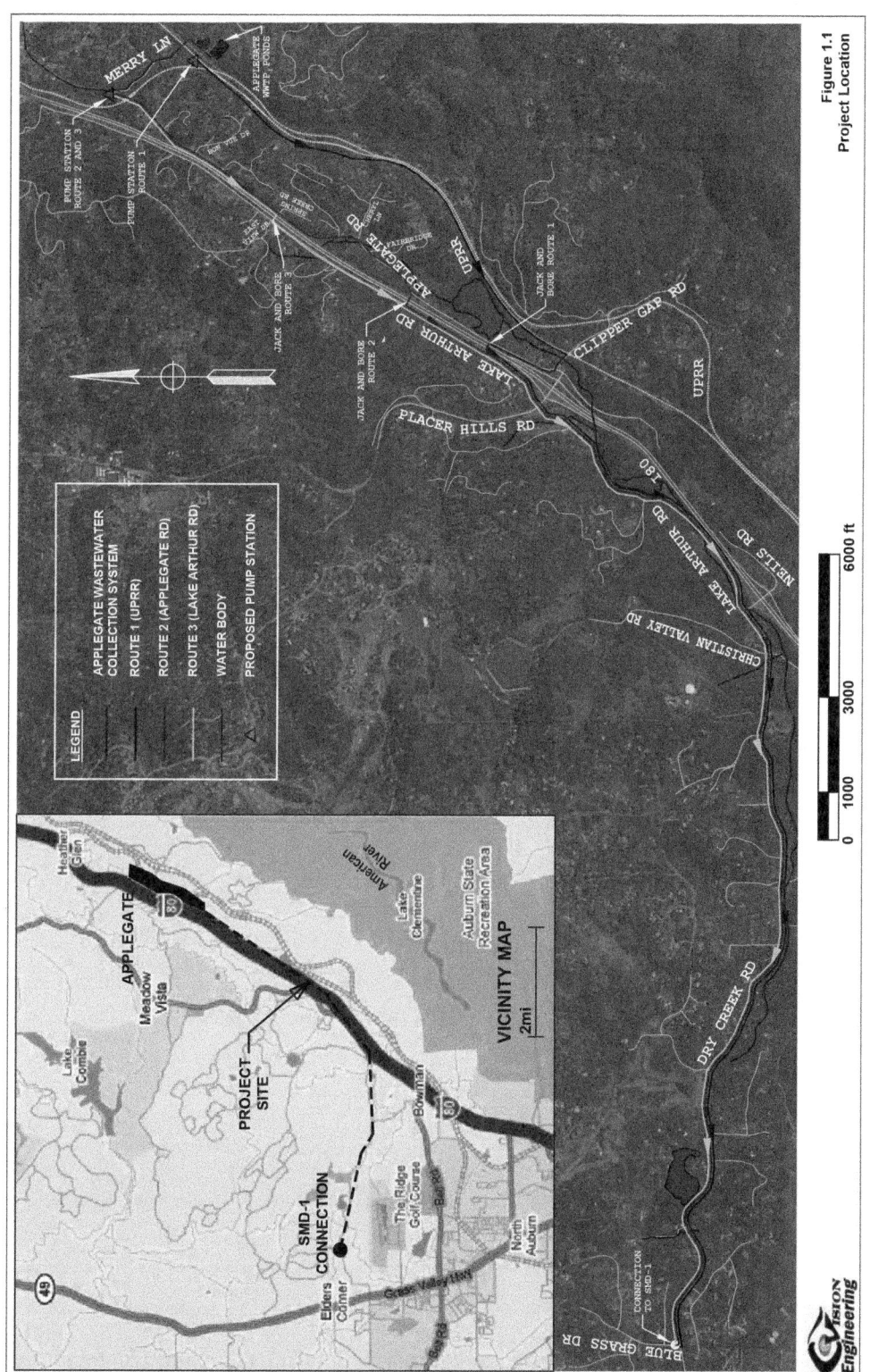

Figure 1.1
Project Location

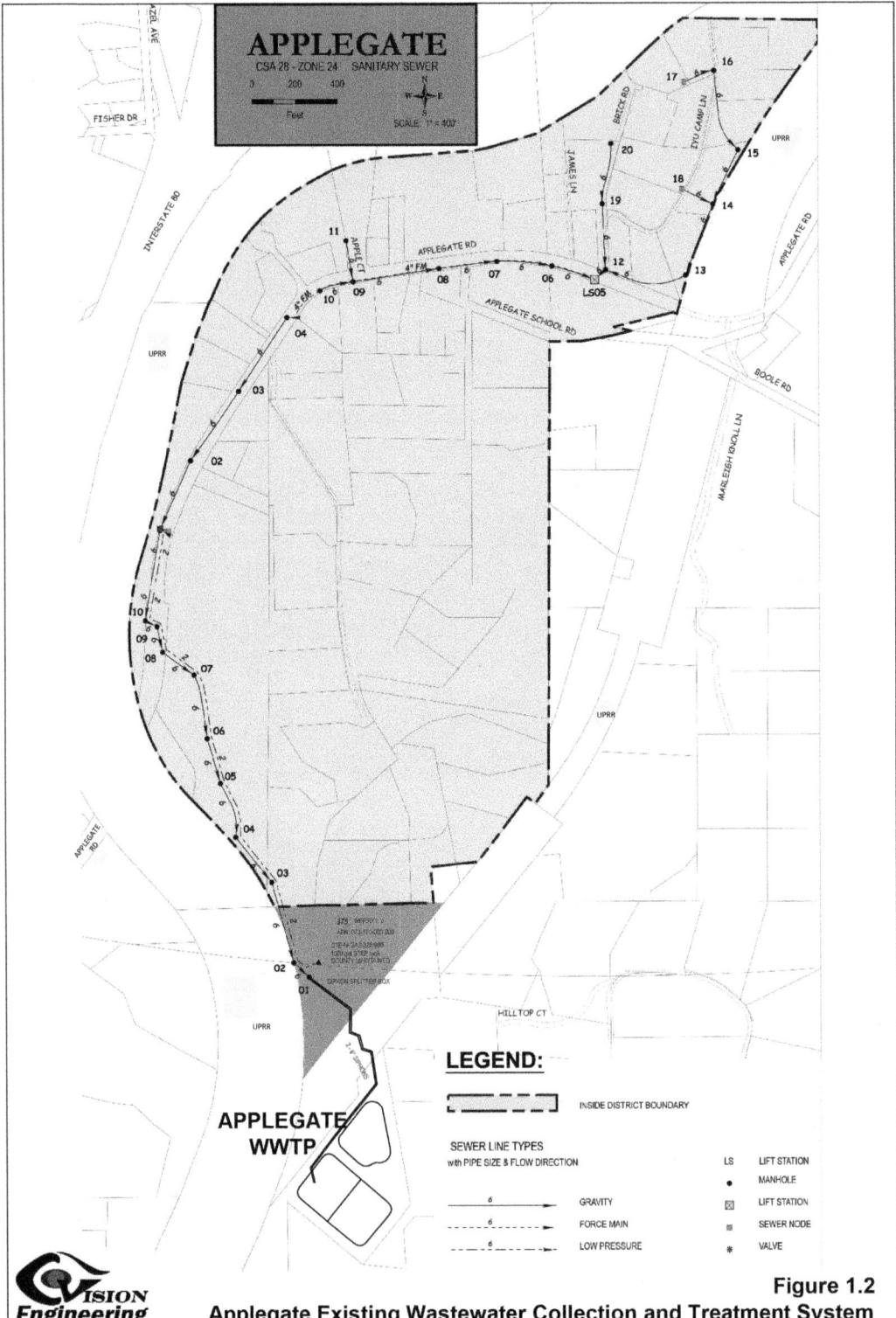

Figure 1.2
Applegate Existing Wastewater Collection and Treatment System

Applegate Road and Brick Road, pumps some of the existing flows over a hill between Apple Court and Brick Road. After overcoming the elevation head between Apple Court and Brick Road, the pipe is again a 6-inch gravity flow system for the remainder of its way to the detention ponds. At the intersection of Applegate Road and Merry Lane, the pipe system leaves the County's right-of-way and turns due south. It follows Merry Lane, which is a private road. Before the sewer pipeline reaches the wastewater treatment plant, it crosses under the Union Pacific Railroad tracks (Figure 1.2). According to the Placer County sanitary sewer record drawings, the last service connection to the existing sewer pipeline is before the pipeline leaves the county right-of-way at the intersection of Applegate Road and Merry Lane.

1.1.2.2 Wastewater Treatment and Disposal Operations

The Applegate WWTP consists of three evaporation and percolation ponds, designed to operate in series. The ponds are only six feet deep and cannot meet the current freeboard requirements, which mandate the pond surface to be a minimum of two feet below the top of the embankment at all times. The anticipated overflow from the ponds into Clipper Creek led Placer County to the decision of implementing a disinfection and chlorination system to achieve the required at the time effluent water characteristics. These discharges are violations of the WWTP permit. Further improvement of the WWTP is unfeasible and hauling away the wastewater during the winter months is only a temporary solution.

1.2 Purpose and Limitations of the Study

The objective of the study is to identify and evaluate a minimum of three feasible pipeline routes to convey the existing wastewater flows from Applegate to the designated connection point with SMD-1 at Dry Creek Road and Blue Grass Drive in Auburn. Alternative routes were considered based on their potential to maximize project benefits and minimize cost and negative impacts to the project area.

The following is a list of the limitations of this study:

- Record drawings of existing utilities were not available at the time of the study. The locations of existing utilities as shown on the design drawings (Appendix B) are approximate and based on field observations. The horizontal and vertical controls of existing facilities in the project area need to be verified by a licensed surveyor prior to further design considerations. (*Appendixes to this report not included.*)

- The profile of the terrain along the selected alternative routes was first approximated using United States Geological Survey (USGS) 20-foot contour data for Placer County. However, field observations showed a discrepancy between the existing terrain conditions and the digitally produced vertical profiles. This was a result of the inability of the 20-foot contour GIS data to account for the cut and fill for the existing roads. To better represent the existing conditions and provide a more realistic pipeline design, adjustments were made to the vertical profiles using a GPS unit. The project site needs to be surveyed prior to further design considerations. Accordingly, adjustments in the alignments and profiles may be further required.

- The research of potential environmental impacts along the proposed alternatives was based on the Placer County Master Conservation Plan and existing Environmental Impact Reports (EIRs) filed in and around the project area. Based on the available data, no significant environmental concerns have been identified at this time. However, the findings of the Final EIR for the project, due in the summer of 2009, may require appropriate mitigation measures.

- This study considers existing flows only. Future connections will not be included in the development of the proposed alternatives.

- The North Auburn SMD-1 system is a Septic Tank Effluent Pumping system (STEP) at the designated for the project connection point. Although Applegate's wastewater flows are relatively low compared to the ones at the tie-in point, backup of the new pipeline at the connection to SMD-1 may occur. This issue could be evaluated through sewer modeling or measurements and observations after the pipeline is operational. Based on the results of this evaluation, the construction of a pretreatment facility for Applegate's wastewater flows, before they enter SMD-1, may be required. Further studies for the treatment of the existing wastewater flows from Applegate are beyond the scope of this routing study.

1.3   Summary of Recommended Alternative

The study rendered Route 3 (Lake Arthur Road) as the recommended alternative (Figure 4.3). The new pipeline would be a force main throughout. Since there are no service connections along the descent between the intersection of Applegate Road and Merry Lane and the detention ponds, the existing sewer pipeline within this stretch would be decommissioned (Appendix B, L3-6). A new lift-station would be constructed at the intersection of Applegate Road and Merry Lane (Appendix B, L3-6). This location avoids the additional 75 feet of elevation head between the wastewater treatment ponds and Applegate Road, which also reduces the size of the required pump. From the lift-station on towards the connection point with SMD-1, the recommended alternative follows Applegate Road and Lake Arthur Road (Figure 4.3) – the only two viable county roads leading from Applegate to Auburn. Both of these county roads are parallel to I-80, with Applegate Road to the East and Lake Arthur Road to the West of I-80. The jack-and-bore crossing of I-80 for Route 3 (Appendix B, L3-5 and X-2) was identified based on an overlay of the vertical profiles of Applegate Road, I-80 and Lake Arthur Road. The selected jack-and-bore location is favorable because of the short distance from Applegate Road to Lake Arthur Road and also because Applegate Road is elevated above Lake Arthur Road, thus eliminating additional elevation head and higher pumping requirements. The only reasonable continuation of the preferred alternative, from Lake Arthur Road to the connection point with SMD-1, is for the route to follow Dry Creek Road. In addition to this, Route 3 has less private residences and existing utilities along its way.

2.0 PROJECT CONSTRAINTS

An extensive research of the possible constraints in the project area was made prior to considering alternative routes. The research included review of relevant standards and regulations, available record drawings, and environmental studies for the project area. Additional information was obtained through contacting regulatory agencies and conducting field visits. The project research determined the following major groups of project constraints:

- Regulatory Compliance
- Physical Constraints
- Environmental Constraints
- Private Stakeholders' Interests

2.1 Design Standards and Permitting Requirements

The project faces regulatory constraints associated with the following agencies:

- Placer County
- California Department of Transportation (Caltrans)
- Union Pacific Railroad (UPRR)
- California Department of Public Health (CDPH)
- Central Valley Regional Water Quality Control Board (CVRWCB)
- Placer County Air Pollution Control District (APCD)
- U.S. Fish and Wildlife Service
- California Department of Fish and Game
- U.S. Army Corps of Engineers
- U.S. Environmental Protection Agency (EPA)
- California Department of Parks and Recreation, Office of Historic Preservation (OHP)

These regulatory constraints include both design standards and required permits that may be either very stringent and/or time consuming to obtain or satisfy. The design of the proposed alternatives is preliminary. Possible delays and design changes to the project may arise because of the complexity of the routes and their close proximity to some environmentally sensitive areas.

2.1.1 Design Standards

The proposed sewer pipeline design must follow the relevant engineering design standards of the regulatory agencies listed above. The following Table 2.1 is a summary of the main design considerations which were accounted for in the present routing study:

December 12, 20XX

Table 2.1- Routing Study Design Considerations

No.	SPECIFICATION NO.	DESCRIPTION
	PLACER COUNTY (REFERENCE [3])	
	SPECIFICATION NO.	DESCRIPTION
1	71-1.02D	Force mains shall only be PVC pipes.
2	71-1.03	Excavation and backfill; trench excavation specifications.
3	71.1.05	Minimum cover of 3 feet; pipe depth may not exceed 20 feet for horizontal directional drilling method of construction.
4	71.1-1.06A	Bracing and shoring requirements.
	CALTRANS (REFERENCE [4])	
	SPECIFICATION NO.	DESCRIPTION
1	606.3	Transverse encroachment is permissible.
2	606.4	Longitudinal encroachments would not be allowed.
3	623.1	No open trenching or overhead sewer crossing is allowed.
		Only jack-and-bore and directional drilling would be allowed.
		Jack-and-bore installation for a 6-inch diameter force main would require a steel casing pipe with a minimum wall thickness of ¼ inch.
		Receiving pits for either jack-and-bore or directional drilling must be outside of the Caltrans right-of-way.
		No pump stations are allowed in the Caltrans right-of-way.
4	623.2D	Minimum 4 feet depth of pipe cover under I-80.
	UPRR (REFERENCE [5])	
No.	DESCRIPTION	
1	Longitudinal utility encroachment within UPRR right-of-way must be a minimum of 35 feet from centerline of the nearest railroad track.	
2	Utilities crossing railroad must be encased in a steel pipe; minimum thickness of a casing pipe with less than a 12-inch diameter is ¼ inches.	
3	Utility crossings must have a minimum cover of 4.5 feet and a maximum cover of 20 feet.	
	CDPH (REFERENCE [6])	
No.	DESCRIPTION	
1	Sewer line must have a minimum horizontal clearance of 10 feet from any drinking. water	
2	Sewer line must cross underneath potable water piping.	

2.1.2 Permitting Requirements

The majority of the construction work for the project would be in the Placer County right-of-way, which requires an Encroachment Permit (Reference [7] & [8]). The Improvement Plans for the new pipeline require the County's approval. Additional permits would need to be obtained from the regulatory agencies discussed above. Table 2.2 below lists the permitting requirements for the project:

Table 2.2- Permitting Requirements

No.	PLACER COUNTY (REFERENCE [7])
1	Improvement Plans approval/Encroachment permit – including approval of stormwater quality, noise pollution, traffic control, tree preservation, road restoration, grading, erosion and sediment control plans and measures.

No.	CALTRANS (REFERENCE [4])
1	Transverse encroachment permit.
2	Jack-and-bore permit.
3	Performing relocation work for utilities (if necessary).

No.	UPRR (REFERENCE [9])
1	Permit to be on railroad property for utility survey.
2	Longitudinal utility encroachment permit.
3	Encased non-flammable pipeline crossing permit.

No.	CVRWQCB (REFERENCE [10])
1	National Pollutant Discharge Elimination System (NPDES) – for construction projects, which encompass more than 5 acres of disturbed soil (Reference [11]).
2	Clean Water Act, Section 401 Certification – regulation of fill and dredged material.

No.	PLACER COUNTY APCD (REFERENCE [12])
1	Authority to construct permit (Reference [13]).

No.	U.S. FISH AND WILDLIFE SERVICE
1	Consultation may be required since the project has a potential to harm protected wildlife and plant species.

No.	CALIFORNIA DEPARTMENT OF FISH AND GAME
1	Consultation may be required since the project has a potential to harm protected wildlife and plant species.

No.	U.S. ARMY CORPS OF ENGINEERS (REFERENCE [14])
1	Any work within the waters of the State, including dredging or discharging sediment laden runoff into a creek requires a CWA Section 404 permit.

No.	CALIFORNIA DEPARTMENT OF PARKS AND RECREATION, OHP (REFERENCE [15])
1	Project will be required to ensure compliance with the National Historic Preservation Act and other regulations pertinent to the protection of cultural resources.

2.2   Physical Constraints

The proposed alternatives would face physical constraints that range from easements from various agencies to unforeseen conditions. Easements from agencies such as Caltrans and Union Pacific Railroad would have to be acquired in order for the sewer line to pass through their right-of-way. The proposed alternatives would require minor land acquisition from private landowners. Additional physical constraints such as Placer County's hilly terrain, deposits of hard rocks, creeks, and lakes, also pose some difficulties. Although the majority of the proposed routes follow existing roads, there is a possibility of uncovering archeological remains.

2.2.1   County Roads

- The narrow County roads along the proposed alternatives serve as the only access to private residences, which would require careful planning, phasing, re-routing and traffic control measures.
- Lane widths may be too narrow to work safely next to live traffic, which may require temporary road closures and/or traffic diversion.
- A proactive approach to public outreach would ensure that the concerns of the interested parties are addressed and possible conflicts that may arise are resolved in a timely manner.

2.2.2   I-80

- Installation of the sewer line would have to be through jack-and-bore or horizontal directional drilling (Table 2.1).
- The vertical and horizontal delineation of Eastbound and Westbound I-80 (Reference [16]) with respect to Applegate Road and Lake Arthur Road limit the reasonable locations for the jack-and-bore crossings. Crossing I-80 at different locations than those in the proposed alternatives would be more expensive due to additional elevation head and/or increased length of the jack-and-bore operation.
- Caltrans may impose time restrictions for construction along the freeway during peak traffic periods such as holidays or commute hours.

2.2.3   Union Pacific Railroad

- Installation of a sewer line along the UPRR right-of-way would require both a minimum of one transverse crossing (through jack-and-bore or directional drilling) and open trenching (Reference [17]).
- Some stretches of UPRR's right-of-way are currently surrounded by dense woods that provide a rich habitat to many species. Utilizing this land would require environmental mitigation.
- The difficult terrain and live train traffic would be major accessibility and safety constraints during construction (Figure 2.1).

Figure 2.1- UPRR Tracks 200 ft South of the Detention Ponds

- Special safety precautions per UPRR and the California Occupational Safety and Health Administration (Cal OSHA) should be taken during construction near the UPRR tracks.
- The lengthy processing time for permits (a minimum of 3 to 6 months) is also a constraint (Reference [9]).
- Depending on the proximity of construction, UPRR may require Railroad Protective Liability Insurance in addition to general liability insurance (Reference [9]).

2.2.4 Existing Utilities

- If properly submitted, the required Notice to Adjacent Utility Owners may take up to 30 calendar days before the County takes a course of action (Reference [7]).
- It may be unfeasible to relocate certain utility installations, such as high voltage PG&E lines, which would require design changes.
- Unmarked utilities pose a great concern for damage during construction.
- Limited space around existing utilities may be an issue during construction.
- Relocation of existing utilities may be required.

Figure 2.2 - Fiber Optic Line and Propane Gas Pipeline along UPRR

2.2.5 Creeks / Lakes / Culverts

- Due to the close proximity of creeks and lakes, care should be taken to prevent chemical contamination and sewer spills that may lead to public health issues, costly fines and time-consuming cleaning (Figure 2.3 – 2.4).
- Culverts that cross the roadway and are near the road surface would require an increased trenching depth (Figure 2.5).
- The installation of the sewer line must provide a minimum of 10 feet clearance to any water bodies or potable water supply lines.

Figure 2.3- Creek Under Applegate Road

Figure 2.4- View of Lake Arthur from Lake Arthur Road

Figure 2.5- Water Supply Pipe to the Helsey Power House on Dry Creek Road

December 12, 20XX

2.2.6 Terrain

- The elevation changes along the alternatives, limit the opportunities to use gravity main and increase the need for force main technology.
- The rocky terrain of the project site (Appendix C), which is part of the Sierra Nevada Foothills, may make excavating trenches to the desired depth extremely difficult and time consuming if hard rock is encountered (Reference [18])
- There is a limited space to for the construction work due to cliffs and dense trees along the County roads.

2.2.7 Archeological

The area of the proposed project is considered to have a high possibility of artifacts from the California Gold Rush era and a moderate possibility of Native American artifacts. Even though the proposed routes mainly follow existing County roads, the depth of excavation still allows for the possibility of uncovering of artifacts with archeological significance. Uncovering such artifacts would affect the project schedule or may cause changes to the proposed routes (Reference [15]).

2.2 Environmental Constraints

Environmental impacts may also be a major constraint in the project area and need to be properly addressed. The following is a list of potential environmental issues:

- Air Pollution
- Sensitive Animal Species
- Sensitive Plant Species
- Naturally Occurring Asbestos
- Sediment Runoff

The Placer County Chapter of the Sierra Club is the most prominent environmental group in the area. The Sierra Club confirmed that they are not aware of any environmental concerns related to the proposed project. However, this might change based on the findings of the Final EIR and the proposed mitigation measures.

2.3.1 Air Pollution

Construction related activities have the potential to conflict with the Placer County APCD regulations, which follow the air quality standards set by the EPA and the California Air Resources Board (Reference [12]). The use of a generator at the proposed pump station would also require adherence to the air quality standards.

2.3.2 Sensitive Animal Species

The proposed alternatives could potentially affect the habitats of protected species (Reference [19]). Possible protected species in this area may include Elderberry Longhorn Beetle, California Red-Legged Frog, Foothill Yellow-Legged Frog and Western Pond Turtle (Figure 2.6 – 2.8). The possibility of these species to be encountered in the area would be addressed in the EIR. Harming of any threatened or endangered species during construction may pose serious consequences to the project, both in time delays and heavy fines.

Figure 2.6- Elderberry Longhorn Beetle.
Source: Jean Landry/Adobe Stock Photos.

Figure 2.7- California Red-Legged Frog.
Source: Nps.gov.

Figure 2.8- Western Pond Turtle.
Source: Yuval Helfman/Adobe Stock Photos.

2.3.3 Sensitive Plant Species

Possible sensitive plant species, such as Blue Oak Tree, Big Scale Balsamroot, Butte County Fritillary, Brandegee's Clarkia, Oval-Leaved Viburnum and Jepson's Onion, could also be affected by the proposed project (Figure 2.9 – 2.12). It is known that Blue Oak Trees are present along the proposed alternatives. The removal of any Blue Oak Tree during construction requires that three Blue Oak Trees are planted for every one that is removed (Reference [20]). The possibility that the rest of the sensitive plant species could be in the project area would also be addressed in the Final EIR. Disturbance or loss of these plants' habitats may pose time delays and heavy fines.

Figure 2.9- Blue Oak Tree.
Source: Wikimedia Commons.

Figure 2.10- Butte County Fritillary.
Source: Maria J.Boudreaux.

Figure 2.11- Brandegee's Clarkia.
Source: Wikimedia Commons.

Figure 2.12- Oval-Leaved Viburnum.
Source: Wikimedia Commons.

2.3.4 Naturally Occurring Asbestos

There is a high probability that a naturally occurring asbestos (NOA) will be encountered during construction because of past project experiences in the Placer County area. A preliminary geotechnical report for the project site has identified rocks, which are likely to contain NOA (Reference [18]). NOA has been previously encountered within metavolcanic and ultramific rock units in the project vicinity (Appendix D). The proximity of the site to fault fractures also suggests the existence of NOA in the project area. The presence of NOA may lead to the implementation of stringent mitigation measures that could delay the project.

2.3.5 Sediment Runoff

The proximity of the proposed alternatives to lakes and the crossing of a number of creeks along each the routes increases the possibility of sediment runoff entering these waterways. This can potentially violate Placer County's erosion control requirements (Reference [7]). Seasonal construction restrictions may be imposed due to high risk of sediment runoff, which could possibly lead to changes in the construction schedule. Construction during the wet season would require trench dewatering operations, thereby increasing project construction costs.

2.3 Private Stakeholders' Interests

- Nuisance complaints concerning noise pollution caused by construction equipment may be filed by residents. Noise from construction may not violate Placer County's noise ordinance of maximum 5 decibels (dB) of exterior ambient sound level or the standards (Reference [7]), whichever is greater.
- Nuisance complaints may also be filed due to loss of business (Figure 2.13) caused by construction activities and re-routing of traffic (Reference [7]). This may cause construction delays or time restrictions may be imposed.
- The minor land acquisition and temporary easements through private property may cause delays. It is desirable to start open communication with the affected property owners as early as possible.

Figure 2.13- Gas Station on Lake Arthur Road

December 12, 20XX

3.0   DEVELOPMENT OF ALTERNATIVES

The flow chart in Figure 3.1 illustrates the general approach to identifying of the proposed feasible alternatives and the selection of the recommended alternative. The specific strategy used in the alternative selection process is discussed in this section.

Figure 3.1- Development of Alternatives Flow Chart

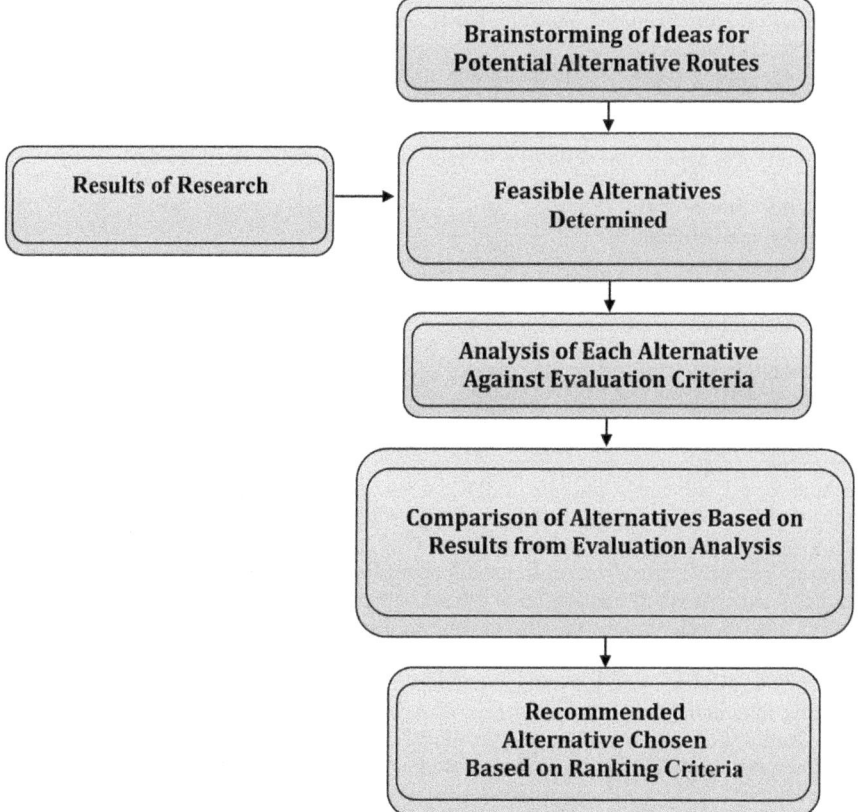

3.1   Strategy for Identifying of Alternatives

In developing a strategy for identifying feasible routes, the factors outlined below were considered. If these factors were not obviously present in an alternative, the route was not pursued as a feasible alternative. The alternatives presented in this study were identified based on the criteria listed below. This section also describes the reasons why the study proposes only three feasible routes.

3.1.1   Delineation Considerations

The path of the alternative is a major factor in the pipeline study. The routing of the alternatives was focused on minimizing the elevation head, which needs to be overcome. Furthermore, the goal was to minimize excavation, pavement replacement, length of pipe, and other high cost items for the new pipeline. Restrictions due to regulatory compliance require a wastewater pipeline route to adhere to standards set forth by Placer County. This limited the number of alternatives to a few possibilities that conform to sanitary sewer design criteria. Once this selection filter was applied, the following delineation considerations were made:

- Public right-of-way
- Existing easements
- Minimal distance
- Obstacles (existing structures, land features and utilities)
- Infrastructure (canals and storm drain)

Each criterion comes with certain challenges that were addressed in the development of each of the proposed alternatives and later evaluated according to weighted criteria.

3.1.2   Required Force Main Technology / Pump Stations

The Applegate Community and the SMD-1 tie-in point are separated by challenging land features characterized by foothill areas, spurs, draws, streams and elevations changes that do not allow a gravity flow system to be considered as an acceptable alternative. A method of minimizing excessive excavation or trenching applications for a gravity pipeline is to use force main technology. At the designated point of connection, the existing Auburn wastewater system is a force main, which requires the flows conveyed by the new pipeline to enter SMD-1 at least at the same pressure. If a partial gravity design / partial force main design is used, the gravity section of the pipeline would require numerous timed-valves and holding tanks in order to achieve the minimum gravity flow velocity of 2 feet per second as specified in the Placer County Design Standards. Also, the initial and operational costs would be increased by a partial gravity design because several pump stations and holding tanks would need to be placed along the pipeline length. Due to the above reasons, each of the proposed alternatives was designed as a force main for the entire length of the pipeline.

Force mains require periodic maintenance which was considered in the study as part of the weighted criteria to suggest a course of action to the County. Pumps, energy consumption, and maintenance personnel, associated with force mains were included as part of the alternatives' evaluation.

The existing Applegate lift station will remain in place with no changes. The project research showed that making upgrades to utilize it as an interception point of Applegate's wastewater flows would require overall improvements to the entire existing Applegate collection system. The reason for this is that only a portion of the service connections are pumped via the 4-inch force main which then outfalls to the 6-inch gravity sewer picking up the rest of the existing service connections (Figure 1.2). Utilizing the existing pump station would bypass a number of service connections, thus requiring their upgrade, which is beyond the scope of this study.

Additionally, each proposed lift station will have two identical pumps, one for normal operation and one for back-up in case of failure of the first pump. Emergency gasoline operated generator will also standby in case of power failure to keep the lift station operational until normal power is restored. Structural enclosures will house the electrical components and pump controls for weather protection and security purposes.

3.1.3 Conflict with Existing Utilities

The alternatives may cross or run next to existing utilities from Applegate to the point of connection to SMD-1. Regulations and improvement standards may provide guidance on the proper construction of wastewater pipeline to mitigate any potential interruption of services or damage to existing utilities. Since record drawings of existing utilities were not available at the time of this study, a consideration for necessary relocation of utilities could only be made after appropriate survey of existing utilities. Relocation of existing utilities may be necessary if it can significantly reduce the conflict with the preferred alternative route.

3.1.4 Land Acquisition / Easements / Right-of-Way

In order to minimize project costs, no land acquisitions are proposed along the pipeline delineation of either one of Route 2 or Route 3 except at the proposed pump station location at the intersection of Applegate Road and Merry Lane. The objective of the proposed alternative routes was to stay as much as possible within the Placer County right-of-way because the County would ultimately own, maintain and operate the proposed facilities. Route 1 would require limited traversing across private property.

3.1.5 Environmental Impacts

Environmental considerations are an important aspect of this project. The proposed alternatives attempt to minimize impact on wildlife or habitat in the project area. Noise, air quality, and other impacts on the environment as identified by regulations and improvement standards associated with the pipeline should be addressed and implemented by the contractor during the planning and construction phases of the project. The study identifies potential environmental impacts along each of the alternative routes. The proposed alternatives attempt to stay along existing right-of-ways to minimize the effects on undeveloped areas.

3.1.6 Cost

For each of the proposed alternatives, the study considered the estimated cost to construct the pipeline. The cost of force mains, pump stations, construction method, and other unique features were estimated to provide a comprehensive comparison between the alternatives presented in the

study. This also includes proposed land acquisition or utility relocation that may be necessary to pursue a feasible alternative.

3.2   Reasons for Disqualification

The following may be potential reasons to disqualify potential alternatives:

- Expected excessive costs due to constraints or limitations, whether regulatory or geographical, were not developed;
- Anticipated schedule delays for any portion or process of a route by external sources, i.e. regulatory agencies, stakeholders, or site conditions;
- Major stakeholder opposition to land use, location, or acquisition necessary for success of an alternative;
- Significant environmental impact due to alternative or construction thereof;
- Significant utility conflicts where a solution is not feasible or acceptable (such as the Kinder Morgan pipeline along the UPRR tracks or PG&E high voltage lines);
- Significant liability to the health or safety of the community;
- Unreasonable maintenance or operations of the proposed pipeline, whether it be access to the facility, dangerous for personnel, or hazardous for equipment (including vehicles).

Examples of routes, which were considered, but not further pursued, are:

- A route intercepting Applegate's sewer flows at the detention ponds and pumping it towards Applegate Road along Bon Vue Drive (Figure 1.1). This route would have required a larger pump due to the elevation head to be overcome. Moreover, Bon Vue Drive serves as the only access road to a cluster of private residences. Closure of this road due to construction would be, if not impossible, than extremely difficult to accommodate.
- Following County roads east of the detention ponds would have increased the length of the pipeline, and ultimately, the construction, operation and maintenance costs.
- Over-ground sewer pipeline route was considered along the UPRR tracks. Such route would have had a lower construction cost. However, safety precautions and maintenance complications deemed this design approach unfeasible.
- The unwillingness of Caltrans to allow the installation of the pipeline along the outside of the overpass at Placer Hills Road mandated the use of jack-and-bore.

4.0 PRIMARY ALTERNATIVES

The present routing study identified three feasible alternatives. As noted in the previous section, the research determined that other routes do not meet the expectation of the Client in terms of cost and schedule, and pose a significant potential for opposition by stakeholders. This section provides details and design information on each one of the proposed alternatives. Each alternative has elements that make it unique and different from the rest. Due to the constraints of the project site, the alternatives share common elements as an attempt to minimize costs related to construction and mitigation of significant impacts. The specific major groups of constraints for the alternative routes are:

- Regulatory Compliance
- Physical Constraints
- Environmental Constraints
- Private Stakeholders' Interests

The focus of this section is the design information related to each of the three alternatives.

4.1 Common Elements / Design Considerations

In all cases, the Applegate Treatment Plant would be decommissioned with the implementation of a new pipeline to convey wastewater flows to SMD-1. Therefore the study finds the following common elements for each alternative (Figures 4.1, 4.2 and 4.3):

- Cut off existing 6-inch gravity sewer to WWTP
- Construct a new pump station to connect the existing pipe to the proposed alternative. For alternative Route 2 (Figure 4.2) and Route 3 (Figure 4.3), this will occur before the UPRR overcrossing at Applegate Road and Merry Lane. The proposed pump station location for Route 1 (Figure 4.1) is at the detention ponds.
- From the point of connection in Applegate to SMD-1 the wastewater flows will be pumped via force main the entire distance.

The approach for construction of the pipeline shall be as follows whenever applicable:

- Open trench installation except for locations where directional drilling or jack-and-bore is required, i.e. crossing at creeks, canals, or the I-80 corridor.
- Minimum design depth of 3 feet per Placer County Standards. The design of the alternatives is based on the more conservative 5 feet minimum depth of cover, which accounts for the known and unknown locations where the pipeline needs to pass under shallow utilities or infrastructure, i.e. dry trench, water supply, drainage culverts, underground communication lines, etc. This approach also accounts for adequate clearance between the pipeline and such shallow utilities.
- There are five distinct creek crossings that all of the alternatives are going to encounter. Crossing underneath these channels will require a fairly deep directional boring operation. The footprint of each of these operations is proposed within the Placer County right-of way and will not require any easements over private land. Should the further field verification of the creek crossings uncover the lack of feasibility of directional

drilling, creek diversions and work in the stream bed will need to be pursued via special permits from the Army Corps of Engineers, the U.S. Fish and Wildlife Service, the California Department of Fish and Game and CVRWCB.

The final section of the pipeline along Dry Creek Road will be the same for all of the proposed Alternatives (Figures 4.1, 4.2 and 4.3):

- Installation of 6-inch force main in the public right-of-way along Dry Creek Road to Bluegrass Drive for a total of 5.1 miles.
- The designated tie-in point is into an existing 10-inch force main at Dry Creek Road and Bluegrass Drive.

For all of the alternatives there are three different construction methods: trenching, directional drilling and jack-and-bore. Trenching would require 15 to 20 feet of construction area centered around the proposed trench delineation in order to accommodate construction equipment and personnel. The depth of trench would dictate the shoring requirements and would not affect the width of the construction area. Directional drilling does not require any surface disturbance along the route except at the locations of the receiving pits. The size of the receiving pits would vary depending on the depth of the pipeline and would not exceed a 20 foot wide by 40 foot long footprint. Jack-and-bore is a method similar to directional drilling. The difference is that jack-and-bore can only be done in a straight line and the steel casing pipe remains in the ground. The jack-and-bore method needs the same construction footprint as the directional drilling method but would require a greater number of receiving pits.

Pump selection for each of the alternatives is challenging due to the low flows produced by the Applegate community and the high total head that needs to be pumped. Another challenge for the pump selection is the solids-handling during low flows. Since Auburn's SMD-1 wastewater system is a STEP system at the designated tie-in point (Reference [21]), Applegate's flows may require a pretreatment facility to avoid backup of the pipeline at the tie-in point. This potential issue would be handled by SMD-1. The specifics of the treatment of the Applegate's sewer flows are beyond the scope of this routing study.

The total operational time for the proposed pumps was selected to be 12 hours per day, which was based on an evaluation of head losses. The effective pumping rate, as shown in the design calculations per Appendix E, is double the assumed peak hour design flow rate, which was assumed to be 78,000 gallons of effluent wastewater per day. The existing effluent wastewater flows from the Applegate community were obtained from Placer County (Appendix E-1).

4.2   Route 1 (UPRR)

The pipeline for Route 1 follows a path along the Union Pacific Railroad (Figure 4.1). The Applegate's wastewater flows will be intercepted at the existing detention ponds. A lift station located at the existing detention ponds will pump the wastewater into the proposed pipeline under the westbound UPRR tracks through an existing tunnel (Reference [1]). From this point, the pipeline will head due south along the westbound UPRR tracks and cross underneath the eastbound UPRR tracks. For the length of the crossing, the proposed sewer pipeline must be incased in a steel pipe (Appendix B, X-1). After the second railroad crossing, the pipeline

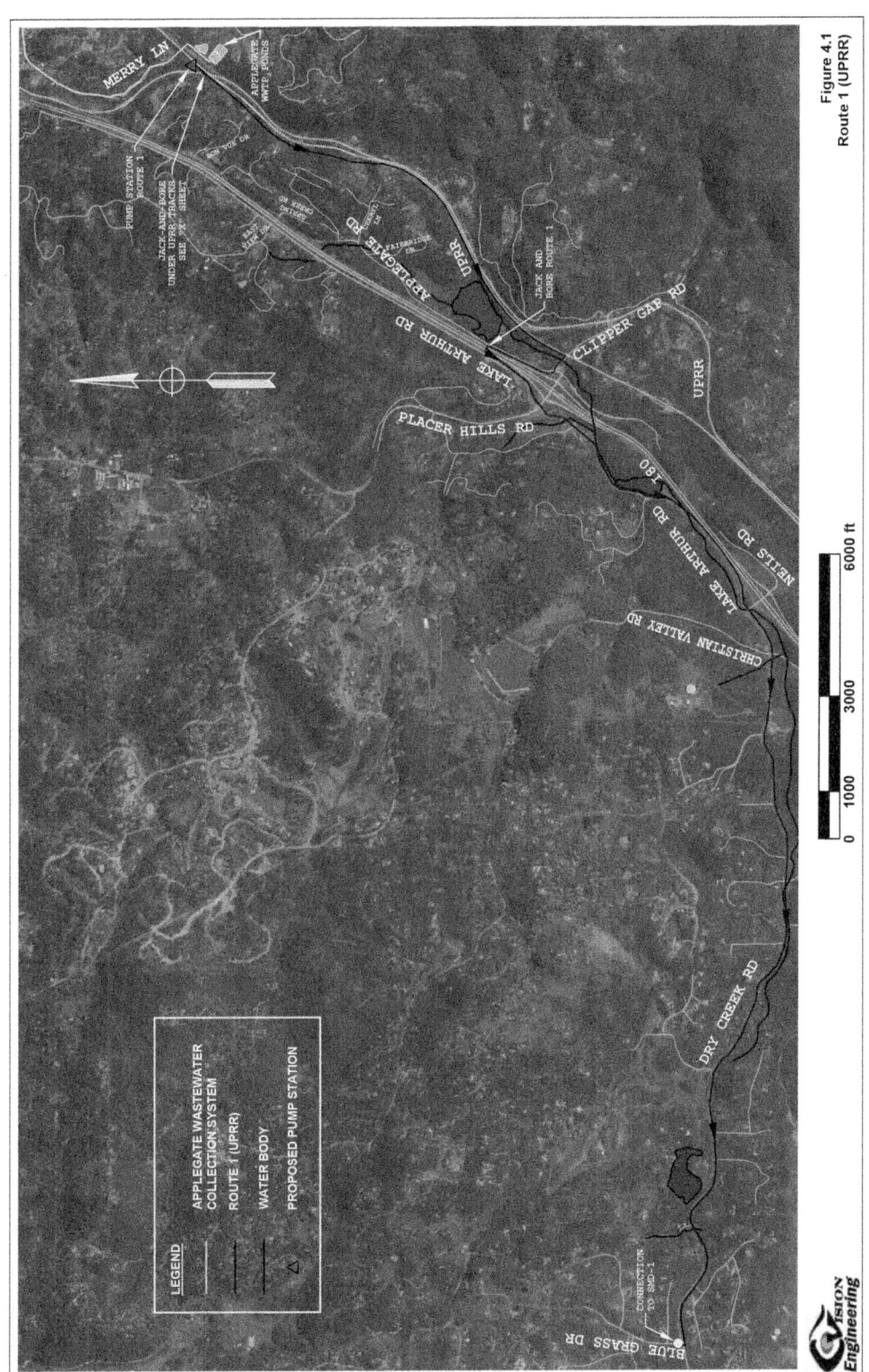

Figure 4.1
Route 1 (UPRR)

LEGEND

APPLEGATE WASTEWATER
COLLECTION SYSTEM

ROUTE 1 (UPRR)

WATER BODY

PROPOSED PUMP STATION

MERRY LN

PUMP STATION 1
ROUTE 1

JACK-AND-BORE
UNDER UPRR TRACKS
SEE 'XX' SHEET

APPLEGATE
WWTP PONDS

APPLEGATE RD

UPRR

JACK AND
BORE ROUTE 1

CLIPPER GAP RD

LAKE ARTHUR RD

PLACER HILLS RD

LAKE ARTHUR RD

I80

NEILS RD

UPRR

CHRISTIAN VALLEY RD

DRY CREEK RD

CONNECTION
TO SRC-1

BLUE GRASS DR

0   1000   3000   6000 ft

Vision
Engineering

follows a maintenance road along the West side of the UPRR right-of-way. The route turns west on Clipper Gap Road until it meets Applegate Road. There, the pipeline heads north to the proposed jack-and-bore crossing of I-80 (Figure 4.1). The Route 1 force main will convey the wastewater flows across the freeway to Lake Arthur Road and then south to Dry Creek Road, which leads to the designated point of connection with SMD-1. The total length of force main is approximately 8.1 miles.

A sewer easement would need to be obtained from UPRR for maintenance and construction of the pipeline, which would be secondary to the UPRR right-of-way easement. Because the pipeline is intended to follow the railroad, the entire construction footprint will remain in the UPRR right-of-way (Reference [21]). Thus, no easements over private land would be required except at the intersection of the UPRR tracks with Clipped Gap Road. There, a small sewer easement would be needed (Appendix B, PM-3).

The remainder of the pipeline, from the intersection of Clipper Gap Road with the UPRR tracks to SMD-1, remains within Placer County's right-of-way. A sewer easement would not be necessary because Placer County is to become the pipeline owner. Crossing underneath I-80 would require either a jack-and-bore or a directional drilling construction and a sewer easement would need to be obtained from Caltrans for this stretch of the pipeline delineation. The receiving pits must be outside of the Caltrans right-of-way.

4.2.1 Design Details

The design drawings for this alternative can be found in Appendix B. This alternative provides the following unique elements to meet the scope of the project:

- Force main route following the maintenance road on the West side of the UPRR tracks up to Clipper Gap Road.
- Force main turns west onto Clipper Gap Road and heads north along Applegate Road to avoid traversing private property before crossing I-80.
- Jack-and-bore crossing of I-80 between Applegate Road and Lake Arthur Road where the pipeline merges with the common for all of the proposed alternatives final stretch of the project (Appendix B, L1-4 and X-1)
- This alternative requires a single pump station at the point of flow interception at the existing detention ponds. The total elevation head that needs to be overcome by the pump is roughly 140 feet. The total dynamic head, accounting for friction losses, minor losses and elevation head, is 196 feet at a pumping rate of 110 gallons per minute identifying the pump operational point. The pump was sized utilizing a published catalog from an industry standard pump manufacturer – Goulds Pumps. Total head losses, system curve and pump selection can be found in Appendix E-2 and Appendix F.
- Route 1 requires one instance of land acquisition. The proposed land acquisition in form of an easement along the pipe delineation is located on parcel 077-130-026-000 which belongs to "Trustee". The easement is projected to be 67 feet wide by 71 feet long, totaling to 290 square feet of the aforementioned property (Appendix B, PM-3). No land acquisition for the pump station and facilities' footprint is needed for Route 1.

- In order to minimize project costs, Route 1 utilizes parcel number 073-120-013-000, which belongs to Placer County, for the construction of the required pump station facilities. The said parcel is the current site of Applegate's detention ponds and already has the necessary access roads in place.

4.2.2   Constraints

Table G.1 in Appendix G show the locations along Route 1 where certain physical, environmental or regulatory constraints may be an issue. Permits that would have to be obtained and their areas of concern along the route are also included. This table was used as an aid for comparing the constraints for each of the alternatives. The following is a summary of the constraints for Route 1:

- Maintenance road along the UPRR tracks

  - Hilly terrain may cause difficulties in installing the force main;
  - Encountering hard rock may have impact on the construction methods;
  - Live train traffic poses construction, accessibility and maintenance challenges;
  - Limited space for construction in some areas;

- Existing infrastructure

  - Construction nearby railroad tunnels;
  - Crossing under the railroad tracks;
  - Crossing drinking water supply lines such as the Bypass Canal near Clipper Gap Road;
  - Creek and canal crossings along Dry Creek Road;
  - Lack of public right-of-way or developed roads for construction crews and equipment along the UPRR easement – only dirt trails and narrow paths;

- Stakeholders

  - Possible delays due to opposition by UPRR to construct in their right-of-way;
  - Possible conflict with the existing high pressure liquefied propane gas pipeline by Kinder Morgan Inc.

- Environmental

  - Existing oak trees along the route that may need to be removed and replaced.

4.2.3   Cost Estimate

Among the proposed alternatives, Route 1 has the longest length of force main off of the existing paved roads. This means a lower cost for surface roadwork repair. Repaving will be the least for this alternative. However, the pump for this alternative (Appendix F) has to overcome nearly twice the total dynamic head and more than twice the elevation head alone. Table 4.1 presents a summary of the preliminary cost estimate for this alternative. A detailed preliminary cost estimate can be found in Appendix H.

December 12, 20XX

Table 4.1- Summary of Preliminary Cost Estimate for Route 1

SUBTOTALS		
FORCE MAIN	$	2,556,060
PUMP STATION	$	475,000
SITE WORK	$	2,762,523
LAND ACQUISITION	$	18,850
ANNUAL MAINTENANCE	$	15,650
TOTAL	$	5,828,084

4.2.4 Alternative Evaluation

This alternative was originally intended to follow immediately next to the UPRR tracks, utilizing the existing, consistently downhill, elevation gradient. However, cobbles, boulders, narrow canyons, train tunnels and passing trains pose safety concerns and construction problems. After a close investigation of these conditions, this alternative had to be routed along the maintenance and access roads parallel to the UPRR tracks. These roads follow the terrain along the West boundary of the UPRR right-of-way, and often drastically change elevation. The rest of the route is along county roads with the exception of the jack-and-bore location to cross under the I-80 corridor. Back-tracking a short distance along Applegate Road was necessary to find a convenient crossing location and to avoid traversing private property. Almost the entire Applegate community is avoided by the delineation of this alternative.

Route 1 has the following benefits:

- Less than significant impact to most of Applegate's residents during construction;
- Minimized road repair since a long stretch of the pipeline is not following paved roads in the public right-of-way;

This alternative was pursued in attempt to mitigate the following:

- Avoid Creek crossings on Applegate Road;
- Avoid utility conflicts on Applegate Road;
- Minimize disruption of transportation and access disruption due to construction on public roads.

4.3 Route 2 (Applegate Road)

The pipeline for Route 2 (Figure 4.2) follows the Placer County's roads of Applegate, Lake Arthur and Dry Creek. It will pick up flows from the existing Applegate sewer system at the intersection of Applegate Road and Merry Lane. A lift station will pump the wastewater south, away from the community, along Applegate Road for approximately 1.8 miles to the location of the jack-and-bore crossing of I-80 (Figure 4.2). The force main will continue to convey flows across I-80 to Lake Arthur Road. Once across, the force main will continue south along Lake Arthur Road for 2.5 miles and west along Dry Creek Road for 2.6 miles before it injects flows into SMD-1. The total length of force main is approximately 6.9 miles.

From the tie-in point at the intersection of Applegate Road and Merry Lane the pipeline is proposed to stay within the Placer County right-of way. A sewer easement would not be necessary because Placer County is to become the pipeline's owner. Crossing underneath I-80 would require either a jack-and-bore or a directional drilling method of construction. A sewer easement would need to be obtained from Caltrans for this part of the pipeline delineation as well as both of the receiving pits. The receiving pits must be outside the Caltrans right-of-way. No surface work is permitted in the Caltrans right-of-way.

4.3.1 Design Details

The design drawings for this alternative can be found in Appendix B. This alternative provides the following unique elements to meet the scope of the project:

- Jack-and-bore crossing of I-80 between Applegate Road and Lake Arthur Road at a location further north than the one proposed in Route 1, north of Lake Theodore and where Applegate Roads veers away from I-80 (Appendix B, L2-4 and X-2)
- This alternative requires a single pump station at the point of Applegate's wastewater flow interception located at the intersection of Applegate Road and Merry Lane. The total elevation head that needs to be overcame by the pump is roughly 68 feet. The total dynamic head is 118 feet at a pumping rate of 110 gallons per minute, which identifies the pump's operational point. The pump was sized utilizing a published catalog from an industry standard pump manufacturer – Goulds Pumps. Total head losses, system curve and pump selection can be found in Appendix E-3 and Appendix F.
- Route 2 requires land acquisition in order to accommodate the footprint of the pump station and its facilities. Parcel 073-120-013-000 owned by "Jane Doe" is a suitable location for the proposed pump station and facilities. The proposed land acquisition is estimated to be 100 feet by 35 feet and totaling to 350 square feet (Appendix B, PM-4).

4.3.2 Constraints

Table G.2 in Appendix G show the locations along Route 2 where certain physical, environmental or regulatory constraints may be an issue. Permits that would have to be obtained and their areas of concern along the route are also included. This table was used as an aid for comparing the constraints for each of the alternatives. The following is a summary of the constraints for Route 2:

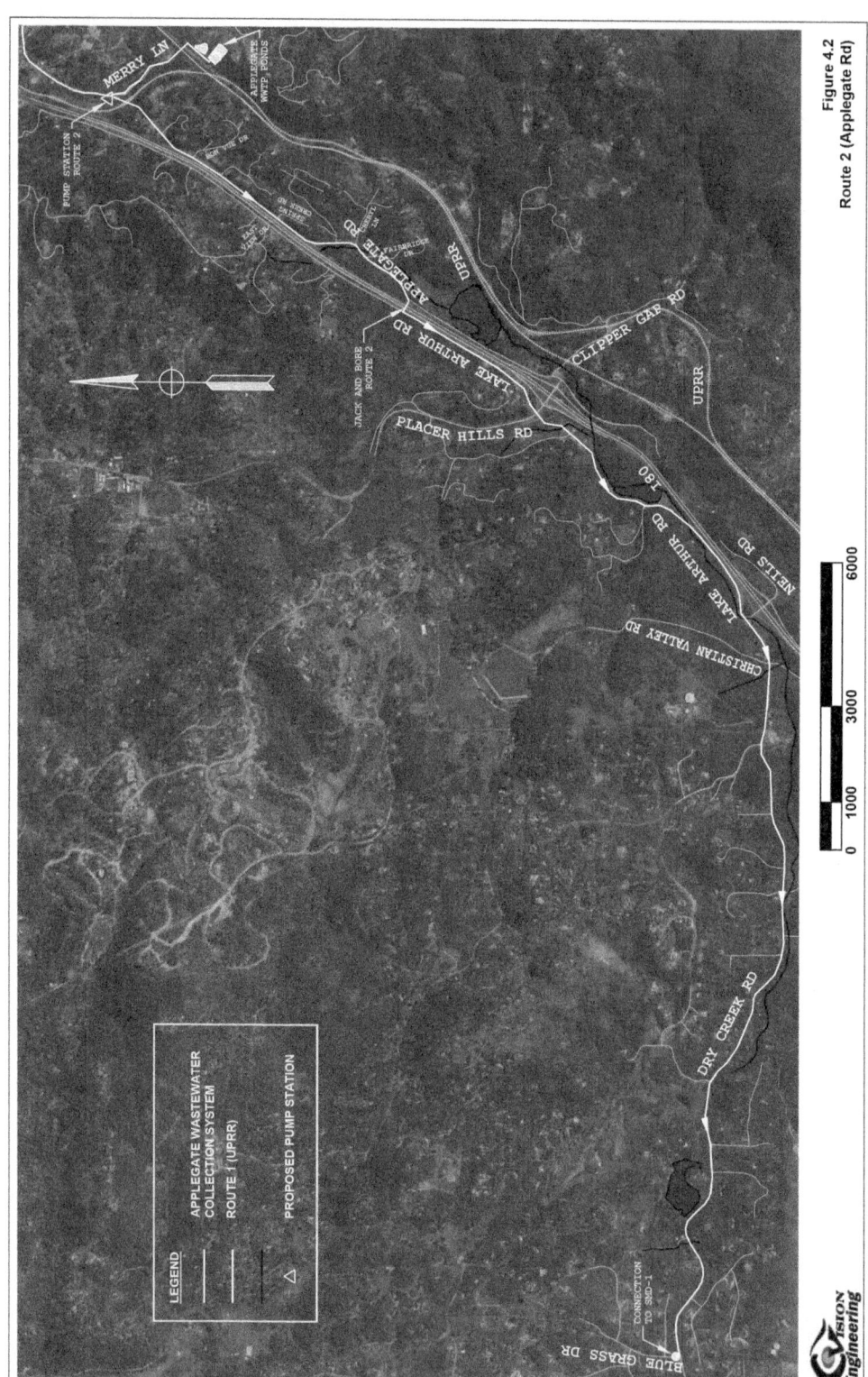

Figure 4.2
Route 2 (Applegate Rd)

LEGEND

APPLEGATE WASTEWATER
COLLECTION SYSTEM

ROUTE 1 (UPRR)

PROPOSED PUMP STATION

Vision
Engineering

- Existing utilities

  o Water mains, gas lines and possibly other underground utilities along Applegate Road;
  o Overhead power and telephone lines along Applegate Road may be too low to provide the necessary clearance for construction vehicles and equipment.

- Existing infrastructure

  o Drainage ditches and culverts;
  o Creek and canal crossings.

- Stakeholders' concerns

  o Air pollution;
  o Construction noise;
  o Road closures;
  o Inconvenience to property owners along the county roads.

4.3.3 Cost Estimate

This route follows Applegate Road for a longer distance and Lake Arthur Road for a shorter distance than Route 3. There will be little difference in cost between Routes 2 and Route 3. There is a high possibility of additional cost for more instances of relocation of utilities along Route 2, due to its path through the community of Applegate. This alternative will pose a greater inconvenience to Applegate's residents since private residence density is higher on Applegate Road than that along Lake Arthur Road. Table 4.2 presents a summary of the preliminary cost estimate for this alternative. A detailed preliminary cost estimate can be found in Appendix H.

Table 4.2- Summary of Preliminary Cost Estimate for Route 2

SUBTOTALS		
FORCE MAIN	$	2,249,820
PUMP STATION	$	445,000
SITE WORK	$	2,667,390
LAND ACQUISITION	$	35,000
ANNUAL MAINTENANCE	$	14,374
TOTAL	$	5,411,584

4.3.4 Alternative Evaluation

This route was developed to stay on County roads to avoid disturbing the environment on undeveloped lands and traversing private property. The route goes through the Applegate community, along the main thoroughfare. Among the three proposed alternatives, Route 2 has

the highest probability of conflicts with existing utilities, particularly from Fairbridge Drive to Spring Creek Road (Appendix B, L2-5). Careful planning, staging and traffic control would need to be performed by contractors to minimize the impact of construction on residents. Local ordinances would need to be observed to lessen the disturbance of construction during specific hours and to avoid generating excessive traffic delays. Staging and observance of ordinances may cause delays during construction.

The jack-and-bore location to cross the I-80 corridor was chosen to keep the pipeline away from Lake Arthur and to avoid additional creek crossings further south on Applegate Road. One of the challenges along this alternative is a creek crossing at Cheryl Lane, which is the Boardman Canal drinking water supply line. The pipeline would need to be incased in a steel pipe and go underneath the channel with a 10 foot minimum clearance. This location also has a difficult terrain and very limited construction space.

Route 2 has the following benefits:

- Less than significant impact on the environment because the pipeline follows a paved road for its entirety;
- Ease of access and maintenance of the new pipeline during the time of construction and operation.

This alternative was pursued in an attempt to mitigate the following:

- Conflict with UPRR;
- Conflict with Kinder Morgan Inc.;
- Minimize the environmental impacts.

4.4 Route 3 (Lake Arthur Road)

The pipeline for Route 3 (Figure 4.3) follows Applegate Road, Lake Arthur Road and Dry Creek Road. It will pick up flows from the existing Applegate sewer pipeline at the same location as Route 2. A lift station will pump the wastewater south, away from the community, along Applegate Road for about 1.4 miles to a jack-and-bore crossing of I-80 between Spring Creek Road and East View Drive (Figure 4.3). The force main will continue to convey flows across I-80 to Lake Arthur Road (Appendix B, X-2). Once across, the force main will continue south along Lake Arthur Road for 3.0 miles and west along Dry Creek Road for 2.5 miles to the connection point with SMD-1. The total length of force main is approximately 6.9 miles.

From the start point at the intersection of Applegate Road and Merry Lane the pipeline will stay within the Placer County right-of way. A sewer easement would not be necessary because Placer County is to become the pipeline's owner. Crossing underneath I-80 would require either a jack-and-bore or a directional drilling method. A sewer easement would need to be obtained from Caltrans for the jack-and-bore as well as both of the receiving pits. The receiving pits must be outside the Caltrans right-of-way. No surface work is to be done in the Caltrans right-of-way.

4.4.1 Design Details

The design drawings for this alternative can be found in Appendix B. This alternative provides the following unique elements to meet the scope of the project:

- Jack-and-bore crossing of I-80 between Applegate Road and Lake Arthur Road is at the location between Spring Creek Road and East View Drive;
- This alternative requires a single pump at the point of flow interception located at the intersection of Applegate Road and Merry Lane. The total elevation head that needs to be overcame by the pump is roughly 68 feet. The total dynamic head is 118 feet at a pumping rate of 110 gallons per minute, which identifies the pump's operational point. The pump was sized utilizing a published catalog from an industry standard pump manufacturer – Goulds Pumps. Total head losses, system curve and pump selection can be found in Appendix E-4 and Appendix F.
- Route 3 requires the same land acquisition for the pump station site as Route 2. The proposed land acquisition is the above mentioned 350 square feet (Appendix B, PM-4) from parcel 073-120-013-000 owned by "Jane Doe".

4.4.2 Constraints

Table G.3 in Appendix G show the locations along Route 3 where certain physical, environmental or regulatory constraints may be an issue. Permits that would have to be obtained and their areas of concern along the route are also included. This table was used as an aid for comparing the constraints for each of the alternatives. The following is a summary of the constraints for Route 3:

- Existing utilities
  - Water mains, gas lines and possibly other underground utilities along Applegate Road and Lake Arthur Road;
  - Overhead power and telephone lines along Lake Arthur Road.

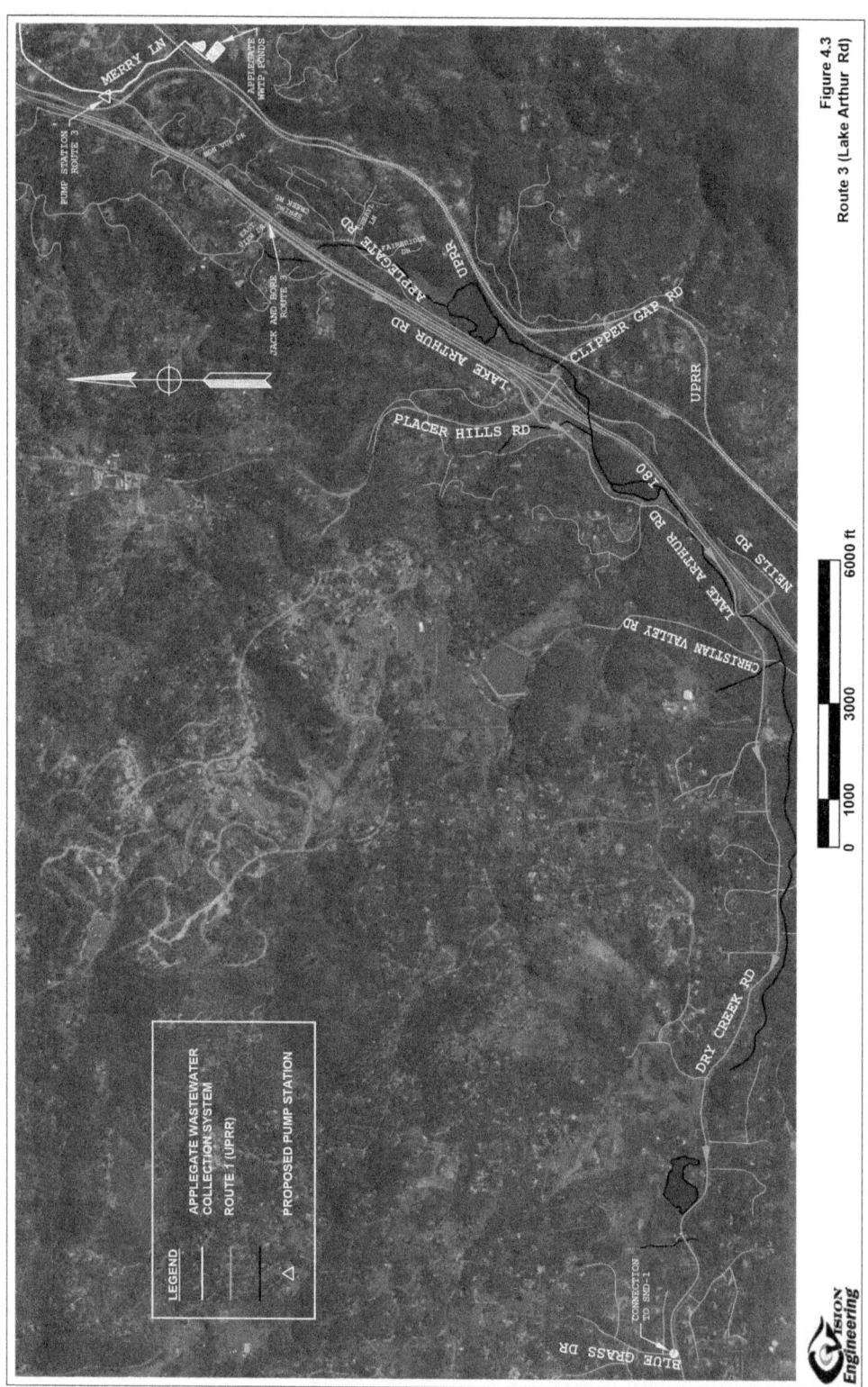

LEGEND

APPLEGATE WASTEWATER
COLLECTION SYSTEM

ROUTE 1 (UPRR)

△ PROPOSED PUMP STATION

Figure 4.3
Route 3 (Lake Arthur Rd)

0   1000   3000   6000 ft

Vision
Engineering

- Existing infrastructure
  - o Drainage ditches and culverts;
  - o Creek and canal crossings.
- Stakeholders' concerns
  - o Air pollution;
  - o Construction noise;
  - o Road closures;
  - o Inconvenience to property owners along Applegate Road and Lake Arthur Road.

4.4.3 Cost Estimate

This route follows Applegate Road for a shorter distance and Lake Arthur Road for a longer distance than Route 3. There will be a little difference in cost between Route 2 and Route 3. However, this route should have less costs associated with utility conflicts since it avoids most of the Applegate community, by crossing I-80 much sooner than Route 2. Table 4.3 presents a summary of the preliminary cost estimate for this alternative. A detailed preliminary cost estimate can be found in Appendix H.

Table 4.3- Summary of Preliminary Cost Estimate for Route 3

SUBTOTALS		
FORCE MAIN	$	2,234,580
PUMP STATION	$	445,000
SITE WORK	$	2,562,310
LAND ACQUISITION	$	35,000
ANNUAL MAINTENANCE	$	14,311
TOTAL	$	5,291,201

4.4.4 Alternative Evaluation

This route was developed to stay on County roads to avoid disturbing the environment on undeveloped lands and traversing across private property. The route follows Applegate Road for a much shorter distance than Route 2 does. Careful planning, staging and traffic control should be performed by contractors to minimize the impact of construction on residents. The route crosses I-80 before going into the populated area along Applegate Road. This delineation will avoid the potential additional utility and private residence conflicts, which are present along Route 2. The location of the jack-and-bore crossing of the I-80 corridor allows for an easier crossing of the Boardman Canal, which makes Route 3 more preferable and easier to construct than Route 2. The crossing of the Boardman Canal drinking water supply line would still require a 10 foot minimum clearance.

Route 3 is proposed with the following benefits:

- Less than significant impact on the environment due to following public right-of-way;
- Significantly less impact on Applegate's residents than Route 2;
- Significantly less conflicts with utilities than Route 2.

This alternative was pursued in an attempt to mitigate the following:

- A difficult creek crossing on Applegate Road;
- Major utility conflicts on Applegate Road;
- Conflict with UPRR;
- Conflict with Kinder Morgan Inc.

5.0 EVALUATION OF ALTERNATIVES

After brainstorming possible alternatives, three feasible alternatives were determined from the results of the gathered research. These three alternatives were then analyzed individually based on the following ranking criteria.

5.1 Ranking Criteria

The evaluation of each alternative was determined by assigning points to each area of constraints which represent a weighted average with respect to each other (Table 5.1). There are a maximum of 100 points that represent 100 percent of the weighted constraint averages. Each area has its own maximum points possible. The higher the points given, the better the alternative is. The alternative with the most points was determined to be the recommended alternative. The following are the nine criteria used for the evaluation of each alternative:

- Physical Constraints
- Permits
- Existing Utility Conflicts
- Environmental Impacts
- Construction Costs
- Impacts on Businesses / Residents
- Safety
- Land Acquisition / Easements
- Maintenance Costs / Accessibility

The weighted averages were chosen for each of the constraints based on their importance to the project and to each other. Since the 6-inch sewer pipeline would follow exiting roads for most of its route, the project would have minimal if any environmental impacts. Therefore, a lower weight was given to the environmental impacts criterion. Maintenance costs, accessibility, construction costs, and impacts to businesses and residents would be of a greater importance to the project. At the same time, maintenance costs and accessibility are long-term conditions so they were rated higher than the construction costs. Conflict with existing utilities is also a major factor for the evaluation of the alternatives, because they may pose significant schedule and budget implications. Since the proposed land acquisitions are minimal, this criterion was assigned lower possible maximum points. Safety should always be a priority, so this area was given a higher importance. Although the working conditions may be difficult along some stretches along the proposed routes, the density and speed of the adjacent traffic is low. This is why the safety criterion was given a lesser importance than some of the other ranking criteria.

December 12, 20XX

Table 5.1- Ranking Criteria with Alternatives' Scores

RANKING CRITERIA	POSSIBLE POINTS	ROUTE 1	ROUTE 2	ROUTE 3
Physical Constraints	10	5	7	9
Permits	8	4	6	7
Existing Utility Conflicts	12	10	6	8
Environmental Impacts	8	3	6	7
Construction Costs	15	12	14	15
Impacts on Businesses / Residents	14	11	9	9
Safety	10	7	8	9
Land Acquisition / Easements	5	5	4	4
Maintenance Costs / Accessibility	18	10	18	18
TOTAL POINTS	100	67	78	86

5.2   Alternative Rankings

Each alternative faces a specific set of constraints. The points given to each alternative are meant to reflect the complexity and the difficulty in constructing that route. Although the value of the points assigned to each area is not supported by precise statistical data, they do serve as a reasonable tool in the comparison of the proposed alternatives to one another and in the selection of the recommended alternative.

5.2.1   Route 1 (UPRR)

The following describes the reasons for which a specific number of points was assigned to Route 1 for each of the ranking criteria (Table 5.1):

- Physical Constraints (Points: 5)
    - The sewer line must overcome significant elevation changes due to the hilly terrain along the UPRR right-of-way.
    - Rocky soil conditions (Appendix C) may make trenching for the sewer pipeline difficult and time consuming.
    - Numerous canal and creek crossings along Applegate Road and Dry Creek Road, including a crossing of the Boardman Canal at Clipper Gap Road.
    - Lack of paved roads for construction equipment to access areas along UPRR.
    - The route encroaches on the UPRR and the Caltrans right-of-way.
    - The route traverses across private property at the intersection of Clipper Gap and UPRR.

- Permits (Points: 4)
  - UPRR has stringent requirements for utilities encroaching on its right-of-way.
  - UPRR has a long permitting process.
- Existing Utility Conflicts (Points: 10)
  - Most of the existing utilities, which would be encountered by the route are along Lake Arthur Road and Dry Creek Road, which is typical for all of the proposed alternatives.
- Environmental Impacts (Points: 3)
  - There is a moderate possibility of removing Blue Oak Trees along the route since it crosses undeveloped areas along the UPRR right-of-way.
  - There is a slight chance of harming other plant and animal species mentioned in Section 2, since a long stretch of the route is along an undeveloped part of the UPRR right-of-way.
  - The route passes near Lake Theodore, the Boardman Canal and other creeks.
- Construction Costs (Points: 12)
  - Highest total cost ($5,828,084).
  - The route would not have as high construction cost associated with excavation, and re-paving as the other two routes.
- Impacts on Business and Residents (Points: 11)
  - Since the majority of the construction would take place in the UPRR right-of-way the impact on residents should be minimal
- Safety (Points: 7)
  - A significant portion of the construction activities for this route would take place in the UPRR right-of-way, but no less than 35 feet away from the railroad tracks. Still, special safety measures should be taken for construction personnel and equipment.
  - Working near live traffic would be done along the County roads, which is typical for all of the routes.
- Land Acquisition / Easements (Points: 5)
  - Easements from UPRR are more stringent and time consuming than those from other regulatory agencies.
  - A limited temporary easement would be required at the intersection of Clipper Gap Road and the railroad tracks.
- Maintenance Costs / Accessibility (Points: 10)
  - The cost of maintenance costs may be greater for this alternative as it requires a larger and more expensive pump. The accessibility would be difficult for the stretch of the route in the UPRR right-of-way.

5.2.2 Route 2 (Applegate Road)

The following describes the reasons for which a specific number of points was assigned to Route 2 for each of the ranking criteria (Table 5.1):

- Physical Constraints (Points: 7)
    - Numerous canal and creek crossings along Applegate Road and Dry Creek Road.
    - The route encroaches on the UPRR and the Caltrans right-of-way.
    - The proposed pump station encroaches on residential property.
- Permits (Points: 6)
    - Placer County does not have permits that are as stringent or time consuming as the UPRR permits.
    - Caltrans required permits for transverse encroachments are not as stringent as those required by UPRR.
    - Route 2 requires a total number of permits which is higher than that for both Route 1 and Route 3.
- Existing Utility Conflicts (Points: 6)
    - Many overhead power lines along Applegate Road and Lake Arthur Road may be a concern for construction equipment.
    - Probable underground utility lines across Applegate Road, Lake Arthur Road and Dry Creek Road.
- Environmental Impacts (Points: 6)
    - There is a slight chance of harming the plant and animal species mentioned in Section 2 since most of the route is on County roads.
    - There are several creek and canal crossings along the route.
- Construction Costs (Points: 14)
    - Total cost of $5,411,584 is lower than the one for Route 1 and higher than the one for Route 3.
    - Compared to Route 1, this route has higher construction costs associated with excavation and re-paving.
- Impacts on Business and Residents (Points: 9)
    - Since most of the construction would take place along County roads passing through densely populated areas, this alternative would cause more inconveniences to residents and businesses in the area than Route 1 and 3.
- Safety (Points: 8)
    - Most of the construction activities for this route would take place on narrow County roads accommodating traffic and serving as the only access to private residences. Safety measures should be taken to prevent traffic accidents and protect construction workers and the general public.
- Land Acquisition / Easements (Points: 4)
    - Easier process of obtaining easements from Caltrans than from UPRR.
    - Minor land acquisition is needed for the footprint of the proposed pump station.
- Maintenance Costs / Accessibility (Points: 18)
    - Lower maintenance cost than Route 1.

   o Better accessibility to the project site than that for Route 1 both during construction and operation.

5.2.3 Route 3 (Lake Arthur Road)

The following describes the reasons for which a specific number of points was assigned to Route 3 for each of the ranking criteria (Table 5.1):

- Physical Constraints (Points: 9)
  - Numerous canal and creek crossings along Applegate Road and Dry Creek Road.
  - The route encroaches on the UPRR and the Caltrans right-of-way.
  - The location of the proposed pump station encroaches on private property.
- Permits (Points: 7)
  - Placer County does not have permits that are as stringent or time consuming as the UPRR permits.
  - Caltrans required permits for transverse encroachments are not as stringent as those required by UPRR.
  - Route 3 requires a total number of permits which is higher than that for Route 1 but less than that for Route 2.
- Existing Utility Conflicts (Points: 8)
  - Probable underground utility lines across Applegate Road, Lake Arthur Road and Dry Creek Road.
  - Lower probability of utility conflicts than Route 2.
- Environmental Impacts (Points: 7)
  - There is a slight chance of harming the plant and animal species mentioned in Section 2 since most of the route is on County roads.
  - There are several creek and canal crossings along the route.
- Construction Costs (Points: 15)
  - Lowest total cost of $5,291,201.
  - Higher construction costs associated with excavation and re-paving than Route 1 and Route 2.
- Impacts on Business and Residents (Points: 9)
  - Less impact on residents and businesses in the area than Route 2.
- Safety (Points: 4)
  - Like Route 2, most of the construction activities for this route will take place on the County roads. Safety measures should be taken to prevent traffic accidents and protect construction workers and the general public.
- Land Acquisition / Easements (Points: 5)
  - Same as Route 2.
- Maintenance Costs / Accessibility (Points: 18)
  - Same as Route 2.

December 12, 20XX

6.0 RECOMMENDATION

This study recommends Route 3, based on the ranking criteria discussed in Section 5 and summarized in Table 5.1. The other two alternatives, though feasible, did not have as many benefits and received a lower score based on the criteria established in this study.

6.1 Reason for Recommendation

Reasons for denying Route 1 (UPRR):

- Permits and timing associated with strict UPRR regulations may go beyond the desired deadline for implementation of the design.

- Environmental concerns due to construction on undeveloped lands are lower as compared to the other alternatives.

- Physical constraints, accessibility along undeveloped land and limited space around the railroad infrastructure are more significant constraints for this alternative than for the other two routes.

- Safety concerns for construction along live train traffic and additional bond insurance required per UPRR standards.

Reasons for denying Route 2 (Applegate Road):

- Complex crossing of the Boardman Canal increases the potential for more delays and costs than Route 3.

- Significant inconveniences to residents along Applegate Road.

- Complicated construction staging to minimize impact on residents.

- Construction costs increased due to delays and staging along a more travelled road.

Route 3 follows a route along the same county roads as Route 2, but would impact the community less by crossing I-80 before going through the populated area on Applegate Road. Since Route 3 stays mostly on county roads, construction equipment would not have to go off road to the extent of Route 1. Finally, potential environmental impacts may be the same as Route 2, however, construction cost is lower. This results in minimizing the following:

- Construction delays

- Conflicts with residents

- Conflict with existing utilities

Although Route 3 did not surpass the other alternatives in each criterion, it established itself as the highest overall scoring alternative. Thus, the study identified Route 3 as the recommended route for the implementation of the new Applegate wastewater conveyance pipeline connecting to North Auburn's SMD-1 at Dry Creek Road and Blue Grass Drive.

6.2   Benefits

Based on the established by the study ranking criteria Route 3 has the following benefits:

- It has the potential to be constructed in the least amount of time.
- It is the least likely to be opposed by stakeholders, residents and environmental groups.
- It is the cheapest of the three proposed alternatives.

6.3   Cost

The preliminary cost for Route 3 is estimated to be $5,291,201.

REFERENCES

1. Landis and Associates. (January 1, 1975). *Placer County Service Area No.24. Improvements to Community Sewer System: Applegate.* (Record Drawings).

2. Central Valley Regional Water Quality Board (CVRWQB). (June 23, 2006). *Administrative Civil Liability Complaint No. R5-2006-0510 Against Placer County Service Area and No. 28, Zone 24 Applegate Wastewater Treatment Facility.*

3. Department of Public Works, Placer County, State of California. (August 2005). *General Specifications.* Section 71: Sewers.

4. State of California, Department of Transportation. (January 19, 2004). *Encroachment Permits Manual.* Chapter 6, Utilities Permits. Available: http://www.dot.ca.gov/hq/traffops/developserv/permits/pdf/manual/Chapter_6.pdf. Retrieved: October, 2008.

5. Union Pacific Railroad. (2008). *Pipeline Installation Engineering Specifications: Specifications for Pipelines with Maximum Casing Diameter Of 48 Inches and Encased Gas Transmission Lines Crossing Under Railroad Tracks.* Available: http://www.uprr.com/reus/pipeline/pipespec.shtml. Retrieved: October, 2008.

6. California Department of Public Health. (May 2006). *Pipeline Separation Design and Installation Reference Guide.*

7. Placer County, California. (October 1, 2008). *Placer County Code.* Available: http://qcode.us/codes/placercounty/. Retrieved: September, 2008.

8. Placer County, California. (2005). *General Plan. Placer County General Plan. Auburn, CA.* Available: <http://www.placer.ca.gov/Departments/CommunityDevelopment/Planning/Documents/CommPlans/GenPlanPC.aspx>;. Retrieved: October, 2008.

9. Union Pacific Railroad. (2008). *Procedures for Encroachments.* Available: http://www.uprr.com/reus/encroach/procedur.shtml. Retrieved: October, 2008.

10. Central Valley Regional Water Quality Board (CVRWQB). (2008). *Laws and Regulations.* Available: http://www.swrcb.ca.gov/centralvalley/laws_regulations/. Retrieved: October, 2008.

11. State Water Resources Control Board. (January 2003). *Water Quality Order 99-08-DWQ. National Pollutant Discharge Elimination System (NPDES). Storm Water Discharges associated with Construction Activity (General Permit).*

12. Placer County Air Pollution Control District. (2008). *Air Quality Standards.* Available: http://www.placer.ca.gov/Departments/Air/airquality.aspx. Retrieved: September 2008.

13. California Air Resources Board. 2008. *Air Districts (APCD or AQMD). Authority to Construct.* Available: http://www.arb.ca.gov/permits/airdisac.htm. Retrieved: October 2008.

14. U.S. Army Corps of Engineers. (2008). *Regulatory Program Overview.* Available: http://www.usace.army.mil/cw/cecwo/reg/oceover.htm. Retrieved: October 2008.

15. California Department of Parks and Recreation. (February 14, 2002). *National Historic Preservation Act.* Available: http://ceres.ca.gov/wetlands/permitting/nhpa.html. Retrieved: October, 2008.

16. State of California, Department of Transportation. (January 19, 2004). *Project Plans for Construction on State Highway in Placer County, from Route 80/193 Separation to Auburn Ravine Undercrossing and from 0.8km West of Auburn Ravine Road Overcrossing to Route 174/80 Separation.* (Record Drawings).

17. Union Pacific Railroad. (July 26, 2006). *Interim Guidelines for Horizontal Directional Drilling (HDD) Under Union Pacific Right-of-Way.*

18. Kleinfelder. (July 18, 2008). *Geotechnical Data Report. Applegate Wastewater Connection to SMD-1 Collection System.*

19. ICF Jones & Stokes. (September, 2008). *Initial Study/Environmental Assessment for Applegate Wastewater Treatment Plant Closure and Pipeline Project.* Available: http://www.placer.ca.gov/Departments/CommunityDevelopment/EnvCoordSvcs/EIR/ApplegateWastewater.aspx. Retrieved: September 2008.

20. Placer County, California. (February 22, 2005). *Placer County Conservation Plan.* Available: http://www.placer.ca.gov/Departments/CommunityDevelopment/Planning/PCCP.aspx. Retrieved: September 2008.

21. Cooper, Rodolf and Associates. (May 27, 1981). *Improvement plans for Low-Pressure Sewer System, Saddleback, California.* (Record Drawings). Sheet 3/10 (Blue Grass Drive).

22. Placer County, California. (1980). *Southern Pacific Transportation Company Right of Way and Track Map.* (Record Drawings).

APPENDICES

A. *American River Watershed Map*

B. *Design Drawings - Route 3*
- o   **Title Sheet (T-1)**
- o   **Route 1 Layouts (L1-1 to L1-6)**
- o   **Route 2 Layouts (L2-1 to L2-6)**
- o   **Route 3 Layouts (L3-1 to L3-6)**
- o   **Cross-Sections (X1 to X2)**
- o   **Details (C1)**
- o   **Parcel Maps (PM-1 to PM-4)**

C. *Site Geologic and Exploration Location Map*

D. *Naturally Occurring Asbestos Map*

E. *Pump Calculations*
- o   **E-1. Applegate's Existing Flows**
- o   **E-2. Route 1 (UPRR)**
- o   **E-3. Route 2 (Applegate Road)**
- o   **E-4. Route 3 (Lake Arthur Road)**

F. *Pump Selections*
- o   **Route 1 – Goulds Model 3771(12 hour)**
- o   **Route 2 and Route 3 – Goulds Model 3655 (12 hour)**

G. *Constraints*
- o   **Table G.1- Constraints along Route 1 (UPRR)**
- o   **Table G.2- Constraints along Route 1 (Applegate Rd)**
- o   **Table G.3- Constraints along Route 1 (Lake Arthur Rd)**

H. *Preliminary Cost Estimates*

APPENDIX D

Example Short Technical Report: The Benefits of Green Roofs

Source: epa.gov.

INTRODUCTION

Green roofs are becoming a popular choice for many cities and corporations. They have many benefits. The purpose of this report is to discuss the benefits of green roofs. The background section contains information on the basic components of a green roof system and explains the two main types of green roofs. The discussion section outlines the benefits of green roofs and briefly explains a few considerations that must be addressed. The conclusion re-iterates the benefits of green roofs and explains who realizes each of those benefits. The recommendations include those who should strongly consider the use of green roofs.

BACKGROUND

Most roof assemblies contain structural support, a waterproof roofing membrane, and thermal insulation. However, a green roof assembly requires additional components not used in conventional roof assemblies. Like a conventional roof, a typical green roof section includes structural support, a waterproof roof membrane, and thermal insulation. In addition, a green roof requires a root barrier to protect the roof membrane, a drainage medium, a filter fabric, a growing medium, and vegetation. There are several products available that can be used to fulfill these functions. Figure 1, Green Roof Assembly, from American Wick Drains (American Wick Drains) and Figure 2, Principle Green Roof Technologies, from National Research Council, Institute for Research in Construction (Green Roofs for Healthy Cities), show the basic components of a green roof assembly.

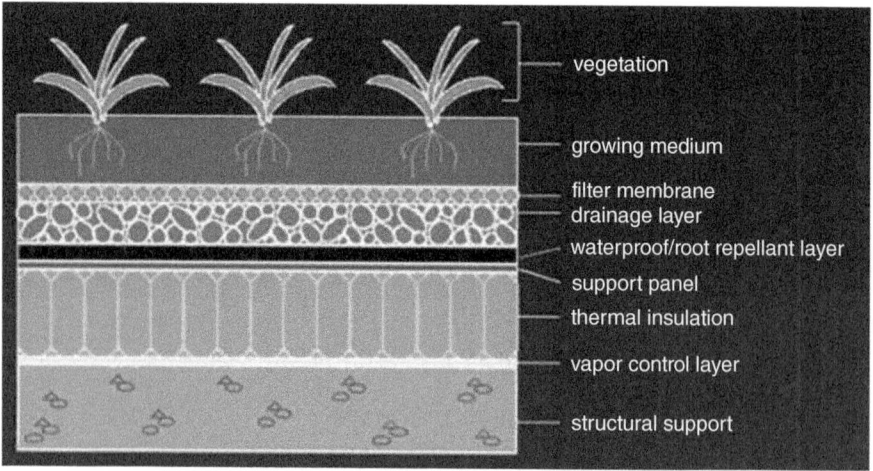

FIGURE 1   Green Roof Technologies.

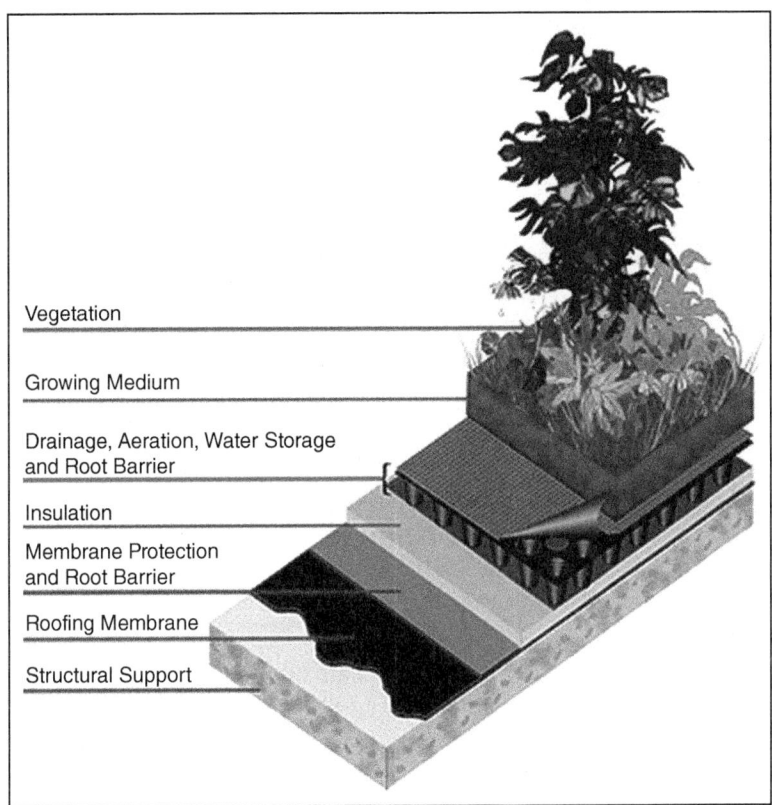

FIGURE 2 Green Roof Assembly.

There are two primary types of green roofs, intensive and extensive. Extensive roofs are characterized by shallow soil depth and low growing plants. Soil depths are usually less than 8 inches with plantings reaching heights up to 36 inches (Garden the Planet). They are lighter and lower maintenance. They provide habitat for flora and fauna. Intensive roofs can resemble park-like spaces. They are characterized by soil depths up to 4 feet and can support a wide range of plant life (Garden the Planet).

DISCUSSION

Green roofs have many advantages over conventional roofing systems. These advantages include:

- Improved storm water runoff management
- Improved air quality
- Reduced heat island effect

- Thermal and sound insulation
- Reduced HVAC costs
- Extended roof life
- Potential food production
- Community social, health, and emotional benefits

One of the most touted benefits of green roofs is the ability to substantially reduce storm water runoff from the buildings they cover. The planting medium absorbs much of the rainwater reducing the load on municipal storm drainage systems. The flow water that is not absorbed is delayed. This delay assists in reducing peak flows and preventing sewage overflows and flooding that often occur during peak flows. A study by North Carolina State University demonstrated reductions in runoff up to 63% for a 3 inch deep green roof. This same study showed peak flow reduction of up to 87% (Tokarz) for .6 inches of rainfall. Milwaukee, Wisconsin, is another city using green roofs to assist in managing storm water runoff. After the 2003 installation of seven green roofs, Milwaukee conducted a modeling study that showed the volume of storm water runoff sent to sewer treatment plants was reduced 31–37% and peak flows were reduced between 5 and 36% (Environmental Protection Agency).

Another benefit of green roofs is improved air quality. Studies have shown that green roofs can absorb particulates. According to Green Roofs for Healthy Cities, a 10 square feet of grass roof can remove as much as 4.4 pounds of particulate matter per year with the proper plantings (Green Roofs for Healthy Cities). Some of the pollutants that can be removed from the air include "nitrogen oxides, sulfur dioxide, carbon monoxide, and ground level ozone" (Massachusettes Department of Environmental Protection). The plant life associated with green roofs also removes carbon dioxide and produces oxygen. Green roofs for Healthy Cities states 10 square feet of uncut grass on a green roof can generate enough oxygen for one person for a year (Green Roofs for Healthy Cities). This also serves to improve air quality, which has become an important issue for many cities facing increased air pollution and its associated health issues.

Green roofs also mitigate the effects of urban heat islands. Green spaces tend to be cooler than typical urban hardscape areas. This is also true of roofing areas. During the summer months, temperatures of conventional roofs can be substantially higher than ambient temperature reaching 130°F (Holladay) or more. Green roofs have been shown to lower the temperature of the roof areas. For instance, the City of Chicago found that the green roof of City Hall measured a range of 91°–119°F, while the adjacent conventional roof measured 169°F on a day with a 90°F temperature during the month of August. Figure 3 below illustrates the temperature differences between the conventional and the green roof (Environmental Protection Agency).

Figure 4, Roof Temperature by Hour, developed by the American Society of Heating, Refrigeration, and Air Conditioning Engineers, Inc., shows how the temperature of a green

FIGURE 3   Temperature Difference between a Green and Conventional Roof.
Source: IOP Publishing.

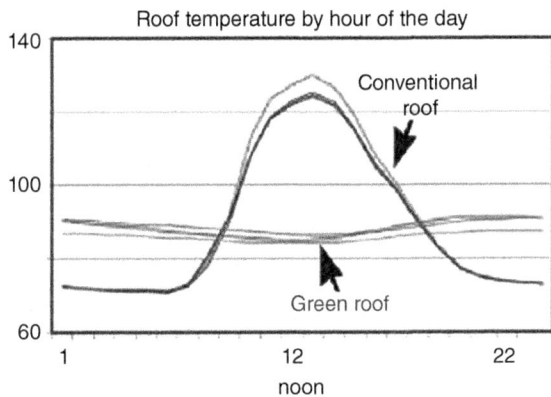

FIGURE 4   Roof Temperature by Hour.

roof compares to a conventional roof throughout a day. The green roof temperature is more consistent and lower overall, reducing the heat radiated into the surrounding area.

Green roofs also act as additional thermal and sound insulation. They stop heat from moving through the roof. In her report titled Energy Efficiency in Green Roofs, Karen Lui states, "The growing medium and the plants enhanced the thermal performance of the rooftop garden by providing shading, insulation and evaporative cooling. It acted as a thermal mass, which effectively damped the thermal fluctuations going through the roofing system (Lui)." In addition to thermal insulation, green roofs provide sound attenuation. According to Green Roofs for healthy Cities, a 4.7 inch substrate has the ability to reduce sound levels 40 decibels, and a 7.9 inch substrate may reduce sound by up to 50 decibels (Green Roofs for Healthy Cities).

The potential for reduced HVAC costs is another benefit of green roofs. By lowering the heat gain and heat loss associated with heat transmission through the roof, the demand for air conditioning can be reduced. In a study performed by Lui, a 6 inch extensive green roof demonstrated a 95% reduction in heat gain and 26% reduction in heat loss when compared to a reference roof (Green Roofs for Healthy Cities). The decreased heat load may allow for a reduction in mechanical equipment for the building. It also decreases the energy consumption used to heat and cool the building. Chicago's City Hall realized savings of up to 30% in their energy costs after the installation of their green roof (Garden the Planet).

The list of benefits continues with prolonged roof life. The plants and planting medium provide protection from UV rays and thermal extremes that deteriorate conventional roofing membranes. It also provides physical protection from wind, hail, fireworks, and vandalism (International Green Roof Association). It's estimated that this protection doubles the life of the roof membrane. This in turn reduces re-roofing expenses and landfill materials.

The potential for food production is another benefit of green roofs. Intensive roofs can support a broad spectrum of plant life, including herbs and vegetables. The Fairmont Waterfront Hotel in Vancouver planted a herb and vegetable garden on their green roof. The roof garden produces $30,000 per year of produce that is used in the hotel's restaurant (Green Roofs for Healthy Cities). The idea can be expanded and has the potential to create food sources close to living centers. This could, in turn, reduce transportation environmental impacts as well by reducing the shipping distances for some food items.

Although much more difficult to quantify, green roofs provide communities with social, emotional, and health benefits. The EPA states, "An increasing number of studies suggest that vegetation and green space - two key components of green infrastructure - can have a positive impact on human health. Recent research has linked the presence of trees, plants, and green space to reduced levels of inner-city crime and violence, a stronger sense of community, improved academic performance, and even reductions in the symptoms associated with attention deficit and hyperactivity disorders. One such study discusses the association between neighborhood greenness and the body mass of children" (Environmental Protection Agency).

The benefits of green roofs continue. However, there are some issues that must be accounted for when considering a green roof. These include:

- Initial costs
- Maintenance costs
- Drainage and irrigation

Green roofs require more materials to build and therefore are more expensive than conventional roofs. In addition, the structural capacity of the roof must be sufficient to support the additional weight of the plants, planting medium, and human activities. According to Green Roofs for Healthy Cities and the EPA, green roofs cost $10–$24 per square foot. While more than conventional roofs, the long-term savings in energy and maintenance costs should be considered.

Maintenance is important consideration for green roofs. The EPA estimates maintenance costs $.75–$1.50 per square foot. Care must be taken to keep woody plants from overgrowing and potentially damaging the roof membrane.

Although the planting medium absorbs much of the rainwater, drainage must still be provided and maintained. Gutters and downspouts must be kept clean and free of debris to avoid roof damage. Provisions for providing irrigation to plant life must be considered as well.

CONCLUSION

The benefits of green roofs cannot be ignored. Green roofs have many long-term benefits. They have the potential to address a myriad of urban issues. Many municipalities are turning to green roofs to address issues of storm water runoff management, urban air quality, and the effects of urban heat islands. Building owners benefit from the reduced energy costs and extended roof life. Community members gain social, health, and emotional benefits. Green roofs are a long-term investment in community well-being.

RECOMMENDATIONS

Given the benefits of green roofs, all cities should promote the use of green roofs. With the initial costs associated with green roofs, they may not be an appropriate choice for speculative developers who are concerned with keeping initial costs low. However, green roofs are an excellent choice for public buildings and corporate entities with long-term commitment to their projects and who will be able to reap the long-term benefits.

REFERENCES

American Wick Drains. *Green Roof.* 14 April 2009 http://www.americanwick.com/applications/detail.cfm?app_id=18.

Environmental Protection Agency. *Managing Wet Weather with Green Infrastructure.* 15 December 2008. 14 April 2009 http://cfpub.epa.gov/npdes/home.cfm?program_id=298.

———. *National Polution Discharge Elimination System: Green Infrastructure Case Studies, Milwaukee Wisconson.* 9 December 2008. 14 April 2009 http://cfpub.epa.gov/npdes/greeninfrastructure/gicasestudies_specific.cfm?case_id=61.

———. "Reducing Urban Heat Islands Compendium of Strategies Green Roofs." 9 February 2009. Green Roofs. 14 April 2009 http://www.epa.gov/heatisland/resources/pdf/GreenRoofsCompendium.pdf.

Garden the Planet. *Green Roofs.* 14 February 2009. 14 April 2009 http://www.gardentheplanet.com/gr_components.htm.

Green Roofs for Healthy Cities. *About Green Roofs*. 2000-2005. 14 April 2009 http://www. greenroofs.org/index.php?option=com_content&task=view&id=26&Itemid=40.

Holladay, April. *Green Roofs Swing Temperatures in Urban Jungles*. 24 April 2006. 14 April 2009 http://www.usatoday.com/tech/columnist/aprilholladay/2006-04-24-green-roofs_x.htm.

International Green Roof Association. *Private Benefits*. 2009. 14 April 2009 http://www.igra-world. com/benefits/private_benefits.php.

Lui, Karen. "Energy Efficiency and Environmental Benefits of Rooftop Gardens." March 2002. *ELT Easy Green*. 14 April 2009 http://www.eltgreenroofs.com/pdf/ELT-NRCC45345.pdf.

Massachusettes Department of Environmental Protection. *Green Roofs and Storm Water Management*. 14 April 2009 http://www.mass.gov/dep/water/wastewater/grnroof.htm.

Tokarz, Erika. "CEER Green Roof Project." May 2006. *Villanova University*. 14 April 2009 http://egrfaculty.villanova.edu/public/Civil_Environmental/WREE/VUSP_Web_Folder/ GR_web_folder/GR_paper.html.

Example Specification: Cast-in-Place Concrete

SECTION 03 30 00

CAST-IN-PLACE CONCRETE

PART 1 – GENERAL

1.01   SECTION INCLUDES

A.   Conveying and placing concrete

B.   Placement under water

C.   Consolidation

D.   Construction joints

E.   Expansion and contraction joints

F.   Curing and protection

1.02   RELATED SECTIONS

A.   Portland cement concrete specified in Section 03 05 15 – Portland Cement Concrete

B.   Finishing and curing of formed and unformed concrete surfaces, including repair and patching of surface defects, are specified in Section 03 35 00 – Concrete Finishing

C.   Shotcrete is specified in Section 03 37 13 – Shotcrete

D.   Vapor barrier under slabs on grade is specified in Section 07 26 00 – Vapor Retarders

RELEASE – R3.0
Issued: January 2013

SECTION 03 30 00
PAGE 1 OF 10

BART FACILITIES STANDARDS
STANDARD SPECIFICATIONS

Civil Engineer's Handbook of Professional Practice, Second Edition. Karen Lee Hansen and Kent E. Zenobia.
© 2025 John Wiley & Sons, Inc. Published 2025 by John Wiley & Sons, Inc.
Companion website: www.wiley.com/go/hansen/CivilEngineersHandbook

1.03   MEASUREMENT AND PAYMENT

A.   General: Measurement and payment for cast-in-place concrete will be either by the lump-sum method or by the unit-price method as determined by the listing of the bid item for cast-in-place concrete indicated in the Bid Schedule of the Bid Form.

B.   Lump Sum: If the Bid Schedule indicates a lump-sum for cast-in-place concrete, the lump-sum method of measurement and payment will be in accordance with Section 01 20 00 – Price and Payment Procedures, Article 1.03.

C.   Unit Prices: If the Bid Schedule indicates a unit price for cast-in-place concrete, the unit-price method of measurement and payment will be as follows:

    1.   Measurement:

        a.   Except as specified otherwise in other Sections of these Specifications or the Contract Specifications, each class of concrete and type of placement of cast-in- place concrete will be measured for payment by the cubic yard, and quantities will be computed, based on the neat lines or pay lines, section profiles, and dimensions indicated on the Contract Drawings, without deduction for chamfers, reinforcing steel and embedded items, and openings and recesses having an area of less than two square feet.

        b.   Additional concrete used to replace overcut or for overbreak, or to repair or replace defective work, will not be measured separately for payment.

    2.   Payment: Cast-in-place concrete will be paid for at the indicated Contract unit prices for the computed quantities as determined by the measurement method specified in Article 1.03.C.1.

1.04   DEFINITIONS

A.   The words and terms used in these Specifications conform with the definitions given in ACI 116R.

1.05   REFERENCES

A.   American Concrete Institute (ACI):

1.	ACI 116R	Cement and Concrete Terminology
2.	ACI 117	Standard Specification for Tolerances for Concrete Construction and Materials
3.	ACI 301	Standard Specifications for Structural Concrete
4.	ACI 302.1R	Guide for Concrete Floor and Slab Construction
5.	ACI 303.1	Standard Specification for Cast-In-Place Architectural Concrete
6.	ACI 304R	Guide for Measuring, Mixing, Transporting, and Placing Concrete
7.	ACI 304.2R	Placing Concrete by Pumping Methods

8.	ACI 305R	Hot Weather Concreting
9.	ACI 306.1	Standard Specification for Cold Weather Concreting
10.	ACI 308	Standard Practice for Curing Concrete
11.	ACI 309R	Guide for Consolidation of Concrete
12.	ACI 318	Building Code Requirements for Structural Concrete
13.	ACI 503.2	Standard Specification for Bonding Plastic Concrete to Hardened Concrete with a Multi-Component Epoxy Adhesive

B. American Society for Testing and Materials (ASTM):

1.	ASTM C31	Standard Practice of Making and Curing Concrete Test Specimens in the Field
2.	ASTM C94	Specification for Ready-Mixed Concrete
3.	ASTM C881	Specification for Epoxy-Resin-Base Bonding Systems for Concrete

1.06 SUBMITTALS

A. General: Refer to Section 01 33 00 – Submittal Procedures, and Section 01 33 23 – Shop Drawings, Product Data, and Samples, for submittal requirements and procedures.

B. Shop Drawings:

1. Submit drawings that indicate the locations of all joints in concrete, including construction joints, expansion joints, isolation joints, and contraction joints. Coordinate with the requirements specified in Section 03 11 00 – Concrete Forming.

2. Submit drawings that indicate concrete placement schedule, method, sequence, location, and boundaries. Include each type and class of concrete, and quantity in cubic yards.

C. Product Data: Submit manufacturer's product data for epoxy adhesive.

D. Records and Reports: Report the location in the finished work of each mix design, and the start and completion times of placement of each batch of concrete placed for each date concrete is placed.

1.07 QUALITY CONTROL

A. Tolerances:

1. Concrete Tolerances: Comply with the requirements of ACI 117 as applicable. Coordinate with the requirements specified in Section 03 11 00 – Concrete Forming.

2. Tolerances for Slabs and Flatwork: Comply with the requirements specified in Section 03 35 00 – Concrete Finishing.

B. Architectural Concrete: Where concrete is indicated as architectural concrete exposed to public view, such concrete shall be produced in accordance with applicable requirements of ACI 301 and ACI 303.1.

C. Site Mock-Ups:

   1. Refer to Section 01 43 38 – Field Samples and Mockups, for mock-up requirements and procedures.

   2. Construct site mock-ups for all architectural concrete work and formed concrete that will be exposed to the public in the finished work, not less than 4 feet by 6 feet in surface area, for review and acceptance by the Engineer, before starting the placement of concrete.

   3. Approved site mock-ups shall set the standard for the various architectural concrete features, formed finishes, and colors of the concrete. Provide as many mock-ups as required to show all the different features and formed surfaces of the concrete.

D. Cold Joints: Cold joints in concrete will not be permitted unless planned and treated properly as construction joints.

E. Monitoring of Formwork: Provide monitoring of forms and embedded items to detect movement, or forms and embedded items out-of-alignment, from pressure of concrete placement.

1.08 ENVIRONMENTAL REQUIREMENTS

A. Delivering and placing of concrete in hot weather and cold weather shall conform with applicable requirements of ACI 305R and ACI 306.1 and Section 03 05 15 – Portland Cement Concrete.

B. Do not place concrete when the rate of evaporation of surface moisture from concrete exceeds 0.2 pounds per square foot per hour as indicated in Figure 2.1.5 of ACI 305R.

C. Do not place concrete in, or adjacent to, any structure where piles are required until all piles in the structure have been driven or installed.

PART 2 – PRODUCTS

2.01 MATERIALS

A. Formwork: Refer to Section 03 11 00 – Concrete Forming, for requirements.

B. Joint Fillers and Sealers: Refer to Section 03 15 00 – Concrete Accessories, for requirements.

C. Waterstops: Refer to Section 03 15 13 – Waterstops, for requirements.

D. Reinforcing Steel: Refer to Section 03 20 00 – Concrete Reinforcing, for requirements.

E. Portland Cement Concrete: Refer to Section 03 05 15 – Portland Cement Concrete, for mix designs and other requirements.

RELEASE – R3.0
Issued: January 2013
SECTION 03 30 00
PAGE 4 OF 10
BART FACILITIES STANDARDS
STANDARD SPECIFICATIONS

F.   Concrete Curing Materials: Refer to Section 03 35 00 – Concrete Finishing, for requirements.

G.   Vapor Barrier Materials: Refer to Section 07 26 00 – Vapor Retarders, for requirements.

H.   Epoxy Adhesive: ASTM C881, Type II for non-load-bearing concrete and Type V for load- bearing concrete, Grade and Class as determined by project conditions and requirements.

PART 3 – EXECUTION

3.01   EXAMINATION

A.   Inspect forms, earth-bearing surfaces, reinforcement, and embedded items, and obtain the Engineer's written approval before placing concrete. Complete and sign a pour card on the form supplied by the Engineer. The Engineer shall countersign the card prior to commencing the pour.

3.02   PREPARATION

A.   Place concrete under the observation of the Engineer and with the Contractor's Quality Control Representative present to document requirements and results of the placement.

B.   Whenever possible, place concrete during normal working hours. When concrete-placement schedules require concrete placement at times other than normal working hours, ensure that the Engineer is notified and is present at the time of placement.

C.   Do not place concrete until conditions and facilities for the storage, handling, and transportation of concrete test specimens are in compliance with the requirements of ASTM C31 and Section 03 05 15 – Portland Cement Concrete, and are approved by the Engineer.

D.   Prior to placement of concrete, the subgrade shall be in a firm, well-drained condition, and of adequate and uniform load-bearing nature to support construction personnel, construction materials, construction equipment, and steel reinforcing mats without tracking, rutting, heaving, or settlement. All weak, soft, saturated, or otherwise unsuitable material shall be removed and replaced with structural backfill or lean concrete.

E.   All structure foundations, including those for Stations and for subway box, shall be inspected and approved, in writing, by a qualified, independent geotechnical engineer prior to placement of footings and base slabs, to confirm the adequacy of the supporting soil for concrete placement.

F.   Earth bottoms or bearing surfaces for footings and slabs shall be dampened but not saturated or muddied just before placing concrete.

3.03   TRANSPORTING

A.   Concrete shall be central-mixed concrete from a central batch plant, transported to the jobsite in a truck mixer, in accordance with the requirements specified in Section 03 05 15 – Portland Cement Concrete, and ASTM C94.

B.   Transport concrete to the jobsite in a manner that will assure efficient delivery of concrete to the point of placement without adversely altering specified properties with regard to water-cement ratio, slump, air entrainment, and homogeneity.

3.04   CONVEYING AND PLACING

A.   Placement Standards: Conveying and placing of concrete shall conform with applicable requirements of ACI 301, ACI 302.1R, ACI 304R, and ACI 318.

B.   Handling and Depositing:

1.   Concrete placing equipment shall have sufficient capacity to provide a placement rate that will preclude cold joints and that shall deposit the concrete without segregation or loss of ingredients.

2.   Concrete placement, once started, shall be carried on as a continuous operation until the section of approved size and shape is completed.

3.   Concrete shall be handled as rapidly as practicable from the mixer to the place of final deposit by methods that prevent the separation or loss of ingredients. Concrete shall be deposited, as nearly as practicable, in its final horizontal position to avoid redistribution or flowing.

4.   Concrete shall not be dropped freely where reinforcing will cause segregation, nor shall it be dropped freely more than 5 feet. Concrete shall be deposited to maintain a plastic surface approximately horizontal.

5.   In placing walls, columns, or thin sections (6 inches or less in thickness) of heights greater than 10 feet, concrete placement rate, lift thickness, and time intervals between lifts shall be as indicated on approved Shop Drawings. Openings in the form, elephant trunk tremies, or other approved devices shall be used that will permit the concrete to be placed without segregation or accumulation of hardened concrete on the forms or metal reinforcement above the level of the fresh concrete.

6.   Concrete that has partially hardened shall not be deposited in the work. The discharge of concrete shall be started not later than 60 minutes after the introduction of mixing water. Placing of concrete shall be completed within 90 minutes after the first introduction of water into the mix.

C.   Pumping:

1.   Concrete may be placed by pumping if the maximum slump can be maintained and if accepted in writing by the Engineer for the location proposed.

2. Placing concrete by pumping methods shall conform with applicable requirements of ACI 304R and ACI 304.2R.

3. Equipment for pumping shall be of such size and design as to ensure a continuous flow of concrete at the delivery end without separation of materials. Concrete from end of hose shall have a free fall of less than 5 feet. Pump hoses shall be supported on horses or similar devices so that reinforcement or post-tensioning ducts or tendons are not moved from their original position.

4. The concrete mix shall be designed to the same requirements as specified in Section 03 05 15 – Portland Cement Concrete, and may be altered for placement purposes with the prior approval of the Engineer.

3.05   PLACEMENT UNDER WATER

A. Placement Standards: Placing of concrete in or under water shall conform with requirements of ACI 304R. All concrete to be placed under water shall be placed by the tremie method or by direct pumping.

B. Placement Requirements: Deposit concrete in water only when indicated or approved in writing by the Engineer, and only under the observation of the Engineer. Use only tremie method and direct pumping with equipment that has been accepted by the Engineer.

3.06   CONSOLIDATION

A. Concrete shall be thoroughly consolidated and compacted by mechanical vibration during placement in accordance with the requirements of ACI 309R.

B. The Engineer will inspect concrete placement to confirm that proper placing methods are being employed, and that special techniques are being used in congested areas and around obstructions such as pipes and other embedded items. Check installation of embedded items for correct location and orientation during concrete placement.

C. Conduct vibration in a systematic manner by competent, skilled, and experienced workers, with regularly maintained vibrators, and with sufficient back-up units at the jobsite. Use the largest and most powerful vibrator that can be effectively operated in the given work, with a minimum frequency of 8,000 vibrations or impulses per minute, and of sufficient amplitude to effectively consolidate the concrete.

D. Insert and withdraw the vibrator vertically at uniform spacing over the entire area of the placement. Space the distance between insertions such that "spheres of influence" of each insertion overlap.

E. Conduct vibration so as to produce concrete that is of uniform texture and appearance, free of honeycombing, air and rock pockets, streaking, cold joints, and visible lift lines.

RELEASE – R3.0
Issued: January 2013

SECTION 03 30 00
PAGE 7 OF 10

BART FACILITIES STANDARDS
STANDARD SPECIFICATIONS

F. On vertical surfaces and on all architectural concrete where an as-cast finish is required, use additional vibration and spading as required to bring a full surface of mortar against the forms, so as to eliminate objectionable air voids, bug holes, and other surface defects. Additional procedures for vibrating concrete shall consist of the following:

1. Reduce the distance between internal vibration insertions and increase the time for each insertion.

2. Insert the vibrator as close to the face of the form as possible, without contacting the form.

3. Use spading as a supplement to vibration at forms to provide fully filled out form surfaces without air holes and rock pockets.

4. Provide vibration of forms only if approved by the Engineer for the location.

3.07   CONSTRUCTION JOINTS

A. Construction joints will be permitted only where indicated or approved by the Engineer.

B. Provide and prepare construction joints and install waterstops in accordance with the applicable requirements of ACI 301 and ACI 304R, and as specified in Section 03 11 00 – Concrete Forming.

C. Make construction joints straight and as inconspicuous as possible, and in exact vertical and horizontal alignment with the structure, as the case may be.

D. Use approved key, at least 1-1/2 inches in depth, at joints unless otherwise indicated or approved by the Engineer.

E. Thoroughly clean the surface of the concrete at construction joints and remove laitance, loose or defective concrete, coatings, sand, sealing compound, and other foreign material. Prepare surfaces of joints by sandblasting or other approved methods to remove laitance and expose aggregate uniformly.

F. Immediately before new concrete is placed, wet the joint surfaces and remove standing water. To allow for shrinkage, do not place new concrete against the hardened concrete side of a construction joint for a minimum of 72 hours.

G. Locate joints that are not indicated so that the strength of the structure is not impaired. Joint types and their locations are subject to prior approval of the Engineer.

H. Ensure that reinforcement is continuous across construction joints.

I. Place waterstops in construction joints where indicated.

J. Where bonding of the joint is required, provide epoxy adhesive hereinbefore specified and apply in accordance with ACI 503.2.

K. Retighten forms and dampen concrete surfaces before concrete placing is continued.

L. Allow at least 72 hours to elapse before continuing concrete placement at a construction joint. Approval for accelerating the minimum time elapsing between adjacent placements will be based on tests and methods that confirm that a minimum moisture loss at a relatively constant temperature will be maintained for the period as necessary to control the heat of hydration and hardening of concrete, and to prevent shrinkage and thermal cracking.

3.08 EXPANSION AND CONTRACTION JOINTS

A. Refer to Section 03 11 00 – Concrete Forming, for slab screeds and for formwork where expansion and contraction joints are indicated as architectural features, such as reveals or rustications.

B. Refer to Section 03 15 00 – Concrete Accessories, for expansion joint filler material and joint sealing compound.

C. Refer to Section 03 35 00 – Concrete Finishing, for finishing of edges of expansion joints in slabs with curved edging tool.

3.09 CURING AND PROTECTION

A. Curing of concrete shall conform with applicable requirements of ACI 301 and ACI 308, except that the curing duration shall be a minimum period of ten days. HVFAC shall be cured a minimum of 28 days including an initial 10 days of moist curing. Curing with earth, sand, sawdust, straw, and hay will not be permitted.

B. Keep concrete in a moist condition from the time it is placed until it has cured for at least ten days. Keep forms damp and cool until removal of forms.

C. Immediately upon removal of forms, exposed concrete surfaces shall be kept moist by applying an approved curing compound or by covering with damp curing materials as specified in Section 03 35 00 – Concrete Finishing.

D. Concrete shall not be permitted to dry during the curing period because of finishing operations.

E. Protect fresh concrete from hot sun, drying winds, rain, damage, or soiling. Fog spray freshly placed slabs after bleed water dissipates and after finishing operations commence. Allow no slabs to become dry at any time until finishing operations are complete.

F. Finishing and curing of slabs are specified in Section 03 35 00 – Concrete Finishing.

G. Protect concrete from injurious action of the elements and defacement of any kind. Protect exposed concrete corners from traffic or use that will damage them in any way.

RELEASE – R3.0
Issued: January 2013

SECTION 03 30 00
PAGE 9 OF 10

BART FACILITIES STANDARDS
STANDARD SPECIFICATIONS

H.   Protect concrete during the curing period from mechanical and physical stresses that may be caused by heavy equipment movement, subjecting the concrete to load stress, load shock, or excessive vibration.

3.10   REPAIR OF SURFACE DEFECTS

A.   Refer to Section 03 35 00 – Concrete Finishing, for requirements.

END OF SECTION 03 30 00

RELEASE – R3.0
Issued: January 2013

SECTION 03 30 00
PAGE 10 OF 10

BART FACILITIES STANDARDS
STANDARD SPECIFICATIONS

EJCDC® Model Contract – Agreement Between Owner and Engineer for Study and Report Professional Services

Civil Engineer's Handbook of Professional Practice, Second Edition. Karen Lee Hansen and Kent E. Zenobia.
© 2025 John Wiley & Sons, Inc. Published 2025 by John Wiley & Sons, Inc.
Companion website: www.wiley.com/go/hansen/CivilEngineersHandbook

This document has important legal consequences; consultation with an attorney is encouraged with respect to its use or modification. This document should be adapted to the particular circumstances of the contemplated Project and the controlling Laws and Regulations.

AGREEMENT BETWEEN OWNER AND ENGINEER
FOR STUDY AND REPORT PROFESSIONAL SERVICES

AMERICAN COUNCIL OF ENGINEERING COMPANIES

AMERICAN SOCIETY OF CIVIL ENGINEERS

 NATIONAL SOCIETY OF
PROFESSIONAL ENGINEERS

Copyright © 2022:

National Society of Professional Engineers

1420 King Street, Alexandria, VA 22314-2794

(703) 684-2882

www.nspe.org

American Council of Engineering Companies

1015 15th Street N.W., Washington, DC 20005

(202) 347-7474

www.acec.org

American Society of Civil Engineers

1801 Alexander Bell Drive, Reston, VA 20191-4400

(800) 548-2723

www.asce.org

The copyright for this EJCDC document is owned jointly by the three sponsoring organizations listed above. The National Society of Professional Engineers is the Copyright Administrator for the EJCDC documents; please direct all inquiries regarding EJCDC copyrights to NSPE.

The use of this document is governed by the terms of the License Agreement for the 2020 EJCDC® Engineering Series Documents.

NOTE: EJCDC publications may be purchased at www.ejcdc.org, or from any of the sponsoring organizations above.

GUIDELINES FOR USE OF EJCDC® E-525, AGREEMENT BETWEEN OWNER AND ENGINEER FOR STUDY AND REPORT PROFESSIONAL SERVICES

1.0   PURPOSE AND INTENDED USE OF THE DOCUMENT

EJCDC® E-525, Agreement Between Owner and Engineer for Study and Report Professional Services, is for use in retaining an engineering firm to prepare a study and report—for example, an evaluation of a project's feasibility, facility needs, treatment options, or siting challenges. E-525 does not include design-phase or construction-phase engineering services, and contains an abbreviated set of terms and conditions that is appropriate for its narrow and preliminary scope of services.

If, after receiving the report prepared by the Engineer under E-525, the Owner decides to move forward with a specific project, the parties may consider using EJCDC® E-500, Agreement between Owner and Engineer for Professional Services (2020), for design, bidding-related, and construction-phase services. E-500 contains comprehensive terms and conditions that account for the more substantial compensation, insurance, indemnity, and liability issues that arise from a broader scope of services.

For notably large, expensive, or complex studies and reports, especially those that are conducted over an extended period of time, or that require the use of numerous subconsultants, the Owner and Engineer may wish to consider using E-500 rather than E-525. As stated above, E-500 contains a more comprehensive set of terms and conditions, which may be appropriate for larger studies and reports. If E-500 is used for such studies and reports, the drafter should delete the majority of the scope suggested in E-500's Exhibit A, Engineer's Services, retaining the study and report obligations but removing design, bidding-related, and construction-phase services.

Owner and Engineer will note that Paragraph 6.05.A.1 of E-525 limits the Engineer's liability to $100,000, or the total amount of the Engineer's compensation, whichever is greater. A limitation of liability is a standard recommended provision for engineering, geotechnical, and architectural services that are preliminary in nature and narrow in scope, and for which the compensation is a small proportion of the project's final, overall cost. The parties are free to revise the cap amount in Paragraph 6.05.A.1, or to remove the clause entirely if a limitation is not appropriate for the specific study and report services.

2.0   OTHER DOCUMENTS

EJCDC documents are intended to be used as a system and changes in one EJCDC document may require a corresponding change in other documents. Other EJCDC documents may also serve as a reference to provide insight or guidance for the preparation of this document.

Other EJCDC documents potentially relevant to a study/report assignment under E-525 include:

EJCDC Doc. No.	Document Title	Edition
E-570	Agreement between Engineer and Subconsultant for Professional Services	2020
E-500	Agreement between Owner and Engineer for Professional Services	2020
E-520	Short Form of Agreement between Owner and Engineer for Professional Services	2020

The current (2017) and pending (2021) editions of EJCDC® E 001, Commentary on the EJCDC Engineering Services Documents, also provide additional information and guidance for the use of this document.

3.0   ORGANIZATION OF INFORMATION

All parties involved in a construction project benefit significantly from a standardized approach in the location of subject matter throughout the documents. Experience confirms the danger of addressing the same subject matter in more than one location; doing so frequently leads to conflicting requirements, confusion, and unanticipated legal consequences. Careful attention should be given to the guidance provided in EJCDC® N-122/AIA® A521™, Uniform Location of Subject Matter (2012 Edition) when preparing construction documents. EJCDC® N-122/AIA® A521™ is available at no charge from the EJCDC website, www.ejcdc.org, and from the websites of EJCDC's sponsoring organizations.

In addition, the current editions of MasterFormat and SectionFormat, published by the Construction Specifications Institute, provide useful guidance on the location of information and requirements in construction documents.

4.0   GUIDANCE NOTES AND NOTES TO USER

EJCDC Documents include Guidance Notes and Notes to User to provide guidance regarding the preparation of Project-specific documents. Guidance Notes and Notes to User are lightly shaded to distinguish them from the proposed text of the document. As Project-specific documents are prepared and made ready to publish, all shaded text (Guidance Notes and Notes to Users) should be deleted. These notes are intended for use by the User in the preparation of the document and are not intended to be included in the completed document.

Guidance Notes provide information regarding the paragraphs which follow, including reasons for the paragraph, discussions of best practices, and alternate approaches for different situations.

Notes to User provide specific information for editing the document. When alternate paragraphs for different situation are presented, explanations on how to select the most appropriate alternate will be provided, with direction to delete those paragraphs not used. Paragraphs will automatically renumber when unused paragraphs are deleted.

5.0   EDITING THIS DOCUMENT

5.1   It is intended that this document be edited for each specific agreement (contract) between the Owner and the Engineer. Guidelines for editing include:

A.   Remove the cover pages, which consist of the title pages and these Guidelines for Use.

B.   Type in required information as indicated by brackets ([ ]). Bracketed text will usually provide instructions for what is to be inserted in place of the brackets. Delete brackets and change formatting to match existing text after project specific text has been added, e.g. change "[Project Name]" to "Peach Street Renovation" (without brackets or bold, or quotation marks).

C.   Fill in blanks, if any. It will be more common for information to be inserted by user to be indicated by a prompt in brackets, as described in Paragraph B above, rather than by an underline-style blank.

D. Most Notes to User are presented before the text to which they apply; some Notes to Users are interspersed in the text, usually within brackets. Delete all "Notes to User" after reviewing each note and taking appropriate action. Delete all associated numbering and brackets.

E. Make Project-specific modifications and supplementations, as appropriate. If such revisions affect any cross-references, revise the cross-references.

F. Complete tables, if any.

G. Address check boxes, if any, by clicking in the appropriate box.

H. Delete Guidance Notes.

6.0 LICENSE AGREEMENT

This document is subject to the terms and conditions of the **License Agreement, 2020 EJCDC Engineering Series Documents**. A copy of the License Agreement was furnished at the time of purchase of this document, and is available for review at www.ejcdc.org and the websites of EJCDC's sponsoring organizations.

AGREEMENT BETWEEN OWNER AND ENGINEER
FOR STUDY AND REPORT PROFESSIONAL SERVICES

This is an Agreement between **[name of Owner]** (Owner) and **[name of Engineer]** (Engineer). Owner's Project, of which Engineer's services under this Agreement are a part, is generally identified as **[name of Project]** (Project).

Owner and Engineer further agree as follows:

ARTICLE 1—ENGINEER'S SERVICES

1.01   Study and Report Services of Engineer

    A.   Engineer's services under this Agreement are generally identified as **[short description of Engineer's study and report services]** ("Study and Report Services").

    B.   Engineer shall perform or furnish the Study and Report Services set forth in this Agreement, expressly including the Basic Services described in Article 1 of Exhibit A, Scope of Engineer's Study and Report Services, and any duly authorized Additional Services described in Article 2 of Exhibit A.

ARTICLE 2—OWNER'S RESPONSIBILITIES

2.01   Owner shall:

    A.   Provide Engineer with all criteria and full information as to Owner's requirements for the Study and Report Services, including but not limited to design objectives and constraints; space, capacity and performance requirements; flexibility and expandability goals; security issues; any anticipated funding sources; and budgetary limitations.

    B.   Furnish to Engineer all existing studies, reports, and other available information pertinent to the Engineer's performance of the Study and Report Services, including reports and data relative to previous investigations, designs, construction, or existing facilities at or adjacent to any Site under consideration.

    C.   Following Engineer's assessment of initially-available Project data and information, and receipt of Engineer's advice regarding the need (if any) for additional Project-related data and information, either (1) authorize Engineer to undertake Additional Services necessary to obtain such additional Project-related data and information, or (2) obtain, furnish, or otherwise make available (if necessary through title searches, or retention of specialists or consultants) such additional Project-related data and information. Such additional data and information would generally include the following:

        1.   Property descriptions.

        2.   Zoning, deed, and other land use restrictions.

        3.   Utility information, reports, and mapping.

        4.   Property, boundary, easement, right-of-way, topographic, and other special surveys or data, including establishing relevant reference points.

5. Explorations and tests of subsurface conditions at or adjacent to a Site; geotechnical reports and investigations; drawings of physical conditions relating to existing surface or subsurface structures at a Site; hydrographic surveys, laboratory tests and inspections of samples, materials, and equipment; with appropriate professional interpretation of such information or data.

6. Environmental assessments, audits, investigations, and impact statements, and other relevant environmental, historical, or cultural studies relevant to the Project, the Site(s), and adjacent areas.

7. Data or consultations as required for the Project but not otherwise identified in this Agreement.

D. Advise Engineer of the identity and scope of services of any independent consultants and contractors retained by Owner to perform or furnish services pertinent to the Study and Report Services.

E. Arrange for safe access to and make all provisions for Engineer to enter upon public and private property as required for Engineer to perform services under the Agreement.

F. Inform Engineer in writing of any specific requirements of safety or security programs that are applicable to Engineer, as a visitor to any Site under study.

G. Examine all Documents submitted by Engineer (and obtain the advice of an attorney, risk manager, financial advisor, insurance counselor, or other advisors or consultants as Owner deems appropriate with respect to such examination), and render in writing timely decisions pertaining to such Document submittals.

H. Inform Engineer regarding any need for assistance in evaluating the possible use of Project Strategies, Technologies, and Techniques, as defined in Exhibit A.

I. Furnish (if necessary by retaining qualified specialists or consultants) accounting services; bond and financial advisory services; independent cost estimating; and insurance, risk management, and legal services, as required in support of Engineer's performance of its Study and Report Services.

2.02 Owner shall be responsible for all requirements and instructions that it furnished to Engineer pursuant to this Agreement, and for the accuracy and completeness of all programs, reports, data, and other information furnished by Owner to Engineer pursuant to this Agreement. Engineer may use and rely upon such requirements, programs, instructions, reports, data, and information in performing or furnishing services under this Agreement, subject to any express limitations or reservations applicable to the furnished items.

2.03 Owner shall give prompt written notice to Engineer whenever Owner observes or otherwise becomes aware of:

A. any development that affects the scope or time of performance of Engineer's services;

B. the presence of any Constituent of Concern at any Site; or

C. any relevant, material defect or nonconformance in Engineer's services or Owner's performance of its responsibilities under this Agreement.

ARTICLE 3—SCHEDULE

Guidance Notes—The schedule in Paragraph 3.01 contemplates two completion dates, the first for submittal of the Report (see Exhibit A Paragraph 1.02.A.17), and the second for submittal of a revised Report after receipt of Owner's comments (see Exhibit A Paragraph 1.02.A.18). If there are other interim schedule requirements (milestones) for the Study and Report Services, such as a date or number of days/weeks/months in which Engineer must submit a specified Document (for example, a first draft of the Report), identify such requirements in a clause added to Paragraph 3.01.

3.01 Schedule for Rendering Services

    A.   Engineer shall furnish the Report and any other Study and Report deliverables to Owner within the following specific time period: **[here insert any specific completion date, or the time for submittal of the required Documents in days, weeks, or months from the Effective Date]**:

        1.   If no specific time periods are indicated in Paragraph 3.01.A, Engineer shall complete its Study and Report Services within a reasonable period of time.

    B.   Owner shall review the Documents submitted by Engineer and provide one set of coordinated comments to Engineer within **[number]** days after Owner receives the Documents from Engineer.

    C.   Engineer shall revise the Report and other deliverables and submit such Documents to Owner within **[number]** days of receipt of Owner's comments.

    D.   If, through no fault of Engineer, such periods of time or dates are changed, or the orderly and continuous progress of Engineer's Study and Report Services is impaired, or such services are delayed or suspended, then the time for completion of Engineer's Study and Report Services, and the rates and amounts of Engineer's compensation, will be adjusted equitably.

ARTICLE 4—ENGINEER'S COMPENSATION

4.01 Invoices and Payments

    A.   Invoices—Engineer shall prepare invoices in accordance with its standard invoicing practices and submit the invoices to Owner on a monthly basis. Invoices are due and payable within 30 days of receipt. Engineer shall also comply with the progress reporting and special invoicing requirements (if any) in Exhibit A Paragraph 1.01.A.

    B.   Payment—As compensation for Engineer providing or furnishing Study and Report Services, Owner shall pay Engineer as set forth in this Paragraph 4.01, Invoices and Payments. If Owner disputes an invoice, either as to amount or entitlement, then Owner shall promptly advise Engineer in writing of the specific basis for doing so, may withhold only that portion so disputed, and must pay the undisputed portion.

    C.   Failure to Pay—If Owner fails to make any payment due Engineer for Study and Report Services or expenses within 30 days after receipt of Engineer's invoice, then (1) the amounts due Engineer will be increased at the rate of 1.0% per month (or the maximum rate of interest permitted by law, if less) from said thirtieth day; (2) in addition Engineer may, after giving 7 days' written notice to Owner, suspend the Study and Report Services under this Agreement until Engineer has been paid in full all amounts due for such services, expenses, and other related charges, and in such case Owner waives any and all claims against Engineer for any such suspension; and (3) if any payment due Engineer remains unpaid after 90 days, Engineer may terminate the Agreement for cause pursuant to Paragraph 5.01.B.

D. Reimbursable Expenses—Engineer is entitled to reimbursement of expenses only if so indicated in Paragraph 4.02.A or 4.02.B. If so entitled, and unless expressly specified otherwise, the amounts payable to Engineer for reimbursement of expenses will be the Project-related internal expenses actually incurred or allocated by Engineer, plus all invoiced external expenses allocable to the Project, including Engineer's subcontractor and subconsultant charges, with the external expenses multiplied by a factor of **[specify numeric factor]**.

4.02   Compensation

Notes to User

1. Choose one of the three compensation methods for Basic Services that follow as Paragraphs 4.02.A.1, 2, and 3. Delete the other two options. Retain Paragraph 4.02.B, Additional Services.

2. Revise the provisions regarding reimbursement of Engineer's expenses to reflect the specific agreement regarding entitlement to reimbursement, or to provide specificity regarding which expenses are reimbursable.

3. Paragraph 4.02.B, regarding Additional Services, indicates as a default assumption that the Additional Services will be compensated using standard hourly rates. If direct labor costs or some other method of compensation is preferred, revise accordingly.

A. Basis of Compensation—Basic Services

1. Lump Sum. Owner shall pay Engineer for Basic Services as follows:

    a. A Lump Sum amount of $**[amount]**.

    b. In addition to the Lump Sum amount, reimbursement of the following expenses: **[specify expenses that Owner will reimburse, or indicate "None."]**

    c. The portion of the compensation amount billed monthly for Engineer's Basic Services will be based upon Engineer's estimate of the percentage of the total Basic Services actually completed during the billing period.

[or]

2. Hourly Rates. Owner shall pay Engineer for Basic Services as follows:

    a. An amount equal to the cumulative hours charged to the Project by Engineer's employees times standard hourly rates for each applicable billing class, plus reimbursement of expenses incurred in connection with providing the Basic Services.

    b. Engineer's Standard Hourly Rates are attached as Appendix 1.

    c. The total compensation for Basic Services and reimbursement of expenses is estimated to be $**[estimated amount]**.

[or]

3. Direct Labor Costs Times a Factor. Owner shall pay Engineer for Basic Services as follows:

a. An amount equal to Engineer's Direct Labor Costs times a factor of **[specify numeric factor]** for services provided by Engineer's employees, plus reimbursement of expenses incurred in connection with providing the Basic Services.

b. Direct Labor Costs means salaries and wages paid to employees but does not include payroll-related costs or benefits.

c. The total compensation for Basic Services and reimbursement of expenses is estimated to be $**[estimated amount]**.

[End of Options for Compensation of Basic Services]

B. Additional Services—For authorized Additional Services, Owner shall pay Engineer an amount equal to the cumulative hours charged by Engineer's employees in providing the Additional Services, times standard hourly rates for each applicable billing class; plus reimbursement of expenses incurred in connection with providing the Additional Services. Engineer's standard hourly rates are attached as Appendix 1.

ARTICLE 5—TERMINATION

5.01   Termination for Cause

A. Either party may terminate the Agreement for cause upon 30 days' written notice in the event of substantial failure by the other party to perform in accordance with the terms of the Agreement, through no fault of the terminating party.

1. Notwithstanding the foregoing, this Agreement will not terminate under Paragraph 5.01.A if the party receiving such notice begins, within 7 days of receipt of such notice, to correct its substantial failure to perform and proceeds diligently to cure such failure within no more than 30 days of receipt thereof; provided, however, that if and to the extent such substantial failure cannot be reasonably cured within such 30-day period, and if such party has diligently attempted to cure the same and thereafter continues diligently to cure the same, then the cure period provided for herein will extend up to, but in no case more than, 60 days after the date of receipt of the notice.

B. In addition to its termination rights in Paragraph 5.01.A, Engineer may terminate this Agreement for cause upon 7 days' written notice (a) if Owner demands that Engineer furnish or perform services contrary to Engineer's responsibilities as a licensed professional, (b) if Engineer's services for the Project are delayed or suspended for more than 90 days for reasons beyond Engineer's control, (c) if payment due Engineer remains unpaid for 90 days, as set forth in Paragraph 4.01.C, or (d) as the result of the presence at the Site of undisclosed Constituents of Concern as set forth in Paragraph 6.06.A.

1. Engineer will have no liability to Owner on account of any termination by Engineer for cause.

5.02   Termination for Convenience—Owner may terminate this Agreement for convenience, effective upon Engineer's receipt of notice from Owner.

5.03   Payments Upon Termination

A. In the event of any termination under this Article 5, Engineer will be entitled to invoice Owner and to receive full payment for all services performed or furnished in accordance with this

Agreement, and to reimbursement of expenses incurred through the effective date of termination. Upon making such payment, Owner will have the limited right to the use of all deliverable Documents, whether completed or under preparation, subject to the provisions of Paragraph 6.04, at Owner's sole risk.

B. If Owner has terminated the Agreement for cause and disputes Engineer's entitlement to compensation for services and reimbursement of expenses, then Engineer's entitlement to payment and Owner's rights to the use of the deliverable documents will be resolved in accordance with the dispute resolution provisions of this Agreement or as otherwise agreed in writing.

C. If Owner has terminated the Agreement for convenience, or if Engineer has terminated the Agreement for cause, then Engineer will be entitled, in addition to the payments identified above, to invoice Owner and receive payment of a reasonable amount for services and expenses directly attributable to termination, both before and after the effective date of termination, such as reassignment of personnel, costs of terminating contracts with Engineer's subcontractors or subconsultants, and other related close-out costs, using methods and rates for Additional Services as set forth in Paragraph 4.02.B.

ARTICLE 6—GENERAL CONSIDERATIONS

6.01 Standard of Care

A. The standard of care for all professional engineering and related services performed or furnished by Engineer under this Agreement will be the care and skill ordinarily used by members of the subject profession practicing under similar circumstances at the same time and in the same locality. Engineer makes no warranties, express or implied, under this Agreement or otherwise, in connection with any services performed or furnished by Engineer. Subject to the foregoing standard of care, Engineer may use or rely upon design elements and information ordinarily or customarily furnished by others, including, but not limited to, specialty contractors, manufacturers, suppliers, and the publishers of technical standards.

6.02 Construction Costs; Project Costs

A. Engineer's opinions (if any) of probable construction costs are to be made on the basis of Engineer's experience, qualifications, and general familiarity with the construction industry. However, because of the limited and preliminary nature (1) of the Study and Report Services and (2) of any capital improvements described in any delivered Document, and because Engineer has no control over the cost of labor, materials, equipment, or services furnished by others, or over contractors' methods of determining prices, or over competitive bidding or market conditions, Engineer cannot and does not guarantee that proposals, bids, or actual construction costs will not vary from opinions of probable construction costs prepared by Engineer. If Owner requires greater assurance as to probable construction costs, then Owner agrees to obtain an independent cost estimate.

B. The services, if any, of Engineer with respect to Total Project Costs will be limited to assisting the Owner in tabulating the various categories that comprise Total Project Costs. Engineer assumes no responsibility for the accuracy of any opinions of Total Project Costs.

6.03 Constructors' Work

A. Engineer shall not at any time supervise, direct, control, or have authority over any Constructor's work, nor will Engineer have authority over or be responsible for the means, methods, techniques, sequences, or procedures of construction selected or used by any Constructor, or the safety precautions and programs incident thereto, for security or safety at any Site, nor for any failure of a Constructor to comply with laws and regulations applicable to that Constructor's furnishing and performing of its work. Engineer shall not be responsible for the acts or omissions of any Constructor.

6.04 Documents

A. All Documents prepared or furnished by Engineer are instruments of service, and Engineer retains an ownership and property interest (including the copyright and the right of reuse) in such Documents, whether or not the Project is completed.

B. Owner may make and retain copies of Documents solely for Owner's information and reference in connection with the specific subject matter of the Documents, subject to receipt by Engineer of full payment for all services relating to preparation of the Documents, and subject to the following limitations:

1. Owner acknowledges that such Documents are not intended or represented to be suitable for use by Owner unless completed by Engineer;

2. If Engineer has completed a Report under this Agreement, and received full payment for such Report, then Owner may furnish copies of the completed Report to Owner's consultants and design professionals for their reference in proceeding with design or similar services, provided that Owner informs such consultants and design professionals of Engineer's ownership interests in the Report, and includes with the Report all Engineer's written statements regarding the purpose, scope, use, and limitations of the Report;

3. Owner acknowledges that the Documents are not design or construction documents;

4. No Document shall be altered, modified, or reused by Owner or any third party for any purpose except with Engineer's express written consent;

5. Any use, reuse, alteration, or modification of the Documents, except as authorized in this Agreement or by Engineer's written consent, will be at Owner's sole risk and without liability or legal exposure to Engineer or to its officers, directors, members, partners, agents, employees, subcontractors, and subconsultants;

6. Owner shall indemnify and hold harmless Engineer and its officers, directors, members, partners, agents, employees, subcontractors, and subconsultants from all claims, damages, losses, and expenses, including attorneys' fees, arising out of or resulting from any unauthorized use, reuse, alteration, or modification of the Documents; and

7. Nothing in this paragraph shall create any rights in third parties.

C. Owner and Engineer agree to transmit, and accept, the Documents and all other Project-related correspondence, text, data, drawings, documents, information, and graphics, in electronic media or digital format, either directly, or through access to a secure Project website, in accordance with a mutually agreeable protocol.

6.05   Waiver of Damages

A.   To the fullest extent permitted by law, Owner and Engineer waive against each other, and the other's officers, directors, members, partners, agents, employees, subcontractors, subconsultants, and insurers, any and all claims for or entitlement to special, incidental, indirect, or consequential damages arising out of, resulting from, or in any way related to this Agreement or the Project, from any cause or causes. In addition:

1.   Limitation of Liability. Owner and Engineer agree that Engineer's total liability to Owner under this Agreement shall be limited to $100,000 or the total amount of compensation received by Engineer, whichever is greater.

6.06   General Provisions

A.   Constituents of Concern—The parties acknowledge that Engineer's Study and Report Services do not include any services related to unknown or undisclosed Constituents of Concern. If Engineer or any other party encounters, uncovers, or reveals an unknown or undisclosed Constituent of Concern, then Engineer may, at its option and without liability for consequential or any other damages, suspend performance of the Study and Report Services on the portion of the Project affected thereby until such portion of the Project is no longer affected, or terminate this Agreement for cause if it is not practical to continue providing Study and Report Services.

B.   Dispute Resolution—Owner and Engineer agree to negotiate each dispute between them in good faith during the 30 days after notice of dispute. If negotiations are unsuccessful in resolving the dispute, then the dispute will be mediated. If mediation is unsuccessful, then the parties may exercise their rights at law.

Notes to User: If necessary, modify Paragraph 6.06.C below to identify a specific controlling jurisdiction if other than the state where the Project is located; if multiple states are involved; or to identify controlling jurisdictions other than a state, such as a U.S. territory, commonwealth, or tribal jurisdiction/domestic dependent nation.

C.   Governing Law—This Agreement is to be governed by the laws of the state in which the Project is located.

D.   Exclusions from Services—Engineer's Study and Report Services do not include: (1) serving as a "municipal advisor" for purposes of the registration requirements of Section 975 of the Dodd-Frank Wall Street Reform and Consumer Protection Act (2010) or the municipal advisor registration rules issued by the Securities and Exchange Commission; (2) advising Owner, or any municipal entity or other person or entity, regarding municipal financial products or the issuance of municipal securities, including advice with respect to the structure, timing, terms, or other similar matters concerning such products or issuances; (3) providing surety bonding or insurance-related advice, recommendations, counseling, or research, or enforcement of construction insurance or surety bonding requirements; or (4) providing legal advice or representation.

Notes to User: The following paragraph establishes basic insurance requirements. If more detailed insurance requirements, such as specific coverage limits, will be part of the Agreement, either state them here or include them in an insurance exhibit. (For reference, see EJCDC® E-500, Agreement between

[Owner and Engineer for Professional Services, Exhibit G, Insurance). If an insurance exhibit is included, identify the exhibit in Paragraph 8.02.]

E.   Insurance—Engineer will maintain insurance coverage for Workers' Compensation, General Liability, Professional Liability, and Automobile Liability, and will provide certificates of insurance to Owner upon request.

F.   Successors and Assigns

1.   Owner and Engineer are hereby bound and the successors, executors, administrators, and legal representatives of Owner and Engineer (and to the extent permitted by Paragraph 6.06.F.2 the assigns of Owner and Engineer) are hereby bound to the other party to this Agreement and to the successors, executors, administrators, and legal representatives (and said assigns) of such other party, in respect of all covenants, agreements, and obligations of this Agreement.

2.   Neither Owner nor Engineer may assign, sublet, or transfer any rights under or interest (including, but without limitation, money that is due or may become due) in this Agreement without the written consent of the other party, except to the extent that any assignment, subletting, or transfer is mandated by law. Unless specifically stated to the contrary in any written consent to an assignment, no assignment will release or discharge the assignor from any duty or responsibility under this Agreement.

G.   Beneficiaries—Unless expressly provided otherwise, nothing in this Agreement shall be construed to create, impose, or give rise to any duty owed by Owner or Engineer to any Constructor, other third-party individual or entity, or to any surety for or employee of any of them. All duties and responsibilities undertaken pursuant to this Agreement will be for the sole and exclusive benefit of Owner and Engineer and not for the benefit of any other party.

ARTICLE 7—DEFINITIONS

7.01   Definitions Used in this Agreement

A.   Constructor—Any person or entity (not including the Engineer, its employees, agents, representatives, subcontractors, or subconsultants), performing or supporting construction activities relating to the Project, including but not limited to contractors, subcontractors, suppliers, Owner's work forces, utility companies, construction managers, testing firms, shippers, and truckers, and the employees, agents, and representatives of any or all of them.

B.   Constituent of Concern—Asbestos, petroleum, radioactive material, polychlorinated biphenyls (PCBs), lead based paint (as defined by the HUD/EPA standard), hazardous waste, and any substance, product, waste, or other material of any nature whatsoever that is or becomes listed, regulated, or addressed pursuant to laws and regulations regulating, relating to, or imposing liability or standards of conduct concerning, any hazardous, toxic, or dangerous waste, substance, or material.

C.   Documents—All documents expressly identified as deliverables in this Agreement, whether in printed or electronic form, required by this Agreement to be provided or furnished by Engineer to Owner. Such specifically required deliverables may include, by way of example, data, studies, models, and reports (including the Report referred to in Exhibit A).

D.   Site—One or more lands or areas that Engineer studies as the location or possible location of the Project.

E.   Total Project Costs—The total cost of planning, studying, designing, constructing, testing, commissioning, and start-up of the Project, including construction costs and all other Project labor, services, materials, equipment, insurance, and bonding costs, allowances for contingencies, and the total costs of services of Engineer or other design professionals and consultants, together with such other Project-related costs that Owner furnishes for inclusion, including but not limited to cost of land, rights-of-way, compensation for damages to properties and private utilities (including relocation if not part of construction costs), Owner's costs for legal, accounting, insurance counseling, and auditing services, interest and financing charges incurred in connection with the Project, and the cost of other services to be provided by others to Owner.

F.   Underground Facilities—All active or not-in-service underground lines, pipelines, conduits, ducts, encasements, cables, wires, manholes, vaults, tanks, tunnels, or other such facilities or systems at a Site, including but not limited to those facilities or systems that produce, transmit, distribute, or convey telephone or other communications, cable television, fiber optic transmissions, power, electricity, light, heat, gases, oil, crude oil products, liquid petroleum products, water, steam, waste, wastewater, storm water, other liquids or chemicals, or traffic or other control systems. An abandoned facility or system is not an Underground Facility.

ARTICLE 8—AGREEMENT, EXHIBITS, ATTACHMENTS

8.01   Total Agreement

A.   This Agreement (including any expressly incorporated attachments), constitutes the entire agreement between Owner and Engineer and supersedes all prior written or oral understandings. This Agreement may only be amended, supplemented, modified, or canceled by a duly executed written instrument.

8.02   Attachments:

A.   Exhibit A, Scope of Engineer's Study and Report Services

B.   Appendix 1, Engineer's Standard Hourly Rates

Notes to User

1.   Always include Exhibit A as part of the Agreement. Modify Exhibit A to establish the specific scope of services that Engineer will provide.

2.   Exclude Appendix 1, Engineer's Standard Hourly Rates, only if hourly rates are not to be used as the basis for either compensation for Basic Services (Paragraph 4.02.A) or for Additional Services (Paragraph 4.02.B).

3.   Itemize any other attachments that will be part of the Agreement, such as a reimbursable expenses schedule.

This Agreement's Effective Date is **[insert date]**.

Owner:

_____
(name of organization)

By: _____
(authorized individual's signature)

Date: _____
(date signed)

Name: _____
(typed or printed)

Title: _____
(typed or printed)

Address for giving notices:

_____

_____

Designated Representative:

Name: _____
(typed or printed)

Title: _____
(typed or printed)

Address:

_____

_____

Phone: _____

Email: _____

Engineer:

_____
(name of organization)

By: _____
(authorized individual's signature)

Date: _____
(date signed)

Name: _____
(typed or printed)

Title: _____
(typed or printed)

Address for giving notices:

_____

_____

Designated Representative:

Name: _____
(typed or printed)

Title: _____
(typed or printed)

Address:

_____

_____

Phone: _____

Email: _____

EXHIBIT A, SCOPE OF ENGINEER'S STUDY AND REPORT SERVICES

Paragraph 1.01 of the Agreement, Study and Report Services of Engineer, is supplemented to include the following provisions:

Guidance Notes—Baseline Information

1. Inherent in establishing the scope of Engineer's Study and Report Services is clearly defining the nature of the facility or facilities to be studied and planned, to the extent known at the time this Agreement is drafted. Providing such information creates a mutual understanding by the parties as to the services that will be needed, as well as the level of effort and time (and thus cost) required to deliver those services. To accurately establish the scope of services, it is almost always in the best interests of both the Owner and the Engineer when the Owner provides as much detailed baseline information regarding the facilities as is reasonably available.

2. If known, the baseline information regarding the physical improvements to be studied or planned should include important controlling features, assumptions, and limitations that may affect Engineer's services. The information may be expressly stated below or by reference to external sources, such as requests for proposal or actual proposals, both of which often contain project descriptions.

Baseline Information: Owner has furnished the following Project information to Engineer as of the Effective Date. Engineer's scope of services has been developed based on this information. As the Project moves forward, some of the information may change or be refined, and additional information will become known, resulting in the possible need to change, refine, or supplement the scope of services.

Notes to User

1. Here insert relevant information about the Project, either directly or by reference, to the extent known and necessary to give context to the Engineer's Study and Report Services that are specified in the remainder of Exhibit A.

2. Below are some suggested or sample categories of information that may have been furnished to Engineer as baseline information about the Project. Other categories may be relevant to the specific engagement.

3. Note that the term "Project" is broadly defined; in most cases it will be sufficient here to describe information relevant to those portions of the Project within Engineer's scope of services.

Project Title:
Type and Size of Facility:
Description of Anticipated
Improvements:
Expected Construction Start:
Prior Studies, Reports, Plans:
Facility Location(s):
Owner's Current Project
Budget:
Funding Sources:
Known Design Standards:

Known Project Limitations:
Project Assumptions:
Other Pertinent Information:

Engineer shall perform or furnish Basic and Additional Services as set forth below.

ARTICLE 1—BASIC SERVICES

Guidance Notes—Management of Engineering Services

1. Proper management of the Engineer's Study and Report Services includes certain administrative, scheduling, personnel coordination, reporting, and other tasks, directly and specifically related to the specific scope of services. These project management tasks are necessary to effectively and successfully complete the required professional and technical services that are more commonly the focus of attention when drafting a Project's engineering scope of services.

2. The following Exhibit A Paragraph 1.01 lists typical Project management tasks that will almost always apply, in some form, to the Engineer's Study and Report Services, whether separately budgeted and tracked to the contract, or simply performed as part of the required professional and technical services.

3. The listed management services are intended to include only those that are directly and solely applicable to the particular Project-specific scope of services, and should not include general business management ("overhead") tasks that apply to or are performed as part of a core business function (finance, accounting, human resources, legal, corporate, facilities, etc.) or to those tasks that are performed in the course of Engineer's management of multiple projects. Some owners and agencies have specific rules and guidance defining what they will allow to be treated as billable project management, and the following scope should be modified for compliance with such applicable rules and guidance.

1.01   Management of Study and Report Services

    A.   Engineer's services will include management of Engineer's Project-specific responsibilities, including but not limited to the following management tasks, whether separately tracked and itemized or included as being incidental to other phase and scope task items.

        1.   Develop and submit an Engineering Services Schedule. The Engineering Services Schedule will:

            a.   be updated on a regular basis, and as required to reflect any programmatic decisions by Owner.

            b.   include, but not be limited to, an anticipated sequence of tasks; estimates of task duration; interrelationships among tasks; milestone meetings and submittals; anticipated schedule of construction; and other pertinent Project events.

        2.   Coordinate services within Engineer's internal team, including Engineer's subcontractors and subconsultants.

Notes to User

1. Paragraph 4.01.A of the Agreement states that Engineer shall prepare invoices in accordance with its standard invoicing practices, and with the progress reporting and special invoicing requirements (if any) in Exhibit A Paragraph 1.01.A. Paragraphs Exhibit A 1.01.A.3 and 4 immediately below are the locations for the specific follow-up to Paragraph 4.01.A.

2. The progress reporting addressed in Exhibit A Paragraph 1.01.A.3 is typically submitted in conjunction with Engineer's periodic invoices. Reporting and invoicing obligations will usually be coordinated.

3. If Owner has progress reporting or invoicing requirements that differ from Engineer's standard and routine reporting and invoicing procedures, modify the requirements in Exhibit A Paragraphs 1.01.A.3 and 4, as appropriate.

> 3. Prepare and submit **[monthly] [other frequency]** engineering services progress reports to Owner. Include summary of services performed in period, expected progress in next period, percent completion of current tasks, and a description of major issues or concerns.
>
> 4. Special Invoicing: In addition to, or as a substitute for, Engineer's standard invoicing, provide the specified additional information or documentation, following the invoicing procedures indicated: **[indicate Not Applicable; or specify required information, documentation, or special invoicing procedures]**.
>
> 5. Establish Project-specific security and health and safety plans (as deemed necessary by Engineer), consistent with Owner's programs and procedures of which Engineer has been made aware in writing. Distribute security and health and safety plans to Engineer's team, and monitor compliance.
>
> 6. Conduct ongoing management tasks, including:
>
>   a. Maintaining communications records and files pertaining to or arising from Engineer's Study and Report Services;
>
>   b. With respect to Engineer's services and other directly relevant parts of the Project, prepare for and participate in periodic progress meetings with Owner to discuss progress, schedule, budget, issues, potential problems and their resolution; and
>
>   c. Preparing agendas prior to and minutes following all Engineer-led meetings.

B. In all phases of Engineer's services, Engineer shall prepare draft and final drawings and other Documents in graphic form in accordance with **[Engineer's CAD standards] [Owner's CAD standards] [other CAD standard (specify)]**, using **[___]** version **[___]** software.

1.02 Study and Report Phase

A. Engineer shall:

> 1. Consult with Owner to define and clarify Owner's requirements for the Project, including design objectives and constraints; space, capacity, and performance requirements; flexibility and expandability; and any budgetary limitations. Identify available data, information, reports, facilities plans, and site evaluations.

a. If Owner has already identified one or more potential solutions to meet its Project requirements, then proceed with the study and evaluation of the Owner-identified potential solutions listed here:

   1)   **[List the specific potential solutions to be studied and evaluated here]**.

b. If Owner has not identified specific potential solutions for study and evaluation, then assist Owner in determining whether Owner's requirements, and available data, reports, plans, and evaluations, point to a single potential solution for Engineer's study and evaluation, or are such that it will be necessary for Engineer to identify, study, and evaluate multiple potential solutions.

c. If it is necessary for Engineer to identify, study, and evaluate multiple potential solutions, then identify **[insert specific number]** alternative solutions potentially available to Owner, unless Owner and Engineer mutually agree that some other specific number of alternatives should be identified, studied, and evaluated.

2. Identify potential solution(s) to meet Owner's Project requirements, as needed.

3. Study and evaluate the potential solution(s) to meet Owner's Project requirements.

4. Visit the Site, or potential Project Sites, to review existing conditions and facilities, unless such visits are not necessary or applicable to meeting the objectives of the Study and Report Phase.

5. Assess initially available Project information and data, including the Baseline Information set forth at the beginning of this Exhibit A.

6. Advise Owner of any need for Owner to obtain, furnish, or otherwise make available to Engineer additional Project-related information and data, for Engineer's use in the study and evaluation of potential solution(s) to Owner's Project requirements, and preparation of a related report.

7. After consultation with Owner, recommend the solution(s) which in Engineer's judgment meet Owner's requirements for the Project.

8. Identify, consult with, and analyze requirements of authorities having jurisdiction to permit or approve construction or operation of the portions of the Project under study, including but not limited to impacts and mitigating measures identified in previously prepared environmental assessments for the Project provided to the Engineer or being concurrently prepared for Owner by others.

9. Advise the Owner of any need for Owner to provide data or services of the types described in Paragraph 2.01 of the Agreement.

10. Assist Owner in evaluating the possible use of building information modeling; civil integrated management; geotechnical baselining of subsurface conditions at the Site; innovative design, contracting, or procurement strategies; project delivery method; or other strategies, technologies, or techniques for assisting in the design, construction, and operation of Owner's facilities. The subject matter of this paragraph will be referred to as "Project Strategies, Technologies, and Techniques."

11. Assist Owner in identifying opportunities for enhancing the sustainability of the Project, and pursuant to Owner's instructions, plan for the inclusion of sustainable features in the design.

12. Review with Owner the thresholds established in applicable codes, standards, and design criteria specifically governing the ability of the proposed facilities or improvements to perform, and to absorb or avoid damage without suffering complete or substantial failure. As part of the review, identify additional risk assessment studies or tools that are available to evaluate the susceptibility of the facilities or improvements to natural and man-made events beyond the applicable established thresholds. Upon Owner request, as an additional service, perform additional risk assessment studies or tools to further evaluate system resiliency beyond the applicable established thresholds.

13. Utilities, including Underground Facilities

   a. Review any utility mapping and surveys and other utilities documentation made available by Owner. Take note of observable utilities during Site visit.

   b. Identify, in a preliminary manner and to the extent determinable by such mapping or other information provided by Owner, and by observations at the Site, those utilities (whether above-ground utilities of any type, or Underground Facilities) likely to be affected by the Project construction and additional utility facilities or extensions that will be needed to serve the Project.

   c. If the impact on existing utilities or the need for additional utility facilities or extensions cannot reasonably be determined in a preliminary manner from mapping or other information provided by Owner, or such information was not available from Owner, then assist Owner in evaluating the need to either obtain additional utility mapping and utility documentation during the Study and Report Phase, or undertake other alternative approaches and contingencies to account for utility uncertainties in this phase.

   d. Advise Owner of additional utility documentation and coordination needed during the design and construction phases to adequately assess, mitigate, and manage the impact of the Project (including any additional utility facilities or extensions needed to serve the Project) on existing utilities.

   e. Use ASCE 38, "Standard Guideline for the Collection and Depiction of Existing Subsurface Utility Data" as a means to advise the Owner regarding the extent and identification and mapping of existing Underground Facilities during the design and construction phases.

      1) If Owner has retained a land surveyor, utility engineer, or utility consultant, collaborate with such individuals or entities regarding the application of ASCE 38.

14. Inquire regarding survey methodologies and technologies that would aid in addressing Owner's Project requirements. Develop a scope of work and survey limits for any topographic and other surveys necessary for design. For recommended survey deliverables, specify a) required technical specifications; b) pertinent datum; c) survey

limits, and d) formats of deliverables. Collaborate with land surveyor, when separately retained by Owner or third party, to develop such scope of work.

15. Prepare a report (the "Report") which will, as appropriate, contain schematic layouts, sketches, and conceptual design criteria with appropriate exhibits to indicate the agreed-to requirements, considerations involved, and Engineer's recommended solution(s).

    a. For each recommended solution, Engineer will separately tabulate Total Project Cost, itemizing those items and services included within the definition of Total Project Costs.

    b. Engineer will meet with Owner to discuss the draft Report and receive Owner's comments.

16. Perform or provide the following other Study and Report Phase tasks or deliverables:

    a. **[List any such tasks or other deliverables here].**

17. Furnish the Report and any other Study and Report Phase deliverables to Owner pursuant to the requirements of the schedule in Paragraph 3.01 of the Agreement, and review the deliverables with Owner.

18. Revise the Report and any other Study and Report Phase deliverables in response to Owner's comments, as appropriate, and submit revised deliverables pursuant to the schedule in Paragraph 3.01.

B. Engineer's services under the Study and Report Phase will be considered complete on the date when Engineer has delivered to Owner the final Report (as revised) and any other Study and Report Phase deliverables.

ARTICLE 2—ADDITIONAL SERVICES

2.01 Additional Services

A. If authorized in writing by Owner, Engineer shall furnish or obtain from others Additional Services of the types listed below.

B. Site-related Additional Services

    1. Services to make measured drawings of existing conditions or facilities, to conduct tests or investigations of existing conditions or facilities, or to verify the accuracy of drawings or other information furnished by Owner or others.

    2. Provide necessary field surveys and topographic and utility mapping to be used for study and design purposes.

    3. Provide additional mapping and documentation services relating to utilities, including Underground Facilities, arising from Engineer's services under Exhibit A Paragraph 1.02.A.13 above, based when applicable on the guidance in ASCE 38, "Standard Guideline for the Collection and Depiction of Existing Subsurface Utility Data."

4. Preparation of environmental assessments and impact statements; and assistance in obtaining approvals of authorities having jurisdiction over the anticipated environmental impact of the Project.

5. Providing assistance in responding to or investigating the presence of any Constituent of Concern at any Site, in compliance with current Laws and Regulations.

C. Additional Study, Investigation, and Report Services

1. Undertaking investigations and studies including, but not limited to:

    a. detailed consideration of operations, maintenance, and overhead expenses;

    b. the preparation of feasibility studies (such as those that include projections of output capacity, utility project rates, project market demand, or project revenues) and cash flow analyses, provided that such services are based on the engineering and technical aspects of the Project, and do not include rendering advice regarding municipal financial products or the issuance of municipal securities;

    c. preparation of appraisals;

    d. evaluating processes available for licensing, and assisting Owner in obtaining process licensing;

    e. detailed quantity surveys of materials, equipment, and labor; and

    f. audits or inventories required in connection with construction performed or furnished by Owner.

2. Services resulting from significant changes in the scope, extent, or character of the Project including, but not limited to, changes in size, complexity, Owner's schedule, character of construction, or method of financing; and revising the Report or other deliverables when such revisions are required by changes in Laws and Regulations enacted subsequent to the Effective Date or are due to any other causes beyond Engineer's control.

3. Services resulting from Owner's request to evaluate additional potential solutions.

D. Additional Services—General

1. Attendance, presentations, and support at public meetings and hearings, including preparation for such meetings and hearings, and drafting related exhibits and handouts.

2. Preparation of applications and supporting documents for private or governmental grants, loans, or advances in connection with the Project.

3. Services required as a result of Owner providing incomplete or incorrect information to Engineer.

4. Preparing for, coordinating with, participating in and responding to structured independent review processes, including, but not limited to, construction management, cost estimating, project peer review, value engineering, and constructability review requested by Owner; and performing or furnishing services required to revise studies, reports, or other documents as a result of such review processes.

5. Providing renderings or models for Owner's use, including services in support of building information modeling or civil integrated management.

6. Services during out-of-town travel required of Engineer, other than for visits to the Site or Owner's office as required in Basic Services (Article 1 of Exhibit A).

7. Preparing to serve or serving as a consultant or witness for Owner in any litigation, arbitration, lien or bond claim, or other legal or administrative proceeding involving the Project.

This is **Appendix 1, Engineer's Standard Hourly Rates**, referred to in and part of the Agreement between Owner and Engineer for Study and Report Professional Services dated **[date]**.

ENGINEER'S STANDARD HOURLY RATES

Notes to User: The categories below (Billing Classes VIII through I) are traditional hourly rate classes for engineering services, but the classes themselves do not currently have widely accepted or understood meanings or definitions. Many approaches are possible for establishing the hourly rates that will be charged. These include defining the categories (for example, "Billing Class VI—Assistant Project Manager"), or using the engineering firm's own professional classifications. If hourly rates are ascribed to specific individuals, the user should ensure that changes in professional personnel and rates are allowable over the Project's course.

A.   Standard Hourly Rates:

   1.   Standard Hourly Rates are set forth in this Appendix 1 and include salaries and wages paid to personnel in each billing class plus the cost of customary and statutory benefits, general and administrative overhead, non-project operating costs, and operating margin or profit.

   2.   The Standard Hourly Rates apply only as specified in Paragraph 4.01 of the Agreement, and are subject to annual review and adjustment.

B.   Schedule of Hourly Rates:

Billing Class	Rate
Billing Class VIII	$ **[enter rate]**/hour
Billing Class VII	$ **[enter rate]**/hour
Billing Class VI	$ **[enter rate]**/hour
Billing Class V	$ **[enter rate]**/hour
Billing Class IV	$ **[enter rate]**/hour
Billing Class III	$ **[enter rate]**/hour
Billing Class II	$ **[enter rate]**/hour
Billing Class I	$ **[enter rate]**/hour

Design and Construction Documents Mosquito Bridge Project

COUNTY OF EL DORADO
DEPARTMENT OF
TRANSPORTATION

Mosquito Road Bridge
At the South Fork of the
American River

**FINAL ENVIRONMENTAL
IMPACT REPORT
(FEIR)**

Board of Supervisors Hearing
August 8, 2017

In Association With:

QUINCY ENGINEERING   ICF INTERNATIONAL

AGENDA
- Project Background
- Project Alternatives
- Proposed Project
- EIR Highlights
- Public Comments
- Conclusion

Project Background
- Highway Bridge Program (HBP)
- Sufficiency Rating = 13.3 (2016), 12.5 (2011), 0.0 (2006) (scale from 0 to 100)
- Extensive Maintenance & Repairs (1990, 2010)
- Most Expensive EDC Bridge to Maintain (approx. $75k/ yr)
- Bridge Replacement Project
- Off System Bridge (Federally Funded, Match with Toll Credits)

Peg Presba

"Swinging Bridge" – History Museum

Events to Date

- 1993 Study
- Programmed in HBP and CIP
- Public Workshop No. 1 – Jan 26, 2013
- Public Workshop No. 2 – Nov 15, 2014
- Alternative Screening 2014/15
- Met w/ Caltrans District 03 – March 12, 2015
- Met w/ HQ and FHWA – June 4, 2015
- Public Workshop No. 3 – July 15, 2015
- Meetings with Contractors – 2015/16
- Meetings with Tribes – 2015/16
- Technical & Environmental Studies - 2016
- Met w/ HQ and FHWA – May 4, 2016
- Value Analysis (VA) – Aug 17–26, 2016
- River Access Feasibility Study –
  Aug 16, 2016
- Draft EIR – Oct 2016
- Public Hearing (WS No. 4) –
  Oct 26, 2016
- Final EIR – Aug 2017

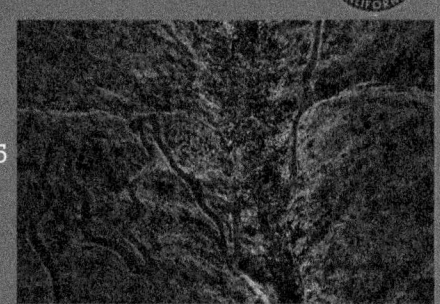

Project Background
PA&ED Schedule

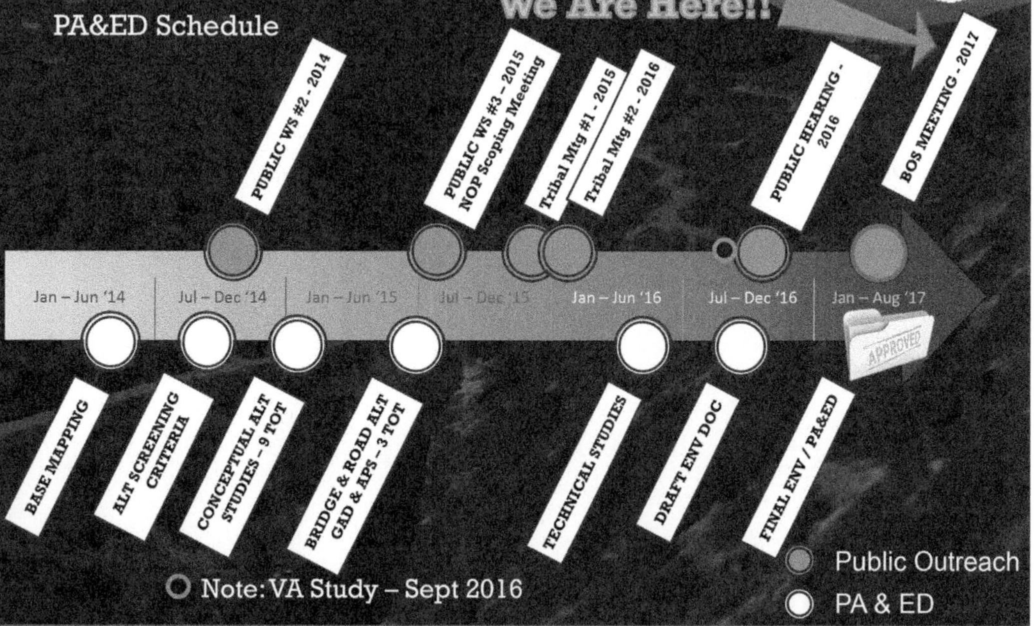

We Are Here!!

PUBLIC WS #2 - 2014

PUBLIC WS #3 - 2015
NOP Scoping Meeting

Tribal Mtg #1 - 2015

Tribal Mtg #2 - 2016

PUBLIC HEARING - 2016

BOS MEETING - 2017

| Jan – Jun '14 | Jul – Dec '14 | Jan – Jun '15 | Jul – Dec '15 | Jan – Jun '16 | Jul – Dec '16 | Jan – Aug '17 |

APPROVED

BASE MAPPING

ALT SCREENING CRITERIA

CONCEPTUAL ALT STUDIES – 9 TOT

BRIDGE & ROAD ALT GAD & APS – 3 TOT

TECHNICAL STUDIES

DRAFT ENV DOC

FINAL ENV / PA&ED

○ Note: VA Study – Sept 2016

● Public Outreach
○ PA & ED

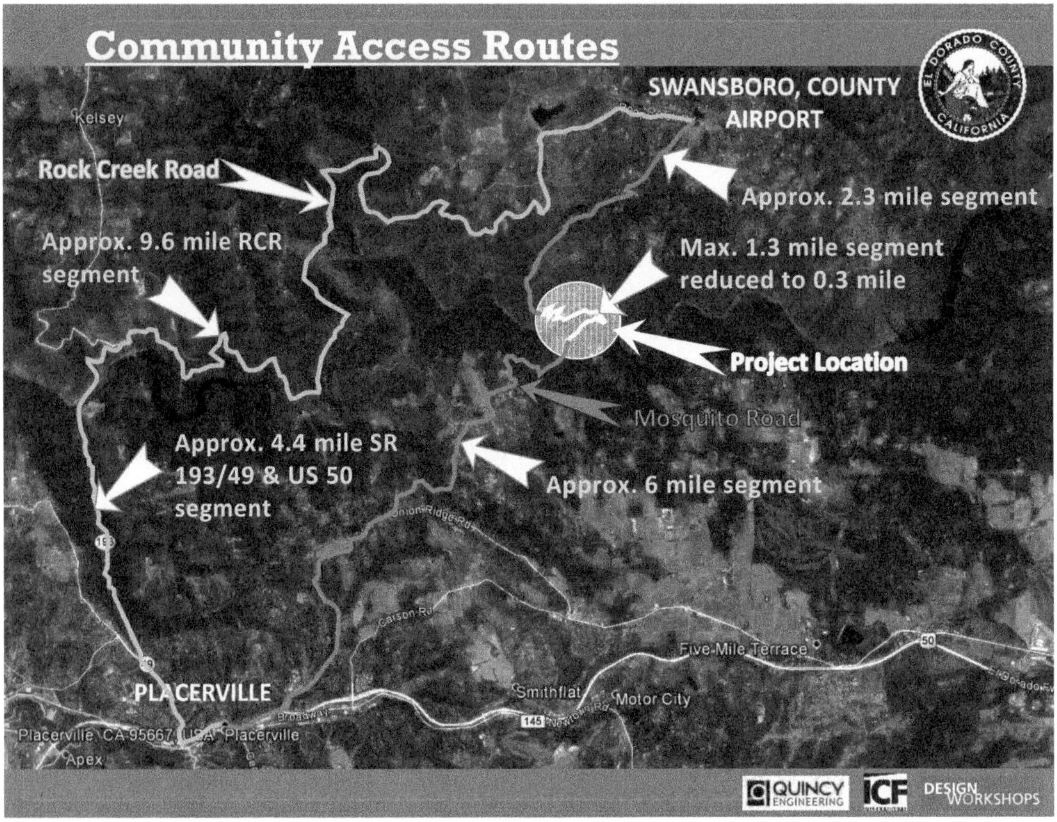
Community Access Routes
SWANSBORO, COUNTY AIRPORT
Kelsey
Rock Creek Road
Approx. 2.3 mile segment
Approx. 9.6 mile RCR segment
Max. 1.3 mile segment reduced to 0.3 mile
Project Location
Mosquito Road
Approx. 4.4 mile SR 193/49 & US 50 segment
Approx. 6 mile segment
Onion Ridge Rd
Carson Rd
Five Mile Terrace
PLACERVILLE
Smithflat
Motor City
Broadway
Placerville, CA 95667, USA Placerville
Apex
El Dorado Fwy
QUINCY ENGINEERING
ICF
DESIGN WORKSHOPS

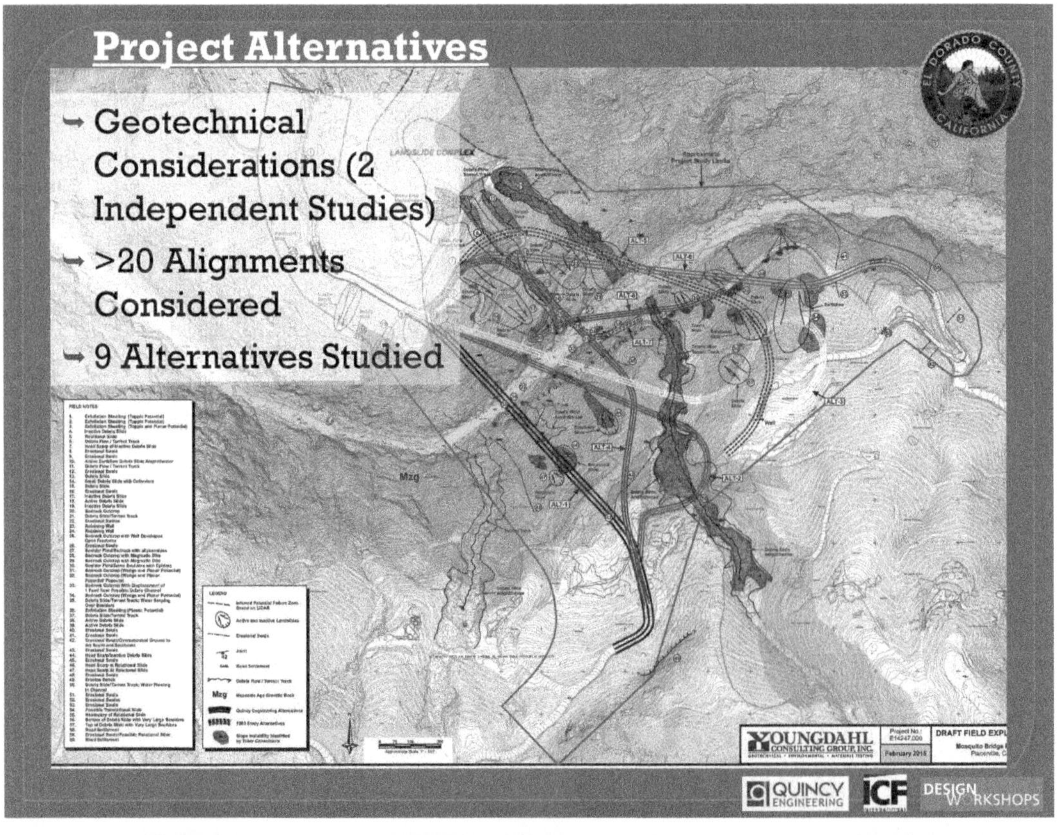
Project Alternatives
Geotechnical Considerations (2 Independent Studies)
>20 Alignments Considered
9 Alternatives Studied
LANDSLIDE COMPLEX
FIELD NOTES
LEGEND
Mzg
YOUNGDAHL CONSULTING GROUP, INC.
February 2016
DRAFT FIELD EXPL
Mosquito Bridge
Placerville, C
QUINCY ENGINEERING
ICF
DESIGN WORKSHOPS

Project Alternatives

- High-, Mid-, & Low-Level Corridors
- Alts 1, 6 & 8 Further Studied
  - Alt 1 Preferred by Public
  - Alt 1 Favored by Contractors
  - Alt 1 Supported by VA

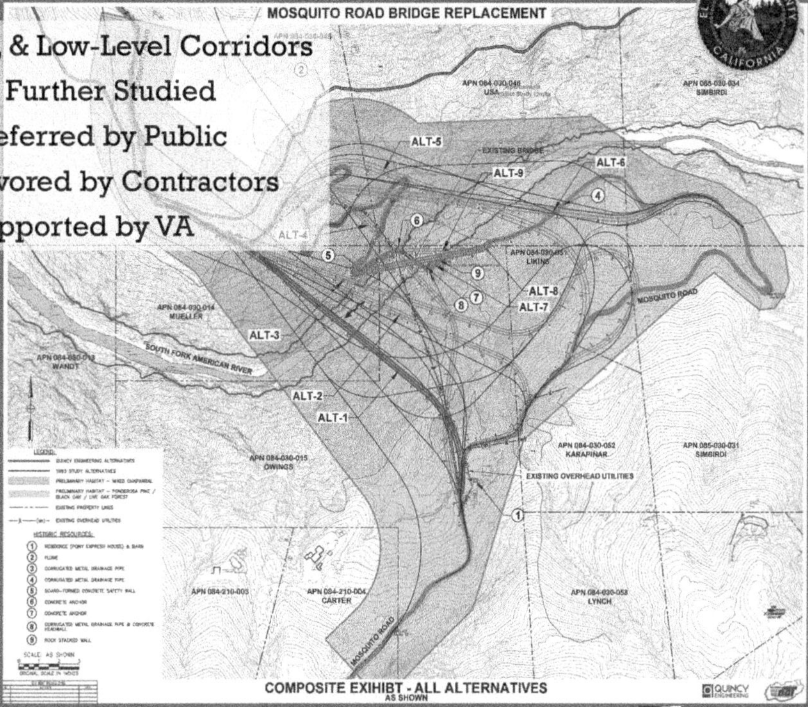

QUINCY ENGINEERING | ICF | DESIGN WORKSHOPS

Project Alternatives

Evaluation in EIR

- No Project
- High-Level Alternative (Proposed Project)
- Mid-Level Alternative
- Low-Level Alternative

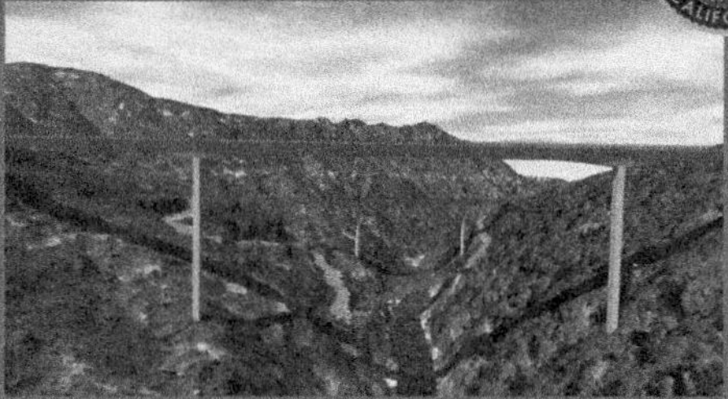

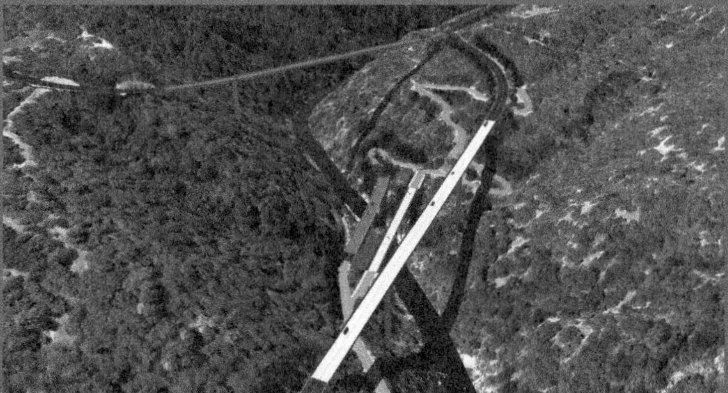

QUINCY ENGINEERING | ICF | DESIGN WORKSHOPS

Proposed Project

High-Level Alternative

- Overview

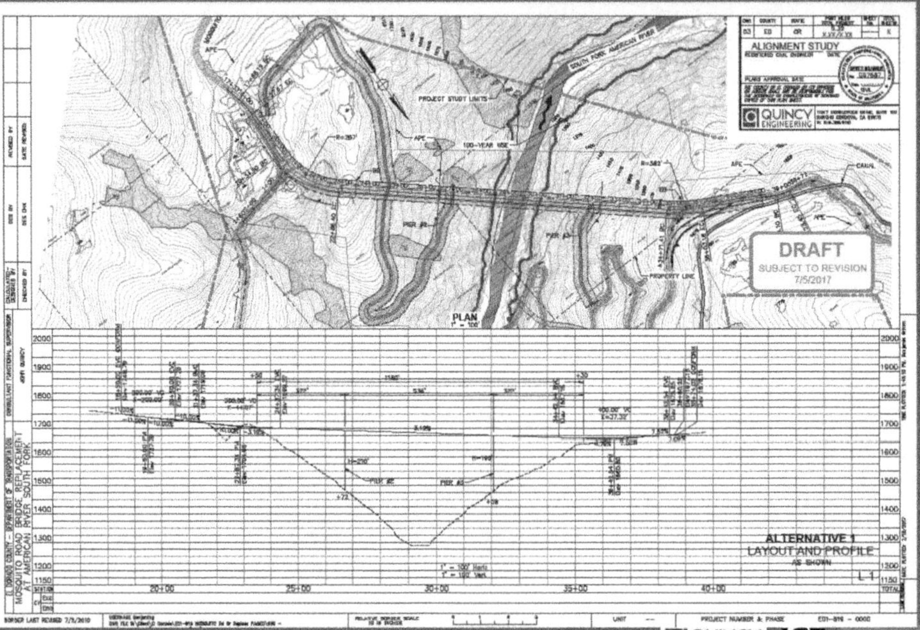

Proposed Project

High-Level Alternative

- Value Analysis (VA)

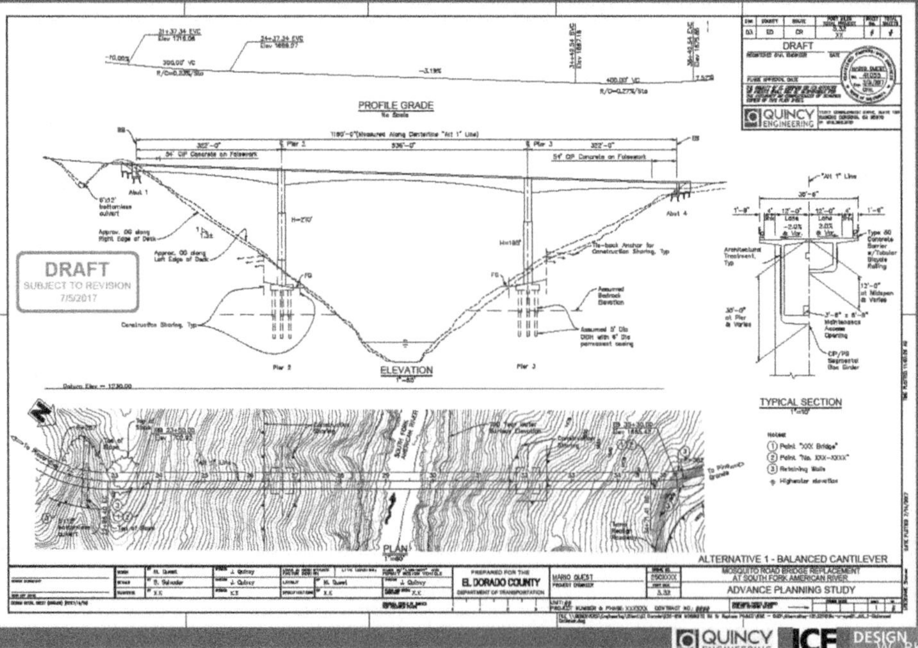

Proposed Project
High-Level Alternative
- Impact Areas

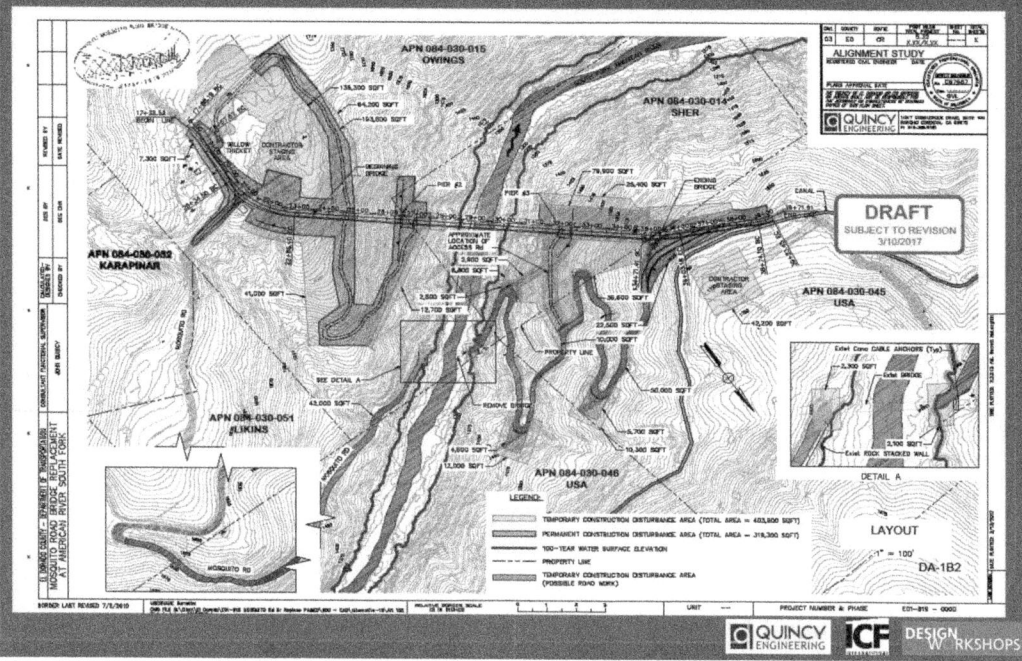

EIR Highlights

- Draft EIR
  - Notice of Availability (NOA) & Public Meeting – Oct. 26, 2016

- Existing Mosquito Road Bridge & Roadway
  - River Access Feasibility Study
  - Remains in Place (Bike & Pedestrian Access across Bridge)

EIR Highlights
Summary of Impacts and Mitigation Measures

Significance Criteria and Significant Impact Summary	Mitigation Measures	Significance after Mitigation
Biological Resources		
Impact BIO-1 • Potential for construction activities to result in the mortality or disturbance of foothill yellow-legged frog, Blainville's horned lizard, nesting bald eagles, nesting California spotted owls, nesting migratory birds, special-status bats and their habitat • Potential for construction activities to result in the loss of willow flycatcher foraging habitat	Mitigation Measure BIO-1 Mitigation Measure BIO-2 Mitigation Measure BIO-3 Mitigation Measure BIO-4 Mitigation Measure BIO-5 Mitigation Measure BIO-6 Mitigation Measure BIO-7 Mitigation Measure BIO-8 Mitigation Measure BIO-9 Mitigation Measure BIO-10	Less than significant
Impact BIO-2 • Permanent and temporary impacts on interior live oak woodland • Permanent and temporary impacts on willow thicket wetland	Mitigation Measure BIO-1 Mitigation Measure BIO-2 Mitigation Measure BIO-3 Mitigation Measure BIO-4 Mitigation Measure BIO-6 Mitigation Measure BIO-7 Mitigation Measure BIO-11	Less than significant
Impact BIO-3 • Potential temporary and indirect effects on intermittent stream	Mitigation Measure BIO-1 Mitigation Measure BIO-2 Mitigation Measure BIO-3 Mitigation Measure BIO-4 Mitigation Measure BIO-6 Mitigation Measure BIO-12	Less than significant
Impact BIO-7 • Potential to create additional disturbed areas for a temporary period and to introduce and spread invasive plant species to uninfected areas within and adjacent to the Project area	Mitigation Measure BIO-1 Mitigation Measure BIO-2 Mitigation Measure BIO-3 Mitigation Measure BIO-12	Less than significant
Geology, Soils, Minerals, and Paleontological Resources		
Impact GEO-3 • Project excavation, grading, and changes in the routing of overland and subsurface flow may reactivate existing failures and initiate failures where none do not presently exist.	Mitigation Measure GEO-1	Less than significant
Hazards and Hazardous Materials		
Impact HAZ-8 • Project construction would involve the use of heavy equipment, welding, and other activities that have potential to ignite fires.	Mitigation Measure HAZ-1	Less than significant
Noise and Vibration		
Impact NOI-4 • Construction equipment noise would increase ambient noise levels at residences located near the southern terminus of the Project, and would potentially result in a substantial temporary or periodic increase in ambient noise levels.	Mitigation Measure NOI-1	Less than significant

EIR Highlights
Comparison of Project Alternatives

Table 4-2. Comparison of Environmental Impacts of Alternatives

Resource Topic	Proposed Project	No-Project Alternative
Aesthetics		
Scenic vistas	LTS	Lesser
Scenic resources	LTS	Lesser
Degrade visual character or quality	LTS with Mitigation	Lesser
New source of light or glare	LTS	Lesser
Air Quality		
Air quality plan conflict	LTS	Lesser
Air quality standard violations	LTS	Lesser
Cumulatively considerable net increase in criteria pollutant	LTS	Lesser
Sensitive receptors	LTS	Lesser
Objectionable odors	LTS	Lesser
Biological Resources		
Special-status species	LTS with Mitigation	Lesser
Sensitive natural communities	LTS with Mitigation	Lesser
Wetlands	LTS with Mitigation	Lesser
Wildlife movement and migration	LTS	Lesser
Local policies and ordinances	LTS	Lesser
Habitat conservation plan	LTS	Lesser
Spread of invasive species	LTS with Mitigation	Lesser
Cultural Resources		
Historical resources	LTS	Lesser
Archaeological resources	LTS	Lesser
Human remains	LTS	Lesser

El Dorado County

Resource Topic	Proposed Project	No-Project Alternative
Geology, Soils, Minerals, and Paleontological Resources		
Seismicity	LTS	Greater
Soil erosion	LTS	Lesser
Unstable geologic unit	LTS with Mitigation	Greater
Expansive soils	LTS	Lesser
Greenhouse Gas Emissions		
Greenhouse gas emissions	LTS	Lesser
Greenhouse gas plan conflict	LTS	Lesser
Hazards and Hazardous Materials		
Use, transport or disposal	LTS	Lesser
Accidental release	LTS	Greater
Emergency response plan	LTS	Greater
Wildland fire	LTS with Mitigation	Lesser
Hydrology/Water Quality		
Water quality standard violations	LTS	Lesser
Alter drainage and result in erosion	LTS	Lesser
Alter drainage and result in flooding	LTS	Lesser
Degrade water quality	LTS	Lesser
Impede floodflows	LTS	Greater
Exposure to flooding	LTS	Greater
Land Use Planning and Agricultural Resources		
Important farmland	LTS	Lesser
Zoning and Williamson Act	LTS	Lesser
Other farmland conversion	LTS	Lesser

El Dorado County — Alternatives

Resource Topic	Proposed Project	No-Project Alternative	Mid-level Alternative	Low-level Alternative
Noise and Vibration				
Noise standards	LTS	Lesser	Greater	Greater
Groundborne vibration/noise	LTS	Lesser	Similar	Similar
Permanent increase	LTS	Lesser	Similar	Similar
Temporary increase	LTS with Mitigation	Lesser	Similar	Similar
Public Services and Utilities				
New/expanded facilities	LTS	Lesser	Similar	Greater
Landfill capacity	LTS	Lesser	Same	Same
Telecommunications disruption	LTS	Lesser	Similar	Similar
Energy	LTS	Lesser	Similar	Similar
Recreation				
Deterioration of existing	LTS	Lesser	Similar	Greater
Traffic and Circulation				
Performance standards conflict	LTS	Same	Same	Same
Design hazards	No Impact	Greater	Greater	Greater
Emergency access	LTS	Greater	Greater	Greater
Alternative modes	No Impact	Greater	Similar	Greater

Impact Level
- No Impact = includes beneficial effects
- LTS = less-than-significant impact
- LTS with Mitigation = less-than-significant impact with mitigation incorporated

Impact Comparisons
- Same = same as the proposed Project
- Similar = similar to proposed Project
- Lesser = lesser than the proposed Project
- Greater = greater than proposed Project

		Lesser	Lesser
Important farmland	LTS	Lesser	Lesser
Zoning and Williamson Act	LTS	Lesser	Lesser
Other farmland conversion	LTS	Lesser	Lesser

Public Comments

- Existing Mosquito Road Bridge & Roadway
    - Remains in Place (Pedestrian and Bicycle Access Across Bridge)
- Development & Traffic Considerations
    - Not capacity increasing project, no changes to planning
- Technical Considerations (& Clarifications)
    - Storm Water
    - Geotechnical Considerations
    - Fire Considerations (Haz-1)
- Biological Considerations
    - Oak Woodland Habitat (Bio-7)
    - Invasive Plant (Bio-12)

Conclusion

- Project meets all Project Objectives
- All Impacts Reduced to Less than Significant with Mitigation Measures
- DOT recommends BOS to:
    - 1) Certify EIR
    - 2) Adopt Findings of Fact and MMRP
    - 3) Approve Mosquito Bridge Replacement Project

COUNTY OF EL DORADO, CALIFORNIA
DEPARTMENT OF TRANSPORTATION

NOTICE TO BIDDERS

The County of El Dorado, State of California, Department of Transportation (Department of Transportation) **HEREBY GIVES NOTICE** that they will receive sealed bids for work in accordance with the Project Plans (Plans) and Contract Documents for:

MOSQUITO ROAD AT SOUTH FORK AMERICAN RIVER
BRIDGE REPLACEMENT
CONTRACT NO. 5084, CIP NO. 36105028 (77126)

through Quest Construction Data Network (Quest). The Department of Transportation will accept bids until **February 11, 2022 at 2:00 PM**, at which time they will publicly open and read all bids. The bid opening will take place virtually through Zoom. Access to the virtual bid meeting will be via the following: https://us06web.zoom.us/j/89272205559 (669)900-9128 US (San Jose), (253)215-8782 US (Tacoma), (346)248-7799 US (Houston).

No Bid withdrawal after the time established for receiving bids or before the award and execution of the Contract, unless there is a delay of award for a period exceeding sixty (60) calendar days. Execution of bids must be in accordance with the instructions, forms, and Contract Documents furnished by the Department of Transportation through Quest Construction Data Network (Quest). **Submittal of the Proposal including the Bidder's Security, Form 590, and Payee Data Record shall be through the Quest website for Project # 7267650:**

"PROPOSAL FOR MOSQUITO ROAD AT SOUTH FORK AMERICAN RIVER - BRIDGE REPLACEMENT"
CONTRACT NO. 5084, CIP NO. 36105028 (77126)
TO BE OPENED AT 2:00 P.M. ON FEBRUARY 11, 2022

LOCATION/DESCRIPTION OF THE WORK: The Project is located along Mosquito Road, 1.3 miles north of the Mosquito Rd/Kona Drive intersection, in the County of El Dorado. Details for the Work are in the Plans and Contract Documents. They generally consists of, but are not limited to:

A. Construction of a cast-in-place pre-stressed concrete segmental box girder bridge over South Fork American River; construction of soil nail, tie-back, soldier pile, and reinforced concrete retaining walls; construction of cast-in-drilled-hole (CIDH) concrete piles and optional micropiles; and grading, paving, and drainage system improvements for the re-aligned roadway. Performance, construction or installation of other items or details not mentioned above, that are in the Plans, Standard Plans, Standard Specifications, or these Special Provisions.

B. Bidding for this Project consists of a Base Bid (Schedule A) with all Pier 2 Alternative Bids (Schedule B through D) and Pier 3 Alternative Bids (Schedule E through H) in accordance with the Proposal, Special Provisions, and Agreement. All bids will be compared by the summation of Proposal Pay Items and Bid Price Schedule for the Base Bid (Schedule A), lowest schedule bid for Pier 2 (Schedules B, C, or D), and lowest schedule bid for Pier 3 (Schedules E, F, or G). In the event of a discrepancy between the unit price bid and the extended unit total as stated on the Proposal, the County uses the amount bid for the unit price in calculating the additive total of the bid items for purposes of award, including revisions by Addenda, and as specified in the Proposal instructions. The All Bidder's Letter and Notice of Award will state the Base Bid (Schedule A) with lowest Pier 2 and Pier 3 Alternative Bid Schedules of Work recommendation for award to the Board and award by the Board respectively.

C. BASE BID (SCHEDULE A): Consists of constructing bridge pile caps, abutments, piers, superstructure, approach slabs, and barrier rails; bridge abutment return walls and Type 1A retaining walls; pier soldier

Mosquito Road at South Fork American River - Bridge Replacement
Contract No. 5084, CIP No. 36105028
December 14, 2021

County of El Dorado
Notice to Bidders
NTB-1

pile, tie-back, and soil-nail retaining walls; roadway approach sections, Type 1 & 1A retaining walls, culvert, drainage structures, and Midwest Guardrail Systems (MGS); all associated excavation, structure backfill, embankment fill, and rock slope protection.

D. PIER 2 ALTERNATIVE BIDS (SCHEDULES B – D): Consists of construction and installation of CIDH piles. All Pier 2 alternatives (Schedules B – D) consisting of 36", 48", and 60" CIDH must be bid. After bid opening, the lowest cost Pier 2 alternative schedule will be the Contract Pier 2 Alternative Bid Schedule for construction.

E. PIER 3 ALTERNATIVE BIDS (SCHEUDLES E – H): Consists of construction and installation of CIDH piles or micro-piles. All Pier 3 alternatives (Schedules E – H) consisting of 36", 48", and 60" CIDH, and 10" micropile alternative schedules must be bid. After bid opening, the lowest cost Pier 3 alternative schedule will be the Contract Pier 3 Alternative Bid Schedule for construction.

F. The amounts bid for Pier 2 Alternative Bids (Schedules B, C, and D) must not calculate to the same value. The amounts bid for Pier 3 Alternative Bids (Schedules E, F, G, and H) must not calculate to the same value. The total values bid for the alternative bid schedules must be independent and calculate to determine the lowest cost alternative bid schedule intended for construction at each Pier.

G. Bids are required for the entire Work described herein.

H. The award of Contract, if the Board awards, will be to the lowest responsive, responsible Bidder whose Proposal complies with all the requirements prescribed. The basis for low bid and Contract award will be the lowest total bid summation for the combination of the Base Bid (Schedule A) with the lowest bid for Pier 2 Alternative Bid (Schedule B, C, or D) and lowest bid for Pier 3 Alternative Bid (Schedule E, F, G, or H). County reserves the right to award BASE BID combined with the lowest Pier 2 Alternative Bid Schedule and lowest Pier 3 Alternative Bid Schedule; or reject all bids.

I. Bids are required for the entire Work described herein.

J. The Contract time is five hundred eighty (580) WORKING DAYS.

K. For bonding purposes, the anticipated Project cost is less than $56,000,000.

L. County will conduct a mandatory pre-bid meeting for this Project on **JANUARY 19, 2022 at 2:00 p.m.** at the County of El Dorado, Department of Transportation, 2441 Headington Road, Placerville, CA. The meeting will occur in the downstairs conference room. County will enforce State requirements for masks and/or social distancing. Staff will conduct a trip to the jobsite after the pre-bid meeting for interested Bidders to attend. **Attendance at the pre-bid meeting is mandatory**.

M. The formal bidding for this Project is in accordance with Public Contract Code 22032 and County of El Dorado Ordinance Code section 3.14.040.

OBTAINING OR VIEWING CONTRACT DOCUMENTS: The Contract Documents, including the Project Plans, are available to view and/or download from the Quest website at http://www.questcdn.com. Interested parties may also access the Quest website by clicking on the link next to the Project Name or entering the Quest Project # on the Department of Transportation's website at http://www.edcgov.us/Government/DOT/pages/BidsHome.aspx.

Interested parties may view the Contract Documents, including the Project Plans, through the Department of Transportations' website at no charge. Download of the digital Contract Documents, including the Project Plans, is available for $30.00 by inputting the Quest Project # 7267650 on the websites' Project Search page. Please contact QuestCDN.com at (952) 233-1632 or info@questcdn.com for assistance in free membership, registration, downloading, and working with this digital project information.

Eligibility for inclusion on the Planholders List, receipt of Addenda notifications, and Bid submittal,

Mosquito Road at South Fork American River - Bridge Replacement
Contract No. 5084, CIP No. 36105028
December 14, 2021

County of El Dorado
Notice to Bidders
NTB-2

requires that interested parties must pay for and download the Contract Documents, including the Project Plans, from Quest. Those downloading the Contract Documents, including the Project Plans, assume responsibility and risk for completeness of the downloaded Contract Documents.

The Contract Documents, including the Project Plans, are available for examination in person at the Department of Transportation's office at 2850 Fairlane Court, Placerville CA. However, the Department of Transportation will no longer sell paper copies of the Contract Documents.

The Department of Transportation will provide cross sections and the following Supplemental Project Information/Information Handout along with the Contract Documents on Quest's website to all planholders who acquire the Contract Documents digitally through Quest:

- Geotechnical Reports
- Approx. Heritage Oaks and ROW Exhibit
- Revised Standard Plans
- Site Video & Project Rendering
- Estimated Earthwork Tables
- Conceptual Staging Area Site Map
- Conceptual Access Road Layouts
- Cross Sections
- Original Ground Survey (CADD File)
- ROW Linework (KMZ File & PDF File)
- Road Closure Detour and Emergency Evacuation Exhibits

The Department of Transportation advise Bidders that electronic files provided as Supplemental Project Information, including the original ground survey, right of way linework, visual rendering videos of project site, and drone flight video of project site are preliminary and approximate. It is the bidders' responsibility to confirm all data including field details, current conditions, and survey information for accuracy and completeness. In the event that Bidders find discrepancies between hard copies and electronic files, the hard copies will govern.

CONTRACTORS LICENSE CLASSIFICATION: Bidders must have proper licensure to perform the Work pursuant to the Contractors' State License Law (Business and Professions Code Section 7000 et seq.) with a **CLASS A** license or equivalent combination of Classes required by the categories and type of Work included in the Contract Documents and Plans at the time of Contract award. Bidder must maintain the valid license(s) through completion and acceptance of the Work, including the guarantee and acceptance period. Failure of the successful Bidder to obtain proper adequate licensing will constitute a failure to execute the Contract and will result in the forfeiture of the Bidder's security.

BUSINESS LICENSE: The County Business License Ordinance provides that it is unlawful for any person to furnish supplies or services, or transact any kind of business in the unincorporated territory of the County of El Dorado without possessing a County business license unless exempt under County Ordinance Code Section 5.08.070. The Bidder to whom an award is made must comply with all of the requirements of the County Business License Ordinance, where applicable, prior to beginning Work under this Contract and at all times during the term of this Contract.

CONTRACTOR REGISTRATION: No contractor or subcontractor may bid on any public works project, be eligible for inclusion in a bid proposal for any public works project, or engage in the performance of any contract for public work unless registered with the Department of Industrial Relations pursuant to Labor Code sections 1725.5 and 1771.1.

An inadvertent error in listing an unregistered subcontractor pursuant to Section 1725.5 in a bid proposal shall not be grounds for filing a bid protest or grounds for considering the bid nonresponsive if the Bidder complies with requirements of Labor Code section 1771.1.

EMISSIONS REDUCTION: Contractor must comply with emission reduction regulations mandated by the

Mosquito Road at South Fork American River - Bridge Replacement
Contract No. 5084, CIP No. 36105028
December 14, 2021

County of El Dorado
Notice to Bidders
NTB-3

California Air Resources Board, sign the certification of knowledge in the Agreement, and provide County a Certificate of Reported Compliance when road legal diesel vehicles with a gross vehicle weight over 14,000 pounds are included in their fleet. Contractor must require all subcontractors to comply with such regulations and provide County a Certificate of Reported Compliance for each subcontractor with road legal diesel vehicles over 14,000-pound gross vehicle weight.

SUBCONTRACTOR LIST: In accordance with the Subletting and Subcontracting Fair Practices Act, commencing with Section 4100 of the Public Contract Code, each Proposal must list therein the name, Contractor's license number, DIR number, and address of each subcontractor to whom the Bidder proposes to subcontract portions of the Work in an amount in excess of 0.5% of the total bid or $10,000, whichever is greater. The Bidder must also list the Bid Items for completion by each subcontractor listed in the Subcontractor List. Show the subcontractor Work by listing the Bid Item number, Bid Item description, and portion in the form of a percentage (not to exceed 100%). Calculate the Bid Item portion by dividing the amount of subcontractor Work by the respective Bid Item amount(s) (not by the total bid price).

Submit the percentage of each subcontractor Bid Item with the Bidder's bid or send via email or fax to Brian Franklin, County of El Dorado, Department of Transportation, email- Brian.Franklin@edcgov.us, Fax-(530) 626-0387 within 24 hours of request. The email or fax must contain the name of each subcontractor submitted with the Bidder's bid along with the Bid Item number, the Bid Item description, and the percentage of each Bid Item subcontracted, as described above. At the time of Contract award, all listed subcontractors must have proper licensure to perform their designated portion of the Work. The Department of Transportation directs the Bidder's attention to other provisions of the Act related to the imposition of penalties for failure to observe its provisions by using unauthorized subcontractors or by making unauthorized substitutions.

An inadvertent error in listing the California Contractor license number on the Subcontractor List will not be grounds for filing a bid protest or grounds for considering the bid non-responsive if the Bidder submits the corrected Contractor's license number to Brian Franklin via fax or email as noted above within 24 hours of request. The corrected Contractor's license number must correspond to the submitted name and location for that subcontractor.

BUY AMERICA: This Project is subject to the "Buy America" provisions of the Surface Transportation Assistance Act of 1982, as amended by the Intermodal Surface Transportation Efficiency Act of 1991, and the Moving Ahead for Progress in the 21st Century Act (MAP-21).

DISADVANTAGED BUSINESS ENTERPRISE (DBE) PARTICIPATION: The Department of Transportation, in accordance with the provisions of Title VI of the Civil Rights Act of 1964 (78 Stat. 252, 42 U.S.C. §§ 2000d to 2000d-4) and the Regulations, hereby notifies all Bidders that it will affirmatively ensure that in any contract entered into pursuant to this advertisement, disadvantaged business enterprises will be afforded full and fair opportunity to submit bids in response to this invitation and will not be discriminated against on the grounds of race, color, or national origin in consideration for an award.

For Federal-aid projects, DBE requirements of Title 49 Part 26 of the Code of Federal Regulations (49 CFR 26) apply. The Department of Transportation advises Bidders that, as Federal law requires, the County of El Dorado implements Disadvantaged Business Enterprise requirements for Disadvantaged Business Enterprises (DBE). Comply with Section 2-1.12 and Section 5-1.13.

In accordance with 49 CFR 26, Bidder will take all necessary affirmative steps to assure that minority firms, women's business enterprises and labor surplus area firms are used when possible.

The Disadvantaged Business Enterprise (DBE) Contract goal is **14%.**

The UDBE Good Faith Effort Submittal Information Handout and the County of El Dorado DBE Training Presentation is available at http://www.edcgov.us/Government/DOT/pages/DBE.aspx. The problems and solutions listed in the Handout apply to DBE Good Faith Efforts Submittals.

NONDISCRIMINATION: Comply with Subchapter 5 of Chapter 5 of Division 4.1 of Title 2, California Code of

Mosquito Road at South Fork American River - Bridge Replacement
Contract No. 5084, CIP No. 36105028
December 14, 2021

County of El Dorado
Notice to Bidders
NTB-4

Regulations and the following.

NOTICE OF REQUIREMENT FOR NONDISCRIMINATION PROGRAM
(GOVERNMENT CODE SECTION 12990)

Comply with Section 7-1.02I(2), "Nondiscrimination," of the Standard Specifications, which is applicable to all nonexempt State contracts and subcontracts, and to the "Standard California Nondiscrimination Construction Contract Specifications" set forth therein. The specifications are applicable to all nonexempt State construction contracts and subcontracts of $5,000 or more.

Comply with the additional nondiscrimination and fair employment practices provisions in the *Draft Agreement* contained in these Contract Documents that will apply to this Federal-aid Contract.

The Department of Transportation hereby notifies all Bidders that it will affirmatively ensure that in any contract entered into pursuant to this advertisement, minority business enterprises will be afforded full opportunity to submit bids in response to this invitation and will not be discriminated against on the grounds of race, color, sex, national origin, religion, age, or disability in consideration for the award.

PREVAILING WAGE REQUIREMENTS: In accordance with the provisions of California Labor Code Sections 1770 et seq., including but not limited to Sections 1773, 1773.1, 1773.2, 1773.6, and 1773.7, the State prevailing wage rates in the county performing the work is subject to the determination by the Director of the California Department of Industrial Relations. Interested parties can obtain the current wage information by submitting their requests to the Department of Industrial Relations, Division of Labor Statistics and Research, PO Box 420603, San Francisco CA 94142-0603, Telephone (415) 703-4708 or by referring to the website at http://www.dir.ca.gov/OPRL/PWD. The rates at the time of the bid advertisement date of a project will remain in effect for the life of the project in accordance with the California Code of Regulations, as modified and effective January 27, 1997.

Copies of the State prevailing wage rates for the county performing the work are on file at the Department of Transportation's principal office, and upon request. In the case of projects involving Federal funds, Federal wage rates as predetermined by the United States Secretary of Labor have been included in the Contract Documents. Issuance of addenda to modify the Federal minimum wage rates, if necessary, will be in accordance with the Project Administration section of this Notice to Bidders.

In accordance with the provisions of Labor Code 1810, eight (8) hours of labor constitutes a legal day's work upon all work done hereunder, and Contractor and any subcontractor employed under this Contract must conform to the binding provisions of Labor Code Sections 1810 through 1815.

This Project is subject to the requirements of Title 8, Chapter 8, Subchapter 4.5 of the California Code of Regulations including the obligation to furnish certified payroll records directly to the Compliance Monitoring Unit under the Labor Commissioner within the Department of Industrial Relations Division of Labor Standards Enforcement in accordance with Section 16461.

In the case of Federally funded projects, where Federal and State prevailing wage requirements apply, compliance with both is a requirement. Funding for this Project is in whole or part by Federal funds. Comply with Exhibit D of the Draft Agreement and the Copeland Act (18 U.S.C. 874 and 29 CFR Part 3), the Davis-Bacon Act (40 U.S.C. 3141-3147 and 29 CFR Part 5), and the Contract Work Hours and Safety Standards Act (40 U.S.C. 3701 and 29 CFR Part 5).

If there is a difference between the Federal wage rates predetermined by the Secretary of Labor and the State prevailing wage rates determined by the Director of the California Department of Industrial Relations for similar classifications of labor, Contractor and subcontractors must pay not less than the higher wage rate. The Department of Transportation will not accept lower State wage rates not specifically included in the Federal minimum wage determinations. This includes "helper" (or other classifications based on hours of experience) or any other classification not appearing in the Federal wage determinations. Where Federal wage determinations do not contain the State wage rate determination otherwise available for use by Contractor and subcontractors,

Mosquito Road at South Fork American River - Bridge Replacement
Contract No. 5084, CIP No. 36105028
December 14, 2021

County of El Dorado
Notice to Bidders
NTB-5

Contractor and subcontractors must pay not less than the Federal minimum wage rate that most closely approximates the duties of the employees in question.

TRAINING: For the Federal training program, the number of trainees or apprentices is 36.

BID SECURITY: Bidder must provide a bid security with each bid. Bid security must be in an amount of not less than ten percent (10%) of the total amount of the Bid. Bid security must be cash, a certified check, or cashier's check drawn to the order of the County of El Dorado or a Bidder's Bond executed by a surety satisfactory to the County of El Dorado **on the form provided in the Proposal section of these Contract Documents**.

BID PROTEST PROCEDURE: The intent of the bid protest procedure is to handle and resolve disputes related to the bid award for this Project pursuant to Title 2 Code of Federal Regulations Part 200.318(k) and County of El Dorado policies and procedures. A protestor must exhaust all administrative remedies with the County of El Dorado before pursuing a protest with a Federal Agency. Reviews of protests by the Federal agency will be limited to:

1. Violations of Federal law or regulations and the standards of 2 CFR Part 200.318(k). Violations of State of California or local law will be under the jurisdiction of the State of California or the County of El Dorado; and

2. Violation of the County of El Dorado's protest procedures for failure to review a complaint or protest. Protests received by the Federal agency other than those specified above will refer back to the County of El Dorado.

The protest procedure is an extension of the formal bid process and allows those who wish to protest the recommendation of an award after bid the opportunity to be heard.

Policy: Upon completion of the bid evaluation, the Department of Transportation will notify all Bidders of the recommendation of award, the basis therefore, and the date and time on which the recommendation for award consideration and action by the Board of Supervisors. All Bidders may attend the Board of Supervisors meeting at the time the agenda item is considered, address the Board of Supervisors, and be heard.

Procedure: If a Bidder wishes to protest the award, this is the procedure:

1. The Department of Transportation will review the bids received in a timely fashion under the terms and conditions of the Notice to Bidders, and notify the Bidders in writing, at the fax number designated in the Proposal, of its recommendation including for award or rejection of bids ("All Bidders Letter").

2. Within five (5) business days from the date of the "All Bidders Letter," the Bidder protesting the recommendation for award must submit a letter of protest to and must be received by the County of El Dorado, Department of Transportation, Attention Brian Franklin, 2850 Fairlane Court, Placerville, CA 95667, and state in detail the basis and reasons for the protest. The Bidder must provide facts to support the protest, including any evidence for consideration, together with the law, rule, regulation, or criteria on which the protest is based.

3. If the Department of Transportation finds the protest to be valid, it may modify its award recommendations and notify all Bidders of that decision. If the Department of Transportation does not agree with the protest, or otherwise fails to resolve the protest, it will notify the bid protestor and all interested parties of its decision and the date and time that the recommendation for award will be agendized for the Board of Supervisors' consideration and action. The Department of Transportation will also include in its report to the Board of Supervisors the details of the bid protest.

4. The Bidder may attend the Board of Supervisors meeting at which the recommendation and bid protest will be considered. The Board of Supervisors will take comment from the Bidder, staff, and members of the public who wish to speak on the item. In the event that the Bidder is not in attendance at that time, the bid protest may be dismissed by the Board of Supervisors without further consideration of the merits; and

Mosquito Road at South Fork American River - Bridge Replacement
Contract No. 5084, CIP No. 36105028
December 14, 2021

County of El Dorado
Notice to Bidders
NTB-6

PRELIMINARY BID SUMMARY

MOSQUITO ROAD AT SOUTH FORK AMERICAN RIVER BRIDGE REPLACEMENT PROJECT

BID DATE/TIME 02/25/2022 02:00 PM PST
OWNER El Dorado County DOT
CONTRACT NO. 5084
PROJECT NO. 36105028 (77126)
PROJECT TYPES Cast in Place Segmental Bridge, CIDH Piles, Micorpiles, Tie-Back Walls, Grading, Paving, Drainage

					Engineer Estimate		Flatiron West Inc.		Shimmick Construction		MCM Construction, Inc.		Golden State Bridge, Inc.		Walsh Construction II, LLC (Concord,CA)	
Line Item	Item Code	Item Description	UofM	Quantity	Unit Price	Unit Total	Unit Price	Unit Total	Unit Price	Unit Total	Unit Price	Unit Total	Unit Price	Unit Total	Unit Price	Unit Total
Base Bid (Schedule A)																
1	072007A	Excavation Safety	LS	1	$150,000.00	$150,000.00	$10,000.00	$10,000.00	$15,000.00	$15,000.00	$250,000.00	$250,000.00	$650,000.00	$650,000.00	$375,000.00	$375,000.00
2	80050	Progress Schedule (Level 3 Critical Path Method)	LS	1	$20,000.00	$20,000.00	$5,000.00	$5,000.00	$50,000.00	$50,000.00	$35,000.00	$35,000.00	$20,000.00	$20,000.00	$250,000.00	$250,000.00
3	100100	Develop Water Supply	LS	1	$20,000.00	$20,000.00	$500,000.00	$500,000.00	$150,000.00	$150,000.00	$65,000.00	$65,000.00	$55,000.00	$55,000.00	$200,000.00	$200,000.00
4	120090	Construction Area Signs	LS	1	$20,000.00	$20,000.00	$7,995.00	$7,995.00	$15,000.00	$15,000.00	$7,000.00	$7,000.00	$5,000.00	$5,000.00	$15,000.00	$15,000.00
5	120100	Traffic Control System	LS	1	$300,000.00	$300,000.00	$45,000.00	$45,000.00	$925,000.00	$925,000.00	$200,000.00	$200,000.00	$75,000.00	$75,000.00	$250,000.00	$250,000.00
6	128652	Portable Changeable Message Sign	LS	1	$50,000.00	$50,000.00	$50,000.00	$50,000.00	$120,000.00	$120,000.00	$60,000.00	$60,000.00	$30,000.00	$30,000.00	$75,000.00	$75,000.00
7	129000	Temporary Railing (Type K)	LF	520	$100.00	$52,000.00	$60.00	$31,200.00	$68.00	$35,360.00	$100.00	$52,000.00	$45.00	$23,400.00	$80.00	$41,600.00
8	129110	Temporary Crash Cushion	EA	5	$7,000.00	$35,000.00	$4,500.00	$22,500.00	$5,000.00	$25,000.00	$6,000.00	$30,000.00	$4,500.00	$22,500.00	$10,000.00	$50,000.00
9	130100	Job Site Management	LS	1	$400,000.00	$400,000.00	$200,000.00	$200,000.00	$700,000.00	$700,000.00	$125,000.00	$125,000.00	$37,000.00	$37,000.00	$660,000.00	$660,000.00
10	130300	Prepare Storm Water Pollution Prevention Plan	LS	1	$5,000.00	$5,000.00	$5,000.00	$5,000.00	$15,000.00	$15,000.00	$3,000.00	$3,000.00	$5,000.00	$5,000.00	$5,000.00	$5,000.00
11	130310	Rain Event Action Plan	EA	20	$500.00	$10,000.00	$500.00	$10,000.00	$500.00	$10,000.00	$500.00	$10,000.00	$500.00	$10,000.00	$500.00	$10,000.00
12	130320	Storm Water Sampling and Analysis Day	EA	20	$500.00	$10,000.00	$500.00	$10,000.00	$500.00	$10,000.00	$500.00	$10,000.00	$500.00	$10,000.00	$500.00	$10,000.00
13	130330	Storm Water Annual Report	EA	2	$2,000.00	$4,000.00	$2,000.00	$4,000.00	$2,000.00	$4,000.00	$2,000.00	$4,000.00	$2,000.00	$4,000.00	$2,000.00	$4,000.00
14	149001A	Prepare Fugitive Dust Control Plan	LS	1	$2,000.00	$2,000.00	$5,000.00	$5,000.00	$10,000.00	$10,000.00	$3,000.00	$3,000.00	$6,000.00	$6,000.00	$5,000.00	$5,000.00
15	160110	Temporary High-Visibility Fence	LF	4700	$4.00	$18,800.00	$2.74	$12,878.00	$4.00	$18,800.00	$6.00	$28,200.00	$7.00	$32,900.00	$5.50	$25,850.00
16	160120	Tree Removal	EA	165	$100.00	$16,500.00	$515.00	$84,975.00	$2,200.00	$363,000.00	$700.00	$115,500.00	$650.00	$107,250.00	$1,000.00	$165,000.00
17	170103	Clearing and Grubbing	LS	1	$181,500.00	$181,500.00	$250,000.00	$250,000.00	$500,000.00	$500,000.00	$225,000.00	$225,000.00	$126,000.00	$126,000.00	$250,000.00	$250,000.00
18	190101	Roadway Excavation (Final Pay)	CY	5788	$100.00	$578,800.00	$80.00	$463,040.00	$55.00	$318,340.00	$310.00	$1,794,280.00	$70.00	$405,160.00	$130.00	$752,440.00
19	190185	Shoulder Backing	TON	60	$200.00	$12,000.00	$90.00	$5,400.00	$250.00	$15,000.00	$305.00	$18,300.00	$77.00	$4,620.00	$130.00	$7,800.00
20	192003	Structure Excavation (Bridge) (Final Pay)	CY	822	$135.00	$110,970.00	$100.00	$82,200.00	$110.00	$90,420.00	$75.00	$61,650.00	$400.00	$328,800.00	$125.00	$102,750.00
21	192004	Supplemental Structure Excavation and Backfill	CY	300	$260.00	$78,000.00	$170.00	$51,000.00	$140.00	$42,000.00	$155.00	$46,500.00	$400.00	$120,000.00	$200.00	$60,000.00
22	192035	Structure Excavation (Abut 4 / Ret Wall 4F) (Rock) (Final Pay)	CY	902	$310.00	$279,620.00	$90.00	$81,180.00	$92.00	$82,984.00	$300.00	$270,600.00	$1,100.00	$992,200.00	$65.00	$58,630.00
23	192049	Structure Excavation (Soldier Pile Wall) (Final Pay)	CY	60	$225.00	$13,500.00	$100.00	$6,000.00	$103.00	$6,180.00	$200.00	$12,000.00	$200.00	$12,000.00	$450.00	$27,000.00
24	192055	Structure Excavation (Soil Nail Wall) (Final Pay)	CY	8180	$225.00	$1,840,500.00	$120.00	$981,600.00	$103.00	$842,540.00	$105.00	$858,900.00	$120.00	$981,600.00	$175.00	$1,431,500.00
25	193000	Backfill (Pier) (Final Pay)	CY	2410	$100.00	$241,000.00	$150.00	$361,500.00	$69.00	$166,290.00	$105.00	$253,050.00	$125.00	$301,250.00	$100.00	$241,000.00
26	193003	Structure Backfill (Bridge) (Final Pay)	CY	1357	$125.00	$169,625.00	$150.00	$203,550.00	$150.00	$203,550.00	$140.00	$189,980.00	$175.00	$237,475.00	$100.00	$135,700.00
27	193028	Structure Backfill (Soil Nail Wall) (Final Pay)	CY	31	$250.00	$7,750.00	$150.00	$4,650.00	$200.00	$6,200.00	$800.00	$24,800.00	$125.00	$3,875.00	$500.00	$15,500.00
28	193029	Structure Backfill (Soldier Pile Wall) (Final Pay)	CY	87	$250.00	$21,750.00	$150.00	$13,050.00	$250.00	$21,750.00	$625.00	$54,375.00	$125.00	$10,875.00	$350.00	$30,450.00
29	193116	Concrete Backfill (Soldier Pile Wall) (Final Pay)	CY	103	$350.00	$36,050.00	$590.00	$60,770.00	$350.00	$36,050.00	$1,000.00	$103,000.00	$750.00	$77,250.00	$550.00	$56,650.00
30	193119	Lean Concrete Backfill (Final Pay)	CY	20	$550.00	$11,000.00	$1,125.00	$22,500.00	$950.00	$19,000.00	$1,000.00	$20,000.00	$500.00	$10,000.00	$1,200.00	$24,000.00
31	194001	Ditch Excavation	CY	89	$100.00	$8,900.00	$80.00	$7,120.00	$65.00	$5,785.00	$130.00	$11,570.00	$125.00	$11,125.00	$95.00	$8,455.00
32	198010	Imported Borrow (Final Pay)	CY	11801	$70.00	$826,070.00	$90.00	$1,062,090.00	$32.00	$377,632.00	$105.00	$1,239,105.00	$65.00	$767,065.00	$150.00	$1,770,150.00
33	198250	Geosynthetic Reinforcement	SQYD	950	$60.00	$57,000.00	$6.00	$5,700.00	$20.00	$19,000.00	$50.00	$47,500.00	$5.00	$4,750.00	$15.00	$14,250.00
34	19XXXX	Construct Access	LS	1	$4,500,000.00	$4,500,000.00	$13,500,000.00	$13,500,000.00	$14,950,000.00	$14,950,000.00	$11,000,000.00	$11,000,000.00	$13,000,000.00	$13,000,000.00	$24,312,050.00	$24,312,050.00
35	210270	Rolled Erosion Control Product (Netting)	SQFT	33000	$0.60	$19,800.00	$0.72	$23,760.00	$2.25	$74,250.00	$1.00	$33,000.00	$2.00	$66,000.00	$1.00	$33,000.00
36	210350	Fiber Rolls	LF	2500	$6.00	$15,000.00	$3.54	$8,850.00	$8.50	$21,250.00	$4.00	$10,000.00	$7.50	$18,750.00	$4.00	$10,000.00
37	210430	Hydroseed (3-Step)	SQFT	60000	$0.20	$12,000.00	$0.18	$10,800.00	$0.30	$18,000.00	$0.25	$15,000.00	$0.15	$9,000.00	$0.30	$18,000.00
38	260203	Class 2 Aggregate Base	CY	1520	$170.00	$258,400.00	$70.00	$106,400.00	$200.00	$304,000.00	$210.00	$319,200.00	$125.00	$190,000.00	$150.00	$228,000.00
39	303000	Pulverize AC	SQYD	690	$2.50	$1,725.00	$8.50	$5,865.00	$22.00	$15,180.00	$35.00	$24,150.00	$7.00	$4,830.00	$1.25	$862.50
40	390132	Hot Mix Asphalt (Type A, PG 64-16)	TON	1400	$220.00	$308,000.00	$150.00	$210,000.00	$200.00	$280,000.00	$155.00	$217,000.00	$160.00	$224,000.00	$170.00	$238,000.00
41	394073	Place Hot Mix Asphalt Dike (Type A)	LF	200	$45.00	$9,000.00	$50.00	$10,000.00	$12.00	$2,400.00	$35.00	$7,000.00	$75.00	$15,000.00	$45.00	$9,000.00
42	398200	Cold Plane Asphalt Concrete Pavement	SQYD	300	$30.00	$9,000.00	$30.00	$9,000.00	$42.50	$12,750.00	$35.00	$10,500.00	$35.00	$10,500.00	$40.00	$12,000.00
43	460220	Ground Anchor (T = 100 Kips) (Final Pay)	EA	121	$6,000.00	$726,000.00	$6,000.00	$726,000.00	$8,700.00	$1,052,700.00	$8,000.00	$968,000.00	$13,000.00	$1,573,000.00	$15,000.00	$1,815,000.00
44	460230	Ground Anchor (T = 210 Kips) (Final Pay)	EA	111	$5,500.00	$610,500.00	$7,000.00	$777,000.00	$9,000.00	$999,000.00	$9,000.00	$999,000.00	$13,000.00	$1,443,000.00	$10,500.00	$1,165,500.00
45	460300	Soil Nail (Final Pay)	LF	5721	$60.00	$343,260.00	$90.00	$514,890.00	$155.00	$886,755.00	$105.00	$600,705.00	$200.00	$1,144,200.00	$175.00	$1,001,175.00
46	490311	Steel Soldier Pile (HP 12 x 74) (Final Pay)	LF	1279	$150.00	$191,850.00	$100.00	$127,900.00	$120.00	$153,480.00	$85.00	$108,715.00	$80.00	$102,320.00	$550.00	$703,450.00
47	490400	24" Drilled Hole (Final Pay)	LF	1058	$250.00	$264,500.00	$380.00	$402,040.00	$750.00	$793,500.00	$450.00	$476,100.00	$900.00	$952,200.00	$770.00	$814,660.00
48	490604	30" Cast-In-Drilled-Hole Concrete Piling	LF	317	$500.00	$158,500.00	$600.00	$190,200.00	$800.00	$253,600.00	$550.00	$174,350.00	$1,230.00	$389,910.00	$600.00	$190,200.00
49	500001	Prestressing Cast-In-Place Concrete	LS	1	$1,780,400.00	$1,780,400.00	$3,800,000.00	$3,800,000.00	$1,900,000.00	$1,900,000.00	$2,200,000.00	$2,200,000.00	$2,000,000.00	$2,000,000.00	$5,000,000.00	$5,000,000.00
50	510051	Structural Concrete, Bridge (Footing) (Final Pay)	CY	2231	$1,200.00	$2,677,200.00	$650.00	$1,450,150.00	$500.00	$1,115,500.00	$1,000.00	$2,231,000.00	$1,800.00	$4,015,800.00	$1,200.00	$2,677,200.00
51	510052	Structural Concrete, Bridge (Pier) (Final Pay)	CY	2396	$1,800.00	$4,312,800.00	$2,550.00	$6,109,800.00	$2,200.00	$5,271,200.00	$1,500.00	$3,594,000.00	$3,500.00	$8,386,000.00	$2,400.00	$5,750,400.00
52	510053	Structural Concrete, Bridge (Abutments) (Final Pay)	CY	282	$1,500.00	$423,000.00	$1,000.00	$282,000.00	$1,700.00	$479,400.00	$1,500.00	$423,000.00	$1,750.00	$493,500.00	$1,800.00	$507,600.00
53	510054	Structural Concrete, Bridge (Polymer Fiber) (Final Pay)	CY	6442	$3,000.00	$19,326,000.00	$1,715.00	$11,048,030.00	$2,200.00	$14,172,400.00	$2,887.00	$18,598,054.00	$3,333.00	$21,471,186.00	$4,000.00	$25,768,000.00
54	510060	Structural Concrete, Retaining Wall (Final Pay)	CY	405	$1,500.00	$607,500.00	$1,300.00	$526,500.00	$2,200.00	$891,000.00	$1,750.00	$708,750.00	$1,450.00	$587,250.00	$2,500.00	$1,012,500.00
55	510085	Structural Concrete, Approach Slab (Type EQ Modified) (Final Pay)	CY	27	$1,800.00	$48,600.00	$1,000.00	$27,000.00	$1,750.00	$47,250.00	$2,000.00	$54,000.00	$1,600.00	$43,200.00	$2,300.00	$62,100.00
56	518051	PTFE Spherical Bearing	EA	4	$25,000.00	$100,000.00	$30,000.00	$120,000.00	$35,000.00	$140,000.00	$40,000.00	$160,000.00	$30,000.00	$120,000.00	$36,000.00	$144,000.00
57	519107	Joint Seal Assembly (MR 10")	LF	71	$3,000.00	$213,000.00	$3,000.00	$213,000.00	$3,500.00	$248,500.00	$5,000.00	$355,000.00	$3,000.00	$213,000.00	$4,400.00	$312,400.00
58	520102	Bar Reinforcing Steel (Bridge) (Final Pay)	LB	2098350	$2.00	$4,196,700.00	$1.85	$3,881,947.50	$1.90	$3,986,865.00	$2.00	$4,196,700.00	$1.83	$3,839,980.50	$2.20	$4,616,370.00
59	520103	Bar Reinforcing Steel (Retaining Wall) (Final Pay)	LB	174767	$1.50	$262,150.50	$1.72	$300,599.24	$1.90	$332,057.30	$2.00	$349,534.00	$1.85	$323,318.95	$2.80	$489,347.60
60	520110	Bar Reinforcing Steel (Epoxy Coated) (Bridge) (Final Pay)	LB	569037	$2.25	$1,280,333.25	$1.13	$643,011.81	$2.70	$1,536,399.90	$3.00	$1,707,111.00	$2.60	$1,479,496.20	$2.60	$1,479,496.20
61	520120	Headed Bar Reinforcement (Final Pay)	EA	12136	$15.00	$182,040.00	$17.57	$213,229.52	$40.00	$485,440.00	$18.00	$218,448.00	$20.00	$242,720.00	$30.00	$364,080.00
62	530050	Sculpted Shotcrete (Final Pay)	SQFT	10355	$30.00	$310,650.00	$5.00	$51,775.00	$8.00	$82,840.00	$7.00	$72,485.00	$18.00	$186,390.00	$50.00	$517,750.00
63	530100	Shotcrete (Final Pay) (Final Pay)	CY	545	$800.00	$436,000.00	$1,700.00	$926,500.00	$2,100.00	$1,144,500.00	$2,200.00	$1,199,000.00	$1,100.00	$599,500.00	$5,500.00	$2,997,500.00
64	530200	Structural Shotcrete (Final Pay)	CY	600	$1,000.00	$600,000.00	$1,030.00	$618,000.00	$1,300.00	$780,000.00	$1,100.00	$660,000.00	$1,600.00	$960,000.00	$3,300.00	$1,980,000.00
65	575004	Timber Lagging (Final Pay)	MFBM	9.2	$7,500.00	$69,000.00	$6,000.00	$55,200.00	$15,000.00	$138,000.00	$7,000.00	$64,400.00	$9,500.00	$87,400.00	$22,000.00	$202,400.00
66	590120	Clean And Paint Steel Soldier Piling	LS	1	$50,000.00	$50,000.00	$100,000.00	$100,000.00	$35,000.00	$35,000.00	$50,000.00	$50,000.00	$33,000.00	$33,000.00	$60,000.00	$60,000.00
67	600037	Prepare Concrete Bridge Deck Surface	SQFT	40120	$1.15	$46,138.00	$0.60	$24,072.00	$0.65	$26,078.00	$0.50	$20,060.00	$0.75	$30,090.00	$0.70	$28,084.00
68	600041	Furnish Polyester Concrete Overlay (Final Pay)	CF	3343	$85.00	$284,155.00	$101.00	$337,643.00	$130.00	$434,590.00	$100.00	$334,300.00	$90.00	$300,870.00	$120.00	$401,160.00
69	600043	Place Polyester Concrete Overlay (Final Pay)	SQFT	40120	$3.60	$144,432.00	$2.10	$84,252.00	$1.25	$50,150.00	$2.50	$100,300.00	$5.00	$200,600.00	$2.40	$96,288.00
70	641107	18" Plastic Pipe	LF	281	$280.00	$78,680.00	$120.00	$33,720.00	$175.00	$49,175.00	$300.00	$84,300.00	$155.00	$43,555.00	$300.00	$84,300.00
71	641131	48" Plastic Pipe	LF	188	$460.00	$86,480.00	$250.00	$47,000.00	$270.00	$50,760.00	$325.00	$61,100.00	$355.00	$66,740.00	$365.00	$68,620.00
72	698100A	18" Plastic Pipe Downdrain (Fusion Welded)	LF	599	$190.00	$113,810.00	$250.00	$149,750.00	$420.00	$251,580.00	$360.00	$215,640.00	$145.00	$86,855.00	$325.00	$194,675.00
73	703450	Welded Steel Pipe Casing (Bridge)	LF	40	$310.00	$12,400.00	$700.00	$28,000.00	$235.00	$9,400.00	$350.00	$14,000.00	$300.00	$12,000.00	$275.00	$11,000.00
74	705011	18" Steel Flared End Section	EA	1	$2,000.00	$2,000.00	$300.00	$300.00	$2,200.00	$2,200.00	$1,500.00	$1,500.00	$1,000.00	$1,000.00	$1,600.00	$1,600.00
75	705031	48" Steel Flared End Section	EA	2	$3,000.00	$6,000.00	$2,500.00	$5,000.00	$9,000.00	$18,000.00	$3,500.00	$7,000.00	$3,000.00	$6,000.00	$5,000.00	$10,000.00
76	707117	36" Precast Concrete Pipe Inlet (Type OCPI or GCP)	EA	3	$5,000.00	$15,000.00	$2,000.00	$6,000.00	$13,000.00	$39,000.00	$17,000.00	$51,000.00	$4,000.00	$12,000.00	$4,200.00	$12,600.00
77	707117A	Drainage Inlet	EA	3	$6,000.00	$18,000.00	$5,000.00	$15,000.00	$12,000.00	$36,000.00	$24,000.00	$72,000.00	$5,000.00	$15,000.00	$9,000.00	$27,000.00

Bid tabulation (rotated landscape table). Columns: Item No. | Spec No. | Description | Unit | Qty | Engineer's Estimate (Unit Price / Amount) | Bidder 1 (Unit / Amount) | Bidder 2 (Unit / Amount) | Bidder 3 (Unit / Amount) | Bidder 4 (Unit / Amount) | Bidder 5 (Unit / Amount)

Item	Spec	Description	Unit	Qty	EE Unit	EE Amount	B1 Unit	B1 Amount	B2 Unit	B2 Amount	B3 Unit	B3 Amount	B4 Unit	B4 Amount	B5 Unit	B5 Amount
78	707125	48" Precast Concrete Pipe Inlet (Type OCPI or GCP)	EA	4	8,000.00	32,000.00	1,500.00	6,000.00	16,000.00	64,000.00	20,000.00	80,000.00	5,000.00	20,000.00	13,000.00	52,000.00
79	707217	36" Precast Concrete Pipe Manhole	EA	1	7,000.00	7,000.00	7,000.00	7,000.00	19,000.00	19,000.00	25,000.00	25,000.00	5,000.00	5,000.00	7,500.00	7,500.00
80	710136	Remove Pipe	LF	32	100.00	3,200.00	100.00	3,200.00	750.00	24,000.00	220.00	7,040.00	55.00	1,760.00	16.00	512.00
81	721420	Concrete (Ditch Lining) (Final Pay)	CY	27	2,500.00	67,500.00	1,200.00	32,400.00	780.00	21,060.00	1,500.00	40,500.00	1,200.00	32,400.00	2,000.00	54,000.00
82	723005A	Rock Slope Protection (1/4 T, Class V, Method A)	TON	76	280.00	21,280.00	100.00	7,600.00	170.00	12,920.00	220.00	16,720.00	400.00	30,400.00	187.00	14,212.00
83	723060A	Rock Slope Protection (300 lb, Class IV, Method A)	TON	644	245.00	157,780.00	100.00	64,400.00	170.00	109,480.00	180.00	115,920.00	165.00	106,260.00	105.00	67,620.00
84	723075	Rock Slope Protection (150 lb, Class III, Method B)	TON	3486	190.00	662,340.00	100.00	348,600.00	120.00	418,320.00	120.00	418,320.00	155.00	540,330.00	170.00	592,620.00
85	723088	Rock Slope Protection (60 lb, Class II, Method B)	TON	299	175.00	52,325.00	100.00	29,900.00	155.00	46,345.00	225.00	67,275.00	165.00	49,335.00	185.00	55,315.00
86	723125	Concreted-Rock Slope Protection (Class III, Method A)	TON	59	350.00	20,650.00	300.00	17,700.00	640.00	37,760.00	400.00	23,600.00	200.00	11,800.00	170.00	10,030.00
87	730045	Minor Concrete (Gutter) (Final Pay)	CY	5	1,700.00	8,500.00	1,200.00	6,000.00	6,000.00	30,000.00	2,500.00	12,500.00	2,000.00	10,000.00	2,400.00	12,000.00
88	750501	Miscellaneous Metal (Bridge) (Final Pay)	LB	13900	12.00	166,800.00	15.00	208,500.00	3.00	41,700.00	14.00	194,600.00	9.00	125,100.00	15.00	208,500.00
89	77000A	Joint Trench	LF	1453	50.00	72,650.00	30.00	43,590.00	475.00	690,175.00	700.00	1,016,400.00	10.00	14,530.00	825.00	1,197,900.00
90	780045	Prepare and Stain Shotcrete (Final Pay)	SQFT	10355	5.00	51,775.00	6.00	62,130.00	6.00	62,130.00	10.00	103,550.00	4.00	41,420.00	13.00	134,615.00
91	780046	Stain Galvanized Surfaces	LS	1	180,000.00	180,000.00	200,000.00	200,000.00	200,000.00	200,000.00	200,000.00	200,000.00	180,000.00	180,000.00	200,000.00	200,000.00
92	780600	Inclinometers	EA	8	7,000.00	56,000.00	10,000.00	80,000.00	50,000.00	400,000.00	17,000.00	136,000.00	10,000.00	80,000.00	26,000.00	208,000.00
93	810180A	Delineator (Type E/Class 2, Barrier Mounted, or Culvert Marker)	EA	29	50.00	1,450.00	100.00	2,900.00	70.00	2,030.00	125.00	3,625.00	65.00	1,885.00	60.00	1,740.00
94	820840A	Roadside Sign - One Post, Two Post, or Barrier Mounted	EA	34	500.00	17,000.00	750.00	25,500.00	500.00	17,000.00	450.00	15,300.00	550.00	18,700.00	315.00	10,710.00
95	832006	Midwest Guardrail System (Steel Post)	EA	290	90.00	26,100.00	95.00	27,550.00	100.00	29,000.00	100.00	29,000.00	100.00	29,000.00	93.00	26,970.00
96	832070	Vegetation Control (Minor Concrete)	SQYD	250	80.00	20,000.00	125.00	31,250.00	110.00	27,500.00	130.00	32,500.00	170.00	42,500.00	91.00	22,750.00
97	833090	Tubular Handrailing (Modified) (Final Pay)	LF	2666	205.00	546,530.00	125.00	333,250.00	110.00	293,260.00	130.00	346,580.00	150.00	399,900.00	135.00	359,910.00
98	839521	Cable Railing (Final Pay)	LF	629	80.00	50,320.00	75.00	47,175.00	220.00	138,380.00	100.00	62,900.00	90.00	56,610.00	170.00	106,930.00
99	839540	Transition Railing (Type STB)	EA	3	8,000.00	24,000.00	12,000.00	36,000.00	10,500.00	31,500.00	12,500.00	37,500.00	8,000.00	24,000.00	9,500.00	28,500.00
100	839543	Transition Railing (Type WB-31)	EA	3	6,000.00	18,000.00	12,000.00	36,000.00	10,500.00	31,500.00	12,500.00	37,500.00	6,000.00	18,000.00	9,200.00	27,600.00
101	839584A	MASH In-Line Terminal System	EA	5	5,000.00	25,000.00	7,500.00	37,500.00	9,000.00	45,000.00	8,000.00	40,000.00	6,000.00	30,000.00	9,500.00	47,500.00
102	837700	Concrete Barrier (Type 85 Modified) (Final Pay)	LF	2678	500.00	1,339,000.00	400.00	1,071,200.00	736.45	1,972,450.00	400.00	1,071,200.00	380.00	1,017,640.00	650.00	1,740,700.00
103	840505	6" Thermoplastic Traffic Stripe	LF	8000	4.00	32,000.00	0.93	7,440.00	1.46	11,680.00	2.00	16,000.00	3.00	24,000.00	1.50	12,000.00
104	840515	Thermoplastic Pavement Markings	SQFT	100	50.00	5,000.00	15.00	1,500.00	20.00	2,000.00	10.00	1,000.00	21.00	2,100.00	28.00	2,800.00
105	999090	Mobilization	LS	1	5,428,473.25	5,428,473.25	2,454,000.00	2,454,000.00	6,600,000.00	6,600,000.00	7,345,000.00	7,345,000.00	5,145,000.00	5,145,000.00	5,999,218.00	5,999,218.00

Base Bid [Schedule A] Total: 59,773,962.00 | 61,780,838.07 | 70,710,452.00 | (—) | 80,826,788.65 | 107,736,885.30

Pier 2 Alternative Bid [Schedule B]:

Item	Spec	Description	Unit	Qty	EE Unit	EE Amount	B1		B2		B3		B4		B5	
P2B1	490605	30" Cast-In-Drilled-Hole Concrete Piling	LF	2389	840.00	2,006,760.00			1,000.00	2,389,000.00	750.00	1,791,750.00	2,000.00	4,778,000.00	1,406.34	
P2B2	520102	Bar Reinforcing Steel (Bridge)	LB	118639	1.75	207,618.25			2.20	261,005.80	5.50	652,514.50	2.30	272,869.70	3.10	
P2B3	500030	HS Steel	LB	54120	3.00	162,360.00			3.59	86,092.80	3.00	162,360.00	3.75	594,710.00	3.98	

Pier 2 Alternative Bid [Schedule B] Total: 2,376,738.25 | 2,464,000.00 | 2,736,098.60 | 2,606,624.50 | 4,009,517.50 | 5,145,579.70

Pier 3 Alternative Bid [Schedule H]:

Item	Spec	Description	Unit	Qty	EE Unit	EE Amount	B1		B2		B3		B4		B5	
P3H1	495000	Micropile	EA	154	16,000.00	2,464,000.00	11,500.00	1,771,000.00	30,000.00	4,620,000.00	18,000.00	2,772,000.00	12,500.00	1,925,000.00	188,239.00	

Pier 3 Alternative Bid [Schedule H] Total: 2,464,000.00 | 1,771,000.00 | 4,620,000.00 | 2,772,000.00 | 1,925,000.00 | 2,808,806.00

Base Bid Total: 64,614,700.25 | 66,237,894.67 | 76,888,214.45 | 77,551,969.50 | 87,997,948.35 | 114,380,376.05

	EE	B1	B2	B3	B4	B5
+ 10% Contingency	6,461,470.03	6,623,789.47	7,688,821.45	7,755,196.85	8,799,734.84	11,438,037.61
+ 15% CM						
Total Con. Cost	80,768,375.31	82,797,368.34	96,110,268.06	96,939,961.88	109,996,685.44	142,975,470.08

The decision of the Board of Supervisors on the bid protest will be final.

AWARD OF CONTRACT: Consideration of bids for award will be by the Board of Supervisors. The County of El Dorado reserves the right after opening bids to reject any or all bids, to waive any irregularity in a bid, or to make award to the lowest responsive, responsible Bidder and reject all other bids, as it may best serve the interests of the County.

As a condition of award, the Department of Transportation requires the successful Bidder to submit bonds and evidence of insurance prior to execution of the Agreement by the County. Failure to meet this requirement constitutes abandonment of the Bid by the Bidder and forfeiture of the Bidder's security. Subsequently, award will be to the next lowest, responsive, responsible Bidder.

The Office Engineer must receive all required documents within ten (10) business days of the date of the Notice of Award of Contract letter.

ESCROW BID DOCUMENTS: Refer to the Special Provisions 3-1.14, "Escrow Bid Documents" for the provisions requiring the successful Bidder to submit all documentary information used to prepare its bid. Bidder must submit Escrow Bid Documents in a sealed lockable container to the Department of Transportation.

RETAINAGE FROM PAYMENTS: The Contractor may elect to receive one hundred percent (100%) of payments due under the Contract from time to time, without retention of any portion of the payment by the County, by depositing securities of equivalent value with the County in accordance with the provisions of Section 22300 of the Public Contract Code. Securities eligible for deposit hereunder are be limited to those listed in Section 16430 of the Government Code, or bank or savings and loan certificates of deposit.

PROJECT ADMINISTRATION: Submit all Requests for Information (RFI) during the bid period to Brian Franklin, County of El Dorado, Department of Transportation, email- Brian.Franklin@edcgov.us. If the response does not require an Addendum, posting of a response will be as a Response to Bidder's Inquiry on the Quest website under Quest Project # 7267650, "Project Q&A". It is the Bidders' responsibility to check this website under "Project Q&A" for responses to Bidders' inquiries during the bid period. Upload of Addenda will be in pdf format to Quest's website and Quest will issue an automatic email notification to all planholders that have acquired the Contract Documents digitally through Quest. The list of planholders will be available on Quest's website under "View Planholders".

Department of Transportation will not give oral responses to any questions concerning the content of the Contract Documents. All responses will be in the form of written Addenda to the Contract Documents or written responses to Bidders' inquiries. Posting of responses to Addenda and Bidders' inquiries will be on the Quest website as described above.

Inquiries or questions based on alleged patent ambiguity of the Plans, Specifications, or estimate must be via Bidder inquiry prior to bid opening. These inquiries or questions, submitted after bid opening will not constitute basis for a bid protest.

BY ORDER OF the Director of the Department of Transportation, County of El Dorado, State of California.

Authorized by the Board of Supervisors on December 14, 2021, at Placerville, California.

By _____
Rafael Martinez, Director
Department of Transportation

Mosquito Road at South Fork American River - Bridge Replacement
Contract No. 5084, CIP No. 36105028
December 14, 2021

County of El Dorado
Notice to Bidders
NTB-7

Index